Radioactivity

Introduction and History

Cover illustration depicts an atomic nucleus and the various types of radiation that may be emitted from decaying nuclei. (Courtesy of Canberra Industries, Meriden, CT, USA.)

Radioactivity

Introduction and History

by

Michael F. L'Annunziata
The Montague Group P.O. Box 5033
Oceanside, CA 92052-5033, USA

with Foreword by
Prof. Dr. Werner Burkart, Deputy Director General
International Atomic Energy Agency, Vienna

ELSEVIER

Amsterdam • Boston • Heidelberg • London • New
Paris • San Diego • San Francisco • Singapore • Sy

Elsevier
Radarweg 29, Box 211, 1000 AE Amsterdam, The Netherlands
Linacre House, Jordan Hill, Oxford OX2 8DP, UK

First edition 2007

Library of Congress Cataloging in Publication Data
A catalog record is available from the Library of Congress

British Library Cataloguing in Publication Data
A catalogue record is available from the British Library

ISBN: 978-0-444-52715-8

Printed and bound in The Netherlands

07 08 09 10 11   10 9 8 7 6 5 4 3 2 1

To Reyna, Michael, and Helen
In memory of Frank

"Through the release of atomic energy, our generation has brought into the world the most revolutionary force since prehistoric man's discovery of fire … We scientists recognize our inescapable responsibility to carry to our fellow citizens an understanding of atomic energy and its implication for society. In this lies our only security and our only hope …"

Albert Einstein

Contents

Foreword   *xi*

Preface   *xiii*

Acronyms, Abbreviations, and Symbols   *xvii*

Introduction: Radioactivity and Our Well-Being   **1**

Radioactivity Hall of Fame—Part I: for Chapter 1

Democritus (c.460–c.370 B.C.), Wilhelm C. Röntgen (1845–1923), Henri Becquerel (1852–1908), Pierre (1859–1906) and Marie Curie (1867–1934), Paul Villard (1860–1934), Ernest Rutherford (1871–1937), Hendrick A. Lorentz (1853–1928), Pieter Zeeman (1865–1943), Joseph John Thomson (1856–1940), and Philipp Lenard (1862–1947)

Chapter 1: Alpha Radiation   **71**

1.1 Introduction   71

1.2 Decay Energy   71

1.3 Alpha-Particle Interactions with Matter   75

1.4 Alpha-Particle Ranges   78

Radioactivity Hall of Fame—Part II: for Chapter 2

Frederick Soddy (1877–1956), C.T.R. Wilson (1869–1959), Frédéric Joliet (1900–1958) and Irène Joliet-Curie (1897–1956), Enrico Fermi (1901–1954), Wolfgang Pauli (1900–1958), Frederick Reines (1918–1998), and Clyde Cowan, Jr. (1919–1974)

Chapter 2: Beta Radiation   **119**

2.1 Introduction   119

2.2 Negatrons   119

2.3 Positrons   124

2.4 Beta-Particle Absorption and Transmission   129

2.5 Stopping Power and Linear Energy Transfer   132

Radioactivity Hall of Fame—Part III: for Chapter 3

Max Planck (1858–1947), Louis de Broglie (1892–1987), Albert Einstein (1879–1955), Arthur H. Compton (1892–1962), Max von Laue (1879–1960), Sir William Henry Bragg (1862–1942) and Sir William Lawrence Bragg (1890–1971), Henry G.J. Moseley (1887–1915), Charles Glover Barkla (1877–1944), Manne Siegbahn (1886–1978), and Robert A. Millikan (1868–1953)

Chapter 3: Gamma- and X-Radiation—Photons   **187**

3.1 Introduction   187

3.2 Dual Nature: Wave and Particle   187

3.3 Gamma Radiation   191

3.4 Annihilation Radiation   195

3.5 Cherenkov Radiation   196

3.6 X-Radiation   196
3.7 Interactions of Electromagnetic Radiation with Matter   201

Radioactivity Hall of Fame—Part IV: for Chapter 4
James Chadwick (1891–1974), Lise Meitner (1878–1968) and Otto Hahn (1879–1968), and
Leo Szilard (1898–1964)

Chapter 4: Neutron Radiation   **253**
4.1 Introduction   253
4.2 Neutron Classification   253
4.3 Sources of Neutrons   260
4.4 Interactions of Neutrons with Matter   276
4.5 Neutron Attenuation and Cross-Sections   281
4.6 Neutron Decay   287

Radioactivity Hall of Fame—Part V: for Chapter 5
Niels Bohr (1885–1962), Gustav Hertz (1887–1975) and James Franck (1882–1964), Werner
Heisenberg (1901–1976), Erwin Schrödinger (1887–1961), Max Born (1882–1970) and Paul A.H.
Dirac (1902–1984), and Clinton Davisson (1881–1958) and George Paget Thomson (1892–1975)

Chapter 5: Atomic Electron Radiation   **341**
5.1 Introduction   341
5.2 Internal-Conversion Electrons   341
5.3 Auger Electrons   344

Radioactivity Hall of Fame—Part VI: for Chapter 6
Victor F. Hess (1883–1964), Carl D. Anderson (1905–1991), Patrick M.S. Blackett (1897–1974),
Hideki Yukawa (1907–1981), Cecil F. Powell (1903–1969), Donald A. Glaser (1926–), and
Pierre Victor Auger (1899–1993)

Chapter 6: Cosmic Radiation   **399**
6.1 Introduction   399
6.2 Classification and Properties   400
6.3 Showers of the Cosmic Radiation   406
6.4 Cosmic Rays Underground   411
6.5 Origins of Cosmic Radiation   412
6.6 Cosmic Background Radiation   412
6.7 Dose from Cosmic Radiation and Other Sources   421

Radioactivity Hall of Fame—Part VII: for Chapter 7
Sergei Ivanovich Vavilov (1891–1951), Pavel Alekseyevich Cherenkov (1904–1990), Il'ja Mikhailovich
Frank (1908–1990), and Igor Yevgenyevich Tamm (1895–1971)

Chapter 7: Cherenkov Radiation   **465**
7.1 Introduction   465
7.2 Theory and Properties   467
7.3 Cherenkov Photons from Gamma-Ray Interactions   482
7.4 Particle Identification (PID)   485
7.5 Applications in Radionuclide Analysis   492

Radioactivity Hall of Fame—Part VIII: for Chapter 8

Ernest O. Lawrence (1901–1958), John Douglas Cockcroft (1897–1967) and Ernest Thomas Sinton Walton (1903–1995), Hans A. Bethe (1906–2005), and Willard F. Libby (1908–1980)

Chapter 8: Radionuclide Decay, Mass, and Radioactivity Units   **529**
8.1 Introduction   529
8.2 Half-Life   529
8.3 General Decay Equations   538
8.4 Secular Equilibrium   539
8.5 Transient Equilibrium   544
8.6 No Equilibrium   547
8.7 More Complex Decay Schemes   548
8.8 Radioactivity Units and Radionuclide Mass   549

Appendix A: Particle Range-Energy Correlations   553
Appendix B: Periodic Table of the Elements   557
References   559
Index   597

Foreword

Over a century has passed since the discovery of radioactivity by Henri Becquerel. His discovery opened the door to a new realm of science where physics, chemistry, and biology would unite and grow, like a tree, spreading its roots into almost all scientific disciplines. Many natural radionuclides discovered and artificially produced have found their way into the disciplines of medicine, agriculture, environment, industry, and power. Although we tend to take radioactivity for granted as it is always out of sight, and, therefore, often out of mind, our lives have been greatly enriched by its many peaceful applications. For example, each and every one of us has applied a bandage to an open wound and received vaccinations through a syringe, each of these being items that were sterilized with ionizing radiation. Those of us unfortunate enough to have been struck by a serious disease, such as cancer, can attest to the numerous nuclear diagnostic and therapy techniques that enable medical science to help us lead longer and healthier lives. We also frequently forget, or fail to realize at all, that the abundant and nutritious foods that we have on our tables are the result of plant breeding, optimized fertilizer and water use, and insect pest control, all of which, to a considerable extent, have been made possible by invisible radioactive allies and their often unique applications.

The International Atomic Energy Agency (the IAEA) promotes scientific and technical cooperation worldwide in the pursuit of development goals by seeking to advance Member States' capabilities in medicine, agriculture, biology, environmental protection, food safety, hydrology, industry, and power generation. The IAEA's efforts have led to the peaceful and safe use of radioactive sources in over a hundred countries around the world. Through research, our understanding of the basic principles of nuclear decay and the properties of nuclear radiation help us to find new and improved nuclear applications that can assist in providing sustainable solutions to basic but also highly refined human needs for all on this globe.

I am pleased that this book brings a fascinating but poorly known world into easy reach of students, teachers, scientists, and lay persons. It describes the origins, properties, and applications of nuclear radiation, and the lives and works of numerous pioneers and Nobel Laureates who have helped us to understand radioactivity and to find peaceful nuclear applications of it for human development. Those who read the book will hopefully come to appreciate from the lives and works of these pioneers, as I have, just how powerful a tool nuclear radiation is and how deeply its applications touch our lives each day.

Professor Dr Werner Burkart
Deputy Director General
Department of Nuclear Sciences and Applications
International Atomic Energy Agency, Vienna, Austria

Preface

This book describes the origins and properties of the various types of radiation emitted by radioactive nuclides, as well as cosmic radiation, cosmic ray showers, and Cherenkov radiation. Included are the principles of radionuclide decay, and an historical narrative of discoveries that revealed the properties of the atomic nucleus, atomic structure, nuclear decay and its radiations. These discoveries revealed, as described in this book, a source of energy vital to our well-being and development. Before each chapter are accounts of the lives and works of early pioneers from the discovery of radioactivity by Henri Becquerel in 1896, which had followed by only a few months the discovery of x-radiation by Wilhelm Röntgen in 1895, to around 1960 when many applications of nuclear energy of benefit to mankind were beginning to take hold in medicine, biology, food and agriculture, hydrology, industry, nuclear power, and other fields vital to our basic needs.

The book is divided into two parts, namely, one part providing the biographical accounts of the lives and works of pioneers in this field and the other part detailing the origins and properties of the various types of radiation and nuclear decay. The biographical accounts are found in the sections entitled *Radioactivity Hall of Fame*, which are included before each chapter of the book. The lives and works of the pioneers are not presented in purely chronological order. Rather, just before each chapter, I included as far as possible the biographical accounts of those, who contributed most to our knowledge of the material presented in the following chapter.

The Introduction includes an overview of some of the many applications of radioactive sources that play an important role in our day-to-day lives. In so doing, it was my objective to highlight how we have become dependent in so many ways on a source of energy, which has enriched our lives. Radioactive sources and the various types of radiation that they emit are invisible to us, and thus, the benefits that these energy sources provide us each day through radiation technology, most often do not cross our minds.

It is my hope that this book will serve as a teaching text for those who are or may become interested in physics and chemistry. It describes in detail how pioneers in these two fields had joined forces to unravel the mysteries of natural and artificial radioactive sources, their invisible radiations, and the numerous ways that this energy source can be applied to improve our lives and well-being.

In the sections entitled *Radioactivity Hall of Fame*, I have included the lives and works of over 60 research pioneers, most of whom were awarded the Nobel Prize. The experiments that they carried out and the reasoning they used to both devise their experiments and arrive at their findings are provided in detail. Many of the trials and tribulations of their lives are also included. It is my hope that they will serve as examples to the student of any field of endeavor. Specifically, I hope the student will extract from this book not only the romance of science but also take note of the tremendous perseverance displayed by these pioneers. They encountered many difficulties and sometimes seemingly insurmountable obstacles in their life and work, and without doubt, fear of failure; nevertheless, these obstacles did not hinder them but only motivated them further. To illustrate my point I will take one example of many from the text of this book. It is taken from the biographical sketch of Ernest Lawrence (1901–1958), who conceived and built the cyclotron, a device found today in many hospitals for the on-site production of artificial radioisotopes of short half-life needed in medical diagnosis and treatment. In the text it is described how Ernest Lawrence one day in 1929 read an article by Rolf Wideröe, a Norwegian physicist, who wrote about the acceleration of positive ions across an increasing voltage potential in a linear device. On that day in 1929 after reading Wideröe's paper, Lawrence invented the principle of the cyclotron. Conceiving the principle of a new concept is one matter, but making it happen may be altogether a seemingly impossible hurdle. At the Nobel Prize Award Ceremony held at the University of California, Berkeley on February 29, 1940, Professor R.T. Birge stated the following:

> "The next morning Dr. Lawrence told his friends that he had found a method for obtaining particles of very high energy, without the use of any high voltage. The idea was surprisingly simple and in principle quite correct – everyone admitted that. Yet everyone said, in effect, "Don't forget that having an idea and making it work are two very different things"… In this connection I can quote with profit some remarks made by Dr. W. D. Coolidge, when he presented to Dr. Lawrence, in 1937, the Comstock Prize of the National Academy of Sciences… "Dr. Lawrence envisioned a radically different course – one which did not have those difficulties attendant upon the use of potential differences of millions of volts. At the start, however, it presented other difficulties and many uncertainties, and it is interesting to speculate on whether an older man, having had the same vision, would have ever attained its actual embodiment and successful conclusion. It called for boldness and faith and persistence to a degree rarely matched."

The above statement is a lesson to us in all fields of work and whatever the pursuit. Many good ideas that conform with known laws of nature may come to us, but to bring these ideas or inventions into reality can be to most an insurmountable burden. Often it is difficult to comprehend the toil, almost endless hours, frustration, and sometimes border despair that one encounters coupled with the needed stubborn determination to find solutions to make an idea become a reality, We can be certain that Lawrence and many persons, who have achieved greatness, have battled these tremendous hurdles. It is my hope that, in relating these stories, the student will find inspiration and examples to follow in the pursuit, not necessarily of greatness, but of achievement and satisfaction in whatever work or endeavor their lives may bring.

While writing I thought how best to portray the greatness of the achievements of the pioneers described in this book. I thought of doing what is common, namely, providing copies of historic photographs, as we have seen in other books, such as Pierre and Marie Curie working together in their Paris laboratory or Lise Meitner and Otto Hahn working in Germany as a team in physics and chemistry. Instead, I thought how best to underscore the significance of their work by illustrating postage stamps issued in many countries commemorating their discoveries or the beneficial applications that their discoveries have brought to mankind. It took time and effort to obtain the needed commemorative stamps, but worth the effort.

There is no better way to draw attention to the significance of the work accomplished by these pioneers than to see that many nations have issued a postage stamp commemorating their achievements.

In the end I hope this book will serve not only to inspire the student with the historical romance of discovery, but also serve as a teaching text of many of the basic principles of radioactivity, its beneficial applications, nuclear decay and radiation physics and chemistry.

I have omitted specific chapters on the detection and measurement of radiation or radionuclide analysis, as it is beyond the scope of this book. This subject matter is covered in detail in other more lengthy books including an earlier text *Handbook of Radioactivity Analysis*, 2nd edition, edited by the author and published by Elsevier Science, Amsterdam in 2003.

Mention of commercial products in this book does not imply recommendation or endorsement by the author. Other and more suitable products may be available. The names of products are included for convenience or information purposes only.

I want to thank Dr Ramkumar Venkataraman for his encouragement and review of some of the text. However, I bear sole responsibility for any errors or omissions. The encouragement of Professor Emeritus Romard Barthel, Ph.D., C.S.C., and his stimulating and dynamic lectures in physics and mathematics at St. Edward's University during 1961–1962 will always be remembered with gratitude, appreciation, and esteem. My appreciation is extended to editorial staff of Elsevier, Amsterdam, including Dr. Andrew D. Gent, Dr. Egbert van Wezenbeek and Joan Anuels for their encouragement, recommendations and patience during the writing of this book and to Joan Anuels and Betsy Lightfoot for their constant attention to every detail of this book. I thank my wife Reyna for her steadfast support, understanding, and patience.

Michael F. L'Annunziata, Ph.D.
February 2007

Acronyms, Abbreviations, and Symbols

A	mass number
a	years (anni)
Å	Angstrom (10^{-10} m)
Ab	antibody
AD	Anno Domini
Ag	antigen
α	alpha particle, internal-conversion coefficient
$\propto$	proportional to
ANL	Argonne National Laboratory
$\sim$	approximately
ARS	agricultural research service
atm	standard atmospheric pressure (760 mmHg, 760 Torr)
B	binding energy
B/A	binding energy per nucleon
barn	10^{-24} cm^2
BP	before present
Bq	Becquerel = 1 disintegration per second (1 dps)
β	particle relative phase velocity, beta particle
β^-	negatron, negative beta particle
β^+	positron, positive beta particle
c	speed of light in vacuum (2.9979×10^8 m/sec)
°C	degrees Celsius
C	Coulomb (2.997×10^9 esu)
cal	calorie (4.186 J)
CAT	computed axial tomography
CCD	charge-coupled device
CEA	Commissariat à l'Energie Atomique
CERN	European Organization for Nuclear Research, Geneva
CFR	Centre des Faibles Radioactivités
Ci	Curie (2.22×10^{12} dpm = 3.7×10^{10} dps = 37 GBq)
CIEMAT	Centro de Investigaciones Energéticas, Medioambientales y Technológicas, Madrid
cm	centimeter (10^{-2} m)
CMB	cosmic microwave background
CN	carbon–nitrogen
CNRS	Centre National de la Recherche Scientifique
CPM, cpm	counts per minute
CPY	counts per year
CT	computed tomography

CTR	controlled thermonuclear reactor
D	deuterium
d	deuteron
D-D	deuterium–deuterium fusion
DDR	German Democratic Republic
DNA	deoxyribonucleic acid
DPM, dpm	disintegrations per minute
DPS, dps	disintegrations per second
D-T	deuterium–tritium fusion
e^-	electron or negatron
e^+	positron
$e^\pm$	positron–negatron pair
E	energy, detection efficiency
E_e	electron energy
E_γ	gamma-ray energy
E_k	kinetic energy
E_{max}	maximum energy
E_{th}	threshold energy
EC	electron capture
erg	10^{-7} J
esu	electrostatic unit (3.335×10^{-10} C)
ETH	Eidgenössische Technische Hochschule (Zurich)
EU	European Union
EUR	Euro
eV	electron volt (1.602×10^{-19} J $= 1.602 \times 10^{-12}$ ergs)
FAO	Food and Agriculture Organization
FIAN	Fizicheskii Institut Akademya Nauk USSR
fm	fermi (10^{-15} m), femtometer
fp	fission products
ft	foot (0.3048 m)
G	gauss (10^{-4} T)
g	gram
γ	gamma radiation
GBq	gigabecquerels (10^9 dps)
GeV	gigaelectron volts (10^9 eV)
Gy	Gray = 1 J/kg $= 6.24 \times 10^{12}$ MeV/kg
H	magnetic field strength
h	hours
h	Planck constant (6.626×10^{-34} J sec or 4.136×10^{-15} eV sec)
$\hbar$	Planck constant, reduced ($h/2\pi = 1.054 \times 10^{-34}$ J sec $= 6.582 \times 10^{-22}$ MeV sec)
HDR	high dose rate
HTGR	High Temperature Gas Cooled Reactor
Hz	hertz ($\sec^{-1}$)
IAEA	International Atomic Energy Agency, Vienna

IC	internal conversion
ICRP	International Commission on Radiation Protection
ICRU	International Commission on Radiation Units and Measurements
IEC	inertial electrostatic confinement
in.	inch = 2.54 cm
ITER	International Thermonuclear Experimental Reactor
J	joule (0.2389 cal or 10^7 ergs)
JET	Joint European Torus
K	degrees Kelvin
K	kinetic energy
K^+, K^-, K^0	positive, negative, and neutral kaon
kBq	kilobecquerel (10^3 dps)
keV	kiloelectron volts (10^3 eV)
kG	kilogauss (10^3 G)
kg	kilogram (10^3 g)
kGy	kilogray (10^3 Gy)
km	kilometer (10^3 m)
km.w.e	km-water-equivalent
kV	kilovolt (10^3 V)
kWh	kilowatt-hour
KWI	Kaiser Wilhelm Institute (Berlin)
l	liters
Λ	hyperon
λ	wavelength, decay constant
LANL	Los Alamos National Laboratory
LET	linear energy transfer
m	particle mass
m_e	electron mass
m_0	particle rest mass
m_r	particle relativistic mass
m	meters
MBq	megabecquerels (10^6 dps)
MCi	megacurie (10^6 Ci)
mCi	millicurie (10^{-3} Ci)
MeV	megaelectron volts (10^6 eV)
mg	milligram (10^{-3} g)
min	minutes
MIT	Massachusetts Institute of Technology
ml	milliliter (10^{-3} l)
mM	millimolar (10^{-3} M)
mm	millimeter (10^{-3} m)
mrad	millirad (10^{-3} rad = 10 μGy), milliradian
mrem	millirem (10^{-3} rem)
MRI	magnetic resonance imaging
msec	milliseconds (10^{-3} sec)

mSv	millisievert (10^{-3} Sv)
μ	linear attenuation coefficient
μ_m	mass attenuation coefficient
μ^+, μ^-	positive muon, negative muon
μCi	microcurie (10^{-6} Ci)
μGy	microgray (10^{-6} Gy)
μl	microliter (10^{-6} l)
μm	micron, micrometer (10^{-6} m)
μsec	microseconds (10^{-6} sec)
MWPC	multiwire proportional chamber
MW	megawatt
n	neutron
$\bar{n}$	antineutron
n	index of refraction
N_A	Avogadro's number (6.022×10^{23} atoms/mol)
nCi	nanocurie (10^{-9} Ci)
NDFF	nitrogen derived from fertilizer
NDFS	nitrogen derived from soil
NDT	non-destructive testing
NIST	National Institute of Standards and Technology
nm	nanometer (10^{-9} m)
NMR	nuclear magnetic resonance
n/p	neutron/proton ratio
NRL	National Radiation Laboratory (New Zealand)
nsec	nanoseconds (10^{-9} sec)
ν	neutrino, photon frequency, particle velocity
$\bar{\nu}$	antineutrino
N/Z	neutron/proton ratio
p	particle momentum
p^+, p	proton
$\bar{p}$	antiproton
PDFF	phosphorus derived from fertilizer
PDFS	phosphorus derived from soil
PET	positron emission tomography
PGA	phosphoglyceric acid
π	3.141592
π^+, π^-, π^0	positive, negative, and neutral pion
PID	particle identification
PMT	photomultiplier tube
Q	disintegration energy
R	roentgen = 2.5×10^{-4} C/kg of air, particle range
RaB	radium B (^{214}Pb)
RaC	radium C (^{214}Bi)
RaD	radium D (^{210}Pb)
rad	radiation absorbed dose (100 erg/g = 10 mGy), radian

rem	roentgen equivalent for man (10 mSv)
ρ	density (g/cm^3), neutron absorption cross-section, particle-path radius of curvature
RIA	radioimmunoassay
RICH	ring imaging Cherenkov (counter)
sec	seconds
SANS	small-angle neutron scattering
SF	spontaneous fission
SI	Système International d'Unités
σ	neutron absorption cross-section
SIT	sterile insect technique
SPA	scintillation proximity assay
SPECT	single-photon emission computed tomography
sr	steradian
SRM	Secondary Modern Reference
STP	standard temperature and pressure
Sv	sievert (1 J/kg)
T	tritium, tesla (10^4 G)
t	time
$t_{1/2}$	half-life
TA-GVHD	transfusion-associated graft versus host disease
τ	particle lifetime
TeV	teraelectron volts (10^{12} eV)
TFTR	Tokamak Fusion Test Reactor
ThA	thorium A (^{216}Po)
ThB	thorium B (^{212}Pb)
ThC	thorium C (^{212}Bi)
ThC$''$	thorium C$''$ (^{208}Tl)
TOA	top of the atmosphere
TOF	time of flight
Tokamak	Russian "toroidal kamera ee magnetnaya katushka" for "torus-shaped magnetic chamber"
TOP	time of propagation
u	atomic mass unit ($(1/12)m$ of ^{12}C = 1.6605402 $\times$ 10^{-27} kg) or 931.494 MeV/c^2
u	particle speed
u_{nr}	non-relativistic particle speed
u_r	relativistic particle speed
UIC	Uranium Information Centre
USDA	United States Department of Agriculture
USSR	Union of Soviet Socialist Republics
UV	ultraviolet
V	volt
W	watt (1 J/sec)
Z	atomic number or nuclear charge

Introduction: Radioactivity and Our Well-Being

Radioactivity is the process of the spontaneous decay and transformation of unstable atomic nuclei accompanied with the emission of nuclear particles and/or electromagnetic radiation (also referred to as nuclear radiation). The chapters in this book focus on the origins and properties of various types of nuclear radiation and an historical account of their discovery. In light of the fundamental nature of the subject matter in this book, the writer felt the need to avail of the introduction to expound briefly on the peaceful applications of radioactivity and its properties upon which we depend dearly. This introduction will provide a very brief sketch of only a very few examples of applications of radioactivity that improve and enrich our lives.

Nuclear radiation and sources of radioactivity, that is, radionuclides, have become a necessary part of our daily lives. The quantity and quality of our food, our health, general well-being,

and consequently our extended life span are due in large part to radioactive sources and their numerous applications in medicine, biology, agriculture, industry, and electric power generation. The significance of the role that radioactivity plays to improve our lives was commemorated with the postage stamp illustrated here issued by France in 1965. The stamp illustrates an artistic depiction of an atom together with drawings representing four fields where radioactivity and nuclear energy play a significant role in development, namely, medicine, agriculture, industry, and nuclear power for electricity.

Our dependence on radioactivity began only a few years after its discovery by Henri Becquerel in 1896. Not long thereafter, during the beginning of the 20th century (Curie, 1905) when Marie and Pierre Curie spearheaded the use of radium for the treatment of cancer. We may consider their work to be the first peaceful application of nuclear energy and the birth of modern nuclear medicine, upon which we now depend for the diagnosis and treatment of cancer and many other infirmities of the human body. Not long afterward Rutherford in 1919 demonstrated the first artificial production of the stable isotope oxygen-17 by bombarding the nucleus of nitrogen-14 with alpha particles. It was not until the 1930s that the example of Rutherford would be followed by Frédéric Joliet and Irene Joliet-Curie, who in 1934 achieved the first artificial production of radioisotopes by bombarding various elements such as Be, B, and Mg with alpha particles from polonium. About the same time, Ernest Lawrence built the first working cyclotron in 1931 capable of accelerating protons, deuterons, or helium ions (alpha particles) to energies capable of penetrating atomic nuclei and thereby producing numerous stable and radioactive isotopes that would find many peaceful applications in improving the well-being of humanity the world over. By 1940 the cyclotron developed by Lawrence and his coworkers would produce artificially as many as 223 radioactive isotopes, many of which would prove to be of immediate and immense value in medicine and studies in the biological sciences. Small cyclotrons are used to this day on-site at hospitals to produce short-lived radioisotopes for the diagnosis and treatment of various cancers and other diseases. The invention of the cyclotron, and the many radioactive isotopes that would be produced as a result, heralded the beginning of an era when peaceful applications of radioactive isotopes would expand worldwide and prove to be of great benefit to humanity.

Within the various chapters of this book, the reader will find information on the origins and properties of the various types of nuclear radiation together with an historical account of the lives and works of many early pioneers and Nobel Laureates, who have opened the doors to the application of radioactive sources in fields of science that have been of great benefit to mankind. In the following paragraphs, a small number of the countless applications of radioactive isotopes and nuclear radiation in medicine, biological research, food and agriculture, water resource management, industry, environmental protection, and nuclear power will be described. Only a few examples of peaceful applications will be provided here. More comprehensive reviews on the current trends and advances in peaceful applications of isotopes and nuclear radiation are found in the "Nuclear Technology Review" published annually by the IAEA, Vienna as well as in numerous books and training manuals published by the IAEA each year. Also, the reader may obtain much from a comprehensive book on *Practical Applications of Radioactivity and Nuclear Radiations* by Gerhart Lowenthal and Peter Airey (2001). A few examples, some of which are cited from the Nuclear Technology Review of the IAEA and other sources, are presented below.

1   HUMAN HEALTH

Among the many applications of nuclear radiation for human health, the technique of medical radiation imaging for cancer diagnosis and treatment is utilized worldwide, and research in its development and applications is making constant advances with each passing day. Imaging by means of radiation medicine techniques is often the first step in clinical management and diagnostic radiology; and nuclear medicine studies play important roles in the screening, staging, monitoring of treatment, and in the long-term surveillance of cancer patients. Numerous imaging techniques are available to the medical doctor. The techniques that include x-radiation and nuclear radiation are x-ray radiography, x-ray computed tomography (x-ray CT), single photon emission computed tomography (SPECT), and positron emission tomography (PET).

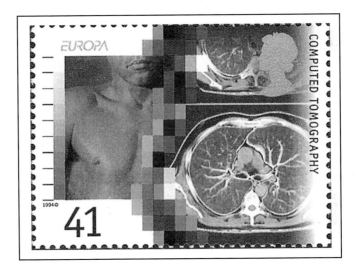

X-radiation, described in this book, is not of nuclear origin, but rather originates from electron energy. For medical purpose, x-rays are artificially produced when needed for diagnosis. It is electromagnetic radiation with properties similar to gamma radiation. The potential of x-rays to medical diagnosis became immediately clear following Wilhelm Röntgen's discovery of x-rays in 1895. The advent of computer technology and more recent digital technology facilitated the development of x-ray computed tomography (CT or x-ray CT), which was originally known as computed axial tomography (CAT or CAT-scan). x-ray CT provides a three-dimensional image of the internal structure and organs of the human body from numerous x-ray images taken around the body in a precise axis of rotation. The important role of computed tomography in medical diagnosis was commemorated in the postage stamp issued by the United Kingdom illustrated here. The various densities of the structures and organs of the human body will absorb the x-radiation to different degrees and thus, with computer data processing, produce an image of the organs and structure within the human body. These images provide the physician with information that might reveal abnormalities such as a lung tumor illustrated in Figure 1. The figure illustrates helical CT scans of the lungs of a patient at four different time intervals providing the physician with information on the progress that a patient is making following radiation therapy. Since its development in the early 1970s (Allan M. Cormack and Godfrey N. Hounsfield shared the Nobel Prize in Physiology and Medicine 1979 for the development of computer-assisted tomography),

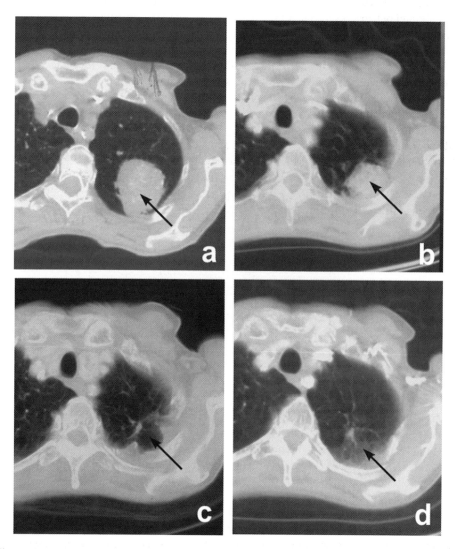

Figure 1 Computerized tomographic scans illustrating (a) bronchial carcinoma, indicated with arrow, before radiotherapy with a single-dose (30 Gy) irradiation; (b) partial remission after 10 months; (c) complete remission 21 months after irradiation leaving a scar-like fibrosis; and (d) dense consolidation after 48 months. (From Fritz *et al.* (2006) with permission © Fritz *et al.*; licensee BioMed Central Ltd.)

computed tomography has become the standard for the evaluation of patients with malignancies, because of its excellent definition of anatomical details.

SPECT is a nuclear diagnostic imaging technique that requires the administration of a radioactive pharmaceutical to the patient. A radiopharmaceutical is a drug labeled with a relatively short-lived radioactive nuclide, such as technetium-99m, gallium-67, iodine-131, or thalium-201. All of these nuclides decay with the concomitant emission of gamma rays. When administered to the patient, the drug and its attached radionuclide will distribute itself throughout the body sometimes settling at higher concentrations where an injury, tumor, or infection may be located. The gamma rays are emitted equally in all directions. Therefore, the patient is normally told to stand or sit in front of a collimator that is in contact with a detector. The collimator permits only the gamma rays emitted in the direction of the detector to be registered providing an image of the radiation intensities emitted from within the body of the patient. The detector may be moved to image the front, rear, or side of a patient, or it may be rotated in a 360° fashion around the patient. Only short-lived radionuclides are used so that the radioactivity administered to the patient will decay in short term and not cause harm. In Chapter 8 of this book, the author provides an example of a calculation illustrating the short duration that ^{99m}Tc remains in the human body after it is administered to a patient. SPECT as well as PET described below are nuclear medicine functional imaging techniques, which have the ability to detect cancerous involvement based on molecular and biochemical processes within the tumor tissue.

The important role of SPECT as a tool in cancer diagnosis was commemorated in the postage stamp illustrated here issued by Germany in 1981. The colors produced by the computerized tomographic image provide the medical doctor with an indication of radiation intensities emitted by the radioactive source in the patient where red > yellow > green > blue. Figure 2 shows colorless front and rear SPECT images taken of two patients administered a ^{99m}Tc radiopharmaceutical via intravenous injection. SPECT images of the skeletal structure of a patient taken a few hours after the administration of the radioactive pharmaceutical is

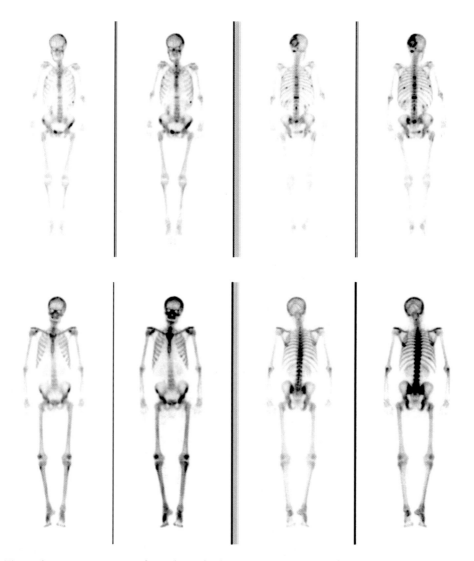

Figure 2 (Top) Bone scan of a patient with lung cancer and metastatic cancer. On the bone scan multiple osseous metastases can be seen (unevenly dark and spotted regions). (Below) Bone scan of a patient with lung cancer and no evidence of metastatic disease. SPECT bone scans were performed on patients 2–4 hours after the intravenous administration of the radiopharmaceutical ^{99m}Tc-DPD or [^{99m}Tc]-3,3-diphosphono-1,2-propandicarbonacid. (From Schoenberger *et al.*, 2004 with permission © 2004 Schoenberger *et al.*, licensee BioMed Central Ltd.)

often used to diagnose cancer metastases such as bone metastases observed in advanced cancers of the lung, prostate, and breast, among others.

PET is yet a more sophisticated form of tomographic imaging with a radioactive source, which can provide three-dimensional images of body organs and a display of the dynamics of radioisotope-labeled compound metabolism in organs. A positron-emitting radionuclide,

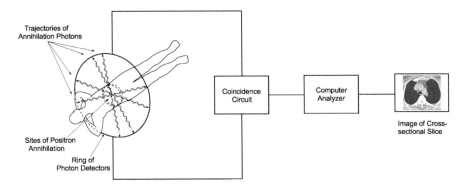

Figure 3   Positron-annihilation detection by a photon-detector ring in a PET instrument illustrating the line segment (cross-sectional slice) in which the positron-emitting nuclides resided. Many thousands of positron annihilations are detected by the ring of photon detectors. The photons arriving at opposite detectors (180°) in coincidence pass the coincidence circuit and are analyzed by the computer. From the data collected, an image of the relative intensities of the radioactive (positron-emitting) sites within the cross-sectional slice are plotted to provide an image of the structure and organs of the patient.

such as fluorine-18, carbon-11, or gallium-68, in the chemical form of a radiopharmaceutical is first administered to the body of a patient through intravenous injection. The patient is placed in a declining position within a chamber containing numerous radiation detectors that surround the body encompassing 360° in a complete circle (see Figure 3).

The positrons emitted by the radionuclide atoms in the patient become annihilated only a few millimeters from their originating atomic nuclei resulting in the emission of annihilation radiation, namely, two 0.511 MeV gamma rays emitted in opposite directions (180° apart) for each positron annihilation. The properties of positron radiation and their annihilation are discussed in detail in this book. The multiple radiation detectors, mounted in a circle, are positioned around the body part of the patient where the labeled organ or radionuclide localization is expected. Two of the many detectors surrounding the body become activated when two gamma-ray photons, originating from one positron–electron annihilation, simultaneously reach detectors 180° apart (see Figure 3). The coincidence detection of annihilation photons accurately determines the line segments in which the radionuclides resided. Many thousands of line segments are analyzed by a computer to reconstruct the distribution of the decayed radionuclides producing a tomographic image in a cross-sectional slice of the organ where the radiopharmaceutical had concentrated. The computer analysis of the line-segment origins of the annihilation radiations yields a tomographic density map or image of the various organs and anomalies (e.g., tumors) in the narrow region (slice) of the patients' body. PET is not a new concept. It was reviewed by the author (L'Annunziata, 1987) 20 years ago; however, with the advances of computer science in the past two decades, PET imaging technology has advanced almost exponentially, and it is now a dominant topic of many medical imaging meetings and nuclear medicine publications. PET serves as a tool in the diagnosis of many cancers. The diagnosis and treatment of other conditions with PET such as cardiovascular and brain degenerative diseases including Alzheimer's and depression are under intensive research.

Other imaging techniques that do not involve radioactivity are available to the medical profession such as magnetic resonance imaging (MRI). Often it is a combination of imaging

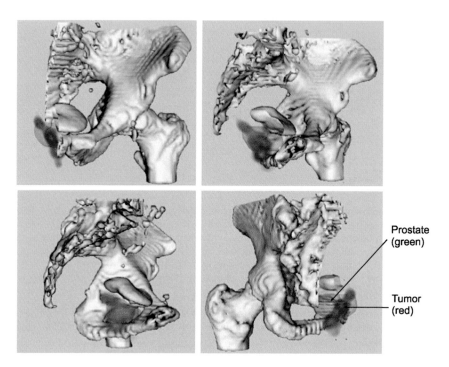

Prostate
(green)

Tumor
(red)

Figure 4   Four views from hybrid rendering of aligned scans from CT (bones), MRI (prostate), and SPECT (tumor, showing seminal vesicle invasion). (From Lee *et al.* (2005), reprinted with permission from Elsevier © 2005.)

techniques that provide the physician with a hybrid image yielding the information needed to arrive at the correct diagnosis. An example is provided by the work of Lee *et al.* (2005), who combined three imaging techniques, namely, MRI, CT, and SPECT into a single image. The hybrid image yielded a three-dimensional picture by blending the image of a prostate tumor distribution obtained by SPECT in the pelvis of a patient with anatomical structures imaged from CT and MRI. An example is provided in Figure 4.

Radiation oncology is a field of science that has a very long history. As described in this book, Marie and Pierre Curie recognized the potential of nuclear radiation for the treatment of cancer. After the death of her husband in 1906 and after receiving a second Nobel Prize in 1911 for her discovery of radium and polonium, the isolation of radium and the study of its properties, Marie Curie spearheaded the application of the nuclear radiation from radium for the treatment of cancer. Since that time countless patients suffering from many types of cancer including brain, neck, lung, breast, cervical, rectal, and prostate cancer have been treated and continue to be treated today with radiation therapy.

One of the common means of radiation therapy is external beam radiotherapy, which involves directing a beam of gamma-ray photons from an external radiation source onto the cancer of a patient or the use of external beams of protons or neutrons produced by a linear accelerator. One common external source of gamma radiation for cancer therapy is cobalt-60, often referred to as cobalt therapy. The importance of cobalt therapy in the treatment of cancer was heralded by the postage stamp issued in Canada illustrated here, which commemorated

the world's first treatment of a cancer patient using cobalt-60 radiation on October 27, 1951, at the Ontario Institute of Radiotherapy, which is known today as the London Regional Cancer Centre. On October 27, 2001, the Cancer Centre celebrated 50 years of cobalt-60 radiotherapy (Battista and van Dyk, 2002). The stamp illustrates a cobalt therapy device, which contains a source of cobalt-60, situated above a patient. The device directs a narrow beam of gamma radiation onto the cancer cells of the patient to kill the malignant cells while causing minimal damage to the healthy cells. The stamp also illustrates the decay scheme of the cobalt-60 radiation source. Cobalt-60 decays with a half-life of 5.3 years to the element nickel-60 by the emission of a 0.32 MeV beta particle. The nickel-60 daughter nuclide is formed at an excited energy state, and it loses this energy immediately with the emission of two gamma rays of energies 1.17 and 1.33 MeV settling at a stable nuclear state. The origins and properties of gamma radiation are discussed in this book. It is the gamma radiation from the cobalt-60 decay that is directed onto the cancer by careful collimation of the beam providing pinpoint irradiation of the cancer cells. Many advances have been made in external beam radiotherapy whereby proton beams are produced by a linear accelerator in the hospital. The protons undergo less scatter than gamma-ray photons, and consequently, with protons, higher dose rates to diseased tissues are sometimes possible while inflicting minimal damage to healthy tissue. With the use of certain target nuclei, as described in this book, neutrons may be produced with a proton accelerator. In this case, external beam irradiation with neutron beams sometimes offers the physician advantages in the treatment of certain tumor types including those in dense tissue where the heavier neutrons can inflict greater damage to malignant cells.

Modern cancer treatment may also involve another form of radiation therapy referred to as brachytherapy, sealed source radiotherapy, or endocurietherapy. This technique entails the insertion of radioactive seed implants into the tumor where the nuclear radiation is directly effective in killing the malignant cells. For example, in the treatment of prostate cancer, the physician may perform brachytherapy by inserting seeds of radioactive palladium-103 or iodine-125 directly into the prostate of the patient using a computer imaging technique to carefully insert the seeds where the radiation would be most effective. Palladium-103 and iodine-125 have 17- and 60-day half-lives and emit gamma radiation of relatively high and low energies, respectively. Both radioactive sources would kill diseased tissue and eventually decay to background levels, whereby the seeds would remain afterwards inside the prostate without causing damage. Alternatively, high dose rate (HDR) radioactive sources have been developed whereby the physician will insert a very small plastic catheter into the prostate gland of the patient, and then provide the patient with a series of HDR radiation treatments by inserting a radioactive source through the catheter. The catheter and radiation source are removed by the physician, after a controlled dose is administered, and no radioactive material is left in the tumor.

There are numerous approaches that a physician may take to the treatment of cancers by external beam radiation therapy or by brachytherapy. The techniques described here are only a few examples. Numerous books are available on nuclear medicine from which thorough treatments on the subject may be obtained. Among these are books by Biersack and Freeman (2007), Bonbardieri *et al.* (2007), Christian and Waterstram (2007), Cook *et al.* (2007), Eary and Brenner (2007), Powsner and Powsner (2006), Treves (2006), Mettler and Guiberteau (2005), and Zeissman *et al.* (2005).

2 BIOLOGICAL RESEARCH

The applications of radioactive isotopes in biological research began with George de Hevesy. He was awarded the Nobel Prize in Chemistry 1943 for, in the words of the Nobel Committee, "his work on the use of isotopes as tracers in the study of chemical processes". De Hevesy was among the first to discover in 1913 that radioactive isotopes of an element could be used to trace the chemical processes of that element in inert and in living systems. Because radioisotopes are chemically inseparable from the stable isotopes of a given element, that is, they possess the same electron shell, de Hevesy demonstrated that a radioactive isotope in trace amounts could be mixed with the stable isotope of the same element and all of the chemical or biological transformations of that element could be followed by detecting and measuring the radiation emitted by the radioactive tracer isotope. George de Hevesy's discovery of the application of isotopes as tracers started when he traveled to Manchester, England in 1910 to study under Ernest Rutherford. At Manchester, Rutherford had a large supply of radioactive-lead containing Radium D (RaD), which had been donated by the Austrian Government. Rutherford gave George de Hevesy the assignment of separating the RaD from lead. The RaD was of little use as a radiation source because the lead would absorb much of the radiation. Thus, one day, as related by de Hevesy (1944), Rutherford addressed de Hevesy in the basement of the Institute of Physics at the University of Manchester where the radioactive lead was stored saying "My boy, if you are worth your salt, you try to

George de Hevesy (1885-1966)

separate RaD from all that lead." George de Hevesy was an enthusiastic budding scientist and made numerous attempts working for almost 2 years at the task of separating chemically the RaD from the lead. Having failed completely at this task, and to make best of a depressing situation, he decided that RaD could not be separated chemically from lead and thus could be used as a radioactive tracer to study the chemical pathways of stable (nonradioactive) lead. De Hevesy was very correct in his conclusion, as it was later discovered that RaD is a radioactive isotope of lead, namely, Pb-210, which cannot be separated chemically from the element lead. De Hevesy then went on to the Vienna Institute of Radium Research, as they had more radium at their disposal than any other institution. There he met with Frederic Paneth, who had also carried out abortive tests to separate RaD from lead.

George de Hevesy's first experiment with radioisotope tracers was carried out with Frederic Paneth in 1913 at the Vienna Institute. Here they were able to determine the solubility of highly insoluble lead compounds with the use of tracer amounts of radioactive lead (de Hevesy and Paneth, 1913a,b). George de Hevesy later demonstrated through numerous works how radioisotopes could be applied to determine the fate of inorganic and organic compounds in living systems such as plant, animal, and the human organism. For example, if we wanted to determine the pathways of a carbon atom in an organic compound within a living system, such as a certain atom of an amino acid, protein, nucleic acid, carbohydrate, lipid, vitamin, or other, the atom in that chemical compound need only be labeled with a radioactive isotope of carbon, namely, carbon-14. From the radiation emissions of the radioisotope tracer, the chemical pathways of the carbon atom could then be followed. Since the pioneering experiments of de Hevesy up to this day, countless biological pathways of carbon, hydrogen, phosphorus, sulfur, and other elements in plants, animals, and microorganisms have been elucidated with the use of radioisotopes as tracers such as carbon-14, tritium (hydrogen-3), phosphorus-32, or sulfur-35, respectively. Where suitable radioactive isotopes are not available, such as is the case with nitrogen, the anabolic and catabolic reactions involving nitrogen in living systems are studied with the stable isotope, nitrogen-15. Since the first demonstration of George de Hevesy in 1913, who is the father of the isotope tracer technique, the examples that may be taken from the scientific literature are innumerable and the knowledge gained from these studies and applied to medicine, agricultural production, and the biological sciences, in general, has been invaluable. One can safely state that almost every biological cycle of carbon compounds that we know today, and that are vital for human health and nutrition, has been elucidated with the use of the aforementioned radioactive isotopes as well as many stable isotopes (e.g., ^{2}H, ^{13}C, and ^{15}N). George de Hevesy also pioneered the use of stable isotopes as tracers in the biological and chemical sciences, after deuterium, the heavy stable isotope of hydrogen, was discovered by Harold Urey in 1932. Biological cycles such as carbohydrate metabolism, the Krebs cycle, lipid metabolism, photosynthesis, biosynthesis and modes of action of nucleic acids and proteins, metabolic regulation, etc. have been studied with the aid of radioactive isotope tracers.

Melvin Calvin (1911-1997)

A classic example of the use of a radioactive tracer in the biological sciences can be taken from the work of Melvin Calvin, who together with coworkers used the radioactive isotope carbon-14 as a tracer for CO_2 to elucidate the initial biochemical pathways in plant photosynthesis. He was awarded the Nobel Prize in Chemistry 1961 for, in the words of the Nobel Committee, "his research on the carbon dioxide assimilation in plants".

It was, of course, known then that plants needed carbon dioxide and light to grow; but knowledge was lacking on the biochemical mechanisms whereby plants could capture CO_2 from the atmosphere and assimilate the carbon from this gaseous substance into the building blocks of plant matter and food upon which we depend dearly. Melvin Calvin and coworkers (Calvin and Benson, 1948; Calvin, 1949, 1953; Bassham *et al.*, 1950; Calvin *et al.*, 1950; Benson *et al.*, 1952) carried out experiments by feeding carbon dioxide labeled with the radioactive tracer carbon-14 (i.e., $^{14}CO_2$) to photosynthesizing plant chloroplasts over different time intervals. During short exposure periods ($\sim$1 sec), the major fraction of the radioisotope label was encountered in the three-carbon (C_3) compound, phosphoglyceric acid (PGA, **3**) as shown below; this was concluded to be the first product of photosynthesis. PGA was found to be a product of the condensation of the $^{14}CO_2$ with ribulose 1,5-biphosphate (**1**) catalyzed by a carboxylase enzyme to produce the 2-carboxy-3-ketoribitol 1,5-biphosphate intermediate (**2**), which then undergoes hydrolysis to PGA (**3**). The asterisks in the illustrated reactions mark the positions of the carbon-14 radioisotope label.

The first step of the above Calvin–Benson–Bassham cycle, producing a C_3 compound (**3**) as a first product of photosynthesis, was considered to be the general pathway for CO_2 fixation in all plants. However, subsequent work of Kortschak, Hartt and coworkers (Kortschak *et al.*, 1965; Kortschak and Hartt, 1966) with sugarcane showed that, for very short periods of photosynthesis with $^{14}CO_2$, the carbon-14 label was found mostly in the four-carbon compounds (C_4) aspartic (**4**) and malic (**6**) acids, with only a small portion of the isotope label residing in the three-carbon PGA (**3**). For example, when an attached sugarcane leaf was exposed to $^{14}CO_2$ for 2 sec and killed with 80% alcohol, the radioactivity intensities of the ^{14}C label in malate, aspartate, PGA, and hexose monophosphates were 54, 37, 7, and 2%, respectively. The pathways of carbon in this newly discovered cycle were elucidated by Hatch and Slack (1966) by counting the radioactivity in the specific carbon atoms of malate, aspartate, and PGA after isolation of the compounds from the plant and selective chemical degradation of their carbon atoms.

The above work on photosynthesis clearly distinguished two types of plants, namely C_3 and C_4 plants and this knowledge together with measurements on the energy requirements for the various biochemical reactions involved were used to improve crop production around the globe as a function of climate variations among other factors. For example, the

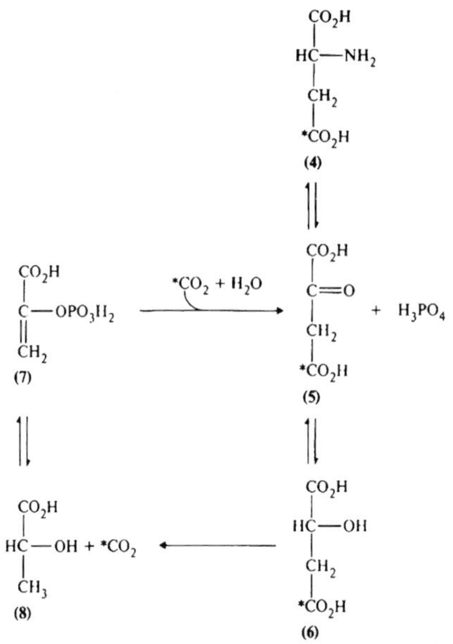

$$
\begin{array}{c}
CO_2H \\
| \\
HC\!-\!NH_2 \\
| \\
CH_2 \\
| \\
{}^{\bullet}CO_2H \\
(4)
\end{array}
$$

$$\Updownarrow$$

$$
\begin{array}{ccccc}
CO_2H & & & CO_2H & \\
| & & {}^{\bullet}CO_2 + H_2O & | & \\
C\!-\!OPO_3H_2 & \xrightarrow{\hspace{2cm}} & & C\!=\!O & +\ \ H_3PO_4 \\
\| & & & | & \\
CH_2 & & & CH_2 & \\
(7) & & & {}^{\bullet}CO_2H & \\
& & & (5) &
\end{array}
$$

$$\Updownarrow \qquad\qquad\qquad \Updownarrow$$

$$
\begin{array}{ccccc}
CO_2H & & & CO_2H & \\
| & & & | & \\
HC\!-\!OH + {}^{\bullet}CO_2 & \longleftarrow & & HC\!-\!OH & \\
| & & & | & \\
CH_3 & & & CH_2 & \\
(8) & & & {}^{\bullet}CO_2H & \\
& & & (6) &
\end{array}
$$

C_4 plants like sugarcane and maize display a higher net photosynthesis in regions of strong light intensity such as in the tropical regions of Brazil where sugarcane and alcohol production for fuel is a viable energy source. In temperate regions, where light intensities are lower, the C_3 plants have higher rates of net photosynthesis and a potential for high crop production.

Radioactive isotopes are sensitive tracers that are helpful in the elucidation of biochemical pathways. For the method to be conclusive, it is necessary to start with a compound precursor labeled with a radioisotope, for example, ^{14}C, ^3H, or ^{32}P. Biological transformations of the radioactive precursor molecule may lead to a number of product compounds, which would also be radioactive, if the atom or atoms that are labeled in the precursor also reside in the product compounds. Thus, if we started with an organic compound A labeled with ^{14}C, which undergoes transformation to product compounds B, C, and D, etc., and if compounds B, C and D are also radioactive, we can conclude that the isotope-labeled carbon atom or atoms in the product compounds originated from compound A. This is illustrated by the following reaction sequence:

$$[^{14}C]\text{-Molecule A} \rightarrow [^{14}C]\text{-Molecule B} \rightarrow [^{14}C]\text{-Molecule C} \rightarrow \ldots$$

The tracer technique is useful in determining very complex sequences of biological reactions such as the intricate carbon transformations that occur in plant photosynthesis described above. The method is complicated by the need to isolate each product compound and determine their molecular structures in addition to measuring the radioactivity intensity emanating from the radioisotope label in each compound. Although the technique may be tedious, the radioisotope tracer method is vital in proving the origin of compounds that

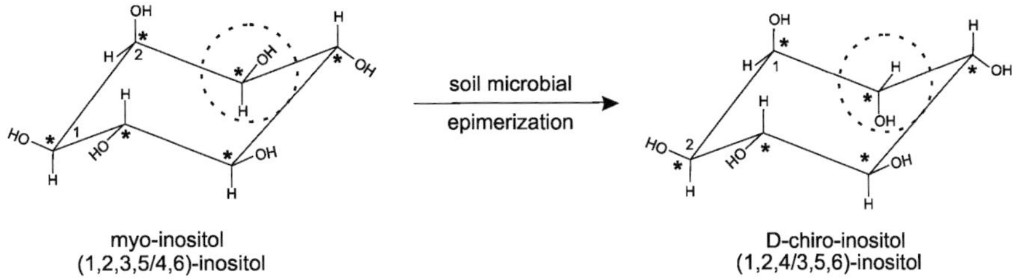

Figure 5 The microbial epimerization of uniformly label [^{14}C]myo-inositol to [^{14}C]chiro-inositol in soil. The asterisks represent the locations of the carbon-14 radioisotope labels and the dotted circles enclose the only carbon atom that differs in the structure of the two molecules. (From L'Annunziata *et al.* (1977).)

are very similar in structure. For example, the writer and coworkers (L'Annunziata *et al.*, 1977) demonstrated the biochemical transformations shown in Figure 5 using carbon-14.

The two molecules illustrated in Figure 5 are stereoisomers. The gross similarities of the two molecules, which share the same chemical formula, differ only in the spatial orientation of the proton and hydroxyl group at one carbon atom. This would make the biological origin of the product compound difficult to ascertain without the aid of the radioisotope label. The intensities of the radioactivity in the precursor and product compounds can also give a quantitative measure of the degree to which the metabolism occurs.

Radioisotopes as tracers in the biological sciences remain a vital tool for the elucidation of anabolic and catabolic reaction pathways. Examples of the application of radioactive isotopes as tracers in the scientific literature since the first demonstration of George de Hevesy in 1913 are numbered in the millions. For additional information on applications of radioactive tracers in the biosciences, the reader may peruse books by Larijani *et al.* (2006), Volkman (2006), Wolfe and Chinkes (2004), Slater (2002), Cobelli *et al.* (2006), Billington (1992), and Wolfe (1992).

3   FOOD AND AGRICULTURE

Scientific research has been vital to improving the quantity and quality of food needed to meet the growing demand of the world population. Radioactive and stable isotopes and nuclear radiation serve as useful tools in research aimed at increasing world food production. In the Foreword of a review book on *Isotopes and Radiation in Agricultural Sciences* (L'Annunziata and Legg, 1984a,b), Hans Blix, then Director General, International Atomic Energy Agency (IAEA), Vienna, wrote the following:

> Nuclear techniques in agricultural research have played a major role in increasing world food production to the level at which it is today. A few areas in which these techniques, largely based on the use of isotopes and radiation, have attained prominence are insect pest control, food preservation, plant nutrition, animal health and production,

crop management and pesticide residue studies. Nuclear methods have contributed in no small way to our understanding of biological processes and, when translated into practice, have made a subsequent impact in various fields of agriculture.... While there have been remarkable advances in the agricultural sciences in recent years, the need for increasing global food production is still critical and will become more so as time goes on. If further advances are to be made in this respect, our need for nuclear techniques will become even greater.

The applications of nuclear techniques aimed at increasing agricultural production are numerous and only a few examples will be cited here.

3.1   Insect pest control

In the field of insect pest control one of the greatest boons to improving agricultural production and human health, while at the same time improving the environment, has been the use of the sterile insect technique (SIT) to control or eradicate insect pests that cause damage to crops, livestock, and human health. The SIT involves the mass rearing of an insect pest and the sterilization of the reared insects by means of high doses of gamma radiation from a sealed source of a radioisotope such as cobalt-60. Following irradiation, the sterile insects, particularly the sterile male insects, are released into the wild by various techniques such as aircraft drops of irradiated insect pupae. The sterile male insects generally mate many times with female insects that they encounter in the wild, whereas the female insects mate only once. Thus, if sufficient male insects of a particular species are reared, sterilized, and released into the wild in a healthy and competitive albeit sterilized state, the population of that insect species can be made to drop. When sufficient numbers of the insect species are reared, sterilized, and dispersed over large regions of land, the insect species can be eradicated from a region. Reinfestation of the insect species could occur, but an SIT program can be maintained to keep the insect species out of a particular region, if not eradicate the species altogether. In an address given at the Opening Ceremony of the Environment Exhibition on the Occasion of the 50th General Conference of the International Atomic Energy Agency (IAEA) on September 19, 2006, in Vienna, Austria, Professor Dr Werner Burkart, Deputy Director General, Department of Nuclear Sciences and Applications noted the following about the sterile insect technique or SIT:

> Many countries worldwide now use the sterile insect technique, a proven nuclear technique, in controlling or even eradicating insect pests. It involves mass breeding of the insects, sterilizing them with gamma radiation and releasing them into the affected areas to mate with wild females, in effect, insect birth control. As no offspring are produced, the populations fall or disappear after repeated applications. How does this have a beneficial effect on the environment? The technique replaces the mass use of pesticides, which are harmful to the environment. With SIT the toxic agent remains in the laboratory, there is no contamination of ground, farm workers or products, and it's good for biodiversity because it is highly targeted towards only one pest, sparing bees and other beneficial insects.

Edward F. Knipling was the father of the SIT. He first developed the sterile male theory of autocidal control of the screwworm pest in 1938 after joining the Agricultural Research Service (ARS) of the United States Department of Agriculture in Menard, TX that year (USDA-ARS, 2006). The screwworm fly is a destructive pest of livestock, which was causing losses of several hundreds of millions of dollars annually in the USA (Kloft, 1984). The USDA-ARS also reports that Edward F. Knipling and coworkers R.C. Bushland, H.J. Muller, and E.D. Hopkins, among many others, began research on the radiation sterilization of the screwworm in 1950 followed with field tests on Sanibel Island, FL during 1951–1953. They then successfully eradicated the screwworm from the island of Curaçao, Dutch East Indies, off the coast of Venezuela in 1954 using radiation sterilized flies reared at the laboratory in Orlando, FL. The sterile flies were shipped and released on the island of Curaçao marking the first successful eradication of an insect pest by SIT. As reported in the Nuclear Technology Review (2004), the screwworm pest was eradicated from all of North and Central America by use of the SIT, and it is estimated that the eradication has provided annual benefits to the livestock industry in the region, through improved livestock productivity and health, that exceed the overall investment in the eradication campaign of over 45 years.

Knipling (1955) was the first to publish the technique of insect control or eradication through the use of sexually sterile males. In his historic paper he demonstrated mathematically that insecticide applications become less efficient and the SIT more efficient as the insect population is reduced. Edward Knipling subsequently published in 1957, 1959, and

E. F. Knipling (1909-2000)
reprinted with permission
from US National Academy
of Sciences, © 2003

1960 more earthshaking papers on SIT with titles such as "Controlled screwworm eradication by atomic radiation", "Sterile-male method of population control", and "Use of insects for their own destruction". He had a very long and successful career in the continued development of his SIT technique for the control and eradication of numerous types of insect pests as well as other methods of insect control, and he published numerous works as recently as 1998, 2 years before his passing. Knipling did not receive the Nobel Prize for his discovery of the SIT, but certainly was deserving of such a prize, as the SIT has proven to be a successful technique for the control and eradication of numerous harmful insect pests around the world. He was admitted to the US National Academy of Sciences, recipient of the Presidential Medal for Merit, the FAO Medal and World Food Prize and Japan Prize as well as numerous honorary doctoral degrees (Adkisson and Tumlinson, 2003). For additional information on the life and work of Knipling the reader is invited to peruse an historical account published by Waldemar Klassen (2003), a former coworker of Knipling and former colleague of the writer.

The SIT has demonstrated to be a useful method for the control of numerous insect pests in addition to the screwworm, including the Mediterranean fruit fly or "Medfly", olive fruit fly, onion fly, pink bollworm, codling moth, cactus moth, boll weevil, stable fly, horn fly, certain species of mosquito, and the tsetse fly. The control and eradication of the tsetse fly is vital to the development of certain countries of Africa where the insect acts as a vector of the "sleeping sickness" affecting the livestock and human population. The Joint FAO/IAEA Division of the United Nations Organizations in Vienna, Austria overseas around the world SIT control and eradication programs against the Medfly, tsetse fly, and malaria mosquito. The use of SIT to control the malaria mosquito is a relatively new and vital challenge for the scientific community. The Joint FAO/IAEA Division initiated a project in 2004 (Nuclear Technology Review, 2006) to assess the feasibility of using the SIT for mosquito control. Research is now focused on the mosquito species *Anopheles arabiensis*, which is the second most important vector of malaria in Africa. As reported in the Nuclear Technology Review (2006) of the IAEA, it is anticipated that by the year 2010, the technique of mass rearing of the mosquito will be improved to the point where sterilized males will be applied to the field in pilot tests. The control or even eventual eradication of malaria in Africa would be one of the greatest achievements of the century for the betterment of mankind.

3.2 Fertilizer and water use efficiency

Another field of agriculture, where radioactivity and nuclear techniques have played vital roles in development, is in research on the optimization of crop nutrition and water use efficiency. The efficient use of fertilizer to maximize crop production is measured with radioactive and stable isotope tracers and the most cost-effective application of water for crop irrigation, upon which optimum fertilizer utilization depends, is measured with neutron radiation. As noted by Menzel and Smith (1984):

> Increasing fertilizer costs and the necessity for minimizing environmental pollution have given added impetus to fertilizer disposition studies … tracer fertilizers [i.e., isotope-labeled fertilizers] provide the only definitive means for determining both

the behavior and fate of applied nutrients. A basic assumption is that the isotopically labeled element behaves in an identical manner to the non-labeled element both physically and chemically, and a plant cannot distinguish between [isotope] labels.

In research studies on fertilizer use efficiency, radioactive isotopes are not applied in the open field; however, greenhouse experiments with radioactive isotopes of plant nutrients (e.g., ^{32}P, ^{33}P, ^{35}S, and ^{65}Zn) can provide research scientists with the information needed to make recommendations to the farmer to help achieve the optimum utilization of plant nutrient elements from soil and fertilizer. For example, a research scientist may use a fertilizer labeled with the radioisotope of phosphorus, ^{32}P, and apply it to the soil to measure percentage utilization of the fertilizer nutrient by plant crops. The radioactive fertilizer phosphorus will have a specific activity defined as the intensity of the radioactivity per unit weight of the element or

$$^{32}P \text{ specific activity} = \frac{DPM}{gP} \tag{1}$$

where DPM is the radioactivity in units of disintegrations per minute, and gP represents the weight of the element phosphorus in grams. The weight of the phosphorus in the fertilizer is the sum of the weights of the radioactive isotope ^{32}P plus the natural stable isotope ^{31}P. However, the weight of the radioisotope tracers used in such studies is so small that the contribution of the radioisotope ^{32}P weight to the total weight of phosphorus can be neglected. (See Eq. (8.59) of Chapter 8, which demonstrates that 1 mCi of ^{32}P comprises only 3.5 ng or 3.5×10^{-9} g.) When the researcher adds the radioactive fertilizer to the soil, the plant will absorb phosphorus from two sources, namely phosphorus derived from the fertilizer (PDFF) and phosphorus derived from the soil (PDFS). The specific activity of the radioactive tracer ^{32}P in the plant will be lower than that of the radioactive fertilizer, because of the absorption of stable ^{31}P from the soil by the plant. The reduction in ^{32}P specific activity is referred to as isotope dilution, and the degree of isotope dilution will serve to measure the proportions of phosphorus in the plant that came from the fertilizer and from the soil from the following:

$$\%PDFF = \frac{^{32}P \text{ specific activity in plant}}{^{32}P \text{ specific activity in fertilizer}} \times 100 \tag{2}$$

where %PDFF is the percent phosphorus in the plant derived from the fertilizer. The remaining phosphorus in the plant would be derived from the soil or

$$\%PDFS = 100\% - \%PDFF \tag{3}$$

In the case of the element N, which is another major plant nutrient, there is no suitable radioactive isotope that could be used by researchers to assess fertilizer use efficiency. In this case, they use the stable isotope of nitrogen or ^{15}N as the tracer. In nature the element nitrogen consists of a mixture of two isotopes, $^{14}N + ^{15}N$ with percent abundances of 99.634 and 0.366%, respectively. There is consequently much more ^{14}N than ^{15}N in nature.

However, researchers have been able to enrich the isotope ^{15}N in samples of natural nitrogen to produce nitrogen with a %^{15}N abundance greater than 0.366% so that this stable isotope may be used as a tracer in scientific experiment such as fertilizer use efficiency studies. (See Radioactivity Hall of Fame—Part V in this book, for the biographical sketch on Gustav Hertz, who was the first to devise a method of isotope enrichment.) The amount of enrichment of ^{15}N in excess of its natural abundance is defined as

$$\%{}^{15}N \text{ atomic excess} = \%{}^{15}N \text{ in the sample} - 0.366\% \qquad (4)$$

Thus, as in the previous example taken in research with fertilizers labeled with a radioactive isotope, scientific researchers can use a fertilizer with a known %^{15}N atomic excess to measure the plant utilization of the nitrogen from the fertilizer according to the following equation:

$$\%NDFF = \frac{\%{}^{15}N \text{ atomic excess in plant}}{\%{}^{15}N \text{ atomic excess in fertilizer}} \times 100 \qquad (5)$$

where %NDFF is the percent nitrogen in the plant derived from the fertilizer. The %^{15}N atomic excess in the plant is always less than that of the fertilizer, because the plant roots will also absorb nitrogen from the soil, which will contain ^{14}N at its natural abundance. Thus, the remaining nitrogen in the plant must be derived from the soil or

$$\%NDFS = 100\% - \%NDFF \qquad (6)$$

While the scientist measures fertilizer use efficiency with the aid of isotope techniques, he or she will also study the conditions that can optimize the plant nutrient use efficiency such as the soil moisture. A source of fast neutron radiation is used by scientists to measure and monitor the water content of soils in the field.

In Chapter 4 of this book, the reader will find much information on the sources and properties of neutron radiation. Neutrons are classified according to their energy, and fast neutrons possess energies in the range of ~200 keV to 10 MeV, whereas slow (thermal) neutrons are lower in energy in the range of ~0.003–0.4 eV. As discussed in Part IV of Chapter 4, collisions of fast neutrons with nuclei can result in the scattering of the neutrons yielding recoil nuclei whereby part of the kinetic energy of the neutrons are transferred to the nuclei with which they collide. The fast neutron will lose more energy when colliding with a nucleus of similar mass, that is, the proton nucleus (^1H), which is similar to the collision of two billiard balls. One billiard ball can transfer almost all of its energy to another billiard ball. Less energy will be lost by collision of a fast neutron with a nucleus of higher mass, for example ^{28}Si, somewhat like the collision of a billiard ball with a bowling ball.

The scientist will lower a source of fast neutrons in a narrow well in the soil as illustrated in Figure 6, and in so doing, the neutrons that are emitted will collide with the various nuclei of the soil matter as well as with nuclei of soil water molecules. A sphere or cloud of slow (thermal) neutrons is created around the source of fast neutrons through numerous collisions of the fast neutrons with the nuclei of water molecules as well as with nuclei of other

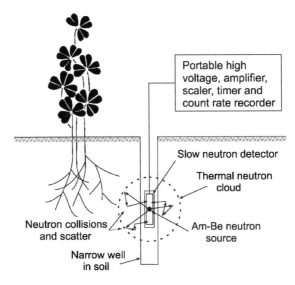

Figure 6 Schematic illustration of a neutron soil moisture gauge. (Adapted from Nielsen and Cassel (1984), reprinted with permission from Elsevier © 1984.)

elements in the soil as illustrated in Figure 6. Because water molecules are high in protons compared with soil minerals, the rate of attenuation of the fast neutrons to slow neutrons will be a function of the moisture content in the soil. The slow neutrons are counted with a gas ionization detector, which is located next to the source of fast neutrons. The rate of production of slow (thermal) neutrons is calibrated with soil moisture content. In this fashion, irrigation can be programmed when needed to maximize crop yields with the most efficient use of fertilizer while conserving water resources.

In an address entitled "Nuclear Technologies for the Environment" given at the Opening Ceremony of the Environment Exhibition on the Occasion of the 50th General Conference of the International Atomic Energy Agency (IAEA) on September 19, 2006, in Vienna, Austria, Professor Dr Werner Burkart stated the following:

> Agriculture is the biggest user of the world's scarce water resources, and as an industry, one of the most wasteful. Any measures that can be taken to reduce the wastage can have dramatic effects on not only people's health and well being, but also on the environment. Neutron moisture gauges can measure the hydrogen component of water in both the plant and the surrounding soil. They are thus ideal instruments to help farmers optimize their irrigation, fulfilling the vision of our Joint FAO/IAEA Division of "MORE CROP PER DROP".

Numerous articles and books are available on the nuclear techniques used to optimize crop nutrient and water use efficiency including a review book edited by L'Annunziata and Legg (1984a) and more recent books published as proceedings of international symposia (IAEA, 2000, 2002, 2005e).

3.3   Animal health and production

The IAEA Nuclear Technology Reviews (2004, 2005, 2006) outline the scope of applications of nuclear techniques in scientific research around the world directed to the improvement of animal productivity and health. Among these applications is the traditional radioimmunoassay (RIA), which is an analytical method capable of measuring very low concentrations (femto-molar or 10^{-15} M) of hormones and other antigens in living systems (e.g., blood or plasma) utilizing antigens labeled with a radioactive isotope. The RIA technique was discovered by Rosalyn Yalow (1921–) and Solomon Berson (1918–1972), which they demonstrated to the scientific community in 1959. Rosalyn Yalow was awarded the Nobel Prize in Physiology and Medicine 1977 for, in the words of the Nobel Committee, "the development of radioim-munoassay of peptide hormones". She is honored by the postage stamp, illustrated here, which was issued by Sierra Leone. Yalow and Berson demonstrated that the binding of a radioisotope-labeled antigen, such as insulin, to a fixed concentration of antibody is a quantitative function of the amount of insulin present. This formed the basis for the quantitative analysis of insulin in human plasma (Yalow and Berson, 1959). RIA was described by Yalow in her Nobel Lecture (Yalow, 1977) given on December 8, 1977, when she stated the following:

> Radioimmunoassay (RIA) is simple in principle.... The concentration of the unknown unlabeled [non-radioactive] antigen is obtained by comparing its inhibitory effect on the binding of radioactively labeled antigen to specific antibody with the inhibitory effect of known standards.

The method is outlined by the competing reactions illustrated in Figure 7.

Rosalyn Yalow (1921-   )

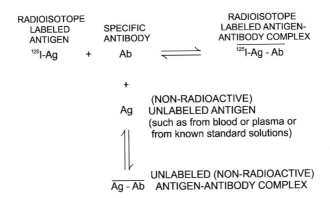

Figure 7 Competing reactions that form the basis of RIA. (Adapted from Yalow (1977) with permission The Nobel Foundation © 1977.) The result of mixing a known quantity of radioactive antigen (e.g., ^{125}I-Ag), such as a hormone, with a known quantity of its antibody (Ab), results in the formation of a specific amount of the radioactive antigen attached to the antibody as a radioactive antigen–antibody complex (e.g., ^{125}I-Ag-Ab). If one then adds a small amount of blood or plasma that contains the antigen (Ag), this newly added antigen will also become attached to the antibodies (Ab) resulting in a competitive removal of a certain amount of the radioactive antigen (^{125}I-Ag) from its antibodies. The amount of radioactive antigen removed from its antibodies, measured by a radioactive detector, is a function of the amount of antigen present in the blood or plasma sample. Following calibration with known standards of antigen, the amount of reduction in radioactivity of the ^{125}I-Ag-Ab complex can be correlated to an exact amount of antigen in the blood or plasma sample.

The radioactivity of antigen–antibody complex is measured in modern scintillation analyzers. The reader may wish to peruse a previous book edited by the writer (L'Annunziata, 2003c) where these techniques are described in detail. The method is very sensitive. In his presentation speech of the Nobel Prize, Professor Rolf Luft stated the following:

Rosalyn Yalow's name is forever associated with her methodology of measuring the presence of hormones in the blood at concentrations as low as one thousand billionths of a gram per milliliter of blood.

Yalow and Berson's first report on the discovery of RIA was concerned with the analysis of insulin (Yalow and Berson, 1959). In the *Mayo Clinic Proceedings* Kyle and Shampo (2002) noted the following concerning the discovery:

Yalow's report was initially rejected by a journal editor, who did not believe that insulin could produce antibodies. The radioimmunoassay technique proved to be so sensitive that a Swedish scientist likened its power to detecting half a lump of sugar in a lake 62 miles square and 30 feet deep.

The IAEA Nuclear Technology Reviews (2004, 2006) point out that RIA, which employs radioisotopes in the measurement of a given molecule in a biological system, remains the dominant technology in the field of animal reproduction and breeding. The radioisotope

^{125}I is commonly employed in RIA to determine the concentrations of hormones and metabolites in the blood or milk of farm animals. For example, RIA of the reproductive hormone progesterone is reported to be an indispensable tool that provides information needed by farmers for breeding management programs, and it can help monitor the utility of the artificial breeding services provided to farmers by government, cooperative, or private organizations. The RIA is also reported to be a very simple and robust technique for the analysis of harmful residues (antigens) in food of animal origin, such as products resulting from the misuse of veterinary drugs.

The Nuclear Technology Reviews (2004, 2006) point to new frontiers where radioactive tracers are applied to studies of animal health and production. Among these are molecular diagnostics, which increase the sensitivity of testing and detecting animal diseases. Technologies such as ^{35}S/^{32}P dual radioisotope phosphor-imaging gene sequencing and the incorporation of radioactive ^{32}P, ^{33}P, or ^{35}S into short DNA synthetic probes to permit the identification of genes that influence animal traits of interest and parentage. Such testing can help determine which animals carry the superior forms of genes of interest to improve accuracy of selection and increase animal productivity (Nuclear Technology Review, 2006). Upham and Englert (2003) provide detailed information on some of the analytical techniques used in this field.

3.4 Plant breeding

Plant breeding is the science of the manipulation of plant species, through pollination, genetic engineering, and selection of progeny, for the purpose of creating new and improved plant genotypes and phenotypes with desirable characteristics such as disease or insect pest resistance, salt or draught tolerance, crop quality, or increased yields, etc. Mutations induced in plants with the aid of gamma rays, x-rays, or fast neutrons, or by means of chemical mutagens aid the plant breeder by increasing germplasm variability.

The significant role that plant breeding can have in the improvement of crop quality and yield was highlighted by the postage stamp issued in Canada in 1988, which is illustrated here. The stamp commemorates the national breeding program that generated the Marquis wheat in 1909. The new variety obtained through plant breeding provided best baking quality, higher yields, earlier maturity, and shorter straw than other wheat varieties (Morrison, 1960; Fedak, 2002).

The impact of plant breeding in alleviating starvation in the world by the creation of new crop varieties with improved crop quality and yields was recognized the world over through the work of geneticist Norman Borlaug (1914–), who was awarded the Nobel Peace Prize 1970 for his "green revolution" born from the impressive results he achieved in wheat improvement through plant breeding, and the organization of the exploitation of the results of this improvement for world agriculture particularly in the developing world. Borlaug's work drew the attention of the world to the importance of plant breeding. Mrs Aase Lionaes (1970), Chairperson of the Nobel Committee, on the presentation of the Nobel Prize on December 19, 1970, underscored the significance of Norman Borlaug's work as follows:

Dr. Borlaug went to the International Maize and Wheat Improvement Center in 1944....
Ever since that day, twenty five years ago, when Dr. Borlaug started his work on the

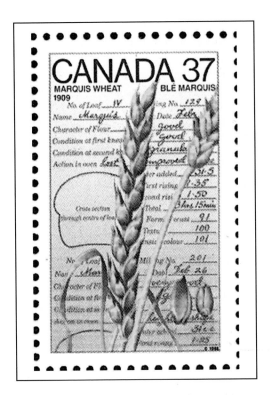

improvement of grain, and right up to the present, he has devoted all his energy to achieving the historical result which today is referred to as the "green revolution". This revolution will make it possible to improve the living conditions of hundreds of millions of people in that part of the globe which today might be called as the "non-affluent world"....

The purpose of this research, Dr. Borlaug continues, was to endeavor to develop a variety of wheat with greater yields, with a great degree of resistance to diseases, and with qualities that rendered it suitable for use in connection with improved agronomic methods, that is to say, the use of artificial fertilizers, improved soil culture, and mechanization. The results of the concerted attack launched by the team of scientists on all these problems was the new Mexican breeds of wheat ..., which produce astonishingly large yields, which are resistant to disease, and which facilitate intensive use of fertilizers. Unlike previously known breeds of wheat, the new types can be transferred to remote parts of the world that differ in climate.

The most important event ... was the development of the so-called "dwarf varieties". After years of research on the part of Dr. Borlaug and his collaborators to develop, by crossing and selection, the so-called Japanese breed of wheat, they evolved the new world-famous "dwarf variety". These are breeds of wheat which, unlike previously known long-bladed varieties, have short blades .... The new dwarf varieties ... provide an increase of yield per decare from the previous maximum of

Norman E. Borlaug (1914-   )

450 kilos to as much as 800 kilos per decare. These varieties can be used in various parts of the world because they are not affected by varying lengths of daylight. They are better than all other kinds in both fertilized and non-fertilized soil, and with and without artificial irrigation. In addition they are highly resistant to the worst enemy of wheat, rust fungus or oromyces. Thanks to these high-yield breeds of wheat, Mexico was self-supporting in this grain in 1956, and in recent years this country has exported several hundred thousand tons annually …

Once the seed corn had been introduced [into Pakistan] and had yielded superb results in the form of increased crops, the triumphal march of the green revolution was ushered in. Pakistan's present-day wheat production amounts to several million tons, and the country is self-supporting in wheat …

The highest results [of wheat production] in the history of India was achieved in 1968 [after the introduction of Borlaug's new wheat] with a crop of seventeen million tons …. After the successful results achieved in Mexico, India, and Pakistan, the new varieties of wheat were introduced into certain parts of Turkey, Afghanistan, [Islamic Republic of] Iran, Iraq, Tunisia, Morocco, and Lebanon …. The new variety of wheat will be able to affect a total transformation of the economic picture in the developing countries.

The induction of mutation in plants by nuclear radiation has proven to be a useful tool for plant breeders for crop improvement. Mutation induction for increasing germplasm variability is carried out with gamma rays, x-rays, and fast neutrons, in addition to chemical mutagens. As reported in the IAEA Nuclear Technology Review (2006) other nuclear techniques used are the radioisotope labeling of nucleic acids used as probes for genetic fingerprinting, mapping and marker-assisted selection, and mutagenesis for the analysis of gene function.

The Joint FAO/IAEA Division in Vienna, Austria carries out extensive coordinated research programs with member states utilizing nuclear and conventional techniques in mutation plant breeding. The induced mutations created by nuclear radiation and chemicals have led to major advances in plant breeding for crop improvement. The beneficial mutants have been selected and used by plant breeders for over 50 years. The Nuclear Technology Review (2006) reports that, to date, approximately 2500 officially registered mutant varieties of more than 160 plant species worldwide are listed in the FAO/IAEA Mutant Variety Database. An example cited is a mutant rice cultivar with high quality and tolerance to salinity, which has been released in Vietnam. It is one of the top five export rice varieties, which occupies 280,000 ha of the export rice area of the Mekong Delta. The target area for the salt tolerant rice cultivar for Bangladesh, India, Philippines, and Vietnam encompass an estimated area of 4.3 million ha.

The IAEA Nuclear Technology Review (2005) reports that mutation induction coupled to selection remains the most "clean" and inexpensive way to create varieties by changing single characteristics without touching the general phenotype. Major successes are reported for mutagenesis-enhanced breeding in the USA (rice, barley, sunflower, grapefruit, peppermint), Pakistan (cotton), India (blackgram), Australia and Canada (linseed), Japan (pear), and China and Australia (rice). Mutant varieties of oil seeds and pulses are being released for commercial cultivation. Other changes created through induced mutation with nuclear radiation and chemicals are disease and pest resistance and crop nutritional and processing quality. There are an increasing number of private enterprises involved in research in mutation-induction breeding for ornamental flowers as well as industrial and food crops.

4   WATER RESOURCES

Water, like the air we breathe, is a valuable and vital resource. It is in great demand for domestic use, agricultural irrigation, and industrial productivity. Many people in many nations do not have adequate water supplies, and scientific research into maintaining and improving the supply and quality of our water is vital to human development and the conservation of the environment. In an address given at the Opening Ceremony of the Environment Exhibition on the Occasion of the 50th General Conference of the International Atomic Energy Agency (IAEA) on September 19, 2006, in Vienna, Austria, Professor Dr Werner Burkart stated the following concerning global water resources and the role of nuclear technology in helping the world meet its future water needs:

> Here in Vienna, there is an abundance of good quality water, sometimes too much pouring from the skies, but more than 1 billion people [worldwide] have no access to clean water, and predictions are that, without intervention, about two-third's of the world's population will face shortages by 2025. It is a goal of the World Summit on Sustainable Development to halve the number of people without access to clean water by the year 2015. Nuclear technology is part of that effort. By using naturally

occurring isotopes, we can trace the origin and movement of water and specific pollutants.... This information is difficult to obtain by other means, but is critical for developing robust and affordable policies for the protection of freshwater resources.

Also, in his opening statement at the International Symposium on Isotope Hydrology and Integrated Water Resources Management (IAEA, 2004a) held in Vienna during May 19–23, 2003, Professor Burkart added

> As many of us here are aware, the global demands for freshwater have been increasing much faster than the ability of many nations to meet these demands.... Currently, surface waters in rivers and lakes are being exploited to the maximum, and ground water resources are getting depleted or contaminated in many parts of the world. Extensive hydrological information is necessary to sustain the current and future levels of human development, to protect available water resources from pollution and over exploitation.... Isotope techniques help to provide rapidly hydrological information for large areas and will undoubtedly be a key component of global efforts in water resource assessment and management.

By tracing the changes in the isotopic composition of our water resources including isotopes of hydrogen (^1H, ^2H, and ^3H), oxygen (^{16}O, ^{17}O, and ^{18}O), carbon (^{12}C, ^{13}C, and ^{14}C), nitrogen (^{14}N and ^{15}N), and others, isotope hydrologists can obtain valuable data on hydrological dynamics that could otherwise be impossible or immensely difficult to obtain. Isotopes are used to measure the processes illustrated in Figure 8 of the hydrologic cycle in

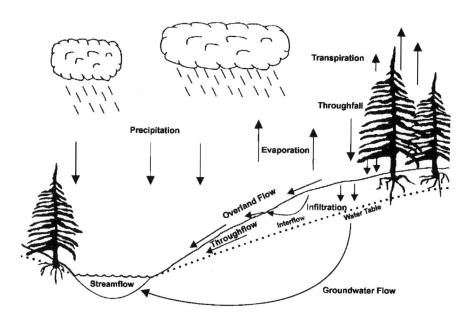

Figure 8 Catchment hydrologic processes composing a local hydrologic cycle. Each of these processes can be assessed by lithogenic and cosmogenic nuclides. (From Nimz (1998), reprinted with permission from Elsevier © 1998.)

catchment, including, the groundwater origins, the dating and renewal rate of groundwater, and sources of pollution or contaminant intrusions into groundwater, etc.

A thorough review of the isotope tracer techniques used in studies of catchment hydrology is provided in the book edited by Kendall and McDonnel (1998). A comprehensive review by Nimz (1998) cites numerous isotopes as tools for measuring the dynamics of catchment hydrolic processes such as variations in ^{87}Sr natural abundance, or $\delta\ ^{87}$Sr, which will vary in waters depending on the water origin (e.g., bedrock, soil leachate, rain, etc.) in the catchment. Precise measurements of variations in isotope natural abundances are made with a mass spectrometric analysis.

Another isotopic technique involves the measurement of ^{36}Cl/Cl, that is, the ratio of the isotope ^{36}Cl to the total chlorine in a sample, because the atmospheric input of ^{36}Cl to the chlorine content of water can be used as a groundwater tracer (Caffee *et al.*, 1992; Nimz *et al.*, 1993, 1997; Nimz, 1998). The thermonuclear atmospheric weapons tests during the 1950s and 1960s enriched the atmosphere with ^{36}Cl, which has a half-life of 3×10^5 years. Water older than 1950 (>50 years), which is water that has not mixed with water exposed to the atmosphere for over 50 years, would be lower in ^{36}Cl radioactivity than younger water. Thus, the ^{36}Cl/Cl ratios can be used to discriminate between older and younger water that may be recharged in a catchment.

The ^{36}Cl/Cl ratios of rainwater and groundwater have been used to determine the amount of water lost from catchments as a result of evapotranspiration (Paul *et al.*, 1986; Nimz, 1998; Magaritz *et al.*, 1990). Rainwater increases in solute, including chloride ion, concentration as it enters the soil profile, which is due to a combination of evapotranspiration and Cl acquired from soil mineral leaching. The evapotranspiration process will increase the chloride concentration, but it does not alter the ^{36}Cl/Cl ratio, whereas the soil mineral contribution of chloride does alter the ratio. The ratio would be reduced by the addition of Cl from soil minerals. Thus, by accurate measurement of the ^{36}Cl isotope and total chloride content of precipitation and groundwater, research hydrologists can factor out the soil mineral contribution of chloride to the soil water and calculate the chloride contribution due to evapotranspiration.

Nimz (1998) gives a detailed account of many applications of lithogenic and cosmogenic stable and radioactive isotopes, including ^{3}H, ^{6}Li, ^{7}Li, ^{7}Be, ^{8}Be, ^{10}Be, ^{10}B, ^{11}B, ^{14}C, ^{22}Na, ^{24}Na, ^{39}Ar, ^{41}Ca, ^{129}I, ^{204}Pb, ^{206}Pb, ^{208}Pb, ^{234}U, and ^{238}U to the measurement of the catchment hydrologic processes illustrated in Figure 8. For additional reading on detailed accounts of the isotope techniques used to assess and improve our water resources see a book chapter by Gonfiantini *et al.* (1998) and books by the IAEA (2004) and Singh and Kumar (2005).

5   MARINE RESOURCES

Radioactive and stable isotopes are vital tools used by the scientific community to study and preserve our marine resources. Relating to nuclear techniques to study and preserve our marine resources the following statement was made by Professor Dr Werner Burkart in the Opening Ceremony of the Environment Exhibition on the Occasion of the 50th General

Conference of the International Atomic Energy Agency (IAEA) on September 19, 2006, in Vienna, Austria:

> If we look to the large picture, climate change and global warming dominate many debates. The oceans are known to play a major part, transporting heat and acting as a sink for the carbon dioxide that we produce, and helping to regulate the earth's climate. The IAEA's Marine Environmental Laboratory in Monaco contributes to the global knowledge of these phenomena, through nuclear techniques to track ocean currents and to measure ocean climate coupling and carbon cycling. Stable and radioactive isotopes are also used to date to reconstruct past temperature, circulation and glacial events, providing the necessary data for the complex predictions that we need today.... Also in the marine environment, one of the more serious problems facing some coastal waters is called Harmful Algal Blooms, or HABs for short. These often grow because of rich nutrient mixes of chemicals from industry, dwellings and agriculture discharged to coastal areas. HABs can cause the entry of toxic substances into the human food chain. Paralytic Shellfish Poisoning is one of the potentially deadly outcomes, and so shellfish have to be monitored and, if necessary, supplies regulated. The traditional methods of testing have been wanting in terms of timeliness and cost, but a nuclear technique called receptor binding assay, which is rapid, more sensitive and cheaper will now ensure sustained and effective shellfish toxicity monitoring worldwide.

The receptor binding assay is a very sensitive analytical technique commonly used in the biological sciences for the measurement of minute concentrations (femtomolar or 10^{-15} M) of antigen in solution. The technique makes use of a microplate of sample wells made of white light-reflecting plastic. The wells are coated with plastic scintillator and can contain small volumes (80–1500 µl) of solutions to be analyzed for antigens. Molecules of a suitable receptor are adhered onto the surface of the plastic scintillator in an immobilized state as illustrated in Figure 9.

Once the receptor molecules are immobilized on the plastic scintillator, a radioisotope-labeled ligand is added. The isotope-labeled ligand will then bind with the receptor molecules and thereby come into close proximity with the plastic scintillator. Generally, a radioisotope that emits low-energy beta particles, such as ^{3}H ($E_{max} = 0.018$ MeV), or a radioisotope that emits Auger electrons, such as ^{125}I, is used as the isotope label. The low-energy beta-particle emissions from ^{3}H decay only travel about 1.5 µm in water before coming to a stop, and the Auger electrons only a little further. The ranges of travel of various types of nuclear radiation in different media are described in this book. The short range of travel of the decay emissions assures that only those emissions resulting from the radioactive decay of radioisotope-labeled ligand molecules bound to the receptor molecules will interact with the plastic scintillator resulting in the emission of visible light. The radiation emissions of radioisotope-labeled ligand molecules in the microplate sample-well solution will not reach the plastic scintillator. The photons of visible light emitted by the plastic scintillator are detected by a photomultiplier tube (PMT) above the sample well (see Figure 9). The PMT converts the light photons to an electric pulse and the pulses counted. If, however, we now add another competing ligand, which is not labeled with radioisotope, it will displace the

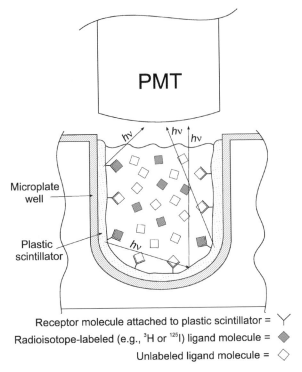

Receptor molecule attached to plastic scintillator = Y

Radioisotope-labeled (e.g., ^{3}H or ^{125}I) ligand molecule = ◆

Unlabeled ligand molecule = ◇

Figure 9 Artistic depiction of a magnified cross-section of a sample well in a microplate. The sample well is coated with plastic scintillator. Onto the surface of the scintillator is attached a layer of receptor molecules. When radioisotope-labeled antigen molecules bind to the receptor molecules, the binding brings the radioisotope into close proximity to the plastic scintillator. The energy of the short-range emissions from radioactive decay of the radioisotope-labeled ligand molecules will excite the plastic scintillator when the radioactive ligands molecules are bound to receptor molecules on the surface of the scintillator-coated microplate wells. The radiation energy absorbed by the plastic scintillator is emitted as photons ($h\nu$) of visible light, which travel through the solution and are detected by a photomultiplier tube (PMT).

radioisotope-labeled ligand on the receptor molecules according to the competitive affinity for the receptor and its concentration in the solution (see Figure 9). The displacement of radioisotope-labeled ligand from the receptor molecules into the medium solution of the sample well and away from the plastic scintillator results in a reduction in the radiation–scintillator interactions and a reduction in the number of photons emitted. The greater the concentration of unlabeled ligand in the solution, the greater will be the reduction in the emission rate of light photons from the plastic scintillator. Thus, with the aid of a standard curve, the unknown concentration of ligand in solution can be determined by the rate of emission of light photons.

The receptor binding assay described here is a type of analysis also referred to as scintillation proximity assay (SPA), because only receptor-bound radioisotope-labeled ligands in close proximity to plastic scintillator can be detected and counted. In countless bioscience laboratories around the world receptor binding assays are performed to analyze very minute concentrations of various types of antigens or ligand molecules. Receptor binding assays and SPAs are described in detail in an earlier book (L'Annunziata, 2003c).

Radioactive and stable isotopes are used to study the physical and chemical processes in our oceans that aid in studies of marine conservation. An interesting example described by Basu (2003) is the measurement of the ratio of the two stable isotopes of oxygen, $^{18}O/^{16}O$, to study the temperature changes of ocean waters that may be correlated with prehistoric climatic changes. Due to the higher mass of ^{18}O, its rates of reaction are slower than those involving ^{16}O in biochemical processes (i.e., isotope effect). Thus, the $^{18}O/^{16}O$ ratio in organic matter of the oceans would be lower than $^{18}O/^{16}O$ ratio of the seawater. At elevated temperatures, the reaction rates of the ^{18}O in living organisms would be faster and the $^{18}O/^{16}O$ ratios would thus increase. Certain types of amoebas, such as the Foraminifera, have been living in our oceans for millions of years. These form a siliceous shell around their bodies, which they have been leaving behind as an ever-accumulating layer of lime and silica at the ocean bottom. Examination of the $^{18}O/^{16}O$ isotopic ratios of the remains of these ancient organisms can provide information on the temperatures during prehistoric times when correlated with information provided by isotope dating (Bornemann et al., 2003; Mora, 2003; Mutterlose et al., 2005).

One dating method of ocean floors is the $^{40}Ar/^{39}Ar$ method, which is based on the decay of ^{40}K to ^{40}Ar (McDougall and Harrison, 1999; Koppers et al., 2000, 2003). The isotope ^{40}K decays with a half-life of 1.26×10^9 years providing a means of dating well beyond that possible by ^{14}C dating.

Numerous other isotope techniques are used to study the physical and chemical processes in our oceans including the application of natural isotope ratios, such as $^{87}Sr/^{86}Sr$, $^{206}Pb/^{207}Pb$, $^{143}Nd/^{144}Nd$, and $^{187}Os/^{186}Os$ to answer geologic questions of plate tectonics based on the isotopic composition of seawater and that of the mineral composition of the sea bedrock (National Academy of Sciences, 1992). Comprehensive treatments of the applications of radioactive and stable isotopes to marine studies can be obtained from recent books such as works by Carroll and Lerche (2003), Povinec and Sanchez-Cabeza (2006), and Schulz and Zabel (2006).

6   INDUSTRIAL APPLICATIONS

The industrial applications of nuclear radiation are widespread throughout the world and vital to our well-being, and only a few will be noted here. Among the many applications are (i) the research reactor- and cyclotron-production of radioisotopes required for radiopharmaceuticals in medical diagnosis and treatment of cancer and other infirmities described earlier in this introduction, (ii) neutron radiation applied to the non-destructive testing of materials, (iii) the sterilization of medical products with high levels of gamma radiation, and (iv) the use of electron beams to process materials such as needed in rubber vulcanization and wood curing.

6.1   Research reactors and accelerators

Research reactors and accelerators are vital to the production of neutrons for industrial applications. This book describes the discovery of neutrons by James Chadwick in 1932, the properties of neutron radiation, and the production of neutrons by various means including controlled nuclear fission in a research reactor and by means of the charged particle accelerator, such as the charged particle accelerator first conceived by Ernest Lawrence in 1929. Research reactors and the cyclotron are major sources of neutrons utilized in the production of radioisotopes, and furthermore, as described in this book, the cyclotron can produce a wide range of radioactive isotopes depending on the particle accelerated (proton, deuteron, or helium nucleus) against different target materials.

Many of the radioactive isotopes produced by either the research reactor or cyclotron are vital to our well-being, as these are used at hospitals and clinics for medical diagnosis and treatment, such as the medical imaging and radiotherapy techniques described earlier in this introduction. Among some of the radioactive isotopes required in medicine are ^{99m}Tc, used mostly in medical diagnosis, ^{67}Ga, ^{131}I, and ^{201}Tl used as seeds in the treatment of prostate cancer, other therapeutic agents containing ^{90}Y, ^{153}Sm, ^{186}Re, ^{165}Dy, and ^{177}Lu, and positron-emitting radionuclides, ^{18}F, ^{11}C, and ^{68}Ga required for the medical imaging of the human body permitting the detailed visualization of the organs within the human body and attached tumors, and ^{60}Co for radiation therapy, just to mention a few. Other radioactive isotopes are needed for important research in the biological and agricultural sciences for which some of the applications were cited previously in this introduction. There are currently 246 research reactors operating in 69 countries of the world and approximately 300 cyclotrons operating worldwide. Small cyclotrons are located at hospital facilities where short-lived radioisotopes may be produced on-site for the diagnosis and treatment of various diseases.

Research reactors also serve as a source of neutron beams applied to research and industrial needs for non-destructive testing and analysis of materials including techniques such as neutron scattering and diffraction for atomic structure determinations, neutron radiography, and neutron activation analysis. These are described briefly in the following paragraphs.

6.2 Neutron diffraction and scattering

The importance of neutron beam technology for human development and well-being was recognized by the Nobel Committee when it awarded the Nobel Prize in Physics 1994 to Bertram N. Brockhouse (1918–2003) and Clifford G. Shull (1915–2001) for, in the words of the Nobel Committee, "pioneering contributions to the development of neutron scattering techniques for studies of condensed matter" and particularly to Bertram Brockhouse "for the development of neutron spectroscopy" and to Clifford Shull "for the development of the neutron diffraction technique".

As background information to the modern applications of neutron diffraction and scattering, we can recall the work of Albert Einstein in 1905, as described in this book, when he demonstrated that a photon of light or a light quantum possess a dual nature, that of a wave and that of a particle. For this work Einstein was awarded the Nobel Prize in Physics 1921. Also described in the book is the work of Louis de Broglie, who demonstrated in 1923 that particles to which we can assign a mass at rest, such as a proton, electron, or neutron, will have also a dual nature, that is, the properties of a particle as well as the properties of a wave, such as light. Furthermore, along these same lines of thought, the book describes the pioneering work of Max von Laue in 1912 when he discovered the diffraction of x-rays by crystalline substances and that of Sir William Henry Bragg and Sir William Lawrence Bragg in 1914 who, as a father and son team, determined the interatomic distances of atoms in a crystal by means of x-ray diffraction. These previous findings lead to the work of Brockhouse and Shull, which is described in the words of Professor Carl Nordling of the Royal Swedish Academy of Sciences, in his Presentation Speech of the Nobel Prize to Brockhouse and Shull on December 8, 1994:

> … they knew that neutrons possessed a dual nature that was characteristic of their tiny world: the ability to behave both as particles and as waves. In their latter guise, neutrons have been reflected against the atomic planes of a crystal in the same way x-rays previously had. This provided a hint that, some day, neutrons might become a tool for studying the microstructure of matter at the atomic level. The door was already ajar, but had not yet been opened wide…. Shull took advantage of the fact that the wavelength of neutrons from a reactor may be roughly equal to the distance between the atoms in a solid body or a liquid. When the neutrons bounce against atomic nuclei, they do not lose energy, but their scattering is concentrated in directions that are determined by the structure in which the atoms are arranged. Shull revealed that neutrons could answer questions that the x-ray diffraction method had failed to answer, such as where the atoms of the light element hydrogen are located in an ice crystal.

Neutrons of specific energy, when directed as a beam toward a crystalline substance, may exit from the substance without any loss of energy (elastic scattering), exit with some loss of energy (inelastic scattering), or exit at specific angles or directions (diffraction). Like in x-ray diffraction, described in this book, a neutron beam from a research reactor may be directed toward a crystalline material. Most of the neutrons exit the crystal without loss of energy and are diffracted by the crystal. A neutron detector may be rotated around the crystal to determine the angles of diffraction to yield a diffraction pattern upon which the positions and distances of the atoms on the crystalline sample may be determined. As discovered by Shull, this would be particularly sensitive in determining the positions of hydrogen atoms, which are similar in mass to the neutron, and which cannot be revealed by x-ray diffraction. This technique has proven to be of great value in the biological sciences where the determination of the structure of proteins, viruses, and other macromolecules and their complexes with DNA and other nucleic acids are vital to our understanding of their modes of action in living systems. An illustrative example is taken from the work of Shu et al. (2000) illustrated in Figure 10, which shows the locations of the hydrogen atoms bound to a residue of myoglobin, the primary oxygen-carrying pigment of muscle tissue. This work demonstrated the enhanced structural information obtained when hydrogen was replaced with deuterium in the molecular structures. As noted in the IAEA Nuclear Technology Review (2006): "The similar scattering signal, using small angle neutron scattering (SANS) from deuterium, carbon, nitrogen and oxygen allows the full determination of the positions and dynamics of the atoms in biological structures. The non-destructive method using neutrons is an ideal complimentary tool of investigation if combined with x-ray scattering and/or nuclear magnetic resonance (NMR) measurements." For an overview of neutron techniques in studies of structural biology see a review paper by the father of neutron protein crystallography Benno Schoenborn (2006).

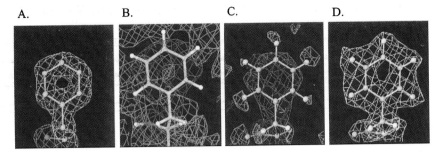

Figure 10   Fourier density maps of a unit of myoglobin from diffraction pattern data obtained by using (A) x-rays where no hydrogen atoms are visible, (B) neutrons where no isotope labels were added, (C) the calculated structure equivalent to (B), and (D) an enhanced image taken with neutrons and a sample where the hydrogen atoms were replaced with deuterium. (From Shu *et al.* (2000) with permission from National Academy of Sciences, U.S.A. ©2000.)

6.3   Neutron radiography

Neutron radiography is an imaging technique similar to x-ray radiography. A neutron beam is directed toward an object and an image is formed onto photographic film sensitive to neutron radiation. The image produced may look like an x-ray image; however, objects often invisible to x-ray radiography are made visible in neutron radiography.

The many properties of neutrons are described in this book. Neutron properties, which make them particularly suitable for radiography, are their neutral charge and their ability to transfer significant amounts of their energy to light elements by inelastic scattering. Because of the lack of charge on the neutron, a neutron beam from a research reactor or other source of neutrons is able to penetrate matter easily somewhat like x-rays are able to pass through materials of varying densities. However, contrary to x-rays, neutrons are sensitive to light elements, such as hydrogen, lithium, boron, and carbon. As described in more detail in this book, elements of high atomic number will attenuate x-rays to the greatest extent, whereas neutrons are attenuated most by elements of low mass. As a consequence, the images that neutrons produce after passing through an object will be particularly revealing of substances containing light elements, which would be invisible with x-rays. Thus, neutron radiography will complement x-ray radiography. This is illustrated by comparison of the neutron and x-ray radiographs illustrated in Figure 11.

Neutron radiography is often used in quality control inspection such as the inspection of components inside manufactured devices that would generally not be visible through x-ray radiography. Some applications of neutron radiography are inspections of manufactured components of many types such as those engineered for the aircraft, aerospace, and automotive industries, including (i) precision in manufacture of turbine engine blades, (ii) corrosion, (iii) flaws in adhesive bindings, (iv) internal flaws, (v) missing or misplaced O-rings or other components, and (vi) cracks, inclusions, and voids or other types of internal defects in materials, etc. Also, neutron radiography is applied in the biomedical and agricultural fields. Examples of applications cited in the IAEA Nuclear Technology Review (2006) are (i) the monitoring of boron distribution needed for boron capture therapy in the treatment

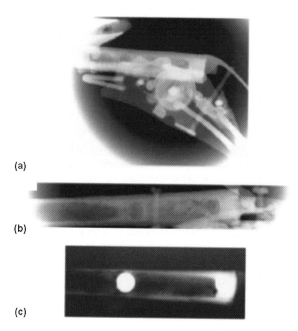

(a)

(b)

(c)

Figure 11   Neutron radiograph and x-ray radiograph of a 16th century wheel-lock pistol discovered in a river in Slovenia. The wheel-lock mechanism and indications of the presence of gunpowder can be observed in the neutron radiographs of the pistol (a) and the barrel (b). The lead bullet cannot be directly observed by the neutron radiograph. Its presence in the barrel is revealed by the detection of the amounts of mud, corrosion products, and gunpowder clogged around the bullet. The position of the lead bullet is clearly revealed by the x-ray radiograph (c). (From Rant *et al.* (2006), reprinted with permission from Elsevier © 2006.)

of cancer, (ii) the inspection of heart valves and materials used in dentistry, and (iii) water movement in plants including the distribution and transport of neutron absorbing elements such as gadolinium, samarium, and cadmium in plant leaves. The neutron absorbing properties of the elements are described in Chapter 4 of this book.

A detailed review of the many applications of neutron beam technology is provided in the Nuclear Technology Review (2006). Also, for further reading, books by Domanus (2004) and Harms and Wyman (2005) provide comprehensive treatments on neutron radiography.

6.4   Neutron activation analysis

Neutron activation analysis is a method that utilizes a neutron beam to determine the concentrations of elements in a given substance. The technique involves irradiating the material to be analyzed with a neutron beam. The irradiation will result in the absorption of neutrons by various nuclei of the elements in the substance yielding many radioactive isotopes. The energies and intensities of the gamma-ray emissions of the radioisotopes produced are then measured with a sensitive semiconductor detector, and the elements in the sample as well as their concentrations are determined by correlating the gamma-ray energy lines and their intensities with standards irradiated simultaneously with the unknown substance.

Neutron activation analysis was first conceived by Nobel Laureate George de Hevesy while working with Hilde Levi (1909–2003) at Niels Bohr's Institute of Theoretical Physics in Copenhagen and first published in 1936 (de Hevesy and Levi, 1936a,b). Hilde Levi received her doctorate degree in physics and chemistry at the University of Berlin in 1934, which was during the early phase of Hitler's dictatorship. Because of her Jewish heritage she left Germany and obtained assistance from the Danish branch of the International Federation of University Woman, which arranged her connections to Denmark and employment at Niels Bohr's Institute (Aaserud, 2003). There she became a collaborator in research with George de Hevesy during the time of his pioneering work on the application of isotope tracers in biological research. In a biography on George de Hevesy, written by Hilde Levi (1985), she related her recollections on this historic discovery as follows:

> The most important discovery from this short period of research was the basic development of neutron activation analysis. As discussed in a previous chapter, the separation and purification of the different elements within the group of rare earths was exceedingly complex. Chemists in many countries were still laboring with this problem using the classical methods of analytical chemistry. Hevesy suggested that—as a means of identification—we make use of the characteristic decay period of each of these elements and of their relative intensities of activation after neutron bombardment. In this way, their presence could be determined in any unknown mixture. In the early phase of this work we observed that, with the neutron sources available, the strongest activity was induced in the rare earth element dysprosium. It was therefore easy to detect even a minute amount of dysprosium compound present as an impurity in the salt of any other rare earth element. This finding was published in 1936.

Frame (1997) also relates an interesting story of de Hevesy's discovery of neutron activation analysis taken in part from de Hevesy's biography (Levi, 1985).

Because of its high level of sensitivity for the trace elements (<0.01%) and accuracy, neutron activation analysis is commonly employed in many fields of study including the biological sciences, agriculture, industry, and food and nutrition (Filby, 1995). It is also commonly used in forensic science because of characteristic signatures that the gamma spectra will display after neutron irradiation of samples (Nuclear Technology Review, 2006), as well as trace element analysis for applications such as the verification of archeological artifacts and precious paintings of the masters. The reader will find thorough treatments on neutron activation in books by Keisch (2003), Baas (2004), and Molnar (2004).

6.5 Radiation processing

Radiation processing is the use of high doses of ionizing radiation to alter the biological, or physical and chemical properties of the irradiated substances. Radiation processing is not a new field; certain applications were reviewed in a book edited by the writer over 20 years ago (Grünewald, 1984; L'Annunziata and Legg, 1984b). It is a field of science out of view and out of mind or thought of the general public, but one to which we owe dearly to our well-being and comfort. The importance of radiation processing in our day-to-day lives was underscored in a publication (IAEA, 2004b) from which the following is quoted:

Radiation Processing: The Invisible Performer

Unknown to the general public, and even to the scientific community, radiation technology has sneaked into our daily routine to enrich the quality of our life in many ways. Today, you may be using radiation processed materials in some form or the other, for example, the car you drive or the television set you watch may be equipped with radiation crosslinked wires or cables, the hot water pipe line bringing warmth to your homes might be made of radiation crosslinked plastic material, the long lasting alkaline battery you use in the camera or watch may use radiation cured membranes to enhance its lifetime, the safe hamburger you ate may have contained meat or spices pasteurized by radiation ... or if you have been unfortunate to have stayed in the hospital, the sterile syringe or the catheter the doctor used might have been sterilized by radiation processing to make it absolutely safe. The examples are numerous. At present, the value of irradiated products produced exceeds several billion dollars and the industry is expanding at a fast rate all over the world ...

Some of the industrial applications of radiation processing are the following:

(i) The sterilization of health care products, such as, bandages, syringes, gloves, surgical blades, sutures, catheters, pharmaceuticals, and many other items used during surgery or outpatient treatment. Even the standard bandages that we often apply at home to minor cuts and bruises are sterilized by high doses of radiation. The advantages are that the sterilization does not require heat or autoclave, and the items are sterilized after they are packed or sealed in wrappings impermeable to air or bacteria, which keeps them sterile until their time of use.

(ii) The irradiation of blood to prevent transfusion-associated graft versus host disease (TA-GVHD), which is a rare but lethal complication of transfusion generally associated with immuno-suppressed patients. Its prevention is generally affected by the irradiation of cellular blood components (red cells and platelets) with suitable doses of gamma radiation before transfusion (Williamson, 1998).

(iii) The sterilization of tissue provided by human donors and used for transplant surgery to reduce infection rate.

(iv) The preparation of sterile hydrogels. The ionizing radiation is used to both prepare and sterilize the hydrogels, which have practical medical applications. Hydrogels are cross-linked polymeric chains created with slight heat and gamma-ray irradiation of synthetic and natural monomers with natural polymers like agar. The preparations are molded according to need in foil impermeable to air or bacteria. Practical applications are contact lenses (silicone hydrogels, polyacrylamides), dressings for the healing of burn victims and other hard-to-heal wounds, disposable diapers, etc.

(v) The irradiation of food and agricultural products for a variety of objectives including disinfestation of seed products, flours, fresh and dried fruit, spices, etc., sprout inhibition, killing of pathogenic bacteria in meat, poultry, and fish, and increasing food shelf life, etc.

(vi)   The synthesis of radiation-cured rubber free of carcinogenic compounds, which may be used in conventional rubber vulcanization, for the manufacture of disposable gloves and other products.

(vii)   The manufacture of radiation-induced heat-shrinkable plastics employed in the packing industry; radiation cross-linking in the manufacture of wire insulations, plastic piping, and other materials to increase their resistance to heat, chemical attack, and/or mechanical damage, and the radiation grafting of polymers to other materials.

(viii) The manufacture of radiation-cured wood polymer composites, resistant to insect attack, moisture, fire, and abrasion, often used for flooring. Also, the radiation-curing of surface coatings of wood, plastics, and metals to increase their potential for use in the industrial manufacture of a wide spectrum of goods and electronic appliances.

(ix)   The irradiation of hospital waste to kill disease-causing bacteria and microorganisms and thus converting hazardous hospital waste to conventional refuse.

(x)   The gamma-ray sterilization of insects that are mass reared for release in the wild to control and eradicate insect pests by the SIT.

(xi)   The use of electron-beam radiation to process the polluting industrial emissions of NO_x and SO_2 and thus, purify the flue gas that is emitted into the atmosphere. The electron-beam radiation converts the oxides of nitrogen and sulfur to a residue that can be utilized as fertilizer. Both nitrogen and sulfur are major nutrients required by crops for optimum growth. Their removal from industrial flue gas helps reduce acid rain while, in turn, leaves behind a valuable by-product.

(xii)   The irradiation of domestic sewage sludge with ^{60}Co gamma radiation or accelerator electron beam radiation to kill pathogenic microorganisms for safe release of the sludge into the environment.

Gamma radiation from cobalt-60 sources are most commonly used for large-scale sterilization of health care products, food, and industrial applications. Accelerated electron beams may also be used as a source of radiation. The electron beam can be converted to a source of penetrating x-radiation when the electron beam is directed toward a high atomic number target such as tungsten. A typical industrial-scale cobalt-60 irradiation facility is illustrated in Figure 12. Although the irradiation plant illustrated here is almost 30 years old, its basic design is identical to modern facilities, which demonstrates that this gamma-irradiation technology has been well developed for considerable time, and was reviewed by Grünewald (1984).

The products to be irradiated are placed in standard boxes, referred to as totes. The boxes are transported on a conveyer belt into the radiation room where a cobalt-60 source of gamma radiation is lifted by a hoist mechanism upwards from its storage pool below ground to a position between the boxes. In the radiation room the boxes will remain exposed to the gamma radiation for a period of time necessary to achieve a calculated dose, measured in units of the gray (Gy) required to achieve the objectives of the radiation, according to the cobalt-60 gamma-ray activity, measured in units of the curie (Ci). Units of radiation activity and radiation dose are described in the book. Typical dose requirements for several radiation processing applications are listed in Table 1.

Data provided by the Nuclear Technology Review (2004) noted that there are over 160 industrial gamma-ray irradiation facilities working worldwide on a service basis; and more

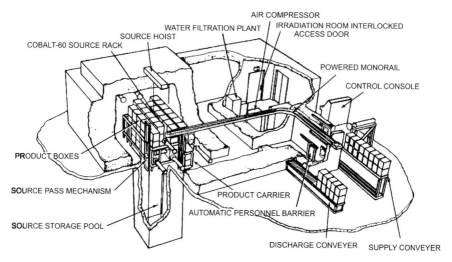

Figure 12 ^{60}Co irradiation facility, tote box system. (From Williams and Dunn (1979), reprinted with permission from Elsevier © 1979.)

Table 1

Some typical radiation processing applications

Product	Intended effect	Typical dose range (kGy)
Blood	Preventing TA-GVHD^a	0.020–0.040
Potatoes, onions, garlic	Inhibiting sprouting	0.05–0.15
Insects	Reproductive sterilization for pest management	0.1–0.5
Strawberries and some other fruit	Extending shelf-life by delaying mould growth and retarding decay	1–4
Meat, poultry, fish	Delaying spoilage, killing certain pathogenic bacteria (e.g., salmonella)	1–7
Spices and other	Killing a variety of microorganisms and insects	1–30
Health care products	Sterilization	15–30
Polymers	Crosslinking	1–250
	Grafting	0.2–30

Source: IAEA (2005c).
^{a}Transfusion-associated graft versus host disease.

than 20% of these gamma irradiators are equipped with activities in excess of 1 MCi of ^{60}Co. There are approximately 1200 electron beam facilities applied to radiation processing worldwide.

As noted through the numerous examples cited in this introduction, radiation processing and radioisotope applications have a great impact on social and industrial development of

Table 2

Economic scale of some industrial applications of radiation and isotopes

Item	Economic scale (billions $)	
	USA	Japan
1. Sterilized medical supplies	4.8	2.3
2. Semiconductors[a]	37.2	28.4
3. Radiographic testing (NDT)[b]	0.65	0.26
4. Radiation cured radial tires	13.5	8.4
Total	56.15	39.36

Source: Nuclear Technology Review (2004).
[a]Utilized in a wide spectrum of industrial applications.
[b]Non-destructive testing of installations, pipes, tanks, etc.

nations worldwide. These benefits are reflected in the betterment of human health, comfort, and environmental protection. In addition, the economic impact of radiation processing and radioisotope applications is substantial as noted from the data provided in Table 2.

For additional reading on radiation processing and technologies, the reader is invited to review informative works by Chmielewski and Haji-Saeid (2004), Masefield (2004), Molins (2001), and Morrissey and Herring (2002).

7 NUCLEAR POWER

There are three peaceful applications of nuclear power, namely (i) nuclear fission for the generation of electricity, (ii) the desalination of seawater, which requires a large source of economically feasible energy from nuclear power, and (iii) nuclear fusion, the source of power for electricity of the future. These applications will be treated subsequently.

7.1 Nuclear fission

The book describes the discovery of nuclear fission in early 1939 by Otto Hahn, Lise Meitner, and Fritz Strassmann. This was followed by the calculations of Robert Anderson, Enrico Fermi, and Leo Szilard in July of the same year with the approximate number of two neutrons produced per each neutron-induced fission of uranium-235 nucleus; and in August of the same year, Walter Zinn and Leo Szilard made more precise calculations arriving at the number of approximately 2.3 neutrons. These historic findings, all within the period of 1 year, illustrated the possibility of a nuclear chain reaction, which could be controlled to release and harness immense amounts of energy for electric power, through nuclear fission. The first successful demonstration of a controlled nuclear chain reaction on December 2, 1942, by Enrico Fermi and collaborators at the University of Chicago is described in the book. Two years later in December of 1944, Enrico Fermi and Leo Szilard jointly filed a highly classified patent as coinventors of the atomic pile, describing how the self-sustained nuclear chain reaction had been achieved.

The invention of the nuclear reactor for the peaceful application of nuclear energy capable of providing electric power to cities was declared by some to be one of the most significant inventions of all time comparable to those of communication, the telegraph and telephone, by Samuel Morse and Alexander Graham Bell. In 1944 the "New Piles Committee" meetings were held with the participation of Enrico Fermi, Leo Szilard, together with Nobel Laureates James Franck and Eugene Wigner among other physicists, chemists, and engineers including Walter Zinn and Alvin Weinberg. The New Piles Committee meetings explored and recommended new designs for peaceful electric power-producing nuclear reactors. Many new reactor concepts were presented which would become the modern power reactors that supply the needed electricity for cities around the world.

It was not until after the Second World War on December 20, 1951, was electricity finally generated from a nuclear power reactor. It was from the Experimental Breeder Reactor-I at the National Reactor Testing Station in Idaho, USA (Nuclear Technology Review, 2004). The breeder reactor, first conceived by Fermi and Szilard, is described in this book. Now, in this day and age, we have already celebrated 50 years of the comforts and benefits of nuclear power. Data from the Nuclear Technology Review (2006) show that, at the end of 2005, there were 441 nuclear power plants in 30 countries of the world generating electricity and supplying approximately 16% of the world's electricity. There are 26 new power reactors under construction in 2006 in nine countries of the world, and it is foreseen that the use of nuclear power will grow worldwide to meet the expanding energy needs of this century. The critical role that nuclear power plays as a source of energy was commemorated in the postage stamp illustrated here issued in Finland, which shows an artistic depiction of a reactor core producing heat from nuclear fission, and a wave of electricity generated from that heat. The forecasted growth of nuclear power worldwide is based on many factors including an international meeting in Paris of 74 government representatives organized by the IAEA in March of 2005 on the future role of nuclear power. The meeting concluded with the affirmation of the vast majority of the participants that nuclear power would contribute to a great extent to help the world meet its energy needs in sustaining development through the 21st century. The foreseen growth in nuclear power is due to several factors including the growing demand for energy to sustain development throughout the world, the environmental constraints placed on the need to reduce global pollution from greenhouse gas emissions by reducing our need to burn fossil fuels, concerns over future energy supplies, and lower generating costs and improvements in nuclear power safety. The Nuclear Technology Review (2006) provides an overview of the current and future advances in fission, which include cost reductions, safety enhancements, and proliferation resistance. The review also concludes that, if the world is to meet even a fraction of the economic aspirations of the developing world, energy supplies must expand significantly; and that oil and natural gas prices and demand will grow faster than supply.

7.2 Desalination

Desalination refers to the removal of the excess salt or other minerals from water, such as seawater, to obtain freshwater suitable for irrigation or human consumption. The desalination process may include several steps such as multistage flash distillation and/or reverse osmosis using semipermeable membranes. Details of the techniques involved are described in books including those by El-Dessouky and Ettouney (2002) and Lauer (2006). Desalination requires high inputs of energy from electric power tapped from either conventional fossil fuel burning plants or nuclear power or a combination of the two.

Current and future global needs for freshwater were underscored by the following data taken from the Nuclear Technology Review (2004): "Of all the earth's water only 2.5% is freshwater and 70% of the freshwater is frozen in the icecaps of the Antarctica and Greenland, and most of the remaining freshwater is found in the form of soil moisture or deep within inaccessible aquifers, or arrives on earth in the form of monsoons or floods that are difficult to contain or exploit. This leaves very little, less than 0.08%, of the earth's freshwater available for human consumption. There are currently about 2.3 billion people living worldwide in water-stressed or water-scarce areas. By the year 2025, it is foreseen that the number of people suffering from water stress or scarcity will increase to 3.5 billion."

There are currently about 7500 desalination plants in operation worldwide and about 60% are located in the Middle East. To meet the growing demands for freshwater, the demand for the desalination of seawater is on the increase, and it is foreseen that the need for freshwater from the sea will increase unabated through the century as countries develop and increase in population with the concomitant need for increased water supply for agricultural production and human consumption. The high-energy inputs for desalination are provided largely by conventional fossil fuel power plants although nuclear power is used in some countries for desalination, and the role of nuclear power in desalination is foreseen to increase for many of the same reasons presented early for the predicted increase in nuclear power, in general. An additional factor for the increased role of nuclear power in desalination was underscored in the Nuclear Technology Review (2004), where it was noted that the High Temperature Gas Cooled Reactor (HTGR), which is one of the leading reactor concepts for future power reactor needs, is particularly suited for desalination. The advantage of the HTGR for desalination is the virtually cost-free waste heat produced by the reactor in the desired temperature range of 100–120°C, which is needed for seawater desalination, and that of the total production costs of seawater desalination, 30–50% is attributed to energy costs. Thus, there exists a high potential of increased use of nuclear power, particularly the HTGR, since it has the advantage of offering increased thermal energy utilization, in addition to electricity, as a safe, clean, and economically competitive heat source. For further reading on the use of nuclear power in seawater desalination see books published by the IAEA (2001, 2005d).

7.3   Nuclear fusion

Nuclear fusion is the process whereby nuclei join together into one nucleus. Such an occurrence is not possible under normal temperature and pressure, because the repulsive coulombic forces between the positive charges of atomic nuclei prevent them from mingling into the required close proximity for them to coalesce into one. However, at temperatures of about 100 million degrees Centigrade, the nuclei of atoms become plasmas in which nuclei and electrons move freely with high kinetic energy sufficient for them to overcome their repulsive forces and combine. Because of the high temperatures required, the process is also referred to as thermonuclear fusion. The theory, principles, and energies released in various fusion reactions are discussed in this book.

The fusion of the light elements, such as deuterium (^{2}H) and tritium (^{3}H), release an energy per nucleon (MeV/u) or energy per unit of material considerably greater than can be achieved by nuclear fission. Thermonuclear fusion requires energy to reach high temperatures and maintain nuclear plasma, but once achieved, the energy output is greater than the input. An additional advantage of the nuclear fusion of deuterium and tritium is that no hazardous radioactive wastes are direct products of the reaction. The products of the thermonuclear fusion of deuterium and tritium nuclei are alpha particles and neutrons. The neutrons can produce some radioactive isotopes in the surrounding reactor shielding material, but most would be short-lived and there would be no need to store radioactive waste in geological deposits for long periods of time as is the custom with nuclear fission.

Nuclear fusion reactors or controlled thermonuclear reactors (CTRs) are under development to achieve nuclear fusion as a practical energy source. The reactors are based on maintaining plasmas through magnetic or inertial confinement. A deuterium–tritium (D-T) plasma burning experiment was performed with 0.2 g of tritium fuel with the Joint European Torus (JET) reactor in the UK in 1991; and a higher power D-T experimental program with 20–30 g of tritium was continued on the Tokamak Fusion Test Reactor (TFTR) at Princeton in 1993. The International Thermonuclear Experimental Reactor (ITER) was established under the auspices of the IAEA to develop a prototype fusion reactor by the year 2030. The prototype reactor has the purpose of demonstrating that fusion can produce useful and relatively safe energy. Fusion energy production via a commercial reactor is assumed to start around the year 2050.

8   SUMMARY

In summary, the peaceful applications of radioactive sources and the energy from these sources are numerous. They encompass a wide spectrum of fields from medicine, biology, agriculture, and the environment to industry and power production. Radioactive sources and nuclear energy applied to peaceful means have enriched our lives in so many ways. The significance of the many peaceful applications of nuclear energy was commemorated by the postage stamp from Canada illustrated here which shows an artist's portrayal of an atom superimposed over a microscope on one side and a nuclear reactor with electric grid on the other. Only a few examples of peaceful applications and how these touch our every-day lives are provided in this introduction, and the writer hopes, that through the examples given, the reader will appreciate the significance of the material presented in the remaining part of the book, if he or she is not already well informed of this realm of science.

Radioactivity Hall of Fame — Part I

Democritus (c.460–c.370 B.C.), Wilhelm C. Röntgen (1845–1923), Henri Becquerel (1852–1908), Pierre (1859–1906) and Marie Curie (1867–1934), Paul Villard (1860–1934), Ernest Rutherford (1871–1937), Hendrick A. Lorentz (1853–1928) and Pieter Zeeman (1865–1943), Joseph John Thomson (1856–1940), and Philipp Lenard (1862–1947).

DEMOCRITUS (c.460–c.370 B.C.)

Democritus was a Greek philosopher born in Abdera in the north of Greece. Democritus was a student of Leucippus, who proposed the atomic theory of matter. There is little documentation on the philosophy of Leucippus; however, it was Democritus, who elaborated extensive works on his theories on the atomic structure of the physical world, of the universe, and the void of space. Although Democritus was a philosopher, he is included here among the list of great pioneers of physics and chemistry of the 19th and 20th centuries, because many of his teachings on the structure of matter were demonstrated finally by scientists over 2000 years after his death.

Democritus (c.460-c.370 B.C.)

Democritus taught the theory of atomism, which held the belief that indivisible and indestructible atoms are the basic components of all matter in the universe. Thus the word atom is derived from the Greek *atomos* meaning indivisible. It was not until 20 centuries after Democritus did Rutherford, Bohr, Soddy, and others demonstrate the atom to be the smallest unit of an element consisting of a positively charged nucleus surrounded by electrons equal to the number of protons in the nucleus. Modern science has demonstrated that atoms remain undivided in matter (in accord with the early philosophical teachings of Democritus) or in chemical reactions with the exception of a limited removal, exchange, or transfer of electrons. The atom is also the basic unit of elements and is the source of nuclear energy. The postage stamp illustrated here was issued by Greece in 1961 to commemorate Democritus' teachings of atomism and the development of peaceful applications of atomic energy in the world.

Democritus was not alone in the teaching of atomism, but his writings on this philosophy were most extensive. He held that atoms were the tiniest of particles, too small to be perceived by the senses, of which all matter was composed, and that the atoms differed in size, shape, and mass. He also argued that atoms were in constant motion and could coalesce to form the larger bodies of matter that we can see, feel, and taste. The properties of matter that we can perceive with the senses such as color, taste, and hardness were the result of the interactions of atoms that constituted a given substance and the interactions of atoms with our body. For example, the taste of a substance would be the result of the interactions of atoms with the atoms of our tongue. Democritus also held to the belief of the existence of the "void" or empty space to which atoms or matter can move into. He argued that the lights of the Milky Way were the lights of distant stars, and that there existed other worlds, some with suns and moons, and others without. Likewise there would be other worlds with animal life, plants, and water and others without.

Much of Democritus' philosophy of atomism was demonstrated by modern science to be true. In honor of Democritus the national institution dedicated to research on peaceful applications of atomic energy for development in Greece is named the Democritus Nuclear Research Center. For additional reading on one of the greatest philosopher–scientists from antiquity see Taylor's book *The Atomists* (1999).

WILHELM C. RÖNTGEN (1845–1923)

Wilhelm Röntgen was born in 1845 in Lennep, Prussia, which is now Remscheid, Germany. He was awarded the very first Nobel Prize in Physics in 1901. The award was granted for the extraordinary services he had rendered by the discovery of Röntgen rays or more commonly referred to as x-rays.

Wilhelm C. Röntgen (1845-1923)

Röntgen discovered the mysterious invisible rays on November 8, 1895 when he was studying cathode rays, that is, the current that would flow through a partially-evacuated glass tube (cathode-ray tube then referred to as a Crookes tube). With such a tube covered in black paper and in a dark room he noticed that a paper plate covered with the chemical barium platinocyanide would become fluorescent (give off light) even at a distance of 2 m from the cathode-ray tube. He was able to demonstrate that the invisible rays came from the collision of the cathode rays (electrons) with the glass surface of the cathode-ray tube or from the collision of the cathode rays with other materials such as aluminum inserted into the glass of the tube. Further studies by Röntgen demonstrated that these rays could travel through various materials to varying extents, when these were placed in the path of the invisible rays, and the transmitted rays could be measured with photographic plates. He gave these mysterious rays the name x-rays, because of their unknown nature and "for the sake of brevity," as he stated in his original paper (Roentgen, 1895, 1896).

A piece of sheet aluminum, 15 mm thick, still allowed the x-rays (as I will call the rays, for the sake of brevity) to pass, but greatly reduce the fluorescence.

Röntgen was a very modest man as evidenced by the fact that after making such a revolutionary discovery he went into isolation for almost 2 months to thoroughly research these mysterious rays. It was not until December 28, 1895 did he make an official report entitled "Über eine neue Art von Strahlen" on his discovery to the Wurzburg Physical-Medical Society (Roentgen, 1895). His original report was translated shortly thereafter and published in *Nature* on January 23, 1896 under the title "On a New Kind of Rays" (see Roentgen, 1896). In his original report Roentgen (1895, 1896) demonstrated the properties of these newly discovered rays and demonstrated that the rays could be attenuated to different degrees according to the densities of the absorber. He showed that platinum, lead, zinc, and aluminum would be transparent to the x-rays to different magnitudes, and he experimentally determined the relative differences of these metals to x-ray transmission with the following data

	Thickness (mm)	Relative thickness	Density
Platinum	0.018	1	21.5
Lead	0.050	3	11.3
Zinc	0.100	6	7.1
Aluminum	3.500	200	2.6

As a result of the transparency of various materials to x-rays, Roentgen reported the images of items enclosed within metal boxes and the images of the bones of the fingers of a human hand when the hand is placed on the surface of photographic film. The varying degrees of transmission of metal and human flesh were illustrated by the x-ray image of his wife's hand with a ring on the third finger. This was the very first x-ray image illustrated in Roentgen's original paper (1895, 1896) and illustrated in the postage stamp above issued in 1995 to commemorate the 100th anniversary of Roentgen's discovery.

Only days after Roentgen's report on December 28, 1895 the news proliferated through the world press, and an "x-ray mania" spread throughout the globe. One can only imagine the thoughts that went through the minds of the public and scientists alike when they saw for the first time the skeletal structure of a living human being or were able to "see through" a closed wood or metal container and observe its contents.

The image that x-rays produced on a photographic plate after traveling through a substance with varying transparencies to x-rays were referred to then as a "röntgenogram," and in 1896 the Wurzburg Physical Medical Society officially named these mysterious rays as "Röntgen rays." Nowadays we call these images as "x-ray images." The medical ramifications of the early x-ray image of bones of the human hand became immediately clear, and Röntgen was awarded an honorary degree of Doctor of Medicine by the medical faculty of his own University of Würzburg. Numerous awards and honorary degrees were deluged upon him, but he nevertheless remained a modest man refusing to take out any patents on his work so that the scientific and medical community could advance without obstacle.

The early cathode-ray tubes used by Röntgen consisted of simple evacuated or partially evacuated glass tubes with a cathode (negative electrode) at one end of the tube and the anode (positive electrode) located off the side or at the other end of the tube. An illustration of such a cathode-ray tube with the anode located off the side of the tube is illustrated in Figure I.4, which is found further on in this Chapter describing the contributions of J.J. Thomson. In these early tubes, with the application of a high voltage to the electrodes, electrons would escape from the cathode (cathode rays) and be accelerated toward the anode located off to the side of the tube. Many of the electrons upon acceleration would acquire sufficient energy to speed by the anode and hit the glass of the tube. Upon colliding with the glass, electron kinetic energy would be lost as x-radiation. These x-rays are also referred to as "Bremmstrahlung" meaning "braking radiation," since the accelerated electrons are forced to divert their path when they approach the nuclei of the glass atoms or target nuclei and, thereby, the accelerated electrons undergo a "slowing-down" or deceleration. The deceleration of the electrons results in the emission of the electron energy as x-radiation.

Not long after Roentgen's discovery, much work went into improving the efficiency of the production of x-rays by the development of x-rays tubes as modified cathode-rays tubes. Such a tube from the early 1900's is illustrated (Figure I.1) in the postage stamp issued by the Czech Republic in 1995 to commemorate 100 years of Röntgen's discovery of x-rays. The stamp illustrates a typical x-ray tube

Electrons (Cathode rays)

Anticathode (+)

Cathode (−)

Anode (+)

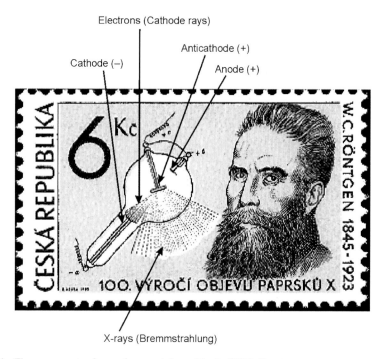

X-rays (Bremmstrahlung)

Figure I.1   The components of an early x-ray tube and bust of W.C. Roentgen on a postage stamp from the Czech Republic issued to commemorate the 100th anniversary of his discovery of x-rays.

of the time. The glass tube was partially evacuated to remove excess gas and maximize the efficiency of x-ray production. Upon application of a high-voltage potential to the electrodes, electrons escaping from the cathode would accelerate towards the positively charged anticathode. The anticathode was the name attributed to a target in the x-ray tube against which the electrons would collide to produce the x-radiation. The name anticathode is derived from the fact that the anode is oppositely charged from that of the cathode. The anticathode was made of a metal of high atomic number such as platinum or tungsten, as x-ray output efficiency was proportional to the target atomic number (see eq. (2.35) of Chapter 2). It was orientated at a 45° angle, as illustrated, to disperse the x-rays away from the electron beam out of the glass tube. The anode, located behind the anticathode, would accelerate the electrons from the cathode. The accelerating electrons would collide with the anticathode placed directly between the cathode and anode.

Röntgen's discovery and his numerous investigations on x-rays revolutionized the fields of medicine and physics. The vital fields of radiology and nuclear medicine, which many of us often take for granted today, were given birth on the day of this revolutionary discovery as commemorated by the postage stamp from Mexico in 1995. We should keep in mind that Röntgen was the first to produce artificial radiation, and it was only shortly after in 1896 did Henri Becquerel discover radioactivity.

HENRI BECQUEREL (1852–1908)

Antoine Henri Becquerel was born in Paris, France in 1852. He was awarded the Nobel Prize in Physics in 1903 for, in the words of the Nobel Committee, "the extraordinary services he has rendered by his discovery of spontaneous radioactivity."

Becquerel (1852-1908)

At the beginning of 1896 on the very day that news reached Paris of the discovery of x-rays Henri Becquerel thought of carrying out research to see whether or not natural phosphorescent materials emitted similar rays. He was then Professor of the École Polytechnic where he went to work on some uranium salts that he had inherited from his father, who had previously studied phosphorescence as professor of applied physics at the Polytechnic. Henri Becquerel placed samples of uranium sulfate onto photographic plates, which were enclosed in black paper or aluminum sheet to protect the plates from exposure to light. After developing the photographic plates he discovered that the uranium salt emitted rays that could pass through the black paper and even a metal sheet or thin glass positioned between the uranium salts and the photographic plates. Becquerel reported his findings to the French Academy of Sciences in February and March of 1896 (Becquerel, 1896a,b) and summarized his discovery in 1901 in the journal *Nature* as follows:

> At the commencement of the year 1896, in carrying out some experiments with the salts of uranium, the exceptional optical properties of which I had been studying for some time, I observed that these salts emitted an invisible radiation, which traversed metals and bodies opaque to light as well as glass and other transparent substances. This radiation impressed a photographic plate and discharged from a distance electrified bodies—properties giving two methods of studying the new rays.

At first he thought the rays were a result of phosphorescence, that is, excitation of the crystals by sunlight forcing the crystals to give off their own rays. However, Henri Becquerel carried out further tests demonstrating that the rays emanating from the uranium salts were independent of any external source of excitation including light, electricity, or heat, and the intensity of the rays did not diminish appreciably with time. "We were thus faced with a spontaneous phenomenon of a new order," which were his words during his Nobel Lecture given on December 11, 1903.

Becquerel provided evidence that all uranium salts emitted the same radiation, and that this was a property of the uranium atom particularly since uranium metal gave off much more intense radiation than the salts of that element. He was also able to demonstrate that the radiation was different from the x-rays discovered by Röntgen. The new radiation produced ionization and the intensity of the radioactivity could be measured by this ionization, as in his words, "the uranium discharged electrically-charged materials located some distance away." Even to this day, some pocket dosimeters used by radiation workers, measure the radiation dose according to the degree of electrical discharge caused by ionizing radiation impinging onto an electrically charged chamber that can fit into the shirt pocket. Not only did these rays produce ionization, but he was also able to demonstrate that a large portion of these rays could be deflected by a magnetic field and were charged particles of property similar to cathode rays. In the journal *Nature* Becquerel (1901) described his findings as follows:

> …in an electric field the radiation of radium undergoes a parabolic inflection in the contrary sense to the [magnetic] field, as would be the case with a flux of negatively charged particles. The comparison of electrostatic and magnetic deviations allows of the determination, like that of Prof. J.J. Thomson for the cathode rays, of the velocity of the particles. For the particular radiations defined by $H\rho = 1600$ [product of magnetic field strength and radius of curvature of the electron, which is proportional to the electron energy; see eq. (VI.24) in Radioactivity Hall of Fame VI], the velocity has been found equal to 1.6×10^{10}—about one half of that of light.

In recognition of his discovery of radioactivity Henri Becquerel shared the Nobel Prize in Physics in 1903 with Pierre and Marie Curie, who were awarded the Prize for their extensive research into this newly discovered radioactivity.

PIERRE CURIE (1859–1906) AND MARIE CURIE (1867–1934)

Pierre Curie and Marie Curie shared the Nobel Prize in Physics in 1903 with Henri Becquerel. The Curies were awarded the prize in recognition of "their joint researches on the radiation phenomena discovered by Professor Henri Becquerel."

Pierre (1859-1906)

Marie (1867-1934)

Pierre Curie was born in Paris in 1859. He received the Doctor of Science degree in 1895 and was then appointed Professor in the School of Physics. His wife Marie Curie was born Maria Sklodowska in Warsaw, Poland in 1867. Maria left Poland in 1891 after years of hard work and saving to finance her own education as well as providing for the education of her sister. When she saved what she thought would be enough for her education, she traveled to Paris to continue her studies of physics and mathematical sciences at the Université Paris "La Sorbonne." It was impossible for her to continue her education to get a university degree under the czarist control of Poland at that time. In France she changed her name from Maria to the more locally customary name of Marie. Although Marie had barely the resources to survive living in the most modest dwelling with barely enough to eat, she studied hard and was the top of her class receiving licentiateships in physics and the mathematical sciences. She was sometimes so engrossed in her studies that she would forget to eat, and it was reported that she once fainted in class for lack of nourishment.

While Marie was studying in 1894 she and Pierre met. Pierre was Professor of the Municipal School of Industrial Physics and Chemistry and already a well-known scientist in the field of crystal chemistry and physics. They fell in love and married in 1895. She encouraged Pierre to complete his doctorate degree that year, and he provided Marie with his laboratory as well as encouragement to pursue her research interests.

Two recent discoveries in physics sparked Madame Curie's interest. The first was the discovery of x-rays by Röntgen in 1895, and the other was the discovery the following year by Becquerel of spontaneous radiation emitted by uranium. She decided to study the mysterious rays emitted by uranium and to apply this work for a doctorate degree. The fact that no woman yet in the world had earned a doctorate degree did not thwart her efforts. Obstacles in life only seemed to motivate Madame Curie. She discovered in 1898 that not only uranium gave off the mysterious rays discovered by Becquerel, but thorium did as well; this was independently discovered by Gerhard Schmidt in Germany the same year.

Madame and Pierre Curie observed that the intensity of the spontaneous rays emitted by uranium or thorium increased as the amount of uranium or thorium increased. They concluded that these rays were a property of the atoms or uranium and thorium; thus they decided to coin these substances as "radioactive." The emanation of such spontaneous rays from atoms would now be referred to as "radioactivity."

Madame Marie Curie pursued her studies with all materials that contained uranium and thorium and discovered that the uranium ore pitchblende exhibited more radioactivity than could be attributed to its content of uranium or thorium. On this basis she made the assumption that there were one or more other radioactive elements in pitchblende other than uranium and thorium. Through tedious chemical separations and analyses Madam and Pierre Curie worked as a team and found that another radioactive element with chemical properties similar to bismuth was present in pitchblende. She named this new element "polonium" in honor of her native country. They found yet a second new radioactive element in the pitchblende ore with chemical properties close to that of barium, and they named that new element "radium" from the Latin word *radius* meaning "ray." The new elements were encountered only as traces in pitchblende and the Curies had to tediously process and fractionate over a ton of pitchblende ore to extract only minute quantities of the new elements. They spent their life savings to procure the pitchblende, and in the words of Madame Curie "it took several years to show unequivocally that pitchblende contains at least one highly-radioactive material which is a new element in a sense that chemistry attaches to the term."

The procedures devised by Pierre and Marie Curie for the extraction of these new radioactive elements were very exhausting physically, and the radiation doses they received from this tedious work and handling of this material surely affected their health. They both suffered from ill health due possibly to a combination of excessive work and radiation sickness. Due in part to ill health, Pierre delivered the Nobel Laureate address on June 6, 1905, 2 years after they were awarded the coveted prize. In 1903, the year Pierre and Marie Curie were awarded the Nobel Prize in Physics, Marie also became the first woman to receive a doctorate degree in France. Pierre was promoted to Professor at La Sorbonne.

The effects of excessive doses of radiation to one's health were then unknown, although Pierre demonstrated that high doses of radiation could damage human tissue. In the Nobel address (Curie, 1905) he stated "Radium rays have been used in the treatment of certain diseases (lupus, cancer, etc.). In certain cases their action may become dangerous. If one leaves a small glass ampoule with several centigrams of a radium salt in one's pocket for a few hours, one will feel absolutely nothing.

But 15 days afterwards a "redness" will appear on the epidermis, and then a sore that will be very difficult to heal. A more prolonged action could lead to death. Radium must be transported in a box of
lead." Over 100 years ago at the Nobel Lecture Pierre Curie had the wisdom to comprehend, like a
double-edged sword, the potential benefits and dangers that radioactivity presented. He noted the
immediate benefits found in using radiation from radium to treat cancer, while on the other hand he
stated "It can even be thought that radium could become very dangerous in criminal hands, and here
the question can be raised whether mankind benefits from knowing the secrets of Nature, whether it
is ready to profit from it or whether this knowledge will not be harmful for it. I am one to believe with
[Alfred] Nobel that mankind will derive more good than harm from the new discoveries." So far to
this day mankind has not betrayed Pierre's belief.

Approximately 10 months after presenting the Nobel address Pierre met with a tragic accident and
died instantly after being struck by a horse-drawn carriage while crossing a street in Paris. Marie suffered from his demise, as they were deeply in love. Like most determined and driven individuals this
tragedy forced her deeper into her work. She was awarded the Nobel Prize in Chemistry in 1911.
Marie Curie then became not only the first woman to receive a Nobel Prize, but also the only person
to ever receive a second Nobel Prize in two distinct sciences. The prize was awarded "in recognition
of her services to the advancement of chemistry by the discovery of the elements radium and polonium,
by the isolation of radium and the study of the nature and compounds of this remarkable element."

The French Academy of Sciences rejected Marie's bid for admission into the Academy. No woman
had yet been admitted into the Academy and she was not of French decent. Nevertheless, following
the death of Pierre she took over his classes, and was appointed professor to become the first woman
ever to teach at La Sorbonne. Much of her work later in life was devoted to finding medical applications of radioactivity including radium therapy for the treatment of illnesses such as cancer, and the
medical applications of x-rays. She played a vital role in the establishment of the Radium Institute in
Paris now known as the Curie Institute, which is recognized for its pioneering work in cancer
research and treatment. The stamp above commemorates the Curie's discovery of radium and Marie's
efforts to apply this radioactive source in the international fight against cancer.

Marie Curie suffered hardship and many obstacles in her life, and these obstacles only seemed to
motivate her further. Also her modesty was unsurpassed. It is reported that she wore the same black dress
to receive both Nobel Prizes in 1903 and 1911. Her life, determination and perseverance to overcome
all obstacles should serve as an example for us all. Many very informative books are written about
Marie Curie including those by Noami Pasachoff (1996), Susan Quinn (1996), Barbara Goldsmith
(2004), and a biography by her daughter Eve Curie (1937).

Marie Curie died of leukemia on July 4, 1934, a disease she likely acquired from excessive exposure to the radioactive materials she studied. Albert Einstein eulogized on her death stating "Marie Curie is of all celebrated beings, the only one whom fame had not corrupted." Her determined work, personal sacrifice, and devoted efforts to help mankind through medical applications of radioactivity had revolutionized the sciences of physics, chemistry, and medicine. In a book Richard Rhodes (1986) eloquently characterized her passing by writing "As if to mark in some distant inhuman ledger the end of one age and the beginning of another, Marie Sklodowska Curie, died that day of [Leo] Szilard's filing, July 4, 1934 [of a patent on the principles of making the atomic bomb]."

In 1995 President François Mitterrand of France transferred the ashes of Marie Curie and those of her husband Pierre to rest under the famous dome of the Panthéon, in Paris alongside the remains of France's greatest dignitaries. The inscription on the Panthéon's ornamental front used to read "To the fatherland's great men, in gratitude." This could be interpreted literally until the remains of Marie Curie were laid to rest among these great men. President Mitterrand conferred the added value of "beings" to the term "men" to set things right and enable the country and world to honor a nonnational among other great French personalities, who had contributed greatly to French prestige. It is said that nobody has sacrificed and contributed so much to scientific research and humankind as Marie Curie.

PAUL VILLARD (1860–1934)

Paul Ulrich Villard discovered gamma radiation in the year 1900. He received very little recognition for his discovery. His work remained largely forgotten by the scientific community until the 100th anniversary of his discovery when Gerward (1999) and Gerward and Rassat (2000) published thorough historical accounts of Villard's life and work. Most of the biographical data presented here was obtained from their reports.

Paul Villard (1860-1934)
© **Académie des sciences de l'Institut de France**

Paul Villard was born in a village near Lyon, France on 28 September 1860. He entered the Ecole Normale Supérieure in Paris in 1881 and upon graduation he taught at various lycées and finally at the Lycée of Montpellier. Villard had a strong interest in scientific research. He thus established a scientific liaison with staff of the local university, which provided him with a small research laboratory. He enjoyed scientific research and preferred to be in Paris, which was then the heart of research in the physical and chemical sciences. Having sufficient financial resources to cover his needs Villard was free to move to Paris where he enjoyed the hospitality of the Chemistry Department of the École Normale at rue d'Ulm. He was provided a small laboratory at the École Normale for his personal scientific research where he remained for his entire scientific career until late in life when he would retire outside of Paris.

His early scientific work during 1888–1896 was devoted to the synthesis and chemistry of gaseous hydrates (1894a,b). Villard (1896) is recognized as the discoverer of the noble gas hydrates including the hydrate of argon. The element was discovered in 1894 by Nobel Laureates Lord Rayleigh and Sir William Ramsey. The gas hydrates are solid compounds of crystalline structure in a water matrix produced under high pressure and low temperature. Villard's pioneering work in this field is cited often in the scientific literature on gaseous hydrates.

The very year that Villard (1896) reported his discovery of argon hydrate Henri Becquerel had made his discovery of radioactivity. The year prior to Henri Becquerel's discovery of radioactivity Wilhelm Röntgen discovery x-radiation, and in 1898 Marie Curie discovered that uranium and thorium also emitted radioactivity. Before the end of 1898 Pierre and Marie Curie had discovered new radioactive elements, which they named polonium and radium. Research in the field of radioactivity became the new frontier, and Villard would take part in the exploration.

In the field of radiation Villard began his research using a Crookes tube to study the effect of magnetic fields on cathode rays and chemical effects of x-rays during 1897–1899. He discovered that x-rays would produce both physical and chemical changes in substances (Villard, 1899, 1900a). Villard next obtained samples of radium from the Curies and at first attempted to study the rays emitted by radium that could be deflected by a magnetic field, referred to as deviable rays, because he considered these rays to be similar to the deviable cathode rays. Although unknown at the time the deviable rays emitted by radioactive atoms were beta particles identical in property to electrons that constituted cathode rays. In his attempt to measure the "refraction" of the deviable rays emitted by radium Villard (1900b) discovered a yet unknown non-deviable highly penetrating radiation, which was to be named years later by Rutherford (1903) as gamma radiation. Villard's discovery of gamma radiation was reported to the French Academy of Sciences (Villard, 1900b,c) and at the Meetings of the French Society of Physics (1900d). The experimental procedures used by Villard are related in detail by Gerward (1999), and only his final and conclusive experiment reported to the French Society of Physics in 1900 is described below. Villard did not provide any diagrams of his experimental arrangements, which led to the discovery of gamma rays; however, the writer sketched Figure I.2, to facilitate the description of an experiment reported by Villard.

Villard placed a sample of barium chloride containing radium sealed in a glass ampoule within a lead shield that contained an opening, which essentially provided a collimated beam of the radiation from the radium source as illustrated in Figure I.2. To the radiation beam he exposed two photographic plates wrapped in black light-tight protective paper. Between the two plates was sandwiched a 0.3 mm thick lead barrier. A magnetic field was applied to the collimated beam to deflect the deviable rays. Alpha particles emitted by the radium are ignored, because these are absorbed by the protective paper wrapping of the photographic plates. The magnetic field caused a deviation of the beta particles whereas a very penetrating radiation remained unaffected or undeviable by the magnetic field as evidenced from the images produced by the radium emanations on the developed photographic emulsions. The developed photographic emulsion A, which was the first to receive the nuclear radiations from the radium, showed two spots produced by two types of radiation, one deviable (marked β) and the other undeviable (marked γ) in the magnetic field. The second photographic emulsion B, which was placed behind a 0.3 mm thick lead barrier, yielded only one spot produced by a highly penetrating radiation unaffected by the magnetic field. The intensity of the spot on emulsion B was the same as that on emulsion A indicating that its intensity remained unaffected to any observable extent by the lead barrier. The spot was also more clearly discernable, because it was not clouded by the deviable

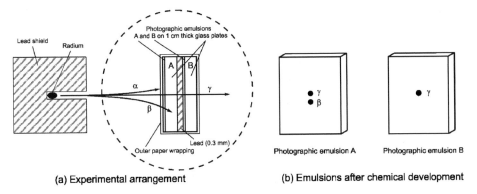

(a) Experimental arrangement   **(b) Emulsions after chemical development**

Figure I.2   Paul Villard's (a) experimental arrangement and (b) experimental results that lead to his discovery of gamma rays in 1900. Photographic plates A and B consisting of emulsion set on 1 cm thick glass supports were separated from each other by a 3 mm thick lead barrier and wrapped in an envelope of light-tight paper. The dashed circle represents the pole of a magnet from which the lines of force are directed into the plane of the page perpendicular to the path of radiation emitted by the radium source.

beta particles. Villard concluded that his experimental evidence demonstrated a radiation of property similar to x-rays, but with a greater penetrating power than x-radiation. In the journal *Nature* Becquerel (1901) acknowledged Villard's discovery of gamma rays, which had not yet been named as such, with the following statement:

> "…there exists two kinds of radiations, one not capable of deviation [in a magnetic field] and of which the nature is still unknown, the other capable of deviation, which later experiments have identified with the cathode rays [later identified as beta-particles]…I might add that recently Mr. Villard has proved the existence in the radium radiation of very penetrating rays, which are not capable of deviation."

Villard did not venture to name this newly discovered highly penetrating radiation. He was a modest person possibly uninterested in fame or in the highly competitive pursuits of his fellow scientists in this new frontier. Gerward (1999) researched the scientific literature and found that it was Rutherford (1903), who was first to name the highly penetrating radiation discovered by Villard, as gamma rays. A few years prior to Villard's discovery Rutherford (1899) has already named two types of nuclear radiation as "alpha" and "beta," which he characterized on the basis of their relative penetrative power in matter, that is, alpha radiation would be more easily absorbed by matter than beta radiation. In harmony with this nomenclature Rutherford (1903) assigned the term gamma rays to the yet more penetrating radiation. In the *Philosophical Magazine* Rutherford (1903) named and characterized the three types of nuclear radiation on the basis of their penetration power in matter as follows:

Radium gives out three distinct types of radiation:

(1)   The α rays, which are very easily absorbed by thin layers of matter, and which give rise to the greater portion of the ionization of the gas observed under the usual experimental conditions.

(2)   The β rays, which consist of negatively charged particles projected with high velocity, and which are similar in all respects to cathode rays produced in a vacuum-tube.

(3)   The γ rays, which are non-deviable by a magnetic field, and which are of a very penetrating character.

These rays differ very widely in their power of penetrating matter. The following approximate numbers, which show the thickness of aluminum traversed before the intensity is reduced to one-half, illustrate this difference:

Radiation	Thickness of aluminum (cm)
α Rays	0.0005
β Rays	0.05
γ Rays	8

As noted by Rutherford in the above quotation, ionization in a gas was one of the principle properties of nuclear radiation, which helped distinguish the radiation type. It was well understood at the time that the three rays possessed different powers of ionization in matter. The number of ion pairs formed in a gas by alpha, beta, and gamma rays of given energy per path length of travel would differ according to the relationship α rays $> \beta$ rays $> \gamma$ rays. Consequently the property of ionization power of a given radiation was used to also measure the intensity of that radiation.

Villard discontinued his research on nuclear radiation and went on to research cathode rays and x-rays. He introduced the concept of the ionization of air by x-rays to help radiologists measure the intensity of x-rays that would be administered to patients (Villard, 1908). Prior to instrumental methods of measuring x-ray intensity, radiologists would practice the long-term harmful technique of using the x-ray image of their hand to measure adequate x-ray exposure times for patients. Van Loon and van Tiggelen (2004) provide an historical account of Villard's contribution to radiation dosimetry in the monitoring of medical exposure. Villard (1908) introduced the Villard unit as a measure of radiation intensity, which was defined as "the quantity of radiation which liberates by ionization 1 esu of electricity per cubic centimeter of air under normal conditions of temperature and pressure." Van Loon and van Tiggelen (2004) report that the Villard unit was adopted 10 years later as the "e" unit by Krönig and Friedrich (1918), and modified by H. Behnken (1927) to become the "R" unit (i.e., Roentgen unit) or "German unit of x-radiation."

In 1908 Villard was appointed a member of the French Academy of Sciences. He retired outside of Paris and died in Bayonne, France on January 13, 1934.

ERNEST RUTHERFORD (1871–1937)

Earnest Rutherford was born on August 10, 1871 in Nelson, New Zealand. He was awarded the Nobel Prize in Chemistry in 1908 "for his investigations into the disintegration of the elements, and the chemistry of radioactive substances." He was still a young man of 37 years at the time of this award, and most of his major achievements and contributions to the science of radioactivity and the field of nuclear physics were made by Rutherford after receiving the Nobel Prize.

Rutherford grew up as a farm boy in New Zealand, raised by parents who had emigrated from Britain with modest means but who had instilled in their children the power of knowledge and the importance of completing all their schoolwork to the best of their abilities. Ernest excelled in science and mathematics, and could only achieve a higher education through financial aid with scholarships. He won a scholarship to Nelson College where he learnt elementary science and mathematics and excelled to win another scholarship to continue his education at Canterbury College, which is now the University of Canterbury in New Zealand. Although Rutherford worked hard and had an inquisitive mind, he credited his teacher, Professor Alexander Bickerton at Canterbury, who displayed an enthusiasm for science, which gave him the stimulus to pursue scientific research on his own. At Canterbury Rutherford won another scholarship as a research student at the Cavendish Laboratory, University of Cambridge under J.J. Thomson, who would later in life also win the Nobel Prize in Physics. At Cambridge Rutherford stood out as a great experimenter and thinker, who could make major discoveries concerning radioactivity and atomic structure without expensive equipment and only limited resources.

Ernest Rutherford (1871-1937)

Rutherford was captivated with the recent discoveries of x-rays by Roentgen and the subsequent discovery by Becquerel of the mysterious natural radiations from uranium. Because Becquerel showed that the radiations from uranium could, as well as x-rays, discharge an electrified body (i.e., cause ionization), he decided to examine the effect of placing successive layers of aluminum foil over uranium oxide on the efficiency of the emanations to cause the electrical discharge. From this experiment he made his first discovery concerning radioactivity stating at his Nobel address "…[I] was led to the conclusion that two types of radiation of very different penetrating power were present—one that is very readily absorbed, which will be termed for convenience the α-radiation, and the other of a more penetrative character, which will be termed the β-radiation, … and when a still more penetrating type of radiation from radium was discovered by Villard, the term γ-rays was applied to them." These terms devised by Rutherford soon took acceptance and came into common use by the scientific community as the convenient nomenclature, which remains today, for identifying these three types of radiation.

In addition to the different penetrating powers of alpha, beta, and gamma radiation, other properties were used to identify these mysterious radiations, such as the differing deflections that the three radiations undergo in electric or magnetic fields. Alpha radiation was known to possess a positive charge, because it would be deflected toward the negative electrode in an electric field potential, while beta particles were known to be negatively charged due to their deflection in the opposite direction toward the anode or positive electrode. Whereas gamma radiation would not undergo any deflection whatsoever as illustrated in Figure I.3. Likewise, the alpha and beta radiations, when traveling in a path perpendicular to the lines of force of a magnetic field will be deflected in opposite directions, which is a characteristic of charged particles. Radiation, which carries no electric charge, would continue along a straight undeviating path in either electric or magnetic fields.

After 3 years at the Cavendish laboratory Rutherford moved in 1898 to McGill University in Montreal at the age of 27 to take on the position of Professor of physics. It is at McGill University where he began to make his major discoveries in the field of nuclear physics. The first of these was the discovery that radioactive atoms that emit alpha particles or beta particles disintegrate into atoms of lighter weight, in other words, atoms of an element such as radium that emit alpha particles undergo transformations to atoms of a lighter and consequently different element. Rutherford and coworkers were able to demonstrate that the alpha particle was an atom of helium (later to be determined to be a nucleus of helium), and that helium gas would accumulate or be entrapped in minerals that contained radium. Furthermore, he demonstrated that the lighter atom produced as a product of the decay of radium would likewise be radioactive, and in turn, decay to another even lighter atom, and so on until the final product atom was stable. It was for this work that Rutherford received the Nobel Prize in Chemistry. Although a great honor for such a young scientist, it was ironic for Rutherford to go through life as a Nobel Laureate of chemistry. He was a physicist and considered physics as the most important science of all. At the Nobel Prize presentation address on December 10, 1908, Prof. K.B. Hasselberg, President of the Royal Academy of Sciences explained

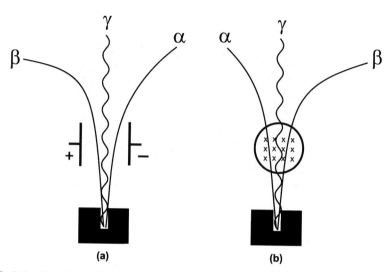

Figure I.3 Paths of travel of collimated beams of alpha, beta, and gamma radiation in (a) electric and (b) magnetic fields. Beam collimation is provided by placing the radiation source in a hole drilled within shielded containers (e.g., lead) of which cross-sections are illustrated. The electric field in (a) is illustrated by positively and negatively charged electrodes separated by a space through which the radiations pass; and the magnetic field in (b) is illustrated by the circle to depict the pole of a magnet through which the lines of force of magnetic flux (marked by the symbol x) are directed into the plane of the page (z-axis) perpendicular to the radiation paths in the xy-axis.

Though Rutherford's work has been carried out by a physicist and with the aid of physical methods, its importance for chemical investigation is so far-reaching and self-evident, that the Royal Academy of Sciences has not hesitated to award to its progenitor the Nobel Prize designed for original work in the domain of chemistry.

Rutherford's work in conjunction with numerous collaborators including F. Soddy, led to the conclusion that one chemical element can transform into other elements, which was previously only a centuries old belief of alchemists, who tried to change lead into gold. Professor Hasselberg in his speech of award presentation, had foresight when he added further

[Rutherford's] disintegration theory [of atoms] and the experimental results upon which it is based, are synonymous with a new department of chemistry.

We can thus give credit to Rutherford for giving birth to the field of radiochemistry.

Rutherford's best work was yet to come after the Nobel Prize. In 1907 he moved to England to fill the position of Professor of Physics at Manchester University. It was at Manchester where Rutherford had a list of research topics to explore, and one of these was the deflection that alpha particles would undergo when passing through thin foils (approximately 5×10^{-5} cm thick) of materials such as mica, aluminum, gold, etc. He knew that the alpha particles, as they travel at high speed, would traverse the foils as if the foil material was not even in the alpha-particle path, and that the flux of alpha-particle radiation would undergo only a very slight dispersion upon exiting the foils. This was understandable to him as strong electrical charges, expected to occur in atoms, could cause slight deflection of the positively charged alpha particles.

The story goes that he was approached by one of his students at Manchester, Dr. Hans Geiger (best known as the person who developed the Geiger counter still used today for monitoring radioactivity). Geiger had asked Rutherford, "What do you suggest we give the new student Ernest Marsden to do?"

Rutherford proposed that they should try to see if any alpha particles would bounce back, that is, not traverse the foil but be deflected by over 90° back towards the particle direction of travel. Rutherford did not expect to see any such deflection, but it had to be investigated. They could observe the alpha-particle path of travel and see and count the deflections by means of a zinc sulfide screen that would produce a microscopic fluorescence (flash of light scintillation) in the dark when each individual alpha particle hit the screen. They used radium as the source of the alpha-particle beam to bombard a thin foil of gold. Geiger later informed Rutherford that they could see the occasional deflection by greater than 90° of one alpha particle for every 8000 particles traversing the gold foil. The commemorative postage stamp above illustrates this phenomenon. The stamp illustrates a central nucleus surrounded by five electrons and the path (arrows) of three alpha particles. One of the alpha particles is illustrated as hitting the nucleus of the gold foil and bouncing back. In a presentation given by Rutherford years later, he described his reaction to this discovery by stating "It was quite the most incredible event that ever happened to me in my life. It was as incredible as if you fired a 15-in. shell at a piece of tissue paper and it came back and hit you." Rutherford grasped this discovery to conclude that there had to be a massive core or nucleus in the atoms of materials that would cause colliding alpha particles to bounce back. By using high-energy (7.7 MeV) alpha particles, which would travel at highest speeds, Rutherford was able to calculate the distance of closest approach and consequently the radius of the atomic nucleus to be approximately 5×10^{-15} m (see Chapter 1).

From here on Rutherford was able to begin to formulate the structure of atoms with a central massive nucleus, the structure that holds today. Rutherford's findings were the initial step that provided the foundation upon which other physicists including Niels Bohr, Werner Heisenberg, and others to elaborate the structure of the atom as we know it today.

In a Letter to the Editor of *Nature*, concerning the structure of the atom, published on December 11, 1913 Rutherford postulated correctly the atomic nucleus as the origin of alpha and beta radiation. In his letter he stated the following:

There appears to me no doubt that the α particle does arise from the nucleus, and I have thought for some time that the evidence points to the conclusion that the beta particle has a similar origin. This point has been discussed in some detail in a recent paper by Bohr (*Phil. Mag.*, September 1913). The strongest evidence in support of this view is, to my mind, (1) that the beta ray, like the alpha ray transformations, are independent of physical and chemical conditions, and (2) that the energy emitted in the form of beta and gamma rays by the transformation of an atom of radium C is much greater than could be expected to be stored up in the external electronic system.

Rutherford's next great discovery came in 1919 when he reported the first evidence of a man-made nuclear reaction, that is, the splitting of the atom. This he was able to demonstrate when a high-speed alpha particle would strike the nucleus of an atom and rearrange it into two different atoms. Rutherford

observed that when alpha particles would strike air, he could detect scintillation on a zinc sulfide screen produced at a distance well beyond the distance of alpha-particle range of travel corresponding to the range of travel of hydrogen atoms (protons). He demonstrated that the production of high-speed hydrogen atoms by collision of alpha particles with air arose from the collision of the alpha particles with nitrogen atoms only, because the effect would not occur with other constituents of air such as oxygen or carbon dioxide. Furthermore, when pure nitrogen was the target, the scintillations produced by the product hydrogen nuclei (protons) were greater then when air was bombarded with alpha particles (air contains only 79% nitrogen). Rutherford was also able to show that the number of swift atoms of oxygen produced by the alpha-particle collisions was about the same as the corresponding number of hydrogen nuclei (protons). This first man-made nuclear reaction is illustrated in the commemorative stamp from New Zealand illustrated above. The stamp above illustrates the historic nuclear reaction where an atom of nitrogen-14, the most common isotope of nitrogen, denoted as $^{14}_{7}N$, because its nucleus contains 7 protons and 7 neutrons, interacts with the colliding alpha particle or helium nucleus ($^{4}_{2}He$). In the collision a proton ($^{1}_{1}H$) is ejected and two protons and two neutrons from the alpha particle can coalesce with the remaining nucleons of the original nitrogen to yield a nucleus having eight protons and nine neutrons, namely, the isotope oxygen-17 denoted above as $^{17}_{8}O$. This was the very first artificial transformation of one element into another, which was the age-old dream of alchemists.

Nobel Laureate Cecil Powell (1903–1969) in his autobiography published in 1987, related that he was only 19 years of age when he arrived at the Cavendish Laboratory as a research student to C.T.R. Wilson, and at this time he recalls Rutherford researching alpha-particle induced transmutations. Powell related his following recollections:

When I arrived in 1922, Rutherford had already, in 1919, demonstrated the artificial disintegration of the light elements by bombarding them with fast alpha-particles from radioactive sources. We used to see him disappearing from time to time into a partitioned corner in his laboratory, with his assistant [George] Crowe, to count the scintillations that recorded the ejected protons. This involved long periods of darkness viewing a zinc sulfide screen under a low-powered microscope, and they used to emerge after an hour or so, blinking in the sudden light, like miners coming out of the pit.

As a result of the above man-made nuclear reaction where protons were emitted as a product, Rutherford has been given the honor of discovering the third elementary particle in matter, the proton. To put this discovery in perspective, the first elementary particle to be discovered was the electron from the work of J.J. Thomson in 1897 (see biographical sketch below), and the second elementary particle to be discovered was the photon from the work of Albert Einstein in 1905 (see Radioactivity Hall of Fame, Part III).

Rutherford contributed much more to the study of radioactivity and nuclear physics including the fact that all radioactive atoms have differing rates of decay, including the concepts of half-life and decay constant that are invariable properties of each radioactive nucleus or radionuclide of given atomic number (proton number) and mass number (number of protons + neutrons) in the nucleus (see Chapter 8).

Another contribution among many others made by Rutherford deserves mention. This is the development, together with Hans Geiger, of the first electronic means of detecting and counting individual alpha-particle emissions from radioactive atoms. The alpha-particle emissions were allowed to travel through a small opening or window into a vessel containing air or other gas exposed to an electric potential. The vessel is referred to today as an ionization chamber. Upon entering the vessel the alpha-particle, which carries a double positive charge, would cause ionization of the gas, and the ions produced by the alpha-particle would accelerate towards electrodes of the chamber thereby magnifying the ionization within the gas. The positive and negative ions produced by the alpha particle would be collected by their apposing electrodes and thereby produce a pulse that would cause a deflection of the electrometer needle. This instrument was the precursor of more modern Geiger counters, but it served its purpose, as Rutherford and Geiger were able to count each alpha-particle emission from a radium sample and calculate its specific activity, in the words of Rutherford on his Nobel Lecture, "In this way it was shown that 3.4×10^{10} alpha-particles are expelled per second from one gram of radium." This was very close to the real value of 3.7×10^{10}, which is the unit used today to define the unit of radioactivity known as the Curie (Ci), where $1\,Ci = 3.7 \times 10^{10}$ disintegrations per second (see Chapter 8). For the entire text of Rutherford's Nobel Lecture as well as those of other Nobel laureates in chemistry and physics the reader may consult books edited by the Nobel Foundation (1967) or Elsevier Science Inc. (1967) entitled *Nobel Lectures, 1901–1970.*

At Manchester many famous nuclear physicists and future Nobel Laureates worked and collaborated with Rutherford including Frederick Soddy, Henry G.J. Moseley, George de Hevesy, and Niels Bohr. At Rutherford's laboratory in Montreal, Otto Hahn, who later discovered nuclear fission with the collaboration of Lise Meitner, also worked with Rutherford. At the Cavendish laboratory other future Nobel Laureates of physics collaborated with Rutherford including James Chadwick, Patrick Blackett, John Cockroft, Ernest Walton, George P. Thomson, Edward V. Appleton, Cecil Powell, and Francis W. Aston, among others. The commemorative stamp from Canada illustrated above celebrated the centennial of Rutherford's birth. The stamp illustrates a reprint of Ray Webber's art work as described by the Canada Post Office as a burst of light symbolizing "the great energy that the harnessing of the atom has given to us and which, unseen, affects so much of all our lives." It is said that Ernest Rutherford was the greatest experimental physicist of the 20th century, while Albert Einstein was the greatest theoretical physicist of the century.

For further reading on the life and accomplishments of Rutherford the reader may peruse books by Boltz (1970), Heilbron (2003), and Pasachoff (2005).

HENDRICK A. LORENTZ (1853–1928)

Hendrik Antoon Lorentz was born on July 18, 1853 at Arnhem, The Netherlands. He shared the Nobel Prize in Physics 1902 with Pieter Zeeman also of The Netherlands "in recognition of the extraordinary service they rendered by their researches into the influence of magnetism upon radiation phenomena."

Lorentz (1853-1928)

Lorentz entered the University of Leyden in 1870, and as a gifted student, obtained his B.Sc. degree in mathematics and physics in only 1 year. He was awarded a doctorate degree at Leyden in 1875 for his work on the reflection and refraction of light. Three years later Lorentz was appointed Chair of Theoretical Physics at Leyden. In 1912 he became Director of the Teyler Institute in Haarlem and Secretary of the Dutch Society of Sciences. Nevertheless, remaining faithful to his alma mater he never relinquished his position as Professor at Leyden where he was well-known for weekly lectures, which he continued from 1912 for the remainder of his life.

For the entire of his scientific career Lorentz did much to develop and extend the theory of James Clerk Maxwell (1831–1879), that electric current oscillations would create alternating electric and magnetic fields, and radiated electromagnetic waves would have the same physical properties of light. His early studies on electromagnetic radiation were focused on defining empirical formulae for the varying velocities of light in differing liquid, solid, and gaseous substances, and hence the index of refraction, as a function of the density and chemical composition of the media. This work has relevance to our understanding of the threshold energies for the production of Cherenkov radiation by charged particles in media of differing refractive indices (see Chapter 7).

Lorentz is best known for his later work, which resulted in the derivation of mathematical formulae that defined the phenomena known today as the Lorentz transformations. These calculations were based on his finding that electromagnetic forces between charges are subject to slight alterations due to their motion, resulting in minute contractions in the sizes of the moving bodies. The Lorentz transformations are formulae that relate the space and time coordinates of two inertial observers moving at relative speeds up to but less than the speed of light (2.99×10^8 m/sec); and these transformations permit the calculations for the solutions of the resulting changes in size, mass, and time for the moving objects. In his Nobel Lecture Lorentz provided the following explanation for these transformations: "Consider for example the case in which water is flowing along a tube and a beam of light is propagated within this water in the direction of flow. If everything that is involved in the light vibrations is subject to the flowing movement, then the propagation of light in the flowing water to the rate of flow of the water will in relation to the latter behave in exactly the same way as in still water. The velocity of propagation relative to the wall of the tube can be found by adding the velocity of propagation in the water to the rate of flow of the water, just as, if a ball is rolling along the deck of a ship in the direction in which it is traveling, the ball moves relative to an observer on the shore as the sum of the two speeds—the speed of the ship and the speed at which the ball is rolling." (*Nobel Lectures, Physics, 1901–1921*). Lorentz's concepts and his mathematical transformations were the precursor to Einstein's theory of relativity. The Lorentz transformations have proven to be very useful in making calculations for the time dilation that cosmic shower radiations are observed to undergo when traveling at very fast (relativistic) speeds towards the earth. A practical example of the use of the Lorentz transformations is provided in Chapter 6 for the case of muons of very short lifetime (2×10^{-6} sec) created in the upper atmosphere and traveling at close to the speed of light towards earth. The muons undergo a time dilation that increase their lifetime and permit an observer to detect their arrival on the earth's surface.

PIETER ZEEMAN (1865–1943)

Pieter Zeeman was born on May 25, 1865 at Zonnemaire, a village on the Isle of Schouwen, The Netherlands. He shared the Nobel Prize in Physics 1902 with Hendrick Lorentz "in recognition of the extraordinary service they rendered by their researches into the influence of magnetism upon radiation phenomena."

RÉPUBLIQUE DE GUINÉE
OFFICE DE LA POSTE GUINÉENNE
750F
1902
PIETER ZEEMAN
Zeeman (1865-1943)

Zeeman studied at Leyden University from 1885 to 1893, the year he was awarded the doctoral degree. At Leyden he was a pupil of two Nobel Laureates, Heike Kamerlingh Onnes (Physics 1913) and Hendrick Lorentz, with whom he shared the Nobel Prize in 1902. In 1896 Pieter Zeeman discovered that a magnetic field would alter the frequency of light emitted by a glowing gas, called the Zeeman effect. The magnetic field would alter the spectral line corresponding to the frequency of the light into a triplet or even a more complex multiplet of lines. This discovery was significant because it was the very first to link the properties of light with electricity and magnetism. Zeeman explained the origin of the light and how the magnetic field would have an effect on its line of emission based on theory of Lorentz. In his Nobel Lecture he stated, "This theory assumes that all bodies contain small electrically charged mass particles, "electrons", and that all electrical and optical processes are based on the position and motion of the "electrons"…as soon as the electron is exposed to the effect of a magnetic field, its motion changes." Zeeman reported his work in the *Proceedings of the Physical Society of Berlin* in 1896, and this was published in English language in *Nature* on February 11, 1897. At the time of Zeeman's discovery not much was known of electrons or of their position in atoms, before it could be explained fully that light emissions originated from electron orbital transitions in atoms, and that electrons had an intrinsic spin and their own magnetic field, the alignment of which could be controlled by external magnetic fields. It was only the year after Zeeman's discovery did Thomson publish his paper in 1897 where he reported the unequivocal discovery of the electron, the first atomic particle to be discovered, and reported its properties including charge and mass.

JOSEPH JOHN THOMSON (1856–1940)

J.J. Thomson was born on December 18, 1856 in Cheetham Hill, a suburb of Manchester, England. He was awarded the Nobel Prize in Physics 1906, as stated by the Nobel Committee "in recognition of the great merits of his theoretical investigations on the conductivity of electricity by gases."

Thomson became a Fellow of Trinity College, Cambridge in 1880 and remained a member of the college for the remainder of his life. He was Cavendish Professor of experimental physics at Cambridge from 1884 to 1918. During Thomson's tenure at the Cavendish Laboratory at Trinity College, Rutherford received a scholarship in 1894 as a research student at the Cavendish Laboratory.

RÉPUBLIQUE DE GUINÉE
OFFICE DE LA POSTE GUINÉENNE

750F

1906
JOSEPH THOMSON

J.J. Thomson (1856-1940)

Rutherford was only 23 years of age then, and Thomson recognized Rutherford as a budding scientist. Rutherford worked with Thomson for a few years up to about 1897 on the behavior of ions in gases produced by x-rays, on the mobility of ions in electric fields, and on the production of ions from metals exposed to light (photoelectric effect).

Thomson's greatest contribution to physics was the discovery of the electron and its properties, the first atomic particle to be discovered. To identify and study the electron Thomson used a cathode-ray tube illustrated in Figure I.4, which is drawn as it appeared in Thomson's original paper (Thomson, 1897) with the exception that the lines illustrating possible paths of travel of the electrons were added to the original sketch. The apparatus was used to study the cathode rays, which were then only unknown rays emitted from the cathode C. The rays were accelerated by positively charged anodes A and B with slits to collimate the beam. The beam of rays would travel across the length of the tube to the far right, which was coated with a fluorescent screen. The fluorescent screen would allow a visual measurement of the location of the beam that hit the glass wall at the far right hand of the tube. Upon striking the fluorescent screen the beam would produce a pinpoint green-colored phosphorescence similar to the phosphorescence we would see on old computer screens. At the time of Thomson's work there was controversy as to the nature of these rays. As explained by Thomson in his Nobel Lecture "some thought the rays were negatively-charged electrified bodies shot off from the cathode with high velocity; the other view was that the rays were some kind of ethereal vibration or waves." Thomson demonstrated that the rays consisted of negatively charged particles by using both electric and magnetic fields to deflect the cathode rays. When a positive potential was applied to electrode D with respect to electrode E the cathode ray would deflect upwards as illustrated and travel onwards to a position G; the degree of deflection depending on the potential applied to the electrode D. Likewise, when the electric field applied to D and E was shut off, and a magnetic field applied perpendicular to the rays, the rays could be seen to be deflected in the opposite direction to a position H on the phosphorescent screen, the degree of deflection again dependent on the strength of the magnetic field. Earlier experimenters could not observe the cathode-ray deflections due to electrical or magnetic fields, because the experiments were carried out with a gas present inside the tube; and as noted by Thomson, "the gas will become a conductor of electricity and the rays will be surrounded by a conductor, which will screen them (the electrons) from the effect of electric force."

Thomson applied both the electric and magnetic fields simultaneously to determine the velocity of the particles, which he called "negative corpuscles." Firstly, when the magnetic field was applied, he defined the force acting upon the particles traveling perpendicular to the lines of magnetic force as Hev where H is the magnetic force, e is the charge on the particle, and v is its velocity. Likewise, he defined the electric force acting upon the particle as Xe, where X was the electric force. Thomson would then adjust the magnetic field to exactly balance the force due to the electric field so that the beam would not undergo any deflection and travel straight onto point F in Figure I.4. In such as case,

$$Hev = Xe$$

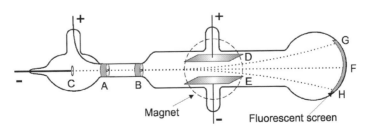

Figure I.4 Thompson's cathode-ray tube. The cathode rays are emitted from the negatively-charged plate C, accelerated and collimated by positively charged anodes A and B. An electric field potential is applied to the cathode rays by plates D and E and the magnetic field is applied by an external magnet illustrated by the circle.

and

$$v = \frac{X}{H}$$

Since the electric and magnetic fields of force, X and H respectively, were applied, these could be eas-ily measured. Thomson, thereby calculated the velocity of the particles at 60,000 miles/sec or 1/3 the speed of light. In Thomson's words "their velocity is much greater than the velocity of any other moving body with which we are acquainted."

Thomson went further now to determine the charge-to-mass ratio (e/m) of this new negative particle. The degree (angle) of deflection of the beam, due to either electric or magnetic force fields, is a function of the charge, mass, and velocity of the particle. From the angle of deflection, Thomson calculated the ratio e/m to be 1.7×10^{11} C/kg. This was very close to the currently accepted value of 1.758820×10^{11} C/kg. The new particle was also the smallest particle to be known at that time, as in Thomson's own words "for the corpuscle in the cathode rays the value e/m is 1,700 times the value of the corresponding quantity for the charged hydrogen atom."

Thomson noted that this new negatively charged particle not only occurred in the cathode rays, but it was widely distributed. As Thomson stated, "They are given off by metals when raised to a red heat; indeed any substance when heated gives off corpuscles to some extent. The corpuscles (electrons) are also given off by alkali metals, when these are exposed to light." (This phenomenon was later on stud-ied by others as the photoelectric effect.) He also added, "They are being continuously given out in large quantities and with very great velocities by radioactive substances." It was not known during Thomson's work; but it was discovered years later that beta particles from radioactive sources and electrons only differed from their origins. Beta particles originate from the nuclei of unstable or radioactive atoms and electrons originate from the outer regions or orbitals of atoms.

The term "electron" was coined by George Johnstone Stoney, Professor of Queen's College (now University College) Galaway, Ireland in a paper in *The Scientific Transactions of the Royal Society*, Dublin in 1891.

PHILIPP LENARD (1862–1947)

Philipp von Lenard was born on June 7, 1862 in Pozsony, a part of the Austro-Hungarian Empire, which is now Bratislava, Slovakia. He received the Nobel Prize in Physics 1905 "for his work on cathode rays."

Lenard completed his Ph.D. degree in physics in 1886 at the University of Heidelberg. He subsequently worked as a privatdozent and assistant to Heinrich Hertz at the University of Bonn from 1892 to 1894, and after a few short stints as professor in other universities, he was appointed Professor Ordinarius at the University of Kiel in 1898. In 1892 at Bonn, Hertz had discovered that when aluminum foil was placed in the path of the cathode rays, within the cathode-ray tube, the rays would pass through the foil and cause uranium glass to glow (Hertz, 1892). A cathode-ray tube of the type used by Hertz is illustrated in the previous biographical sketch on J.J. Thomson. Lenard took it upon himself to test aluminum foil as a substance that could be used as a possible window through which the cathode rays might be directed out of the tube. After several experiments in 1894 he was successful in finding a proper thickness of aluminum foil that could be used to close an opening in the cathode-ray tube to assure the maintenance of a vacuum inside the tube (the cathode rays, i.e., electrons traveled best under maximum vacuum) while permitting the cathode rays to travel through the thin aluminum layer (Lenard, 1894). The aluminum-sealed opening became known as the "Lenard window." The window, for the first time, allowed cathode rays to exit the tube and come out into the "open air" although the rays could not travel very far in air (approximately 8 cm). Previously these rays, (i.e., electrons) were limited to the confines of the cathode-ray tube, the electrons generated were now able to be studied "out in the open."

Lenard extended another finding of Hertz. In 1887 Heinrich Hertz was the first to observe that metal plates would give off negative electricity (electrons) when exposed to UV light (Hertz, 1887).

Centenario de los Premios Nobel

Philipp E. A. von Lenard (Alemania) Física, 1905

CORREOS 1995

NICARAGUA ₡ 2.50

Lenard (1862-1947)

This phenomenon is known as the photoelectric effect (see Chapter 3). To study the photoelectric effect further Lenard constructed a special cathode-ray tube, illustrated in Figure I.5. The tube was equipped with a quartz window B through which UV light could pass. The UV light originating from position L would pass through the quartz window and strike a metal plate U within the evacuated tube. The impact of the UV photons onto the metal would force electrons to be ejected from the metal surface. These electrons, referred to as photoelectrons would then travel in the tube straight on toward the collector plate labeled α. The path of the photoelectrons could be deflected upwards toward the collector labeled β when the poles of a magnet, depicted by a dashed circle in the diagram, were placed perpendicular to the path of the electrons. The tube and external magnet (not visible) provided evidence that the particles emitted from the metal plate U were indeed electrons from data collected including angle of deflection, particle velocity, and magnetic field strength, from which the e/m ratio could be calculated (see bio-sketch on J.J. Thomson). The ion collectors α and β were connected to electrometers that would provide data on the photoelectron intensity via the photoelectric current produced at the collectors (Lenard, 1899).

Lenard was able to measure the photoelectron intensity from the photoelectric current, and the speed or kinetic energy of the photoelectrons as a function of the intensity and frequency of the light hitting the metal target (Lenard, 1902). He concluded that the electron velocity (kinetic energy) was independent of the light intensity, but that the velocity would vary only according to the wavelength of the light, and that highest velocities were attained for shorter wavelengths of light. Lenard provided excellent experimental data on the photoelectric effect, but erred in his interpretation of the facts, which is reflected in his misinterpretation stated as follows in his Nobel Lecture "I have found that the velocity [of the electrons] is independent of the UV light intensity, and thus concluded that the energy of escape [of the electrons from the metal surface] does not come from the light at all,..." It was only a few years later in 1905 when Albert Einstein fully explained the photoelectric effect after grasping Max Planck's calculations and showing that light not only traveled as waves but also as packets of energy or particles, which he named "energy quanta" and that the energy of the photoelectron did indeed come from the light and depended on the energy of the photon. Einstein was awarded the Nobel Prize in Physics 1921 for his discovery of the new law of the photoelectric effect (see Albert Einstein in *Radioactivity Hall of Fame, Part III* for a more thorough treatment of the photoelectric effect and Einstein's explanation of it.).

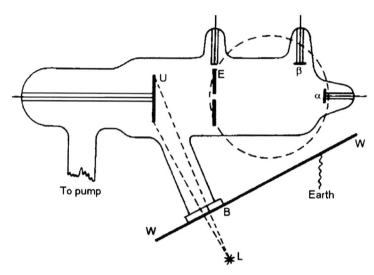

Figure I.5   Ray tube used by Philipp Lenard to produce photoelectrons and demonstrate that the photoelectron kinetic energy (velocity) was independent of the light intensity, but dependent only on the wavelength of the light producing the photoelectrons.

The following interpretation of Lenard is provided in *Nobel Laureates, Physics 1901–1921*, Elsevier Publ., "Lenard never forgave Einstein for discovering and attaching his own name to this law. Lenard was an experimentalist of genius, but more doubtful as a theorist. Some of his discoveries were great ones and others were very important, but he claimed for his discoveries more than their true value. He believed that he was disregarded and this probably explains why he attacked other physicists in many countries. He became a convinced member of Hitler's National Socialist Party and maintained unreserved adherence to it. The party responded by making him Chief of Aryan or German Physics."

– 1 –

Alpha Radiation

1.1 INTRODUCTION

The alpha particle, structurally equivalent to the nucleus of a helium atom and denoted by the Greek letter α, consists of two protons and two neutrons. It is emitted as a decay product of many radionuclides predominantly of atomic number greater than 82. For example, the radionuclide americium-241 (^{241}Am) decays by alpha-particle emission to yield the daughter nuclide ^{237}Np according to the following equation:

$$^{241}_{95}\text{Am} \rightarrow\, ^{237}_{93}\text{Np} + {}^{4}_{2}\text{He} + 5.63\,\text{MeV} \tag{1.1}$$

The loss of two protons and two neutrons from the americium nucleus results in a mass reduction of four and a charge reduction of two on the nucleus. In nuclear equations such as the preceding one, the subscript denotes the charge on the nucleus (i.e., the number of protons or atomic number, also referred to as the Z number) and the superscript denotes the mass number (i.e., the number of protons plus neutrons, also referred to as the A number). The 5.63 MeV of eq. (1.1) is the decay energy, which is described subsequently.

1.2 DECAY ENERGY

The energy liberated during nuclear decay is referred to as decay energy. Many reference books report the precise decay energies of radioisotopes. The value reported by Holden (1997a) in the Table of Isotopes for the decay energy of ^{241}Am illustrated in eq. (1.1) is 5.63 MeV. Energy and mass are conserved in the process; that is, the energy liberated in radioactive decay is equivalent to the loss of mass by the parent radionuclide (e.g., ^{241}Am) or, in other words, the difference in masses between the parent radionuclide and the product nuclide and particle.

We can calculate the energy liberated in the decay of ^{241}Am, as well as for any radioisotope decay, by accounting for the mass loss in the decay equation. Using Einstein's equation for equivalence of mass and energy

$$E = mc^2 \tag{1.2}$$

71

we can write the expression for the energy equivalence to mass loss in the decay of ^{241}Am as

$$Q = (M_{^{241}\text{Am}} - M_{^{237}\text{Np}} - M_\alpha) c^2 \qquad (1.3)$$

where Q is the disintegration energy released in joules, $M_{^{241}\text{Am}}$, $M_{^{237}\text{Np}}$, and M_α are the masses of ^{241}Am, ^{237}Np, and the alpha particle in kg and c is the speed of light in a vacuum, 3.00×10^8 m/sec). When the nuclide masses are expressed in the more convenient atomic mass units (u) the energy liberated in decay equations can be calculated in units of mega-electron volts according to the equation

$$Q = (M_{^{241}\text{Am}} - M_{^{237}\text{Np}} - M_\alpha)(931.494\,\text{MeV/u}) \qquad (1.4)$$

The precise atomic mass units obtained from reference tables (Holden, 1997a) can be inserted into eq. (1.4) to obtain

$$\begin{aligned} Q &= (241.056822\,\text{u} - 237.048166\,\text{u} - 4.00260325\,\text{u})(931.494\,\text{MeV/u}) \\ &= (0.00605275\,\text{u})(931.494\,\text{MeV/u}) \\ &= 5.63\,\text{MeV} \end{aligned}$$

The energy liberated is shared between the daughter nucleus and the alpha particle. If the parent nuclide (e.g., ^{241}Am) is at rest when it decays, most of the decay energy will appear as kinetic energy of the liberated less-massive alpha particle and only a small fraction of the kinetic energy remaining with the recoiling massive daughter nucleus (e.g., ^{237}Np). The kinetic energy of the recoiling daughter nuclide is comparable to that of a recoiling canon after a shell is fired; the shell being analogous to that of the alpha particle shooting out of the nucleus (Figure 1.1).

Figure 1.2 illustrates the transitions involved in the decay of ^{241}Am. The interpretation of this figure is given in the following paragraph. There are four possible alpha-particle transitions in the decay of ^{241}Am each involving an alpha-particle emission at different energies

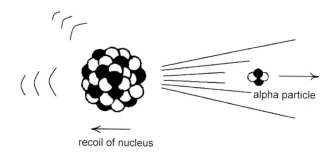

alpha particle

recoil of nucleus

Figure 1.1   The alpha particle shoots out of a nucleus. In the decay of nuclides by alpha-particle emission the alpha particle will shoot out of the nucleus, and the daughter nucleus will recoil similar to that of a canon that recoils upon firing a shell from its barrel.

and relative abundances. These are illustrated in Figure 1.2. The decay energy of 5.63 MeV for ^{241}Am calculated above and reported in the literature is slightly higher than any of the alpha-particle energies provided in Figure 1.2. This is because there remains also the recoil energy of the daughter nucleus and any gamma-ray energy that may be emitted by the daughter, when its nucleus remains at an excited state.

The recoil energy, E_{recoil}, of the daughter nucleus is a very small part of the total energy of transition from parent to daughter nuclide, as it is inversely proportional to the mass of the nucleus as described by the following equation derived by Ehman and Vance (1991)

$$E_{recoil} = \left(\frac{M_\alpha}{M_{recoil}} \right) E_\alpha \tag{1.5}$$

where M_α is the mass of the alpha particle as defined in eq. (1.3), M_{recoil} is the mass of the recoil nucleus and E_α is the alpha-particle energy. For example, the recoil energy of

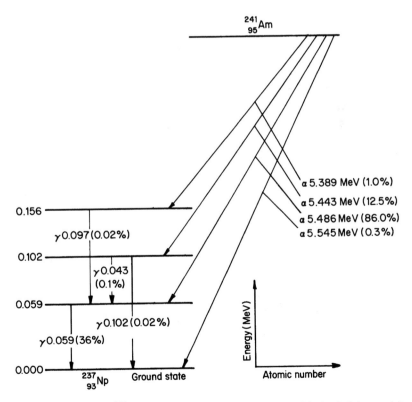

Figure 1.2 Decay scheme of ^{241}Am. The relative abundances (intensities) of alpha-particle and gamma-ray emissions are expressed in percent beside the radiation energy values in MeV.

the ^{237}Np daughter nucleus for the transition of the 5.545 MeV alpha particle (Figure 1.2) can be calculated according to eq. (1.5) as

$$E_{recoil} = \left(\frac{4.00260325 \, u}{237.0481 \, u} \right) 5.545 \, MeV$$

$$= (0.0168851)(5.545 \, MeV)$$

$$= 0.0936 \, MeV$$

The transition energy, E_{trans}, for the above alpha-particle emission is the sum of the alpha particle and recoil nuclear energies or

$$E_{trans} = E_\alpha + E_{recoil}$$

$$= 5.545 \, MeV + 0.0936 \, MeV \qquad (1.6)$$

$$= 5.63 \, MeV$$

In the above case, the transition energy turns out to be equal to the reported and calculated decay energy, because as illustrated in Figure 1.2 the ^{241}Am radionuclides decay directly to the ground state whenever 5.545 MeV alpha particles are emitted. This is not the case when alpha particles of other energies are emitted from ^{241}Am. If we take, for example, the 5.486 MeV alpha-particle transition of Figure 1.2, the decay energy, E_{decay}, would be the sum of the transition energy plus gamma-ray energy, E_γ, emitted from the daughter nucleus or

$$E_{decay} = E_{trans} + E_\gamma$$

$$= E_\alpha + E_{recoil} + E_\gamma$$

$$= E_\alpha + \left(\frac{M_\alpha}{M_{recoil}} \right) E_\alpha + E_\gamma \qquad (1.7)$$

$$= 5.486 \, MeV + (0.0168851)(5.486 \, MeV) + 0.059 \, MeV$$

$$= 5.486 \, MeV + 0.0926 \, MeV + 0.059 \, MeV$$

$$= 5.63 \, MeV$$

The gamma-ray energy emitted from the daughter nucleus for the 5.486 MeV alpha-particle transition in ^{241}Am decay is found in Figure 1.2. Gamma-ray energy values of other radionuclides are available from reference tables (Browne et al., 1986; Firestone et al., 1996; L'Annunziata, 2003d). The emission of gamma radiation often accompanies radionuclide decay processes that occur by alpha-particle emission. The properties and interactions of gamma radiation are discussed in Chapter 3.

As described in the previous paragraphs alpha particles are emitted with a certain quantum of energy as the parent nuclide decays to a lower energy state. The energy emitted from radionuclides as nuclear radiation can be described by a decay scheme such as that given in Figure 1.2. Decay schemes are written such that the energy levels of the nuclides are plateaus along the ordinate, and these energy plateaus are distributed along the abscissa according to atomic number. The alpha particles, as the example shows (Figure 1.2), are emitted with certain magnitudes of kinetic energy, which is most often expressed in units of megaelectron volts. The energies of alpha particles from most nuclear decay reactions fall within the range 1–10.5 MeV.

Alpha particles are emitted from unstable nuclei with discrete quanta of energy, often leaving the daughter nuclide at an excited energy state. In such cases, when the daughter nuclide occurs at an elevated energy state, it may reach the ground state via the emission of energy in the form of electromagnetic gamma radiation as illustrated in Figure 1.2.

The nuclei of daughter atoms of alpha-particle-emitting nuclides are often unstable themselves and may also decay by further alpha-particle emission. Thus, alpha-particle-emitting nuclides may consist of a mixture of radionuclides, all part of a decay chain, as illustrated in Figure 8.7 of Chapter 8. Additional reading on radionuclide alpha decay is available from Das and Ferbel (1994).

1.3   ALPHA-PARTICLE INTERACTIONS WITH MATTER

Now consider what happens to an alpha particle that dissipates its kinetic energy by interaction with matter. Alpha particles possess a double positive charge due to the two protons present. This permits ionization to occur within a given substance (solid, liquid, or gas) by the formation of ion pairs due to coulombic attraction between a traversing alpha particle and atomic electrons of the atoms within the material the alpha particle travels. The two neutrons of the alpha particle give it additional mass, which further facilitates ionization by coulombic interaction or even direct collision of the alpha particle with atomic electrons. The much greater mass of the alpha particle, 4 u, in comparison with the electron (5×10^{-4} u) facilitates the ejection of atomic electrons of atoms through which it passes, either by direct collision with the electron or by passing close enough to it to cause its ejection by coulombic attraction. The ion pairs formed consist of the positively charged atoms and the negatively charged ejected electrons. The alpha particle continues along its path suffering, for the most part, negligible deflection by these collisions or coulombic interactions because of the large difference in mass between the particle and the electron. Thus, an alpha particle travels through matter producing thousands of ion pairs (see the following calculation) in such a fashion until its kinetic energy has been completely dissipated within the substance it traverses.

In air, an alpha particle dissipates an average of 35 eV of energy per ion pair formed. Before it stops, having lost its energy, an alpha particle produces many ion pairs. For example, as a rough estimate, a 5 MeV alpha particle will produce 1.4×10^5 ion pairs in air before coming to a stop:

$$\frac{5{,}000{,}000 \, \text{eV}}{35 \, \text{eV/ion pair}} = 1.4 \times 10^5 \text{ ion pairs in air}$$

The thousands of interactions between a traveling alpha particle and atomic electrons can be abstractly compared with a traveling bowling ball colliding with stationary ping-pong balls. Because of the large mass difference of the two, it will take thousands of ping-pong balls to stop a bowling ball. The additional stopping force of electrons is the binding energy of the atomic electrons.

The amount of energy required to produce ion pairs is a function of the absorbing medium. For example, argon gas absorbs approximately 25 eV per ion pair formed and a semiconductor material requires only 2–3 eV to produce an ion pair. Ionization is one of the principal phenomena utilized to detect and measure radionuclides. The energy threshold for ion pair formation in semiconductor materials is approximately 10 times lower than in gases, which gives semiconductor materials an important advantage as radiation detectors when energy resolution in radioactivity analysis is an important factor.

In addition to ionization, another principle mechanism by which alpha particles and charged particles, in general, may impart their energy in matter is via electron excitation. This occurs when the alpha particle fails to impart sufficient energy to an atomic electron to cause it to be ejected from the atom. Rather, the atoms or molecules of a given material may absorb a portion of the alpha-particle energy and become elevated to a higher energy state. Depending on the absorbing material, the excited atoms or molecules of the material may immediately fall back to a lower energy state or ground state by dissipating the absorbed energy as photons of visible light. This process, referred as fluorescence, was first observed by Sir William Crookes in London in 1903 and soon confirmed by Julius Elster and Hans Geitel the same year in Wolfenbüttel, Germany. They observed fluorescence when alpha particles emitted from radium bombarded a zinc sulfide screen. In darkness, individual flashes of light were observed and counted on the screen with a magnifying glass with the screen positioned a few millimeters from the radium source. The phenomenon of fluorescence is very significant and utilized in the detection and measurement of radionuclides. Thus, as described in the previous paragraphs, alpha particles as well as other types of charged particles dissipate their energy in matter mainly by two mechanisms, ionization, and electron excitation.

Because the atomic "radius" is much bigger ($\approx 10^{-10}$ m) than the "radius" of the nucleus ($\approx 10^{-14}$ m), the interactions of alpha particles with matter via direct collision with an atomic nucleus are few and far between. In this case, though, the large mass of the nucleus causes deflection or ricocheting of the alpha particle via coulombic repulsion without generating any change within the atom. Such deflection was discovered in the early part of this century by Ernest Rutherford and his students Hans Geiger and Ernest Marsden (Rutherford, 1906; Geiger, 1908; Geiger and Marsden, 1909), who bombarded very thin gold foil (only 4×10^{-5} cm thick) with alpha particles and observed the occasional deflection of an alpha particle (approximately 1 in 20,000) by more than 90°, even directly backwards toward the alpha-particle source. Lord Rutherford took advantage of this discovery to provide evidence that the greater mass of an atom existed in a minute nucleus. In his own words at a seminar Rutherford expressed his reaction to the observed alpha-particle backscattering with the following statement, related by N. Feather (1940) in a biographical essay a few years following his death:

It was quite the most incredible event that ever happened to me in my life. It was almost as incredible as if you fired a 15-inch shell at a piece of tissue paper and it came back and hit you. On consideration, I realized that the scattering backwards must be the result of a single collision, and when I made calculations I saw that it was impossible to get anything of that order of magnitude unless you took a system in which the greater part of the mass of the atom was concentrated in a minute nucleus.

The discovery of the deflection of alpha particles by matter in the direction backward from which they came was, in the opinion of Rutherford, an opportunity to explore the structure of the atom. As Rutherford (1911) reported

It seems reasonable to suppose that the [α-particle] deflection of a large angle is due to a single atomic encounter...it should be possible from a close study of the nature of the deflection to form some idea of the constitution of the atom to produce the effects observed.

There was at that time little concrete evidence on the structure of the atom. To explain and predict the deflections he considered the atom as containing a large positive charge at its center surrounded by the distribution of negative electricity (electrons). By considering that the velocity of a massive positively charged particle (alpha particle) would not be changed appreciably by its passage through the atom, Rutherford derived the path that the alpha particle would take as it approached close to the center of an atom. He stated

the path of the α-particle under the influence of a repulsive force [coming from the nucleus] varying inversely as the square of the distance will be a hyperbola with the center of the atom S as the external focus. (The original figure of the hyperbola, taken from Rutherford's paper of 1911, is illustrated here on a stamp, that was issued in 1971 commemorating 100 years since Rutherford's birth). Suppose the particle to enter the atom in the direction PO (O is the position of the circle in the hyperbola illustrated on the commemorative postage stamp), and that the direction of motion on escaping the atom is OP′. OP and OP′ make equal angles with the line SA, where A is the apse of the hyperbola. p = SN = perpendicular distance from the [atom] center on direction of initial motion of the α-particle.

By assigning two velocity variables to the alpha particle entering the atom to position A and departing or backscattering from position A, and considering angular momentum and conservation of energy, Rutherford derived the equation $b = 2p \cot \theta$ for the calculation of the factor b, which would be the distance of closest approach of the alpha particle to the atom center. Initial estimates from his paper of 1911 gave the value of b for an alpha particle of velocity 2.09×10^9 cm/sec to be 3.4×10^{-12} cm.

Rutherford went even further to make use of this interaction to determine the nuclear radius of aluminum. By selecting a metal foil of low Z (aluminum, $Z = 13$) and thus low Coulomb barrier to alpha penetration, and applying alpha particles of high energy (7.7 MeV) whereby defined alpha-particle scattering at acute angle due to coulombic repulsion would begin to fail, Rutherford (1919, 1920) was able to demonstrate that the distance of closest approach of these alpha particles to the atom center according to Coulomb's law was equivalent to the nuclear radius of aluminum, $\sim 5 \times 10^{-15}$ m.

Scattering of alpha particles at angles of less than 90° may occur by coulombic repulsion between a nucleus and a particle that passes in close proximity to the nucleus. These deflected particles continue traveling until sufficient energy is lost via the formation of ion pairs. The formation of ion pairs remains, therefore, the principal interaction between alpha particles and matter.

1.4 ALPHA-PARTICLE RANGES

The high mass and charge of the alpha particle, relative to other forms of nuclear radiation, give it greater ionization power but a poorer ability to penetrate matter. In air, alpha particles may travel only a few centimeters. This short range of travel varies depending on the initial energy of the particle. For example, a 5.5 MeV alpha particle, such as that emitted by the radionuclide ^{241}Am previously described, has a range of approximately 4 cm in dry air at standard temperature and pressure, as estimated by empirical formulae, such as eqs. (1.8) and (1.9) provided below

$$R_{air} = (0.005E + 0.285)E^{3/2} \tag{1.8}$$

where R is the average linear range in cm of the alpha particle in air and E the energy of the particle in MeV. The empirical formula is applied for alpha particles in the energy range 4–15 MeV. According to calculations of Fenyves and Haiman (1969), the ranges of alpha particles with energies between 4 and 7 MeV can be estimated by using a simplified version of eq. (1.8) as follows:

$$R_{air} = 0.3E^{3/2} \tag{1.9}$$

Ranges of alpha particles in air over a more wide range of alpha-particle energy can be obtained from Figures A.1 and A.2 of Appendix A. A thorough treatment of range calculations for charged particles is available from Fenyves and Haiman (1969). The approximate 4 cm range of 5.5 MeV alpha particles in air is illustrated in Figure 1.3. There is no abrupt drop in the number of alpha particles detected at the calculated range of 4 cm owing to statistical variations in the number of collisions that the particles may have with air molecules and to variations in the amount of energy loss by the particles for each ion pair formed. After being halted, an alpha particle acquires two free electrons through coulombic attraction and is converted to helium gas.

In materials other than air, such as liquids and solids, the range of alpha particles is obviously much shorter owing to their higher densities, which enhance the number of collisions

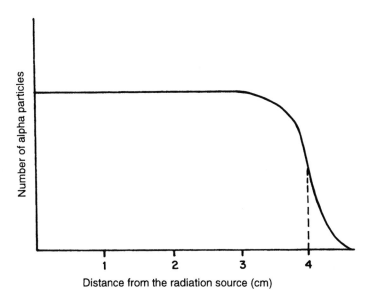

Figure 1.3   Range of 5.5 MeV alpha particles in air.

a particle may undergo per path length of travel. The range of alpha particles in liquids and solids may be approximated by comparison with ranges in air according to the following formula described in a previous text (L'Annunziata, 1987)

$$R_{cm} = 0.00032 \left(\frac{A^{1/2}}{\rho} \right) R_{air} \qquad (1.10)$$

where R_{cm} is the average range in cm of the alpha particle in an absorber other than air, A the atomic weight of the absorber, ρ the absorber density in g/cm^3, and R_{air} the calculated average linear range of the alpha particle in air (from eq. (1.8) or (1.9)). For example, the 5.5 MeV alpha particles emitted by ^{241}Am have a calculated linear range of only 2.4×10^{-3} cm or 24 μm in aluminum ($A = 27$ and $\rho = 2.69$ g/cm^3).

The linear ranges of alpha particles in liquids and solid materials are too short to measure with conventional laboratory instrumentation. The alternative is to express range in units of weight of absorber material per unit area, such as mg/cm^2, which is a measure of milligrams of absorber per square centimeter in the absorption path, or in other words, a measure of absorber thickness. If we multiply the linear range of the alpha particle measured in cm of absorber material by the density of the absorber in units of mg/cm^3, the range of the alpha particle in an absorber will be expressed in terms of the weight of absorber per unit area (mg/cm^2) as described by eq. (1.11), as follows

$$R_{mg/cm^2} = (R_{cm})(\rho) \qquad (1.11)$$

where R_{mg/cm^2} is the range of alpha particles of a given energy in units of mg/cm^2, also referred to as mass thickness units or material surface density, R_{cm} the linear range of the alpha particles, and ρ the absorber density. For example, the linear range of the 5.5 MeV alpha particles in aluminum calculated above with eq. (1.10) is converted to range in mass thickness units according to eq. (1.11) as follows

$$R_{\text{mg/cm}^2} = (2.4 \times 10^{-3} \text{ cm})(2690 \text{ mg/cm}^3) = 6.4 \text{ mg/cm}^2$$

Therefore, the mass thickness of 6.4 mg/cm^2 of aluminum absorber is sufficient to absorb alpha particles of 5.5 MeV energy.

Ranges of alpha particles as well as other charged particles such as protons and deuterons of a given energy in absorber elements of atomic number $Z > 10$ in units of absorber mass thickness can be calculated directly by comparison to the calculated range of the same charged particles of the same energy in air according to the following formula described by Friedlander *et al.* (1964)

$$\frac{R_Z}{R_{\text{air}}} = 0.90 + 0.0275Z + (0.06 - 0.0086Z) \log \frac{E}{M} \tag{1.12}$$

where R_Z is the range of the charged particle in mass thickness units, mg/cm^2, R_{air} the range of the charged particle in air in the same mass thickness units, Z the atomic number of the absorber element, E the particle energy in MeV, and M the mass number of the particle (i.e., 1 for protons, 2 for deuterons, and 4 for alpha particles). For example, if we use the empirical formula provided above (eq. (1.12)) to calculate the range of 5.5 MeV alpha particles ($M = 4$) in aluminum ($Z = 13$), we obtain the value of $R_Z = 6.1$ mg/cm^2, which is in close agreement to the mass thickness range calculated previously. In this example, eq. (1.12) requires the value of R_{air} for 5.5 MeV alpha particles, which is determined according to eq. (1.11) as the product of the 5.5 MeV alpha-particle linear range in air (previously calculated) and the density of air at STP ($\rho = 1.226$ mg/cm^3), that is, $R_{\text{air}} = (4 \text{ cm})$ $(1.226 \text{ mg/cm}^3) = 4.90$ mg/cm^2. The formula provided by eq. (1.12) is applicable to charged particles over a wide range of energies (approximately over the range 0.1–1000 MeV) and for absorber elements of $Z > 10$. For lighter absorber elements the term $0.90 + 0.0275Z$ is replaced by the value 1.00 with the exception of hydrogen and helium, where the value of 0.30 and 0.82 are used, respectively (Friedlander *et al.*, 1964).

Where alpha particles alone are concerned, the range in mass thickness units can be calculated according to eq. (1.13) described by Ehman and Vance (1991), as follows

$$R_{\text{mg/cm}^2} = 0.173E^{3/2}A^{1/3} \tag{1.13}$$

where E is the energy of the alpha particle in MeV and A the atomic weight of the absorber. If we continue to use the 5.5 MeV alpha particles emitted from ^{241}Am as an example, we can calculate their range in mass thickness units in aluminum according to eq. (1.13) as follows

$$R_{\text{mg/cm}^2} = 0.173(5.5)^{3/2}(27)^{1/3} = 6.6 \text{ mg/cm}^2$$

Ranges reported in mass thickness units (mg/cm^2) of absorber can be converted to linear range (cm) in that same absorber material from the absorber density (ρ) from the relationship described in eq. (1.11) or

$$R_{cm} = \frac{R_{mg/cm^2}}{\rho} \tag{1.14}$$

For example, the linear range of the 5.5 MeV alpha particles in aluminum ($\rho = 2.69\,g/cm^3$) is calculated as

$$R_{cm} = \frac{6.6\,mg/cm^2}{2690\,mg/cm^3} = 0.0024\,cm = 24\,\mu m$$

When the absorber material is not a pure element, but a molecular compound (e.g., water, paper, polyethylene, etc.) or mixture of elements, such as an alloy, the ranges of alpha particles in the absorber are calculated according to eq. (1.15) on the basis of the atomic weights of the elements and their percent composition in the absorber material or, in other words, the weight fraction of each element in the complex material. Thus, the range in mass thickness units for alpha particles in absorbers consisting of compounds or mixtures of elements is calculated according to the equation

$$\frac{1}{R_{mg/cm^2}} = \frac{w_1}{R_1} + \frac{w_2}{R_2} + \frac{w_3}{R_3} + \cdots + \frac{w_n}{R_n} \tag{1.15}$$

where R_{mg/cm^2} is the range of the alpha particles in mass thickness of the complex absorber material, and $w_1, w_2, w_3, \ldots w_n$ the weight fractions of each element in the absorber, and $R_1, R_2, R_3, \ldots R_n$ the ranges in mg/cm^2 of the alpha particle of defined energy in each element of the absorber. For example, the range of 5.5 MeV alpha particles in Mylar (polyethylene terephthalate) in units of mass thickness are calculated as follows

$$\frac{1}{R_{mg/cm^2}} = \frac{w_C}{R_C} + \frac{w_H}{R_H} + \frac{w_O}{R_O}$$

where w_C, w_H, and w_O are the weight fractions of carbon, hydrogen, and oxygen, respectively, in Mylar and R_C, R_H, and R_O are the mass thickness ranges of the alpha particles in pure carbon, hydrogen, and oxygen, respectively. The ranges of 5.5 MeV alpha particles in carbon, hydrogen, and oxygen are calculated according to eq. (1.13) as

$$R_C = 0.173(5.5)^{3/2}\,(12)^{1/3} = 5.10\,mg/cm^2$$

$$R_H = 0.173(5.5)^{3/2}\,(1)^{1/3} = 2.23\,mg/cm^2$$

$$R_O = 0.173(5.5)^{3/2}\,(16)^{1/3} = 5.62\,mg/cm^2$$

The weight fractions of the carbon, hydrogen, and oxygen in Mylar [-(C$_{10}$H$_8$O$_4$)$_n$-] are calculated as

$$w_C = \frac{(12 \times 10)}{192} = 0.625$$

$$w_H = \frac{(1 \times 8)}{192} = 0.042$$

$$w_O = \frac{(16 \times 4)}{192} = 0.333$$

The calculated ranges of the 5.5 MeV alpha particles in each element and the values of the weight fractions of each element in Mylar can now be used to calculate the alpha-particle range in Mylar in mass thickness units according to eq. (1.15) as

$$\frac{1}{R_{Mylar}} = \frac{0.625}{5.10} + \frac{0.042}{2.23} + \frac{0.333}{5.62} = 0.200$$

$$R_{Mylar} = \frac{1}{0.200} = 5.0 \text{ mg/cm}^2$$

The linear range of these alpha particles in Mylar are obtained from range in mass thickness units and the density of Mylar ($\rho = 1.38$ g/cm^3) as

$$R_{cm} = \frac{5.0 \text{ mg/cm}^2}{1380 \text{ mg/cm}^3} = 0.0036 \text{ cm} = 36 \,\mu\text{m}$$

To provide illustrative examples the values of the ranges of 5.5 MeV alpha particles in units of mass thickness of various absorber materials are given in Table 1.1. These values represent the milligrams of absorber per square centimeter in the alpha-particle absorption path required to stop the particle. It can be difficult to envisage alpha-particle distance of travel from the values of range when expressed in units of mass thickness. However, it is intuitively obvious that, the greater the charge on the nucleus of the absorber (i.e., absorber atomic number, Z), the greater the atomic weight of the absorber (A), and the greater the absorber density (ρ), the shorter will be the path length of travel of the alpha particle through the absorber. This is more evident from the calculated values of linear range of 5.5 MeV alpha particles in various gaseous, liquid, and solid absorbers provided in Table 1.2. From the linear ranges we can see that 5.5 MeV alpha particles could not pass through fine commercial aluminum foil 0.0025 cm thick. Although commercial paper varies in thickness and density, the linear range in paper calculated in Table 1.2 illustrates that 5.5 MeV alpha particles would not pass through 0.0034 cm thick paper, which has an average density value of 1.45 g/cm^3.

Table 1.1

Ranges of 5.5 MeV alpha particles in various absorbers in units of surface density or mass thickness

Water[a]	Paper[a,b]	Aluminum[c]	Copper[c]	Gold[c]
$4.8 \, mg/cm^2$	$4.9 \, mg/cm^2$	$6.6 \, mg/cm^2$	$8.9 \, mg/cm^2$	$12.9 \, mg/cm^2$

[a] Calculated with empirical formula provided by eq. (1.15) on the basis of the weight fraction of each element in the absorber.
[b] Cellulose, $(C_6H_{10}O_5)_n$ calculated on the basis of the weight fraction of each element in the monomer.
[c] Calculated with empirical formula provided by eq. (1.13).

Table 1.2

Linear ranges of 5.5 MeV alpha particles in various absorbers in units of cm and μm or 10^{-6} m

Air[a]	Water[b]	Mylar[b,c]	Paper[b,d]	Aluminum[b]	Copper[b]	Gold[b]
4 cm	0.0048 cm	0.0036 cm	0.0034 cm	0.0024 cm	0.001 cm	0.00075 cm
40,000 μm	48 μm	36 μm	34 μm	24 μm	10 μm	7.5 μm

[a] Calculated with empirical formula provided by eqs. (1.8) and (1.9).
[b] Calculated by dividing the range in mass thickness by the absorber density according to eq. (1.14).
[c] Polyethylene terephthalate, $\rho = 1.38 \, g/cm^3$.
[d] Cellulose, $(C_6H_{10}O_5)_n$, $\rho = 1.45 \, g/cm^3$.

Also, the alpha particles of the same energy would not pass through a layer of Mylar only 0.0036 cm thick. Mylar is a polymer sometimes used as a window for gas ionization detectors. From our previous calculations in this chapter we can see that a Mylar window of mass thickness 5 mg/cm^2 would not allow 5.5 MeV alpha particles to pass into the gas ionization chamber. A sample emitting such alpha particles would have to be placed directly into the chamber in a windowless fashion to be detected and counted.

From the above treatment it is clear that the range of alpha-particle travel depends on several variables including (i) the energy of the alpha particle, (ii) the atomic number and atomic weight of the absorber, and (iii) the density of the absorber. The higher the alpha-particle energy, the greater will be its penetration power into or through a given substance as more coulombic interactions of the alpha particle with the electrons of the absorber will be required to dissipate its energy before coming to rest. Also, if we consider an alpha particle of given energy, their ranges will be shorter in absorbers of higher atomic number or atomic weight, as the absorber atoms will contain a higher number of atomic electrons, and consequently increase the number of coulombic interactions of the alpha particle per path length of travel.

As the alpha particle travels through air and undergoes energy loss via numerous collisions, the velocity of the particle obviously diminishes. At reduced velocity and consequently reduced momentum, an alpha particle is more affected by coulombic attraction within the vicinity of a given atom. Progressive reduction in the velocity of travel of the alpha particle therefore results in an increase in the number of ion pairs produced per millimeter of path

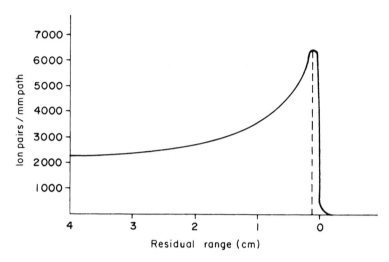

Figure 1.4   Specific ionization of an alpha particle in air along its range of travel.

length of travel. The increase in ionization per path length of travel of an alpha particle is illustrated in Figure 1.4. The highest specific ionization (number of ion pairs formed per millimeter of path) occurs shortly before termination of the alpha particle's travel, some 2 or 3 mm before the end of its range.

Radioactivity Hall of Fame—Part II

Frederick Soddy (1877–1956), C.T.R. Wilson (1869–1959), Frédéric Joliet (1900–1958) and Irène Joliet-Curie (1897–1956), Enrico Fermi (1901–1954), Wolfgang Pauli (1900–1958), Frederick Reines (1918–1998), and Clyde Cowan, Jr. (1919–1974)

FREDERICK SODDY (1877–1956)

Frederick Soddy was born in Eastbourne, Sussex, England on September 2, 1877. He was awarded the Nobel Prize in Chemistry 1922, in the words of the Nobel Committee, "for his contributions to our knowledge of the chemistry of radioactive substances, and his investigations into the origin and nature of isotopes".

Frederick Soddy (1877-1956)

Soddy studied at Eastbourne College and the University College of Wales, Aberystwyth and, after obtaining a scholarship at Merton College, Oxford, he graduated with highest honors in chemistry in 1898. After a couple of years of research at Oxford, Frederick Soddy joined the Chemistry Department at McGill University, Montreal, Canada where he worked with Rutherford during the period of 1900–1902. It is with Rutherford at McGill University where Soddy began to do pioneering work in the field of radioactivity. Rutherford (1900) was the first to define the concept of radioactive decay in terms of half-life, that is, the time it takes for a radioactive sample to lose one-half of its original radioactivity. In 1900 Rutherford had also determined the equations that define the rates of decay of radioactive atoms (see Chapter 8 and eq. (II.1) below). Subsequently Rutherford and Soddy (1902) reported the rates of transformation of one radioactive element (parent nuclide) into another element (daughter nuclide). They demonstrated that the rate of decay of the parent nuclide would govern the rate of formation of the daughter nuclide. The terminology "parent" and "daughter" nuclides were not used by them in this initial report, because such nomenclature had not yet been adopted at this early stage of development. They defined the exponential decay equation for radioactive nuclides in accord with Rutherford's initial mathematical findings (Rutherford, 1900), which are now used universally in this field of science, namely,

$$\frac{I}{I_0} = e^{-\lambda t} \tag{II.1}$$

where I is the intensity of the radioactivity after a given time t measured in sec, min, days, or years, I_0 the initial intensity of radioactivity or the intensity at the time observations are initiated, λ the decay constant with units of reciprocal time, and e the base to the natural logarithm (see eq. (8.6) and Chapter 8 for a more detailed treatment of radionuclide decay). They also defined another equation, now widely used, that defines the rate of production of daughter element (nuclide) from the parent as

$$\frac{I}{I_{max}} = 1 - e^{-\lambda t} \tag{II.2}$$

where I is the intensity of radioactivity of the daughter after a given period of time t, and I_{max} the intensity of the radioactivity of the daughter nuclide when its maximum value is reached. See eq. (8.33) in Chapter 8 for the practical applications of this equation. The important conclusions, made by Rutherford and Soddy from their work, were concerned with the nature of radioactivity. They stated in their paper:

> this property (radioactivity) is the function of the atom and not of the molecule. Uranium and thorium possess the property (radioactivity) in whatever molecular condition they occur…the intensity of the radiation appears to depend only on the quantity of active element present.

After completing the above work with Rutherford, Soddy moved to the University College, London where he worked with Sir William Ramsay. In their joint work Ramsay and Soddy (1903) demonstrated that the alpha particles that constitute the radioactive emanations from radium turned into helium gas. They trapped the emanations from 50 mg of radium bromide into a glass tube and studied the contents by spectroscopic means. At first they could not recognize the spectrum as that of any known element, but after a few days the contents of the tube produced spectroscopic lines characteristic of helium. This was an important discovery, because it provided evidence that alpha particles consisted of helium nuclei. It is now known that alpha particles, when they lose their energy traveling through matter, they come to a stop and acquire two free electrons and form helium gas. The following year Sir William Ramsay was awarded the Nobel Prize in Chemistry 1904 for his discovery of the inert gases, such as helium, and their location in the periodic table.

After two short stints of work under the influence of the two Nobel Laureates Ernest Rutherford and William Ramsay, Frederick Soddy went on at the young age of 27 to the University of Glasgow where he became lecturer in physical chemistry and radioactivity. He remained at Glasgow University from 1904 to 1914 during which time he would conduct yet more pioneering work on radioactivity. While at Glasgow Soddy carried out studies that established the "displacement law" or "periodic law" of radioactive elements (radionuclides). In the journal *Nature*, Soddy (1913a) stated the law clearly in the following words

> The successive expulsion of one α and two β particles in three radioactive changes in any order brings the inter-atomic charge of the element back to its initial value, and the element back to its original place in the (Periodic) table, though its atomic mass is reduced by four units.

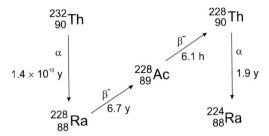

Figure II.1   Soddy's displacement or periodic law as illustrated with thorium and radium isotopes.

An example of Soddy's displacement law can be taken from a part of Figure 8.7 in Chapter 8, which is illustrated above in Figure II.1.

As illustrated in Figure II.1, Soddy demonstrated that the emission of an alpha particle would produce an element that is two atomic numbers lower, while the emission of a beta particle would displace an atom one atomic number higher. In the illustration above, we see that the radioactive atom of thorium-232 emits an alpha particle to become an atom of radium-228, which in turn decays by the emission of a beta particle to become radioactive Actinium-228. The Actinium atom also decays by the emission of a beta particle to become thorium, the same element of atomic number 90 from which it had originated, but with a mass of four units less. The emission of an alpha particle from the thorium-228 brings the atom back to the element radium with four mass units less than the atom of radium-228. The parent and product atoms of atomic number 90 are of the same element thorium, but of different mass. These two distinct atoms of thorium and of radium illustrated above are today known as "isotopes", because Soddy (1913a,b) coined the word "isotopes" to refer to such atoms that have the same atomic number (i.e., the number of protons in the nucleus) but different mass number (i.e., number of protons plus neutrons in the nucleus). Soddy recognized that these are atoms of the same element in the periodic table but of different mass. In his words Soddy (1913a) explained

> I call (them) "isotopes" or "isotopic elements", because they occupy the same place (from the Greek) in the periodic table. They are chemically identical, and save only as regards the relatively few physical properties which depend on atomic mass directly, physically identical also.

Soddy went further by alluding to the fact that not only can we consider the radioactive atoms as "isotopes" of known elements, but that many, if not most, of the stable elements may actually consist of a mixture of isotopes. In his Nobel Lecture, Soddy (1922) gave credit to the Swedish scientists Strömholm and Svedberg (1909a,b) by quoting them as follows:

> Perhaps, one can see, as an indication in this direction, the fact that the Mendeleev scheme (Periodic Table of Elements) is only an approximate rule as concerns the atomic weight, but does not possess the exactitude of a natural law; this would not be surprising if the elements of the scheme were mixtures of several homogeneous elements of similar but not completely identical atomic weight.

In a subsequent paper (Soddy, 1913c) and in his Nobel Lecture (1922) Soddy mapped out his periodic law, which shed much light on the identification of isotopes. Figure II.2 is the original Figure 2, taken from Soddy's Nobel Lecture (1922), which illustrates his charting of the radioactive elements and the periodic law.

This work advanced considerable knowledge as to the identities of radioactive atoms and where these resided in the periodic table of elements. The radioactive atoms were then identified, just as they are to this day, by their decay rate (i.e., half-life), energies and types of radiation they emitted, or from their parent atoms. But in the early part of the 20th century, the radioactive atoms were named after the parent atom, for one of the two reasons: it was thought that the daughter atom was of the same element as the

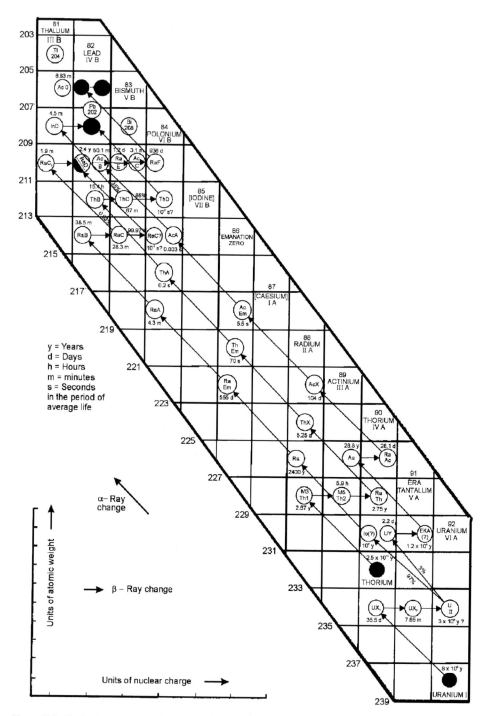

Figure II.2 Radioelements and periodic law. All elements in the same vertical column are isotopes…are chemically non-separable. (From Soddy (1922) with permission © The Nobel Foundation 1922.)

parent or it was simply not known what element the daughter atom was. For example, as illustrated in Figure II.2, the radioactive atoms previously referred to as thorium A, thorium B, and thorium C, originated from thorium, but were not thorium at all. Rather, the three radioactive atoms were isotopes of the elements polonium, lead, and bismuth, respectively. Soddy's work was the precursor to the Chart of the Nuclides, which is the current chart of all known isotopes of the atomic elements.

In 1914 at the age of 37 Soddy was appointed to the position of Professor of Chemistry and Chair at the University of Aberdeen where he remained until 1919 to take on a position of Professor of Chemistry at Oxford University. He remained at Oxford until 1936 retiring after the death of his wife.

It is interesting to note that Soddy's greatest contribution to science, the work for which he won the Nobel Prize, was confined to the short period of 1902–1914. After that period he diverted his attention from radioactivity to other interests, which were summarized succinctly in *Nobel Prize Lectures in Chemistry, Vol. 1, 1901–1921* (1967) as follows: "After his period at Glasgow (1914) he did no further work in radioactivity and allowed the later developments to pass him by. His interest was diverted to economic, social and political theories which gained no general acceptance, and to unusual mathematical and mechanical problems".

C.T.R. WILSON (1869–1959)

Charles Thomson Rees Wilson was born in Glencorse Parish near Edinburgh, Scotland on February 14, 1869. He received the Nobel Prize in Physics 1927 "for his method of making the paths of electrically charged particles visible by condensation of vapor".

Wilson initially studied biology at the University of Manchester with intentions of earning a degree in medicine. He won a scholarship to Cambridge in 1888 where his interests changed from medicine to physics and chemistry, and graduated in 1892.

**C.T.R Wilson
(1869-1959) with permission
© The Nobel Foundation 1927**

In his Nobel Lecture Wilson explained how he became interested in cloud and mist formation and eventually discovered a method of making visible the paths of travel of subatomic particles. During September of 1894 he spent a few weeks at an observatory located on the summit of Ben Nevis, the highest of the Scottish hills. The beautiful optical phenomena that the sun produced when its rays passed through the clouds surrounding the hilltop and the shadows cast on the mist and cloud glories excited his interest in the possibility of reproducing this effect in the laboratory. In early 1895 Wilson first tried to make clouds in the laboratory by expanding moist air. It was known then that dust particles in the air served as nuclei for the formation of clouds, and during his experimentation he discovered that even if all dust particles were removed whereby no cloud formation should occur, he could still achieve a cloud formation if the air expansion and supersaturation would exceed a certain limit. His studies indicated that a cloud would form in dust-free air that had an expansion ratio of $v_2/v_1 = 1.25$ where v_2 and v_1 are the expanded and initial volumes of air, respectively, resulting in an approximately fourfold supersaturation of air moisture. He also observed that an even more rapid critical dust-free air expansion yielded an approximately eightfold supersaturation of the water vapor, and expansions exceeding this limit produced dense cloud formation. It was clear to Wilson at this point in his work that a critical supersaturation of water vapor could be reached where cloud formation would occur with nothing more than the gas molecules of air.

The news of Röntgen's discovery of x-rays reached Scotland in a short time during late 1895, and by early 1896 J.J. Thomson was already studying the ionization (conductivity) of air by the newly discovered x-rays. That very year Wilson got hold of one of the x-ray tubes manufactured in Thompson's lab to see what effect these rays would have on his new synthetic clouds. When exposing the expanding moist air to the x-rays Wilson discovered that a dense long-lasting fog would be produced when the expansion ratio exceeded 1.25. The x-rays were apparently producing condensation nuclei in the super-saturated air to enable the formation of the long-lasting fog or mist. During the following 2 years Wilson studied the effects of several forms of radiation on the cloud formations including the effects of x-rays, the newly discovered Becquerel rays from uranium and UV light.

During the first decade of the 20th century the corpuscular nature of radioactivity, such as alpha and beta particles and the ionization that these particles produced in air, was known (see Rutherford in Radioactivity Hall of Fame—Part I). Sometime around 1910 Wilson decided to build his cloud chamber with the idea of possibly visualizing and even photographing the tracks or paths of travel of the various radiations by the ionization and consequent condensation or miniature clouds these might form. The expansion chamber or cloud chamber designed by Wilson in 1911 for these studies is illustrated in Figure II.3.

The diameter of the cylindrical cloud chamber (A) is 16.5 cm and the expansion depth (height) is 3.4 cm. The chamber contains a hollow piston or "plunger" (M) that moves up or down within a fixed outer cylinder according to the pressure differential within and outside of the piston. The chamber is sealed with a pool of water within which the piston is able to move. A vacuum is applied to the round glass chamber (C) with a suitable pump, and rapid expansion of the air within the chamber (region A) is achieved by opening the valve (B) to the vacuum chamber (C). The vacuum will suck air out of the hollow piston forcing it to move in the downward direction thereby rapidly expanding the moist air in the cloud chamber (A), which is the expansion space above the piston. The floor (I) of the cloud chamber drops suddenly when valve (B) is opened to the evacuated chamber (C) until it is brought to a sudden stop when the plunger hits the rubber-covered base (H). The expansion of the chamber would yield the supersaturation of moisture required for particle track formation while simultaneously triggering the flash of a camera. Pinch cocks (F) and (G) on rubber tubing connections allowed for opening communication with the atmosphere to control the air in the cloud chamber and adjustment of the plunger (piston) providing the desired volume to the chamber (approximately 750 ml). A hollow cylinder of wood (D) is enclosed inside the inner piston, which reduces the volume of air passing through the connecting tubes at each expansion. The top (L), walls (K), and floor (I) of the cloud chamber were made of glass. To provide for visualization and photography of the charged-particle tracks in the cloud chamber a dark background was provided by painting the base of the expansion chamber black and coating the walls with gelatin. Wilson found that an expansion ratio of at least 1.25, yielding a fourfold supersaturation of water vapor, was sufficient for negative particles such as Compton electrons produced by x-rays to condense water along the electron tracks in ionized air.

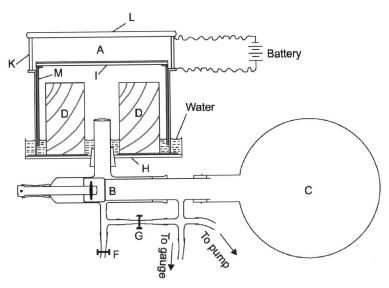

Figure II.3 Wilson Cloud Chamber. The charged-particle tracks were produced in the expansion space above the piston and these were observed and photographed with proper illumination as described in the text. An electric field potential was applied to the cloud chamber expansion space to remove any ions that may be present in the water vapor prior to the entry of ionizing charged particles under observation. The original Wilson Cloud Chamber is on exhibit at The Cavendish Laboratory Museum, University of Cambridge. (See Wilson (1911), (1912), and (1923) for details.)

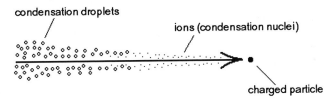

Figure II.4 The formation of condensation droplets onto ions formed along the path of travel of a charged particle in a Wilson Cloud Chamber.

However, for positively charged particles, such as alpha particles, the expansion ratio had to exceed 1.31 corresponding to an approximately sixfold supersaturation of water vapor to yield the condensation of the water vapor along the particle tracks. The water vapor would condense along the charged-particle path of travel as illustrated in Figure II.4.

The ions produced by the charged particle as it traveled through the air of the chamber would act as seeds or condensation nuclei on which very minute droplets of the size of large molecules would form. To achieve the proper conditions so that the charge particle tracked could be observed and photographed the following conditions had to be met in the words of C.T.R. Wilson in his Nobel Lecture:

> The expansion must be effected without stirring up the gas (air); this condition is secured by using a wide, shallow cloud chamber of which the floor can be made to drop suddenly and so produce the desired increase in volume. The cloud chamber must be freed not only from "dust" particles, but from ions other than those produced by the ionizing particles under observation.

It was not only desirable to observe the particle tracks, but also photograph these for subsequent detailed study. With this in mind Wilson further explained

For the purpose of obtaining sharp pictures of the tracks, the order of operations has to be firstly, the production of the necessary super-saturation by sudden expansion of the gas; secondly, the passage of the ionizing particles through the super-saturated gas (air); and finally, the illumination of the cloud condensed on the ions along the track. (From Nobel Lectures, Physics, 1922–1941.)

The commemorative postage stamp above illustrates numerous straight tracks produced in a Wilson Cloud Chamber by alpha particles emitted in all directions from a radium source. Other interesting particle-track photographs taken by Wilson with his cloud chamber and described at his Nobel Lecture on December 12, 1927 are illustrated in Figures II.5–II.8. The track in Figure II.5 is that of an alpha particle. Alpha-particle tracks are generally straight and wider than those produced by beta particles or electrons. This is because alpha particles produce a higher specific ionization (ion pairs per mm of path traveled) and they have a much higher mass than the electron. The usual straight path of an alpha particle may bend when it comes into close proximity to an atomic nucleus such as point A in Figure II.5; or the alpha particle may deflect sharply if it collides with a nucleus as illustrated by point B in Figure II.5. In Figure II.6 Wilson illustrates how the beta particle of much lower mass than the alpha particle produces tracks that are thinner than those produced by alpha particles, and they will also bend or undergo numerous deflections by collisions with atomic electrons. The collisions of electrons with other electrons is illustrated in Wilson's photograph of Figure II.7. This photograph

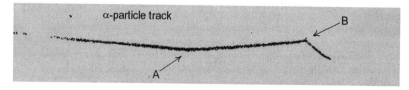

Figure II.5   A track produced by an alpha particle. The bend (A) and deflection (B) in the track were interpreted by Wilson as a close encounter of the alpha particle with an atomic nucleus and a collision with a nucleus, respectively. (From Wilson (1927) with permission © The Nobel Foundation 1927.)

β-particle track

Figure II.6   A track of a beta particle. Beta particles and electrons yielded thinner tracks than alpha particles. The tracks would be straight for high-energy electrons, but low-energy electrons would have more tortuous paths as illustrated above. (From Wilson (1927) with permission © The Nobel Foundation 1927.)

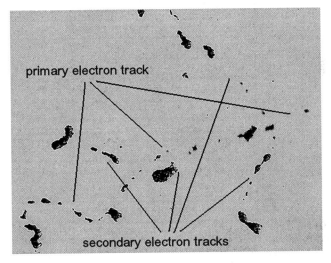

primary electron track

secondary electron tracks

Figure II.7   A track produced by a primary electron and four branching tracks produced by secondary electrons after collision with the primary electron. (From Wilson (1927) with permission © The Nobel Foundation 1927.)

shows a long winding track of a primary electron and four branches of tracks made by secondary electrons. The secondary electrons are the result of collisions of the primary electron with four atomic electrons in the cloud chamber. One can also see that the electron tracks become much wider at the very end of each track. This is a common trait in Wilson Cloud Chamber track images. He explained that as the electron looses energy its path becomes more tortuous, and as it slows down it accumulates more condensation. The tracks of high-energy electrons and beta particles generally yielded straight thin tracks. The two tracks illustrated in Figure II.8 are interpreted by Wilson as one produced by a Compton electron possibly ejected from a K shell of an atom when irradiating the cloud chamber with x-rays, and the other track arising from the same atom or origin would be that of an Auger electron ejected from an outer shell of the atom after absorbing the K x-ray.

The significance of Wilson's work in the development and application of the cloud chamber cannot be underestimated. For the first time he made it possible to observe the interactions of charged particles with matter by photographing their tracks before and after collision with atomic particles. At the time of his work he was able to put into clear view in black and white photographs many interactions of atomic particles, such as, the existence of Compton scatter electrons, Rutherford's alpha-particle interactions with atomic nuclei, the production of secondary electrons from primary (fast) electron collisions, and the existence of delta rays, which are very fast high-energy electrons produced by head-on collision of heavy particles with electrons, among other interactions.

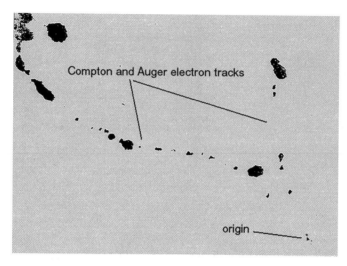

Figure II.8 Two branching tracks produced by Compton and Auger electrons originating from the same atom. (From Wilson (1927) with permission © The Nobel Foundation 1927.)

FRÉDÉRIC JOLIET (1900–1958) AND IRÈNE JOLIET-CURIE (1897–1956)

Frédéric Joliet and Irène Joliet-Curie shared the Nobel Prize in Chemistry 1935 "in recognition of their synthesis of new radioactive elements".

Jean Frédéric Joliet was born in Paris on March 19, 1900. He studied at the Ecole de Physique et Chemie, Paris, and in 1925 became a laboratory assistant to Marie Curie at the Radium Institute in Paris. Then Irène Curie, a daughter of Marie and Pierre Curie, was given the task of teaching Frédéric laboratory methods used when working with radioactive materials. Frédéric and Irène married in 1926. Frédéric completed his doctorate degree in 1930 with a thesis on the electrochemistry of the radioactive elements and became lecturer of the faculty of Science in 1935. Frédéric Joliet and Irène

Frédéric and Irène Joliet-Curie

Joliet-Curie had two children and, like Pierre and Marie Curie, devoted their lives to their work as a team in laboratory research on radioactivity and to the education of their children.

Irène Curie was born in Paris on September 12, 1897. She was elder of the two daughters born to Pierre and Marie Curie. The year that she met Frédéric, Irène received her doctorate degree in 1925 with a thesis on polonium alpha radiation, the element discovered by her mother.

Their work yielded crucial evidence that led to the discovery of the neutron, but the discovery of this elementary particle eluded their grasp. In early 1932, the Joliet-Curies reported that the highly penetrating radiation emitted when the alpha-particle emitter polonium was in contact with beryllium could knock out high-energy protons from paraffin and other hydrogenous materials and that the radiation responsible for this was gamma rays (Joliot-Curie and Joliot, 1932). At the Cavendish Laboratory James Chadwick recognized the findings of the Joliet-Curies to be startling and very significant, but the radiation capable of knocking out protons could not be gamma rays, but another source of unknown radiation consisting of a particle with a mass or energy equivalent to that of the proton. Chadwick then set about in a matter of weeks to discover this new particle as the neutron (see Radioactivity Hall of Fame—Part IV).

Although they missed the opportunity to discover the neutron, the Joliot-Curies were the first to synthesize new radioactive elements (isotopes) for which they were awarded the Nobel Prize. By the time the Joliot-Curies began their work together it was well understood, following the discoveries of Rutherford and Soddy, that one element would transform itself into another element by radioactive decay. In their jointly prepared Nobel Lecture presented on December 12, 1935 (They split their lecture so that Irène Joliot-Curie would present the first half of the lecture followed by Frédéric Joliot, who gave the remaining part.), the Joliot-Curies explained that, when writing these transformations "the atomic weight is placed above and to the left of the chemical symbol, and the atomic number to the left and below". The following transformation of radium into the radioactive radon gas, served as an example:

$$\ce{^{226}_{88}Ra} \rightarrow \ce{^{222}_{86}Rn} + \ce{^{4}_{2}He}(\text{alpha particle}) \qquad \text{(II.3)}$$

We should keep in mind that the neutron was discovered only in 1932 and, the term mass number (representing the sum of protons and neutrons in the nucleus) did not yet exist. Hence, the Joliot-Curies referred to the superscript mass number as the atomic weight.

Much had occurred in 1932, first the discovery of the neutron by James Chadwick, which the Joliot-Curies missed. This was followed by the discovery of the positron (positively charged electron) in cosmic-ray showers by Carl Anderson from Caltech later that year. The Joliot-Curies now searched for the positron, which they thought they had also missed. They not only found the missed positron (the first artificially produced positron), but they also made a new discovery, the first artificially produced radioactive isotopes.

The discovery of the Joliot-Curies of artificially produced radioisotopes came about when they began their search for their missed positrons. They decided to irradiate various metals with alpha particles of different energies to see if the result would be the production of a predominance of neutrons or positrons. It was already known that neutrons would result from the irradiation of beryllium metal with alpha particles according to the following transformation:

$$\ce{^{9}_{4}Be} + \ce{^{4}_{2}He} \rightarrow \ce{^{12}_{6}C} + \ce{^{1}_{0}n} \qquad \text{(II.4)}$$

They irradiated aluminum foil with alpha particles from a polonium source. The energy of the alpha particles colliding with the aluminum could be controlled by placing the polonium at various distances from the aluminum. The irradiation of aluminum foil by the polonium alpha particles produced neutrons and positrons. When the polonium source was moved away from the aluminum to a distance beyond the reach of the alpha particles (the alpha particles from polonium-216 have an energy of 6.78 MeV and these cannot travel beyond 5.3 cm in dry air as calculated according to eq. (1.9) of Chapter 1), the neutron emission from the alpha-irradiated aluminum would cease; however, to their amazement, the Joliot-Curies discovered that the positron emission would continue. They measured the positron emission with a Geiger counter and found that the number of positrons emitted would diminish with time with an average lifetime of less than 5 min. The interaction of the alpha particles with the aluminum foil produced a new radioactive isotope, yet unknown to mankind.

The Joliot-Curies were able to demonstrate that the new isotope was radioactive phosphorus using the following tests and reasoning:

(1) By balancing the transformation of the nuclear reaction between alpha particles and aluminum yielding neutrons they could write the reaction

$$\ce{^{27}_{13}Al} + \ce{^{4}_{2}He} = \ce{^{30}_{15}P} + \ce{^{1}_{0}n}$$ (II.5)

The Joliot-Curies used the equal sign rather than an arrow for the above reaction to demonstrate that the sum of the mass numbers and atomic numbers on one side of the equation should equal those on the other side, i.e., $27 + 4 = 30 + 1$ for the mass numbers and $13 + 2 = 15 + 0$ for the atomic numbers. Expressing nuclear reactions was not a practice at that time. The Joliot-Curies were one of the first to introduce the practice.

(2) The Joliot-Curies took a piece of aluminum foil that had been irradiated with alpha particles and placed it into a test tube containing hydrochloric acid. The aluminum would react to produce hydrogen gas as expected according to the reaction

$$2Al + 6HCl \rightarrow 2AlCl_3 + 3H_2 \uparrow$$ (II.6)

They collected the hydrogen gas in an inverted test tube as illustrated in Figure II.9. The new radioactive substance emitting positrons was carried over with the hydrogen gas into the test tube as illustrated in the figure.

CHEMICAL EVIDENCE OF TRANSMUTATION

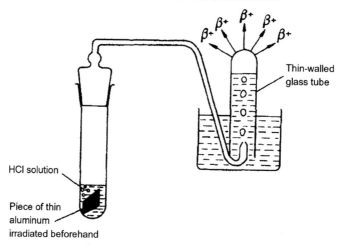

Figure II.9 Acid treatment of aluminum foil irradiated with alpha particles. (From Joliot and Joliot-Curie (1935) with permission © The Nobel Foundation 1935.)

(3)   Finally by taking another piece of aluminum foil, which had previously been irradiated with alpha particles and dissolving this metal in a mixture of acid and chemical oxidant to which was added a small quantity of sodium phosphate and a zirconium salt, the Joliot-Curies found that zirconium phosphate formed and its precipitates carried with it the new radioactive substance.

From the above tests the Joliot-Curies could conclude that the new radioactive isotope was that of phosphorus, which emits positrons. The new radioisotope, phosphorus-30, decays to stable silicon-30 with positron emission and a short half-life of 2.5 min according to the following:

$$\ce{^{30}_{15}P} \rightarrow \ce{^{30}_{14}Si} + \beta^+ \qquad (II.7)$$

Thus, the Joliot-Curies not only produced a previously non-existent radioisotope, but they also were the first to produce positrons in the laboratory.

By irradiating other metals, such as boron and magnesium, with alpha particles the Joliot-Curies were able to produce radioactive nitrogen and radioactive silicon according to the following reactions (Joliot-Curie and Joliot, 1934a–c):

$$\ce{^{10}_{5}B} + \ce{^{4}_{2}He} \rightarrow \ce{^{13}_{7}N} + \ce{^{1}_{0}n} \qquad (II.8)$$

$$\ce{^{24}_{12}Mg} + \ce{^{4}_{2}He} \rightarrow \ce{^{27}_{14}Si} + \ce{^{1}_{0}n} \qquad (II.9)$$

The radioisotopes of nitrogen and silicon in eqs. II.8 and II.9 decay by positron emission.

The Joliot-Curies were awarded the Nobel Prize in Chemistry the very year following the report of these discoveries. The rapid award of the Nobel Prize underscored the impact this discovery had on the scientific community and the advancement of science. Their work established firmly a new branch of chemistry called "radiochemistry". As explained by Frédéric Joliot in his part of the Nobel Lecture

The succession of radioactive transformations provides numerous examples in which the quantities of radioelements are extremely small and not capable of being weighed, yet nevertheless, by the methods of radiochemistry it has been possible to examine correctly their chemical properties, and identify some of them as being isotopes of elements…This special kind of chemistry in which one handles unweighable quantities, sometimes of the order of 10^{-16} g, is made possible thanks to the fact that one can determine and follow by measuring the radiation emitted, infinitesimal traces of radioactive matter dispersed in the midst of other matter.

While this work established firmly "radiochemistry" as a new branch of chemistry, the Joliot-Curies were the first to coin the terminology, which we are very accustomed to, when referring to radioactive elements. In their Nobel Lecture Frédéric Joliot stated

We have proposed that these new radio-elements (isotopes, not found in nature, of known elements) be called radio-nitrogen (for radioactive isotopes of nitrogen), radio-phosphorus (for radioactive isotopes of phosphorus), …

At the time of this great discovery Irene's mother, Marie Curie, was ailing from leukemia after many years of extensive overexposure to radiation and would not live to see her daughter and son-in-law receive the Nobel Prize. This would be the second Nobel Prize to be granted to a woman and the daughter, who she and her late husband Pierre had helped form into a pioneering scientist. Nevertheless, the Joliot-Curies related the satisfaction that the ailing Marie Curie got from knowing of this great discovery. This was conveyed in Frédéric's words taken from an historical account by Rhodes (1986):

> Marie Curie saw our research work and I will never forget the expression of intense joy which came over her when Irène and I showed her the first artificially radioactive element in a little glass tube. I can still see her taking in her fingers (which were already burnt with radium) this little tube containing the radioactive compound…To verify what we had told her she held it near a Geiger-Müller counter and she could hear the rate meter giving off a great many "clicks". This was doubtless the last great satisfaction of her life.

At reporting their discovery the Joliot-Curies suggested that possibly more artificial radioisotopes might be produced by collisions of other particles with atomic nuclei, particles such as protons, neutrons, deuterons, etc. It did not take long for this to happen. Many researchers all over the world started to make radioisotopes. In just 2 years time since the discovery of the first artificial transmutation Frédéric Joliot related in his Nobel Lecture (Nobel Lectures, Chemistry, 1922–1941)

> In England and the United States, where physicists have at their disposal equipment of very high voltages, several new elements were prepared using protons and deuterons as projectiles. In Italy first, and then in other countries, research workers, in particular, Fermi and his co-workers, used neutrons, projectiles which are outstandingly suitable to cause transmutations…among which were radio-phosphorus (^{32}P), and more than fifty new radio-elements…It was indeed a great source of satisfaction for our lamented teacher Marie Curie to have witnessed this lengthening of the list of radioelements, which she had had the glory, in company with Pierre Curie, of beginning.

After receiving the Nobel Prize Irène Joliot-Curie together with Paul Savitch continued research into the formation of radioelements, this time by irradiation of uranium and thorium with neutrons. They came close to discovering the phenomena of nuclear fission (Joliot-Curie and Savitch, 1938a,b, 1939); but the discovery of fission was reported by Lise Meitner and Otto Frisch (1939) and Otto Hahn and Fritz Strassmann (1939b). Their contributions to advances in nuclear fission and its eventual application to nuclear power for electricity were commemorated in the above postage stamp from France shown on page 94. The postage stamp illustrates in the foreground, superimposed with the busts of the Joliot-Curies, an artistic depiction of a nucleus, followed by the liquid-drop effect after it absorbs a neutron, and its fission into two smaller nuclides. The chain reaction of subsequent fission reactions

Frédéric and Irène Joliet-Curie

caused by the emission of more than one neutron on the average per fission is illustrated in the left-hand portion of the stamp.

The synthesis of new man-made radioisotopes by the Joliot-Curies had a tremendous impact on advances in the biological and medical sciences. Following their discoveries, scientists in many parts of the world began to make artificially many more new radioactive isotopes that could be used as tracers in research. Most of the biological transformations including catabolic and anabolic transformations in living systems that we understand today are due to research using artificially made radioisotopes such as ^{3}H, ^{14}C, ^{32}P, ^{35}S, etc. Medical applications of man-made radioisotopes that have resulted include techniques such as radiotherapy with radioactive sources such as cobalt-60, computerized tomography, positron emission tomography (PET), and the use of radiopharmaceuticals for the diagnosis and treatment of cancer. The stamp illustrated above from Mauritania commemorates the significance of the Joliot-Curie discovery of man-made radioisotopes, as countless lives have been saved by the applications of radiopharmaceuticals in the treatment of cancer. The stamp from Mauritania illustrates a pharmaceutical containing a short-lived radioisotope which, when administered to a cancer patient, will travel and concentrate directly at the location of a particular tumor and provide data for the imaging, diagnosis, and treatment of cancer before the radioactivity dissipates to a level harmless to the patient.

Frédéric Joliot was admitted to the French Academy of Science; however, Irène Joliot-Curie was not admitted. Like her mother, women were not yet granted that privilege. She was, however, appointed to Undersecretary of State for Scientific Research, and her husband the first High Commissioner for Atomic Energy in France. They both passed away before reaching the age of 60. Two years before Frédéric's demise, Irène died of leukemia in 1956; a disease she may have contracted like her mother from overexposure to radioactivity from childhood to her pioneering years of research.

ENRICO FERMI (1901–1954)

Enrico Fermi was born in Rome, Italy on September 29, 1901. He received the Nobel Prize in Physics 1938 "for his demonstrations of the existence of new radioactive elements (radioisotopes) produced by neutron radiation, and for his related discovery of nuclear reactions brought about by slow neutrons".

Enrico Fermi
(1901-1954)

Fermi had a natural aptitude for physics and mathematics. According to historical accounts by Rhodes (1986, 1999), Enrico Fermi discovered physics at the age of 14 after he encountered two old volumes on elementary physics in the book stacks of Rome's Campo dei Fiori. He read these books thoroughly with such intense interest, sometimes correcting the mathematics that he had not even noticed that the books were written in Latin. He earned his doctorate degree in physics at the University of Pisa at 21 years of age. After short fellowships in Gottingen and Leyden, Fermi became lecturer in Physics at the University of Florence during 1924–1926. In 1927 he was elected Professor of theoretical physics at the University of Rome, the youngest scientist to be elected professor since Galileo. Fermi retained this post until he was awarded the Nobel Prize in 1938; an opportunity he took to escape from Mussolini's fascist regime and emigrate with his family to the United States of America.

Enrico Fermi is well-known for his elaboration in 1934 of the beta decay process of radioisotopes often referred to as "Fermi beta decay". The mystery was how beta particles, which are equivalent to electrons, could originate and be expelled from an atomic nucleus. Fermi proposed that within the nucleus of a radioactive atom, the neutron may decay to a proton and vice versa, a proton into a neutron. He proposed the existence of a weak interaction as a new force, within the nucleus, that permits a neutron to decay into a proton and a negatively charged electron ($n \rightarrow p^+ + e^-$) or vice versa, a proton could decay into a neutron and a positively charged electron ($p^+ \rightarrow n + e^+$). The weak interaction would allow a nucleon (proton or neutron) to transform itself into the other to achieve nuclear stability. This force would be weaker than the strong nuclear force or strong interaction that binds the nucleons together in the nucleus. The weak interaction proposed by Fermi has advanced into a unified electroweak theory (Nobel Prize in Physics 1979 awarded to Sheldon L. Glashow, Abdus Salam, and Steven Weinberg "for their contributions to the theory of unified weak and electromagnetic interaction between elementary particles..."), as it has subsequently been discovered that the weak interactions and electromagnetic interactions are of similar strengths at short distances typical of nucleons (10^{-18} m). These two forces can thus be combined into what is called the electroweak force.

The beta-decay theory of Fermi explains the origins for either a negative beta particle or positive beta particle from a decaying nucleus. However, an additional mystery to beta-particle radiation was the broad range of energies that a beta particle could possess from near zero to a maximum energy when these were emitted from a particular radioactive atom. When an unstable nucleus emits radiation it decays from a higher energy state to a nucleus with a lower energy state, and the energy difference in these two states should exactly correspond to the sum of the energies of the radiations emitted by the decaying nucleus. This could easily be observed when alpha particles or gamma radiation were emitted from nuclides with discrete energies that corresponded to the energy transitions of the parent and daughter nuclides. The emission of a beta particle from a nucleus that could possess any energy between zero and a maximum value would appear to contradict the laws of conservation of energy, except that Wolfgang Pauli had proposed in 1930 the existence of a neutral particle of near-zero rest mass that would share the decay energy with the beta particle. Pauli called this particle the "neutron", as this was prior to the time when the neutron was discovered (Chadwick, 1932a,b) and prior to Fermi's beta-decay theory (1934). Fermi coined Pauli's neutral particle as the "neutrino" from the Italian meaning "little neutral one". The Fermi beta decay including the accompanying neutrino (v) can be written as follows:

$$n \rightarrow p^+ + \beta^- + \bar{v} \qquad (II.10)$$

$$p^+ \rightarrow n + \beta^+ + v \qquad (II.11)$$

When the neutron transforms itself into a proton in a decaying nucleus, a negative beta particle (negatron) would be ejected from the nucleus together with an antineutrino. The opposite would occur when a proton transforms itself into a neutron in a decaying nucleus, that is, a positive beta particle (positron) would be ejected from the decaying nucleus together with a neutrino. As the two transformations involve the inverse or opposite of the other, the beta particle and neutrino emitted in one of the decay processes are the antiparticles of the beta and neutrino radiations emitted in the other decay process. Also, a decay process that involves the emission of a negative beta particle would result in a daughter nucleus with one unit higher in atomic number (an additional proton in the nucleus) whereas nuclear decay with the emission of a positive beta particle would produce a daughter nucleus one unit lower in atomic number (one less proton in the nucleus). Examples of the nuclear transitions that occur in negative beta decay and positive beta decay are illustrated in Figures II.10 and II.11.

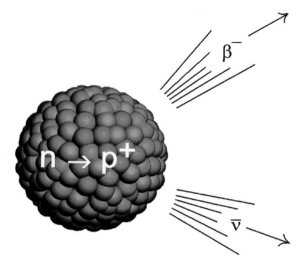

Figure II.10 Negative beta decay. A neutron transforms itself into a proton in the nucleus with the production and emission of a negative beta particle (negatron) and antineutrino from the nucleus. The energy of decay is carried off by the negatron and antineutrino.

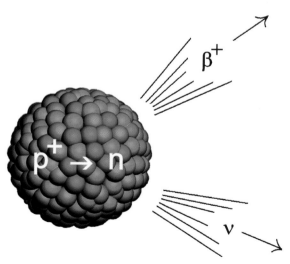

Figure II.11 Positive beta particle decay. A proton transforms itself into a neutron in the nucleus with the production and emission of a positive beta particle (positron) and neutrino. The energy of decay is carried off by the positron and neutrino.

Example: Decay of Carbon-14

$$^{14}_{6}C \rightarrow {}^{14}_{7}N + \beta^- + \bar{\nu} + 0.156\,\text{MeV} \tag{II.12}$$

Example: Decay of Nitrogen-13

$$^{13}_{7}N \rightarrow {}^{13}_{6}C + \beta^+ + \nu + 1.190\,\text{MeV} \tag{II.13}$$

Fermi's original manuscript on the transformations of nucleons within the nucleus was rejected by the editors of the journal *Nature* because of the its theoretical nature, and therefore his first reports on beta-decay theory were published in the Italian and German languages (Fermi, 1934a,b). Evidence for these transformations was provided years later by Snell and Miller (1948), who demonstrated that the neutron when outside the confines of a stable nucleus would decay to a proton and an electron with a lifetime of 15 min (see Chapter 4) according to the scheme

$$n \rightarrow p^+ + e^- + \bar{\nu} + 0.782\,\text{MeV} \tag{II.14}$$

Although Fermi is well remembered for his beta-decay theory, the Nobel Prize was awarded to him in 1938 for his contribution to the application of neutron radiation to the production of many new radioactive isotopes and to his discovery that certain nuclear reactions are brought about by slow neutrons. In his Nobel Lecture (December 12, 1938) Fermi acknowledged that Rutherford was the first to start the technique of nuclear bombardments using high-energy alpha particles to enable the transmutation of one element into another in 1919 (see Rutherford, Radioactivity Hall of Fame—Part I). Fermi further noted that immediately after the discoveries of the artificial production of radioactive elements (radioisotopes) by the Joliot-Curies in 1934 by alpha particle induced nuclear reactions, he thought that alpha particles might not be the only type of bombarding projectiles that could produce artificial radioactivity, particularly for atoms with atomic number greater than 20, and then began to investigate the effects of bombarding various elements with neutrons.

Fermi observed that neutrons, having no electric charge, should be able to reach the nuclei of all atoms without having to overcome the coulombic barrier of the surrounding atomic electrons. He knew that the neutron would not be hindered by the atomic electrons and that the probability of nuclear collisions by neutrons would be higher than that achievable with charged projectiles such as alpha particles or protons. As a source of neutrons he used a small glass bulb of beryllium powder together with 800 mCi of radon, which provided a source of neutrons with an emission rate of 2×10^7 neutrons per sec. The alpha particles from the radon interact with the beryllium nucleus to produce neutrons. (see eq. (4.19) of Chapter 4). The initial results obtained by Fermi in early 1934 (Fermi, 1934c,d; Fermi *et al.*, 1934) and his coworkers during 1934–1940 including H. Anderson, E. Amaldi, O. D'Agostino, F. Rasetti, E. Segré and L. Szilard (Fermi *et al.*, 1934, 1938; Amaldi and Fermi, 1935, 1936a,b; Fermi and Amaldi, 1936; Amaldi *et al.*, 1937; Anderson *et al.*, 1939a,b; Fermi, 1940a,b;) yielded 63 artificial radioisotopes produced by the following three types of nuclear reactions using Fermi's original notation:

$$(1) \quad {}^{M}_{Z}A + {}^{1}_{0}n \rightarrow {}^{M-3}_{Z-2}A + {}^{4}_{2}He \tag{II.15}$$

$$(2) \quad {}^{M}_{Z}A + {}^{1}_{0}n \rightarrow {}_{Z-1}^{M}A + {}^{1}_{1}H \tag{II.16}$$

$$(3) \quad {}^{M}_{Z}A + {}^{1}_{0}n \rightarrow {}^{M+1}_{Z}A \tag{II.17}$$

where ${}^{M}_{Z}A$ is a symbol for any element A with atomic number Z and mass number M, and n is the symbol for the neutron. Fermi found reactions (1) and (2) to occur chiefly among the light elements (there were 92 known elements at the time), and the reaction of type (3) were found to occur most often among the heavy elements. More than one of the above reactions would sometimes occur when elements were bombarded with neutrons. In Fermi's words at his Nobel Lecture

> In many cases the three processes are found at the same time in a single element. For instance, neutron bombardment of aluminum that has a single isotope ^{27}Al, gives rise to three radioactive products: ^{24}Na, with a half-life of 15 h by process (1); ^{27}Mg, with a period [half-life] of 10 min by process (2); and ^{28}Al with a period [half-life] of 2 to 3 min by process (3).

In addition to the above three neutron reactions described by Fermi, he and his coworkers were also able to observe the production of artificial radioactive isotopes by a fourth process that he described:

> The primary neutron does not remain bound in the nucleus, but knocks off instead, one of the nuclear neutrons out of the nucleus; the result is a new nucleus, that is isotopic with the original one and has an atomic weight less by one unit.

This fourth neutron interaction, where the mass number is reduced by one (loss of a neutron) and the atomic number remains unaltered, can be written as follows:

$$(4) \quad {}^{M}_{Z}A + {}^{1}_{0}n \rightarrow {}^{M-1}_{Z}A + {}^{1}_{0}n \qquad\qquad (II.18)$$

Only about 6 months after their first experiment with neutron bombardment Fermi made a great discovery. He found that the neutron irradiation of elements became more effective in producing new radioisotopes when the neutrons were made to slow down, by allowing the neutron beam to pass through a given barrier of water or paraffin. By doing so the fast neutrons are slowed down to become thermal neutrons or in other words, neutrons corresponding to thermal agitation. In his Nobel Lecture Fermi explained the properties of thermal neutrons:

> After their energy is reduced to a value corresponding to thermal agitation, the neutrons go on diffusing without further change of their average energy. All of the processes of capture of slow neutrons by any nucleus are generally accompanied by the emission of gamma rays: Immediately after the capture of the neutron, the nucleus remains in a state of high excitation and emits one or more gamma-quanta, before reaching the ground state.

The idea of slowing down neutrons to improve the efficiency of neutron capture was contrary to what one would believe at that time, because previously only charged particles were used to bombard atomic nuclei and the higher the speed of the charged particle, the greater the probability of penetrating the coulombic barrier of the atomic electrons to yield a reaction. Because neutrons have no difficulty diffusing into close proximity of atomic nuclei, Fermi and his coworkers were able to produce more than 400 new radioisotopes using slow neutron bombardment of all known elements with the exception of hydrogen and helium.

Fermi also discovered a neutron absorption anomaly whereby the strongest absorption of a neutron by a nucleus could occur at a precise speed, referred to as a resonance effect. He noted "elevated neutron absorption by nuclei at precise neutron speeds was explained by Bohr, Breit, and Wigner as due to resonance with a virtual energy level of the compound nucleus (i.e., the nucleus composed of the bombarded nucleus and the neutron". Resonance effects are illustrated in Figure 4.9 of Chapter 4.

Upon receiving the Nobel Prize in 1938 Enrico Fermi immigrated with his family to the USA and was appointed Professor of Physics at Columbia University in 1939, a post that he held until 1942. His work up to 1938 on neutron reactions was the final precursor needed for the discovery of nuclear fission by Lise Meitner, Otto Hahn, and Fritz Strassmann in 1939 and Fermi's next great achievement, the construction of the first nuclear reactor.

Enrico Fermi collaborated with Leo Szilard in the construction of the first nuclear reactor. Leo Szilard was a renowned theoretical physicist and lifelong friend of Albert Einstein (see Radioactivity Hall of Fame—Part IV). Fermi's great knowledge of neutron physics and Szilard's determination to beat Germany at the production of the atomic bomb was an excellent combination for the successful working reactor. In the words of Richard Rhodes (1999)

In 1939, still official enemy aliens, Fermi and Szilard co-invented the nuclear reactor at Columbia University, sketching out a three-dimensional lattice of uranium slugs dropped into holes in black, greasy blocks of graphite moderator, with sliding neutron-absorbing cadmium control rods to regulate the chain reaction.

As soon as it became known that neutron-induced fission of uranium-235 would yield more than one neutron per fission (2.4 neutrons on the average for ^{235}U fission and a high amount of energy, ~194 MeV) Fermi, Szilard, and others knew that a fission chain reaction would be possible, which would release large amounts of energy. The reactor they built was given the name of an "atomic pile", coined by Fermi, because their unique apparatus had the appearance of a pile of uranium fuel, graphite moderator, and cadmium control elements. Fermi knew from his previous work with neutrons that cadmium has a very high efficiency for absorbing slow (thermal) neutrons and should serve to control the chain reaction since fission in ^{235}U is optimal at thermal incident neutron energies. They assembled the pile in such a way that neutrons would be produced with the correct energy needed to create fission in the uranium whereby the splitting of uranium nuclei would create additional neutrons and keep the chain reaction going to produce energy with the application of cadmium rods to control the reaction. The reactor "pile" was assembled by Fermi and his team on a squash court under the stands of the University of Chicago's Stagg (football) Field in late 1942. On December 2, 1942 under the direction of Enrico Fermi the reactor went critical, that is, the chain reaction was sustained with the production of a neutron for every neutron absorbed by the uranium nucleus. Fermi allowed the reactor to continue as a carefully controlled chain reaction for 4.5 min at 0.5 W power. A plaque, at the site of the first "atomic pile", commemorates this historic event with the words, "On December 2, 1942, man achieved here the first self-sustaining chain reaction and thereby initiated the controlled release of nuclear energy".

The first nuclear reactor was a giant step for mankind, as it has had since Fermi's "atomic pile" and continues to have a vital role in development for peaceful applications including (1) the production of isotopes for medical, biological, and other vital sciences, (2) the production of heat that can drive steam turbines and generate enormous amounts of electricity, and (3) the production of neutrons for research. A fourth ominous application of the nuclear reactor is, of course, for the production of plutonium (see Chapter 4) for nuclear weapons. The peaceful applications of nuclear power have an enormous impact on our daily lives. This is underscored in an article by ElBaradei (2004) where he notes

At the end of last year there were 440 nuclear power units operating worldwide. Together, they supply about 16% of the world's electricity. That percentage has remained relatively steady for almost 20 years—meaning that nuclear electricity generation has grown at essentially the same rate as total electricity use worldwide.

After overseeing the assembly and operation of the first self-sustaining nuclear chain reaction Fermi joined a group of eminent physicists and chemists in the Manhattan Project at Los Alamos, New Mexico, which was established in mid 1943 under the direction of J. Robert Openheimer, as part of Americas efforts during the Second World War. Fermi played a vital role in this project. In an address marking the centennial celebration of the birth of Enrico Fermi on November 27, 2001, Dr. John H. Marburger III, the Science Advisor to the President of the United States, noted that in the preparations for testing the bomb, Fermi's help was invaluable and quoted Emilio Segré (Fermi's lifelong friend and collaborator, and a Nobel Laureate in Physics, 1959), as follows:

This was one of those occasions when Fermi's dominion over all physics, one of his most startling characteristics, came into its own. The problems involved in the Trinity test ranged from hydrodynamics to nuclear physics, from optics to thermodynamics, from geophysics to nuclear chemistry. Often they were closely interrelated, and to solve one it was necessary to understand all the others. Even though the purpose was grim and terrifying, it was one of the greatest physics experiments of all time. Fermi completely immersed himself in the task. At the time of the test he was one of the very few persons (or perhaps the only one) who understood all the technical ramifications of the activities at Alamogordo.

Enrico Fermi became an American citizen in 1944 and held the position of Professor at the University of Chicago's Institute for Nuclear Studies from 1946 until his untimely death in 1954 after succumbing to stomach cancer. During his last years at the University of Chicago he concentrated his efforts on high-energy physics and the origins of cosmic radiation, which remain yet the source of high-speed particles with energies not yet achieved by any man-made accelerator.

WOLFGANG PAULI (1900–1958)

Wolfgang Pauli was born in Vienna on April 25, 1900. He received the Nobel Prize in Physics 1945 for "his discovery of the Exclusion Principle also called the Pauli Principle". Also Pauli made a major contribution to the understanding of beta-particle emissions from radioactive atoms by proposing the existence of the neutrino, a particle of zero charge and near-zero rest mass that is emitted from the nucleus during beta decay.

Wolfgang Pauli (1900-1958)

Wolfgang Pauli graduated in 1918 from the Döblinger Gymnasium in Vienna after comprehensive studies in the fields of physics and mathematics. Pauli was a gifted and ambitious student often preferring to study Einstein's works on relativity during his classes at the Döblinger Gymnasium, which he sometimes considered boring. Having graduated with distinction Pauli became a pupil of Arnold

Sommerfeld at the Ludwig-Maximilian University of Munich, where he received his doctorate degree in 1921. During his budding years as a student at the University of Munich Pauli wrote a most comprehensive and knowledgeable book on Einstein's theory of relativity, which illustrated in Einstein's words Pauli's "sureness of mathematical deduction and profound insight". Einstein recognized the genius in Pauli and considered Pauli his intellectual successor, and it was Einstein who, later in life, would nominate Pauli for the Nobel Prize. Pauli continued to study with great physicists of the time including a year with Max Born at the University of Göttingen followed by a year with Niels Bohr at the University of Copenhagen. After a few years as lecturer at the University of Hamburg (1923–1928) Pauli was appointed Professor of Theoretical Physics at the Swiss Federal Institute of Technology in Zurich (Eidgenössische Technische Hochschule or ETH, Zurich). The German annexation of Austria in 1938 placed a threatening cloud over Pauli. He was of part-Jewish descent, and the possibility of German occupation of Switzerland forced him to flee. He was invited to the United States in 1939 by Albert Einstein as visiting professor of the Institute of Advanced Studies at Princeton and elected as Chair of theoretical physics at Princeton in 1940.

Wolfgang Pauli was awarded the Nobel Prize in Physics 1946 for his discovery of the exclusion principle proposed in 1925, which states that no two electrons in an atom could possess the same four quantum numbers, that is, no two electrons in an atom could exist at the same quantum or energy state. In his Nobel Lecture (December 13, 1946) Pauli explained that he began to formulate his thoughts on the exclusion principle after listening to a series of lectures given by Niels Bohr in 1922 at the University of Göttingen where, in the words of Pauli (1946) Bohr explained

…the spherically symmetric atomic model, the formation of the intermediate shells of the atom and the general properties of the rare earths. The question, as to why all electrons for an atom in its ground state were not bound in the innermost shell, had been emphasized by Bohr as a fundamental problem….

Pauli's Exclusion Principle solved the problem by providing "a general explanation for the closing of every electron shell [of an atom]…" (Nobel Lectures, Physics, 1942–1962). The Pauli Exclusion Principle can be visualized by first considering the definition of the four quantum numbers. Three quantum numbers originate from the space geometry of the electron while the fourth arises from the electron spin, and these are defined as follows:

(1) *Principal quantum number*, assigned values of $n = 1, 2, 3 \ldots$, correspond to the principal energy level of the electron. For example, $n = 1$ for a K electron (an electron in the innermost K shell or lowest energy level), $n = 2$ for an L electron, $n = 3$ for an M electron, and $n = 4$ for an N electron, etc.

(2) *Orbital quantum number*, assigned values of $\ell = 0, 1, 2, 3 \ldots n - 1$, correspond to values of the angular momentum of the electron.

(3) *Magnetic quantum number*, assigned values of whole numbers from $m_\ell = -\ell$ to $+\ell$, correspond to the possible properties of an electron in a magnetic field. For example, when $\ell = 2$, m_ℓ can have values of $-2, -1, 0, +1$, and $+2$.

(4) *Spin quantum number*, $m_s = +\frac{1}{2}$ or $-\frac{1}{2}$, corresponds to the two possible spin vectors or orientations of an electron in a magnetic field with spin vectors classified as spin up ($+$ or $\uparrow$) and spin down ($-$ or $\downarrow$).

An example can be taken by arbitrarily selecting an element and applying the rules of the Pauli Exclusion Principle to demonstrate how the Principle can explain the intermediate and main electron energy levels of an atom. For example, zinc has an atomic number of 30 (i.e., 30 protons); and its 30 orbital electrons may possess the quantum numbers listed in Table II.1.

According to the Pauli Exclusion Principle, the maximum number of electrons allowable in the sublevels of the energy levels K, L, M, N, and O are illustrated in Figure II.12 with increasing energy and distance from the nucleus. The nomenclature used to specify the electron configurations of the elements depicts both the numbers of electrons in the main and sublevels such as the following using the

Table II.1

Possible quantum numbers for electrons in atomic orbitals for zinc (atomic number 30)
starting with the lowest K energy level

Main quantum level[a]	Quantum number				Electrons in sublevels or subshells[f]	Orbital type[g]	Electrons in main levels[h]
	Principal[b] (n)	Orbital[c] (ℓ)	Magnetic[d] (m_ℓ)	Spin[e] (m_s)			
K	$n = 1$	$\ell = 0$	$m_\ell = 0$	$m_s = +$			
				$m_s = -$	2	s	2
L	$n = 2$	$\ell = 0$	$m_\ell = 0$	$m_s = +$			
				$m_s = -$	2	s	
		$\ell = 1$	$m_\ell = -1$	$m_s = +$			
				$m_s = -$			
			$m_\ell = 0$	$m_s = +$			
				$m_s = -$			
			$m_\ell = +1$	$m_s = +$			
				$m_s = -$	6	p	8
M	$n = 3$	$\ell = 0$	$m_\ell = 0$	$m_s = +$			
				$m_s = -$	2	s	
		$\ell = 1$	$m_\ell = -1$	$m_s = +$			
				$m_s = -$			
			$m_\ell = 0$	$m_s = +$			
				$m_s = -$			
			$m_\ell = +1$	$m_s = +$			
				$m_s = -$	6	p	
		$\ell = 2$	$m_\ell = -2$	$m_s = +$			
				$m_s = -$			
			$m_\ell = -1$	$m_s = +$			
				$m_s = -$			
			$m_\ell = 0$	$m_s = +$			
				$m_s = -$			
			$m_\ell = +1$	$m_s = +$			
				$m_s = -$			
			$m_\ell = +2$	$m_s = +$			
				$m_s = -$	10	d	18
N	$n = 4$	$\ell = 0$	$m_\ell = 0$	$m_s = +$			
				$m_s = -$	2	s	2

[a] The electron energy level K is the closest to the nucleus and contains no sublevel. Other levels, namely, L, M, N, O, etc. are of increasing energy and distance from the nucleus, and these contain sublevels.
[b] The principal quantum number is assigned values of $n = 1, 2, 3...$
[c] The orbital quantum number is assigned values according to $\ell = 0, 1, 2, 3...n - 1$.
[d] The magnetic quantum number is assigned values in the range of $m_\ell = -\ell$ to ℓ.
[e] The spin quantum number can have only two possibilities, $+\frac{1}{2}$ or $-\frac{1}{2}$. The $+$ or $-$ signs are used above to denote the possible spin orientations.
[f] The maximum number of electrons in the sublevel is $2(2\ell + 1)$. When a subshell is filled, any additional electrons that an atom would possess, must go into the next and higher-energy sublevel.
[g] The term "orbital" does not refer to fixed orbits or circular paths about the nucleus. Following the work of Schrödinger and others, a certain electron cannot be located at any fixed distance from the nucleus. The term "orbital" thus refers to a region in space about the nucleus where an electron has the highest probability of being found.
[h] The number of electrons in the main levels are the sum of the electrons in the sublevels.

above case of zinc as an example: the electron configuration of zinc would be described as $1s^2$, $2s^2$, $2p^6$, $3s^2$, $3p^6$, $3d^{10}$, $4s^2$, which would read as $2s$ electrons in the first main level, $2s$ electrons in the second main level, 6 p electrons in the second mail level, and $2s$ electrons in the third main level, etc. To shorten the notations in the nomenclature, it is possible to add a configuration to the known configuration of another element, such as one of the inert gases, He, Ne, Ar, or Kr. For example, the electron

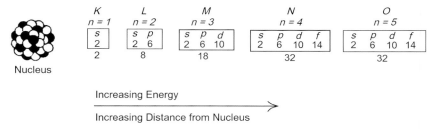

Figure II.12 Maximum number of electrons in the main (K, L, M, N, and O) and sublevels (s, p, d, and f) of electron energies of neutral atoms in the ground (unexcited) state.

configuration for zinc may be simplified to Ar-$3d^{10}$, $4s^2$, because the configuration of the eighteen electrons of argon is $1s^2$, $2s^2$, $2p^6$, $3s^2$, $3p^6$. There are some irregularities in the electron configurations of the elements, and the exact configurations are found in reference tables such as those provided by Weise and Martin (1989) and Lide (1997). However, the Pauli Exclusion Principle is precise in that it excludes, without exception, the number of electrons from exceeding 2, 6, 10, and 14 in the s, p, d, and f sublevels or 2, 8, 18, 32, and 32, in the K, L, M, N, and O main levels, respectively.

The Pauli Exclusion Principle applies to all fermions (odd half-integer spin particles) in a given atom, which includes protons and neutrons in addition to the electrons. The principle solved the electronic structure of the atom, which was clearly emphasized in the following excerpt from an obituary that appeared in *The Times* on the eve of Pauli's death on December 15, 1958:

> The quantum theory of the atoms accounts in some detail for many of the properties of the 100 or so chemical elements, and in this feat of generalization, Pauli's Exclusion Principle played a necessary part. Although the theory of the nucleus of the atom is in a less satisfactory state than that of the electronic or outer structure of the atom, it appears that in the nucleus too, Pauli's Exclusion Principle operates.

It might be said that Wolfgang Pauli is equally famous for his bold hypothesis in 1930 for the existence of the neutrino as for his monumental Exclusion Principle for which he received the Nobel Prize. In 1930 physicists were perplexed over the fact that beta particles were emitted from radioactive atoms with a broad range of energies. That is, the energy of the beta particle could not be predicted on the basis of the radionuclide decay energy. The emissions of alpha particles and gamma rays from the nuclei of atoms could be easily explained. The energy of an alpha particle emitted from a given radionuclide always possessed a discrete energy, which represented the decay energy of the radionuclide less the recoil energy of the nucleus. Likewise gamma rays were of discrete energies representing the energy lost by the nucleus as it decayed from an excited to more stable state. Beta-particle emission presented a totally different and perplexing picture. The energy of a beta particle could not be predicted on the basis of the decay energy of the radionuclide. For example, a radionuclide that decays by beta-particle emission, such as ^{32}P, has a decay energy of 1.7106 MeV, which is the energy lost by the radioactive ^{32}P when it decays by beta-particle emission to the stable nuclide ^{32}S. However, a beta particle emitted from the nucleus of ^{32}P would have an energy ranging from close to zero to a maximum energy close to the decay energy of the nucleus (see Figure 2.1 of Chapter 2). The law of energy conservation required that the energy of the radiation emitted from the radioactive atom should equal its decay energy.

Some leading scientists, including Niels Bohr, postulated that possibly the laws of conservation of energy might have some restricted validity in certain circumstances, such as in beta decay. Pauli, however, concluded that there can never be any alternatives or limitations to the fundamental law of energy conservation. Therefore, in a bold move, he postulated the neutrino, as a small neutral particle of zero charge and near-zero mass that was simultaneously emitted from the nucleus together with the beta particle. The neutrino and beta particle would share the decay energy of the nucleus. While Pauli was convinced of his hypothesis, it would have been too bold to publish. He, therefore, put his hypothesis in a letter addressed to the participants of a conference on radioactivity at Tübingen on December 4, 1930. The letter was addressed to all present "Liebe Radioactive Damen und Herren", and is translated as follows:

Open Letter to the Radioactivity Group
at the Regional Meeting at Tübingen

Zurich, 4 Dec. 1930
Gloriastrasse

From
Physics Institute
of the Federal Institute of Technology (ETH)
Zurich

Dear Radioactive Ladies[1] and Gentleman,

I beg you to most favorably listen to the carrier of this letter. He will tell you that, in view of the wrong statistics of the N and the Li^6 nuclei and of the continuous beta spectrum, I have hit upon a desperate remedy to save the exchange theorem of statistics and the law of conservation of energy. This is the possibility that electrically neutral particles exist which I will call neutrons[2], which exist in nuclei, which have a spin ½ and obey the exclusion principle, and which differ from the photons also in that they do not move with the velocity of light. The mass of the neutrons [i.e., neutrinos][3] should be of the same order as those of the electrons and should in no case exceed 0.01 proton masses. The continuous beta spectrum would then be understandable if one assumes that during beta decay with each electron a neutron [i.e., neutrino][3] is emitted in such a way that the sum of the energies of neutron [i.e., neutrino][3] and electron [i.e., beta-particle][4] is constant.

I admit that my remedy may seem incredible, because one should have seen these neutrons [i.e., neutrinos][3] long ago, if they really exist.[5] But only he who dares can win and the difficult situation caused by the continuous beta spectrum is illuminated by a remark of my honored predecessor, Mr. Debye, who told me recently in Brussels: Oh, it is best not to think at all, just as with the new taxes. Henceforth every possible solution must be discussed. So, dear radioactive people, because I am indispensable here due to a ball which will take place in Zurich during the night from December 6 to 7. With my best regards to you and also to Mr. Back.

Your humble servant,
W. Pauli

[1] A lady at this meeting was Lise Meitner, a key figure later in the discovery of nuclear fission.
[2] Pauli called this particle the neutron in this letter; however, Enrico Fermi, changed the name of this neutral particle of near-zero mass to the "neutrino" from the Italian meaning "little neutral one". The neutron, a particle of relatively high mass and neutral charge in the nucleus was discovered by Chadwick in 1932.
[3] The actual name of neutrino, while not included in Pauli's letter, is included here for clarification with the reader as to the real particle referred to by Pauli.
[4] Pauli used the word electron in his letter while referring to the beta particle, because the beta particle is an electron originating from nuclear decay.
[5] Pauli was being too courteous or playing up to his audience. It would really have been difficult or impossible to detect these particles during the time of Pauli's letter or earlier, because of the particle's neutral charge and near-zero rest mass. So difficult were the neutrinos to detect that it was not until over 25 years later (June 1956) did technology permit Frederick Reines and Clyde Cowan, Jr. to finally achieve the detection of the neutrino.

Pauli's hypothesis was taken by Enrico Fermi to elaborate the beta decay process in 1934 when Fermi (1934a, b) also adopted the name "neutrino" for this particle, since another particle of neutral charge and mass similar to the proton was discovered by Chadwick in 1932 and named the neutron. The beta decay process and examples of how the beta particle and neutrino share the nuclide decay energy is described in detail in this section in the biographical sketch of Enrico Fermi and in Chapter 2.

FREDERICK REINES (1918–1998) AND CLYDE L. COWAN, JR. (1919–1974)

Frederick Reines was awarded the Nobel Prize in Physics 1995 "for the detection of the neutrino", and it is said that Clyde Cowan would have shared that Nobel Prize for his collaboration with Frederick Reines, if the Nobel Prize were awarded posthumously. It was Wolfgang Pauli, who in 1930, first postulated the neutrino in a desperate attempt to explain the wide spectrum of beta-particle energies emitted from radioactive atoms and the need to postulate, a neutral elusive particle therefore, of near-zero rest mass to conform to the laws of conservation of energy in radionuclide decay (see Wolfgang Pauli in this section). Not until 1956, some 26 years later, was the neutrino demonstrated to exist after the extraordinary scientific collaboration and determination of Frederick Reines and Clyde Cowan, Jr.

Frederick Reines (1918-1998)[a]

Frederick Reines was born in Paterson, NJ on March 16, 1918 of Russian immigrant parents. During his first 2 years of high school he was more interested in literature and did not excel in his science studies. However, during his junior and senior years, his science teacher at Union High School, New Jersey took an interest in young Frederick, and encouraged him to pursue science. By the time he graduated from high school Frederick Reines was determined to be a pioneering physicist. Upon graduation, his ambition as noted in the high school yearbook was "to be a physicist extraordinaire". This is another example as in the life of Ernest Rutherford (see Rutherford, Chapter 1), how one teacher can have a pivotal influence on the life achievements of a student. Frederick Reines was also a gifted singer; and he was encouraged by others to pursue a career as a professional singer. So good was his talent that he was awarded cost-free coaching at the Metropolitan Opera. Nevertheless, he decided against a singing career. Frederick Reines' determination to become a "physicist extraordinaire" was set in stone.

After high school Frederick Reines gained admission to the renowned Massachusetts Institute of Technology (MIT) and had promise of a scholarship. However, he selected the Stevens Institute of Technology all because of the enthusiasm for Stevens displayed by an admissions officer in a chance encounter. At Stevens he received his undergraduate degree in Engineering in 1939 and a Master of Science degree in mathematical physics in 1941. While working on his Ph.D. dissertation on

---

[a]Photo from Los Alamos Science (1997) with permission © Los Alamos National Security, LLC.

The Liquid Drop Model for Nuclear Fission, Frederick Reines was recruited in 1944 to work under Richard Feynman on the Manhattan Project at the Los Alamos National Laboratory (LANL). He found the LANL to be the most stimulating environment for scientific advancement. In his own words Frederick Reines noted, "During my participation in the Manhattan project and subsequent research at Los Alamos, encompassing a period of 15 years, I worked in the company of perhaps the greatest collection of scientific talent the world has ever known" (Barany, 2000). It was at Los Alamos that Frederick Reines began a fruitful collaboration with Clyde Cowan that led to the detection of the neutrino.

Clyde Cowan, Jr. (1919-1974)[b]

Clyde L. Cowan, Jr. was born in Detroit, MI on December 6, 1919. He received a Bachelor of Science degree in 1940 at the Missouri School of Mines and Metallurgy, which is now part of the University of Missouri. He was awarded his Master of Science and Ph.D. degrees in physics at Washington University, St. Louis in 1947 and 1949, respectively. During World War II Cowan joined the US Army and was a 2nd lieutenant with the 51st Troop Carrier Wing. During the war he was stationed in England and was assigned to work on the newly developed radar. He was awarded the Bronze Star for his significant contributions to the improvement of radar. In 1949 Cowan joined the scientific team at the Los Alamos National Laboratory and in 1951 became group leader of the Nuclear Weapons Test Division at Los Alamos.

Frederick Reines and Clyde Cowan began their historic collaboration in 1951 at Los Alamos to detect the neutrino. Reines primarily wanted to do something big, deserving of a "physicist extraordinaire", and personally decided to accomplish the detection of the neutrino postulated by Pauli in 1930. He first consulted with Enrico Fermi in 1951 at Los Alamos about the idea. Reines decided that a nuclear weapon test would serve as the best source of neutrinos, because of the immense number of fissions that would occur and the concomitant numerous beta-particle emissions from fission products that would result. He decided that he would also need a very large detector, because of the elusive nature of the neutrino. Enrico Fermi agreed with Reines that a weapon test would be the best source of neutrinos, but that the design of a detector large enough and capable of detecting a neutrino interaction would be a tremendous hurdle. Later that year Reines proposed the idea to Clyde Cowan, who was a good experimentalist with a sense of daring, and they set out to solve this challenging problem even without having in mind the design of a detector.

During their initial assessment of the problem of the detection of the neutrino Frederick Reines and Clyde Cowan first had to consider the following questions: (1) What features of the neutrino interaction would they use for signals? (2) How large a detector would be needed? (3) How many counts would they expect to obtain from neutrino interactions? To select a possible neutrino interaction that could provide a signal to a detector, they considered two processes of reverse beta decay that would be possible according to the Pauli–Fermi beta-decay theory. The two possible reverse beta decay processes are the following:

$$(1) \quad \bar{\nu} + p^+ \rightarrow n + \beta^+, \quad \text{i.e.,} \quad \bar{\nu} + {}_Z^A X \rightarrow {}_{Z-1}^A Y + \beta^+ \tag{II.19}$$

where an antineutrino would interact with a proton to yield a neutron plus a positron (positive electron), that is, the antineutrino would cause a nuclide X of charge Z and mass number A to convert to a nuclide Y of charge Z–1 with the emission of a positron, or

$$(2) \quad \nu + n \rightarrow p^+ + \beta^-, \quad \text{i.e.,} \quad \nu + {}_Z^A X \rightarrow {}_{Z+1}^A Y + \beta^- \tag{II.20}$$

where a neutrino would interact with a neutron of a nucleus to yield a proton and a beta particle (negative electron). They selected the first possible neutrino interaction (eq. (II.19)), because the positron produced by the first process could provide a signal from positron annihilation (i.e., the emission of two gamma rays of 0.55 MeV in opposite directions). Liquid scintillation detection, which uses organic fluor solutions, had been newly discovered by Kallman (1950) and Reynolds *et al.* (1950), and Reines and Cowan decided that a detector composed of liquid scintillation fluor surrounded by numerous photomultiplier tubes to convert light emissions into electrical pulse signals would serve their purpose. The largest detector ever manufactured at that time was only $\sim 1 \, m^3$ in volume, which was thousands of orders of magnitude smaller than what would be needed to detect a neutrino interaction. A liquid detector provided the versatility of manufacturing a detector to almost any size and geometry.

Because of the neutral charge and near-zero mass of the neutrino, they knew that a very large detector would be required. Their initial estimations based on a neutrino of a few MeV in energy yielded a cross section of $\sim 10^{-44} \, cm^2$, which represents a miniscule interaction probability proportional to a mean free path of ~ 1000 light years of liquid hydrogen. Such estimates made the idea of detecting a neutrino seem almost impossible, and the comments from some pioneering physicists were not encouraging. In his Nobel Laureate address Frederick Reines noted that Wolfgang Pauli remarked during a visit to Caltech "I have done a terrible thing. I have postulated a particle that cannot be detected". Also Hans Bethe and Rudolf Peierls (1934) wrote in the journal *Nature* "there is no practically possible way of observing the neutrino". In 1954 during Reines' search for the neutrino, he confronted Bethe concerning his pronouncement in *Nature* and Bethe replied in good humor "Well, you shouldn't believe everything you read in the papers". A major obstacle that Reines and Cowan had to overcome was the general belief that the neutrino was undetectable. Also, there was no evidence of any kind that the neutrino existed. Nevertheless, the existence of the neutrino was generally accepted by the scientific community. The elusive and then theoretical particle provided the only explanation for the wide spectrum of energies encountered in beta decay.

Knowing the difficult task at hand Reines and Cowan nevertheless forged ahead in their quest. They discarded their plans to use a nuclear weapon's test as a source of neutrinos, and decided to make an initial test at one of the nuclear reactors of the Hanford Engineering Works in Hanford, Washington, which would serve as a suitable source of neutrinos, and where the detector could be adequately shielded to reduce gamma and neutron background radiation (Figure II.13). They constructed a 300 liter liquid scintillator surrounded by 90, 5 cm diameter photomultiplier tubes. The positron produced by the neutrino interaction (eq. (II.19)) would excite the liquid scintillator resulting in a flash of light that

Figure II.13 Frederick Reines (left) and Clyde Cowan, Jr. (right) with some of the equipment used at the Hanford experiment. (From Reines (1995) Nobel Lecture with permission © The Nobel Foundaton 1995.)

would be detected as a signal by the photomultiplier tubes. The neutron produced by the neutrino interaction would produce a delayed gamma pulse as a result of neutron capture. To reduce backgrounds even further they decided to count only the delayed coincidence between the positron and neutron capture pulses. After months of testing at the Hanford reactor, results were inconclusive because of high backgrounds from reactor-independent signals arising from cosmic radiation.

In spite of the difficulties encountered they continued to forge ahead with the design of a larger detector that would be yet more discriminating in its rejection of background, because of the very specific signals that would be used as the antineutrino signature. The newly designed detector would be taken to a new 700 MW nuclear reactor at the Savannah River Plant in Aiken, SC. The modified detection scheme is illustrated in Figure II.14 and a sketch of the new detector is provided in Figure II.15. The target chamber would consist of 200 liters of water containing 40 kg of dissolved $CdCl_2$ sandwiched between two tanks of 1400 liters of liquid scintillator solution. Each end of the scintillator tanks were viewed by 55 photomultiplier tubes. The water would provide target protons for antineutrinos. As illustrated in Figure II.14 the interaction of an antineutrino with a water proton would create a

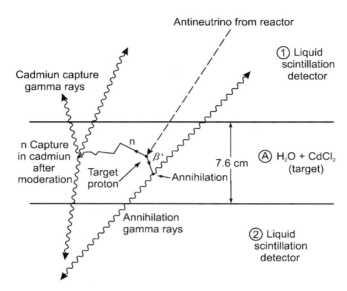

Figure II.14   Detection scheme used by Reines and Cowan for the antineutrino signature signal. (From Reines (1995) Nobel Lecture with permission © The Nobel Foundation 1995.) An antineutrino is illustrated entering the tank of aqueous $CdCl_2$ solution and striking a target proton. The proton converts to a neutron and positron. The positron annihilates on an electron with the emission of two 0.511 MeV gamma rays in opposite directions detected by the liquid scintillator in tanks above and below the water target tank. The neutron produced by the antineutrino interaction slows down in the water and is captured by a cadmium nucleus and the resulting gamma rays are detected by the liquid scintillator in the adjacent tanks approximately 10 μsec after the positron annihilation.

neutron (n) and a positron (β^+). The positron would be annihilated when coming to rest and in contact with an electron, and the resulting annihilation radiation (two 0.51 MeV gamma rays emitted in opposite directions) would be detected in coincidence by the two liquid scintillation detectors above and below the water tank. The neutron produced by the antineutrino interaction would slow down quickly (~10 μsec) in the water and be captured by a cadmium nucleus in the water target chamber. The characteristic multiple gamma rays following the neutron capture would be detected in coincidence by the liquid scintillation detectors. The antineutrino signature would therefore consist of a delayed coincidence between the prompt pulses produced by the β^+ annihilation and the pulses produced microseconds later by the neutron capture in cadmium.

With the unique signature for the antineutrino provided by the detector design devised by Reines and Cowan, the high neutron flux produced by the Savannah River reactor ($1.2 \times 10^{13}/cm^2$ sec), and reduced cosmic ray backgrounds from massive shielding 11 m from the reactor and 12 m underground were essential to the success of their experiment. Nevertheless, a detector running time of 100 days over a period of about 1 year was required to provide sufficient conclusive signals from the antineutrino signature. Frederick Reines and Clyde Cowan, Jr. and their coworkers provided conclusive evidence for the antineutrino signature signals by demonstrating, among other evidence, the following: (i) the first signal pulse of the delayed coincidence pair (two gamma rays) was demonstrated to be due to the positron annihilation by varying the thickness of a lead sheet interposed between the water target and one of the liquid scintillators, thereby reducing the positron detection efficiency in one of the detector triads and not in the others; (ii) the second signal pulse was shown to be due to a neutron by varying the cadmium concentration in the target water. Removal of the cadmium, as expected for antineutrinos, totally removed the correlated count rate; and (iii) the time interval spectrum between the first and second pulse signals agreed with that expected for annihilation and neutron-capture gamma rays. For additional reading on the history of this magnificent feat see papers by Cowan and

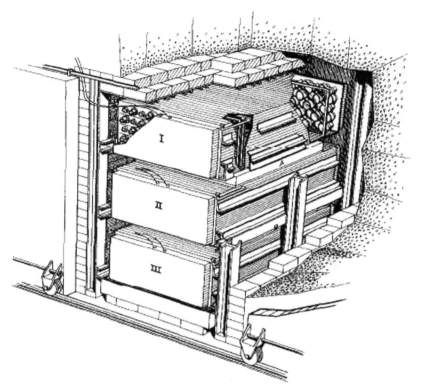

Figure II.15   A sketch of the equipment used by Reines and Cowan for the detection of the antineutrino. (From Reines (1995) Nobel Lecture with permission © The Nobel Foundation 1995.) Tanks marked I, II, and III, with combined heights of 2 meters, contained 1400 liters of liquid scintillator solution and are viewed on each end with 55 photomultiplier tubes. Thinner tanks marked A and B were filled with 200 liters of water providing target protons for the antineutrinos, and 40 kg of CdCl$_2$ were dissolved in each water target solution to provide Cd nuclei that would capture the product neutrons.

Reines (Cowan, 1964; Cowan *et al.*, 1953, 1956; Reines and Cowan, 1953; Reines *et al.*, 1960; Reines, 1979, 1982) and a review of this work in *Los Alamos Science* (1997).

After the discovery of the neutrino Reines and Cowan sent a telegram on 14 June of 1956 to Wolfgang Pauli at Zurich University to inform him of their successful detection of the neutrino. It was Pauli, who as a young man in 1930, had postulated the existence of the neutrino as a desperate measure to explain the broad spectrum of beta-particle energies emitted from decaying nuclides. The telegram is illustrated in Figure II.16 and reads as follows: "We are happy to inform you that we have definitely detected neutrinos from fission fragments by observing inverse beta decay of protons. Observed cross section agrees well with expected six times ten to minus forty four square centimeters".

It is reported that Pauli and friends celebrated in Zurich that evening, and Pauli had immediately sent a reply letter illustrated in Figure II.17 to Reines and Cowan, which reads as follows: "Thanks for message. Everything comes to him who knows how to wait".

Pauli died exactly 18 months later on December 15, 1958 in Zurich. He waited 26 years for the neutrino to be detected, and many of the most renowned physicists of the time did not believe that the neutrino could ever be detected and would remain a postulated particle. One can only imagine the immense satisfaction that Pauli could cherish during the remaining months of his life and the

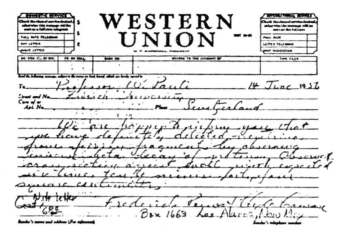

Figure II.16   Telegram from Reines and Cowan to Wolfgang Pauli informing him of their discovery of the neutrino. (From Reines (1995) Nobel Lecture with permission © The Nobel Foundation 1995.)

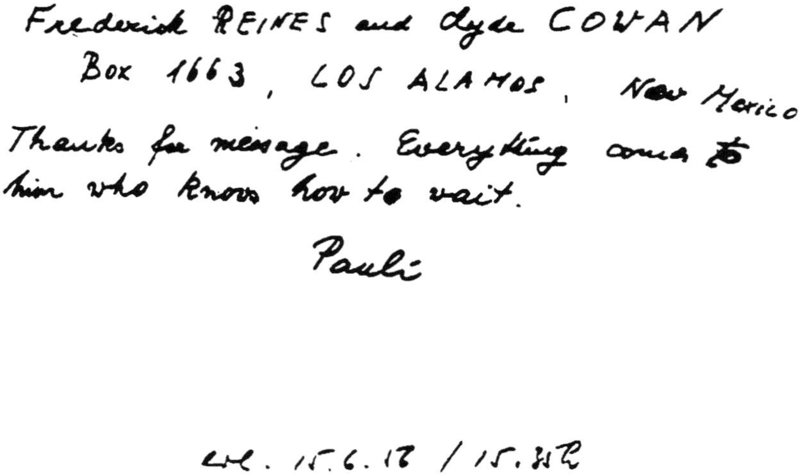

Frederick REINES and Clyde COWAN
Box 1663, LOS ALAMOS, New Mexico
Thanks for message. Everything comes to him who knows how to wait.
                                    Pauli

Figure II.17   Reply letter to Reines and Cowan from Wolfgang Pauli thanking them for informing him of the discovery of the elusive neutrino that he had postulated 26 years earlier. (From Reines (1995) Nobel Lecture with permission © The Nobel Foundation 1995.)

immeasurable sense of accomplishment felt by Frederick Reines and Clyde Cowan, Jr. after completing this colossal achievement.

A very practical outcome of the work of Reines and Cowan was the development of large radioactivity detectors. Among these, is the whole-body counter, which is now a very practical means of measuring low levels of natural radiation or internal contamination in the human body. A person can step into a chamber or be lying down on a bed with surrounding detectors capable of measuring body radioactivity. One of the first whole-body counters was demonstrated to the public in 1958 at the Atoms for Peace Conference in Geneva.

Frederick Reines went on to head a research group at the University of California, Irvine (UCI) and became Dean of the School of Physical Sciences until 1974. He continued at UCI as Professor of Physics and became Professor Emeritus there in 1988. He was awarded the Bruno Rossi Prize in 1989 for the joint observations of his group at UCI and the Kamiokande Experiment in Japan of neutrinos from supernova 1987A. His research group at UCI had large detectors that they used to detect neutrinos originating from stellar collapse and these detectors were adorned with signs identifying each as "Supernova Early Warning System". He died on August 26, 1998.

After the discovery of the neutrino Clyde Cowan, Jr. was awarded a Guggenheim fellowship for studies on the physics of neutrino and neutrino interactions. He went on to become Professor of Physics at the Catholic University of America in 1958 and an inspiration to many of his students until his early death in 1974.

– 2 –

Beta Radiation

2.1 INTRODUCTION

The beta particle has the same mass and charge as an electron. It differs from the electron in its origin. The beta particle, which may be either negatively charged (negatrons) or positively charged (positrons), originates from the nucleus of an atom. A beta particle is emitted from the nucleus of an atom during radioactive decay. The electron, however, occupies regions outside the nucleus of an atom. The beta particle, like the electron, has a very small mass compared to the proton or neutron. Its mass is 1/1836 that of the proton mass or 1/1838 that of the neutron mass. The beta particle has an origin in addition to that of the decaying nucleus. The beta particle may be produced in pairs, one positive and the other negative in charge, by the conversion of gamma radiation energy into the mass of two beta particles in the vicinity of a nucleus. This phenomenon is referred to as pair production. The origins, properties, and interaction of beta particles with matter are discussed in this chapter.

2.2 NEGATRONS

A negatron or negative beta particle (β^-) is an electron emitted by the nucleus of a decaying radionuclide that possesses an excess of neutrons or, in other words, a neutron/proton (n/p) imbalance. (see Section 2.3.1 for a brief discussion of n/p ratios and nuclear stability). The nuclear instability caused by the n/p imbalance results in the conversion of a neutron to a proton within the nucleus, where the balance of charge is conserved by the simultaneous formation of an electron (negatron) according to the equation

$$n \rightarrow p^+ + \beta^- + \bar{\nu} \tag{2.1}$$

A neutrino (ν), which is a particle of zero charge, accompanies beta-particle emission. The neutrino can be identified further as two types with opposite spin, namely, the antineutrino ($\bar{\nu}$), which accompanies negative beta-particle (negatron) emission and the neutrino (ν), which accompanies positive beta-particle (positron) emission (see Section 2.3 of this chapter). Because the neutrino and antineutrino have similar properties with the exception of spin, it is common to use the word "neutrino" to simplify references to both particles. The explanation for the neutrino and its properties, also emitted from the decaying nucleus, is given further on in this section. The electron formed cannot exist within the nucleus and is thus ejected as a negatron or negative beta particle, β^-, as a product of the nuclear decay with a maximum energy equivalent to the slight mass difference between the parent and daughter atoms less the mass of the beta particle, antineutrino or neutrino in the case of positron emission, and any gamma-ray energy that may be emitted by the daughter nucleus if it is left in an excited energy state (see Chapter 3). Tritium (^{3}H), for example, decays with β^- emission according to the following:

$$^3_1\text{H} \rightarrow {}^3_2\text{He} + \beta^- + \bar{\nu} + 0.0186\,\text{MeV} \tag{2.2}$$

The value of 0.0186 MeV (megaelectron volts) is the maximum energy the beta particle may possess. The unstable tritium nucleus contains two neutrons and one proton. The transformation of a neutron to a proton within the tritium nucleus results in a charge transfer on the nucleus from $+1$ to $+2$ without any change in the mass number. Although there is no change in the mass number, the mass of the stable helium isotope produced is slightly less than that of its parent tritium atom. Eqs. (2.3)–(2.8) illustrate other examples of β^- decay.

$$^{14}_{6}\text{C} \rightarrow {}^{14}_{7}\text{N} + \beta^- + \bar{\nu} + 0.156\,\text{MeV} \tag{2.3}$$

$$^{32}_{15}\text{P} \rightarrow {}^{32}_{16}\text{S} + \beta^- + \bar{\nu} + 1.710\,\text{MeV} \tag{2.4}$$

$$^{35}_{16}\text{S} \rightarrow {}^{35}_{17}\text{Cl} + \beta^- + \bar{\nu} + 0.167\,\text{MeV} \tag{2.5}$$

$$^{36}_{17}\text{Cl} \rightarrow {}^{36}_{18}\text{Ar} + \beta^- + \bar{\nu} + 0.714\,\text{MeV} \tag{2.6}$$

$$^{45}_{20}\text{Ca} \rightarrow {}^{45}_{21}\text{Sc} + \beta^- + \bar{\nu} + 0.258\,\text{MeV} \tag{2.7}$$

$$^{89}_{38}\text{Sr} \rightarrow {}^{89}_{39}\text{Y} + \beta^- + \bar{\nu} + 1.490\,\text{MeV} \tag{2.8}$$

The energies of beta-particle decay processes are usually reported as the maximum energy, E_{max}, that the emitted beta particle or antineutrino may possess. The maximum energy is reported because beta particles are emitted from radionuclides with a broad spectrum of energies. A typical spectrum is illustrated in Figure 2.1. Unlike alpha particles, which have a discrete energy, beta particles have a wide spectrum of energies ranging from zero to E_{max}.

The majority of beta particles emitted have energies of approximately $\frac{1}{3}E_{\text{max}}$. Only a very small portion of the beta particles is emitted with the maximum possible energy from any radionuclide sample. Wolfgang Pauli was the first to postulate why beta particles were not emitted with fixed quanta of energy, quite the contrary to what is observed in alpha-particle emission. He proposed the existence of an elusive, neutral, and almost massless

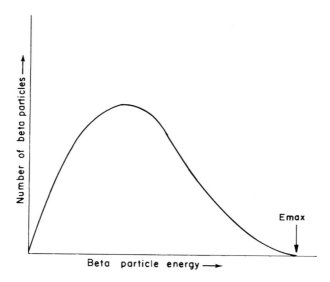

Figure 2.1   General energy spectrum of beta particles.

particle in a letter to the participants of a radioactivity conference at Tübingen on December 4, 1930. The neutrino was considered elusive, because if it existed, its zero charge and near zero rest mass would make the neutrino undetectable by conventional means and allow a neutrino to penetrate matter easily and even pass through the entire earth without causing a single interaction. It is the neutrino that would be emitted simultaneously with the beta particle from the decaying nucleus and share the energy of decay with the beta particle. For example, if a beta particle was emitted from tritium (decay energy = 0.0186 MeV) with an energy of 0.0086 MeV, the accompanying neutrino would possess the remaining energy of 0.01 MeV, that is, the decay energy less the beta-particle energy (0.0186 MeV − 0.0086 MeV). Consequently, if we observe any number of beta particles emitted from a tritium sample or other beta-emitting nuclide sample (e.g., ^{14}C, ^{32}P, ^{90}Sr), they would possess different energies and display an energy spectrum from zero to E_{max} as illustrated in Figure 2.1.

With Pauli's postulation of the neutral particle, Enrico Fermi elaborated the beta-decay theory in 1934 and coined the neutrino from the Italian language meaning "little neutral one". The particle remained elusive until the observation of the neutrino was demonstrated by Reines and Cowan in 1956 (see Cowan et al., 1956; Reines and Cowan 1956, 1957; Reines, 1960, 1979, 1994). They confirmed the existence of the neutrino by demonstrating inverse beta decay where an antineutrino interacts with a proton to yield a neutron and positron

$$\bar{\nu} + p \rightarrow n + \beta^+ \tag{2.9}$$

They used a tank of water containing a solution of ^{113}CdCl$_2$. Neutrinos interacted with the protons of the water to produce neutrons and positrons. Some of the neutrons produced would be absorbed by the ^{113}Cd with the concomitant emission of characteristic gamma radiation. In coincidence, they observed two 511 keV gamma rays, which originate when a positron comes to rest in the vicinity of an electron, its antiparticle, resulting in the annihilation of two electrons into two gamma-ray photons of energy equivalent to the electron masses, 0.511 MeV. In the same year Lee and Yang (1956) proposed that neutrinos and antineutrinos possessed left-handed and right-handed spins, respectively. Reverse beta decay remains an important nuclear process utilized in the measurement of solar neutrinos today (Gratta and Wang, 1999).

Since its inception by Pauli in 1930 up to recent years, the neutrino or antineutrino had been thought to be almost massless or to possess a near-zero rest mass. It was not until June 5, 1998 that the Super-Kamiokande Collaboration, including scientists from 23 institutions in Japan and the United States, announced at the "Neutrino 98" International Physics Conference in Takayama, Japan, that neutrinos possessed a definite mass (Gibbs, 1998; Kesterbaum, 1998; Kearns et al., 1999; Nakahata, 2000). The mass was not reported, but evidence was provided that the neutrino did possess mass although it was considered to be "very small", at least 0.07 eV, which would be less than a millionth of the electron mass. Evidence for the neutrino mass was provided by demonstrating that neutrinos can "oscillate" from one type into another (i.e., electron-, muon-, and tau-neutrinos) as they travel through space and matter. Oscillation is the changing of neutrino types back and forth from one type to another, and this could occur only if the neutrino possessed mass.

More recently at the "Neutrino 2000" Conference held at Sudbury, Canada in June 16–21, 2000, groups from the University of Mainz, Germany (Bonn et al., 2005), and Institute for Nuclear Research, Moscow (Lobashev et al., 2005) reported the mass of the neutrino to be 2.2 and 2.5 eV/c^2, respectively at 95% confidence levels. It is common to express subatomic particle mass in units of energy on the basis of equivalence of mass and energy ($E = mc^2$), so that the particle mass m is measured in units of E/c^2 or eV/c^2. To put the mass of the neutrino in perspective, we can take the experimental value of the neutrino rest mass, $m_\nu = 2.2$ eV/c^2, from the University of Mainz Group and convert this to kilograms as follows:

By definition 1 eV $= 1.60 \times 10^{-19}$ J, and from the equation $E = mc^2$

$$m_\nu = \frac{E}{c^2} = \frac{(2.2\,\text{eV})(1.60 \times 10^{-19}\,\text{J/eV})}{(3.00 \times 10^8\,\text{m/s})^2} = 3.9 \times 10^{-36}\,\text{kg}$$

If we compare the rest mass of the neutrino, m_ν, to that of the miniscule electron, m_e, we see that the neutrino mass is approximately four millionths that of the electron or

$$\frac{m_\nu}{m_e} = \frac{3.9 \times 10^{-36}\,\text{kg}}{9.1 \times 10^{-31}\,\text{kg}} = 4.2 \times 10^{-6}$$

Owing to the very low mass of the β^- particle compared with the alpha particle, it travels at a much higher velocity than an alpha particle of equivalent energy. Because of its greater velocity, lower mass, and lower charge, the specific ionization produced in air by the traveling beta particle is much lower (by a factor of a thousand) than that of an alpha particle of equivalent energy.

Like the alpha particle, the beta particle interacts with matter via (i) ionization and (ii) electron orbital excitation as it dissipates its kinetic energy. A third mechanism of interaction with matter, which distinguishes the beta particle, is radiative energy dissipation via Bremsstrahlung production (see Chapter 3, Section 3.6.2). Thus as described by Turner (1995) the stopping power for beta particles (β^- or β^+) is the sum of the collisional and radiative contributions or

$$\left(-\frac{dE}{dx} \right)_{tot}^{\pm} = \left(-\frac{dE}{dx} \right)_{col}^{\pm} + \left(-\frac{dE}{dx} \right)_{rad}^{\pm} \tag{2.10}$$

where the superscript $\pm$ refers to positively or negatively charged electrons. The radiative contribution, that is, the absorption of beta-particle energy with the concomitant emission of Bremsstrahlung radiation is significant with high-energy beta particles (e.g., ^{32}P or ^{80}Y beta-particle emissions) in absorbers of high atomic number (e.g., Pb-glass). Bremsstrahlung radiation is discussed in Chapter 3.

Collision interactions of beta particles are somewhat different than those that occur with alpha particles. A beta particle may collide with an orbital electron or come into close proximity to it and cause the electron to be ejected, resulting in the formation of an ion pair. Considerable scattering of beta particles occurs in such collisions because the mass of the beta particle is equivalent to that of an atomic electron. This is in direct contrast to the alpha particle, which, for the most part, retains a relatively undeviating path while passing through matter and interacting with atomic electrons. The mass equivalence of beta particles and electrons is an important factor that gives bombarding beta particles the power to impart a major portion of their kinetic energy to atomic electrons in a single collision. The atomic electrons ejected upon beta particle collisions themselves cause ionization in a similar fashion. This is referred to as secondary ionization, and the ionization caused by the first beta particle–electron collisions is classified as primary ionization. Because the major portion of beta particle energy may be imparted to an atomic electron upon collision, secondary ionization may account for as much as 80% of the total ionization produced in a given material bombarded by beta particles.

The probability of beta-particle interactions with atomic electrons increases with the density of the absorbing material. Beta-particle absorption is consequently proportional to the density and thickness of an absorber. When we compare substances of similar atomic composition, we find that the range of beta particles (β^- or β^+) expressed in mass thickness units (mg/cm^2) is approximately the same. For example, Figure A.3 of Appendix A provides a curve where the range in units of g/cm^2 in substances of low atomic number can be estimated for beta particles of energies from 0.01 to 10 MeV. The range of beta particles expressed in terms of surface density or mass thickness (g/cm^2) of absorber can be converted to absorber thickness (cm) when the absorber density (g/cm^3) is known. Several empirical formulas exist for calculating beta particle ranges and are solved on the basis of the E_{max} of the beta particle. The formulas reported by Glendenin (1948) are

$$R = 0.542E - 0.133 \quad \text{for } E > 0.8\,\text{MeV} \tag{2.11}$$

and

$$R = 0.407E^{1.38} \quad \text{for } 0.15\,\text{MeV} < E < 0.8\,\text{MeV} \tag{2.12}$$

where R is the beta particle range in g/cm^2 and E the energy of the beta particle (i.e., E_{max}) in MeV. Also, the following empirical formula of Flammersfeld (1946) described by Paul and Steinwedel (1955) can be used:

$$R = 0.11\left(\sqrt{1 + 22.4E^2} - 1\right) \quad \text{for } 0 < E < 3\,\text{MeV} \tag{2.13}$$

This formula provides calculated ranges in units of g/cm^2 in close agreement to those obtained from eqs. (2.11) and (2.12) or those found from Figure A.3 in Appendix A. The above empirical formula and the graph in Figure A.3 are suitable for low atomic number absorbers such as aluminum ($Z = 13$). Semi-empirical equations developed by Tabata and Okabe (1972) and described in detail by Tsoulfanidis (1995) yield beta particle or electron ranges in absorber elements and compounds of a wide range of atomic number.

According to eq. (2.11), a 1.0 MeV beta particle has a calculated range of 0.409 g/cm^2. This value may be divided by the density, ρ, of the absorber material to provide the range in centimeters of absorber thickness. Thus, it can be estimated that a 1.0 MeV beta particle travels approximately 334 cm in dry air ($\rho = 0.001226$ g/cm^3 at STP), 0.40 cm in water ($\rho = 1.00$ g/cm^3), and 0.15 cm in aluminum ($\rho = 2.7$ g/cm^3). The effect of absorber density on beta particle range is obvious from the foregoing examples, which demonstrate that 1 cm of dry air has about the same stopping power as 0.004 mm of aluminum.

The range of beta particles in matter is considerably greater than that of alpha particles of the same energy. Again, this is due to the lower mass, lower charge, and higher velocity of travel of the beta particle in comparison with an alpha particle of equivalent energy. The significance of this difference may be appreciated by reference to Table 2.1, in which the alpha particle and beta particle and/or electron ranges in air as a function of particle energy are compared. To put this data into historical perspective, it is interesting to recall the origin of the names "alpha and beta radiation". Before alpha and beta particles were characterized fully, Ernest Rutherford carried out experiments in 1898 that demonstrated the existence of two types of radiation; one radiation that was most easily absorbed by matter and another that possessed a greater penetrating power. Out of convenience, he named these radiations as "alpha" and "beta". Not much later P. U. Villard in France discovered in 1900 a yet more penetrating radiation that was named "gamma" in harmony with the nomenclature coined by Rutherford.

It is important to emphasize that, although all beta particles can be completely absorbed by matter, the shields we select can be of great consequence. Hazardous Bremsstrahlung radiation can be significant when high-energy beta particles interact with shields of high atomic number. The phenomenon of Bremsstrahlung production is discussed further in Chapter 3.

2.3 POSITRONS

In contrast to negatron emission from nuclei having n/p ratios too large for stability, positrons, which consist of positively charged electrons (positive beta particles), are emitted from nuclei

Table 2.1

Ranges of alpha and beta particles (or electrons) of various energies in air

Energy (MeV)	Range (mg/cm^2)		Range (cm)^c	
	Alpha particle^a	Beta particle^b	Alpha particle	Beta particle
0.1	0.013^d	13	0.010	11
0.5	0.4	163	0.3	133
1.0	0.6	412	0.5	336
1.5	0.9	678	0.7	553
2.0	1.2	946	1.0	772
2.5	1.6	1217	1.3	993
3.0	2.1	1484	1.7	1210
4.0	3.1	2014	2.5	1643
5.0	4.2	2544	3.5	2075
6.0	5.6	3074	4.6	2507
7.0	7.2	3604	5.9	2940
8.0	8.7	4134	7.1	3372

^{a}From the curve provided in Figure A.1 of Appendix A with the exception of 0.1 MeV particle energy.
^{b}Calculated from the formulas for range (R) in units of g/cm^2, $R = 0.412E^{1.27-0.0945\ln E}$ for $0.01 \leq E \leq 2.5$ MeV and $R = 0.530E - 0.106$ for $E > 2.5$ MeV. (see Figure A.3 of Appendix A.)
^{c}Calculated from the range in mass thickness units (mg/cm^2) and the density of dry air at STP, $\rho_{air} = 1.226$ mg/cm^3 according to eq. (1.14).
^{d}Calculated from eqs. (1.13) and (1.15) using weight averages of elements in air according to the following: 78.06% N, 21% O, 0.93% Ar, and 0.011% C.

having n/p ratios too small for stability, that is, those which have an excess of protons. (See Section 2.3.1 for a brief discussion of n/p ratios and nuclear stability.)

To attain nuclear stability, the n/p ratio is increased. This is realized by a transformation of a proton to a neutron within the nucleus. The previously discussed alteration of a neutron to a proton in a negatron-emitting nuclide (eq. (2.1)) may now be considered in reverse for the emission of positrons. Eq. (2.14) illustrates such a transformation

$$p^+ \rightarrow n + \beta^+ + \nu \tag{2.14}$$

^{58}Co may be cited as an example of a nuclide that decays by positron emission:

$$^{58}_{27}\text{Co} \rightarrow {}^{58}_{26}\text{Fe} + \beta^+ + \nu \tag{2.15}$$

Note that the mass number does not change but the charge on the nucleus (Z number) decreases by 1. As in negatron emission, a neutrino, ν, is emitted simultaneously with the positron (beta particle) and shares the decay energy with the positron. Thus, positrons, like negatrons emitted from a given radionuclide sample, may possess a broad spectrum of energies from near zero to E_{max} as illustrated in Figure 2.1.

Decay by positron emission can occur only when the decay energy is significantly above 1.02 MeV. This is because two electrons of opposite charge are produced (β^+, β^-) within the nucleus, and the energy equivalence of the electron mass is 0.51 MeV (see Section 3.7.4 of Chapter 3). The positive electron, β^+, is ejected from the nucleus and the negative electron, β^-, combines with a proton to form a neutron:

$$\beta^- + p^+ \rightarrow n \tag{2.16}$$

Thus, the E_{max} of a positron emitted from a nucleus is equivalent to the mass difference of the parent and daughter nuclides, less the mass of the positron and neutrino (albeit, the neutrino mass is very small compared to the mass of the positron) emitted from the nucleus (see equivalence of mass and energy, Chapter 3, Section 3.7.4) and less any gamma-ray energy of the daughter nuclide if left in an excited state (see Chapter 3, Section 3.3).

From the Chart of Nuclides it is possible to cite specific examples of the n/p imbalance in relation to negatron and positron emission. Figure 2.2 illustrates the relative positions of the stable nuclides ^{12}C, ^{13}C, ^{14}N, and ^{15}N, etc. and of their neighboring radionuclides. The nuclides are positioned as a function of the number of protons, Z, and the number of neutrons, N, in their respective nuclei. Dashed arrows are placed through the blocks that segregate radionuclides interrelated with common daughter nuclides resulting from β^- or β^+ decay processes. For example, the stable nuclide ^{12}C of atomic number 6 has a nucleus with an n/p ratio of 6/6. However, the nuclide ^{12}N of atomic number 7 has an unstable n/p ratio of 5/7, an excess of protons. Thus, this nuclide decays via positron emission according to the equation

$$^{12}_{7}\text{N} \rightarrow {}^{12}_{6}\text{C} + \beta^+ + \nu \tag{2.17}$$

to ^{12}C by positron emission as indicated by a dashed arrow of Figure 2.2.

The nuclide ^{12}B of atomic number 5 has the unstable n/p ratio of 7/5, an excess of neutrons. This nuclide thus decays to ^{12}C by negatron emission according to the equation

$$^{12}_{5}\text{B} \rightarrow {}^{12}_{6}\text{C} + \beta^- + \bar{\nu} \tag{2.18}$$

Similar reasoning may be used to explain positron and negatron decay of the unstable nuclides shown in Figure 2.2 to the stable products ^{13}C, ^{14}N, and ^{15}N. The interrelationship between β^- and β^+ decay leading to the formation of stable nuclides is to be found throughout the chart of the nuclides.

Positrons dissipate their energy in matter via the same mechanisms as previously described for negatrons, which is understandable, as both are electrons. The stopping powers and ranges of positrons are virtually identical to negatrons and electrons over the broad energy range of 0.03–10^3 MeV (Turner, 1995). Although equations in Section 2.5.1 of this chapter are cited for calculating the ionization–excitation stopping powers for negatrons and positrons due to collision interactions with absorbers, their difference as noted by Tsoulfanidis (1995) is due only to the second term in the brackets of these two equations, which is much smaller then the logarithmic term, and consequently the differences between negatron and positron stopping powers do not exceed 10%. However, positrons are unique in that these particles produce annihilation gamma radiation in matter discussed in the following chapter.

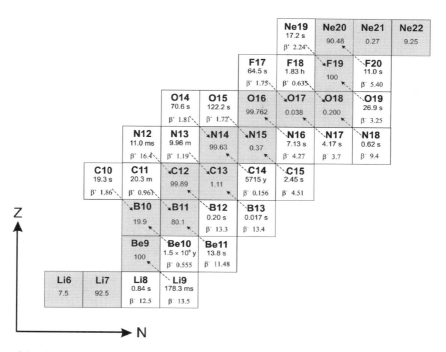

Figure 2.2 A segment of the Chart of the Nuclides showing the relative positions of some stable (shaded) and unstable nuclides. The ordinate Z and abscissa N represent the number of protons (atomic number) and the number of neutrons within the nucleus, respectively. The mass number (number of protons + neutrons) in the nucleus of an isotope is provided alongside the abbreviation of the element. For example, the isotope ^{14}C (carbon-14) is written in the Chart as C14. Radionuclides that undergo positron (β^+) emission are illustrated with arrows pointing downward to stable nuclide products of lower atomic number; and radionuclides undergoing negatron (β^-) emission are illustrated with arrows pointing upward to stable nuclide products of higher atomic number. The maximum beta-particle energy in units of MeV is provided alongside the positron or negatron symbol. Numbers in shaded areas are the percent natural abundances of the stable isotopes, which for each element should add up to 100%. For example, the element lithium (Li) exists in nature as two stable isotopes, namely, ^6Li and ^7Li with abundances of 7.5 and 92.5%, respectively. When only one stable isotope of an element exists, its natural abundance will be expressed as 100%, such as is the case for ^9Be and ^{19}F illustrated above. The half-lives of the unstable nuclides are given in units of seconds (s), milliseconds (ms), minutes (m), hours (h) and years (y).

2.3.1 *N/Z* ratios and nuclear stability

In Sections 2.2 and 2.3 of this chapter we discussed negatron and positron decay as processes whereby unstable nuclei may achieve stability via neutron or proton transformations, respectively. These processes in the nucleus of the radionuclide result in a change in the n/p or *N/Z* ratio of the nucleus.

 If we look throughout the Chart of the Nuclides we will notice that the stable nuclides of low atomic number will have a *N/Z* ratio of approximately 1. However, as the atomic number increases ($Z > 20$), the *N/Z* ratio of the stable nuclides increases gradually and reaches as high as approximately 1.5 (e.g., $^{209}_{83}$Bi, $Z = 83$, $N/Z = 1.518$). Furthermore, there are no stable nuclides of atomic number greater than 83.

The nature of nuclear forces and the relationship of N/Z ratio to nuclear stability are discussed in detail by Serway *et al.* (1997) and Sundareson (2001). In brief, the importance of N/Z ratio to nuclear stability is explained by the fact that there exists a short-range attractive nuclear force, which extends to a distance of ≈ 2 fm (2 fermi or 2×10^{-15} m). This attractive force has charge independence and is a consequence of the relative spins of the protons and neutrons and their relative positions in the nucleus. These binding exchange forces exist therefore, regardless of charge on the particles, between two protons, two neutrons, and a proton and neutron. While the attractive nuclear forces will tend to hold the nucleus together there exists, at the same time, repelling coulombic forces between the positively charged protons that act to force them apart. For nuclides of low Z, the attractive nuclear forces exceed the repelling coulombic forces when $N \approx Z$. However, increasing the number of protons (e.g., $Z > 20$) further increases the strength of the repelling coulombic forces over a larger nucleus, which will tend to force the nucleus apart. Therefore, additional neutrons, $N > Z$, provide additional attractive nuclear forces needed to overcome the repelling forces of the larger proton population. As the atomic number increases further, $Z > 83$, all nuclides are unstable. Even though N/Z ratios reach 1.5, nuclear stability is not achieved when the number of protons in the nucleus exceeds 83.

2.3.2   Positron emission versus electron capture

Another mechanism by which an unstable nucleus can increase its n/p ratio is via the capture by the nucleus of a proximate atomic electron (e.g., K- or L-shell electron). The absorbed electron combines with a proton to yield a neutron within the nucleus as follows:

$$e^- + p^+ \rightarrow n + \nu + \text{inner Bremsstrahlung} + \text{x-rays} + \text{Auger electrons} + (\gamma) \quad (2.19)$$

The decay process is known as electron capture (EC), or sometimes referred to as K capture, because most of the electrons are captured from the K shell, which is closest to the nucleus. A neutrino, ν, is emitted and this is accompanied by the emission of internal Bremsstrahlung, which is a continuous spectrum of electromagnetic radiation that originates from the atomic electron as it undergoes acceleration toward the nucleus. Unlike the beta-decay process, which results in the emission of a neutrino from the nucleus with a broad spectrum of energies, the neutrino emitted from the EC decay process does not share the transition energy with another particle and, therefore, it is emitted with a single quantum of energy equal to the transition energy less the atomic electron binding energy. The capture of an atomic electron by the nucleus leaves a vacancy in an electron shell, and this is usually filled by an electron from an outer shell, resulting in the production of x-radiation (see Chapter 3). The electron that fills the vacancy leaves yet another vacancy at a more distant shell. A cascade of electron vacancies and subsequent filling of vacancies from outer electron shells occurs with the production of x rays characteristic of the daughter atom. The x rays will either travel out of the atom or interact with orbital electrons to eject these as Auger electrons. Gamma radiation is illustrated in the above eq. (2.19), because it is emitted only when the daughter nuclide is left at an unstable elevated energy state (see Figure 3.4 in Chapter 3).

The EC decay process may compete with β^+ emission. That is, some radionuclides may decay by either EC or, β^+ emission. As discussed previously, positron emission requires a transition energy of at least 1.02 MeV, the minimum energy required for pair production in the nucleus (i.e., two electron rest mass energies or 2×0.511 MeV). Positron emission, therefore, will not compete with EC for decay transitions less than 1.02 MeV. In general, positron emission will predominate when the transition energy is high (well above 1.02 MeV) and for nuclides of low atomic number, while the EC decay process will predominate for low transition energies and nuclides of higher atomic number. The decay transitions of ^{22}Na and ^{65}Zn serve as examples. In the case of ^{22}Na, decay by β^+ emission predominates (90%) as compared with decay via EC (10%),

$$^{22}_{11}\text{Na} \rightarrow {}^{22}_{10}\text{Ne} + \beta^+ + \nu \quad (90\%) \tag{2.20}$$

and

$$^{22}_{11}\text{Na} \xrightarrow{\text{EC}} {}^{22}_{10}\text{Ne} + \nu \quad (10\%) \tag{2.21}$$

The transition energy of ^{22}Na is 2.842 MeV (Holden, 1997a), well above the 1.02 MeV minimum required for positron emission. On the other hand, taking the example of the nuclide ^{65}Zn, we see that EC predominates over β^+ emission

$$^{65}_{30}\text{Zn} \rightarrow {}^{65}_{29}\text{Cu} + \beta^+ + \nu \quad (1.5\%) \tag{2.22}$$

and

$$^{65}_{30}\text{Zn} \xrightarrow{\text{EC}} {}^{65}_{29}\text{Cu} + \nu \quad (98\%) \tag{2.23}$$

In the case of ^{65}Zn, the transition energy is only 1.35 MeV (Holden, 1997a). It is not much above the minimum energy of 1.02 MeV required for positron emission. Consequently, EC decay predominates.

It is generally known that, chemical factors do not control nuclear decay processes. However, because the EC decay process involves the capture of an orbital electron by the nucleus, atomic or molecular binding effects, which vary with chemical structure can influence the EC decay process. Ehman and Vance (1991) cite the interesting examples of ^{7}Be and ^{90m}Nb, which display different electron-capture decay rates depending on the chemical state of the nuclides. ^{7}Be as a free metal and in the form $^7\text{BeF}_2$ salt display a 0.08% difference in EC decay rates, while ^{90m}Nb as a free metal and the salt form $^{90m}\text{NbF}_3$ exhibit an even greater 3.6% difference in EC decay rates.

2.4 BETA-PARTICLE ABSORPTION AND TRANSMISSION

Early research work on measuring the range of beta particles involved placing absorbers of increasing thickness between the radioactive source and the detector. The detector would measure the beta particles transmitted through the absorber. Increasing the absorber thickness would increasingly diminish the number of beta particles transmitted on to the detector. The transmission of beta particles was then plotted against absorber thickness as illustrated in Figure 2.3 in an attempt to determine the thickness of absorber required to fully stop the beta particles. Unfortunately, the plots could not be used directly to accurately determine beta-particle ranges; rather they had to be compared to an absorption curve of a beta emitter of known range by what became known as Feather analysis (Feather, 1938; Glendenin, 1948). An auspicious outcome of this work was the observation that the plots of beta-particle absorption had more or less an exponential character. When plotted logarithmically against distance the beta-particle absorption and/or transmission through the absorber were linear or near linear when plotted against absorber thickness as illustrated in Figure 2.3. This was a fortuitous outcome of the continuous energy spectrum of beta particles emitted from any given source. These findings are quite the contrast to the absorption curve of alpha particles discussed previously (Figure 1.3), where the alpha particle intensity remains constant and then comes to an almost abrupt stop.

The curve illustrated in Figure 2.3 is characteristic of beta particles. The somewhat linear segment of the semi-logarithmic plot of activity transmitted versus absorber thickness levels off horizontally due to a background of Bremsstrahlung radiation. Negatrons and positrons both display a somewhat linear semi-logarithmic plot with the exception that, in the case of positrons, the horizontal portion of the plot has an added background due to annihilation radiation (Glendenin, 1948). Because beta particles have a definite range in matter, beta-particle transmission is not a purely logarithmic one as we shall see is the case for gamma radiation (see Section 3.7.5 of Chapter 3). The curves may not display a purely exponential character and the plots may have a degree of concavity to them depending on the distance of the source and detector to the absorber and on the shape of the beta particle continuous

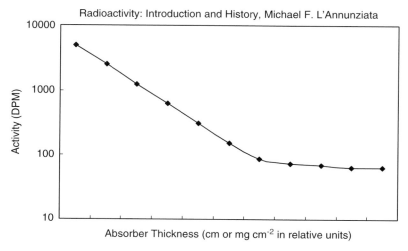

Figure 2.3   The transmission of beta particles through absorber material of increasing thickness. The semi-logarithmic plot is linear over a specific range of absorber thickness and then levels off horizontally due to a background of Bremsstrahlung radiation.

energy spectrum. The greater the atomic number of the beta-particle emitter, and the more the beta spectrum is displaced toward the lower energies, the more nearly exponential (linear) will be the absorption curve (Glendenin, 1948). It is common to express the amount of absorber in mass thickness units, that is, mass per unit area (e.g., g/cm^2), which is the product of absorber thickness and density, as it is easier to measure accurately very thin absorbers simply from their weight.

On the basis of the exponential character of beta-particle absorption we can describe the transmission of beta particles through the absorber as

$$I = I_0 e^{-\mu x} \tag{2.24}$$

where I is intensity of the beta particles (DPM) transmitted through the absorber, I_0 the initial DPM incident on the absorber, μ the linear absorption coefficient in units of cm^{-1}, and x the absorber thickness in cm. If we express absorber thickness in mass thickness units (e.g., mg/cm^2 or g/cm^2) we can rewrite eq. (2.24) as

$$I = I_0 e^{-(\mu/\rho)(\rho x)} \tag{2.25}$$

or

$$\frac{I}{I_0} = e^{-(\mu/\rho)(\rho x)} \tag{2.26}$$

and

$$\ln \frac{I}{I_0} = -(\mu/\rho)(\rho x) \tag{2.27}$$

where μ/ρ is the mass absorption coefficient (also referred to as mass attenuation coefficient) in units of cm^2/g, that is, the linear absorption coefficient divided by the absorber density, and ρx is the absorber thickness in mass thickness units g/cm^2, that is, the product of the absorber density and absorber thickness.

Eq. (2.27) can be used to determine experimentally the unknown thickness of absorber materials. A standard curve is plotted with the ratio I/I_0 on a logarithmic scale versus mass thickness (ρx) of the absorber on a linear scale. A value for I in units of DPM are determined with a detector by measuring the beta-particle intensity transmitted through a given absorber thickness. This is repeated with absorbers of different thickness. The magnitude of the incident beta-particle intensity, I_0, is a constant value and determined with the detector in the absence of absorber. The linear portion of the plot has a negative slope, such as that illustrated in Figure 2.3, and from least squares analysis the mass attenuation coefficient μ/ρ is determined (Yi et al., 1999). Consequently, the thickness of an unknown similar material can be determined from measured intensity, I, of the transmitted beta-particle radiation after placing the material between the beta-particle source and detector without altering the counting geometry. The sample thickness is calculated or determined directly from the aforementioned curve (Tumul'kan, 1991; Clapp et al., 1995).

Beta-particle transmission has many practical applications today in industrial manufacturing. Beta-particle sources and detectors are placed on the production line to test for thickness, uniformity, and defects in the manufacture of paper, metal, and plastic films as well as on-line inspection of sewn seams in the textile industry (Ogando, 1993; Clapp *et al.*, 1995; Mapleston, 1997; Titus *et al.*, 1997) and in agronomic research to measure leaf water content (Mederski, 1961; Mederski and Alles, 1968; Nakayama and Alles, 1964; Obregewitsch *et al.*, 1975). These are commonly referred to as beta transmission thickness gauges. The beta-particle sources used depend on the absorber thickness to be measured and the E_{max} of the beta particles. Three sources commonly used are ^{147}Pm (E_{max} = 0.224 MeV), ^{85}Kr (E_{max} = 0.672 MeV), and ^{90}Sr(^{90}Y) in secular equilibrium (E_{max} of ^{90}Sr and ^{90}Y = 0.546 and 2.280 MeV, respectively). The source with the lowest beta-particle E_{max} (e.g., ^{147}Pr) is used to measure the finest thickness of material (Balasubramanian, 1997, 1998), and the sources are changed according to beta-particle energy, penetration power and thickness of material to be tested.

2.5   STOPPING POWER AND LINEAR ENERGY TRANSFER

Throughout the text of this book information is provided on the mechanisms of interaction of radiation with matter. In summary, we can state that the principle mechanisms of interaction of charged particles (e.g., alpha particles, protons, deuterons, and electrons or beta particles) with matter, which result in significant charged-particle energy loss are (i) ionization via coulombic interactions of the charged particles with atomic electrons of the absorbing medium, (ii) electron orbital excitation of the medium, which occurs when the energy transfer through coulombic interaction is not sufficient to actually eject an electron from an atom, and (iii) the radial emission of energy as Bremsstrahlung when an electron or beta particle decelerates as it approaches an atomic nucleus. Release of particle energy by Bremsstrahlung radiation becomes increasingly significant as the beta-particle energy and absorber atomic number increase. On the other hand, electromagnetic radiation (i.e., the photon) dissipates its energy in matter via three possible mechanisms, namely, (i) the photoelectric effect, (ii) Compton scattering, and (iii) pair production. The photoelectric effect and Compton scattering generate ion pairs directly within the absorbing medium, whereas, pair production results in the creation of charged particles (positrons and negatrons) that will subsequently dissipate their energy via ionization, electron excitation and, in the case of positrons, annihilation. Also, as described in Chapter IV, energetic neutrons will dissipate their energy in matter through elastic collisions with atomic nuclei of the absorbing medium. When hydrogen is present in the absorbing material, the bulk of the fast neutron energy is passed on to the hydrogen nuclei. In turn, the kinetic energy of these protons is absorbed in the medium via ionization and excitation processes. Low- and high-energy neutrons are absorbed principally via inelastic neutron collisions, which can result in the production of charged particles and gamma radiation.

The radiation properties (e.g., charge, mass, and energy) and mechanisms of interaction previously described govern the rate of dissipation of energy and consequently the range of travel of the nuclear radiation in the absorber. This brings to bare the concepts of stopping power and linear energy transfer (LET), which are described subsequently.

2.5.1 Stopping power

Stopping power is defined by The International Commission on Radiation Units and Measurements or ICRU (Taylor *et al.*, 1970) as the average energy dissipated by ionizing radiation in a medium per unit path length of travel of the radiation in the medium. It is, of course, impossible to predict how a given charged particle will interact with any given atom of the absorber medium. Also, when we consider that the coulombic forces of charged particles will interact simultaneously with many atoms as it travels through the absorbed medium, we can only predict an average effect of energy loss per particle distance of travel. Taking into account the charge, mass, and speed (energy) of the particle and the density and atomic number of the absorbing medium, Hans Bethe (1933) and Bethe and Ashkin (1953) derived the formula for calculating the stopping power resulting from coulombic interactions of heavy charged particles (e.g., alpha particles, protons, and deuterons) traveling through absorber media. Rohrlich and Carlson (1954) have refined the calculations to include energy losses via Bremsstrahlung radiation, significant when high-energy electrons and beta particles interact with absorbers of high atomic number. Also, refinements to the stopping-power formulae in the low energy ranges of heavy particles have been made by several researchers including Bohr and Lindhard (1954), Lindhard and Scharff (1960, 1961), Northcliffe (1963), and Mozumder *et al.* (1968). Derivations of the stopping-power formulas can be obtained from texts by Friedlander *et al.* (1964), Roy and Reed (1968), Segré (1968), and Evans (1972). The formulas for the stopping power of charged particles due to coulombic interactions (i.e., ionization and electron orbital excitation) are most clearly defined by Tsoulfanidis (1995) as the following:

(i) for heavy charged particles (e.g., protons, deuterons, and alpha particles),

$$\frac{dE}{dx} = 4\pi r_0^2 z^2 \frac{mc^2}{\beta^2} NZ \left[\ln\left(\frac{2mc^2}{I}\beta^2\gamma^2\right) - \beta^2 \right] \tag{2.28}$$

(ii) for electrons or negatrons (negative beta particles), and

$$\frac{dE}{dx} = 4\pi r_0^2 \frac{mc^2}{\beta^2} NZ \left\{ \ln\left(\frac{\beta\gamma\sqrt{\gamma-1}}{I}mc^2\right) + \frac{1}{2\gamma^2}\left[\frac{(\gamma-1)^2}{8} + 1 - (\gamma^2+2\gamma-1)\ln 2\right] \right\} \tag{2.29}$$

(iii) positrons (positive beta particles),

$$\frac{dE}{dx} = 4\pi r_0^2 \frac{mc^2}{\beta^2} NZ \left\{ \begin{array}{l} \ln\left(\dfrac{\beta\gamma\sqrt{\gamma-1}}{I}mc^2\right) - \dfrac{\beta^2}{24}\left[23 + \dfrac{14}{\gamma+1} + \dfrac{10}{(\gamma+1)^2}\right. \\ \left. + \dfrac{4}{(\gamma+1)^3}\right] + \dfrac{\ln 2}{2} \end{array} \right\} \tag{2.30}$$

where dE/dx is the particle stopping power in units of MeV/m, r_0 the classical electron radius $= 2.818 \times 10^{-15}$ m, z the charge on the particle ($z = 1$ for p, d, β^-, β^+ and $z = 2$ for α), mc^2 is the rest energy of the electron $= 0.511$ MeV (see eqs. (3.29) and (3.30) of Chapter 3.), N is the number of atoms per m^3 in the absorber material through which the charged particle travels ($N = \rho(N_A/A)$) where ρ is the absorber density in units of g/cm^3, N_A is Avogadro's number $= 6.022 \times 10^{23}$ atoms per mol, A and Z are the atomic weight and atomic number, respectively, of the absorber, $\gamma = (T + mc^2)/mc^2 = 1/\sqrt{1 - \beta^2}$ where T is the particle kinetic energy in MeV and M the particle rest mass (e.g., proton $= 938.2$ MeV/c^2, deuteron $= 1875.6$ MeV/c^2, alpha particle $= 3727.3$ MeV/c^2, and β^- or $\beta^+ = 0.511$ MeV/c^2 and β is the relative phase velocity of the particle $= v/c$, the velocity of the particle in the medium divided by the speed of light in a vacuum $= \sqrt{1 - (1/\gamma^2)}$ (see also Chapter 7 for a treatment on β), and I is the mean excitation potential of the absorber in units of eV approximated by the equation

$$I = (9.76 + 58.8Z^{-1.19})Z, \quad \text{when } Z > 12 \tag{2.31}$$

where pure elements are involved as described by Tsoulfanidis (1995). However, when a compound or mixture of elements is concerned, a mean excitation energy, $\langle I \rangle$, must be calculated according to Bethe theory as follows

$$\langle I \rangle = \exp \left\{ \frac{[\Sigma_j w_j (Z_j/A_j) \ln I_j]}{\Sigma_j w_j (Z_j/A_j)} \right\} \tag{2.32}$$

where w_j, Z_j, A_j, and I_j are the weight fraction, atomic number, atomic weight, and mean excitation energy, respectively, of the jth element (Seltzer and Berger, 1982a). See Anderson $et\ al.$ (1969), Snow $et\ al.$ (1973), Janni (1982), Seltzer and Berger (1982a,b, 1984), Berger and Seltzer (1982), and Tsoulfanidis (1995) for experimentally determined values of I for various elements and thorough treatments of stopping-power calculations. Values of mean excitation potentials, I, for 100 elements and many inorganic and organic compounds are provided by Seltzer and Berger (1982a, 1984).

An example of the application of one of the above equations would be the following calculation of the stopping power for a 2.280 MeV beta particle (E_{max}) emitted from ^{90}Y traveling through an aluminum absorber. Of course, the beta particles emitted from ^{90}Y encompass a wide spectrum of energies between zero and a maximum value or E_{max}. It is the E_{max} that will be used in this example. The solution is as follows:

Firstly, the calculation of relevant variables are

$$\gamma = \frac{2.280\,\text{MeV} + 0.511\,\text{MeV}}{0.511\,\text{MeV}} = 5.462$$

$$\beta = \sqrt{1 - \frac{1}{\gamma^2}} = 0.9831 \quad \text{and} \quad \beta^2 = 0.9665$$

The atomic weight A for aluminum is 27, atomic number Z is 13, and its density ρ is 2.70 g/cm^3. The mean excitation potential, I, for aluminum is 166 eV, which is obtained from Seltzer and Berger (1982a). From eq. (2.29) the stopping power for the 2.280 MeV beta particle traveling through aluminum is calculated as

$$
\begin{aligned}
\frac{dE}{dx} = {} & 4(3.14)(2.818 \times 10^{-15} \, \text{m})^2 \left(\frac{0.511 \text{MeV}}{0.9665} \right)(2.70 \, \text{g/cm}^3) \\
& \times \left(\frac{6.022 \times 10^{23} \, \text{atoms/mol}}{27 \, \text{g/mol}} \right) \left(\frac{10^6 \, \text{cm}^3}{\text{m}^3} \right) \\
& \times (13) \left\{ \begin{array}{l} \ln \left(\dfrac{(0.9831)(5.462)\sqrt{4.462}}{166 \, \text{eV}} (0.511 \, \text{MeV})(10^6 \, \text{eV/Mev}) \right) \\[2mm] + \dfrac{1}{2(5.462)^2} \left[\dfrac{(4.462)^2}{8} + 1 - (5.462^2 + 2 \cdot 5.462 - 1) \right] \ln 2 \end{array} \right\} \\
= {} & 431.4 \, \text{MeV/m}
\end{aligned}
$$

The term $(10^6 \, \text{cm}^3/\text{m}^3)$ is included in the above calculation to maintain consistency of units, thus converting the units of density expressed as g/cm^3 to units of g/m^3. In SI units the stopping power can be expressed in units of J/m or

$$(431.4 \, \text{MeV/m})(1.602 \times 10^{-13} \, \text{J/MeV}) = 6.91 \times 10^{-11} \, \text{J/m}$$

The stopping power is often expressed in units of $\text{MeV cm}^2/\text{g}$ or $\text{J m}^2/\text{kg}$, which provides values for stopping power without defining the density of the absorber medium (Taylor *et al.*, 1970; Tsoulfanidis, 1995). In these units the above calculation can also be expressed as

$$\frac{1}{\rho} \left(\frac{dE}{dx} \right) = \frac{4.314 \, \text{MeV/cm}}{2.70 \, \text{g/cm}^3} = 1.60 \, \text{MeV cm}^2/\text{g} \tag{2.33}$$

Eq. (2.29) used above to calculate the stopping power for the 2.280 MeV beta particle from ^{90}Y in aluminum accounts only for energy of the beta particle lost via collision interactions resulting in ionization and electron-orbital excitations. The equation does not account for radial energy loss via the production of Bremsstrahlung radiation, which can be very significant with beta particles of high energy and absorber materials of high atomic number. Thus, a complete calculation of the stopping power must include also the radial energy loss via Bremsstrahlung as described by eq. (2.10), that is,

$$\left(\frac{dE}{dx} \right)_{\text{total}} = \left(\frac{dE}{dx} \right)_{\text{ion.}} + \left(\frac{dE}{dx} \right)_{\text{rad.}} \tag{2.34}$$

where $(dE/dx)_{\text{total}}$ is the rate of total beta particle or electron energy loss in an absorber, $(dE/dx)_{\text{ion.}}$ is the rate of energy loss due to ionization, and $(dE/dx)_{\text{rad.}}$ is the rate of energy loss due to Bremsstrahlung radiation production. We must keep in mind that each stopping-power calculation, such as the above example, provides values for only one beta particle or electron energy. Beta particles, on the other hand, are emitted with a broad spectrum of energies from zero to E_{max}, the majority of which may posses an average energy, E_{av}, of approximately one-third of E_{max}.

The ratio of beta-particle energy loss via Bremsstrahlung emission to energy loss via collision interactions causing ionization and excitation is described by the relation

$$\frac{E_{Brems.}}{E_{ioniz.}} = \frac{EZ}{750} \tag{2.35}$$

where E is the beta-particle energy in MeV and Z the atomic number of the absorber material (Friedlander *et al.*, 1964; Evans, 1972; Tsoulfanidis, 1995). From eqs. (2.34) and (2.35), we can write

$$\left(\frac{dE}{dx}\right)_{rad.} = \frac{ZE}{750}\left(\frac{dE}{dx}\right)_{ion.}$$
$$= \frac{(13)(2.280)}{750}(4.314\,\text{MeV/cm}) = 0.170\,\text{MeV/cm} \tag{2.36}$$

The total stopping power of the 2.280 MeV beta particle in aluminum according to eqs. (2.34) and (2.36) is thus

$$\left(\frac{dE}{dx}\right)_{total} = \left(\frac{dE}{dx}\right)_{ion.} + \frac{EZ}{750}\left(\frac{dE}{dx}\right)_{ion.} \tag{2.37}$$
$$= 4.314\,\text{MeV/cm} + 0.170\,\text{MeV/cm} = 4.48\,\text{MeV/cm}$$

Beta-particle loss via Bremsstrahlung radiation of the 2.280 MeV beta particles from ^{90}Y is low in aluminum, namely, 0.170/4.314 or 3.9% of the total energy loss. However, Bremsstrahlung production will increase as the absorber atomic number (Z) and beta particle or electron energy increase. Bremmstrahlung is highly penetrating electromagnetic radiation, and thus low atomic number absorbers such as aluminum or plastic (high in carbon and hydrogen) should be used as shields against beta particles and electron radiation. Shielding material of high atomic number elements, such as lead, will yield high levels of Bremsstrahlung radiation. The effect of atomic number (Z) and beta particle or electron energy (E) on the ratio of energy loss via Bremstrahlung production to ionization energy is illustrated in Figure 2.4.

The effective atomic numbers of compounds such as polyethylene ($-C_2-H_4)_n$, $Z_{ef} = 5.28$), water ($Z_{ef} = 6.60$), and air ($Z_{ef} = 7.3$) provided in Figure 2.4 above are a function of the weight fraction of each element in the compound as is calculated according to the equation

$$Z_{ef} = \frac{\sum_{i=1}^{L} w_i(Z_i^2/A_i)}{\sum_{i=1}^{L} w_i(Z_i/A_i)} \tag{2.38}$$

where L is the number of elements in the compound, w_i the weight or mass fraction of element i, Z_i the atomic number of element i, and A_i the atomic weight of element i (Tsoulfanidis, 1995; Andreo *et al.*, 2005). As illustrated in Figure 2.4 absorbers of low atomic number as protective shields will yield less harmful highly penetrating Bremsstrahlung radiation particularly when absorbers of high-energy (MeV) beta particles or electrons are needed.

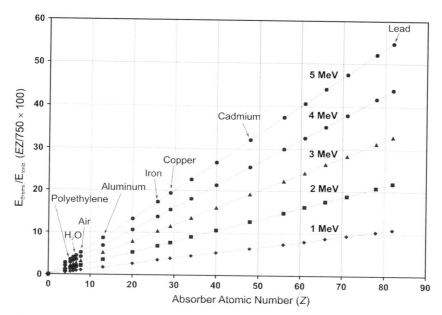

Figure 2.4 Ratio of energy loss via Bremsstrahlung production to ionization expressed as a percent ($EZ/750 \times 100$) for beta particles or electrons of energies (E) ranging from 1 to 5 MeV as a function of absorber atomic number (Z). Some common absorbers are identified above data points corresponding to their atomic number (Z) or effective atomic number (Z_{ef}) for observers that are compounds.

2.5.2 Linear energy transfer

The International Commission on Radiation Units and Measurements or ICRU (Taylor *et al.*, 1970) defines LET (L) of charged particles in a medium as

$$L = \frac{dE_L}{dl} \tag{2.39}$$

where dE_L is the average energy locally imparted to the medium by a charged particle of specified energy in traversing a distance dl. The term "locally imparted" refers either to a maximum distance from the particle track or to a maximum value of discrete energy loss by the particle beyond which losses are no longer considered as local. LET is generally measured in units of keV/μm. The ICRU recommends, when a restricted form of LET is desired, that the energy cut-off form of LET be applied because this can be evaluated using restricted stopping-power formulae (Taylor *et al.*, 1970). The energy-restricted form of LET or L_Δ is therefore defined as that part of the total energy loss of a charged particle which is due to energy transfers up to a specified energy cut-off value

$$L_\Delta = \left(\frac{dE}{dl} \right)_\Delta \tag{2.40}$$

where the cut-off energy (Δ) in eV units must be defined or stated. If no cut-off energy is applied, then the subscript ∞ is used in place of Δ, where L_∞ would signify the value of LET, which includes all energy losses and would therefore be equal to the total mass stopping power.

Figure 2.5 illustrates charged particle interactions within an absorber involved in the measurement of LET. The possible types of energy loss, ΔE, of a charged particle of specified energy, E, traversing an absorber over a track length Δl is illustrated, where O represents a particle traversing the absorber without any energy loss, U is the energy transferred to a localized interaction site, q the energy transferred to a short-range secondary particle when $q \leq \Delta$, and Δ a selected cut-off energy level (e.g., 100 eV), Q' is the energy transferred to a long-range secondary particle (e.g., formation of delta rays) for which $Q' > \Delta$, γ the energy transferred to photons (e.g., excitation fluorescence, Cherenkov photons, etc.), r a selected cut-off distance from the particle's initial trajectory or path of travel, and θ the angle of particle scatter. The interactions q, Q, and γ are subdivided in Figure 2.5 when these fall into different compartments of the absorber medium. See Taylor *et al.*, (1970) for methods used for the precise calculations of LET. Some examples of LET in water for various radiation types are given in Table 2.2. The table clearly illustrates that radiation of a given energy with shorter range in a medium will yield higher values of LET than radiations of the same energy with longer ranges in the same medium. This may be intuitively obvious, because the shorter the range of the radiation, the greater is the energy dissipated per unit path length of travel. We can take this further and generalize that the following radiation types will yield LET values of decreasing orders of magnitude (the heavier charged particles are considered here to be of the same energy for purposes of comparison) according to the sequence:

Decreasing LET:

Fission Products
Alpha Particles
Deuterons
Protons
Low-energy x Rays and Beta Particles
High-energy x Rays and Beta Particles
Gamma Radiation and High-energy Beta Particles

Although electromagnetic x and gamma radiations are not charged particles, these radiations do have the characteristics of particles (photons) that produce ionization in matter. They are, therefore, included in the above sequence and among the radiations listed in Table 2.2.

The term delta rays, referred to in the previous paragraph, is used to identify energetic electrons that produce secondary ionization. When a charged particle, such as an alpha particle, travels through matter, ionization occurs principally through coulombic attraction of orbital electrons to the positive charge on the alpha particle with the ejection of electrons of such low energy that these electrons do not produce further ionization. However, direct head-on collisions of the primary ionizing particle with an electron does occur occasionally whereby a large amount of energy is transferred to the electron. The energetic electron will then travel on in the absorbing matter to produce secondary ionization. These energetic electrons are referred to as delta rays. Delta rays form ionization tracks away from the track produced by

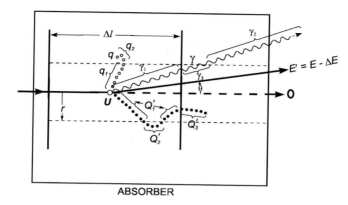

Figure 2.5   Diagram of the passage of particle of energy E through a thickness Δl of material illustrating the several types of energy loss that may occur. (From Taylor *et al.* (1970) reprinted with permission from ICRU © 1970.)

Table 2.2

Track-average values of LET (L_Δ) in water irradiated with various radiations

Radiation	Cut-off energy, Δ (eV)	L_Δ (keV/μm)
^{60}Co gamma rays	Unrestricted	0.239
	10,000	0.232
	1,000	0.230
	100	0.229
22 MeV x-rays	100	0.19
2 MeV electrons (whole track)	100	0.20
200 kV x-rays	100	1.7
^{3}H beta particles	100	4.7
50 kV x-rays	100	6.3
5.3 MeV alpha particles (whole track)	100	43

Source: (Taylor *et al.* (1970) reprinted with permission from ICRU © 1970.)

the primary ionizing particle. The occurrence and effects of delta rays in radiation absorption are applied to studies of radiation dosimetry (Casnati *et al.*, 1998; Cucinotta *et al.*, 1998).

When we compare particles of similar energy, we can state that, the ranges of particles of greater mass and charge will obviously be shorter and the magnitude of their LET values would be consequently higher in any given medium. The relationship between mass, charge, energy, range of particles, and their corresponding LET values can be appreciated from Table 2.3. The LET values in Table 2.3 are estimated by dividing the radiation energy by its range or path length in the medium. Such a calculation provides only an estimate of the LET, because the energy dissipated by the radiation will vary along its path of travel, particularly in the case of charged particles. More energy is released when the particle slows down before it comes to a stop as illustrated in Figure 1.4 of Chapter 1, when energy liberated in ion-pair formation is the highest. Nevertheless, the LET values provided in Table 2.3 give good orders of magnitude for comparative purposes.

Table 2.3

Range and LET value for various charged-particle radiations in water in order of decreasing mass[a]

Nuclide	Radiation energy (MeV)	Range in water (mm)	Average LET in water (keV/μm)
Thorium-232	α, 4.0	0.029[b]	138
Americium-241	α, 5.5	0.048[b]	114
Thorium-227	α, 6.0	0.055[b]	109
Polonium-211	α, 7.4	0.075[b]	98
----------	d, 4.0	0.219[c]	18.3
----------	d, 5.5	0.377[c]	14.6
----------	d, 6.0	0.440[c]	13.6
----------	d, 7.4	0.611[c]	12.1
----------	p, 4.0	0.355[d]	11.3
----------	p, 5.5	0.613[e]	9.0
----------	p, 6.0	0.699[f]	8.6
----------	p, 7.4	1.009[g]	7.3
Tritium	β^-, 0.0186(E_{max})	0.00575[h]	3.2[h]
Carbon-14	β^-, 0.156(E_{max})	0.280[h]	0.56[h]
Phosphorus-32	β^-, 1.710(E_{max})	7.92[h]	0.22[h]
Yttrium-90	β^-, 2.280(E_{max})	10.99[h]	0.21[h]

[a]The deuteron (d) and proton (p) energies were arbitrarily selected to correspond to the alpha practicle (α) energies to facilitate the comparison of the effects of particle mass and charge on range and LET.
[b]Calculated according to eqs. (1.14) and (1.15).
[c]The deuteron range is calculated from the equation $R_{Z,M,E} = M/Z^2 R_{p,E/M}$. The equation provides the range of a particle of charge Z, mass M, and energy E, where $R_{p,E/M}$ is the range in the same absorber of a proton of energy E/M (Friedlander et al., 1964).
[d]Calculated according to eqs. (1.12), (1.14), and (1.15), R_{air} = 28.5 mg/cm² (Figure A.1, Appendix A).
[e]Calculated according to eqs. (1.12), (1.14), and (1.15), R_{air} = 49.5 mg/cm² (Figure A.1).
[f]Calculated according to eqs. (1.12), (1.14), and (1.15), R_{air} = 56.5 mg/cm² (Figure A.1).
[g]Calculated according to eqs. (1.12), (1.14), and (1.15), R_{air} = 82.0 mg/cm² (Figure A.1).
[h]Calculations are based on the maximum energy (E_{max}) of the beta particles. When the lower value of average beta-particle energy (E_{av}) is used, the calculated value of range would be shorter and LET higher. The range was calculated according to the empirical formula $R = 0.412E^{1.27-0.0954\ln E}$ available from the curve provided in Figure A.3, Appendix A.

The concept of LET and the calculated values of LET for different radiation types and energies can help us interpret and sometimes even predict the effects of ionizing radiation on matter. For example, we can predict that heavy charged particles, such as alpha radiation, will dissipate their energy at shorter distances within a given absorber body than the more penetrating beta or gamma radiations. Also, low-energy x-radiation can produce a similar effect as certain beta radiations. The order of magnitude of the LET will help us predict the penetration power and degree of energy dissipation in an absorber body, which is critical information in studies of radiation chemistry, radiation therapy, and dosimetry, among others. For additional information, the reader is referred to works by Ehman and Vance (1991), Farhataziz and Rodgers (1987), and Spinks and Woods (1990).

Radioactivity Hall of Fame—Part III

Max Planck (1858–1947), Louis de Broglie (1892–1987), Albert Einstein (1879–1955), Arthur H. Compton (1892–1962), Max von Laue (1879–1960), Sir William Henry Bragg (1862–1942) and Sir William Lawrence Bragg (1890–1971), Henry G. J. Moseley (1887–1915), Charles Glover Barkla (1877–1944), Manne Siegbahn (1886–1978), and Robert A. Millikan (1868–1953)

MAX PLANCK (1858–1947)

Max Karl Ernst Ludwig Planck was born in Kiel, Germany on April 23, 1858. He was awarded the Nobel Prize in Physics 1918 "in recognition of the services he rendered to the advancement of physics by his discovery of energy quanta".

Max Planck (1858-1947)

Max Planck entered the University of Munich in 1874 to study physics and mathematics and transferred to the University of Berlin in 1877 to broaden his studies. In Berlin he was able to attend lectures by the prominent physicists Hermann von Helmholtz and Gustav Kirchhoff. At the age of 21, he received the doctor of philosophy degree in physics after defending his dissertation in Munich *On the Second Law of the Mechanical Theory of Heat*. He submitted his habilitation thesis the following year to become a university teacher, and began to serve at the university as a lecturer without salary.

The lack of remunerative compensation for his university lectures forced him to live with his parents in Munich, which he considered to be an unfavorable burden on them. He felt the need to demonstrate to the scientific community his worth as a physicist in order to gain a professorship. For the next 5 years (1880–1885) Max Planck established himself as a private lecturer or Privatdozent in physics at the University of Munich until he was able to get an appointment to the position of Associate Professor of Theoretical Physics at the University of Kiel in 1885 where he remained until 1889. His appointment at the University of Kiel at the age of 27 provided him with the financial resources necessary to marry Marie Merck (1861–1909) on March 31, 1887. After the death of Gustav Kirchhoff in 1887, Max Planck was able to succeed him as Chair of Theoretical Physics at the University of Berlin in 1888 starting as Associate Professor and Director of the Institute of Theoretical Physics, and then advancing to full professorship in 1892. He remained at the University of Berlin until his retirement in 1926.

Upon his arrival in Berlin, Max Planck joined the German Physical Society and remained a very active and devoted member for three decades serving in various years on the Society's Board as Treasurer, Committee Member, and Chairman. The year 1929 marked the 50th anniversary of his doctorate degree, and the German Physical Society established on that occasion the Max Planck Medal. Albert Einstein was a recipient of the medal in 1929. The medal is bestowed annually by the German Physical Society, and it remains the reward of highest honor granted by the Society.

Max Planck's main contributions originated from his work on thermal radiation, the results of which opened the field of quantum physics and provided the groundwork for subsequent discoveries on the properties of light waves and other classes of electromagnetic radiation including x-rays and gamma rays. Several key discoveries in the properties of radiation were made prior to Planck's and a brief description of this earlier work is provided here. Very early observations on the properties of thermal radiation were documented in 1792 by Thomas Wedgewood, the famous British manufacturer of Wedgewood porcelain. Wedgewood observed that all objects in his ovens, where he prepared fine china, would glow the same red color when they reached a certain temperature, regardless of their size, shape, or composition. It was not until 1859 did Gustav Kirchhoff demonstrate that, any body in thermal equilibrium with radiation, the emitted energy is proportional to the energy absorbed by the object. An example of such a case would be the heated walls of a kiln with its door closed and at a constant temperature. The radiation within the walls of the kiln would be in thermal equilibrium when the radiation energy within the kiln is absorbed, exchanged, and reemitted many times over until the entire walls of the cavity of the kiln are in thermal equilibrium. The radiation in thermal equilibrium within the walls of a kiln is similar to the radiation emitted by a black body, which is an object that absorbs radiation of all wavelengths or frequencies and therefore would appear black. A black body emits the energy it absorbs in accord with Kirchhoff's observations; and the energy emitted is a function of the temperature of the black body and frequency of the emitted light, and independent of the size, shape, and chemical nature of the black body. William Wien, in 1893, mathematically defined the spectral density of a black-body cavity, that is, the energy per unit volume per unit frequency within a black-body cavity, as a function of black-body temperature. The equation derived by Wien became known as Wien's exponential law, because the energy density was an exponential function of the radiation frequency and black-body temperature. Radiation spectroscopists at the time determined experimentally that Wien's law fit well for the short wavelengths of radiation (0–4 μm) over a wide range of temperatures (400–1600 K), but that the law failed for longer radiation wavelengths (Serway *et al.*, 1997).

Over a period of 6 years Planck labored to find a general formula that would describe the energy density of black-body radiation that would fit for both low and high radiation frequencies. In October of 1900, Planck found such a formula that included a constant $h = 6.626 \times 10^{-34}$ J sec, which is called Planck's constant in his honor. The postage stamp illustrated below was issued by the German Democratic Republic in honor of Max Planck and his discovery of the Planck constant h. While this was a great achievement it did not satisfy Plank's efforts to find more meaning to his formula. In his Nobel Lecture he expressed his struggles toward this goal as follows:

> However, even if the radiation formula should prove itself to be absolutely accurate, it would still only have, within the significance of a happily chosen interpolation formula, a strictly limited value. For this reason, I buried myself, from then on, that is, from the day of its establishment, with the task of elucidating a true physical character for the formula…after some weeks

of the most strenuous work of my life, light came into the darkness, and a new undreamed of perspective opened up for me.

Max Planck was referring above to his discovery of the quantum theory, which we have yet to describe, and the role of Plank's constant h in the new field of quantum physics that he was about to open. In the course of his efforts, Max Planck correlated black-body radiation to radiation produced by many oscillators, which he called resonators, within the cavity surrounded by a sphere of reflecting walls from which he developed the quantum theory. In his Nobel Lecture of June 2, 1920 he described his thoughts at the time as follows:

> Heinrich Hertz's linear oscillator, whose laws of emission, for a given frequency, …seemed to me to be a particularly suitable device for this purpose. If a number of such Hertzian oscillators are set up within a cavity surrounded by a sphere of reflecting walls, then by analogy with audio oscillators and vibrating resonators, energy will be exchanged between them by the output and absorption of electromagnetic waves, and finally stationary radiation corresponding to Kirchhoff's Law, the so-called black-body radiation, should be set up within the cavity.

Planck expected that the vibrating resonators would emit radiation within the reflecting walls of a cavity over a continuous spectrum of wavelengths with frequencies corresponding to the varying frequencies of the vibrating resonators. However, to fit experimental observations, Planck had to envisage that the vibrating resonator could not possess energies corresponding to all frequencies, but that the energy levels would correspond to discrete multiples of $h\nu$ (Serway *et al.*, 1996) or

$$E_{\text{resonator}} = nh\nu \qquad \text{(III.1)}$$

where $n = 1,2,3…$ and ν is the resonator frequency. Therefore, as illustrated in Figure III.1 the radiation emitted by the resonator could only occur as "packets" of energy when the energy of the resonator would drop from a high energy level to a lower one or

$$\Delta E = h\nu \qquad \text{(III.2)}$$

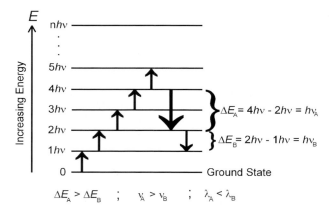

Figure III.1   Planck's quantum theory. The energy levels of a vibrating resonator or black-body would not correspond to energies of all frequencies v, but that the energy levels would correspond to multiples of hv. The energy emitted by the resonator or body could only occur as "packets" of energy when the energy of the body would drop from a higher energy level to a lower one or $\Delta E = hv$. The arrows pointed upward correspond to increasing energy levels, and the two arrows pointed downward illustrate two distinct emissions of energy and radiation frequencies.

It was found that radiation is emitted at discrete energies corresponding to differences in energy states of a body, and the energies of emitted radiations would be found as the product of Planck's constant, h, having units of energy × time and the radiation frequency, v. Planck named the discrete radiation energy as "the quantum" from the Latin *quantus* meaning "how great". Eq. (III.2) is referred to as the Planck–Einstein equation, as Einstein in 1905 was able to demonstrate that light not only traveled as waves of electromagnetic radiation but also discrete packets of energy, which he named "energy quanta" or photons.

To commemorate the significance of the quantum theory, the postage stamp illustrated here was issued in Germany in 1994. It pictures a black-body cavity or kiln in which an opening is created and from which radiation is emitted. The energy of that radiation is illustrated as equivalent to Planck's constant h times the radiation frequency.

After the death of his wife Marie in 1909 Planck married her cousin Marga von Hösslin. He remained in Germany throughout the Second World War and had to cope with the power and intervention of the National Socialists in Germany. He was President of the Kaiser Wilhelm Society and thus headed the

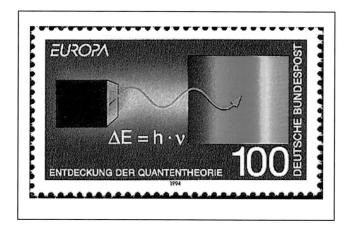

primary research institution in Germany during 1930–1938. His leadership was notable in hindering the Nazis from engulfing the Society, and to maintaining high scientific standards, but he was unable to prevent it (Max-Planck Society). During the last few years of his life he suffered great hardship and loss. In 1944 his home in the Berlin suburb of Grunewald was totally destroyed together with most of his possessions and scientific papers after an Allied air-bombing raid. That same year on July 20 his son Erwin, who was also his close friend and trusted advisor, was arrested for participating in an assassination attempt on Hitler's life. After the war in 1945, American colleagues escorted the world famous physicist to Göttingen, where he spent the remaining years of his life until his death in October of 1947. His wife Marga wrote Lise Meitner on June 16, 1947 "But I can assure you of one thing: without the help of our friends abroad, my husband would very certainly, not have survived the winter". (Max-Planck Society).

LOUIS DE BROGLIE (1892–1987)

Prince Louis-Victor de Broglie was born at Dieppe, Seine Inférieure, France on August 15, 1892. He was awarded the Nobel Prize in Physics 1929 "for his discovery of the wave nature of electrons".

After completing his secondary school education at the Lycée Janson de Sailly in 1909 Louis de Broglie decided to pursue studies in literature and history and was awarded a degree in history at the Sorbonne in 1910 contemplating a career in the diplomatic service; but he also had an interest in science. When given a research topic in history he gave serious thought to his goals in life and his interest in science prevailed. He decided to pursue studies in physics and mathematics and obtain a degree in theoretical physics. In 1913 he was awarded his License ès Sciences at the Sorbonne. De Broglie was then conscripted into military service for the duration of the war of 1914–1918, and was stationed at the Eifel Tower in the wireless section of the army. After the war de Broglie pursued studies in theoretical physics taking an interest in the x-ray research work of his elder brother Maurice de Broglie, which strengthened his interest in science while working for his doctorate degree in theoretical physics in the early 1920s. During his Nobel Lecture on December 12, 1929, de Broglie elucidated his thoughts on physics at the time of his graduate studies for the doctorate degree as follows:

> I was attracted to theoretical physics by the mystery enshrouding the structure of matter and the structure of radiations, a mystery which deepened as the strange quantum concept introduced by Planck in 1900 in his research on black-body radiation continued to encroach on the whole domain of physics.

While pursuing his university studies in theoretical physics, de Broglie focused his interest in problems related to quanta. The year prior to presenting his dissertation de Broglie (1923a–c) presented his thoughts on the dual nature of matter and light, that is, matter could be interpreted as having the properties of light with wavelength and frequency or the properties of a particle with mass. For example, de Broglie (1923a) referred to the wavelength and associated frequency of a particular matter as follows:

> An observer bound to the portion of matter will associate with it a frequency determined by its internal energy, namely by its "mass at rest". An observer for whom a portion of matter is in steady motion with velocity βc, will see this frequency lower in consequence of the Lorentz-Einstein time transformation.

The term β in de Broglie's statement refers to the relative phase velocity of a particle of matter, that is, the velocity of the particle in a medium (e.g., air) divided by the velocity of light in a vacuum, that is, $c = 2.99 \times 10^{10}$ cm per sec. (See Chapter 6 for an example of a time transformation calculation based on an observer on earth detecting the arrival of a cosmic particle.) De Broglie (1923a), in turn, referred to light as having the properties of particles with mass, when he wrote the following:

> A radiation of frequency v has to be considered as divided into atoms of light of very small internal mass ($\sim 10^{-50}$ gm.) which move with a velocity very nearly equal to c given by $m_0 c^2 / \sqrt{1 - \beta^2} = hv$.

In 1924 at Paris University he delivered his doctoral dissertation on *Recherches sur la Théorie des Quanta* (Research on the Quantum Theory). What is very astonishing is that de Broglie received the Nobel Prize mainly for the work he did on his doctoral dissertation presented to the Faculty of Sciences only 5 years prior to his receiving the Nobel Prize in Physics. His dissertation expounded on the dual particle and wave nature of the electron and other matter. His ideas, based on previous work of Planck and Einstein, were first received with astonishment, but subsequently accepted universally following demonstrations by experimental work on electron diffraction by crystals carried out by Davisson and Gerner at Bell Laboratories, New York in 1927 and others.

The significance of de Broglie's findings was expounded by Professor C. W. Oseen, Chairman of the Nobel Committee for Physics of the Royal Swedish Academy of Sciences, on December 10, 1929 as follows:

> Louis de Broglie had the boldness to maintain that not all properties of matter can be explained by the theory that it consists of corpuscles...there are others, according to him, which can be explained only by assuming that matter is, by its nature, a wave motion. At the time when no single known fact supported this theory, Louis de Broglie asserted that a stream of electrons, which passed through a very small hole in an opaque screen, must exhibit the same phenomena as a light ray under the same conditions. It was not quite in this way that Louis de Broglie's experimental investigation concerning his theory took place. Instead, the phenomena arising when beams of electrons are reflected by crystalline surfaces, or when they penetrate thin sheets, etc. were turned to account. The experimental results obtained by these various methods have fully substantiated Louis de Broglie's theory. It is thus a fact that matter has properties, which can be interpreted only by assuming that matter is of a wave nature.

Professor Orseen concluded his remarks on the work of de Broglie with the following statement, which underscored the impact of de Broglie's findings:

> Hence there are not two worlds, one of light and waves, one of matter and corpuscles. There is only a single universe. Some of its properties can be accounted for by the wave theory, others by the corpuscular theory.

Louis de Broglie expounded on his approach to the findings of his theoretical work as follows:

> I thus arrived at the following overall concept which guided my studies: for both matter and radiations, light in particular, it is necessary to introduce the corpuscle concept and the wave concept at the same time. In other words, the existence of corpuscles accompanied by waves has to be assumed in all cases…it must be possible to establish a certain parallelism between the motion of a corpuscle and the propagation of the associated wave.

He derived the fundamental relation of his theory as the wavelength of the electron or other particle in motion as

$$\lambda = \frac{h}{p} \qquad \text{(III.3)}$$

where λ is the wavelength of the particle, that is, the distance between two consecutive peaks of the particle wave, h the Planck's constant ($6.626 \times 10^{-34}\,\mathrm{J\,s}$), and p the particle momentum. De Broglie noted that for electrons that are not extremely fast (not relativistic), the electron momentum could be expressed as

$$p = m_0 v \qquad \text{(III.4)}$$

where m_0 is the rest mass of the electron, and v its velocity. The wavelength of the electron could then be calculated as

$$\lambda = \frac{h}{m_0 v} \qquad \text{(III.5)}$$

This eq. (III.5) is the basic formula for calculating the de Broglie wavelength of a particle. The discovery of de Broglie in his doctoral dissertation of 1924 is commemorated in the postage stamp illustrated above that was issued in France in 1994.

Louis de Broglie calculated the wavelength of an electron that is accelerated through a specific voltage potential after deriving the following:

For the non-relativistic case the energy of the electron can be expressed as:

$$eV = \frac{1}{2} m_0 v^2 \tag{III.6}$$

and since $p = m_0 v$ we can write

$$\frac{p^2}{2m_0} = \frac{m_0^2 v^2}{2m_0} = \frac{1}{2} m_0 v^2 = eV \tag{III.7}$$

Thus

$$p = \sqrt{2m_0 \, eV} \tag{III.8}$$

and

$$\lambda = \frac{h}{p} = \frac{h}{\sqrt{2m_0 \, eV}} \tag{III.9}$$

He demonstrated that the wavelength of the electron that had been accelerated some tens of volts would have a wavelength of the order of the Angström unit. For example, an electron that is accelerated across a potential of 100 V would have a wavelength calculated according to eq. (III.9) as

$$\lambda = \frac{h}{\sqrt{2m_0 \, eV}} = \frac{6.626 \times 10^{-34} \, Jsec}{\sqrt{2(9.11 \times 10^{-31} kg)(1.6 \times 10^{-19} C)(100 \, V)}} = 1.2 \times 10^{-10} m = 1.2 \, \text{Å}$$

De Broglie concluded that with a wavelength of the order of magnitude of the Angström unit, it would be expected that crystals, which have similar atomic dimensions, could cause the diffraction of these electron waves similar to the diffraction of x-rays in crystals. In his Nobel lecture he noted the following:

A natural crystal such as rock salt, for example, contains nodes composed of the atoms of the substances making up the crystal and which are regularly spaced at distances of the order of an Angström. These nodes act as diffusion centers for the waves, and if the crystal is impinged upon by a wave, the wavelength of which is also of the order of an Angström, the waves diffracted by the various nodes are in phase agreement in certain well-defined directions and in these directions the total diffracted intensity is a pronounced maximum. The arrangement of these diffraction maxima is given by the nowadays well-known mathematical theory developed by von Laue and Bragg, which defines the position of the maxima as a function of the spacing of the nodes in the crystal and of the wavelength of the incident wave...However, if the electron is assumed to be associated with a wave and the density of an electron cloud is measured by the intensity of the associated wave, than a phenomenon analogous to the Laue phenomenon ought to be expected for electrons. The electron wave will actually be diffracted intensely in the directions which can be calculated by means of the Laue-Bragg theory from the wavelength $\lambda = h/mv$, which corresponds to the known velocity v of the electrons imping- ing on the crystals.

Proof of Louis de Broglie's theory of the wave nature of the electron was provided by Davisson and Germer at the Bell Laboratories, New York in 1927 when they demonstrated that the phenome- non of electron diffraction by crystals was possible, and that the diffraction obeys exactly and quan- titatively the laws of wave mechanics. Davisson and Germer shared the honor of being the first to observe the phenomenon of electron diffraction by a method analogous to that of von Laue for x-rays.

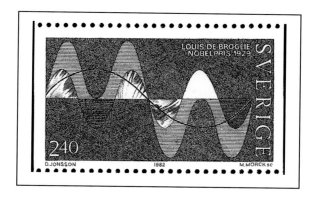

The postage stamp illustrated here was issued in Sweden in 1982 to commemorate de Broglie's discovery of the wave nature of matter. The impact of de Broglie's findings on modern physics was underscored when he concluded his Nobel lecture with the following:

Thus to describe the properties of matter as well as those of light, waves and corpuscles have to be referred to at one and the same time. The electron can no longer be conceived as a single, small granule of electricity; it must be associated with a wave and this wave is no myth; its wavelength can be measured and its interferences predicted…And it is on this concept of the duality of waves and corpuscles in Nature, expressed in a more or less abstract form, that the whole recent development of theoretical physics has been founded and that all future develop- ment of this science will apparently have to be founded.

After presenting his doctoral dissertation in 1924 de Broglie completed 2 years of compensation- free lectures at the Sorbonne after which he was appointed to the faculty of the Institut Henri Poincaré in Paris, which was newly established for the purpose of teaching and development of mathematical and theoretical physics. In 1932, he was appointed to the chair of theoretical physics at the Faculty of Sciences at the University of Paris only 8 years after receiving his doctorate at the same university. During 1930 to 1952 de Broglie continued studies on wave mechanics, Dirac's electron theory, the new theory of light photons, etc., during which time he was the author of 25 books. Many of these books were based on his courses; and many students traveled from many parts of the world and within France to work with him. Many doctorate dissertations were prepared by students under his guidance.

Louis de Broglie received many awards including the first Kalinga Prize awarded by UNESCO for his efforts in explaining modern physics to the layperson. He was an elected member of the French Academy of Sciences and its Permanent Secretary for the mathematical sciences since 1942. Louis de Broglie died on March 19, 1987.

ALBERT EINSTEIN (1879–1955)

Albert Einstein was born at Ulm in Württemberg, Germany, on March 14, 1879. He was awarded the Nobel Prize in Physics 1921 "for his services to Theoretical Physics and especially for his discovery of the law of the photoelectric effect". Einstein was the greatest theoretical physicist of the 20th century. He is known for making many contributions to science and mankind including his interpretation of quantum physics and the photoelectric effect for which he received the Nobel Prize, his theory of relativity including the demonstration of the equivalence of mass and energy, his correspondence with President Franklin Delano Roosevelt warning of the potential of Germany to build a neutron fission bomb, his pacifism, and his philosophy. This biographical sketch will, in addition to providing information on his life and the impact it has made on our lives, underscore mostly Einstein's scientific contributions to the field of our understanding of the properties of radioactivity.

Albert Einstein (1879-1955)

When Einstein was 6 weeks of age his family moved to Munich where he began his elementary education at Luitpold Gymnasium. His family moved to Italy in 1894, but Einstein remained in Munich. In 1895 he tried to gain admission to the Eidgenössische Technische Hochschule (ETH) or Swiss Federal Polytechnic School in Zurich, but failed the entrance exam. In 1896 he renounced his German citizenship and remained stateless until being granted Swiss citizenship in 1901. Also in 1896 Einstein gained admission to the ETH and graduated in 1900 with a diploma as a teacher in mathematics and physics. After graduation, he was unsuccessful in getting a job as a teacher in any university, and at the time gave up his ambitions to do so. Instead Einstein had to take temporary teaching employment in High Schools in Winterthur and Schaffhausen during 1901. With the help of Marcel Grossman, a friend and classmate at the ETH, Einstein was appointed in 1902 to work as a clerk at the Swiss Patent Office in Bern where he worked until 1909.

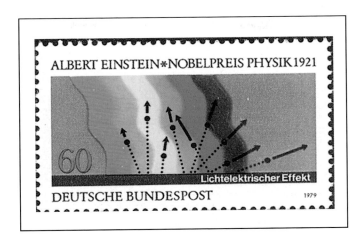

During the time of his employment at the Patent Office and during his spare time Einstein produced a great amount of his most remarkable work without any close association with other scientific colleagues or scientific literature. At the age of 26, he wrote three historic papers in his spare time and submitted these to the prestigious journal *Annalen der Physik* requesting publication "provided there was room" in the journal. All papers were published in the same journal issue in 1905. The first paper was Einstein's interpretation of Max Planck's quantum theory of radiation. He applied Planck's quantum theory to the interpretation to the photoelectric effect and explained the dual nature of light, that is, light may behave as a particle (i.e., photon) or as a wave. The phenomenon of the photoelectric effect was already well known. It was for Einstein's interpretation of the photoelectric effect that he was awarded the Nobel Prize in Physics 1921.

The photoelectric effect is the process whereby the energy from electromagnetic radiation, such as visible light, gamma radiation or other, hits an atomic electron whereby the energy of the radiation is transferred in its entirety to the electron causing the electron to be ejected from the atom. An artists' depiction of the photoelectric effect is illustrated above in the postage stamp from Germany issued in 1979. The stamp illustrates light of various wavelengths (colors) striking the surface of a substance (e.g., metal plate) causing the ejection of electrons from the surface of the plate.

Credit for the discovery of the photoelectric effect has been given to Heinrich Hertz, who in 1887 found that an electrical spark passing between two spheres would occur more readily, if its path were illuminated with the light from another electrical discharge. A more thorough study by the German physicist Wilhelm Hallwachs in 1888 demonstrated that under certain conditions a negatively charged body, for example, a metal plate, illuminated with light of a particular color, loses its negative charge and ultimately assumes a positive charge (Arrhenius, 1922). Philipp Lenard (Lenard, 1902) demonstrated that the photoelectron velocity (i.e., its' kinetic energy) was independent of the light intensity, but that the electron velocity would vary only according to the wavelength of the light, and the highest velocities were attained when shorter wavelengths of light would hit a metal electrode (See Lenard, Radioactivity Hall of Fame, Part I). Increasing the light intensity would increase the number of photoelectrons, but not their energies.

Einstein was able to demonstrate in his first paper submitted to *Annalen der Physik* (Einstein, 1905) that when a quantum of light hits a substance such as a metal plate, it can at most yield all of its energy to an electron. A part of the energy of the light quantum is consumed in knocking the electron out of its atom and the remaining energy will stay with the electron as kinetic energy of the electron. Einstein was awarded the Nobel Prize for his findings described in his first publication of 1905, and Professor S Arrhenius, Chairman of the Nobel Committee for Physics, announced that the Nobel Committee named these findings as Einstein's Law of the Photoelectric Effect. The law states that the light quantum must have sufficient energy for the electron to become detached or released from its structure (electron binding energy) and provide the remaining energy to the electron as kinetic energy.

Only light having a frequency greater than a certain limit [greater than the electron binding energy] is capable of inducing a photoelectric effect, however high the intensity of the irradiating light. If this limit is exceeded the effect is proportional to the light intensity at a constant frequency (Arrhenius, 1922).

This effect can occur in any substance gas, liquid, or solid, provided the frequency of the light quantum is of sufficient energy to liberate an electron from its atom and thereby cause ionization. A postage stamp from Bosnia and Hertzegovina issued in 2001 commemorates Einstein's findings. The stamp illustrates a quantum of light (hv) striking a substance, such as a metal plate, and an electron is ejected. The stamp provides the equation describing the photoelectric effect where

$$hv = A + \frac{mv^2}{2} \tag{III.10}$$

The equation states that the light quantum (hv) is proportional to the electron binding energy (A) plus the kinetic energy of the electron ($E_K = 1/2\,mv^2$). In Chapter 3, the equation is described in reverse, that is, the kinetic energy of the photoelectron is proportional to the light quantum less the electron binding energy (see eq. (3.19)). In summary, Einstein grasped the calculations of Planck to explain that light not only traveled as waves but also as particles of energy (energy quanta). Today we refer to these energy quanta as photons, which are classified by particle physicists as one of the elementary particles of nature. The photoelectric effect is also one of the key interactions now used in gamma-ray spectrometry to identify radioactive sources when the energy imparted to an electron of a radiation detector is proportional to the energy of the gamma-ray photon.

The significance of Einstein's Law of the Photoelectric Effect was emphasized by the Chairman of the Nobel Committee (Gullstrand, 1925) when he stated

A logical consequence of Planck's theory is that a transition from one [energy] state to another can only take place in such a way that an integral number of energy quanta is emitted or absorbed. An exchange of energy between matter and radiation, therefore—that is to say an emission or absorption of radiation—can be effected only by the transmission of an integral number of energy quanta. It was not Planck, however, but Einstein, that drew this conclusion, which involves the law of the photoelectric effect—a law that now, especially thanks to Millikan's work, has been verified in a brilliant manner. It is through Einstein's law that the Planck's constant and the whole quanta theory has attained their greatest importance.

Another paper written by Einstein, when he was a patent clerk, explained Brownian motion, which was first observed by botanist Robert Brown in 1828. Most students of biology have witnessed Brownian motion when they observed in the lens of an optical microscope the erratic and jittery movement of microscopic substances in a liquid, such as pollen grains suspended in water on a microscope slide. Einstein provided calculations that could account for Brownian motion as that due to random collisions of molecules with the microscopic particle in the fluid, and his calculations provided an accurate measurement of the dimensions of the yet hypothetical molecules. His findings provided the first evidence for the existence of atom-sized molecules in matter and that the kinetic-energy theory could explain how heat would have an effect on the motion of atoms and molecules. From this paper Einstein wrote a thesis titled *Eine neue Bestimmung der Molekuldimensionen* (A New Determination of Molecular Dimensions), which earned him a doctorate degree from the University of Zurich in 1905. He dedicated the thesis to Marcel Grossman, his friend and former classmate at the ETH.

A third of Einsteins' papers submitted to *Annalen der Physik* in 1905 proposed his theory of special relativity. Einstein's *Principle of Relativity* is based on two postulates, which are namely: (i) that in making observations or measurements in physics, there should be no preferred inertial (non-accelerating) frame. In other words, for all inertial observers, the laws of physics are the same. (ii) The speed of light in a vacuum is a constant ($c = 2.99 \times 10^8$ m/sec) for all inertial observers, regardless of the velocity of travel of the source. Einstein developed his principle of relativity into what would be referred to later as the special theory of relativity that would provide an analysis of space and time. The special theory of relativity can be used to explain changes of length, time, and mass when an observer in one inertial frame makes measurements of length, time, and mass of another body or clock in another inertial frame. A person in one inertial frame of reference can make measurements of length, time, and mass on another inertial frame and vice verse. Space, time, mass, and speed are all relative depending on which inertial frame of reference is used. In a lecture on the theory of relativity he underscored the two principles upon which the special theory of relativity is based (Einstein, 1923) as follows:

> The special theory of relativity is an adaption of physical principles to Maxwell-Lorentz electrodynamics. From earlier physics it takes the assumption that Euclidian geometry is valid for the laws governing the position of rigid bodies, the inertial frame and the law of inertia. The postulate of the equivalence of inertial frames for the formulation of the laws of Nature is assumed to be valid for the whole of physics (special relativity principle). From Maxwell-Lorentz electrodynamics it takes the postulate of invariance of the velocity of light in a vacuum (light principle).

As a consequence of his special theory of relativity Einstein concluded that energy was equivalent to mass, that is, if a body emitted energy, the mass of the body would decrease proportionally to the amount of energy lost. In his explanation of the fundamental ideas and problems of the theory of relativity he stated (Einstein, 1923) the following:

> The special relativity theory…united the momentum and energy principle, and demonstrated the like nature of mass and energy.

He expressed this relationship of mass and energy with the equation

$$E = mc^2 \tag{III.11}$$

where E is energy, m the mass, and c the velocity of light in a vacuum. Einstein's equation of equivalence of mass and energy has been demonstrated many times over, and it is an equation of standard use in the study of radioactivity and nuclear decay processes. For example, it is known that a quantum of gamma radiation must possess a minimum of 1.02 MeV of energy for it to be converted to two electrons (positron and negatron pair) when the gamma-ray photon interacts with the coulombic field of a nucleus. This phenomenon is referred to as pair production, because the mass of two electrons is

created from the energy of the gamma-ray photon. Using Einstein's eq. III.11, the equivalent energy of one electron can be calculated from the electron mass of 9.109×10^{-31} kg as

$$E = mc^2 = (9.109 \times 10^{-31}\,\text{kg})(2.997 \times 10^8\,\text{m/sec})^2 = 8.182 \times 10^{-14}\,\text{J}$$

The joule (J) as a unit of energy can be converted mathematically to other units of energy. The amount of energy in joules calculated above, as equivalent to one electron mass, can be converted to another unit of energy, the electron-volt as follows:

$$\frac{8.182 \times 10^{-14}\,\text{J}}{1.602 \times 10^{-19}\,\text{J/eV}} = 0.511\,\text{MeV}$$

Therefore, the mass of one electron is equivalent to 0.511 MeV, and the mass of two electrons would be twice that amount or 1.02 MeV, which is exactly the amount of energy that a gamma-ray must possess when its energy is converted to the mass of two electrons in the process of pair production (see Pair Production in Chapter 3). Likewise, we can consider this calculation in the reverse. The mass of the positively charged electron (positron) will be converted to energy as gamma radiation when it loses kinetic energy, slows down, and encounters a negatively charged electron, its antiparticle. In such an encounter, both electrons become annihilated and the masses of the two electrons are converted to energy in the form of two gamma-ray photons of 0.511 MeV energy that are emitted in opposite directions from their site of encounter (see Annihilation in Chapter 3). Einstein's equation of equivalence of mass and energy is used often to calculate the energy lost by nuclides that undergo radioactive decay transitions from the differences in mass of parent and daughter nuclides. See the example calculations for ^{241}Am decay given in Chapter 1. The stamp illustrated here, issued in China in 1979, is one example of many issued from many countries that commemorate Einstein's work and his equation of the equivalence of mass and energy.

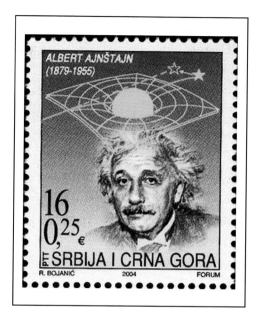

Einstein extended his special theory of relativity to phenomena involving acceleration into his general theory of relativity, which he wrote in 1915 (Einstein, 1916). He proposed that as mass was equivalent to energy, as described above, the same principle of equivalence would require that gravitational mass would interact with the mass of light photons (electromagnetic radiation). From this reasoning, Einstein predicted the deflection of the light from stars as the light would pass near a massive body such as the sun (Einstein, 1911). The gravitational pull of the sun would attract and bend the light from a distant star as that light passed near the body of the sun. Einstein concluded that this deflection of light toward the body of the sun could be observed from earth when the light of the sun would be blocked by a total eclipse. In 1913 prior to the total eclipse of May 1919 Einstein diagramed a sketch illustrating how the gravity of the sun would deflect light near the sun making stars to appear to observers on earth that they have shifted their position in space. Einstein's prediction was found to be true when the British astronomers in May of 1919 took photographs of the total eclipse of the sun. The British astronomer Arthur Eddington demonstrated Einstein's prediction to be true. He spotted a star that should have been hidden behind the sun. Photographs of the total eclipse illustrated how the positions of some stars deviated from their positions when the stars were photographed on other occasions with the sun in a different location in the sky. This finding made Einstein an instant celebrity. *The London Times* on November 7, 1919 ran the headline "Revolution in Science, New Theory of the Universe, Newtonian Ideas Overthrown". The demonstration of Einstein's theory is commemorated in the stamp issued by Serbia in 2004. The stamp illustrates how the bending of the light near the sun gave the appearance that the distant star was located to the side rather than directly behind the sun. Eddington commented that it was the greatest moment of his life when he measured the image of a star and found that the sun's gravity warped the space through which the light had traveled. The effect was reconfirmed in photographs taken during the Solar Eclipse of 1922. As noted by the American Institute of Physics

The eclipse experiments, like most important new science, were done at the very limit of available techniques. It was not until the 1960s, with vastly improved methods, that the gravitational bending of light could be demonstrated beyond reasonable doubt. Until then one could almost say that the logic and beauty of Einstein's theory did as much to confirm the observations as the observations did to confirm his theory.

In addition to the deflection of light by the sun's gravity described above Einstein made several predictions from general relativity theory including the perihelic orbit of the planet Mercury (Einstein, 1915). In his Nobel Lecture presented on July 11, 1923 (see clarification below for the lateness and subject matter of the lecture), Einstein underscored some of the predictions of his theory when he stated:

> The considerations mentioned led to the theory of gravity which yields the Newtonian theory as a first approximation and furthermore it yields the motion of the perihelion of Mercury, the deflection of light by the sun, and the red shift of spectral lines [due to an expanding universe]... (Einstein, 1923).

Einstein's publications in *Annalen der Physik* in 1905 while a clerk in the Swiss Patent Office put him in great demand by universities and research institutes, which previously knew nothing of him. In 1908 he became Privatdozent in Bern and in 1909 became Professor Extraordinary at the University of Zurich. In 1911 he was appointed Professor of Theoretical Physics at the Karl-Ferdinand University in Prague; and returned to Zurich in 1912 to take up a chair at the ETH, the prestigious polytechnic school of Switzerland, which had rejected his admission as a student in 1895. In 1914, Einstein was appointed Director of the Kaiser Wilhelm Institute of Physics in Berlin and Professor of the University of Berlin. At the outbreak of World War I in 1914 Einstein's pacifist sympathies were brought into public view. Ninety-three leading German intellectuals, including physicists such as Max Planck, signed a manifesto defending Germanys' war conduct. Einstein and three others signed an anti-war counter-manifesto (American Institute of Physics). He became a citizen of Germany in 1914 and remained in Berlin until 1933 when he renounced his citizenship during the rise of facism and immigrated to the USA to take on a position of Professor of Theoretical Physics at Princeton. He earned American citizenship in 1940 and retired from his post at Princeton in 1945.

Some wonder why Einstein was awarded the Nobel Prize in Physics 1921 for his contributions to the understanding of the photoelectric effect and not for his extensive works on relativity for which he is best known. Einstein's theories of relativity remained controversial among the less flexibly minded physicists of the Nobel Committee at the time of the awarding of the Nobel Prize. He therefore was awarded the prize not for his contributions to relativity, but for a paper he wrote on the photoelectric effect in 1905. A possible indication of the hard feelings at the time may be manifested by the fact that Einstein did not present himself to receive the Nobel Prize, as he was on a voyage to Japan, and moreover, he presented his Nobel Lecture not on the subject matter for which he was awarded the Nobel Prize but rather on the *Fundamental Ideas and Problems of the Theory of Relativity*. In addition, he did not deliver his Nobel Lecture in Sweden, but to the Nordic Assembly of Naturalists at Gothenburg on July 11, 1923 (Einstein, 1923). Einstein had been nominated many times over for the Nobel Prize and, according to a report by Schultz (2005), he was nominated 11 different years before finally getting the Prize. As reported by Schultz (2005) one Nobel Committee member wrote "Einstein must never receive a Nobel Prize even if the entire world demands it". And, when finally awarded the prize for his interpretation of the photoelectric effect attributing to light the dual properties of wave and particle (photon), the Nobel Committee directed Einstein not to mention relativity in his acceptance lecture. Well, Einstein ignored the request of the Committee and only discussed relativity.

During the Second World War, nuclear physicists approached Einstein for help in drawing the attention of the American Government to the threat of German research on nuclear fission and the possibility that Germany would develop a nuclear weapon. Although he was a pacifist against war of any kind, Einstein agreed to help urge the American Government to pursue research into the making of a nuclear weapon in light of Hitler's aggression. With the assistance of his longtime friend and colleague, Leo Szilard, Einstein sent a letter to Franklin Delano Roosevelt, President of the United States, on August 2, 1939. A portion of the letter is provided here

Albert Einstein
Old Grove Rd.
Nassau point
Paconic, Long Island
August 2nd, 1939

F. D. Roosevelt
President of the United States
White House
Washington, D.C.

Sir:

Some recent work by E. Fermi and Leo Szilard, which has been communicated to me in manuscript, leads me to expect that the element uranium may be turned into a new and important source of energy in the immediate future. Certain aspects of the situation which has arisen seem to call for watchfulness and, if necessary, quick action on the part of the Administration. I believe therefore that it is my duty to bring to your attention the following facts and recommendations:

In the course of the last four months it has been made possible—through the work of Joliot in France as well as Fermi and Szilard in America—that it may become possible to set up a nuclear chain reaction in a large mass of uranium, by which vast amounts of power and large quantities of new radium-like elements would be generated. Now it appears almost certain that this would be achieved in the immediate future.

This new phenomenon would also lead to the construction of bombs, and it is conceivable—though much less certain—that extremely powerful bombs of a new type may thus be constructed. A single bomb of this type, carried by boat and exploded in a port, might very well destroy the whole port together with some of the surrounding territory...

A part of a second letter by Einstein to the President dated March 7, 1940 reads as follows:

...Since the outbreak of the war, interest in uranium has intensified in Germany. I have now learned that research there is carried out in great secrecy and that it has been extended to another of the Kaiser Wilhelm Institutes, the Institute of Physics. The latter has been taken over by the government and a group of physicists, under the leadership of C. F. von Wetzsäcker, who is now working there on uranium in collaboration with the Institute of Chemistry. The former director was sent away on leave of absence, apparently for the duration of the war.... Dr. Szilard has shown me the manuscript which he is sending to the Physics Review in which he describes in detail a method of setting up a chain reaction in uranium. The papers will appear in print unless they are held up, and the question arises whether something ought to be done to withhold publication. (Clark, 1970).

In a third letter to the President dated April 25, Einstein urges that the efforts to carry out the research needed to create the bomb move at a faster and more organized manner possibly under the auspices of a Special Advisory Committee (Clark, 1970).

Einstein explained his position in urging the American President to pursue the creation of an atomic bomb when he stated the following (American Institute of Physics):

Because of the danger that Hitler might be the first to have the bomb, I signed a letter to the President, which had been drafted by [Leo] Szilard. Had I known that the fear was not justified, I would not have participated in opening this Pandora's box, nor would Szilard. For my distrust of governments was not limited to Germany.

Aside from these letters Einstein contributed in no other way to the creation of the atomic bomb. Einstein was a true pacifist. One week before his death he signed his last letter to Bertrand Russel whereby he agreed that his name be included in a manifesto, The Russell-Einstein Manifesto, urging all nations to abolish nuclear weapons. The entire text of the Manifesto is provided in Radioactivity Hall of Fame, Part IV.

Einstein was also a philosopher with a strong kinship to mankind, which is evidenced by his writings particularly his essay "The World as I See It" published in *Living Philosophies* (Einstein, 1931) and in a book edited by Seelig (1954).

For a long period of his life starting before 1920 up to his death in 1955 Einstein struggled without success to define a unified field theory that would unite electromagnetism and gravity and space and time. Einstein died on April 18, 1955 at Princeton, NJ.

ARTHUR H. COMPTON (1892–1962)

Arthur Holly Compton was born in Wooster, OH on September 10, 1892. He was awarded the Nobel Prize in Physics 1927 "for his discovery of the phenomenon named after him the Compton Effect".

Arthur Compton graduated in 1913 from the College of Wooster with a Bachelor of Science degree. He then went to Princeton University where he received his M.A. degree in 1914 and his Ph.D. in physics in 1916. His doctoral dissertation was on the angular distribution of x-rays reflected by crystals. Compton then took on a year as physics instructor at the University of Minnesota before taking on a position as a research engineer for the Westinghouse Company until 1919. He left industrial engineering to do research of his choosing when he received a research fellowship from the National Research Council, which enabled him to continue research with x-ray and gamma-ray scattering at the Cavendish Laboratory, Cambridge University where he carried out his first tests on wavelength variations of x- and gamma-rays as a function of scattering angle. After a year at Cambridge he became Professor and Head of the Department of Physics at Washington University, St. Louis in 1920 and subsequently succeeded

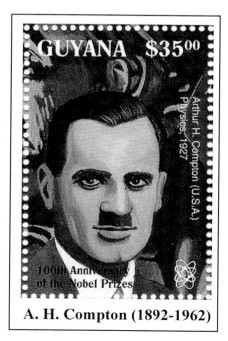

A. H. Compton (1892-1962)

Robert A. Millikan as Professor of Physics at the University of Chicago in 1923, a post he occupied until the end of the Second World War in 1945 (Nobel Lectures, Physics 1922–1941).

Arthur Compton is best known for his work on the scattering of x-rays and gamma rays when he discovered the increase in wavelength of an x-ray or gamma ray after the incident radiation was scattered by an electron in matter, demonstrating that the scattered radiation had lost energy to an atomic electron (Compton, 1923a) known as the Compton effect. It was for this work that Compton received the Nobel Prize for Physics 1927. He studied the x-radiation scattered by low atomic weight materials. By irradiating substances with homogeneous or monochromatic x-rays, that is, x-rays of a specific wavelength, he found that the rays would be divided into two lines or wavelengths, one line would be of one and the same wavelength as the incident radiation, and the other would be the scattered rays of a longer wavelength. The change in wavelength would be independent of the irradiated material, but dependent on the angle between the incident and scattered rays.

Compton explained this phenomenon on the basis of the quantum theory of light. In his account for the quantum theory of the scattering of x-rays and gamma rays by light elements Compton (1923a) stated

> The hypothesis is suggested that when an x-ray quantum is scattered it spends all of its energy and momentum upon some particular electron. This electron, in turn, scatters the ray in some definite direction. The change in momentum of the x-ray quantum, due to the change in its direction of propagation, results in a recoil of the scattering electron. The energy in the scattered quantum is thus less than the energy of the primary quantum by the kinetic energy of recoil of the scattering electron.

Compton elucidated that, of the total energy of the primary beam incident on a material, a fraction of the energy reappears as scattered radiation, while the remainder is absorbed and transformed into kinetic energy of recoil of the scattering electrons. The Compton effect is illustrated in Figure III.2. The scattered quantum is of longer wavelength (λ') and thus lower in energy than the incident quantum

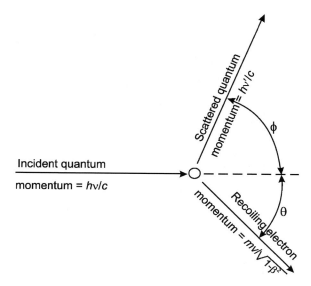

Figure III.2   The Compton effect. An incident x-ray or gamma-ray photon is deflected through an angle ϕ by an electron, which, in turn, recoils at an angle θ, taking part of the energy of the photon. The symbol β represents the relative phase velocity of the recoil electron, that is, the ratio of the velocity of the electron to the velocity of light. (Adapted from Compton (1927) with permission from The Nobel Foundation © 1927.)

of shorter wavelength (λ). The change in wavelength can be calculated as a function of the angle of deflection of the scattered ray as derived by Compton according to the equation

$$\lambda' - \lambda = \frac{h}{mc}(1 - \cos \phi) \tag{III.12}$$

where h is Planck's constant, m the mass of the scattering electron, c the velocity of light in a vacuum, and ϕ the angle between the incident and scattered ray. The difference in energy of the two wavelengths is manifested as the kinetic energy of the recoil electron less the electron binding energy. As the electron binding energy is relatively small, it can be ignored. Compton (1923a,b) thus described the energy of the scattered quantum ($h\nu_\phi$) as being equal to that of the incident quantum ($h\nu_0$) less the kinetic energy of recoil of the scattering electron or

$$h\nu_\phi = h\nu_0 - mc^2 \left(\frac{1}{\sqrt{1 - \beta^2}} - 1 \right) \tag{III.13}$$

where β is the relative phase velocity of the electron or the ratio of the velocity of recoil of the electron to the velocity of light.

When Compton (1923a,b) first presented his theory the scientific community reacted with both amazement and some opposition. However, Compton and others were able to provide evidence by several methods including: (i) exact spectrometric analysis of the incident and scattered x-ray or gamma-ray lines, (ii) visualization of the directions of the scattered photon and recoil electron tracks by means of the Wilson Cloud Chamber (Compton, 1927), and (iii) by coincident detection methods developed by Bothe and Geiger (1924) and Compton and Simon (1925a,b), by which it could be established that the scattered x-ray photon and recoil electron appear at the same instant.

In his Nobel Lecture, Compton (1927) explained that an examination of the spectrum of the scattered (secondary) x-rays and incident (primary) x-rays showed that the primary beam would be split into two parts, as illustrated in Figure III.3, one of the same wavelength and the other of an increased wavelength. When different primary wavelengths are used, the same difference in wavelength between the two components is encountered, but the relative intensities of the two components change. For the longer wavelengths of incident radiation, the unmodified ray has the greatest intensity, whereas for shorter wavelengths of incident radiation, the scattered (modified) ray is predominant; and when very hard gamma rays are used (very short wavelengths) as incident rays, all of the rays end up as scattered (modified) radiation.

Very conclusive evidence for the Compton effect was provided by visualization of the directions of emission of the scattered photon and recoil electron by means of the Wilson Expansion Cloud Chamber. The experimental setup used by Compton and coworkers is illustrated in Figure III.4. A photograph taken by Compton in the Wilson Cloud Chamber showing the electron recoil track and associated secondary β-track produced by the scattered x-ray photon is illustrated in Figure III.5. In his Nobel lecture on December 12, 1927, Compton explained his method as follows:

A narrow beam of x-rays enters a Wilson Expansion Cloud Chamber. Here it produces a recoil electron. If the photon theory [Compton Effect] is correct, associated with this recoil electron, a photon is scattered in the direction ϕ. If it should happen to eject a β-ray, the origin of the β-ray tells the direction in which the photon was scattered. [The term β-ray was adopted by Compton for convenience and it refers to a photoelectron produced by the scattered x-ray photon]. A measurement of the angle θ at which the recoil electron is ejected and the angle ϕ of the origin of the secondary β-particle, shows close agreement with the photon formula. This experiment is of special significance, since it shows that, for each recoil electron there is a scattered photon, and that the energy and momentum of the system photon plus electron are conserved in the scattering process.

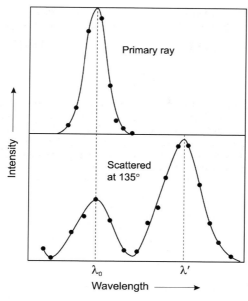

Figure III.3   A typical spectrum of scattered x-rays, showing the splitting of the primary (incident) ray (λ_0) into a modified (scattered, λ') and unmodified ray. (Adapted from Compton (1927) with permission from The Nobel Foundation © 1927.)

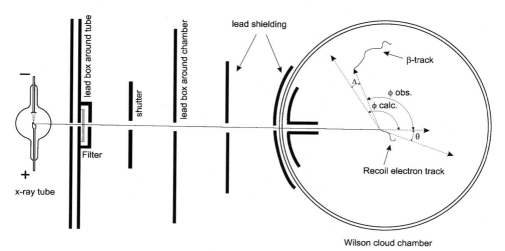

Figure III.4   A schematic illustration of the experimental setup used by Compton to observe and measure the recoil electron and x-ray scatter angles relative to the incident x-ray photon. The circular cloud chamber measured 18 cm in diameter and 4 cm deep. It is traversed by a carefully shielded narrow beam of homogeneous (monochromatic) x-rays. A photograph is taken of the electron recoil track as well as the track of a photoelectron produced by the scattered x-ray (referred to by Compton as a β-track for convenience). A line is manually drawn on the photograph from the beginning of the electron recoil track to the beginning of the β-track, which provides the direction of the x-ray after scattering. An electron recoiling at an angle θ is associated with an x-ray photon scattered simultaneously through an angle ϕ. The angle Δ represents the difference between the observed and calculated angles ϕ of Compton photon scatter. (Adapted from Compton (1927) with permission from The Nobel Foundation © 1927.)

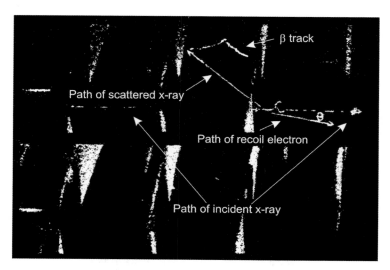

Figure III.5 Photograph taken in the Wilson Cloud Chamber showing an electron recoil track and the β-track produced by the scattered x-ray photon. The direction of travel of the scattered photon is found by drawing a line on the photograph from the beginning of the electron recoil track to the beginning of the β-track. The photograph was modified with lettering. (From Compton (1927) with permission from The Nobel Foundation © 1927.)

It was Professor Manne Siegbahn, Nobel Laureate in Physics 1924 and member of the Nobel Committee in Physics, who gave the presentation speech during the awarding of the Nobel Prize to Arthur Compton on December 10, 1927. Manne Siegbahn summarized the Compton effect and its significance as follows

> Compton deduced a new kind of corpuscular theory [It reconfirmed Einstein's theory of the photon as an elementary particle.], with which all experimental results showed perfect agreement within the limits of experimental error. According to this theory, a quantum of radiation is re-emitted in a definite direction by a single electron, which in so doing must recoil in a direction forming an acute angle with that of the incident radiation. In its mathematical dress this theory leads to an augmentation of the wavelength of the incident radiation and implies a velocity of the recoil electron that varies between zero and about 80% of the velocity of light, when the angle between the incident and the scattered radiation varies between zero and 180°. This theory predicts recoil electrons with a velocity generally much smaller than that of the electrons which correspond to [Einstein's] photoelectric effect…Finally, the fact deserves to be emphasized that the Compton effect has proved to be of decisive influence upon the absorption of short-wave electromagnetic—especially radioactive—radiation…

During 1930–1940 Compton became a leading research physicist on cosmic radiation (Compton, 1932, 1933). Leading a world-wide study of the geographic variations of cosmic rays intensities, he confirmed the findings of J. Clay from Amsterdam, who observed geographic variations in the intensity of cosmic radiation and a consistent lower intensity near the equator as a result of a Dutch cosmic ray expedition in 1933 (Clay, 1934). Compton confirmed these findings from data on the cosmic ray intensities obtained by eight different expeditions at 69 stations distributed at representative points over the earth's surface (Compton, 1933).

While at the University of Chicago during World War II, Compton collaborated with Enrico Fermi, Leo Szilard, Eugene P. Wigner, and others on the first uranium fission reactor (see Enrico Fermi in

Radioactivity Hall of Fame—Part II) and subsequently in the establishment of the large plutonium-producing reactors in Hanford, Washington (Nobel Lectures, Physics 1922–1941).

Arthur Compton died on March 15, 1962 in Berkeley, CA.

MAX VON LAUE (1879–1960)

Max von Laue was born in Pfaffendorf (near Koblenz), Germany on October 9, 1879. He was awarded the Nobel Prize in Physics 1914 "for his discovery of the diffraction of x-rays by crystals".

Von Laue remained keenly interested in physics since he was a young man, and as he noted in his Nobel Lecture "that goes back to the days in which I was privileged to receive first-class instruction at the famous old Protestant Gymnasium in Strassburg, Alsatia—my particular attention was drawn to the field of optics, and within that field the wave theory of light" (von Laue, 1913). After a year in military service in 1898 he went to the University of Strassburg to study physics, chemistry, and mathematics. Soon he moved on to study for short periods at the University of Göttingen and a semester at the University of Munich until 1902 and at the University of Berlin to work under Max Planck. Von Laue recalled that

RÉPUBLIQUE DE GUINÉE
OFFICE DE LA POSTE GUINÉENNE
750F
1914
MAX VON LAUE

von Laue (1879-1960)

the lectures, which I heard during my student days from Profs. W. Voigt, Max Planck, and O. Lummer provided me at that time…with thorough experimental and theoretical knowledge; and as I had been finally able to cultivate what one could almost term a special feeling or intuition for wave processes. (von Laue, 1915).

Max von Laue received his doctorate from the University of Berlin in 1903 after which he returned for 2 years to the University of Göttingen until receiving a post as assistant to Max Planck in 1905 at the Institute for Theoretical Physics at Berlin where he continued research on radiation fields and light waves.

In 1909, Max von Laue became Privatdozent at the University of Munich lecturing on optics, thermodynamics, and relativity theory. It was upon his arrival at Munich that von Laue began to concentrate his thoughts on the true nature of x-rays. It was not that long before had Röntgen discovery x-rays in 1895, and there was much debate in the scientific community about the true nature of x-rays. The debate surrounded the question of whether x-rays had the properties of light waves or were the rays of particulate nature? Since the discovery of x-rays Röntgen and others made efforts to observe the diffraction or interference of the rays that could provide evidence as to whether the x-rays represented a wave phenomenon or the emission of small particles. Some argued that, according to Maxwell–Lorentz electrodynamics, electromagnetic waves must be produced whenever electricity carriers alter their velocity, which is precisely what occurs in the creation of x-rays. Others experts maintained at the time that x-rays were corpuscular in nature, because x-rays would, upon striking a substance, excite the creation of secondary x-rays and, at the same time, they would free electrons from the substance. The velocities of the electrons emitted were not dependent on temperature of the substance bombarded or on the intensity of the x-radiation, but rather on the energy of the x-radiation. The velocities of the electrons knocked out of the bombarded substance would increase with radiation hardness or energy. From the quantum theory of Planck and Einstein's then recent revelations on the photoelectric effect, Max von Laue and others could see that the x-rays had properties similar to light rays in the photoelectric effect. Nevertheless, there was the need to demonstrate the true nature of x-rays. On the assumption that x-rays were of wave nature, Wilhelm Wien and Arnold Sommerfeld used Planck's new quantum theory to arrive at an estimate of the x-ray wavelengths to lie between 10^{-10} and 10^{-9} cm, which would give them a wavelength shorter than visible light.

In 1912, von Laue became Professor of Physics at the University of Zurich and in February of that year he was approached by P. P. Ewald about a research problem concerning crystal optics. Von Laue related (von Laue, 1915)

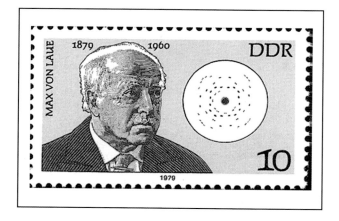

During the conversation I was suddenly struck by the obvious question of the behavior of waves which are short by comparison with the lattice constants of the [crystal] space lattice. And it was at that point that my intuition for optics suddenly gave me the answer:...The fact that the lattice constant in crystals is of an order of 10^{-8} cm was sufficiently known from the analogy with other interatomic distances in solid and liquid substances, and,...The order of x-ray wavelengths was estimated by Wien and Sommerfeld to be 10^{-9} cm. Thus the ratio of wavelengths and lattice constants was extremely favorable if x-rays were to be transmitted through a crystal.

Shortly thereafter at the University of Munich Max von Laue and his associates W. Friedrich and P. Knipping carried out an experiment whereby they irradiated a crystal of copper sulfate with x-rays.

MAX VON LAUE * NOBELPREIS PHYSIK 1914
Röntgenstrahl-Beugung
am Kristallgitter
60
DEUTSCHE BUNDESPOST   1979

A photographic plate was placed behind the crystal. Right from the outset they observed on the photographic plate a considerable number of deflected x-rays together with an image of the primary ray. The deflection maxima left images on the photographic plate as spots at defined distances from a larger center spot produced by the primary x-ray (Friedrich *et al.*, 1912; von Laue, 1912). The deflected maxima represented the crystal lattice spectra as von Laue anticipated. One of the original crystal x-ray diffraction images taken by von Laue and his coworkers is illustrated as spots on a circular photographic plate in the postage stamp from the former German Democratic Republic (DDR) issued in 1979. A more intricate crystal diffraction image, such as those produced by more complex crystalline substances, is illustrated in another stamp from the Federal Republic of German, also issued in 1979 commemorating 100 years of the birth of Max von Laue.

Max von Laue was able to demonstrate that the spots on the photographic plates represented x-ray diffraction maxima produced by crystal lattice points of interference, corresponding with three integral numbers that originated from three space-lattice periodicities, the ratios of which determine the position of each point of interference (von Laue, 1915). This discovery provided evidence for the wave nature of x-rays and also the lattice theory of crystals. The wavelengths of the x-rays could then be determined by the dimensions of the crystal space lattice. However, for this to be done, it would be necessary to know the exact molecular structure of the crystals. Such work was soon to follow with the discoveries of William Henry Bragg and his son William Lawrence Bragg. The significance of Max von Laue's discovery was underscored by Professor G. Granqvist, Chair of the Nobel Committee for Physics when he stated

As a result of von Laue's discovery of the diffraction of x-rays in crystals proof was thus established that these light waves are of very small wavelengths. However, this discovery also resulted in the most important discoveries in the field of crystallography. It is now possible to determine the position of atoms in crystals. (Granqvist, 1915).

In 1914, von Laue moved as professor of Theoretical Physics at the Goethe University in Frankfurt on Main and in 1916 worked at the University of Würzburg on wireless communication for the war effort. In 1917 he worked under Albert Einstein as Second Director to the Institute for Physics at Berlin–Dahlem. From 1919 to 1943 von Laue was Professor of Physics at the University of Berlin. During this period, von Laue made significant contributions to the understanding of superconductivity.

During the reign of Hitler and the National Socialist Power in Germany, Max von Laue defended scientific views, such as Einstein's theory of relativity, that were not approved by the party or by party loyalists such as Philipp Lenard. Von Laue was the only member of the Berlin Academy to protest a statement by the President of the Academy, who claimed that the resignation of Albert Einstein from the Academy was no loss (Nobel Lectures, Physics, 1901–1921). He welcomed the arrival of the French troops and he was taken by an Anglo-American mission to England with other German scientists until 1946 when he returned to Göttingen as Acting Director of the Max Planck Institute and Professor of the University. In 1951, von Laue was appointed Director of the Fritz Haber Institute for Physical Chemistry in Berlin–Dahlem where he continued work on x-ray optics. He retired in 1958 and on April 24, 1960 succumbed to injuries suffered in an auto accident while driving alone to his laboratory on the Berlin autobahn at the age of 80.

SIR WILLIAM HENRY BRAGG (1862–1942) AND
SIR WILLIAM LAWRENCE BRAGG (1890–1971)

Sir William Henry Bragg shared the Nobel Prize in Physics 1915 with his son Sir William Lawrence Bragg "for their services in the analysis of crystal structure by means of x-rays".

Sir William Henry Bragg was born in Westward, Cumberland, England on July 2, 1862. He was educated at Market Harborough Grammar School and King William's College, Isle of Man. William Henry was elected a scholar of mathematics at Trinity College, Cambridge in 1882 where he studied for 2 years achieving high honors in the final mathematics examinations. He then studied physics at the Cavendish Laboratory during part of 1885 being elected that year as Professor of Physics and Mathematics at the University of Adelaide, South Australia. He returned to England in 1909 to become Cavendish Professor of Physics at the University of Leeds where he remained until 1915. During his tenure at Leeds William Henry focused his work on the study of x-rays, which he at first thought were elementary particles until the work of Max von Laue of 1912 demonstrated that x-rays had wavelengths of the order of magnitude equivalent to the distances of crystal lattices (10^{-8}cm) and thus the crystal lattices could serve as molecular x-ray diffraction gratings. At Leeds, Willam Henry designed the Bragg x-ray spectrometer and began work with his son William Lawrence on

W. H. (1862-1942) and W. L. Bragg (1890-1971)

studies of x-ray diffraction by crystalline materials. Their joint studies provided information as soon as 1914 on the interatomic distances and structure of crystals as well as x-ray wavelengths, whereby father and son were awarded jointly the Nobel Prize in Physics 1915. During 1915–1925, William Henry was Quain Professor of Physics at the University College London. In 1923 William Henry became Head of the Royal Institution where he remained in service until his death in 1942.

Sir William Lawrence Bragg was born in Adelaide, South Australia on March 31, 1890 the son of Sir William Henry Bragg. He attended St. Peter's College at Adelaide and thereafter studied mathematics at Adelaide University graduating in honors in 1908. When his father returned to England in 1909 with his son, William Lawrence entered Trinity College, Cambridge and graduated with honors in Natural Science in 1912. William Lawrence started work on x-ray diffraction in 1912 and published his first paper on the diffraction of short electromagnetic waves in the Proceedings of the Cambridge Philosophical Society (Bragg, 1913). This was followed by another paper on the analysis of crystals by the x-ray spectrometer the following year (Bragg, 1914a). In 1912, William Lawrence started a collaboration in the field of x-ray diffraction and crystal structure with his father, who then developed the very first x-ray diffraction spectrometer.

The combined efforts of father and son during 1912–1914 earned them the joint Nobel Prize in 1915. William Lawrence, who delivered the Nobel Lecture in 1915, is yet the youngest person (25 years) to receive the Nobel Prize. Prior to winning the Nobel Prize young William Lawrence wrote a thorough text that gave birth to x-ray diffraction as a standard method for the determination of crystalline structures (Bragg, 1914b). During 1915–1919 William Lawrence served as Technical Advisor on Sound Ranging for the General Headquarters, British Troops in France. From 1919–1937 he held the post of Langworthy Professor of Physics, University of Manchester (Nobel Lectures, Physics, 1901–1922). During 1937 to 1938 he was Director of the National Physical Laboratory, but despised pure administration. He served also in World War II and, after the war succeeded Rutherford as Cavendish Professor of Experimental Physics at Cambridge. In 1948 William Lawrence became interested in the structure of protein and nucleic acids. He did not play any direct part in the x-ray diffraction study and 1953 discovery of the helical structure of DNA derived from the work of Rosalind Franklin, James Watson, Francis Crick, and Maurice Wilkins; however, the x-ray diffraction technique developed by William Lawrence Bragg was a vital tool in the discovery of the nucleic acid structure. James Watson admitted that the x-ray method developed by Bragg 40 years before was at the heart of this profound insight into the nature of life itself (University of Cambridge, Cavendish Laboratory). In 1953, William Lawrence became resident professor of the Royal Institution, London and remained in that position until his retirement in 1966. He passed away in Waldringfield on July 1, 1971.

When awarded the Nobel Prize, William Lawrence was only 25 years of age, the youngest person ever to receive the Nobel Prize. The fact that the joint research and Nobel Prize was the result of collaboration between father and son led many to believe that the father had initiated the research, a fact that upset the son (University of Cambridge, Cavendish Laboratory). The credit for this joint work was clarified by Professor G. Granqvist, Chairman of the Nobel Committee, in the Nobel Presentation Speech (Granqvist, 1915), when he stated

The problem of calculating the crystal structures was a discovery of epoch-making significance when W. L. Bragg [the son] found out that the phenomenon could be treated mathematically as a reflection by the successive parallel planes [i.e., the atomic planes of a crystal] that may be placed so as to pass through the lattice points, and that in this way the ratio between the wavelengths and the distances of the said planes from each other can be calculated by a simple formula from the angle of reflection…The instrument requisite for the said purpose, the so-called x-ray spectrometer, was constructed by Professor W. H. Bragg, W. L. Bragg's father, and it has been with the aid of that instrument that father and son have carried out, in part conjointly, in part each on his own account, a series of extremely important investigations respecting the structure of crystals.

Giving much deliberation on the work of Max von Laue (see von Laue in this Section), who demonstrated the wave nature of x-rays and that the diffracted wave pattern was a function of the symmetry of the underlying crystal structure, W. L. Bragg considered the diffraction effects as a reflection of the x-rays by the planes of the crystal structure. In his Nobel Lecture given at the Royal Swedish Academy of Sciences in Stockholm on September 6, 1922 W. L. Bragg stated how he approached the mathematical solution to the application of x-ray diffraction to the determination of crystal structure (Bragg, 1922). In the lecture he stated

> The points of a space lattice may be arranged in series of [atomic] planes, parallel and equidistant from each other. As a pulse passes over each diffracting point, it scatters a wave, and if the number of points are arranged on a plane the diffracted wavelets will combine together to form a reflected wave front, according to the well known Huygens construction. The pulses reflected by successive [atomic] planes build up a wave train given by the formula

$$n\lambda = 2d \sin\theta \qquad\qquad \text{(III.14)}$$

> In this impression, n is an integer, λ is the wavelength of the x-rays, d the spacing of the [atomic] planes, and θ the glancing angle at which the x-rays are reflected.

Eq. (III.14) has become to be known as Bragg's Law, and the diffraction of x-rays according to Bragg's Law is illustrated in Figure III.6. The figure illustrates two atomic planes in a crystal separated by the distance d. The diffracted rays are those reflected at an angle θ to the atomic planes, and such reflection requires a coherence of rays that are in-phase yielding a reinforced beam, which can be detected by the spectrometer at a given angle θ by an ionization detector or with photographic film in a circular Bragg camera. The reflection of the rays in-phase as a coherent beam requires incident x-ray wavelengths that are integer multiples along the x-ray path of travel between the two atomic planes. Only those x-rays with wavelengths of 1λ, 2λ, 3λ over the path length of travel ($2d \sin\theta$)

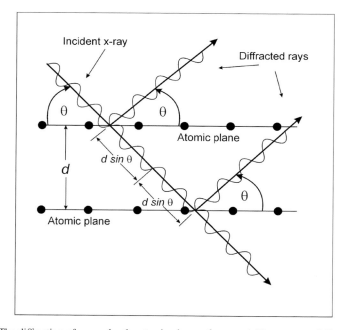

Figure III.6 The diffraction of x-rays by the atomic planes of a crystal. The rays are diffracted according to Bragg's Law, $n\lambda = 2d \sin\theta$, where λ is the x-ray wavelength, d the spacing of the atomic planes, and θ the glancing angle at which the x-rays are reflected.

between each successive crystal atomic plane will produce coherent reflected x-ray beams. By measurement of the angle of reflection θ of an incident monochromatic x-ray of known wavelength, the spacing d between each atomic plane can be measured.

In addition to the measurements of the d spacing of the atomic planes, W. H. and W. L. Bragg extended their diffraction experiments of various crystalline materials to determine their structure or atomic arrangements. As explained by William Lawrence in the Nobel lecture

> In examining the effect for varying angles of incidence my father discovered that a very strong reflection appeared when a given crystal face was set at certain definite angles. Since the relationship [diffraction] must hold between glancing angle and wavelength, this constituted the first evidence of the existence of characteristic [x-ray] lines in the radiation given off by the anticathode [of the x-ray tube]. These same lines could be recognized in the reflections from other faces, and the measurement of the angles at which they appeared proved a most powerful method of finding the arrangement of the atoms in the crystal.

The method used by the father and son team to analyze crystal structure by x-rays as described in the Nobel Lecture was firstly to find the dimensions of a unit cell of the crystal space lattice, which has for its sides the basic atomic structure. To achieve this they would measure the angle θ by which a monochromatic beam of x-rays would be reflected by various faces of the crystal (see a diagram of their x-ray spectrometer in Figure III.7). A small crystal would be set at the center of the instrument as illustrated, adjusted so that a zone of the crystal is parallel to the axis, and then the crystal is rotated to permit the reflections of various faces to be measured by the ionization detector in turn. The d spacing of the atomic planes parallel to any crystal face under examination could then be calculated according to Eq. (III.14). The dimensions of a unit cell are then found by the measurement of d for several crystal faces. When the dimension of the unit cell is known, the number of atoms in the unit cell is calculated from the crystal density and mass of each atom. The second step in the structural analysis is to determine how the atoms are grouped to form the unit cell. The structure or the manner in which the atoms are grouped in the unit cell will influence the strengths of the various x-ray reflections. This is so, because the x-rays are diffracted by the electrons grouped about the nucleus of each atom. In some directions of the x-ray beam the atoms give a strong scattered beam, whereas in other directions the effects of the electron densities annul each other by interference. The exact arrangement of the atoms indicated by the electron

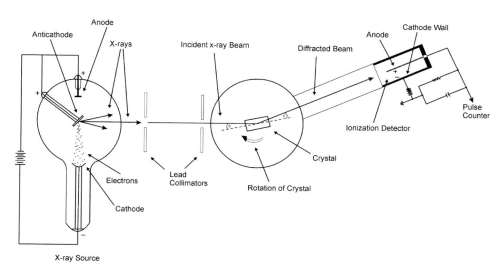

Figure III.7   Bragg's x-ray spectrometer.

density distributions are deduced by comparing the strengths of the x-ray reflections from different faces and in different orders. When placed in the spectrometer, the crystal was set at the center of the instrument and adjusted so that a zone of the crystal was parallel to the axis; and then the crystal would be rotated round so that the reflections of various faces would be observed in turn. The x-ray spectrometer measures the angles between planes of the crystal structure, and the spacings of the planes.

W. H. and W. L. Bragg analyzed several crystal types reporting the atomic arrangements of these and the spacings between the atoms. One of theses was the cubic crystalline lattice of sodium chloride (rock salt), whereby they reported for the first time, that the arrangement of atoms in sodium chloride were alternate Na and Cl atoms and not associated in the crystal as molecular NaCl. In other words, a Na atom was surrounded by six Cl atoms, and vice versa. The atomic arrangement of the sodium chloride crystal is illustrated in the postage stamp issued by the United Kingdom, which commemorates the achievement of W. H. and W. L. Bragg. The stamp illustrates a unit cell of sodium chloride crystal, where the central atom (e.g., sodium ion, blue ball) is surrounded by six atoms of the other ion (e.g., chloride, green ball). As stated by William Lawrence Bragg in his Nobel Lecture

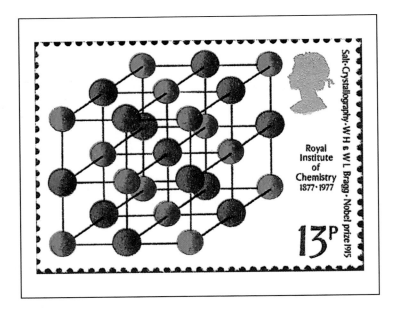

For the first time the exact arrangement of the atoms in solids has become known, we can see how far the atoms are apart and how far they are grouped...One type of chemical binding has already been confirmed, In a structure of sodium chloride the atoms of sodium and chlorine are not associated into chemical molecules, but are arranged alternately equidistant from several neighbors.

Prior to their discovery of the atomic arrangements of the crystalline structure of NaCl in 1914, it was always imagined that crystalline materials such as NaCl, KCl, KI, KBr, etc. were molecular associations not associations of the separate ions or atoms in the crystal. The importance of this discovery was underscored by Professor G. Granqvist (1915), Chairman of the Nobel Committee, when he stated in the Nobel Presentation Speech

From these investigations it follows that a metal atom in the crystals of the alkaloid salts is situated at one and the same distance from the six haloid atoms nearest to it, and vice versa—a relationship that was found to prevail, mutatis mutandis, in all the crystals examined. That means the exceedingly important discovery, both for molecular physics and chemistry, is that the crystals consist of atomic lattices and not, as has been always imagined, of molecular ones.

The above statement of Professor Granqvist can give some insight into the impact of the findings, as when this discovery was first reported (Bragg, 1914a,b) some members of the scientific community were initially in disbelief. I personally had the good fortune to meet Sir William Lawrence Bragg around 1968 during my graduate student days. I recall him relating that, he had saved correspondence, received soon after his publications of the discovery in 1914, from some of the scientific community stating that he must be out of his mind.

By the time of the Nobel Lecture given by William Lawrence Bragg on September 6, 1922, 7 years after the Nobel Prize was awarded to Sir Lawrence and his father, they had determined the atomic structures of several crystalline materials including diamond, magnetite, calcite, graphite, ice, and the organic crystalline structure of naphthalene, among others.

In conclusion, the discovery of x-ray diffraction by crystalline compounds was the beginning of a new era. It would provide us with a new tool that could produce images of the small atomic world, the dimensions of which the light microscope could not provide. W. L. Bragg clearly pointed this out when he stated in the Nobel Lecture

> The examination of crystal structure, with the aid of x-rays has given us for the first time an insight into the actual arrangement of the atoms in solid bodies. The study of structure by means of a microscope is limited by the coarseness of the light which illuminates the object, for we can never hope to see details smaller then the wavelength of the light. By using x-rays with their very short wavelengths, this limit of minuteness has at one step been decreased ten thousand times, for the wavelength of the x-rays is of a smaller order than the dimensions of the atomic structure. We are actually looking into the interior of the molecule and the atom with this fine-grained form of light.

From the original work of W. H. and W. L. Bragg the field of x-ray crystallography, that is, the study of crystal structures by x-ray diffraction, was given birth and remains a vital analytical technique today. The x-ray diffraction data obtained by Rosalind Franklin (1920–1958) was a critical component in the elucidation of the double helical structure of DNA, which was one of the greatest discoveries of the century. Franklin's contribution to the discovery of the structure of DNA with x-ray crystallography was underscored in a publication of the American Institute of Physics (Elkin, 2003)

> Watson and Crick could not have proposed their celebrated structure for DNA as early in 1953 as they did without access to experimental results [including x-ray crystallography data] obtained by King's College scientist Rosalind Franklin. Franklin had died of cancer in 1958 at age 37, and so was ineligible to share the honor [Nobel Prize in Physiology and Medicine 1962]. Her conspicuous absence from the awards ceremony—the dramatic culmination of the struggle to determine the structure of DNA—probably contributed to the neglect, for several decades, of Franklin's role in the DNA story.

Another historic example of the vital role of x-ray diffraction to scientific development was the determination of the crystal structure of myoglobin, a protein of 153 amino acids, which is the primary oxygen-carrying pigment of muscle tissue. In 1957 John Kendrew (1917–1997) and Max Perutz (1914–2002) determined the structure of myoglobin by high-resolution x-ray crystallography and were awarded the Nobel Prize in Chemistry 1962. Two years later Dorothy Hodgkin (1910–1994) was awarded the Nobel Prize in Chemistry 1964 "for her determination by x-ray techniques of the structures of important biochemical substances [including penicillin and Vitamin B12]".

With the aid of modern computers the diffraction images, that is the images of the x-rays diffracted off the electrons in a crystal structure, are recorded on photographic film and scanned into a computer. From this data the electron density maps of the molecular structure in a crystalline material are constructed with the aid of computer processing and thereby the arrangement of the atoms of a molecular structure can be determined. The molecular structures of innumerable inorganic, organic, and biological compounds have been determined by x-ray diffraction.

HENRY G. J. MOSELEY (1887–1915)

Henry Gwyn Jeffreys Moseley was born in Weymouth, England on November 23, 1887. He did not receive the Nobel Prize, because of his untimely death at the age of 27. His contributions to atomic physics and x-ray spectrometry of the elements were of great importance; and it was stated by the Chairman of the Nobel Committee (Gullstrand, 1925), that Moseley would have been awarded the Nobel Prize, if he had survived.

In 1906, Moseley entered Trinity College of the University of Oxford and graduated in 1910 with a Second degree in Physics. Upon graduation he went to Manchester University to work under Ernest Rutherford. At Manchester he was given a full-time teaching position, but he had ambitions to become a research physicist. After a year of teaching, Moseley was relieved of his duties and allowed to do scientific research. It was in 1911 that W. H. and W. L. Bragg at the University of Leeds were carrying out research in x-ray diffraction, which caught his interest. By the time Moseley started his work it was known that the diffraction of x-rays by a crystal made it possible to accurately determine the frequencies and wavelengths of x-radiation. In addition, it had been demonstrated that an element would emit its characteristic x-radiation, if excited by a beam of sufficiently fast cathode rays. Moseley devised an apparatus whereby he could bombard individually all of the known and available elements with cathode rays and measure the wavelengths of the x-rays emitted by each element (Moseley, 1913, 1914). To measure the wavelengths (λ) Moseley directed the x-rays at a crystal of potassium ferrocyanide and measured the angle of diffraction (θ) according to the Bragg Law

$$n\lambda = 2d \sin \theta \qquad \qquad (\text{III.15})$$

where d is the spacing of the interatomic planes of the crystal ($d = 8.454 \times 10^{-8}$ cm). He determined the order of the diffraction ($n = 1, 2,$ or $3\ldots$) by imaging the diffracted light on photographic film.

**H. G. J. Moseley (1887-1915), University of Oxford,
Museum of the History of Science, courtesy AIP Emilio
Segre Visual Archives, Physics Today Collection**

Diffraction of the first order ($n = 1$) was more intense than that of the second and third orders. In his first report, Moseley (1913) measured the x-ray emission spectra of 21 elements of increasing numbers of the Periodic Table of the Elements from aluminum to silver and determined the wavelengths of the α and lower intensity β lines that belong to Barkla's K series. Charles Glover Barkla (1877–1944) identified the x-radiation resulting from atomic electron transitions to the K shell as K x-rays, and x-rays emitted as a result of atomic electron transitions to the L shell as L x-rays. In a subsequent paper, Moseley (1914) measured the wavelengths of the L series of x-ray emissions of 24 elements of increasing numbers of the Periodic Table of the Elements from zirconium to gold. Of the L series he measured the wavelengths of α, β, γ, and δ lines of decreasing wavelengths and decreasing intensities. Moseley then plotted the square root of the radiation frequencies or the logarithm of the radiation wavelengths against a characteristic integer (N) assigned to each element. Starting with aluminum, being the 13th element in the periodic table, he assigned it the number $N = 13$. The next element, silicon, was assigned the next highest integer $N = 14$, phosphorus $N = 15$, etc. A copy of his plots are illustrated in Figure III.8. The plots showed an incredibly straight line relationship between the assigned integer N and the x-radiation frequencies for each element. From this relationship he concluded the following (Moseley, 1914):

> Now if either the elements were not characterized by these integers or any mistake had been made in the order chosen, or in the number of places left for unknown elements [places for three unknown elements in the Periodic Table were found, See Fig. III.8], these irregularities would at once disappear. We can therefore conclude from the evidence of the x-ray spectra alone, without using any theory of atomic structure, that these integers are really characteristic of the elements. Further, as it is improbable that two different stable elements should have the same integer, three, and only three, more elements are likely to exist between Al and Au. As the x-ray spectra of these elements can be confidently predicted, they should not be difficult to find…Now Rutherford has proved that the most important constituent of an atom is its central positively charged nucleus, and van der Broek has put forward the view that the charge carried by this nucleus is in all cases an integral multiple of the charge on the hydrogen nucleus. There is ever reason to suppose that the integer which controls the x-ray spectrum is the same number of electrical units in the nucleus…

From Moseley's discovery as stated above, it was clear that his number N could be identified with the atomic number Z, which is the number of protons in the nucleus. The atomic number, therefore, was a meaningful number with respect to atomic structure, and not just an arbitrary number assigned to an element in the Periodic Table of Elements. Also, this number was related to the electron distributions about the atom according to Bohr Theory. Another outcome of this work was Moseley's confident identification of three gaps or missing elements in the Periodic Table, namely elements of atomic number 43, 61, and 75. Moseley could predict that there were three elements missing from the Periodic Table and yet to be discovered. Two of these elements were Technetium ($Z = 43$), Promethium ($Z = 61$), both of which are radioactive and made artificially. Technetium was the very first element to be made artificially and discovered in Italy by Perrier and Segre in 1937. Promethium was identified as a product of uranium fission in 1945. The third element predicted by Moseley ($Z = 75$) was the stable natural element Rhenium, which was discovered eventually in 1925 in platinum ores. Another outcome from Moseley's work was the potential of identifying or analyzing any element in the Periodic Table from its x-ray emissions.

The same month of Moseley's first publication of 1913 relating x-ray emission frequencies to the atomic number of the element, Rutherford wrote a Letter to the Editor of Nature on December 11, 1913 (Rutherford, 1913), which underscored the importance of Moseley's findings as related to the structure and properties of the atom. In his letter Rutherford stated the following:

> The original suggestion of van der Broek that the charge of the nucleus is equal to the atomic number and not to half the atomic weight seems to me very promising. This idea has already

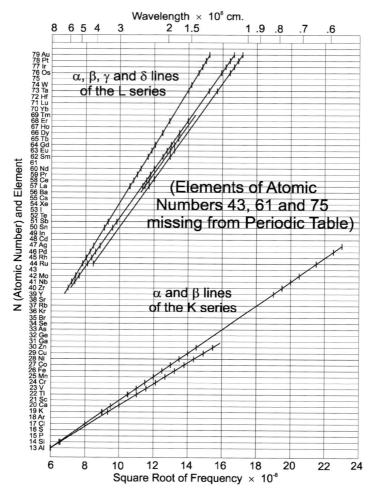

Figure III.8   Artist's portrayal of Moseley's original 1914 graph of the frequencies and wavelengths of x-rays emitted by all of the then known elements plotted against each element and an integer N assigned to each element starting at Aluminum (Al, assigned $N = 13$) to Gold (Au, assigned $N = 79$). Three missing and yet unknown elements are listed with blank spaces adjacent to integers $N = 43, 61$, and 75. The two lines of the lower curves (elements $N = 13$–47) are x-rays resulting from α and β lines of the K series of electron transitions; and the four lines of the upper curves (elements $N = 40$–79) are x-rays resulting from α, β, γ, and δ lines of the L series of electron transitions. (From Moseley, 1914 also the original Moseley graph, drawn by his own hand, may be seen in the Moseley Room of the Cavendish Laboratory, University of Oxford).

been used by Bohr in his theory of the constitution of atoms. The strongest and most convincing evidence in support of this hypothesis will be found in a paper by Moseley in *The Philosophical Magazine* of this month. He there shows that the frequency of the X radiations from a number of elements can be simply explained if the number of unit charges on the nucleus is equal to the atomic number. It would appear that the charge on the nucleus is the fundamental constant which determines the physical and chemical properties of the atom,…

In his above letter Rutherford came close to predicting the proton as the source of the positive charge on the nucleus. It would be only 6 years later that Rutherford would discover the proton. (See Rutherford in Radioactivity Hall of Fame—Part I.)

In 1914, Henry Moseley had planned to return to Oxford to continue his research, but the First World War broke out, and he volunteered for active service against the advice of family and of the Army itself (Clarendon Laboratory Archive, Oxford Physics). He served under the Royal Engineers and was killed on the Gallipoli Peninsula on August 10, 1915 at the age of 27 years. It is said that World War I was an unjustified war solely on the basis of the loss of Henry Moseley. As evidenced from his work as a budding scientist, we can only imagine the additional contributions and advancements to science that he could have made, had he lived. He had accomplished more in about 3 years of research than most could accomplish in a lifetime, and his findings were worthy of the Nobel Prize. The significance of Moseley's work was underscored by the Chairman of the Nobel Committee on December 10, 1925 (Gullstrand, 1925) when he stated

> He [Moseley] discovered the simple mathematical law by means of which the frequencies determined by the position of the lines—and consequently the corresponding wavelengths—can be obtained by what is known as the atomic number, i.e. the number of the element in a series in which all the elements are arranged with a generally increasing atomic weight. As the atomic number has proved to distinguish the elements better than the atomic weight, it has now attained the very greatest importance for atomic physics of the present day. Moseley fell at the Dardanelles before he could be awarded the prize, but his researches had directed attention to the merits of Barkla, who consequently in 1918 was proposed for the Nobel Prize, which was awarded to him without delay.

The Museum of the History of Science in Oxford has some of Moseley's original experimental apparatus. The original graph, artistically approximated by Figure III.8, published by Moseley in the *Philosophical Magazine* in 1914 is in the Moseley Room of the Clarendon Laboratory.

CHARLES GLOVER BARKLA (1877–1944)

Charles Glover Barkla was born in Widness, Lancashire, England on June 7, 1877. He was awarded the Nobel Prize in Physics 1917 "for his discovery of the characteristic Röntgen radiation of the elements".

Charles Glover Barkla (1877-1944)

Charles Barkla was educated at the Liverpool Institute and he received his university degree in Physics in 1898 with First Class Honors at the University College Liverpool followed by a Masters degree the following year. In 1899 he received a scholarship, which enabled him to work in the Cavendish laboratory with J. J. Thomson at Trinity College, Cambridge. In 1900 Barkla moved to King's College, and in 1902 he returned to the University College Liverpool. He was promoted at the University in Liverpool during 1905–1909 from demonstrator, assistant lecturer in physics, and special lecturer in advanced electricity. In 1909 he became Wheatstone Professor of Physics at the University of London. In 1913, Barkla was awarded the Chair in Natural Philosophy at the University of Edinburgh, which he occupied for the remainder of his life.

Barkla devoted most of his scientific career to the study of x-rays. Among his many contributions is the finding that when x-rays (e.g., from an x-ray tube) traverse matter of any kind, the matter becomes a source of x-radiation similar to the primary radiation used to bombard the matter. He found that this secondary radiation, also referred to as scattered x-radiation, displayed a definite degree of polarization with respect to the primary x-rays originating from an x-ray tube. He thus demonstrated that x-rays consist of oscillating waves in different directions and these could be separated or polarized. Thus x-rays had properties similar to light.

His most significant contribution was the discovery that each element has it's own line spectrum in the x-ray region. That is, when any substance, is irradiated by x-rays, it emits x-radiation characteristic of the elements in the substance. He found that, upon bombardment by an x-ray beam, the line spectra or x-ray energies emitted by any element are characteristic of that element irrespective of the chemical composition of the substance within which the element may be found. This formed the basis for the analytical technique known as x-ray fluorescence. In his Nobel Lecture presented on June 3, 1920 Barkla explained his findings as follows:

Each element when traversed by x-rays emits x-radiations characteristic of the element; each characteristic radiation is unaffected by changes in the physical condition or state of chemical combination of the radiating element, and its quality is independent of that of the exciting primary radiation. But only primary radiations of shorter wavelength are able to excite the characteristic x-radiations... all of the radiations hitherto definitely observed have fallen into three series, the K-, L-, and M-series (the M-series was discovered by Siegbahn and his collaborators)

The K-, L-, and M-series referred to by Barkla are x-ray lines emitted by atoms of elements as a result of atomic electron energy transitions from outer to inner shells. Electron transitions to the K shell belong to the K series, transitions to the L shell belong to the L series, etc. Barkla's findings provided important information regarding the concept of the atomic structure as it provided evidence for electron energy levels. Professor G. Granqvist, Chairman of the Nobel Committee, noted in his presentation speech

Barkla's discovery, that two domains of differing hardness are to be differentiated in the characteristic radiation, is of fundamental importance as regards the modern conception of the structure of atoms. Barkla has named the two domains the K-series and the L-series respectively. Thus every chemical element, when irradiated by x-rays, emits two rays of different penetrability, that is to say, every element can by fluorescence emit an x-ray spectrum of two lines or line-groups, the so-called K-series and L-series. Of these the K-series have the greater penetrability.

In addition to grasping the x-ray fluorescence data to provide information concerning energy levels of electrons, Barkla went further to obtain experimental data on x-ray absorption to obtain information on the number of electrons per atom. Firstly he determined that the secondary radiation resulting from x-ray bombardment of matter consisted of two types, namely, electrons and x-rays of the K- and L-series. X-rays of the primary radiation were colliding with atomic electrons and liberating them from the atom by mechanisms such as the photoelectric effect. This caused electron vacancies in the atoms, which would be filled by electrons from outer shells to yield the secondary K- and L-series x-rays. He then utilized the following equation derived by J. J. Thomson which describes the fraction (f) of a primary x-ray beam lost by scattering per centimeter of substance traversed to determine the number of electrons per atom of element bombarded by x-rays:

$$f = \frac{8\pi}{3} N \frac{e^4}{m^2} \mu^2 \tag{III.16}$$

where N is the number per cubic centimeter of particles of charge e and mass m. With precise measurements of N, e/m, and e, the calculated values for the number of electrons per atom were as noted in his Nobel Lecture

> one electron per atom of hydrogen, 6 per atom of carbon, 7 per atom of nitrogen, 8 for oxygen, 15 or 16 for sulfur, and so on. As these conclusions regarding the number of electrons (outer electrons) within the atom have been confirmed by the researches of Rutherford, Bohr and Moseley, it is perfectly legitimate to use the agreement as evidence in support of the theory of radiation upon which it was based (Nobel Lectures, Physics, 1901–1921).

Barkla died at Braidwood, Edinburgh on October 23, 1944 the year after the death of his youngest son Flight Lieutenant Michael Barkla, also a brilliant scholar, who was killed in action in World War II.

MANNE SIEGBAHN (1886–1978)

Karl Manne Georg Siegbahn was born in Örebro, Sweden on December 3, 1886. He was awarded the Nobel Prize in Physics 1924 "for his discoveries and research in the field of x-ray spectroscopy".

Siegbahn (1886-1978)

Manne Siegbahn received his doctorate in physics at the University of Lund in 1911. During his years as a graduate student he served as assistant to Johannes (Janne) Robert Rydberg in the Physics

Institute of Lund University. After receiving his doctorate degree he was appointed lecturer and in 1915 Deputy Professor of Physics. Following the death of Rydberg in 1919 Manne Siegbahn was appointed full professor at Lund in 1920. He became Professor of Physics at Uppsala University in 1923 and Research Professor of Experimental Physics at the Royal Swedish Academy of Sciences in 1937. During 1937–1964 he served as Director of the Nobel Institute of Physics in Stockholm.

Siegbahn's major discoveries were obtained from research in the field of x-ray spectroscopy during the years 1912 to 1937. Research on x-radiation at this time, in addition to applications in medicine, were focused in two directions, namely, (i) the use of x-rays to research the structures of crystalline lattices as pioneered by Sir William and Sir Lawrence Bragg, and (ii) spectral investigation of the x-radiation itself pioneered by Manne Siegbahn, which would provide much information on the electronic structure of the atom. He firstly made many improvements that would facilitate the study of the x-ray spectral lines including the design of improved vacuum pumps and x-ray tubes that yielded more intense x-radiation together with the construction of improved crystal and linear gratings, which provided increased accuracy of measuring x-ray spectral frequencies or wavelengths. Siegbahn thus brought x-ray spectroscopy to a new level, which would provide new information on the electron within the structure of the atom, and the characteristic x-ray emissions of the elements. The contributions of Siegbahn were highlighted in the Nobel Presentation Speech of Professor A. Gullstrand, Chairman of the Nobel Committee for Physics on December 10, 1925 (Gullstrand, 1925) when he stated

> It had already become clear that the x-radiation must arise in the inner parts of the atoms, and that consequently exact spectroscopic investigations form the only means for an experimental research of those parts. Clearly perceiving this fact, Siegbahn has in the course of ten years' assiduous and systematic labor devised a series of improvements and new designs dealing with almost every detail of the various apparatus and so constantly increased the exactitude of his measurements…The high level to which he has brought x-ray spectroscopy can perhaps be best defined by the statement that the exactitude with which wavelengths can now be measured by his methods is a thousand times greater than that attained by Moseley. It was only to be expected that these more accurate means would in his hands be used for a series of new discoveries. Thus to begin with, he has found a large number of new lines in the K- and L-series. Moreover he has made the experimental discovery of a new characteristic x-radiation, the M-series; and another such radiation, the N-series…

Moseley had originally identified two K lines and four L lines (see Moseley in this Part III of the Radioactivity Hall of Fame). Manne Siegbahn 10 years later identified many new lines of x-ray frequencies or energy quanta within the different series for the known 92 elements. He subjected 42 elements to a fresh investigation of the K series and identified four lines of the K series for 27 elements and even eight fainter lines for the lighter elements. Of the L series he identified 28 lines for some 50 elements. A new M series was discovered by Siegbahn for 16 elements and another new N series was measured for three of the heaviest elements, which included five lines belonging to uranium and thorium.

Manne Siegbahn's work provided much insight into the role of the electron in the structure of the atom. The wavelengths or frequencies of the various lines reflected the energy quanta emitted by the x-rays and consequently the energy states of the electrons in the K-, L-, M-, and N-orbitals of the elements. The impact of the findings of Siegbahn on our knowledge of the electron configurations in the atom was emphasized in the Nobel Prize Presentation Speech (Gullstrand, 1925) with the statement

> In an element that can emit both K- and L-rays, the former radiation has much shorter wavelengths and consequently greater frequencies than the latter. As the energy quanta are proportional to the frequencies, therefore, the K-radiation involves a larger change in the energy of the atom than the L-radiation; and in the atomic theory this is as much as to say that an orbit into which an electron falls on emission of a K-line must lie nearer the nucleus than an orbit to which an electron falls on emission of an L-line. In this way it was inferred that there is a K-level nearest the nucleus, outside that and L-level, and after that an M-level and an N-level, all these four being experimentally determined [by Manne Siegbahn].

In his Nobel Lecture Manne Siegbahn explained the various lines of x-ray emission by the elements as electron rearrangements within the atom. He stated

Let us assume that at a given moment a certain atom is in such state that its total energy has the value E_t. At that moment a rearrangement occurs within the atom which has the effect of reducing the total energy to E_0. In the course of the process of rearrangement the quantity of energy liberated is therefore

$$hv = E_t - E_0 \qquad (\text{III.17})$$

According to the Einstein–Bohr formula there is thus a relationship between the frequency v_1 for the wave and the energy liberated from the atom

$$v_1 = \frac{E_1 - E_0}{h} \qquad (\text{III.18})$$

where h is Planck's constant. We can therefore imagine that the energy of the atom during [another] rearrangement of this nature, is then changed from E_2 to E_0 during which a wave frequency v_2 is emitted. Thereby

$$v_2 = \frac{E_2 - E_0}{h} \qquad (\text{III.19})$$

The series can be continued further:

$$v_3 = \frac{E_3 - E_0}{h}, \text{ and so on.} \qquad (\text{III.20})$$

By measuring the wavelengths it is possible experimentally to determine a series of v-values: v_1, v_2, v_3,... It would, however, require a special and not always simple analysis of the wavelengths available in order to select and bring together those frequency values which belong to one and the same series.

Manne Siegbahn measured x-ray spectra lines for the K-, L-, M-, N-series for the 92 elements and, even for the heaviest elements, the O-, and P-series (x-ray lines from electron transitions to the O- and P-orbitals). He confirmed that x-ray emissions from atoms are the result of electron transitions in atomic orbitals. In his Nobel Lecture he explained the mechanism of x-ray emission by the elements and the relevance of these findings to our understanding of the atom as follows

For the x-ray spectra, too, we must assume that light emission takes place when an electron, but in this instance one belonging to the inner electron system of the atom, moves from one quantum orbit to another while the state of all the remaining electrons is not altered to any significant extent...x-ray spectroscopy measurements provide us with a quantitative and thorough knowledge of the energy content and energy relationships within each particular atom. Any further work on the structure of the atom must rest upon this firm, empirical foundation.

Later in life, as Director of the Nobel Institute of Physics in Stockholm (1937–1962), Manne Siegbahn oversaw much scientific research toward problems in nuclear physics, and he lectured extensively in the most prestigious scientific research institutes of Europe, United States, and Canada. His son Kai Siegbahn (b.1918) followed him in his footsteps by occupying the Chair of the Physics Department of Uppsala University (since 1954), and by winning the Nobel Prize in Physics 1981.
Manne Siegbahn died in Stockholm, Sweden on September 26, 1978.

ROBERT A. MILLIKAN (1868–1953)

Robert Andrews Millikan was born in Morrison, Illinois, USA on March 22, 1868. He was awarded the Nobel Prize in Physics 1923 "for his work on the elementary charge of electricity and on the photo-electric effect".

Milliken (1868-1963)

Millikan was a descendant of New England settlers, who had come to America before 1750 and who were pioneer settlers in the Middle West. He was brought up in a rural environment attending Maquoketa High School in Iowa. In 1891, he graduated from Oberlin College in Ohio with major studies in Greek and Mathematics. At Oberlin he taught himself physics, as he was disappointed with the content or scope of the short physics course that the college offered at the time. After graduation Millikan accepted a teaching post in physics for 2 years mostly for the need of money, and during this time he developed an interest in physics, which would become his lifelong passion. Millikan continued his education at Columbia University, and was appointed a Fellow in Physics after receiving the Masters degree at the university. He was awarded the Ph.D. degree in physics at Columbia in 1895. The following year Millikan studied at the Universities of Berlin and Göttingen, as it was popular and thought advisable then for Americans to study advanced physics in Germany. Upon invitation of A. A. Michelson he returned to take on a position as a teaching and research assistant at the University of Chicago in 1896. He taught class with a passion becoming an eminent educator and author of several textbooks on physics for both high school and college students. As noted by the American Institute of Physics Millikan became only Associate Professor at the age of 38 when, on the average, an American physicist with published research would become professor at the age of 32. He later recalled

Although I had for ten years spent on research every hour I could spare from my other pressing duties, by 1906 I knew that I had not yet published results of outstanding importance, and certainly had not attained a position of much distinction as a research physicist.

Millikan thus decided to curtail his writing of textbooks and make one great effort to make his mark in physics by accurately measuring the charge carried by an electron, which he successfully accomplished by 1910 (Millikan, 1911) with greater precision two years later (Millikan, 1913)

demonstrating once and for all the elementary particulate nature of the electron and electricity. He was awarded full professorship at the University of Chicago shortly after his first publication on the elementary electrical charge. He subsequently went to work earnestly to disprove Einstein's concept of the photon as an elementary particle in the photoelectric effect, and after 10 years of effort by 1921 (Millikan, 1914; 1916a,b; 1921) all of his research in this direction proved to be a great contribution to physics, because it gave more credence to Einstein's light quantum or photon. Millikan was thus awarded the Nobel Prize in Physics 1923 for his dual contribution on his measurement of the elementary charge of electricity and the photoelectric effect.

The significance of Millikan's work on the charge of the electron was highlighted by Professor A. Gullstrand, Chairman of the Nobel Committee (Gullstrand, 1923) when he stated

> Millikan's aim was to prove that electricity really has the atomic structure, which, on the base of theoretical evidence, it was supposed to have. To prove this it was necessary to ascertain, not only that electricity, from whatever source it may come, always appears as a unit of charge or as an exact multiple of units, but also that the unit is not a statistical mean, as, for instance, has of late been shown to be the case with atomic weights. In other words it was necessary to measure the charge of a single ion with such a degree of accuracy as would enable him to ascertain that this charge is always the same, and it was necessary to furnish the same proofs in the case of free electrons. By a brilliant method of investigation and by extraordinary exact experimental technique Millikan reached his goal.

To prove the nature and charge of the electron Millikan devised his famous oil-drop experiment. He devised the apparatus, illustrated in Figure III.9, which contains two horizontal metal plates separated by a short distance and joined only by an electric field of about 6000 V/cm. The air between the plates was ionized by means of x-rays or suitable radioactive source.

The positively charged upper plate contained a pin-hole in its' very center over which an atomizer was arranged to spray minute oil droplets with a radius of approximately 1/1000 of a millimeter. Eventually one of the oil droplets would fall through the pin-hole and enter the region between the plates. The space between the plates was illuminated so that Millikan could observe the oil droplet with a telescope. The oil drop would appear in the telescope as a bright star with a dark background.

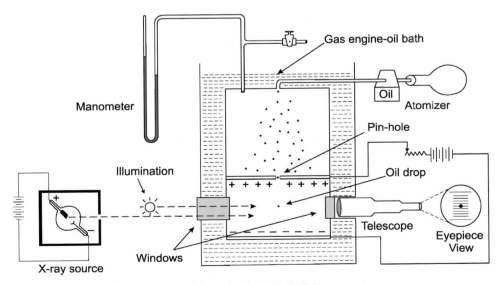

Figure III.9   Schematic of Millikan's oil-drop apparatus.

The eyepiece of the telescope contained crosshairs so that Millikan could measure the time that the droplet needed to pass between each cross hair and thereby measure the velocity of fall of the oil drop, which would be only a fraction of a millimeter per second. The oil droplet would be charged with electricity from the friction produced during the droplet formation by the atomizer. When the drop would fall and approach the lower plate Millikan would switch on the electric field so that the drop would rise through the attractive forces of the positively charged upper plate. The velocity of rise of the droplet was measured with the telescope. After its rise, the charge to the plates would be turned off, and the oil drop would begin to fall again. The fall and rise of the drop could be repeated many times during a period of several hours by simply connecting or short-circuiting the charge to the metal plates, and each time Millikan could repeat his measurements of it's velocity of fall or velocity of rise by use of a stopwatch or chronometer. The velocity of fall of the oil droplet was always constant, but it's velocity of rise would vary and increase according to the amount of charge that the droplet had acquired from the ionized air between the plates. x-rays directed into the space between the plates would produce ionized air, and the oil droplet would rise at a velocity that would increase proportionally to the charge it had acquired. The droplet would increase in velocity upwards by a certain value or an exact multiple of that value. Thus Millikan could demonstrate that the oil droplet had acquired one or more units of electrical charge, each charge of exact equal strength. He was thus able to measure the charge of a single ion with a precision of one in a thousand. Some experimental data presented by Millikan in his Nobel Lecture on May 23, 1924, provided in Table III.1, illustrate how exact were his measurements of the multiples of velocity of rise.

With reference to his data provided in Table III.1 Millikan commented

> ...speeds exactly two times, three times, four times, five times, etc. (always within the limits of observational error—still less than a percent) could be communicated to the droplet, but never any fraction of these speeds. He who has seen that experiment, and hundreds of investigators have observed it, has literally seen the electron. For he has measured (in terms of speed) the smallest of the electrical forces, which a given electrical field ever exerts...by its aid [the experiment] he can count the number of electrons in a given small electrical charge with exactly as much certainty as he can attain in counting his fingers and his toes....but the electron itself, which man has measured, as in the case shown in the table, is neither an uncertainty nor an hypothesis. It is a new experimental fact that this generation in which we live has for the first time seen, but which anyone who wills may henceforth see.

Table III.1

Velocities of rise (electric field on) and fall (electric field off) of oil droplets

Time of fall 1.303 cm under gravity (sec)	Time of rise 1.303 cm in field (sec)	Mean times of rise in field (sec)	Divisors for speeds due to field	The electron in terms of a speed
120.8	26.2			
121.0	11.9			
121.2	16.5	67.73	1	3.007
120.1	16.3	26.40	2	3.009
120.2	26.4	16.50	3	2.993
119.8	67.4	11.90	4	3.008
120.1	26.6			
–	16.6			
120.2	16.6	Mean time of fall under gravity = 120.35		
–	16.4			
120.2	68.0			
119.9	67.8			
–	26.4			

Source: From Millikan (1924) with permission from The Nobel Foundation ©1924.

Millikan applied Stokes' law and the oil-drop velocity of free-fall under gravity and oil-drop velocity of rise in an electric field to calculate the exact charge on the electron in terms of electrostatic units, which can be converted to coulombs the SI unit for electric charge. Stokes' law, which was discovered by George Stokes (1819–1903), predicts the drag or frictional force D that is exerted on a spherical ball moving through a viscous medium. It is defined by the equation

$$D = 6\pi a\eta v \tag{III.21}$$

where variable a is the radius of the sphere or droplet, η the viscosity of the medium, and v the velocity of the ball or droplet. Millikan had to apply a correction factor to the equation, because of the very small radius of the oil droplets he created and the fact that the droplet would fall or rise is an almost frictionless evacuated chamber. He then defined the forces acting on the oil droplet during its free-fall when the electric field was turned off and during its rise when the electric field was turned on. These forces are illustrated in Figure III.10, where b according to Stoke's law is the drag force proportional to the velocity of the droplet, m the droplet mass, g the acceleration due to gravity, q_n the charge on the droplet, and E the electric field strength.

When the electric field of Millikan's apparatus is turned off, the oil drop will fall with the velocity, v_f, under the force of gravity, mg, against the drag force defined by Stoke's law $D = bv_f$; and when the electric field is turned on, the oil droplet would rise with the velocity v_r, proportional to the electric charge on the droplet, q_n and the electric field strength, E, and against the gravitational and drag forces. Millikan derived the equation for calculating the charge on the electron by defining the velocities of fall and rise of the droplet in the absence and presence of an electric field according to the following:

(a)   With the electric field off, the oil drop is static when

$$bv_f = mg \tag{III.22}$$

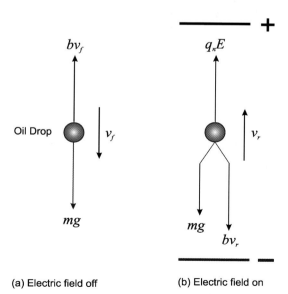

(a) Electric field off   (b) Electric field on

Figure III.10   Forces defining (a) the velocity of free-fall, v_f, of the oil drop in the absence of an electric field, and (b) the velocity if rise, v_r, of the oil drop in the presence of an electric field.

and the drop will fall with a velocity

$$v_f = \frac{mg}{b} \qquad \text{(III.23)}$$

(b)   With the electric field on, the oil drop remains static when

$$q_n E = mg + bv_r \qquad \text{(III.24)}$$

and the drop rises with the velocity

$$v_r = \frac{(q_n E - mg)}{b} \qquad \text{(III.25)}$$

Millikan solved eqs. (III.23) and (III.25) to provide the working equations defining the charge on the electron as

$$q_n = \frac{mg}{Ev_f}(v_f + v_r) \qquad \text{(III.26)}$$

or

$$q_n = \frac{mgt_f}{E}\left(\frac{1}{t_f} + \frac{1}{t_r}\right) \qquad \text{(III.27)}$$

where q_n is the unit charge on the electron, and t_f and t_r are the respective times required for the oil drop to fall under gravity and to rise under the electric field E. If the electron acquires more than one unit of charge in the ionized air of Millikan's apparatus, the oil drop would rise at a higher velocity while its rate of fall in the absence of the electric field always remained constant. Therefore, Millikan defined the equation for the charge on the electron that rose at the faster rate, due to additional units of charges acquired, as

$$q_n = \frac{mgt_f}{E}\left(\frac{1}{t_f} + \frac{1}{t_r'}\right) \qquad \text{(III.28)}$$

The above eq. (III.28) would yield various values for the charge on the electron that would be successive multiples of the unit charge on the electron as defined by eq. (III.27). This was explained by Millikan in his classic paper in *The Physical Review* (Millikan, 1913) as follows:

Now equations 11 and 12 (eq. III.27) and (III.28) above) show, since mgt_f/E remains constant, that, as the drop changes charge, the successive values of its charge are proportional to the successive values assumed by the quantity $(1/t_f + 1/t_r)$ and the elementary charge itself is obviously this same constant factor mgt_f/E multiplied by the *greatest common divisor* of all these successive values.

From the experimental data measured from 23 oil drops Millikan calculated the electron charge, q, which he represented with the symbol e as follows:

$$e = 4.774 \pm 0.009 \times 10^{-10} \text{ electrostatic units,}$$

which is expressed currently in the SI unit of coulombs (C) as

$$e = 1.602 \times 10^{-19} \text{C}$$

As Millikan was able to provide a most precise measurement of the electron charge, he was also able to report (Millikan, 1913) new and more precise calculated values of several other physical constants including the Avogadro constant (N_A), which he recalculated from the Faraday constant ($F = N_A e$) and Planck's constant (h).

Millikan could not accept at first Einstein's interpretation of the photoelectric effect, that is, Einstein's concept of the photon as an elementary particle or "quantum of light" that could collide with an atomic electron in a substance, such as a metal surface, and thereby transfer its entire energy to the electron to cause the electron to be emitted as a free electron with the same energy as the photon (less the small binding energy of the electron in the atom from which it came). The photoelectric effect was a proven phenomenon (See Lenard, Radioactivity Hall of Fame—Part I); however, for over a decade Millikan could not come to terms with Einstein's concept of the corpuscular nature of light. He thus devoted about 10 years of research to disprove Einstein's concept. In his Nobel Lecture Millikan commented

> This…semicorpuscular concepton of localized radiant energy was taken up in 1905 by Einstein who, by combining it with the facts of quanta discovered by Planck through his analysis of black-body radiation, obtained an equation which should govern, from his viewpoint, the interchange of energy between waves and electrons, viz. $\frac{1}{2}mv^2 = h\nu - A$ (Einstein equation), the first term representing the energy with which the electron escapes, the second term Planck's energy quantum for the particular light employed, and the last the work necessary to get the electron out of the metal. After 10 years of testing and changing and learning and sometimes blundering, all efforts being directed from the first toward the accurate experimental measurement of the energies of emission of photoelectrons, now as a function of temperature, now of wavelength, now of material, this work resulted, contrary to my own expectation, in the first direct experimental proof in 1914 of the exact validity, within narrow limits of experimental error, of the Einstein equation, and the first direct photoelectric determination of Planck's h…In view of all these methods and experiments the general validity of Einstein's equation is, I think, now universally conceded, and *to that extent the reality of Einstein's light-quanta may be considered as experimentally established.* But the conception of localized light-quanta out of which Einstein got his equation must still be regarded as far from being established.

From the last sentence of Millikan's statement taken from his Nobel Lecture given on May 23, 1924, he still could not accept Einstein's concept of the photon, even after proving many times over the validity of Einstein's equation and Planck's constant. However, in his autobiography published in 1950, Millikan admitted to eventually yielding to universal thinking and accepted Einstein's conception of the photon, as a chapter in his autobiography (Millikan, 1950) is titled "The Experimental Proof of the Existence of the Photon—Einstein's Photoelectric Equation".

Because of the extensive experimental research that Millikan put into the photoelectric effect the Nobel Committee decided to award the Nobel Prize in Physics 1921 to Albert Einstein in the words of the Nobel Committee "especially for his discovery of the law of the photoelectric effect". The Nobel Committee acknowledged that Millikan's extensive experimental research on the photoelectric effect convinced them to award the prize to Einstein. The was clearly admitted when Professor A. Gulstrand, Chairman of the Nobel Committee, stated in the Nobel Presentation Speech of 1923

> In justifying the reward of Millikan the Academy has not omitted to refer also to his investigations of the photoelectric effect. Without going into details I will only state that, if these researches of Millikan had given a different result, the law of Einstein would have been without value, and the theory of Bohr without support. After Millikan's results both (Einstein and Bohr) were awarded a Nobel Prize for Physics last year.

Millikan remained as Professor at the University of Chicago until 1921. In 1921 he was appointed Director of the Norman Bridge Laboratory of Physics at the California Institute of Technology (Caltech), Pasadena and made Chairman of the Executive Council of the Institute, a post he held until his retirement in 1946 (Nobel Lectures, Physics, 1922–1941, 1965). It was George Ellery Hale (1868–1938) the famed astronomer and astrophysicist of the time, who lured Millikan to join him at the new California Institute of Technology. Under Millikan's guidance, Caltech soon became one of the top American centers of research. At Caltech Millikan pioneered research on the phenomena, which he named "cosmic rays". He designed many small electroscopes and Geiger-Müller detectors of cosmic radiation that could be taken on aircraft and high-altitude balloons, and contributed much to our knowledge of the latitude effect of cosmic radiation at high altitudes (Millikan and Bowen, 1926; Millikan and Cameron, 1928; Bowen *et al.* (1934, 1936, 1937, see also Chapter 6); Millikan and Neher, 1935;). He stubbornly insisted that cosmic rays consisted in whole or in large part of electromagnetic radiation (American Institute of Physics), contrary to the belief of others, such as Arthur Compton, who considered cosmic rays to consist mostly of charged particles. Even to this day some tables of electromagnetic radiation wavelengths and frequencies still equate cosmic rays with gamma rays. (See Chapter 6 for information on cosmic rays.)

Millikan died on December 19, 1953 in San Marino, CA. During the early 1980s his reputation was tarnished with accusations of having swindled his graduate student, Harvey Fletcher (1884–1981) out of credit for the oil-drop experiment (*ScienceWeek*, 2001). They possibly could have shared the Nobel Prize. Harvey Fletcher wrote a manuscript autobiography that included an account of his work in the famous oil-drop experiment for which his thesis advisor, Robert A. Millikan, won the Nobel Prize in 1923. Fletcher wrote the manuscript with instructions that it be published only posthumously, so there would be no doubt that Fletcher had no personal interest motivating its publication. Harvey Fletcher's account of his work with Millikan on the oil-drop experiment is published in *Physics Today* (Fletcher, 1982). Fletcher accounts that after Millikan assigned him the thesis work of finding a liquid substance, other than water, that could be used to determine the charge of the electron (*e*), Millikan went immediately to design a crude oil-drop apparatus. In his autobiography Fletcher relates the following:

> I went immediately to find Millikan [after having set up the initial crude oil-drop apparatus], but could not find him so I spent the rest of the day playing with these oil droplets and got a fairly reasonable value of *e* before the day ended. The next day I found him. He was very much surprised to learn that I had a setup that was working. He came down to the laboratory and looked through the telescope and saw the same beautiful sight of the starlets jumping around that I had already seen and have described above. He was very much excited, especially after turning on the field. After watching for some time he was sure we could get an accurate value of *e* by this method. He stopped working with Begeman [another graduate student with whom Millikan was working on the possibility of utilizing water droplets and an expansion cloud chamber to determine *e*] and started to work with me. We were together nearly every afternoon for the next two years…My name ran right along with Professor Millikan's in the newspapers…This was all great publicity, but I began to wonder if this work was to be my thesis as Millikan had promised at that first conference in December 1909. However, during the spring of 1910 we started together writing a paper to be published about this new research. I wrote more of it than he did, particularly about the modification of Stoke's law and the arrangements of the data…All the time I thought we were to be joint authors.

A defense of Millikan was written in *American Scientist* (Goldstein, 2001). A quotation from the defense article is taken from *ScienceWeek* February 23, 2001 "No doubt Millikan understood that the measurement of [electron-charge] would establish his reputation, and he wanted full credit. Fletcher understood this too, and he was somewhat disappointed, but Millikan had been his protector and champion throughout his graduate career, so he had little choice but to accept".

– 3 –

Gamma- and X-Radiation—Photons

3.1 INTRODUCTION

Gamma radiation is electromagnetic radiation that is emitted by an unstable nucleus of an atom during radioactive decay. A nucleus in an unstable state may fall to a more stable state by the emission of energy as gamma radiation. The radiation has a dual nature, that of a wave and a particle with zero mass at rest. It was Albert Einstein who first demonstrated the dual nature of electromagnetic radiation when he explained the photoelectric effect. He thus discovered the elementary particle known as the photon, which is electromagnetic radiation with its particulate nature. Another origin of gamma radiation is via the phenomenon referred to as annihilation whereby a positron coming to rest encounters an electron, its antiparticle, and the two particles are annihilated. Their annihilation results in the conversion of their mass into energy as gamma radiation. The amount of energy produced by this process is equivalent to the mass of the two electrons annihilated according to Einstein's equation of equivalence of mass and energy ($E = mc^2$). x-radiation is electromagnetic radiation with a dual nature of wave and particle, like gamma radiation; however, x-radiation may originate from electron energy level transitions in an unstable atom. The transition of electrons from higher to lower energy states may result in the emission of x-radiation. Also x-radiation may be emitted when a charged particle decelerates in a series of collisions with atomic particles. For example, when a high-energy electron or beta particle traveling through matter approaches a nucleus, the electron may be deflected and caused to decelerate with the emission of x-radiation referred to as Bremmstrahlung or "braking radiation". This chapter deals with the origins and properties of gamma- and x-radiation and their mechanisms of interaction with matter.

3.2 DUAL NATURE: WAVE AND PARTICLE

In the latter part of the 19th century, Heinrich Hertz carried out a series of experiments demonstrating that an oscillating electric current sends out electromagnetic waves similar to light waves, but of different wavelength. Hertz proved, thereby, the earlier theory of James Clerk Maxwell that electric current oscillations would create alternating electric and magnetic fields, and radiated electromagnetic waves would have the same physical properties of light.

A subsequent discovery by Pieter Zeeman in 1896 further linked the properties of light with electricity and magnetism when he discovered that a magnetic field would alter the frequency of light emitted by a glowing gas, known as the Zeeman effect (Serway *et al.*, 1997).

Not long after the discoveries of Hertz and Zeeman came the work of Max Planck, who in 1900 proposed a formula to explain that the vibrating particles in the heated walls of a kiln could radiate light only at certain energies. These energies would be defined by the product of a constant having the units of energy × time and the radiation frequency. The constant, which he calculated became known as the universal Planck's constant, $h = 6.626 \times 10^{-34}$ J sec. Therefore, radiation would be emitted at discrete energies, which were multiples of Planck's constant and the radiation frequency, ν. Planck named the discrete radiation energy as the quantum from the Latin *quantus* meaning "how great".

In 1905, Einstein grasped the calculations of Planck to explain and provide evidence that light not only traveled as waves but also existed as discrete packets of energy or particles, which he named "energy quanta". Today we refer to these energy quanta as photons. Einstein demonstrated the existence of the photon in his explanation of the photoelectric effect (see Section 3.7.2 of this chapter). It was known from the work of Philipp Lenard (1902) that the energy of an electron (photoelectron) ejected from its atomic orbital after being struck by light was not dependent on the light intensity, but rather on the wavelength or frequency of the light. In other words, increasing the light intensity would increase the number of photoelectrons, but not their energy. Whereas, altering the frequency, thus energy, of the light would alter the energy of the photoelectron. In summary, Einstein demonstrated that the energy of the photoelectron depended on the energy of the photon that collided with the electron or, the product of Planck's constant times the light frequency according to the formula

$$E = h\nu = \frac{hc}{\lambda} \tag{3.1}$$

Eq. (3.1) is referred to as the Planck–Einstein relation (Woan, 2000). Notice from eq. (3.1) that the product of the photon frequency, ν, and wavelength, λ, always yields the velocity, c, the speed of light. The photon always travels at a constant speed is vacuum, $c = 2.9979 \times 10^8$ m/sec; it cannot travel at a speed less than c in vacuum.

From our previous treatment we see that the photon behaves as a particle, which could knock out an electron from its atomic orbit provided it possessed sufficient energy to do so, that is, an energy in excess of the electron binding energy. Therefore, the photon can be considered also as another elementary particle. In his explanation of the photoelectric effect Einstein was the first to demonstrate the particulate nature of light, and it is for this work that he won the Nobel Prize. Since these findings of Einstein, electromagnetic radiation is known to have a dual nature as energy that travels as a wave and a particle.

Electromagnetic radiation may be classified according to its wavelength or origin. For example, we will see in this section of the chapter that gamma rays and x-rays are similar, but have different origins. Gamma rays arise from the nucleus of an atom, while x-rays come from extranuclear electrons. The classification of electromagnetic radiation according to wavelength and frequency is illustrated in Figure 3.1.

Since electromagnetic radiations or photons have properties of particles, they should also possess momentum. We calculate momentum as the product of mass and velocity.

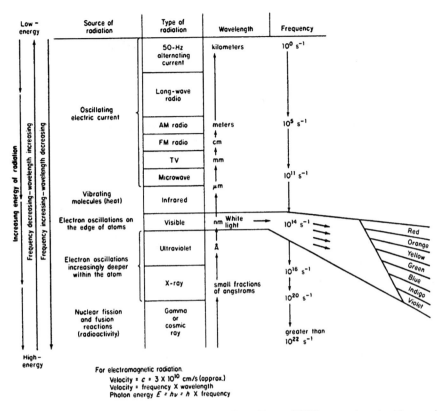

Figure 3.1   Electromagnetic radiation spectrum. (From Dean (1995), reproduced with permission from The McGraw-Hill Companies © 1995.)

For relativistic conditions, the mass of a particle is a function of its speed according to eq. (4.6) of Chapter 4 or

$$m = \frac{m_0}{\sqrt{1 - (u^2/c^2)}} \qquad (3.2)$$

where m and m_0 are the particle relativistic and rest masses, u the particle speed, and c the speed of light. Gautreau and Savin (1999) transform eq. (3.2) by squaring both sides and then multiplying each side by $c^2[1 - (u^2/c^2)]$ to yield the equation

$$m^2c^4 - m^2u^2c^2 = m_0^2c^4 \qquad (3.3)$$

Using $E = mc^2$ and $E_0 = m_0c^2$ to define the relativistic and rest energies and $p = mu$ to define the particle momentum together with the fact that the rest energy of the photon is always zero, that is, $m_0 = 0$, eq. (3.3) becomes

$$E^2 - p^2c^2 = (m_0c^2)^2 \qquad (3.4)$$

$$E^2 - p^2c^2 = 0 \tag{3.5}$$

and

$$p = \frac{E}{c} \tag{3.6}$$

From eqs. (3.1), (3.5), and (3.6) we can further describe the photon momentum as

$$p = \frac{E}{c} = \frac{h\nu}{c} = \frac{h}{\lambda} \tag{3.7}$$

To illustrate the use of the above equations defining the relationships of photon properties, let us calculate the wavelength, frequency, and momentum of a 2-MeV gamma-ray photon. From eq. (3.1) we can write the equation for calculating the wavelength as

$$\lambda = \frac{hc}{E} \tag{3.8}$$

Planck's constant, h, can be converted from units of J sec to eV sec as

$$h = \frac{6.626 \times 10^{-34} \text{ J sec}}{1.602 \times 10^{-19} \text{ J/eV}} \tag{3.9}$$
$$= 4.136 \times 10^{-15} \text{ eV sec}$$

and hc is calculated as

$$hc = (4.136 \times 10^{-15} \text{ eV sec})(2.9979 \times 10^8 \text{ m/sec})$$
$$= 12.399 \times 10^{-7} \text{ eV m} \tag{3.10}$$
$$= 12.4 \text{ keV Å}$$

The wavelength according to eq. (3.8) becomes

$$\lambda = \frac{12.4 \text{ keV Å}}{2 \times 10^3 \text{ keV}} = 0.0062 \text{ Å}$$

The frequency is calculated according to eq. (3.1) as

$$\nu = \frac{c}{\lambda} = \frac{2.9979 \times 10^8 \text{ m/sec}}{0.0062 \times 10^{-10} \text{ m}} = 484 \times 10^{18} \text{ sec}^{-1} = 4.84 \times 10^{20} \text{ Hz}$$

The momentum is expressed according to eq. (3.6) as

$$p = \frac{E}{c} = 2.0 \text{ MeV}/c$$

Notice that relativistic calculations of momentum have units of MeV/c, while conventional units of momentum are derived from mass times velocity or kg m/sec. Units of MeV/c can be converted to the conventional units with the conversion factor 1 MeV/c = 0.534 × 10^{-21} kg m/sec (Gautreau and Savin, 1999).

3.3 GAMMA RADIATION

Radionuclide decay processes often leave the product nuclide in an excited energy state. The product nuclide in such an excited state either falls directly to the ground state or descends in steps to lower energy states through the dissipation of energy as gamma radiation.

A nuclide in an excited energy state is referred to as a nuclear isomer, and the transition (or decay) from a higher to a lower energy state is referred to as isomeric transition. Gamma rays are emitted in discrete energies corresponding to the energy state transitions a nuclide may undergo when in an excited state. The energy, E_γ, of a gamma ray may be described as the difference in energy states of the nuclear isomers:

$$E_\gamma = h\nu = E_1 - E_2 \tag{3.11}$$

where $h\nu$ is the energy of the electromagnetic radiation described previously in Section 3.2, and E_1 and E_2 the energy levels of the nuclear isomers.

Let us consider the decay schemes of some radionuclides to illustrate the process in more detail.

Figure 3.2 shows the decay scheme of $^{86}_{37}$Rb with a half-life of 18.8 days. This nuclide decays by β^- emission with an increase in atomic number to $^{86}_{38}$Sr Eighty-eight percent of the beta particles emitted have a maximum energy of 1.77 MeV; the remaining 11% have a maximum energy of 0.70 MeV. The percentages cited and illustrated in the figure are referred to as transition probabilities or intensities. Obviously, a greater quantum of energy is released by the 1.77 MeV, beta-decay process. As a consequence, the ^{86}Sr product nuclides that result from β^- emission of 0.70 MeV (11%) are at a higher energy state than those that result from β^- emission of 1.77 MeV. The energy difference of the two ^{86}Sr product nuclide isomers, $E_1 - E_2$, is equivalent to the difference of the two β^- energies, 1.77 − 0.70 = 1.07 MeV. Consequently, the ^{86}Sr nuclide isomers, which are products of the 0.70-MeV beta-decay process, can emit the remaining energy as 1.07-MeV gamma-ray photons.

As illustrated in Figure 3.2, 11% of the parent ^{86}Rb nuclides decay to a ^{86}Sr nuclear isomer at an elevated energy state. Not all of these isomers immediately decay to the ground state. Only 8.8% of the ^{86}Rb → ^{86}Sr disintegrations result in the emission of a gamma-ray photon of 1.07 MeV. For example, a 37-kBq sample of ^{86}Rb by definition would emit 2.22 × 10^6 beta particles in 1 min (37,000 dps × 60 sec/min). However, only (2.22 × 10^6)(0.088) = 1.95 × 10^5 gamma-ray photons of 1.07 MeV can be expected to be emitted in 1min from this sample.

Figure 3.3 shows the somewhat more complicated decay scheme of $^{144}_{58}$Ce, which has a half-life of 284.5 days. This nuclide decays by β^- emission with an increase in atomic number to $^{144}_{59}$Pr. In this case, three distinct beta-decay processes produce three nuclear isomers of the daughter ^{144}Pr. Seventy-five percent of the beta particles emitted have a maximum energy of 0.31 MeV, 20% have a maximum energy of 0.18 MeV, and the remaining 5% have a maximum energy of 0.23 MeV. Obviously, a greater amount of energy is released by the

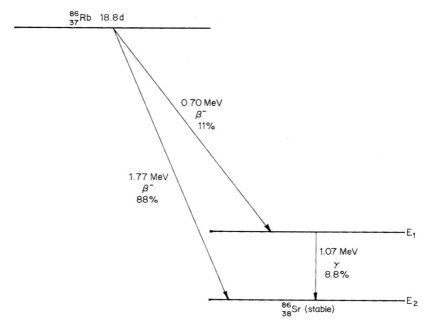

Figure 3.2   Decay scheme of $^{86}_{37}$ Rb.

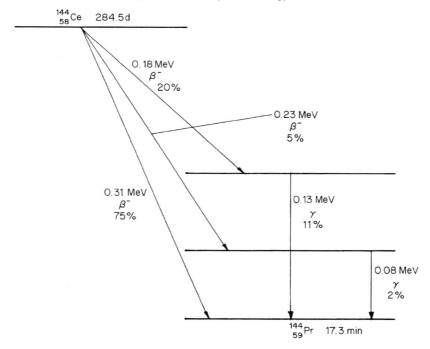

Figure 3.3   Decay scheme of $^{144}_{58}$ Ce.

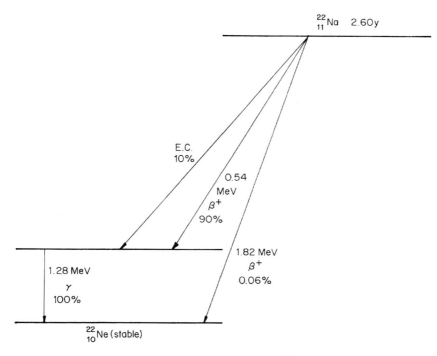

Figure 3.4   Decay scheme of $^{22}_{11}$Na.

0.31-MeV beta-decay process. As a consequence, ^{144}Pr nuclides that result from β^- emission of 0.23 MeV can decay to the ground state with the emission of gamma-ray photons with an energy equivalent to 0.08 MeV, that is, 0.31–0.23 MeV. Likewise, ^{144}Pr isomers at an even higher energy state are products of the 0.18-MeV beta-decay process. These can decay to the ground state with the emission of gamma-ray photons of energy 0.13 MeV, that is, 0.31–0.18 MeV. Not all of the product isomers decay with the immediate emission of gamma radiation, and the abundance of these transitions is given in Figure 3.3.

It is also possible that essentially all of the product nuclides of a decay reaction will be at an excited or elevated energy state and subsequently fall to a lower energy state by the emission of gamma radiation.

The decay scheme of the nuclide $^{22}_{11}$Na with a 2.6-year half-life serves as an example (see Figure 3.4). The $^{22}_{11}$Na nuclides decay by both electron capture and β^+ emission, at relative proportions of 10 and 90%, respectively, to yield immediate $^{22}_{10}$Ne product nuclides in an elevated energy state. Only a trace of the ^{22}Na nuclides (0.06%) decay directly to the ground state. All of the $^{22}_{10}$Ne isomers in the excited energy state decay immediately with the emission of gamma-ray photons of 1.28 MeV energy, which is equivalent to the difference of the energy levels of the two $^{22}_{10}$Ne isomers and also equivalent to the difference in energies released by the two β^+-decay processes (1.82–0.54 MeV).

Isomeric transition, as described earlier, is a decay process in which gamma emission is the sole process of eliminating energy from an excited nucleus. This mode of decay is referred to as isomeric transition because neither the mass number, A, nor the atomic number, Z, of a nuclide (^A_ZX) changes in the decay process, and the nuclides are considered to be in isomeric energy states.

In the previous examples (Figures 3.2–3.4) the isomeric energy state transitions are short-lived; that is, they occur virtually immediately after the other decay processes (e.g., β^-, β^+, and EC) and the half-life of the parent nuclide is dependent on these initial processes. If, however, the isomeric transitions are long-lived, the nuclide is considered to be in a metastable state. These nuclides are denoted by a superscript m beside the mass number of the nuclide. The radionuclide $^{119m}_{50}$Sn with a 250-day half-life is an example. Its decay scheme, shown in Figure 3.5, illustrates the emission of two gamma photons of 0.065 and 0.024 MeV energies falling from the 0.089-MeV excited state to the ground (stable) state.

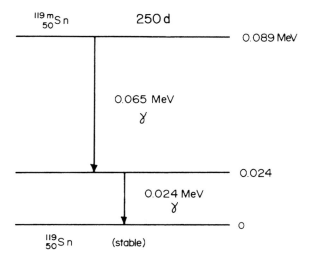

Figure 3.5   Decay scheme of $^{119m}_{50}$Sn.

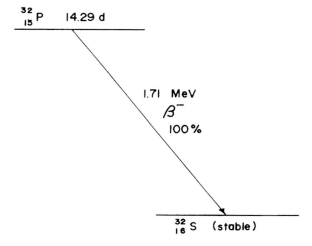

Figure 3.6   Decay scheme of $^{32}_{15}$P.

Gamma radiation is not produced in all radionuclide decay processes. Instead, some radionuclides decay by emitting only particulate radiation to yield a product nuclide at an unexcited ground state. An example is the commonly used radionuclide ^{32}P, which decays by the scheme shown in Figure 3.6.

3.4 ANNIHILATION RADIATION

The negatron or negative beta particle, produced by beta decay or by pair production (see Section 3.7.4 of this chapter), will travel through matter until it has completely dissipated its kinetic energy via ionization, electron excitation, or Bremsstrahlung. The negatron then at rest acts as an atomic or free electron in matter.

A positron or positive beta particle, however, may be considered an "antiparticle" of an electron and consequently, in the electron environment of atoms, has a definite instability. A given positron emitted by pair production or by β^+ decay will also dissipate its kinetic energy in matter via interactions described previously for the case of the negatron. However, as the positron loses its kinetic energy and comes to a near stop, it comes into contact with an electron (Figure 3.7) with nearly simultaneous annihilation of the positron and the electron masses and their conversion into energy. The annihilation involves the formation of positronium, which is a short-lived association of the positron and electron. Its lifetime is only approximately 10^{-10} or 10^{-7} sec, depending on whether the spin states of the associated particles are parallel (ortho-positronium) or opposed (para-positronium). The para-positronium is the shorter lived spin state. The energy released in this annihilation appears as two photons emitted in opposite directions. This transformation of mass into energy, considered as the reverse of pair production, is described as

$$e^+ + e^- = 2h\nu = 2E_\gamma \qquad (3.12)$$

where a positron, e^+, and an electron, e^-, combine to form two gamma-ray photons of energy E_γ. To maintain the equivalence of mass and energy (see eq. (3.28)), the equivalent of two

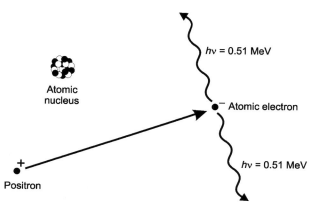

Figure 3.7   Annihilation. The interaction between a positron and an electron, and the conversion of their mass into two photons of 0.51 MeV energy.

electron rest masses (0.51 MeV) must appear as photon energies (see Section 3.7.4 of this chapter). In agreement with eq. (3.31), the annihilation results in the emission of two 0.51-MeV photons in opposite directions.

3.5 CHERENKOV RADIATION

Charged particles, when they possess sufficient energy, may travel through a transparent medium at a speed greater than the speed of light in that medium. This occurrence causes emission of photons of light. These photons extend over a spectrum of wavelengths from the UV into the visible portion of the electromagnetic radiation spectrum.

The photon emission is a result of a coherent disturbance of adjacent molecules in matter caused by the traveling charged particle, which must possess a certain threshold energy. This phenomenon has practical applications in the measurement and detection of radionuclides that emit relatively high-energy beta particles (L'Annunziata and Passo, 2002). The theory and applications of Cherenkov photons are discussed in detail in Chapter 7.

3.6 X-RADIATION

3.6.1 X-radiation following nuclide decay processes

Mention has been made of the electron capture decay process whereby an electron from one of the atomic shells (generally the innermost K shell) is absorbed by the nucleus, where it combines with a proton to form a neutron. No particle emission results from this decay process. However, the vacancy left by the electron from the K shell is filled by an electron from an outer shell (generally the adjacent L shell). Transitions produced in electron shell energy levels result in the emission of energy as x-radiation. This radiation consists of photons of electromagnetic radiation similar to gamma radiation. x-radiation and gamma-radiation differ in their origin. x-rays arise from atomic electron energy transitions and gamma rays from transitions between nuclei of different energy states. The production of x-radiation from atomic electron transitions is illustrated in Figures 3.8 and 5.1.

The electron transitions that ensue in the filling of vacancies are deexcitation processes, and the energy lost by the atom as x-radiation is equivalent to the difference of the electron energies of the outer or excited state, E_{outer}, and its new inner ground state, E_{inner}, as described by

$$h\nu = E_{outer} - E_{inner} \qquad (3.13)$$

The radiation emitted consists of a discrete line of energy characteristic of the electron shell and, consequently, of the atom from which it arises.

The production of x-rays in radionuclide decay is, however, more complex. The filling of one electron vacancy in an inner shell is followed by a series of electron transitions in an overall adjustment of electrons in outer shells. This gives rise to further x-rays with lines characteristic of outer shells. Such electron transitions, each resulting in the emission of discrete lines of characteristic x-rays, are illustrated in Figure 3.9. The transitions are identified by a letter corresponding to the shell (K, L, M, etc.) with vacancy giving rise to the x-ray

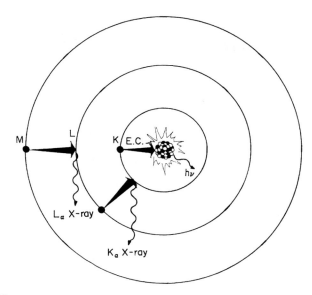

Figure 3.8 Electron capture (EC) decay and the accompanying gamma ($h\nu$) and x-radiation.

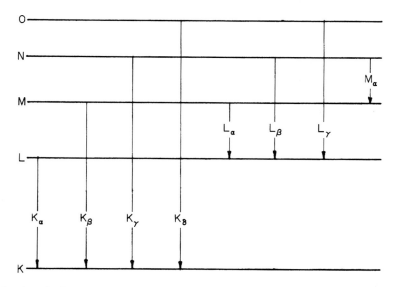

Figure 3.9 Atomic electron energy levels or shells (K, L, M, etc.) and lines of transition corresponding to characteristic x-rays (K_α, K_β, K_γ, etc.).

photon and a subscript (α, β, γ, etc.) to identify, from among a series of outer electron shells of the atom, the shell from which the electron vacancy is filled. For example, an x-ray arising from an electron transition from the L to the K shell is denoted as K_α and that arising from a transition from the M to the K shell as K_β. Transitions involving the filling of electron vacancies in the L shell from outer M, N, and O shells are denoted by L_α, L_β, and L_γ, etc.

Because x-radiation is characteristic of the atom from which it arises, it is customary to identify the element along with the x-ray photon (e.g., Cr K x-rays, Hg L x-rays, etc.). In these examples, the fine structure of the x-ray emissions is not given and the lines are grouped together as K and L x-rays.

The complexity of x-ray lines emitted and their abundances of emission are compounded by the existence of other mechanisms of x-ray production in unstable atoms. One of these mechanisms is the production of Auger electrons. An x-ray emitted from an atom may produce an Auger electron via an internal photoelectric effect (see Chapter 5), which results in the emission of an atomic electron from a shell farther away from the nucleus. The vacancy left by the Auger electron gives rise to additional x-rays characteristic of outer shells following the electron readjustments that ensue. Auger electrons can be emitted from a variety of electron shells, followed by an equal variety of characteristic x-rays from subsequent electron adjustments in outer shells.

Any process that would cause the ejection of an atomic electron of an inner shell can result in the production of x-radiation. Other processes not yet mentioned in this section that involve the ejection of atomic electrons are the emission of internal-conversion electrons (see Chapter 5) and radiation-induced ionization (see Section 3.7 of this chapter).

3.6.2 Bremsstrahlung from high-energy beta particles

Bremsstrahlung is electromagnetic radiation similar to x-radiation. It is emitted by a charged particle as it decelerates in a series of collisions with atomic particles. This mechanism is illustrated in Figure 3.10, where a beta particle traveling through matter approaches a nucleus and is deflected by it. This deflection causes a deceleration of the beta particle and consequently a reduction in its kinetic energy with the emission of energy as a photon of Bremsstrahlung or "braking radiation." The phenomenon is described by

$$h\nu = E_i - E_f \qquad (3.14)$$

where $h\nu$ is the energy of the photon of Bremsstrahlung, E_i the initial kinetic energy of the beta particle prior to collision or deflection, producing a final kinetic energy E_f of the electron. When beta particles from a particular radionuclide source strike an absorber material, a wide spectrum of Bremsstrahlung photon wavelengths (or energies) will be produced. The broad spectrum of Bremsstrahlung is due to the broad possibilities of different interactions, that is, deflections or collisions that the beta particles can have with atomic nuclei of the absorber and the broad spectrum of beta-particle energies emitted from any given radionuclide. In a given spectrum of Bremsstrahlung the shortest wavelength, λ_{min}, is observed when a beta particle or electron undergoes a direct collision with the nucleus of an atom and loses all of its kinetic energy, $h\nu_{max}$, as Bremsstrahlung or x-radiation according to the relation

$$h\nu_{max} = \frac{hc}{\lambda_{min}} \qquad (3.15)$$

which follows the energy–wavelength relation previously described by eq. 3.1.

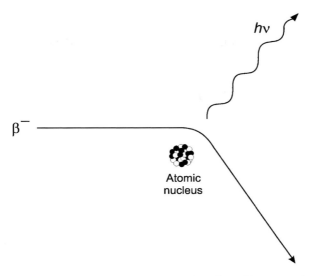

Figure 3.10   Bremsstrahlung production. A beta particle is deflected by an atomic nucleus and loses kinetic energy with the emission of a photon of x-radiation.

Let us consider an example of a 1710 keV beta particle from ^{32}P ($E_{\text{max}} = 1.71\,\text{MeV}$) striking a nucleus of Pb in a lead-glass shield. If the beta particle loses all of its energy in the collision, the wavelength of the Bremsstrahlung emitted from this interaction according to eq. (3.15) would be

$$\lambda = \frac{hc}{h\nu_{\text{max}}} = \frac{12.4\,\text{keV \AA}}{1710\,\text{keV}} = 0.00725\,\text{\AA}$$

See eq. (3.10) for the conversion of the constant hc to convenient units of eV m or eV Å. Bremsstrahlung production by high-energy beta particles in absorber material of high atomic number is significant (see Section 2.5.1 of Chapter 2). Consequently, to avoid the production of Bremsstrahlung in radiation shielding against the harmful effects of high-energy beta particles, an absorber of low atomic number (e.g., plastic) may be preferred over one of high atomic number (e.g., Pb-glass).

3.6.3   Inner bremsstrahlung

Bremsstrahlung of very low intensity also results from the transforming nucleus in electron capture decay processes (see Section 2.3.2 of Chapter 2). This is referred to as internal or inner Bremsstrahlung. Because a neutrino is emitted in these decay processes, the quantum of energy not carried away by the neutrino is emitted as internal Bremsstrahlung. Thus, in electron capture decay, internal Bremsstrahlung may possess energies between zero and the maximum, or transition energy of a radionuclide. When gamma radiation is also emitted, the internal Bremsstrahlung may be masked by the more intense gamma rays and go undetected. In such cases, internal Bremsstrahlung may be of insufficient intensity to lend itself to radionu-clide detection. However, in the absence of gamma radiation, the upper limit of the internal Bremsstrahlung can be used to determine the transition energy of a nuclide in electron capture decay. Some examples of radionuclides that decay by electron capture without the emission of gamma radiation are as follows:

$$^{55}_{26}\text{Fe} \rightarrow {}^{55}_{25}\text{Mn} + \nu + h\nu \quad (0.23\,\text{MeV}) \tag{3.16}$$

$$^{37}_{18}\text{Ar} \rightarrow {}^{37}_{17}\text{Cl} + \nu + h\nu \quad (0.81\,\text{MeV}) \tag{3.17}$$

$$^{49}_{23}\text{V} \rightarrow {}^{49}_{22}\text{Ti} + \nu + h\nu \quad (0.60\,\text{MeV}) \tag{3.18}$$

where $h\nu$ is the internal Bremsstrahlung, the upper energy limits of which are expressed in MeV.

3.6.4 Artificially created x-radiation

An apparatus used to artificially produce x-rays, such as those employed in medical diagno-sis or x-ray diffraction, functions on a similar principle of Bremsstrahlung described previ-ously. The x-ray tube consists of an evacuated tube containing a cathode filament and a metal anode target such as tungsten ($A = 74$). A voltage potential is applied to the tube so that electrons emitted from the cathode accelerate toward the anode. Upon colliding with the tungsten anode, the accelerated electrons lose energy as Bremsstrahlung x-radiation. For example, an electron accelerated in an x-ray tube to an energy of 40 keV, which loses all of its energy upon impact with a tungsten nucleus would produce a single x-ray photon of wave-length calculated as

$$\lambda = \frac{hc}{h\nu} = \frac{12.4\,\text{keV}\,\text{Å}}{40\,\text{keV}} = 0.31\,\text{Å} = 0.031\,\text{nm}$$

Ionization and electron excitation were previously described as predominant mechanisms by which a traveling beta particle may lose its kinetic energy in matter (see Section 2.5 of Chapter 2). However, the production of Bremsstrahlung may also be another significant mech-anism for the dissipation of beta-particle energy, particularly as the beta-particle energy and the atomic number of the absorber increase (Kudo, 1995). A more thorough treatment is found in Section 2.5 of Chapter 2, which includes examples of calculations involved to determine the degree of Bremsstrahlung production as a function of beta-particle energy and absorber atomic number. In general terms we can state that for a high-energy beta particle such as the "strongest" beta particle emitted from ^{32}P ($E_{\text{max}} = 1.7\,\text{MeV}$) in a high-atomic-number material such as lead ($Z = 82$ for Pb), Bremsstrahlung production is significant. In a substance of low atomic number such as aluminum ($Z = 13$ for Al), Bremsstrahlung occurs at a low and often insignificant level.

In view of the wide spectrum of beta-particle energies emitted from radionuclides and the wide variations of degree of beta-particle interactions with atomic nuclei, the production of a broad spectrum, or smear, of photon energies of Bremsstrahlung is characteristic. This contrasts with x-radiation, which is emitted in atomic electron deexcitation processes as discrete lines of energy. We have excluded Bremsstrahlung production by charged particles other than beta particles or electrons, because other charged particles are of much greater mass than the beta particle or electron, and consequently they do not undergo such a rapid deceleration and energy loss as they travel through absorber material.

3.7 INTERACTIONS OF ELECTROMAGNETIC RADIATION WITH MATTER

3.7.1 Introduction

The lack of charge or rest mass of electromagnetic gamma- and x-radiation hinder their interaction with, and dissipation of their energy in, matter. Consequently, gamma radiation and x-rays have greater penetration power and longer ranges in matter than the massive and charged alpha and less-massive but charged beta particles of the same energy. Nevertheless, gamma- and x-radiation are absorbed by matter, and the principal mechanisms by which this type of radiation interacts with matter are discussed in this section.

3.7.2 Photoelectric effect

The energy of a photon may be completely absorbed by an atom. Under such circumstances, the entire absorbed photon energy is transferred to an electron of the atom and the electron is released, resulting in the formation of an ion pair (see also Section 3.2 of this chapter). Consequently, the energy of the emitted electron is equal to the energy of the impinging photon minus the binding energy of the electron. This is described by the photoelectric equation of Einstein

$$E_e = h\nu - \phi \tag{3.19}$$

where E_e is the energy of the ejected electron, $h\nu$ the energy of the incident photon, and ϕ the binding energy of the electron or the energy required to remove the electron from the atom. The ejected electron is identical in property to a beta particle and produces ionization (secondary ionization in this case) as it travels through matter as previously described for beta particles.

When an electron from an inner atomic K or L shell is ejected, electrons from outer shells fall from their higher energy states to fill the resulting gap. These transitions in electron energy states require a release of energy by the atomic electrons, which appears as soft (low-energy) x-rays. x-radiation is identical in properties to gamma radiation. The essential difference between x- and gamma-radiation lies in their origin. As previously described, gamma radiation originates from energy state transformations of the nucleus of an atom, whereas x-radiation originates from energy state transformations of atomic electrons.

3.7.3 Compton effect

There is a second mechanism by which a photon (e.g., x-ray or gamma ray, etc.) transfers its energy to an atomic electron. In this interaction, illustrated in Figure 3.11, the photon, E_γ, imparts only a fraction of its energy to the electron and in doing so is deflected with energy E'_γ at an angle Θ, while the bombarded electron is ejected at an angle θ to the trajectory of the primary photon. This interaction is known as the Compton effect and also as Compton scattering. The result of this interaction is the formation of an ion pair as in the case of the photoelectric effect. However, the deflected photon continues traveling through matter until it dissipates its entire kinetic energy by interacting with other electrons in a similar fashion or via other mechanisms of interaction with matter discussed in this section. The ejected electron, being identical in property to a beta particle, loses its energy through the secondary ionization it causes according to mechanisms previously described.

Our understanding of the Compton effect comes from the original work of Compton (1923a,b), who discovered that x-ray photons scattered by thin foils underwent a wavelength shift. The shift in wavelength of the scattered photon with respect to the incident photon was a function of the angle of scatter Θ. To interpret this effect he treated the x-radiation as photon particles or quanta according to the Einstein–Plank relation $E = h\nu$ (see eq. (3.1)) and the scattering to occur as photon–electron collisions somewhat like billiard-ball collisions as illustrated in Figure 3.11. Compton derived the equation, which describes the

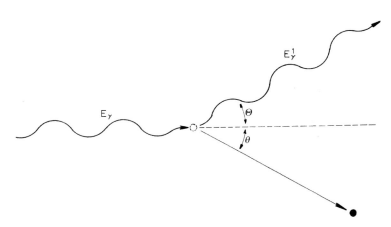

Figure 3.11 The Compton effect. An incident photon collides with an atomic electron and imparts energy to it, the photon and electron being deflected at angles Θ and θ, respectively, to the trajectory of the incident photon.

wavelength shift between the incident and scattered photons and angle of scatter as

$$\lambda' - \lambda = \frac{h}{m_0 c}(1 - \cos\Theta) \tag{3.20}$$

where λ' and λ are the wavelengths of the deflected and incident photons, respectively, h the Planck's constant, m_0 the rest mass of the electron, c the speed of light, and Θ the angle of scatter of the photon relative to its original direction of travel.

The Compton scatter photon will always be of longer wavelength (lower energy) than the incident photon because of energy lost in the collision with the electron. For example, let us calculate the wavelength shift and energy loss by an incident photon of wavelength 0.300 nm that collides with a free electron, and where the photon is scattered at an angle of 70°. The wavelength of the scattered photon is calculated according to eq. (3.20) as

$$\lambda' = \lambda + \frac{h}{m_0 c}(1 - \cos\Theta)$$

$$= 3.0 \times 10^{-10}\,\text{m} + \frac{6.626 \times 10^{-34}\,\text{J sec}}{(9.109 \times 10^{-31}\,\text{kg})(2.997 \times 10^8\,\text{m/sec})}(1 - \cos 70°)$$

$$= 3.0 \times 10^{-10}\,\text{m} + 2.43 \times 10^{-12}\,\text{m}(1 - 0.342)$$

$$= 0.3016\,\text{nm}$$

The energy lost by the incident photon according to the Einstein–Planck relation (eq. (3.1)) is given by

$$\Delta E = E_\gamma - E'_\gamma$$

$$= \frac{hc}{\lambda} - \frac{hc}{\lambda'}$$

$$= \frac{12.4\,\text{keV Å}}{3.00\,\text{Å}} - \frac{12.4\,\text{keV Å}}{3.016\,\text{Å}}$$

$$= 4.133\,\text{keV} - 4.111\,\text{keV} = 0.022\,\text{keV}$$

and the fraction of photon energy lost becomes

$$\frac{\Delta E}{E} = \frac{0.022\,\text{keV}}{4.133\,\text{keV}} = 0.0053 = 0.53\%$$

We can calculate directly the energy of the Compton scatter photon, λ', if we know the incident x-ray or gamma-ray photon energy and angle of scatter of the photon according to the equation

$$E'_\gamma = \frac{E_\gamma}{1 + (E_\gamma/mc^2)(1 - \cos\Theta)} \tag{3.21}$$

where E'_γ is the energy of the Compton scatter photon, E_γ the incident photon energy, mc^2 the rest energy of the electron (511 keV or 0.511 MeV, see Section 3.7.4 of this chapter), and Θ the Compton photon angle of scatter (Tait, 1980). If we take the data from the previous example where the incident photon energy was 4.133 keV and the angle of scatter was 70°, we can calculate the energy of the Compton photon according to eq. (3.21) to be

$$E'_\gamma = \frac{4.133\,\text{keV}}{1 + (4.133\,\text{keV}/511\,\text{keV})(1 - \cos 70°)} = 4.111\,\text{keV}$$

The result is in agreement with the above calculations using eq. (3.20) derived by Compton.

It has been shown by Compton that the angle of deflection of the photon is a function of the energy imparted to the electron. This angle may vary from just above $\Theta = 0°$ for low Compton electron energies to a maximum $\Theta = 180°$ for the highest Compton electron energy. Compton electrons are thus emitted with energies ranging between zero and a maximum energy referred to as the Compton edge. The Compton edge is the Compton electron energy corresponding to complete backscattering of the gamma-ray photon. With $\Theta = 180°$ or $\cos \Theta = -1$, eq. (3.21) is reduced to the following equation describing the energy, E'_γ, of the gamma-ray photon at the Compton edge in MeV units:

$$E'_\gamma = \frac{E_\gamma}{1 + (E_\gamma/0.511\,\text{MeV})(1 - \cos 180°)} \tag{3.22}$$

or

$$E'_\gamma = \frac{E_\gamma}{1 + (2E_\gamma/0.511)} \tag{3.23}$$

or

$$E'_\gamma = \frac{E_\gamma}{1 + 3.914\,E_\gamma} \tag{3.24}$$

As an example, the energy of the gamma-ray photon in MeV at the Compton edge for an incident gamma ray from ^{137}Cs ($E_\gamma = 0.662\,\text{MeV}$) is calculated according to eq. (3.24) to be

$$E'_\gamma = \frac{0.662}{1 + 3.914(0.662)} = 0.184\,\text{MeV}$$

A Compton scatter photon is of longer wavelength and lower energy than the incident photon. Deflected Compton photons occur with a broad spectrum of energies. Spectra of Compton scattered photon energies contain a peak known as the backscatter peak. The backscatter peak arises from Compton scattering into a gamma photon detector (e.g., NaI(Tl) crystal) from the surrounding detector shielding and housing materials. The backscatter peak occurs at increasing values of energy (MeV) in proportion to the incident photon energy and approaches a constant value of 0.25 MeV, according to eq. (3.24), for incident photon energies greater than 1 MeV (Tait, 1980). The energy of the Compton electron, E_e, may be described by

$$E_e = E_\gamma - E'_\gamma - \phi \qquad (3.25)$$

where E_γ and E'_γ are the energies of the incident and deflected photons, respectively, and ϕ the binding energy of the electron. As the binding energy of the atomic electron is relatively small, the energy of the ejected electron is essentially the difference between the incident and deflected photon energies. Substituting the value of E'_γ from eq. (3.21) and ignoring the electron binding energy, the Compton electron energy can be expressed as

$$E_e = E_\gamma - \frac{E_\gamma}{1 + (E_\gamma/mc^2)(1 - \cos \Theta)} \qquad (3.26)$$

$$E_e = E_\gamma - \frac{E_\gamma}{1 + (E_\gamma/0.511\,\text{MeV})(1 - \cos \Theta)} \qquad (3.27)$$

where the electron energies are given in MeV. For example, the energy of a Compton electron, E_e, scattered at 180° (Compton edge: $\cos \Theta = -1$) and originating from an incident gamma-ray photon from ^{137}Cs ($E_\gamma = 0.662\,\text{MeV}$) is calculated according to eq. (3.27) as

$$E_e = 0.662 - \frac{0.662}{1 + (0.662/0.511)(1 - \cos 180°)} = 0.478\,\text{MeV}$$

Alternatively, if we ignore the negligible electron binding energy and know the incident photon energy and Compton scatter photon energy, we can calculate the Compton electron energy by difference according to eq. (3.25)

$$E_e = 0.662\,\text{MeV} - 0.184\,\text{MeV} = 0.478\,\text{MeV}$$

which is in agreement with the electron energy calculated above.

3.7.4 Pair production

The interactions of gamma radiation with matter considered earlier involve the transfer of gamma energy, in whole or in part, to atomic electrons of the irradiated material. Pair production, as another mechanism of gamma-energy dissipation in matter, results in the creation of atomic particles (i.e., electrons) from the gamma energy. The electrons produced are a negatron and a positron from an individual gamma-ray photon that interacts with the coulombic field of a nucleus (see Figure 3.12). Consequently, this phenomenon involves the creation of mass from energy. The creation of an electron requires a certain quantum of energy of a gamma-ray photon, which may be calculated according to Einstein's equation for the equivalence of mass and energy

$$E = m_e c^2 \qquad (3.28)$$

where E is energy, m_e the electron rest mass, and c the speed of light in a vacuum.

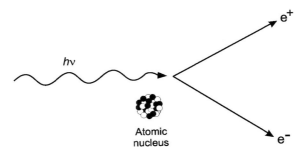

Figure 3.12   Pair production. The conversion of a gamma-ray photon into a negatron and positron pair.

According to eq. (3.28) the rest energy of the electron (negatron or positron) is calculated as

$$E = (9.109 \times 10^{-31}\,\text{kg})(2.997 \times 10^{8}\,\text{m/sec})^2 = 8.182 \times 10^{-14}\,\text{J} \tag{3.29}$$

Since by definition, $1\,\text{eV} = 1.602 \times 10^{-19}\,\text{J}$, the electron rest energy in joules is converted to electron volts as

$$8.182 \times 10^{-14}\,\text{J}/1.602 \times 10^{-19}\,\text{J/eV} = 0.511\,\text{MeV} \tag{3.30}$$

Thus, the creation of an electron (negatron) requires a minimum energy of 0.511 MeV. However, a gamma ray of 0.511 MeV energy cannot alone create a negatron, as there must also be the simultaneous creation of its antiparticle, the positron of equal mass and opposite charge. The minimum gamma-ray photon energy required for the creation of the negatron-positron pair is

$$E_{\text{pair}} = m_{e^-}c^2 + m_{e^+}c^2 = 2mc^2 = 2(0.511\,\text{MeV}) = 1.022\,\text{MeV} \tag{3.31}$$

where m_{e^-} and m_{e^+} are the rest masses of the negatron and positron, respectively. Thus, the absorption by matter of gamma radiation greater than 1.02 MeV may result in pair production. The probability of pair production increases in proportion to the magnitude of gamma-ray photon energy above 1.02 MeV, and pair production is the predominant mechanism for absorption of photons of energies of 5 MeV and above (see Figures 3.14 and 3.15). In pair production, gamma-ray energy in excess of 1.02 MeV appears as kinetic energy of the negatron and positron produced, or

$$h\nu = 2mc^2 + E_{e^-} + E_{e^+} \tag{3.32}$$

where $h\nu$ is the energy of the gamma-ray photon, $2mc^2$ the 1.02 MeV required for pair production, and E_{e^-} and E_{e^+} the kinetic energies of the negatron and positron produced. As discussed previously in Section 3.4, positrons will produce annihilation radiation when they come to rest in the proximity of a negative electron, that is their antiparticle, resulting in the simultaneous conversion of two electron masses into two gamma-ray photons of 0.511 MeV energy.

Table 3.1

Examples of nuclides that exhibit internal pair production, their gamma radiations and relative intensities of the positron–negatron pairs

Nuclide	Gamma radiations		Pair/gamma ratio ($e^{\pm}/\gamma$)
	Energy (MeV)	Abundance (%)	
^{24}Na	1.369	100	6×10^{-5}
	2.754	100	7×10^{-4}
^{56}Mn	1.81	29	5.6×10^{-4}
	2.11	15	4.6×10^{-4}
^{59}Fe	1.099	57	1.4×10^{-4}
	1.292	43	1.1×10^{-4}
^{60}Co	1.17	100	3.7×10^{-5}
	1.33	100	Combined^a
^{142}Pr	1.576	4	1.1×10^{-4}
^{144}Pr	1.489	0.3	1.9×10^{-4}
	2.186	0.7	6.7×10^{-4}
^{154}Eu	1.274	37	8.0×10^{-5}

^{a}The value of ($e^{\pm}/\gamma$) of 3.7×10^{-5} for ^{60}Co represents the pair/gamma ratio for both gamma emissions combined.

Pair production does not only occur in the vicinity of atomic nuclei bombarded by gamma radiation. It may also originate from nuclei that emit gamma radiation with transition energies greater than 1.02 MeV. This is referred to as internal pair production, and the mechanism competes to a small extent with the emission of gamma radiation. The degree to which this competition occurs is measured by the ratio of intensities of positron-negatron pairs to gamma radiation or ($e^{\pm}/\gamma$). Some examples of nuclides that emit such positron-negatron pairs and the intensities of these pairs relative to gamma radiation are given in Table 3.1

3.7.5 Combined photon interactions

Because of its zero rest mass and zero charge, gamma radiation has an extremely high penetration power in matter in comparison with alpha and beta particles.

Materials of high density and atomic number (such as lead) are used most often as absorbers to reduce x- or gamma-radiation intensity. Radiation intensity, I, is defined here as the number of photons of a radiation beam that traverse a given area per second, the units of which can be photons cm^{-2} sec^{-1}. Suppose a given absorber material of thickness x attenuates or reduces the intensity of incident gamma radiation by one-half. Placing a similar barrier of the same thickness along the path of the transmitted gamma radiation would reduce the intensity again by one-half. With three barriers each of thickness x and an initial gamma-ray intensity I_0 there is a progressive drop in the transmitted gamma-ray intensities: $I_1 = 1/2I_0$, $I_2 = 1/2I_1$, $I_3 = 1/2I_2$, and $I_n = 1/2I_{n-1}$. Obviously, incident x- or gamma-radiation may be reduced from I_0 to I_3 by using a $3x$ thickness of the same material as an absorber. Consequently, the intensity of the transmitted electromagnetic radiation is proportional to the thickness of the absorber material and to the initial intensity of the radiation. An increasing absorber thickness increases the probability of photon removal because there is a corresponding increase of absorber atoms that may attenuate the incident photons via the photoelectric effect, the Compton effect, and pair production mechanisms.

If gamma-ray attenuation with respect to absorber thickness is considered, the change in gamma-ray intensity, ΔI, with respect to the absorber thickness, Δx, is proportional to the initial gamma-ray photon intensity, I. This may be written as

$$\frac{\Delta I}{\Delta x} = -\mu I \qquad (3.33)$$

where μ is the proportionality constant, referred to as the linear attenuation coefficient or linear absorption coefficient. Its value is dependent on the atomic composition and density of the absorber material. The change in intensity over an infinitely thin section of a given absorber material may be expressed as

$$\frac{dI}{dx} = -\mu I \qquad (3.34)$$

or

$$\frac{dI}{I} = -\mu \, dx \qquad (3.35)$$

Integrating eq. (3.35) over the limits defined by the initial intensity, I_0, to the transmitted intensity, I, and over the limits of absorber thickness from zero to a finite value x, such as

$$\int_{I_0}^{I} \frac{dI}{I} = -\mu \int_{0}^{x} dx \qquad (3.36)$$

gives

$$\ln I - \ln I_0 = -\mu x \qquad (3.37)$$

or

$$\ln \left(\frac{I_0}{I} \right) = \mu x \qquad (3.38)$$

Eq. (3.38) may be written in exponential form as

$$I = I_0 \, e^{-\mu x} \qquad (3.39)$$

which is somewhat similar to the exponential attenuation of neutrons discussed in Chapter 4.

Because gamma-ray absorption is exponential, the term half-value thickness, $x_{1/2}$, is used to define the attenuation of gamma radiation by matter. Half-value thickness is the thickness of a given material of defined density that can reduce the intensity of incident gamma radiation by one-half. The half-value thickness may also be defined according to eq. (3.38), in which the initial gamma-ray intensity, I_0, is given an arbitrary value of 1 and the transmitted intensity must, by definition, have a value of 1/2, or

$$\ln\left(\frac{1}{0.5}\right) = \mu x_{1/2} \tag{3.40}$$

or

$$\ln 2 = \mu x_{1/2} \tag{3.41}$$

and

$$x_{1/2} = \frac{0.693}{\mu} \tag{3.42}$$

From the linear attenuation coefficient, μ, of a given material and gamma-ray photon energy, it is possible to calculate the half-value thickness, $x_{1/2}$. The linear attenuation coefficient has units of cm^{-1}, so that calculated half-value thickness is provided in units of material thickness (cm). Linear attenuation coefficients for some materials as a function of photon energy are provided in Table 3.2. The table refers to these as total linear attenuation coefficients, because they constitute the sum of coefficients due to Compton, photoelectric, and pair production interactions. Calculated half-value thicknesses of various absorber materials as a function of gamma-ray energy are illustrated in Figure 3.13 to illustrate some examples of the varying amounts of absorber material required to attenuate gamma-ray photons. The linear attenuation coefficient is a constant for a given absorber material and gamma-ray photon energy and has units of reciprocal length such as cm^{-1}. It is, however, dependent on the state of the absorber or the number of atoms per unit volume of absorber. A more popular coefficient is the mass attenuation coefficient, μ_m, which is independent of the physical state of the absorber material and is defined as

$$\mu_m = \frac{\mu}{\rho} \tag{3.43}$$

where ρ is the density of the absorber in units of g/cm^3, and μ_m has units of cm^2/g. Some examples of mass attenuation coefficients according to x- and gamma-ray photon energy are provided in Table 3.3. Using the mass attenuation coefficient, eq. (3.39) changes to

$$I = I_0 e^{-\mu_m \rho x} \tag{3.44}$$

and the half-value thickness is calculated according to eq. (3.42) as

$$x_{1/2} = \frac{0.693}{\mu_m \rho} \tag{3.45}$$

Mass attenuation coefficients for x- or gamma-ray photons over a wide range of energies from 1 keV to 1000 MeV in 100 elements are available from Berger and Hubbell (1997).

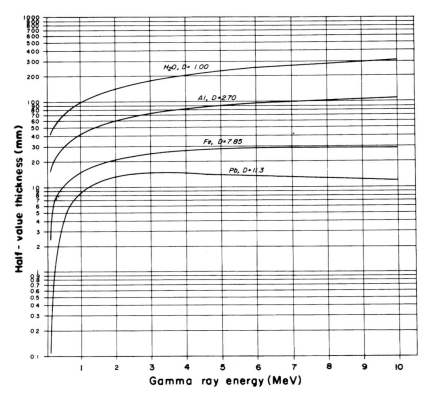

Figure 3.13   Half-value thicknesses of various materials as a function of gamma-ray energy. D is the density of each material.

Table 3.2

Total linear attenuation coefficients (cm^{-1}) for gamma-ray photons in various materials

Photon energy (MeV)	Water	Aluminum	Iron	Lead
0.1	0.167	0.435	2.704	59.99
0.2	0.136	0.324	1.085	10.16
0.4	0.106	0.2489	0.7223	2.359
0.8	0.0786	0.1844	0.5219	0.9480
1.0	0.0706	0.1658	0.4677	0.7757
1.5	0.0575	0.1350	0.3812	0.5806
2.0	0.0493	0.1166	0.3333	0.5182
4.0	0.0339	0.0837	0.2594	0.4763
8.0	0.0240	0.0651	0.2319	0.5205
10.0	0.0219	0.0618	0.2311	0.5545

Source: Argonne National Laboratory (1963), Hubbell (1969), and Serway *et al.* (1997).

Table 3.3

Total mass attenuation coefficients (cm^2/g) for x- or gamma-ray photon energy in various materials

Photon energy (MeV)	Air	Water	Aluminum	Iron	Lead
0.005			193	140	730
0.01			26.2	171	131
0.05			0.368	1.96	8.04
0.1	0.151	0.167	0.170	0.372	5.55
0.2	0.123	0.136	0.122	0.146	0.999
0.4	0.0953	0.106	0.0922	0.0919	0.208
0.8	0.0706	0.0786	0.0683	0.0664	0.0836
1.0	0.0655	0.0706	0.0614	0.0595	0.0684
1.5	0.0517	0.0575	0.0500	0.0485	0.0512
2.0	0.0445	0.0493	0.0432	0.0424	0.0457
4.0	0.0307	0.0339	0.0310	0.0330	0.0420
8.0	0.0220	0.0240	0.0241	0.0295	0.0459
10.0	0.0202	0.0219	0.0229	0.0294	0.0489

Source: Argonne National Laboratory (1963), Hubbell (1969), and Berger and Hubbell (1997).

A sample of mass attenuation coefficients over the range of 5 keV–10 MeV in a few materials are listed in Table 3.3.

The following calculation illustrates the use of the data from Tables 3.2 and 3.3 to calculate half-value thickness and radiation attenuation.

Let us calculate the half-value thickness of lead ($\rho = 11.3$ g/cm^3) for 2.0 MeV gamma radiation, and further calculate what the reduction in radiation intensity would result if we positioned four times the half-value thickness of lead in the path of the radiation beam. First, the linear attenuation coefficient, μ, or mass attenuation coefficient, μ_m, for 2.0 MeV photons in lead are obtained from either Table 3.2 or 3.3 and the half-value thickness of lead for 2.0 MeV photons is calculated as

$$x_{1/2} = \frac{0.693}{\mu} \quad \text{or} \quad \frac{0.693}{\mu_m \rho} \tag{3.46}$$

or

$$x_{1/2} = \frac{0.693}{0.5182 \, \text{cm}^{-1}} \quad \text{or} \quad \frac{0.693}{(0.0457 \, \text{cm}^2/\text{g})(11.3 \, \text{g/cm}^3)}$$

$$x_{1/2} = 1.34 \, \text{cm}$$

Thus, a barrier of 1.34 cm thickness of lead is sufficient to reduce the radiation intensity of 2.0 MeV photons by 1/2 or 50%. According to eq. (3.39) the relation between the initial radiation intensity, I_0, and the transmitted intensity, I, is

$$\frac{I}{I_0} = e^{-\mu x} \tag{3.47}$$

and for $x = 1.34$ cm, if the initial radiation intensity is given an arbitrary value of 2, the transmitted intensity would be 50% of the initial intensity or equal to 1. We can then write

$$\frac{I}{I_0} = \frac{1}{2} = e^{-1.34\mu} \tag{3.48}$$

If we employ four times the half-value thickness of lead or 4×1.34 cm $= 5.36$ cm, we can calculate that the transmitted radiation would be reduced to the following:

$$\frac{I}{I_0} = (e^{-1.34\mu})^4 = \left(\frac{1}{2}\right)^4$$

or

$$e^{-5.36\mu} = \frac{1}{16} = 0.0625 = 6.25\% \text{ transmitted}$$

The remaining 15/16 or 93.75% of the initial radiation is attenuated by the 5.36 cm lead barrier. In general, we need not know the half-value thickness of the material or shield, but simply obtain the linear or mass attenuation coefficient for a given energy of x- or gamma-radiation from reference tables and use eq. (3.39) or (3.44) to calculate the degree of radiation attenuation for any thickness of the absorber material. For example, if we used only 2.5 cm of lead barrier, the attenuation of 2.0 MeV gamma rays could be calculated as

$$\frac{I}{I_0} = e^{-\mu x} = e^{-\mu_m \rho x} \tag{3.49}$$

and

$$\frac{I}{I_0} = e^{-(0.5182\,\text{cm}^{-1})(2.5\,\text{cm})} = e^{-(0.0457\,\text{cm}^2/\text{g})(11.3\,\text{g/cm}^3)(2.5\,\text{cm})}$$

$$= e^{-1.29} = 0.275 = 27.5\%$$

Thus the 2.0 MeV radiation transmitted through a shield of 2.5 cm of lead would be 27.5% of the initial radiation intensity.

As previously discussed, the absorption of gamma radiation is a process that principally involves three mechanisms of gamma-ray attenuation: the Compton effect, the photoelectric effect, and pair production. The attenuation coefficients just discussed above are also referred to as total attenuation coefficients because they consist of the sum of three independent coefficients or

$$\mu = \mu_c + \mu_e + \mu_p \tag{3.50}$$

where μ_c, μ_e, and μ_p are attenuation coefficients for Compton, photoelectric, and pair production processes. The attenuation coefficients are proportional to the probabilities of occurrence of these radiation attenuation processes and can be used as a measure of the relative roles these processes play in the absorption of gamma-ray photons. Accordingly, the total and partial mass attenuation coefficients can be written as

$$\mu_{\mathrm{m}} = \frac{\mu}{\rho} = \frac{\mu_{\mathrm{c}}}{\rho} + \frac{\mu_{\mathrm{e}}}{\rho} + \frac{\mu_{\mathrm{p}}}{\rho} \tag{3.51}$$

Figures 3.14 and 3.15 provide a graphic representation of the relative frequency of occurrence of the Compton, photoelectric, and pair production processes in aluminum and sodium iodide

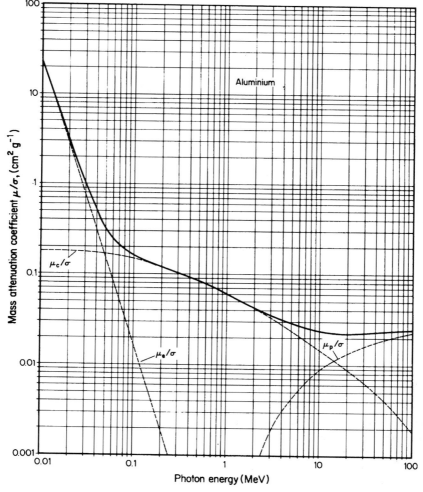

Figure 3.14 Mass attenuation coefficients for photons in aluminum. The total attenuation is given by the solid line, which is the sum of the partial attenuations due to the Compton effect, μ_{c}/ρ; the photoelectric effect, μ_{e}/ρ; and pair production, μ_{p}/ρ. Linear attenuation coefficients are obtained from these values by multiplying by the density of aluminum, $\rho = 2.70\,\mathrm{g/cm^3}$. The symbol σ in the figure is synonymous to the symbol ρ used in this text for absorber density. (From Evans (1955), reproduced with permission from The McGraw-Hill Companies © 1995.)

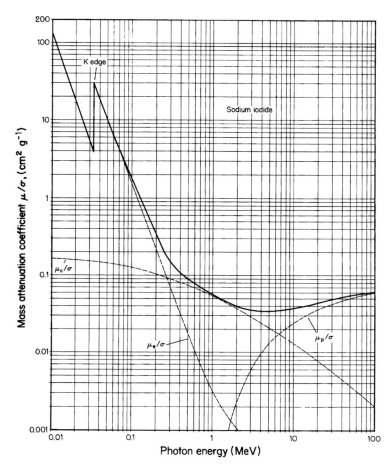

Figure 3.15 Mass attenuation coefficients for photons in sodium iodide. The total attenuation is given by the solid line, which is the sum of the partial attenuations due to the Compton effect, μ_c/ρ; the photoelectric effect, μ_e/ρ; and pair production, μ_p/ρ. Linear attenuation coefficients are obtained from these values by multiplying by the density of sodium iodide, $\rho = 3.67\,\mathrm{g/cm^3}$. The symbol σ in the figure is synonymous to the symbol ρ used in this text for absorber density. (From Evans (1955), reproduced with permission from The McGraw-Hill Companies © 1995.)

absorbers as a function of photon energy. These serve as two examples of absorber material, which differ in atomic number. It is seen from the two figures that the photoelectric effect plays an increasing role in total gamma-ray attenuation at lower gamma-ray energies and with absorber materials of higher atomic number.

In some absorber materials of relatively high density, absorption edges can be measured for low photon energies such as the K edge illustrated in Figure 3.15. The absorption edge is a discontinuity in the attenuation coefficient curve for the photoelectric effect that is caused when photon energies are less than the binding energies of electrons of a certain shell (e.g., K shell) and that reduces the number of electrons that may be ejected by the photoelectric effect.

When the photon energy is in excess of the K-electron binding energy of the absorber, the photoelectric effect will occur primarily in K-electron shell. At such photon energies ($>K$-shell binding energy) the probability of the photoelectric effect occurring in the L shell is only approximately 20% and even less probable for outer shells (Friedlander $et\ al.$, 1964). Such a phenomenon may appear strange, considering that the binding energies of the electrons are lower in shells further away from the nucleus. If one considers only the binding energies of the electrons, it would appear, although incorrectly that for photon energies in excess of the K-shell binding energy, electrons further away from the nucleus (e.g., L, M, N shells) would be ejected as photoelectrons more easily. This is not the case because energy and momentum must be conserved in the process; and the atom, because of its much higher mass, will carry off the excess momentum while taking away negligible energy in the process. The entire photon energy is transferred to the photoelectron minus its binding energy. Consequently, the photoelectric effect cannot occur with a free electron because there is no atom to carry off excess momentum, as demonstrated by Gautreau and Savin (1999). In the same sense, when the photon energy is much greater than the electron binding energy, the electrons, such as those in the outer shells with very low binding energies, act more like free electrons. These electrons contribute less to photon attenuation by the photoelectric effect. As explained by Lilley (2001), the conservation of energy and momentum is the reason why the most tightly bound electrons contribute most to photoelectric absorption for $E_\gamma > B_K$, where E_γ is the photon energy and B_K the K-shell electron binding energy, and also why there is a marked dependence of photon attenuation on photon energy and absorber atomic number. The photon absorption in NaI, illustrated in Figure 3.15, shows a K edge at approximately 0.033 MeV or 33 keV. This corresponds to the electron binding energy in the K shell of iodine, which from reference tables (Lide, 2001) is reported to be 33,169 eV or 33.1 keV. Photon attenuation by the photoelectric effect for photon energies in excess of 33 keV occurs predominantly with the K-shell electrons. There is no K edge observed for sodium in Figure 3.15 for the NaI absorber because the binding energy of the K shell of sodium is only 1070 eV or 1.07 keV, a factor of 10 less than the photon energy scale provided in the figure. At much lower photon energies, corresponding to L- and M-shell electron binding energies, characteristic L and M edges can be observed. The binding energies of electrons in the various atomic electron shells of the elements are listed in reference tables, such as those provided by Lide (2001).

As illustrated in Figures 3.14 and 3.15, the pair production process does not occur at gamma-ray energies below the threshold value of 1.02 MeV, as expected, in accordance with the combined positron and negatron rest energies (2×0.511 MeV) required for pair production. A thorough treatment of the attenuation and absorption of gamma radiation in matter is available from Hubbell (1969), Krane (1988), Lilley (2001), Serway $et\ al.$ (1997), and Turner (1995).

Radioactivity Hall of Fame—Part IV

James Chadwick (1891–1974), Lise Meitner (1878–1968) and Otto Hahn (1879–1968), and Leo Szilard (1898–1964)

JAMES CHADWICK (1891–1974)

James Chadwick was born in Cheshire, England on October 20, 1891. He was awarded the Nobel Prize in Physics 1935 "for the discovery of the neutron".

James Chadwick (1891-1974)

After attending Manchester High School Chadwick entered Manchester University in 1908 and graduated from the Honours School of Physics in 1911. He then went to study under Ernest Rutherford at the Physics Laboratory in Manchester on research studies related to radioactivity and graduated with a Master of Science degree in 1913. A stint with Rutherford was the springboard for the successful careers of many famous physicists and future Nobel Laureates interested in pursuing studies on radioactivity and the structure of the atom. After receiving the M.Sc. degree in 1913 Chadwick received a scholarship to study under Professor Hans Geiger, also a former student of Rutherford, at the Physikalisch Technische Reichsanstalt in Charlottenburg, Berlin. During the war Chadwick was interred in the Zivilgefangenenlager in Ruhleben until 1919. He then returned to England to work again under Ernest Rutherford, who was then at the Cavendish Laboratory, Cambridge, and to continue his studies in physics at the Gonville and Caius College in Cambridge. At this time Rutherford had already accomplished the very first artificial nuclear transmutation by splitting atoms of nitrogen with alpha particles yielding oxygen atoms and protons (see Rutherford, Radioactivity Hall of Fame—Part I). Chadwick worked with Rutherford on studies related to the structure of the nucleus and on the successful nuclear transmutation of other elements (Nobel Lectures, Physics, 1922–1941).

While at the Cavendish Laboratory in 1932 James Chadwick made a discovery for which he received the Nobel Prize. He demonstrated the existence of the neutron, accurately determined its mass, and thereby explained the structure of the atomic nucleus. The existence of the neutron was first suggested by Rutherford in his Bakerian Lecture delivered on June 3, 1920 (Rutherford, 1920), but because of the particle's neutral charge its detection was difficult and the particle remained elusive until Chadwick's discovery in 1932. As noted by Chadwick in his Nobel lecture on December 12, 1935

> He [Rutherford in 1920] thought that a proton and an electron might unite in a much more intimate way than they do in the hydrogen atom, and so form a particle of no net charge and with a mass nearly the same at that of the hydrogen atom. His view was that with such a particle as the first step in the formation of atomic nuclei from the two elementary units in the structure of matter—the proton and the electron.

Two years prior to Chadwick's discovery of the neutron it had been shown by Bothe and Becker (1930) that some light elements particularly beryllium, when bombarded by alpha particles from polonium, would emit an unknown very penetrating radiation resembling gamma radiation. At the time, the unknown radiation originating from the interaction of alpha particles with light elements, such as boron and beryllium, was referred to as "beryllium radiation", because the radiation was more intense with beryllium. Subsequently Irène Joliot-Curie (1931) and Webster (1932) demonstrated that the "beryllium radiation" had a penetration power much greater than any gamma radiation known to exist. The radiation was reported to be able to pass through a brass plate, several centimeters thick, without any noteworthy loss of velocity or several kilometers of air, before losing its energy of motion. Also Irène Joliot-Curie and Frédéric Joliot (1932) made a key observation that the unknown radiation from beryllium and from boron would eject protons when interacting with materials containing hydrogen such as paraffin. They showed that the protons were ejected with a high velocity of approximately 3×10^9 cm/sec and estimated the quantum energy of the presumed gamma radiation to be extremely high, about 50 MeV. They had thought that the protons were ejected by the energy quanta of the mysterious radiation by a process similar to the Compton Effect. The Joliot-Curie's came very close to discovering the neutron, but it eluded them. They unfortunately were fixated on their conviction that the radiation coming from beryllium was gamma. The discovery of the neutron was almost at their grasp, but they failed to consider that the radiation could be a particle of relatively high mass and neutral charge. Chadwick shortly thereafter, after a series of experiments including repeating the paraffin experiment of the Joliot-Curies, identified the mysterious radiation as neutron radiation and not a quantum of gamma radiation.

On the basis of the findings described above and by a combination of intuition, logical thought, and experiment, James Chadwick (1932a,b) was able to prove that the "beryllium radiation" was not gamma radiation, but the elusive neutron. To demonstrate the existence of the neutron and its properties Chadwick made several systematic observations and experiments including (i) disproving on the basis of conservation of energy that the high energy emitted by the "beryllium radiation" was a quantum of gamma radiation; (ii) demonstrating the particulate nature of the neutral radiation by observing recoil atoms of light elements including lithium, beryllium, boron, carbon, and nitrogen; and (iii) measuring the approximate mass of the neutral particle on the basis of recoil atom velocities, and (iv) deducing the precise mass of the neutron from the mass–energy relations of the nuclei involved in the emission of the neutron.

Chadwick's first consideration that led to his discovery of the neutron was the improbability that a nuclear reaction involving an alpha particle with kinetic energy of only 5 MeV interacting with a beryllium nucleus could yield a high-energy gamma quantum of 50 MeV. In line with the laws of conservation and energy and momentum Chadwick was convinced that there was another explanation for the identity of this high-penetrating radiation. He then proceeded to study the interaction of the "beryllium radiation" with protons from paraffin as well as atoms of other light elements. Such interactions via collisions of the unknown "beryllium radiation" would produce recoil atoms not only with hydrogen of paraffin (producing free protons), but also recoil atoms of the light elements. Figure IV.1 illustrates the basic experimental setup Chadwick used to observe the recoil atoms produced by the mysterious "beryllium radiation".

As illustrated in Figure IV.1 Chadwick produced proton recoil atoms by placing a paraffin barrier in the path of the "beryllium radiation". He could then measure the ranges and hence velocities of the protons by counting the protons that reached the ionization counter after interposing aluminum foil of different thicknesses between the wax and the counter. He measured the maximum range of the protons to be equivalent to 22 cm of air corresponding to a velocity of about 3.3×10^9 cm/sec (Chadwick, 1932a,b). He concluded that such a velocity could arise from a collision of a particle of similar mass as the proton, that is, the neutron, because a head-on collision between particles of equal mass could entail the transfer of practically all of the energy of one particle to the other, such as the head-on collision of one billiard ball with another. Chadwick arrived at the following "neutron hypothesis" prior to presenting yet further evidence in support of his hypothesis (Chadwick, 1932b):

> It is evident that we must either relinquish the application of conservation of energy and momentum in these collisions or adopt another hypothesis about the nature of the radiation. If we suppose that the radiation is not a quantum radiation, but consists of particles of mass very nearly equal to that of the proton, all the difficulties connected with the collisions disappear, both with regard to their frequency and to the energy transfer to different masses. In order to explain the great penetrating power of the radiation we must further assume that the particle has no net charge. We must suppose it to consist of a proton and an electron in close combination, the "neutron" discussed by Rutherford in his Bakerian Lecture of 1920…The experiments showed that the maximum velocity of the protons ejected from paraffin wax was about 3.3×10^9 cm per second. This is therefore the maximum velocity of the neutrons emitted from beryllium bombarded by α-particles of polonium.

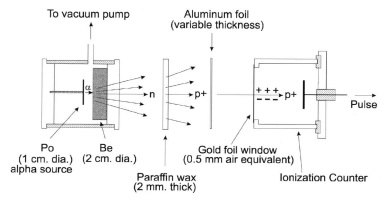

To vacuum pump   Aluminum foil
(variable thickness)

→n   →p+   +++→p+   Pulse

Po   Be   Gold foil window   Ionization Counter
(1 cm. dia.) (2 cm. dia.)   (0.5 mm air equivalent)
alpha source

Paraffin wax
(2 mm. thick)

Figure IV.1 Schematic illustration of Chadwick's experimental setup of 1932 used in the discovery of the neutron and its properties. The aluminum foil of various thicknesses was used to determine the ranges of the recoil protons ejected from the paraffin. Also, Chadwick removed the paraffin layer and aluminum foil and inserted a brass plate coated with one of the following light elements: Li, Be, B, C, and N very close to the ionization counter to measure the recoil atoms of the elements from direct neutron collisions. In another variation of the setup Chadwick removed all paraffin and metal foils or plates between the neutron source and ionization counter and filled the ionization counter with hydrogen, helium, nitrogen, oxygen, or argon gas to observe the number of recoil atoms produced by neutron collisions with atoms of the different gases (see Chadwick, 1932b).

Chadwick made additional measurements of recoil atoms of other light elements including lithium, beryllium, boron, carbon, and nitrogen, which provided additional evidence that the "beryllium radiation" was a nuclear particle of mass similar to that of the proton. In one of these studies he removed the paraffin layer and aluminum foil barriers to the neutron source (see Figure IV.1) and inserted a brass plate coated with one of the following light elements: Li, Be, B, C, and N very close to the ionization counter to measure the recoil atoms of the elements from direct neutron collisions. The recoil atoms would be liberated from the brass plate and would, in turn, produce ionization and a consequent pulse count for each ion entering into the counter. The brass plate coated with any one of the light elements had to be placed very close to the ionization counter, because the recoil atoms of these elements could only travel a few millimeters in air. The counter was equipped with a very thin gold foil window equivalent to an absorbing power of only 0.5 mm of air. In each case, Chadwick found that the number of counts registered would increase when one of the light elements was bombarded with the neutron radiation. The magnitude of the pulses, measured by oscillograph deflections, produced by the light-element recoil atoms was larger than those produced by proton recoil atoms. Chadwick concluded that this was due to the greater atomic size and consequent greater ionization power that the light-element recoil atoms would have over the relatively smaller proton recoil atoms. The fact that the "beryllium radiation" could produce recoil atoms from collisions with such elements considerably larger than hydrogen provided Chadwick with additional more conclusive evidence that the "beryllium radiation" was the particulate neutron radiation and not quanta of gamma radiation. He made this clear in the simple statement

In general, the experiment results show that if the recoil atoms are to be explained by collision with a quantum, we must assume a larger and larger energy for the quantum [e.g., 90 MeV for nitrogen recoil] as the mass of the struck atom increases (Chadwick, 1932b).

Chadwick proceeded to measure the mass of the neutron as conclusive proof that the neutron was indeed a fundamental nuclear particle. He used three approaches to this problem. The first approach was to get an approximate mass of the neutron by measuring and comparing the relative velocities and masses of the recoil nitrogen and proton atoms. With the assistance of Norman Feather (see Feather, 1932) Chadwick measured the ranges of the nitrogen and proton recoil atoms by photographing these from tracks of the recoil atoms produced in a Wilson Cloud Chamber (see C. T. R. Wilson in Radioactivity Hall of Fame—Part II). Chadwick calculated the maximum velocity that could be attributed to a hydrogen atom (proton recoil) from collision with a neutron as

$$U_H = \frac{2M}{M+1} V \tag{IV.1}$$

Where U_H is the hydrogen atom (proton recoil) velocity and M and V the mass and velocity of the neutron, respectively. Likewise, the maximum velocity, U_N, that could be given to a nitrogen atom (nitrogen recoil) upon collision with a neutron was calculated by Chadwick as

$$U_N = \frac{2M}{M+14} V \tag{IV.2}$$

Then the ratio of the two velocities of the recoil atoms could be written as

$$\frac{M+14}{M+1} = \frac{U_H}{U_N} \tag{IV.3}$$

The maximum range of the nitrogen recoil atom was measured at 3.5 mm in air, which corresponded to a velocity, U_N, of 4.7×10^8 cm/sec. The previously deduced maximum velocity for the proton (hydrogen recoil) atom, U_H, was taken by Chadwick as 3.3×10^9 cm/sec. Therefore, Chadwick (1932b) estimated the mass, M, of the neutron as

$$\frac{M + 14}{M + 1} = \frac{U_H}{U_N} = \frac{3.3 \times 10^9}{4.7 \times 10^8}$$

$$M = 1.15,$$

which was very close to the mass of the proton.

Because the ranges and velocities of the recoil atoms could not be so accurately determined, the above calculation of the neutron mass was only an approximation. Chadwick went further to a second approach to getting a more accurate measurement of the neutron mass. For this, he turned to consider the energy relations in one of the disintegration processes in which a neutron is liberated from an atomic nucleus and where the masses of the atomic nuclei concerned in the process were already accurately known. He took the process where boron (^{11}B) would liberate neutrons when bombarded with alpha particles (^4He) from polonium, that is

$$^{11}_{5}B + ^4_2He \rightarrow ^{14}_{7}N + ^1_0n \tag{IV.4}$$

From the calculated maximum velocities of the particles concerned, namely the alpha particle from polonium, the recoil atom of nitrogen, and the neutron, Chadwick could calculate the kinetic energy (K.E.) of each particle concerned, and considering that energy and momentum were conserved in the collisions, he wrote the following energy equation:

$$\text{mass of } ^{11}B + \text{mass of } ^4He + \text{K.E. of } ^4He = \text{mass of } ^{14}N + \text{mass of } ^1n$$
$$+ \text{ K.E. of } ^{14}N + \text{K.E. of } ^1n \tag{IV.5}$$

or

$$11.00825\,u + 4.00106\,u + 0.00565\,u = 14.0042\,u + \text{mass of } ^1n$$
$$+ 0.00061\,u + 0.0035\,u \tag{IV.6}$$

and

$$\text{mass of } ^1n = 1.0066\,u \tag{IV.7}$$

Chadwick estimated an error of ± 0.003 and the mass of the neutron to be somewhere between 1.005 and 1.008 u. This was very close to the actual value of 1.00866 u taken from tables of physical constants (Lide, 1997) and the neutron/proton mass ratio from current tables is 1.001378. In his calculations Chadwick converted the kinetic energy of the particles from units of MeV to atomic mass units (u) according to Einstein's equation of equivalence of energy and mass and the calculated conversion unit of 931.494 MeV/u (see Chapter 1). For example, the kinetic energy of the alpha particle (K.E. of ^4He) from polonium in eq. (IV.5) above has an average energy of 5.26 MeV. Its energy was converted to atomic mass units in eq. (IV.6) as follows:

$$\frac{5.26\,\text{MeV}}{931.494\,\text{MeV/u}} = 0.00565\,u \tag{IV.8}$$

A yet third and most accurate approach to the problem of the mass of the neutron was taken by Chadwick and Goldhaber (1934). The method they designed involved the disintegration of a nucleus into its component nucleons by means of gamma-ray bombardment. They selected a radioactive source that emitted gamma rays of sufficient energy to break apart a deuterium nucleus into its component proton and neutron. The necessary condition for this was that the gamma-ray energy of the source should be greater than the binding energy of the proton and neutron, that is, an energy greater than that which holds the proton and neutron together forming the deuterium nucleus. In the words of Chadwick and Goldhaber (1934)

> Heavy hydrogen was chosen as the element first to be examined, because the diplon has a small mass defect and also because it is the simplest of all nuclear systems and its properties are as important in nuclear theory as the hydrogen atom is in atomic theory. The disintegration to be expected is

$$\,^2_1\mathrm{H} + h\nu \rightarrow \,^1_1\mathrm{H} + \,^1_0\mathrm{n} \tag{IV.9}$$

> Since the momentum of the quantum is small and the masses of the proton and neutron are nearly the same, the available energy, $h\nu - W$, where W is the binding energy of the particles, will be divided nearly equally between the proton and the neutron.

The term "diplon" used by Chadwick and Goldhaber refers to the nucleus of the recently discovered deuterium (heavy hydrogen nucleus) consisting of only one proton and one neutron. By "mass defect" of the nucleus, Chadwick and Goldhaber refer to the fact that the mass of the nucleus is always less than the combined masses of the separate nucleons. This difference in mass of the deuterium nucleus and the combined masses of the proton and neutron would be the binding energy of the deuterium nucleus. According to Einstein's equation of equivalence of mass and energy ($E = mc^2$) we can equate this difference in mass to the energy required to bind the proton and neutron together in the nucleus of deuterium, that is, the "binding energy". This binding energy is also the energy released when the deuterium nucleus is created by nuclear fusion, and it is the amount of energy needed to break apart the deuterium nucleus into its component proton and neutron. They therefore selected the isotope of thorium C″, which is known today as $^{208}\mathrm{Tl}$. This isotope of thallium decays by beta-particle emission with maximum energies of 1.8 MeV followed by the emission of a gamma ray of 2.62 MeV energy. Chadwick and Goldhaber estimated that the gamma ray from thorium C″, if imparted to a nucleus of deuterium, would be more than sufficient to break it apart into its constituent proton and neutron. The result was exactly as they had hoped. An artist's depiction of the experimental arrangement they used is illustrated in Figure IV.2

Gamma rays from the thorium C″ were sufficient to produce pulse events in the ionization chamber due to the proton produced by the splitting of the deuterium nucleus. From the magnitude of the pulse events, the energy of the proton could be determined to be 0.25 MeV. Since the proton and neutron have masses of similar magnitude Chadwick and Goldhaber deduced that they would be ejected from the nuclear disintegration with equal energies, that is, 0.25 MeV each. They provided further evidence that the 2.62 MeV gamma ray from the thorium C″ was indeed causing the splitting of the deuterium nucleus by replacing the thorium C″ source with radium C, that is, $^{212}\mathrm{Bi}$. The 1.6 MeV gamma rays from radium C would produce no pulse events over background, as this gamma radiation was not of sufficient energy to exceed the binding energy to break apart the deuterium nucleus.

From the results obtained and from the then-known atomic masses of deuterium (2.0136 u) and proton (i.e., hydrogen atom = 1.0078 u) and by use of Einstein's equation of mass and energy equivalence ($E = mc^2$) Chadwick and Goldhaber could calculate the mass of the neutron. The following reaction of their photon-induced splitting of deuterium can be written

$$\,^2_1\mathrm{H} + h\nu \rightarrow \,^1_1\mathrm{H} + \text{proton energy} + \,^1_0\mathrm{n} + \text{neutron energy} \tag{IV.10}$$

Because the mass of deuterium includes the mass of an electron, we must use atomic mass units. The above eq. (IV.10) in units of mass or energy can be written as

$$2.0136\,\mathrm{u} + 2.62\,\mathrm{MeV} \rightarrow 1.0078\,\mathrm{u} + 0.25\,\mathrm{MeV} + \,^1_0\mathrm{n} + 0.25\,\mathrm{MeV} \tag{IV.11}$$

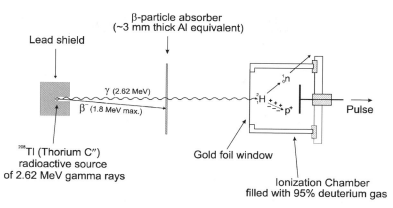

Figure IV.2   An artist's depiction of Chadwick's and Goldhaber's experimental arrangement for the determination of the mass of the neutron. Gamma radiation of 2.62 MeV from a source of thorium C″ was directed toward an ionization chamber filled with 95% pure deuterium gas. An absorber equivalent to 3 mm of aluminum would be required to prevent the beta particles from entering the chamber. Upon impact with a deuterium nucleus in the ionization chamber the 2.62 MeV gamma ray would break the nucleus into its constituent neutron and proton. The neutron would escape from the chamber without effect, and the proton would be fully stopped producing ion pairs during its travel in the deuterium gas. The magnitude of the pulse produced by the proton could be calibrated to measure electron volts of energy absorbed by the ionization chamber.

The gamma-ray energy of 2.62 MeV and the proton and neutron energies of 0.25 MeV can be converted to atomic mass units with Einstein's equation. His equation may be used to determine the factor for converting energy in eV to atomic mass units as follows:

$$E = mc^2 \tag{IV.12}$$

where m is mass in kg and c the constant velocity of light in a vacuum (2.9979×10^8 m/sec). If by definition, the mass of 1 u $= 1.6605 \times 10^{-27}$ kg, the energy equivalent of 1 u could be calculated as

$$E = (1.6605 \times 10^{-27} \text{ kg})(2.9979 \times 10^8 \text{ m/sec})^2$$
$$= 14.924 \times 10^{-11} \text{ J} \tag{IV.13}$$

The calculated energy in units of joules (J) is then converted to units of electron volts according to the conversion factor 1 eV $= 1.602 \times 10^{-19}$ J or

$$E = \frac{14.924 \times 10^{-11} \text{ J}}{1.602 \times 10^{-19} \text{ J/eV}} = 9.315 \times 10^8 \text{ eV} = 931.5 \text{ MeV} \tag{IV.14}$$

Thus 1 u $= 931.5$ MeV is established as a factor for converting energy into mass and vice versa. With this conversion factor the energies in MeV of the gamma-ray photon and of the proton and neutron of eq. (IV.11) can be converted to atomic mass units providing all of the values in mass units as follows:

$$2.0136 \text{ u} + \left(\frac{2.62 \text{ MeV}}{931.5 \text{ MeV/u}} \right) \rightarrow 1.0078 \text{ u} + \left(\frac{0.25 \text{ MeV}}{931.5 \text{ MeV/u}} \right)$$
$$+ {}_0^1\text{n} + \left(\frac{0.25 \text{ MeV}}{931.5 \text{ MeV/u}} \right) \tag{IV.15}$$

or

$$2.0136\,u + 0.0028\,u \rightarrow 1.0078\,u + 0.00027\,u + {}_{0}^{1}n + 0.00027\,u \qquad \text{(IV.16)}$$

And the mass of the neutron would be calculated as

$$
\begin{aligned}
{}_{0}^{1}n &= 2.0136\,u + 0.0028\,u - 1.0078\,u - 0.00027\,u - 0.00027\,u \\
&= 1.0081\,u
\end{aligned}
\qquad \text{(IV.17)}
$$

With an estimate of errors Chadwick and Goldhaber (1934) reported the mass of the neutron as 1.0081 u $\pm$ 0.0005. This was yet the closest to the current value for the mass of the neutron of 1.00866 u taken from tables of physical constants (Lide, 1997). They then were able to report the binding energy of the deuterium nucleus as 2.1 MeV by taking the difference of the mass of deuterium and that of the sum of its proton and neutron or

$$
\begin{aligned}
\text{Binding energy of } {}_{1}^{2}H &= (\text{mass of } {}_{1}^{1}H + \text{mass of } {}_{0}^{1}n) - \text{mass of } {}_{1}^{2}H \\
&= (1.0078\,u + 1.0081\,u) - 2.0136\,u \\
&= 0.0023\,u
\end{aligned}
\qquad \text{(IV.18)}
$$

Converting the 0.0023 u to units of MeV yields

$$\text{Binding energy of } {}_{1}^{2}H = (0.0023\,u)(931.5\,\text{MeV/u}) = 2.1\,\text{MeV} \qquad \text{(IV.19)}$$

Chadwick's discovery of the neutron, its mass and its charge provided science finally with what was missing to formulate the basic structure of the nucleus and the atom. The importance of this milestone was emphasized by Professor H. Pleijel, Chairman of the Nobel Committee for Physics on December 10, 1935 when he stated

> The existence of the neutron having thus been proved.... The nucleus of atoms is nowadays considered to be composed of a number of protons and neutrons...about the nucleus there circle in the atom electrons...Isotopes are formed by surplus or lack of the number of neutrons in the solid atom...The existence of the neutron having been fully established, scientists have come to a new conception of the structure of atoms which agrees better with the distributions of energy within the nuclei of atoms. It has proved obvious that the neutron forms one of the building stones of atoms and molecules and thus also of the material universe.

In 1935 Chadwick left Cambridge to occupy the Lyon Jones Chair of Physics at the University of Liverpool. During 1943–1946 he worked in the United States as Head of the British team on the Manhattan Project for the development of the atomic bomb. Afterwards he returned to England and in 1948 retired from his post at the University of Liverpool to become Master of Gonville and Caius College at Cambridge until his retirement in 1959. Chadwick served as a part-time member of the United Kingdom Atomic Energy Authority from 1957 to 1962 (Nobel Lectures, Physics, 1922–1941). James Chadwick died on July 24, 1974.

LISE MEITNER (1878–1968) AND OTTO HAHN (1879–1968)

The biographical sketches of Lise Meitner and Otto Hahn are included jointly here because they worked closely together and collaborated professionally for over 30 years. Many of their major contributions to science, including the discovery of nuclear fission, for which they deserve equal credit, were the result of their mutual collaboration. Austria has honored Lise Meitner with a postage stamp illustrated here, which was issued to commemorate the tenth death anniversary.

Lise Meitner was born in Vienna, Austria on November 7, 1878 to Jewish parents. Her father was a successful Jewish lawyer. Although of Jewish heritage, Lise Meitner did not practice the Jewish faith. When she and her two sisters were young adults Lise converted to Protestantism and her sisters to Catholicism, but she was not outwardly religious.

Lise Meitner (1878-1968)

When Lise Meitner was a young girl her parents provided her with private tutoring, as a good formal education for women at that time was difficult. She was gifted in the subjects of physics and mathematics. In a biography by Ruth Lewin Sime, 1996 *Lise Meitner: A Life in Physics* we can appreciate the adverse social and cultural conditions she had to surmount. She passed the entrance exam to the University of Vienna in 1901 after the restrictions on female students were lifted. At the prestigious University of Vienna, she majored in physics and studied under Ludwig Boltzmann. Science, at that time, was the domain of men. Consequently her performance had to be outstanding to overcome bias of some unsympathetic professors and students. She could often find herself as the only woman in a class of over a hundred men. Lise Meitner graduated with a doctorate degree in physics from the University of Vienna in 1905. She wanted to continue her studies in theoretical physics at the University of Vienna, but to her great distress, Boltzmann had taken his life in 1906 during one of his attacks of depression (Frisch, 1978).

Lise Meitner's doctoral dissertation was titled *Wärmeleitung in inhomogenen Körpern* (*Heat Conduction in Inhomogeneous Solids*). After graduation she worked at the Institute for Theoretical Physics in Vienna and wrote her first paper on radioactivity in 1907 *Über die Absorption von α- und β-Strahlen* (*Concerning the Absorption of α- and β-rays*). Meitner was able to demonstrate that alpha particles do not travel in straight lines, but that in passing through matter, they undergo slight irregular deflections (Frisch, 1978). Ernest Rutherford later discovered the backscattering of alpha particles and the existence of the atomic nucleus in 1911. There were then few opportunities in Vienna for women in physics so, with financial assistance from her parents, she moved to Berlin to pursue studies and research on nuclear physics. In Germany she met with the great physicists Max Planck and Albert Einstein and began to study and work under Max Planck, the leading theoretical physicist at the time. Planck let her attend lectures and assist in evaluating the essays of students and to organize seminars. Planck was puzzled by her quest for knowledge and need to carry out research in theoretical physics, possibly because she was a woman, and asked, "You have the doctorate, what more do you want" (Frisch, 1978). Eventually Lise Meitner teamed up with a young chemist Otto Hahn at the Kaiser Wilhelm Institute (KWI), with whom she carried on a very successful research collaboration in the field of radioactivity for over thirty years and most of her career. At the start of their collaboration Otto Hahn had suggested that the young woman physicist help him install and operate the new physical instruments used to measure radioactive sources. As related by Otto Frisch (1978) the suggestion of Hahn alarmed his boss Emil Fischer, who did not want women in his laboratory for fear—it was said—"their fuzzy hair might catch fire from a Bunsen burner". Frisch (1978) relates that women's education was regularized in two years time and the ban lifted. Hahn and Meitner were given, in the meantime, an old carpentry shop to set up their equipment.

Otto Hahn (1879-1968)

Otto Hahn was born on March 8, 1879 in Frankfurt on Main, Germany. He studied chemistry at universities in Marburg and Munich and graduated with his doctorate degree in organic chemistry at Marburg in 1901. After receiving the doctorate degree he remained in Marburg as an assistant at the Chemical Institute for two years before transferring to the University College, London to work under Sir William Ramsay (Nobel Laureate in Chemistry, 1904) where he started his research work on radioactivity. From the autumn of 1905 to the summer of the following year Hahn worked with Ernest Rutherford at the Physical Institute of McGill University in Montreal, Canada. Here he carried out research with Rutherford on alpha radiation from thorium and actinium. Otto Hahn returned to Germany in 1906 and moved to Berlin as university lecturer at the Chemical Institute of the university, and also Head of the Radiochemistry Department of the Kaiser Wilhelm Institute. At the end of

1907 Otto Hahn began his long and fruitful collaboration with Lise Meitner after her arrival from Vienna. Their collaboration would last over thirty years until July 1938. Hahn helped Meitner, as it was not easy for a woman to work at the institute and they had a mutual interest in research in the field of radioactivity. They also made a perfect team; he as the chemist with the knowledge and ability to separate and isolate the many radioactive elements they would encounter, and she as the physicist with the theoretical knowledge needed to interpret the physical properties of radiation and its interaction with matter. Their collaboration was interrupted during 1914–1918 when Hahn was recruited into the war effort where he could apply his scientific knowledge as a chemical-warfare specialist. Lise Meitner was also recruited into the war effort, but used her expertise in x-ray physics to save lives of injured soldiers as an x-ray nurse. They continued their research collaboration at the KWI in 1918.

When Lise Meitner arrived at the KWI in Berlin in 1907 she was a woman in a man's world, and was not immediately welcomed with the exception of a few who knew her talents. The previously noted biography of Lise Meitner by Ruth Lewin Sime (1996), a more recent biography by Patricia Rife, 1999, *Lise Meitner and the Dawn of the Nuclear Age*, and a critique on the book by Goldstein (2001), as well as an article on Lise Meitner's life by Bartusiak (1996) provide much insight into the obstacles that she as a woman and an accomplished physicist at that time encountered, and the perseverance, hard work, and life-long battle for the ultimate truth in nuclear physics, which enabled her to overcome most obstacles. Some of their impressions are summarized herein. It was an uphill battle for her to gain recognition as she was the first among women in both Austria and Germany to be allowed into the institutions of higher learning and research. When she started to work at the KWI in 1907, Lise Meitner was compelled to set up her lab in a converted carpentry shop in the building basement as it was unimaginable to some in management to have a woman work in a laboratory that belonged to men. However, her knowledge and perceptive abilities in nuclear physics quickly established her as an outstanding scientist at the KWI. Lise Meitner and Otto Hahn collaborated as equal partners and Hahn always insisted that she be provided with due credit and be included as coauthor on all of their collaborative research at least during the years she remained at the KWI. For many years Lise Meitner worked with Otto Hahn without salary, as women had no official status at the institute. She received financial assistance from her parents and a meager stipend from her friend, Max Planck, while working at the institute. Eventually by the year 1916, she received a salary comparable to that of Otto Hahn even though she had already for several years carried out equal responsibilities as Hahn. In 1918 she became Head of the Radiophysics Department at the KWI.

During the early stages of their collaboration one of the only precise instruments available to them for measuring radioactive sources was the gold-leaf electroscope. With the aid of a stopwatch they could measure an ionization current by the rate of movement of the gold leaf of the electroscope across a reference scale (see Figure VI.1 and a treatment on the electroscope in Radioactivity Hall of Fame—Part VI). In a biographical sketch Otto Frisch (1978) relates a story about Lise Meitner using the gold-leaf electroscope at the time a mailman entered the room. His story goes as follows: "It [gold-leaf electroscope] was surprisingly sensitive. On one occasion Lise Meitner peering through the eyepiece, greatly startled the postman who had just entered the room by telling him that he had a letter from Manchester in his bag! Rutherford occasionally sent her some radon in a small glass tube, just wrapped in cardboard; unthinkable under today's safety rules, which would demand a leaden box with inch-thick walls. Hahn and Meitner, to work fast and unencumbered by gloves often risked and got small burns on their fingers, sometimes slow to heal; but neither of them suffered serious injury."

As in the case of Marie Curie, Lise Meitner volunteered her expertise in the new x-ray technology as an x-ray nurse (Röntgenschwester) during the war. She worked to exhaustion to help in the diagnosis of many severely injured soldiers for the Austrian army in a military hospital on the East Front. By 1922 Lise Meitner became the first woman to be appointed as extraordinary professor of experimental nuclear physics at the University of Berlin. This was a more prestigious position that would draw more attention both nationally and internationally than that of a researcher at the KWI. During these times such accomplishments for a woman were remarkable. For example, when the name of Professor Meitner had become well known, a book publisher enquired about getting the collaboration of Professor Meitner in the production of an encyclopedia. After finding out that Professor Meitner's first name was Lise, he had nothing to do with a lady as author to any chapter. Lise Meitner remained in Berlin as long as possible, but she had to flee for her life to Sweden in 1938. The German *Anschluss*

or annexation of Hitler's regime with Austria made Lise Meitner a German citizen and thereby subject to all of the anti-Semitic laws of the Nazi state. Her long career in Berlin and collaboration with
Otto Hahn resulted in several historic discoveries. She was eventually honored by Germany with the
postage stamp issued during 1986–1991. The stamp, shown here, illustrates how she appeared as a
young girl, when she embarked on a career of discovery in nuclear physics.

One of the first major outcomes of her collaboration with Otto Hahn at the KWI was the discovery of element 91, protactinium. The element is radioactive, as all natural elements of atomic number
greater than 83 are radioactive. They identified the element as a radioactive isotope and the parent
substance of actinium (Hahn and Meitner, 1918). They named it "protoactinium" from the Greek
"prōtos" meaning first, because it held the first place after uranium in the uranium–actinium decay
series, better known today as the uranium-235 decay series illustrated in Figure IV.3. The name of the
element was shortened in 1949 to protactinium, and the isotope (^{231}Pa) in the decay series identified
by Hahn and Meitner is highlighted in Figure IV.3. Soddy and Cranston (1918) also independently
identified the isotope as the parent of actinium. The two papers reporting the discovery were published almost simultaneously in two different journals, and consequently both parties shared the
credit for the discovery. Another isotope of protactinium (^{234m}Pa) of very short half-life (1.18 min)
was discovered earlier by Kasimir Fajans and Otto H. Göring of Karlsruhe in 1913. They called the
isotope "brevium" meaning brief (Fajans and Goring, 1913). Hahn and Meitner discovered the more
stable isotope (^{231}Pa) of the element, which has a half-life of 3.3×10^4 years).

Lise Meitner (1923) discovered Auger (pronounced OH-ZHAY) electrons. The French physicist
Pierre Victor Auger (1899–1993) also independently discovered this process and carried out studies
on the electron emissions from atoms (Auger, 1923, 1925a,b, 1926). Although both Lise Meitner and
Pierre Auger independently discovered the atomic electron emissions about the same time only
Auger's name has been attached to this phenomena. Auger electrons are atomic electrons that are
emitted from atoms after acquiring energy from an atomic electron transition within the atom. For a
detailed discussion of Auger electrons and their origins see the biographical sketch on Pierre Victor
Auger in Radioactivity Hall of Fame—Part VI.

By 1937 Lise Meitner and Otto Hahn together with a German chemist Fritz Strassmann at the
KWI had identified about ten previously unknown radioactive isotopes of the elements in the uranium

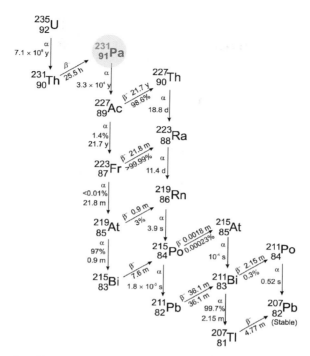

Figure IV.3 The uranium-235 decay series. The protactinium-231 discovered by Hahn and Meitner is high-lighted with a gray circle. They named the element "protoactinium" from the Greek "prōtos" meaning first and the parent of actinium-227. In the figure we can see that it is the first element after uranium in the decay series accord-ing to atomic number. Uranium has atomic number 92, and protactinium has the next lower atomic number of 91. The decay modes are identified as α or β^- and percentages included for branching decay. The abbreviations s, m, h, d, or y represent the half-lives of the radioisotopes in seconds, minutes, hours, days, or years.

decay series. They, as well as others, including Enrico Fermi's group in Rome and Irene Joliot-Curie and Paul Savitch in Paris were researching the transmutation of uranium under the effect of neutron bombardment. This work would in short time lead to Lise Meitner's greatest discovery. Lise Meitner was very productive at the KWI in Germany, well known internationally by 1937 and happy with her work. Albert Einstein had once said "She is our German Marie Curie." The most significant scientific contribution made by Lise Meitner was her discovery of nuclear fission jointly with Otto Hahn. This discovery was not made until February of 1939 after her escape from Nazi Germany in July of 1938.

Lise Meitner was of Jewish decent and, as a consequence of the annexation of Austria by Germany in 1938, she became a German citizen. With her job and life in peril Lise Meitner escaped to Sweden on July 17, 1938 by way of The Netherlands in the company of Dirk Coster (1889–1950), the Dutch physicist, who together with George de Hevesy in Copenhagen discovered the element hafnium (Hf) taken from the word *Hafnia*, Latin for Copenhagen. Lise Meitner's escape from Germany was a dan-gerous one, and the details of her preparation and escape can be found in a detailed account by Rhodes (1986).

During Lise Meitner's escape to Sweden Otto Hahn and Fritz Strassmann continued their research on the transmutations that occur upon the neutron bombardment of uranium. Similar work by Joliot-Curie and Savitch in Paris also continued. In Sweden Lise Meitner had been provided a position in Manne Siegbahn's Institute of Physics in Stockholm. Here she was provided with a place to work, but with insufficient equipment, staff or funds to carry out significant work. Nevertheless, by correspondence Otto Hahn kept Lise Meitner informed of the details of their work in Germany on the transmutations resulting from the neutron bombardment of uranium, and he asked Lise Meitner for guidance and input into the interpretation of their findings. By December 1938, Hahn and Strassmann had confirmed through fractionation and recrystallization of the numerous transmutation products with barium carrier that radioactive barium (atomic number 56) was one of the products of neutron bombardment of uranium (atomic number 92). Such a result was inconceivable to him, as he or anyone else could not then imagine how uranium with its large nucleus containing 92 protons, could yield such a small atomic nucleus as barium with only 56 protons. As related in Rhodes' (1986) detailed account, Hahn wrote Meitner on December 19, 1938 requesting advice as follows

> Perhaps you can suggest some fantastic explanation,"…We understand that it *can't* break up into barium…. So try to think of some other possibility. Barium isotopes with much higher atomic weights than 137? If you can think of anything that might be publishable, then the three of us would be together in this work after all. We don't believe this is foolishness or that contaminations are playing tricks on us.

Lise Meitner wrote Hahn back the same week of his letter accepting the fact that barium was a product of slow-neutron bombardment of uranium and informing him that, although it is difficult to believe, anything is possible. As related by Richard Rhodes' (1996) Hahn and Strassmann on December 23 had reconfirmed that the barium isotope was the product because it would decay to the element lanthanum only one atomic number higher. Hahn was anxious to find an understanding as to what type of phenomenon was occurring, and he was fearful that others, such as Joliot-Curie and Savitch, might soon make the same discovery—a race was on and the matter was urgent. As related by Rhodes (1986) Hahn wrote Lise Meitner again in late December desperate for an explanation of the results

> We cannot hush up the results, even though they may be absurd in physical terms. You can see that you will be performing a good deed if you find an explanation. When we finish tomorrow or the day after I will send you a copy of the manuscript.

Hahn's letter to Stockholm did not reach Meitner's hands immediately as she was away from Stockholm to meet her nephew physicist Otto Frisch in the village of Kungälv, Sweden. Frisch had arrived from Copenhagen. While hiking in Kungälv on December 19, Lise Meitner discussed with Otto Frisch the problem of barium—that is, how could barium of such low atomic number (56) be a product of neutron bombardment of uranium of high atomic number (92). As related by Frisch (1978), they thought of Bohr's theory that the nucleus of an atom was like that of a liquid drop. It then occurred to Lise Meitner that the large uranium nucleus made unstable by its high electric charge, might burst and divide itself into two smaller nuclei when shaken up by a neutron, such as a living cell becomes two smaller cells by fission. In the outdoors of that winter day Lise Meitner sketched the form of a liquid drop to simulate the uranium nucleus, which after capturing a neutron, could become so unstable as to elongate into a dumbbell shape with opposing positive forces of the nuclear protons and break apart into two smaller nuclei. Meitner and Frisch then and there on that winter hike made calculations of the immense energy that would be liberated upon splitting the uranium nucleus—approximately 200 MeV. Nuclear fission had been discovered by Meitner and Frisch while hiking in December of 1938. They coined the word "fission" by borrowing the word from biology, a term used to describe splitting bacteria cells (Michaudon, 2000). Rhodes (1986) relates, in his detailed historical account, within a few days of her discovery of fission, Meitner wrote Hahn the following on January 1, 1939:

> We have read your work very thoroughly and consider it *perhaps* possible energetically after all that such a heavy nucleus bursts.

On January 3 after reviewing Hahn's paper (Hahn and Strassmann, 1939a) proving barium to be a product of neutron bombardment of uranium, she wrote Hahn explaining to him that what was occurring in the production of barium from uranium was definitely a splitting of the nucleus, but she did not go further with providing him with any further information concerning a theory on how the nucleus of uranium could split. She wrote the following:

> I am fairly *certain* now that you really have a splitting towards barium and I consider it a wonderful result for which I congratulate you and Strassmann very warmly.

Now that Hahn and Strassmann (1939a) would publish their experimental results proving that barium was a product of neutron bombardment of uranium (their paper was published in January 1939), Meitner and Frisch intended to publish the interpretation of these results. In an article published in the journal *Nature* on February 11, 1939 entitled *Disintegration of Uranium by Neutrons: A New Type of Nuclear Reaction*, Meitner and Frisch were the first to describe and name the phenomenon of nuclear fission. An excerpt from Meitner and Frisch (1939) describing fission for the first time is the following:

> ...Hahn and Strassmann (1939a) were forced to conclude that isotopes of barium ($Z = 56$) are formed as a consequence of the bombardment of uranium ($Z = 92$) with neutrons...In the basis, however of present ideas about the behaviour of heavy nuclei (Bohr, 1936), an entirely different and essentially classical picture of these new disintegration processes suggests itself. On account of their close packing and strong energy exchange, the particles of a heavy nucleus would be expected to move in a collective way which has some resemblance to the movement of a liquid drop. If the movement is made sufficiently violent by adding energy, such a drop may divide itself into two smaller drops...It seems therefore possible that the uranium nucleus has only small stability of form, and may, after neutron capture, divide itself into two nuclei of roughly equal size (the precise ratio of sizes depending on final structural features and perhaps partly on chance). These two nuclei will repel each other and should gain a total kinetic energy of approximately 200 MeV, as calculated from nuclear radius and charge. This amount of energy may actually be expected to be available from the difference in packing fraction between uranium and the elements in the middle of the periodic system. The whole fission process can be described in an essentially classical way...

An illustration of neutron capture of uranium, as currently known today, is illustrated in Figure IV.4. The figure also illustrates uranium fission as first described by Meitner and Frisch (1939).

Meitner and Frisch (February 11, 1939) went even further to explain in their historic paper reporting nuclear fission for the first time that, if one of the fission products (e.g., X of Figure IV.4) was barium ($Z = 56$), then the other fission product (e.g., Y of Figure IV.4) must be krypton ($Z = 36$), as the atomic number of the two fission products must equal 92, the atomic number of the parent element uranium. In their paper Meitner and Frisch (1939) stated

> After division, the high neutron/proton ratio of uranium will tend to readjust itself by beta decay to the lower value suitable for lighter elements. Probably each part will give rise to a chain of disintegrations. If one of the parts is an isotope of barium (Meitner, Strassmann, and Hahn, 1938), the other will be krypton ($Z = 92-56$),...

In the above statement Meitner and Frisch gave one specific example of two fission fragments of uranium-236; however, they noted that numerous different fission fragments are possible. The example provided in the above quotation can be written as follows:

$$\ce{^{235}_{92}U + ^{1}_{0}n \rightarrow ^{236}_{92}U} \xrightarrow{\text{fission}} \ce{_{56}Ba + _{36}Kr} + x\, \ce{^{1}_{0}n} \tag{IV.20}$$

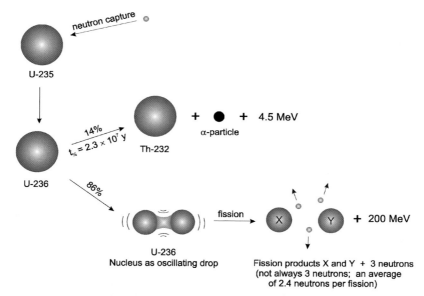

Figure IV.4 Uranium-235 neutron capture. The capture of a slow neutron by uranium-235 yields the nuclide uranium-236. A small number of the neutrons captured (14%) yield a U-236 nucleus that decays with a long half-life ($t_{1/2}$) to thorium-232 with the emission of an alpha particle and 4.5 MeV energy. A larger number of the neutrons captured by U-235 (86%) yield a U-236 nucleus, that is highly unstable, capable of oscillating like a liquid drop and breaking into two smaller nuclides X and Y of different masses and of atomic numbers that add up to 92 (the atomic number of the parent uranium) plus the emission of an average of 2.4 neutrons and a lot of energy (approximately 200 MeV). The emission of three neutrons is illustrated above.

The x in the last term of the equation represents an indeterminant number of neutrons. The number of neutrons emitted from uranium-236 fission will vary, and the average number is 2.4. The fission reaction first explained by Meitner and Frisch was commemorated by a postage stamp shown here issued by the German Democratic Republic (DDR) in 1979. The stamp pictures only Otto Hahn together with the historic fission reaction. Lise Meitner had not been given equal credit for the work leading to the discovery or for her part in the discovery of fission. Due recognition did not come until the award of the Fermi Prize to Meitner, Hahn, and Strassmann in 1966 for the discovery of nuclear fission only a couple of years prior to the deaths of Hahn and Meitner in 1968.

Meitner and Frisch also tried to explain the relative stability of some of the uranium isotopes that is, why they all do not undergo fission (e.g., U-236 with a long half-life illustrated in Figure IV.4). In their paper announcing fission they provided the following explanation:

> The formation of this body [uranium nucleus] is a typical resonance process [Bethe and Placzec, 1937], the compound state must have a life-time of a million times longer than the time it would take the nucleus to divide itself. Perhaps this state corresponds to some highly symmetrical type of nuclear matter, which does not favor fission of the nucleus.

In this paper Meitner and Frisch (1939) also described fission of thorium for the first time. The following is an excerpt of that statement:

> By bombarding thorium with neutrons, activities are which have been ascribed to radium and actinium isotopes (Meitner, Strassmann, and Hahn, 1938). Some of these periods [half-lives] are approximately equal to periods of barium and lanthanum isotopes resulting from the bombardment of uranium. We should therefore like to suggest that these periods [half-lives] are due to a fission of thorium, which is like that of uranium and results partly in the same products.

Hahn and Strassmann (1939b) published their paper providing the analytical chemistry proof of uranium fission as well as the fission of thorium on February 10, 1939, 1 month after the publication of Meitner and Frisch's paper describing nuclear fission. They also identified the fission fragments of radioactive strontium ($Z = 38$) and yttrium ($Z = 39$) for the first time. Although Hahn and Strassmann did not include Lise Meitner as coauthor of the paper, they did acknowledge her contribution to the work at the end of the paper as follows:

> In a rather short time it has been possible to identify numerous new reaction products described above—with considerable certainty, we believe—only because of the previous experience we had gathered, in association with L. Meitner, from the systematic study of uranium and thorium reaction products.

Nuclear Fission

The postage stamp shown here was issued in Germany in 1979. It illustrates the fission of a uranium nucleus as initially proposed by Meitner and Frisch in 1939. The stamp illustrates from left to right a neutron being captured by a uranium nucleus, which forms an unstable "liquid-drop" or dumbbell shape and then undergoes fission to two separate and smaller atomic nuclei (fission products) and three neutrons. The stamp does not cite Lise Meitner for her role in the discovery of nuclear fission, but only Otto Hahn, who received the Nobel Prize in Chemistry 1944 for the discovery of nuclear fission. This was an apparent injustice or mistake on the part of the Nobel Committee and particularly on the part of Otto Hahn for not giving her sufficient credit to share the Nobel Prize. The truth about this unfortunate discrimination has recently come to public light with the detailed biographies of Lise Meitner written by Sime (1996) and Rife (1999). The questions about the unfairness, which Lise Meitner was dealt concerning the credit due her were highlighted by Goldstein (2001), and a relevant excerpt of his comments is the following:

> The history of fission's discovery raises many questions. Why was Meitner not included as co-author on Hahn and Strassmann's paper? Rife (1999) doesn't consider this point, but the fact that Meitner had been an active partner in this particular research until her escape in July reasonably should have led to her inclusion on the paper. Sime's book (1996) speculated that it might have been dangerous for Hahn to include this recently escaped non-Aryan as a co-author. Most of the scientists involved in this line of research knew that Meitner had been an equal partner in the work anyway. Unfortunately, though, after the war Hahn continued to downplay Meitner's role in the research, as Rife (1999) shows. During the war "he refused to take a stand on the politics in and out of his Institute: the Third Reich was blinding Hahn and he began to discount Meitner's insights and contributions he had frantically *sought out* months earlier. We witness here appeasement, professional cowardice, and worse" (Rife, 1999, p. 213). Perhaps Hahn's initial lapse is understandable, given that the editor or some other authorities may have rejected the groundbreaking paper for political reasons, although Hahn knew the editor personally. More likely he would have had to stand up to the Nazi dictatorship of the Institute and the official scientific establishment, thereby jeopardizing his position. And when he finally received recognition for the paper and the acclaim from his German colleagues that followed, he was not about to share credit with Meitner. There is no question that his behavior was scurrilous. He must have constructed a tortuous self-justification to assuage his guilt for deserting his decades-long collaborator and friend. He never adequately acknowledged her essential role in the discovery, even long after he was awarded the Nobel Prize and was a leading figure in post-war German science.

The Nobel Committee overlooked Lise Meitner when they awarded the Nobel Prize in Chemistry 1945 to only Otto Hahn even though she had collaborated with Otto Hahn for three decades right up to July 1938, when she escaped from Nazi Germany, and the work that led to the discovery of fission was concluded only a few months later. The absence of Lise Meitner's name as coauthor to Hahn and Strassmann's paper of February 1939 may have contributed to this mistake of the Nobel Committee, although she was never acknowledged by the Royal Swedish Academy of Sciences. Niels Bohr tried to assure priority credit to Lise Meitner for her interpretation of fission, but this was not sufficient to prevent the 1944 Nobel Prize in Chemistry to be awarded exclusively to Otto Hahn (Goldstein, 2001), which was announced after the war on November 15, 1945. This mistake certainly had been gnawing at the consciences of the scientific community for at least two decades since the awarding of the Nobel Prize to only Otto Hahn. Some rectification of the error was made with the award of the United States Fermi Prize in 1966 to Hahn, Meitner, and Strassmann. The Enrico Fermi Award was established in 1954. It is the most prestigious science and technology award given by the U.S. Government. It recognizes scientists of international stature for their lifetimes of exceptional achievement in the development, use, or production of energy. The Fermi Award in 1966 was bestowed equally to Lise Meitner, Otto Hahn, and Fritz Strassmann "For contributions to nuclear chemistry and extensive experimental studies contributing to the discovery of fission." A yet greater honor was bestowed posthumously to Lise Meitner when element 109 was named meitnerium in 1997 in honor of Lise Meitner, the Austrian physicist who first conceived the idea of nuclear fission. Finally, following the detailed historical accounts of Lise Meiner's life and work by Sime (1996) and Rife (1999) the general public has come to realize the mistake, and consequently the very first postage stamp that honors both Lise Meitner and Otto Hahn for the discovery of fission was issued by St. Vincent & The Grenadines in the year 2000. The stamp shown here is that of a famous photo taken of Lise Meitner and Otto Hahn in their laboratory

Lise Meitner and Otto Hahn

at the KWI in Berlin-Dahlem prior to her escape from Germany in 1938. The fitting statement of the stamp reads "1938: The discovery of Fission".

Albert Einstein called her "Our Madame Curie" (Frisch, 1978). The life of Lise Meitner, like that of Marie Curie, should serve as an example for all on how hard work and dedication can overcome many obstacles no matter how insurmountable they may appear. Goldstein (2001) put it very clearly when he wrote

> Though she was never given the same level of recognition and reward as her collaborator Hahn, Lise Meitner's single-minded dedication to physics while faced with a myriad of obstacles is proof of an extraordinary person. And with such extraordinary people we want to know how they came to be.

Her life was difficult, as she apparently encountered obstacles at almost every turn; nevertheless, Lise Meitner had a life full of satisfaction and fulfillment. We can be certain of this from the following quotation of Lise Meitner, five years before her death, from a biography by Rosner and Strohmaier (2003) and a review article in the CERN Courier (2004):

> I believe that all young people think about how they would like their lives to develop. When I did so, I always arrived at the conclusion that life need not be easy; what is important is that it not be empty. And this wish I have been granted.

Lise Meitner retired in 1960 in Cambridge, England. She continued to lecture and visit friends up to 1964 when she suffered a heart attack after a strenuous trip to the United States. She recovered from the heart attack and the following year was awarded the Enrico Fermi Prize with Otto Hahn and Fritz Strassmann for contributions to nuclear chemistry and the discovery of nuclear fission.

Glenn T. Seaborg, then Chairman of the US Atomic Energy Commission went to Cambridge to present her share of the award (Frisch, 1978). Lise Meitner died on October 27, 1968. Otto Hahn, after receiving the Nobel Prize in 1945, continued his research on the identification and separation of many radioactive isotopes that arise through fission. He died a few months before Lise Meitner on July 28, 1968 (Nobel Lectures, Chemistry, 1901–1970).

LEO SZILARD (1898–1964)

Leo Szilard was born Leo Spitz on February 11, 1898 to an affluent Jewish family in Budapest, Austro-Hungary. The family name was changed to Szilard in 1900. He was one of the world's greatest thinkers among the physicists of the 20th century. His mind was never idle and was always conceiving new scientific theories and inventions. He is recognized mostly for his pioneering thought and contributions to nuclear physics, his numerous patents of instruments that have revolutionized 20th century nuclear physics, his patent for the atomic bomb and his spearheading of the formation of a program for its development, the design of the first nuclear reactor (with Enrico Fermi), and his efforts to prevent the nuclear arms race after the Second World War. Szilard did not win the Nobel Prize for any of his achievements; however, two of his patents, which were for the design and operation of the cyclotron (Rhodes, 1986) and electron microscope, would be developed subsequently by Ernest Lawrence and Ernst Ruska, respectively, providing them with the instrumentation to make discoveries and merit the Nobel Prize. Other inventions, patents, and scientific discoveries of Leo Szilard, which are described subsequently, established his legacy forever among the greatest physicists of the century.

Leo Szilard (1898-1964)

Leo Szilard graduated from the University of Budapest in 1916 winning the Eötvös Prize, the Hungarian National Prize in mathematics (Rhodes, 1986). His main interest was physics, but at that time in Hungary there were no future professional opportunities in physics. Thus, he decided to begin his higher education in engineering at the Budapest Technical University. However, his studies in Hungary were cut short as he was drafted into the Austro-Hungarian Army in 1917 and honorably discharged in 1918 while on leave after recovering from influenza. He was fortunate to escape death, as most of his fellow troops were killed. It was a war that stole so many potentially productive young souls from either side of the trenches including the budding physicist Henry G. J. Moseley (see Radioactivity Hall of Fame—Part III).

In 1919, Szilard left his homeland both to escape the anti-Semitic regime there and to pursue studies in Berlin, which was one of the leading centers of the world for scientific learning and research. He enrolled in the Technische Hochschule (Institute of Technology) in Berlin-Charlottenburg as an engineering student in 1920, but only shortly thereafter, the following year, he began a career in physics transferring to the University of Berlin where some of the greatest physicists of the time were located including Nobel Laureates Albert Einstein, Max Planck, and Max von Laue.

It was at the University of Berlin where Leo Szilard at the age of 23 would demonstrate his amazing intellect by completing his doctoral thesis or dissertation in physics in 1922 only one year after enrolling in the university. A detailed account of Szilard's early accomplishment as a university student is given by Rhodes (1986), only a brief summary will be given here. The award of a university doctoral degree in the sciences at the University of Berlin required the satisfactory completion of four years of course work and a dissertation describing a truly original work. The dissertation could describe either a noteworthy discovery or research results that would contribute significant knowledge to a particular field of science. The topic of the dissertation could be of the students' choice or one selected by his or her supervisor. Following approval of the dissertation by a committee of professors, the final requirement is the completion of a rigorous oral examination. These requirements for a doctoral degree in the sciences still exist today in the major universities of the world. Max van Laue would accept only outstanding students to supervise in their doctoral work and Leo Szilard was one of those. Szilard went to von Laue during his first year as a student to request a topic for his dissertation. Von Laue gave Szilard a very perplexing problem on relativity, possibly to put Leo Szilard in a humble position or just to challenge him and observe his reaction.

For six months Szilard worked on the thesis problem that von Laue had given him and was convinced that it could not be solved. He then abandoned the thesis topic and just put his mind to thought for three weeks to whatever idea would come to mind. A topic on thermodynamics came to him and from time-to-time he would put to paper whatever ideas materialized in his mind. Over a period of only three weeks he was convinced that he had a manuscript that was truly original. Szilard was afraid to take the manuscript to Max von Laue, because it was not the topic given to him to work on. Instead he cornered Einstein after a seminar and asked him if he would look at what he was working on. Einstein asked him specifically what he was doing. When Szilard told Einstein the topic, Einstein replied that it could not be done. However, after Szilard gave Einstein a 5- to 10-min detailed explanation of what he had done, Einstein understood. This boosted Szilard's confidence to take his manuscript to von Laue as his completed dissertation. The manuscript was entitled *Uber die thermodynamischen Schwankungserscheinungen*. In this work Szilard mathematically demonstrated that the second law of thermodynamics covers not only the mean values, as was up to then believed, but also determines the general form of the law that governs the fluctuating values. The dissertation presented ideas that would be the foundation of modern information theory. (UCSD, 2005). The following morning Max von Laue called Leo Szilard to inform him that his thesis was accepted for his doctorate degree, which he earned *"cum laude"* in 1922. Leo Szilard would then begin a lifelong friendship and collaboration with Albert Einstein. Subsequently Leo Szilard published additional work on the topic of his thesis in the journal *Zeitschrift für Physik* (Szilard, 1925, 1929).

Szilard did his postdoctoral work on x-ray diffraction in crystals at the KWI in Berlin-Dahlem until 1925 when his paper on thermodynamics (Szilard, 1925) was accepted for his *Habilitationsschrift*. He was thereafter appointed as a private lecturer *Privatdozent* in 1927 at the University of Berlin. While remaining in Berlin during 1925–1933 Szilard applied for numerous German patents, some of these with Albert Einstein. Some of the most prominent German patents and patent applications submitted by Szilard are described by Telegdi (2000) with diagrams. Among these are the patent applications for particle accelerators in great detail including the linac or linear accelerator, cyclotron, and betatron, which Szilard submitted January 17, 1929. Ernest Lawrence, who won the Nobel Prize in Physics 1939 "for the invention and development of the cyclotron and for the results obtained with it…" did not get the idea for the cyclotron until the summer of 1929, when he came across a publication of the Norwegian engineer Rolf Wideröe (Wideröe, 1928) concerning his idea for the linac or linear accelerator; although the linear accelerator was first reported by the Swedish scientist Gustaf Ising (1924). The year following Wideröe's paper, Ernest Lawrence (1930) published the cyclotron principle in the journal *Science*. It is recognized clearly that Leo Szilard was the first to conceive of the principle of the cyclotron for accelerating charged nuclear particles, which was very instrumental in producing many artificial radioactive elements. However, the Nobel Prize would go to Ernest Lawrence, who independently conceived and developed the cyclotron and demonstrated its utility.

Gene Dannen (1997, 1998, 2001) has thoroughly reviewed the life and work including the many patents of Leo Szilard. Dannen (1998) provides the following comments of Dennis Gabor, one of Szilard's close associates in Berlin:

Dennis Gabor (1900-1979)

Dennis Gabor, who won the 1971 Nobel Prize for the invention of holography, was one of Szilard's best friends in Berlin. "He used to discuss all his inventions with me," Gabor later recalled. "I was so full of admiration that I felt stupid in his presence. Of all the many men I have met in my life, he was, by far, the most brilliant." Gabor also said of Szilard. "Had he pushed through to success all his inventions, we would now talk of him as the Edison of the 20th century."

Gene Dannen (1998) also relates the following circumstances around Leo Szilard's patent on the electron microscope and his urging that Dennis Gabor build it:

It is not as well known that Szilard also foresaw, and filed for patent on, the electron microscope. Dennis Gabor later recalled how, in 1928 at the Café Wien in Berlin, Szilard tried to convince him to build such a microscope. Gabor told Szilard that such a device would be useless. Gabor told Szilard, and I quote, "one cannot put living matter into a vacuum and everything will burn away to a cinder under the electron beam." Gabor later regretted that he had not taken Szilard's advice. It was only a few years later that work on developing electron microscopes began at several different laboratories. Szilard filed for patent on the electron microscope on July 4, 1931.

Historians have pondered over the question of why Szilard did not pursue his patent applications onto the development and application of the instruments he conceived. Gene Dannen (1997, 1998) has carefully studied this question and points out the following:

Why, in fact, were so many of Szilard's inventions not developed further? It is often said that Szilard was psychologically incapable of pursuing his ideas to completion. I would like to disagree with that answer. It is certainly true that Szilard was willing to drop one idea when a more important idea came along. And he was a person with so many important ideas. But when you are so far ahead of your time as Szilard often was, the obstacles of the acceptance of your ideas can be almost insurmountable. In the case of each of Szilard's inventions, there were specific reasons why they were not developed. Above all other reasons loomed the largest: the lack of a peaceful world in which to pursue them.

It is true that Leo Szilard could not pursue the development of most of his patents, because he became overwhelmed with Hitler's rise to power and the need to escape his anti-Semitic regime. However, we will see further on in this biosketch that, Leo Szilard did pursue the development of his patent for the atomic bomb. He would pursue this out of fear for Hitler's regime, to beat German scientists in the development of the bomb, and to keep the bomb in the hands of the free world.

A number of patents were filed by Leo Szilard together with Albert Einstein in the development of refrigerators without moving parts. This work was outside the scope of nuclear physics, as it involved thermodynamics. They developed the first cooling system that involved liquid sodium metal as the refrigerant whereby the metal was pumped by a variable magnetic field. The Szilard–Einstein pump was developed into a prototype refrigerator by the German Electric Company (A.E.G) for which Einstein and Szilard were paid handsomely for their patent. The subsequent discovery of Freon as a cheap refrigerant changed the direction of commercial production away from the Einstein–Szilard refrigerator. Dannen (1997) gives an excellent review of the history of the Einstein–Szilard refrigerators. He points out that had Freon not been implemented as a coolant, the Einstein–Szilard refrigerator might have been a common household appliance today and the problem of Freon diminishing the ozone layer less of a concern. Also Dannen (1997) points out that the Einstein–Szilard pump did find use eventually in the cooling of the nuclear breeder reactor.

After Hitler became Chancellor of Germany in January of 1933, it soon became unsafe for Leo Szilard to remain in the country. He escaped by train from Berlin to Vienna in April one day before it would have become impossible or very difficult for him to leave the country. Richard Rhodes (1986) writes that on the weekend of April 1, Julius Streicher directed a national boycott of Jewish businesses and Jews were beaten in the streets. Rhodes then quotes Szilard concerning his escape as follows:

I took a train from Berlin to Vienna on a certain date, close to the first of April, 1933. The train was empty. The same train the next day was overcrowded, was stopped at the frontier, the people had to get out, and everybody was interrogated by the Nazis. This just goes to show that if you want to succeed in this world you don't have to be much cleverer than other people, you just have to be one day earlier.

Leo Szilard tried for months to help other Jewish scholars who were evicted from their professional posts in Germany to emigrate and find jobs in other countries of the world. Later in 1933 Szilard moved to London. On September 12, 1933 Szilard read a news report in the *The Times* where Rutherford was interviewed. The reporter had asked Rutherford what were his thoughts concerning nuclear transmutations and nuclear energy. Rutherford replied that charged particles, such as the proton, could eventually be accelerated with sufficient high voltage to eventually transform all of the elements. He added that the energy released would be too small to produce energy on a large scale for power and anyone who looked for transmutation as a source of power was talking "moonshine". Szilard agreed with Rutherford's comments concerning accelerated protons, that these would not be a useful source of energy. He knew that the proton had to overcome the coulombic barrier of its repelling positive charge with that of the atomic nucleus. However, he thought, that the neutron, because of its neutral charge, should be able to penetrate the coulombic barrier of the electron shells of an atom and collide with the nucleus without any external force and cause a reaction or transmutation of the nucleus. Szilard's mind was restless, while standing at a street corner in London in the following month of October and during the time that it took a traffic light to change, the idea of the possibility of a neutron-induced nuclear chain reaction hit him. Szilard knew, that in chemistry, there existed chain reactions where one reaction could yield two products that could further interact with other reactants to produce four products and these could react to produce eight products, etc. resulting in an exponential increase in reactions. The chain reactions could be very numerous and limited only by the quantity of reactants available. He thought, that if a chain reaction with neutrons could occur, that is, if a neutron would be able to interact with an atomic nucleus to produce more than one neutron and the additional neutrons created would interact with other atomic nuclei to produce yet more neutrons, a tremendous amount of energy could be liberated. Szilard could foresee that the initial neutron reaction and the resultant multiplication of neutrons should occur in a very short period of time, in microseconds, as nuclear reactions would be fast, producing an explosion of unforeseeable magnitude. Leo Szilard was the first to conceive the possibility of obtaining large amounts of energy and even an explosion from nuclear transmutations.

Having formulated further his initial idea of the potential energy that could be released from neutron-induced chain reactions on March 12, 1934 Szilard filed for a patent, which contained the basic concepts of a nuclear chain reaction, the concept of "critical mass", that is, the minimum mass of a material required for a sustained nuclear chain reaction, and the "explosion" that could result. He was awarded British Patent No. 630,726 entitled *Improvements in or Relating to the Transformation of Chemical Elements*. In his patent Szilard described the concept of critical mass in his statement "If the thickness is larger than the critical value…I can produce an explosion." (Feld and Weiss-Szilard, 1972; Loeber, 2002). Lise Meitner had not discovered nuclear fission until 1939. Consequently, the fissile materials required to produce neutron-induced nuclear chain reactions had not yet been discovered and there is no record of any such nuclear chain reactions ever been discussed prior to Szilard's patent. Szilard was clearly ahead of his time. The patent as written by Szilard stated "This invention has for its object the production of radioactive bodies and the liberation of nuclear energy for power production and other purposes through nuclear transmutation." (Feld and Weiss-Szilard, 1972). Szilard was a pacifist and idealist, and the objective of his patent was not personal gain but for control over the harmful use of nuclear weapons. After a year's time he convinced the British Government to accept the patent in order to keep the concepts of the patent secret and inaccessible to the world. Later in life Leo Szilard joined the Manhattan Project, which was the secret American project for the production of the atomic bomb, out of fear that Germany could become the first to acquire the weapon. Rhodes (1986) and Loeber (2002) point out that Szilard tried to use his patent as a means to gain control of the decision-making process for the atomic bomb. He felt that the control of the bomb and its use should be in the hands of the wisest scientists and not in the hands of government officials. Rhodes (1986) noted that the U.S. Government rejected Szilard's claim to the atomic bomb on simple legalistic grounds, because he did not disclose the existence of his patent before joining the Manhattan Project.

When Szilard first conceived of neutron-induced fission in 1933, he did not know what element or isotope would undergo such fission. His patent had made a slight mention of uranium and thorium, but his thoughts were mistakenly on beryllium. He therefore, started a search for the chain-reacting element. With this in mind Szilard began a collaboration of nuclear research with T. H. Chalmers at St. Bartholomew's Hospital. This research did not lead to the discovery of the neutron-induced chain reaction, but lead to other important discoveries including (i) the discovery of a photoneutron (γ, n) source, that is, the induction of neutron emission by gamma- or x-radiation, and (ii) the Szilard–Chalmers effect, which is the rupture of a chemical bond between and atom and a molecule of which the atom is a constituent, as a result of a nuclear reaction or radioactive decay of that atom. The Szilard–Chalmers effect led to the Szilard–Chalmers process or reaction, which has enabled the isolation of high-specific activities of radionuclides. The two discoveries and their applications will be discussed subsequently in brief.

In his search for neutron emission that might initiate a neutron-induced chain reaction, Leo Szilard and T. H. Chalmers irradiated beryllium with radium gamma rays. They discovered a radiation from the beryllium that would induce radioactivity in iodine and concluded that neutrons were emitted from the beryllium by the gamma rays (Szilard and Chalmers, 1934a,b, 1935). Their report was the first where neutron emission was initiated by gamma radiation, and the reaction that they observed can be abbreviated as

$$^9Be(\gamma, n)^8Be \qquad\qquad\qquad (IV.21)$$

which reads:

^9Be target nuclide(gamma ray projectile, neutron as the detected particle)^8Be product nucleus.

This discovery was significant, because today photoneutron (γ, n) sources are useful portable sources of neutrons. A very common neutron source of this nature used today is the mixture of ^{124}Sb + Be where the gamma rays from ^{124}Sb yield neutrons via a photonuclear reaction (see Section 4.3.4 of Chapter 4).

Szilard and Chalmers (1934b) used iodine as the indicator for neutrons as the neutrons would induce radioactivity in iodine by creating radioiodine via neutron capture. A subsequent work was reported shortly thereafter by Brasch and coworkers together with Szilard and Chalmers and the assistance of Lise Meitner in Berlin (Brasch et al., 1934) whereby the neutron emission from beryllium was induced by x-ray photons and neutron-capture by bromine and the isolation of the product nuclide, radiobromine, was used as the neutron indicator.

An important outcome of this work was the discovery of the Szilard–Chalmers process whereby radionuclides may be easily separated and isolated from their parent atoms and consequently high specific activities of radionuclides may be obtained which otherwise would be difficult when both the target and product atoms are isotopes. For example, Szilard and Chalmers (1934a,b) surrounded 150 mg of radium (gamma-ray source) with 25 g of beryllium in a sealed container of 1 mm thick platinum. The Ra–Be neutron source was immersed in 100 ml of ethyl iodide. The neutrons produced radioactive iodine-128 from stable iodine-127 by neutron capture. The ^{128}I would precipitate as radioactive silver iodide ($Ag^{128}I$) after the addition of water containing silver ions (Ag^+). However, a control experiment without the neutron source produced no radioactivity over background in the silver iodide precipitate. This experiment indicated that the chemical bond between the carbon atom of the ethyl group and the ^{128}I was broken after ^{127}I neutron capture, because free unbound ionic $^{128}I^-$ would precipitate with the Ag^+ to produce radioactive $Ag^{128}I$ precipitate in the aqueous phase according to the sequence illustrated in Figure IV.5

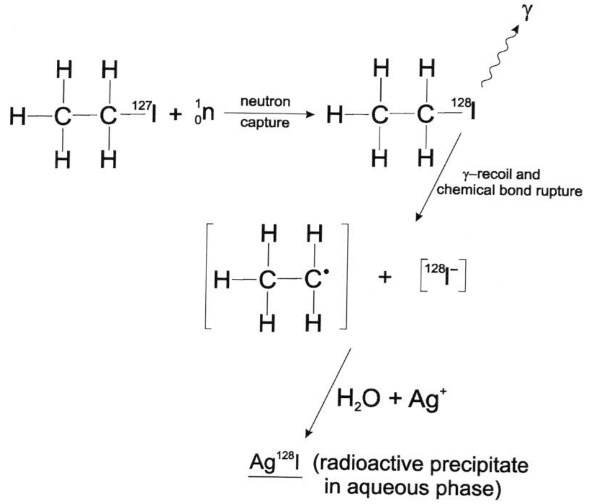

Figure IV.5 Szilard–Chalmers process. High specific activity ^{128}I is separated from the stable ^{127}I as a result of the Szilard–Chalmers effect. The capture of a neutron by the stable isotope ^{127}I in the ethyl iodide molecule created radioactive ^{128}I. The neutron capture process leaves the ^{128}I nucleus in an excited state, which emits a gamma ray upon deexcitation. The gamma recoil causes a rupture of the chemical bond between the carbon atom of the ethyl group and the ^{128}I thereby liberating the ^{128}I as an anion. The addition of water containing silver cations (Ag^+) to the organic solution enables the ionic $^{128}I^-$ to enter the aqueous phase and separate out as precipitated $Ag^{128}I$. The radioactivity of ^{128}I is detected by its beta-particle emissions, and it decays with a half-life of 25 min.

The energy of bonding between the carbon and iodine atoms is about 2 eV, which is higher than the neutron recoil energy. Consequently, the rupture of this bond is due to the gamma recoil when the ^{128}I undergoes gamma emission upon deexcitation after neutron capture. Some atoms of ^{128}I could recombine with the free ethyl group and even exchange with some stable atoms of ^{127}I, but if these processes are slow and, if these processes are further reduced by the addition of water or even alcohol to dilute the organic phase, the precipitate of radioactive inorganic iodide proceeds yielding a highly enriched radioactive ^{128}I with minimal ^{127}I.

The Szilard–Chalmers process remains to this day a very practical application to the isolation of high specific activity radioactive sources from the medium in which the radionuclide sources were synthesized. A few examples taken from the literature are the isolation of high specific activity radioisotopes after cyclotron production (Birattari *et al.*, 2001; Bonardi *et al.*, 2004;), the preparation of high specific activity ^{64}Cu for medical diagnosis (Hetherington *et al.*, 1986), high specific activity radiohafnium complexes (Abbe and Marques-Netto, 1975; Marques-Netto and Abbe, 1975), high specific activity ^{51}Cr (Green and Maddock, 1949; Harbottle, 1954), high specific activity ^{56}Mn (Zahn, 1967a,b), high specific activity ^{156}Re (Jia and Ehrhardt, 1997), high specific activity ^{166}Ho (Zeisler and Weber, 1998), high specific activity radioisotopes of tin (Spano and Kahn, 1952), and the concentration of high specific activity of the radioisotopes of Groups IV and V elements (Murin and Nefedov, 1955), etc.

On September 30, 1938 Germany, Italy, France, and Britain signed the Munich Pact, whereby Premier Édouard Daladier of France and Prime Minister Neville Chamberlain of Britain, to avoid confrontation with German Chancellor Adolf Hitler, agreed to his demands that Germany occupy the Sudetenland, a region of Czechoslovakia bordered by Germany. Italy was represented at the Munich Pact by Benito Mussolini, who colluded with Hitler and had anti-Semitic laws enforced. At the time of the Munich Pact Leo Szilard was a visiting professor in the United States. He had foresight to see the impending danger of the weakening policies of France and Britain toward Germany and moved to New York where he carried on a collaboration from Columbia University with Walter Zinn. The foresight that Szilard displayed in escaping from Berlin and the Nazis in 1933 would again prove to be vital. It would be in the United States where Leo Szilard could safely carry out his efforts to beat Germany or any other aggressive nation in the development of an atomic bomb. On September 1st of 1939, less than a year after the signing of the Munich Pact, Hitler invaded Poland; and France and England declared war on Germany marking the beginning of World War II.

After the first report of neutron-induced fission by Lise Meitner and Otto Frisch on February 11, 1939 following the work of Otto Hahn and Fritz Strassman (1939a,b) it became clear to Leo Szilard that uranium would be the element that could sustain the nuclear chain reaction that he had patented over 5 years prior to the discovery of neutron-induced fission. In July of the same year Herbert Anderson, Enrico Fermi, and Leo Szilard determined that there are about two neutrons produced for every neutron consumed in the fission of uranium-235 (Anderson *et al.*, 1939b). Leo Szilard finally found the isotope that could sustain a neutron-induced chain reaction that could yield immense amounts of energy and even the atomic bomb that he had patented 6 years prior to this discovery. Szilard knew that the neutron yield had to be greater than one neutron produced per neutron consumed to maintain the chain reaction, and that the production of two neutrons for each neutron consumed left no doubt in his mind that the self-sustaining chain reaction was in his grasp. In the following month (August 14) Walter Zinn at City College, New York, NY together with Leo Szilard from Columbia University, New York, NY reported a more precise number of 2.3 neutrons produced on average for each thermal neutron absorbed in uranium (Zinn and Szilard, 1939). This is very close to the current more precise average number of 2.5 neutrons produced per thermal neutron absorbed in uranium-235. They also determined the general energy spectrum of the neutrons emitted in fission. The neutrons were classified as fast neutrons with an upper energy limit of about 3.5 MeV; however, Szilard knew that the fast neutrons could easily be thermalized (slowed down) with a low atomic number element or moderator such as graphite to sustain the chain reaction. From this work it was clear to Szilard that there was little doubt that the conditions needed for the self-sustaining neutron-induced chain reaction in uranium-235 could be made and its demonstration would soon be a reality.

In Szilard's mind time was of the essence. Szilard urged his lifelong friend Albert Einstein, then Theoretical Physicist at Princeton, to send a letter to the President of the United States, Franklin Delano Roosevelt, to inform him of the threat of Germany's nuclear research program and the danger that Germany could develop a nuclear weapon urging the president to initiate a program for the weapon development. A series of letters were sent by Einstein, the first of which was dated on August 2, 1939. A portion of the letter reads as follows:

> In the course of the last four months it has been made possible—through the work of Joliot of France as well as Fermi and Szilard in America—that it may become possible to set up a nuclear chain reaction in a large mass of uranium, by which vast amounts of power … would be generated … that extremely powerful bombs of a new type may thus be constructed …

In a subsequent letter dated March 7, 1940 Einstein wrote

> …Since the outbreak of the war, interest in uranium has intensified in Germany … research there is carried out in great secrecy … Dr. Szilard has shown me the manuscript, which he is sending to the Physics Review in which he describes in detail a method of setting up a chain reaction in uranium. The papers will appear in print unless they are held up, and the question arises whether something ought to be done to withhold publication.

In a third letter dated April 25, 1940 Einstein urged President Roosevelt to carry out the research needed to develop the bomb at a faster and more organized manner.

The secret American project to develop atomic weapons, which became known as the Manhattan Project (derived from its official name of the Manhattan Engineering District), was authorized by President Franklin D. Roosevelt on October 9, 1941. While working on this project Rudolf Peierls (1907–1995) made initial calculations on the number of neutron-chain reactions or neutron generations that would occur if a critical mass of uranium-235 could be confined in a small sphere about the size of a golf ball. See Figure IV.6 for an illustration of the first four generations of a nuclear chain reaction. Rhodes (1986) relates that Peierls based his calculations on the assumption that the time between each neutron generation would be short, only a few microseconds. In reality, the neutron chain reactions are faster. The time it takes for one neutron to travel from one nucleus and hit another nucleus is about 10 nsec. If the velocity of the approximately 2 MeV fission neutrons reach about 20 million m/sec (see Figure 4.1) Peierls estimated that only about 80 neutron generations would be achieved in the nuclear chain reaction before the explosion produced an expansion and dispersion of the U-235 thereby terminating the chain reaction.

We can make a simple calculation of the energy released by 80 neutron generations of the chain reaction based simply on the reaction as propagated by two neutrons per fission with the release of 200 MeV energy per fission (Table IV.1). The energy released after each generation (n) is exponential (2^n) as illustrated in Table IV.1 and Figure IV.6, and the chain reaction culminates in an estimated energy release of 1.26×10^{26} MeV on the 80th neutron generation alone. The estimated energy of 1.21×10^{26} MeV provided in Table IV.1 is equivalent to approximately 4.6×10^{12} calories of heat calculated by converting MeV energy units to joules (J), which is a unit of power, that is, 1 J = 1 W sec (watt second) and then to calories (cal), a unit of heat as follows:

By definition,

$$1\,MeV = 1.60 \times 10^{-13}\,J \tag{IV.22}$$

and

$$1J = 0.239\,cal$$

Then the energy released from the 80th neutron generation in joules and calories may are calculated as

$$(1.21 \times 10^{26}\,MeV)(1.60 \times 10^{-13}\,J/MeV) = 1.94 \times 10^{13}\,J \tag{IV.23}$$

$$(1.94 \times 10^{13}\,J)(0.239\,cal/J) = 4.63 \times 10^{12}\,cal \tag{IV.24}$$

If, by definition, 1 cal would be sufficient heat to raise the temperature of 1 g of water by 1°C, the four million-million calories of heat produced in the 80th neutron generation of the U-235 fission chain reaction would yield a staggering temperature hotter than the center of the sun (approximately 15,000,000 °C) and, if the chain reactions all occur in a period of time shorter than one-millionth of a second [(10 nsec/neutron generation)(80 generations) = 800 nsec = 0.8 μsec], the pressures from the explosion would be staggering. The exponential energy release after only 14 neutron generations is illustrated in Figure IV.7.

The nuclear chain reaction had first to be demonstrated in a controlled fashion. In 1939 Leo Szilard and Enrico Fermi coinvented the nuclear reactor at Columbia University by jointly sketching

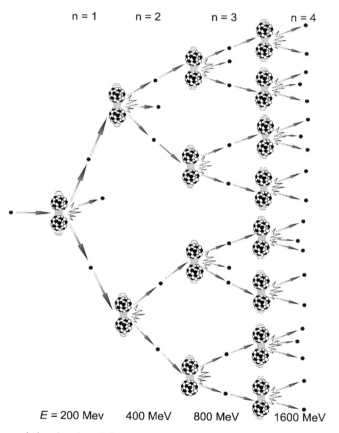

E = 200 Mev   400 MeV   800 MeV   1600 MeV

Figure IV.6   Neutron-induced nuclear chain reaction with uranium-235. On the far left of the figure is illustrated a neutron interacting with a U-235 nucleus, which becomes an unstable nucleus of U-236 oscillating as a liquid droplet. The U-236 nucleus is illustrated as bursting apart into two smaller nuclides (fission) with the emission of three neutrons. Two of the neutrons interact with two other nuclei of U-235 to cause nuclear fission of each with the concomitant emission of two or more neutrons from each nucleus. These neutrons then interact with four U-235 nuclei, which in turn produce yet more neutrons, etc. that propagate the exponential chain reaction. The chain reaction continues producing increasing energy after each neutron generation (n). Only four neutron generations are illustrated, with the energy (E) emitted from each generation increasing exponentially according to 2^n or 200, 400, 800, 1600, 3200 ... n. The actual number of neutrons emitted per nuclear fission will vary with the average being 2.5.

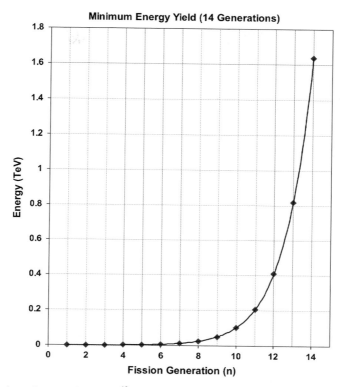

Figure IV.7   Estimated energy (TeV or 10^{12} eV) released from a neutron-induced fission chain reaction with uranium-235. The graph illustrates the energy released only up to the 14th generation of the chain reaction; however, the exponential increase in energy after each generation can be visualized.

Table IV.1   Minimum energy yield from various generations of ^{235}U fission chain reactions[a]

Fission generation (n)	Minimum energy yield (MeV)
1	2.00×10^2
5	3.20×10^3
10	1.02×10^5
20	104,857,600 or 1.05×10^8
30	107,300,000,000 or 1.07×10^{11}
40	109,900,000,000,000 or 1.10×10^{14}
50	112,500,000,000,000,000 or 1.12×10^{17}
60	115,200,000,000,000,000,000 or 1.15×10^{20}
70	118,000,000,000,000,000,000,000 or 1.18×10^{23}
80	120,800,000,000,000,000,000,000,000 or 1.21×10^{26}

[a] Energy yield is estimated on the basis of only two neutrons per fission and an energy emission of 200 MeV per fission, or $\sum_{n=1}^{x}(2^n)(100)$ where n is the neutron generation and the energy emitted after each generation undergoes an exponential multiplication (2^n) or 200, 400, 800, 1600, 3200 ... n.

out a three-dimensional lattice of uranium slugs dropped into holes in black, greasy blocks of graphite moderator, with sliding neutron-absorbing cadmium control rods to regulate the chain reaction referred to by Enrico Fermi as an "atomic pile". The first demonstration of the nuclear chain reaction in uranium was carried out by Enrico Fermi with the collaboration of Leo Szilard and others at the University of

Chicago on December 2, 1942 (see Enrico Fermi in Radioactivity Hall of Fame—Part II). In December of 1944 Enrico Fermi and Leo Szilard jointly filed for a highly classified patent as coinventors of the atomic pile, describing how the self-sustained nuclear chain reaction had been achieved. The patent (No. 2,708,656) was not issued until May 18, 1955. The invention of the nuclear reactor for the peaceful application of nuclear energy capable of providing electric power to cities, as patented by Fermi and Szilard, was declared by some to be one of the most significant inventions of all time comparable to those of communication, the telegraph and telephone, by Samuel Morse and Alexander Graham Bell. In 1944 around the time of writing the patent for the nuclear power reactor Leo Szilard and Enrico Fermi participated in the "New Piles Committee" meetings together with Nobel Laureates James Frank and Eugene Wigner among other physicists, chemists, and engineers including Walter Zinn and Alvin Weinberg. The Committee meetings were held at the University of Chicago Metallurgical Laboratory to explore and recommend new designs for peaceful electric power-producing nuclear reactors. Many new reactor concepts were presented which would become the modern power reactors that supply the needed electricity for our cities today. The history and minutes of the New Piles Committee meeting are reviewed by Lawson and Krause (2004). Among the new reactor concepts and designs was that of the sodium-cooled "breeder reactor" proposed by Szilard and Fermi. The term "breeder reactor" was first proposed by Leo Szilard, as it refers to the type of reactor that produces more nuclear fuel or fissile material than they consume. Although ^{235}U is the only naturally occurring fissile radionuclide, it stood to reason that if an excess of neutrons were produced in a thermal reactor (an average of 2.5 neutrons per ^{235}U fission) it would be possible to produce fissile ^{239}Pu or ^{233}U fuel (see eqs. (4.38–4.43) of Chapter 4) in a reactor in excess of the fuel actually consumed in the reactor. This concept formed the basis for the new generation of "breeder" reactors.

The successful demonstration of the nuclear chain reaction in 1942 led to the successful construction and testing of the first nuclear weapon under the Manhattan Project on July 16, 1945.

In light of the official surrender of Germany to the allied forces in May of 1945 Leo Szilard began opposition to the use of the bomb on moral grounds. Szilard's efforts, first in patenting the atomic bomb and then promoting its development, were driven by the threat that Germany under the rule of Hitler would be first to develop the bomb. Now that the threat of Germany was a matter of the past and a peace treaty, other than an unconditional surrender of Japan appeared possible, he saw no justification to use the bomb against Japan. It was, however, believed by many that countless lives would have been lost on both sides over a prolonged war with Japan, if the bomb had not been used. Over a decade prior to the German surrender he had patented his original idea for the bomb in 1934 in order to maintain control over the eventual discovery of the nuclear chain reaction and then conferred his patent to the British War Office in 1935 to keep the patent a secret and out of the hands of those who might use the information for malicious objectives. It may appear ironic that Leo Szilard was a pacifist. We may say that Leo Szilard was the founder of nuclear arms control or what we now call nuclear non-proliferation (NNP). This is evidenced by his tireless efforts to reach the American President and prevent the deployment of the atomic bomb against Japan.

In March of 1945 Leo Szilard prepared a memorandum to President Roosevelt to avert the use of the bomb against Japan. He warned that the use of the bomb would start an atomic arms race with Russia, and he raised the question on whether avoiding such an arms race might not be more important than the short-term goal of knocking Japan out of the war. Szilard's efforts to get to the President were related by him in an interview with a news reporter (U.S. News & World Report, 1960). Szilard suspected that his memorandum would not reach the President if it went through official channels. He consulted first with Nobel Laureate Arthur H. Compton (Compton Effect), who was in-charge of the secret work of the University of Chicago Metallurgical Laboratory in the Manhattan Project, and Arthur Compton agreed with Szilard and wished him success in getting the letter to the President. Szilard therefore requested and obtained an appointment to see the First Lady, Mrs. Eleanor Roosevelt, so that his letter could be handed over to the President in a sealed envelope. However, President Roosevelt died on April 12, 1945. Now Szilard was back to square one and had to find a way to get his letter to President Truman. Szilard and Walter Bartky, Associate Director of the Met Lab, met with Truman's Appointments Secretary, who agreed that the matter was serious and suggested that they meet with Secretary of State, Jimmy Byrnes. They agreed to meet with Byrnes and asked permission that Nobel Laureate Harold C. Urey accompany them to the meeting. Harold Urey was awarded the Nobel Prize

GUYANA $100

Harold C. Urey U.S.A., 1934

Chemistry

H. C. Urey (1893-1981)

in Chemistry 1933 for his discovery of deuterium, an isotope of hydrogen also referred to as "heavy hydrogen". He had expertise in isotope separation and was part of the team that worked on the separation and enrichment of U-235 in the Manhattan project. Urey's presence would add authority and prestige to the meeting. The three Szilard, Bartky, and Urey took the overnight train to Spartanburg, South Carolina to meet with Byrnes. Szilard related that after having read the memorandum Byrnes commented that General Groves, who had headed the Manhattan Project and the development of the atomic bomb, informed him that Russia had no uranium and thus would not be able to participate in any atomic-arms race and that Byrnes was concerned over the Russian acquisition of Poland, Rumania, and Hungary and he thought that the possession of the bomb by America would render the Russians manageable in Europe. When Leo Szilard returned to Chicago after the meeting he learned that Byrnes had been appointed Secretary of State and consequently concluded that his arguments would be receiving no consideration. Consequently Szilard next considered addressing his concerns to the Secretary of War, Henry L. Stimson.

On June 11, 1945 Leo Szilard coauthored The Franck Report, which was entitled Report on Political and Social Problems Manhattan Project "Metallurgical Laboratory" University of Chicago, addressed to Secretary Stimson. The report was prepared by a committee consisting of Nobel Laureate James Franck, Donald J. Hughes, J. J. Nickson, Eugene Rabinowitch, Nobel Laureate Glenn T. Seaborg, J. C. Stearns, and Leo Szilard. The eight-page report in summary urged that "the use of nuclear bombs in this war be considered as a problem of long-range national policy rather than military expediency, and that this policy be directed primarily to the achievement of an agreement permitting an effective international control of the means of nuclear warfare". As to the outcome of the Franck Report Szilard commented "This report was addressed to Secretary Stimson, but none of those who participated in the writing of the report, including Professor James Franck, had an opportunity to see Mr. Stimson."

On July 4, 1945 Leo Szilard drafted a Petition to the President of the United States opposing the use of the bomb, and by July 17, 1945, weeks before the bomb was dropped on Hiroshima on August 6, 1945, he got altogether 70 signatures on his petition including that of Nobel Laureate Eugene

Wigner and other members of the Metallurgical Laboratory (Met Lab), the secret laboratory at the University of Chicago, developed for the production of the atomic bomb. Szilard sent copies of his petition on July 4, 1945 to the laboratories at Oak Ridge and Los Alamos where the secret work on the development of the atomic bomb was also carried out and appealed that the scientists take a moral stand on the issue. Excerpts of his cover letter to obtain signatures on his petition are the following:

> Enclosed is the text of a petition which will be submitted to the President of the United States. As you will see, this petition is based on purely moral considerations…On the basis of expediency, many arguments could be put forward both for and against our use of atomic bombs against Japan…However small the chance might be that our petition may influence the course of events, I personally feel that it would be a matter of importance if a large number of scientists who have worked in this field went clearly and unmistakably on record as to their opposition on moral grounds to the use of these bombs in the present phase of the war…The fact that the people of the United States are unaware of the choice which faces us increases our responsibility in this matter since those who have worked on "atomic power" represent a sample of the population and they alone are in a position to form an opinion and declare their stand.

It is unknown whether the petition ever got to the President, as Leo Szilard commented the following in an interview on August 15, 1960 with the *U.S. News & World Report*:

> Some of those who signed insisted that the petition be transmitted to the President through official channels. To this I reluctantly agreed. I was, at this point, mainly concerned that the members of the project had an opportunity to go on record on this issue, and I didn't think that the petition would be likely to have an effect on the course of events. The petition was sent to the President through official channels, and I should not be too surprised if it were discovered one of these days that it hadn't ever reached him.

After the war Leo Szilard abandoned physics and entered into research in the biosciences. Maurice Fox (1998) recounts his recollections of the time when Leo Szilard, after a meeting of atomic scientists in Chicago invited Aaron Novick, a physical organic chemist, to join him on "an adventure in biology". Both scientist joined forces by first taking a phage course at Cold Spring Harbor, New York and a microbiology course at the Hopkins Marine Station of Stanford University at Pacific Grove, California. They invented the chemostat, a device for growing bacteria in continuous culture (Novick and Szilard, 1950). With the chemostat they were able to measure bacterial mutation rates (Novick and Szilard, 1951, 1952). Szilard could never rest, first with inventions and patents in nuclear physics, and later, in the field of biological sciences and tireless efforts to promote peace and nuclear arms control.

In his efforts to promote nuclear arms control and world peace Szilard participated in the first three Pugwash Conferences on Science and World Affairs during 1957–1959. The objective of the Pugwash Conferences is to bring together, from around the world, influential scholars and public figures concerned with reducing the danger of armed conflict and seeking cooperative solutions to global problems. The name "Pugwash Conference" originated from the very first meeting, which took place in Pugwash, Nova Scotia in 1957. The Conferences from the very start have derived their inspiration from the *Russell–Einstein Manifesto* issued in London on July 9, 1955 by Nobel Laureates Bertrand Russell and Albert Einstein and signed by 10 other Nobel Laureates and eminent scientists. The Manifesto reads as follows:

THE RUSSELL–EINSTEIN MANIFESTO
LONDON 9 JULY 1955

*I*ₙ *the tragic situation which confronts humanity, we feel that scientists should assemble in conference to appraise the perils that have arisen as a result of the development of weapons of mass destruction, and to discuss a resolution in the spirit of the appended draft.*

We are speaking on this occasion, not as members of this or that nation, continent, or creed, but as human beings, members of the species Man, whose continued existence is in doubt. The world is full of conflicts; and, overshadowing all minor conflicts, the titanic struggle between Communism and anti-Communism.

Almost everybody who is politically conscious has strong feelings about one or more of these issues; but we want you, if you can, to set aside such feelings and consider yourselves only as members of a biological species which has had a remarkable history, and whose disappearance none of us can desire.

Bertrand Russell (1872-1970)

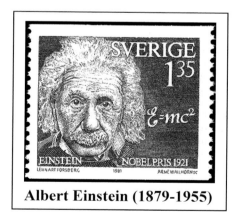

Albert Einstein (1879-1955)

We shall try to say no single word which should appeal to one group rather than to another. All, equally are in peril, and, if the peril is understood, there is hope that they may collectively avert it.

We have to learn to think in a new way. We have to learn to ask ourselves, not what steps can be taken to give military victory to whatever group we prefer, for there no longer are such steps; the question we have to ask ourselves is: what steps can be taken to prevent a military contest of which the issue must be disastrous to all parties?

The general public, and even many men in positions of authority, have not realized what would be involved in a war with nuclear bombs. The general public still thinks in terms of the obliteration of cities. It is understood that the new bombs are more powerful than the old, and that, while one A-bomb could obliterate Hiroshima, one H-bomb could obliterate the largest cities, such as London, New York, and Moscow.

No doubt in an H-bomb war great cities could be obliterated. But this is one of the minor disasters that would have to be faced. If everybody in London, New York, and Moscow were exterminated, the world might, in the course of a few centuries, recover from the blow. But we now know, especially since the bikini test, that nuclear bombs can gradually spread destruction over a very much wider area than had been supposed.

It is stated on very good authority that a bomb can now be manufactured which will be 2,500 times as powerful as that which destroyed Hiroshima. Such a bomb, if exploded near the ground or under water, sends radioactive particles into the upper air. They sink gradually and reach the surface of the earth in the form of a deadly dust or rain. It was this dust which infected the Japanese fishermen and their catch of fish. No one knows how widely such lethal radioactive particles might be diffused, but the best authorities are unanimous in saying that a war with H-bombs might possibly put an end to the human race. It is feared that if many H-bombs are used there will be universal death, sudden only for a minority but for the majority a slow torture of disease and disintegration.

Many warnings have been uttered by eminent men of science and by authorities in military strategy. None of them will say that the worst results are certain. What they do say is that these results are possible, and no one can be sure that they will not be realized. We have not yet found that the views of experts on this question depend in any degree upon their politics or prejudices. They depend only, so far as our researches have revealed, upon the extent of the particular expert's knowledge. We have found that the men who know most are the most gloomy.

Here, then, is the problem which we present to you, stark and dreadful and inescapable: Shall we put an end to the human race; or shall mankind renounce war? People will not face this alternative because it is so difficult to abolish war.

The abolition of war will demand distasteful limitations of national sovereignty. But what perhaps impedes understanding of the situation more than anything else is that the term "mankind" feels vague and abstract. People scarcely realize in imagination that the danger is to themselves and their children and their grandchildren, and not only to a dimly apprehended humanity. They can scarcely bring themselves to grasp that they, individually and those whom they love are in imminent danger of perishing agonizingly. And so they hope that perhaps war may be allowed to continue provided modern weapons are prohibited.

This hope is illusory. Whatever arguments not to use H-bombs had been reached in time of peace, they would no longer be considered binding in time of war, and both sides would set to work to manufacture H-bombs as soon as war broke out, for, if one side manufactured the bombs and the other did not, the side that manufactured them would inevitably be victorious.

Although an agreement to renounce nuclear weapons as part of a general reduction of armaments would not afford an ultimate solution, it would serve certain important purposes. First, any agreement between East and West is to the good in so far as it tends to diminish tension. Second, the abolition of thermonuclear weapons, if each side believed that the other had carried it out sincerely, would lessen the fear of a sudden attack in the style of Pearl Harbor, which at present keeps both sides in a state of nervous apprehension. We should, therefore, welcome such an agreement though only as a first step.

Most of us are not neutral in feeling, but, as human beings, we have to remember that, if the issues between East and West are to be decided in any manner that can give any possible satisfaction to anybody, whether Communist or anti-Communist, whether Asian or European or American, whether White or Black, than these issues must not be decided by war. We should wish this to be understood, both in the East and in the West.

There lies before us, if we choose, continual progress in happiness, knowledge and wisdom. Shall we instead, choose death, because we cannot forget our quarrels? We appeal as human beings to human beings: Remember your humanity, and forget the rest. If you can do so, the way lies open to a new Paradise; if we cannot, there lies before you the risk of universal death.

Resolution:

WE invite this Congress, and through it the scientists of the world and the general public, to subscribe to the following resolution:

> *In view of the fact that in any future world war nuclear weapons will certainly be employed, and that such weapons threaten the continued existence of mankind, we urge the governments of the world to realize, and to acknowledge publicly, that their purpose cannot be furthered by a world war, and we urge them, consequently, to find peaceful means for the settlement of all matters of dispute between them.*

Signed:

Max Born, Percy W. Bridgman, Albert Einstein, Leopold Infeld, Frederic Joliot-Curie, Herman J. Muller, Linus Pauling, Cecil F. Powell, Joseph Rotblot, Bertrand Russell, and Hideki Yukawa

Again, by participating in the first three Pugwash Conferences, Leo Szilard was among the first who stood out among most of his colleagues in working toward a solution to the most difficult and vital challenge facing mankind, nuclear disarmament, and world peace. Since the first meeting at Pugwash, in 1957, the Conferences have evolved to include a continuing series of meetings held at many locations throughout the world. Over 275 Pugwash Conferences, Symposia, and Workshops have been held all over the globe.

Leo Szilard had the foresight to know the important role that the Pugwash Conferences would play in saving the world from nuclear war and annihilation of the human race. The participants of the Pugwash Conferences are individuals including distinguished scientists and eminent scholars, who are not representatives of governments. This provides them with the liberty of flexible and frank discussions not normally possible in official meetings between governments. Because of the eminent stature of many of the participants as advisors to governments and some of whom become holders of government positions the important insights reached by the Pugwash Conferences rapidly penetrate into the appropriate levels of the policy-making branches of governments. As to underscore the important efforts that the Pugwash Conferences have made and the vital need for the Conferences to continue, the Nobel Committee awarded the Nobel Peace Prize for 1995 in two equal parts to Joseph Rotblat (1908–2005) and to the Pugwash Conferences in Science and World Affairs. The Nobel Prize was granted to both parties, in the words of the Nobel Committee, "for their efforts to diminish the part played by nuclear arms in international politics and, in the longer run, to eliminate such arms." The rationale for the award was highlighted by the Nobel Committee in the following statement:

> The Pugwash Conferences are based on the recognition of the responsibility of scientists for their inventions. They have underlined the catastrophic consequences of the use of the new weapons. They have brought together scientists and decision-makers to collaborate across political divides on constructive proposals for reducing the nuclear threat. The Pugwash Conferences are founded in the desire to see all nuclear arms destroyed and, ultimately, in a vision of other solutions to international disputes than war. The Pugwash Conference in Hiroshima in July this year declared that we have the opportunity today of approaching these goals. It is the Committee's hope that the award of the Nobel Peace Prize for 1995 to Rotblat and to Pugwash will encourage world leaders to intensify their efforts to rid the world of nuclear weapons.

The tireless efforts of Leo Szilard toward nuclear disarmament and world peace were summarized by Crow (2004) in his Addendum to *Szilard, A Personal Remembrance* by Werner Maas (2004) as follows:

> After the war, he began devoting his creative imagination, his boundless energy, and his dogged persistence to social causes, especially attempts to assure that nuclear bombs would not be used. He concocted various schemes, scientific and political, to this end. He was instrumental in

putting the development of nuclear energy into civilian rather than military hands at the Pugwash Conferences, in founding the *Bulletin of Atomic Scientists*, and, later, the Council for a Livable World.

In 1959, Szilard was diagnosed with cancer of the bladder. He rejected the advice of his medical doctors, which included the conventional surgery and *in situ* use of a tubercular vaccine to stimulate the immune system (Allen, 2001). Rather, he designed his own radiation therapy (Crow, 2004). Szilard consulted with Jonas Salk, famous for developing the polio vaccine, and he had seeds of radioactive silver implanted into his tumor. Again, Szilard was ahead of his time by designing a nonconventional method of nuclear medicine to treat cancer. Sealed sources of radium-226 had been used to treat external surface lesions caused by lupus since the early work of Marie and Pierre Curie (see Radioactivity Hall of Fame— Part I); however, the use of radioactive sources surgically implanted into tumors in body organs was yet to become recognized as an effective method to treat cancer until the 1980s, and by the year 2000, approximately 10% of the 120,000 patients treated each year for prostate cancer, would be treated with radioactive seeds. The number of patients treated with radioactive seeds is expected to increase in the future (Blasko *et al.*, 1998; Coursey and Nath, 2000).

Leo Szilard's cancer was either cured or abated into remission until his death from a heart attack on May 30, 1964 in La Jolla, CA, the very year he was appointed a Resident Fellow of the Salk Institute in La Jolla. A biography included with *The Register of Leo Szilard Papers 1898–1998* at the Mandeville Special Collections Library of the University of California at La Jolla, San Diego, CA (UCSD, 2005) summarizes his life as follows:

> Szilard lived a peripatetic life. After leaving Budapest in 1919 he had no true permanent residence. He stayed mostly in hotels, and his associations with various universities were usually tenuous. Because he had no long-term institutional affiliations, Szilard had difficulty in marshalling the material forces—such as clerical and laboratory staff—needed to follow through on many of his important ideas. Szilard was essentially a thinker, and he preferred to leave for others the tasks involved in implementing his ideas.

– 4 –

Neutron Radiation

4.1 INTRODUCTION

The neutron is a neutral particle, which is stable only in the confines of the nucleus of the atom. Outside the nucleus the neutron decays with a mean lifetime of about 15 min. Its mass, like that of the proton, is equivalent to 1 amu (atomic mass unit). Unlike the particulate alpha and beta nuclear radiations previously discussed, neutron radiation is not emitted in any significant quantities from radionuclides that undergo the traditional nuclear decay processes with the exception of a few radionuclides such as ^{252}Cf and ^{248}Cm, which decay to a significant extent by spontaneous fission (see Section 4.3.2). Significant quantities of neutron radiation occur when neutrons are ejected from the nuclei of atoms following reactions between the nuclei and particulate radiation. The lack of charge of the neutron also makes it unable to cause directly any ionization in matter, again unlike alpha and beta radiations. The various sources, properties, and mechanisms of interaction of neutrons with matter are described subsequently.

4.2 NEUTRON CLASSIFICATION

Neutrons are generally classified according to their kinetic energies. There is no sharp division or energy line of demarcation between the various classes of neutrons; however, the following is an approximate categorization according to neutron energy:

- Cold neutrons: <0.003 eV
- Slow (thermal) neutrons: 0.003–0.4 eV
- Slow (epithermal) neutrons: 0.4–100 eV
- Intermediate neutrons: 100 eV to 200 keV
- Fast neutrons: 200 keV to 10 MeV
- High-energy (relativistic) neutrons: >10 MeV

The energies of neutrons are also expressed in terms of velocity (m/sec) as depicted in the terminology used to classify neutrons. A neutron of specific energy and velocity is also described in terms of wavelength, because particles in motion also have wave properties. It is the wavelength of the neutron that becomes important in studies of neutron diffraction.

The values of energy, velocity, and wavelength of the neutron, as with all particles in motion, are interrelated. The velocity of neutrons increases according to the square root of the energy, and the wavelength of the neutron is inversely proportional to its velocity. Knowing only one of the properties, either the energy, velocity, or wavelength of a neutron, we can calculate the other two. We can relate the neutron energy and velocity using the kinetic energy equation

$$E = \frac{1}{2} mv^2 \quad \text{or} \quad v = \sqrt{\frac{2E}{m}} \tag{4.1}$$

where E is the particle energy in joules ($1 \text{ eV} = 1.6 \times 10^{-19} \text{J}$), m the mass of the neutron (1.67×10^{-27} kg), and v the particle velocity in meters per second. The wavelength is obtained from the particle mass and velocity according to

$$\lambda = \frac{h}{p} = \frac{h}{mv} \tag{4.2}$$

where λ is the particle wavelength in meters, h the Planck's constant (6.63×10^{-34} J sec), p the particle momentum, and m and v the particle mass and velocity, respectively, as previously defined. The correlation between neutron energy, velocity, and wavelength is provided in Figure 4.1, which is constructed from the classical eqs. (4.1) and (4.2) relating particle mass, energy, velocity, and wavelength. However, calculations involving high-energy particles that approach the speed of light will contain a certain degree of error unless relativistic calculations are used, as the mass of the particle will increase according to the particle speed. In Section 3.7.4 of Chapter 3, we used the Einstein equation $E = mc^2$ to convert the rest mass of the positron or negatron to its rest energy (0.51 MeV). When gauging particles in motion, the *total energy* (E) of the particle is the sum of its kinetic (K) and rest energies (mc^2) or

$$E = K + mc^2 = \gamma mc^2 \tag{4.3}$$

And thus, the kinetic energy of the particle would be the difference between the *total energy* of the particle and its rest energy or

$$K = \gamma mc^2 - mc^2 \tag{4.4}$$

where

$$\gamma = \frac{1}{\sqrt{1 - (u^2/c^2)}}, \tag{4.5}$$

u is the particle speed, and $u < c$. If we call the particle rest mass m_0, then the relativistic mass, m_r, which is the speed-dependent mass of the particle is calculated as

$$m_r = \frac{m_0}{\sqrt{1 - (u^2/c^2)}} \tag{4.6}$$

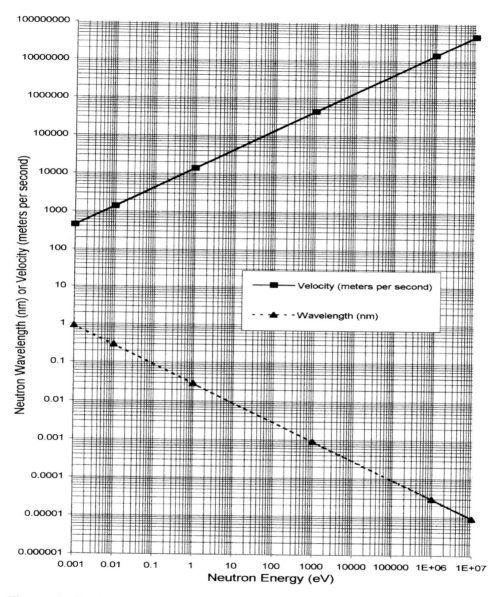

Figure 4.1   Correlation between neutron energy in electron volts (eV), velocity (m/sec), and wave-length (nm). From the energy of the neutron in eV on the abscissa, a line is drawn vertically to cross the wavelength and velocity curves. The values of neutron wavelength and velocity are obtained directly from the ordinate. For example, to determine the wavelength and velocity of 0.025 eV thermal neutrons, the value of 0.025 eV is found on the abscissa. A line is then drawn vertically from the point of 0.025 eV to cross the values of 0.18 nm wavelength and 2200 m/sec velocity.

And from eq. (4.4) the kinetic energy of the particle can be written as

$$K = \frac{m_0}{\sqrt{1-(u^2/c^2)}} c^2 - m_0 c^2 \tag{4.7}$$

Eq. (4.7) can be transformed to read

$$K + m_0 c^2 = \frac{m_0 c^2}{\sqrt{1-(u^2/c^2)}} \tag{4.8}$$

and

$$\frac{K + m_0 c^2}{m_0 c^2} = \frac{1}{\sqrt{1-(u^2/c^2)}} \tag{4.9}$$

and

$$\left(\frac{K}{m_0 c^2} + 1\right)^2 = \frac{1}{1-(u^2/c^2)} \tag{4.10}$$

which can be transformed to read

$$\frac{u^2}{c^2} = 1 - \frac{1}{(K/m_0 c^2 + 1)^2} \tag{4.11}$$

and from eq. (4.11) the relativistic speed of the particle can be defined as

$$u_r = c\sqrt{1 - \left(\frac{K}{m_0 c^2} + 1\right)^{-2}} \tag{4.12}$$

where K is the kinetic energy, and the particle speed u is always less than c (Serway *et al.*, 1997). The non-relativistic speed is that described by eq. (4.1) or $u_{nr} = \sqrt{2E/m_0}$.

To confirm the validity of the use of non-relativistic calculations of particle speed for the construction of Figure 4.1, let us use eqs. (4.1) and (4.12) to compare the differences between the non-relativistic and relativistic speeds of a neutron of 10 MeV kinetic energy. This energy was selected, because it is the highest neutron energy included in Figure 4.1, and differences between non-relativistic and relativistic calculations increase with particle energy. The difference between the two calculated speeds is defined by the ratio of the two or

$$\frac{u_{nr}}{u_r} = \frac{\sqrt{2E/m_0}}{c\sqrt{1-(K/m_0 c^2 + 1)^{-2}}} \tag{4.13}$$

The rest energy of the neutron, mc^2, is first calculated as

$$mc^2 = (1.6749 \times 10^{-27} \text{ kg})(2.9979 \times 10^8 \text{ m/sec})^2 = 1.505 \times 10^{-10} \text{ J}$$

and

$$\frac{1.505 \times 10^{-10} \text{ J}}{1.602 \times 10^{-19} \text{ J/eV}} = 939.5 \text{ MeV}$$

since by definition, $1 \text{ eV} = 1.602 \times 10^{-19} \text{ J}$. From eq. (4.13) the ratio of the non-relativistic and relativistic speeds are calculated as

$$\frac{u_{nr}}{u_r} = \frac{\sqrt{2(10 \text{ MeV})(1.602 \times 10^{-13} \text{ J/MeV})/1.6749 \times 10^{-27} \text{ kg}}}{c\sqrt{1 - ((10 \text{ MeV}/939.5 \text{ MeV}) + 1)^{-2}}}$$

$$= \frac{4.3737 \times 10^7 \text{ m/sec}}{0.144751c} = \frac{4.3737 \times 10^7 \text{ m/sec}}{(0.144751)(2.9979 \times 10^8 \text{ m/sec})}$$

$$= \frac{4.3737 \times 10^7 \text{ m/sec}}{4.339 \times 10^7 \text{ m/sec}} = 1.0079 = 0.79\% \text{ error}$$

The error between the non-relativistic and relativistic calculations is small at this high neutron energy. However, if we consider higher neutron energies in excess of 10 MeV the error of making non-relativistic calculations increases.

As we observed above in the case of particle speed, we will see that particle wavelength will also differ for non-relativistic and relativistic calculations. In 1923 Louis Victor de Broglie first postulated that all particles or matter in motion should have wave characteristics just as photons display both a wave and particle character. We therefore attribute the wavelength of particles in motion as de Broglie wavelengths. Let us then compare calculated non-relativistic and relativistic wavelengths. From eq. (4.2), we can describe the non-relativistic wavelength, λ_{nr}, as

$$\lambda_{nr} = \frac{h}{p} = \frac{hc}{pc} = \frac{hc}{cmv} = \frac{hc}{cm\sqrt{2E/m}} = \frac{hc}{\sqrt{2mc^2E}} \qquad (4.14)$$

where $p = mv = m\sqrt{2E/m}$. For relativistic calculations, the value of pc is calculated according to the following equation derived by Halpern (1988):

$$pc = \left[2m_0 c^2 K \left(1 + \frac{K}{2m_0 c^2} \right) \right]^{1/2} \qquad (4.15)$$

and the calculation for the relativistic de Broglie wavelength, λ_r, then becomes

$$\lambda_r = \frac{hc}{pc} = \frac{hc}{[2m_0c^2K(1+(K/2m_0c^2))]^{1/2}} \tag{4.16}$$

We can then compare the difference between the non-relativistic and relativistic wavelengths for the 10 MeV neutron as follows:

$$\frac{\lambda_{nr}}{\lambda_r} = \frac{hc/\sqrt{2mc^2E}}{hc/[2m_0c^2K(1+(K/2m_0c^2))]^{1/2}}$$

$$= \frac{\left\{\begin{array}{l}[(6.626\times10^{-34}\text{ J sec})(2.9979\times10^8\text{ m/sec})/\\1.602\times10^{-13}\text{ J/MeV}]/\sqrt{2(939.5\text{ MeV})(10\text{ MeV})}\end{array}\right\}}{\left\{\begin{array}{l}[(6.626\times10^{-34}\text{ J sec})(2.9979\times10^8\text{ m/sec})/1.602\times10^{-13}\text{ J/MeV}]/\\\sqrt{2(939.5\text{ MeV})(10\text{ MeV})[1+(10\text{ MeV}/2(939.5\text{ MeV}))]}\end{array}\right\}} \tag{4.17}$$

$$= \frac{12.3598\times10^{-4}\text{ MeV nm}/\sqrt{18790\text{ MeV}^2}}{12.3598\times10^{-4}\text{ MeV nm}/\sqrt{18890\text{ MeV}^2}}$$

$$= \frac{9.0430\times10^{-6}\text{ nm}}{9.0190\times10^{-6}\text{ nm}} = 1.0026 = 0.26\%\text{ error}$$

From the above comparison of non-relativistic and relativistic calculations of neutron wavelength and velocity, we see that the data provided in Figure 4.1 based on non-relativistic calculations are valid with less than 1% error for the highest energy neutron included in that figure. However, if we consider higher energies beyond 10 MeV, where we classify the neutron as relativistic, the errors in making non-relativistic calculations will increase with neutron energy. It will be clearly obvious to the reader that factors in eq. (4.17) can be cancelled out readily and the equation simplified to the following, which provides a quick evaluation of the effect of particle energy on the error in non-relativistic calculation of the de Broglie wavelength:

$$\frac{\lambda_{nr}}{\lambda_r} = \sqrt{1+\frac{K}{2m_0c^2}} \tag{4.18}$$

where K is the particle kinetic energy in MeV and m_0c^2 the rest energy of the particle (e.g., 939.5 MeV for the neutron and 0.511 MeV for the beta particle). For example, a non-relativistic calculation of the wavelength of a 50-MeV neutron would have the following error:

$$\frac{\lambda_{nr}}{\lambda_r} = \sqrt{1+\frac{50\text{ MeV}}{2(939.5\text{ MeV})}} = 1.0132 = 1.32\%\text{ error}$$

Note that the above-computed errors in non-relativistic calculations of the de Broglie wavelength increased from 0.26% for a 10-MeV neutron to 1.31% for a 50-MeV neutron, and the error will increase with particle energy. Errors in non-relativistic calculations are yet greater for particles of smaller mass (e.g., beta particles) of a given energy compared to neutrons of the same energy. This is due obviously to the fact that particles of lower mass and a given energy will travel at higher speeds than particles of the same energy but higher mass. This is illustrated in Figure 4.2 where the particle speed, u, is a function of the particle kinetic energy, K, and its mass or rest energy, m_0c^2. The particle energy in Figure 4.2 is expressed as K/m_0c^2 to permit the reader to apply the curves for non-relativistic and relativistic calculations to particles of different mass. For example, from the abscissa of Figure 4.2, the values of K/m_0c^2 for a 2-MeV beta particle is 2 MeV/0.51 MeV = 3.9 and that for a 2-MeV neutron is 2 MeV/939.5 MeV = 0.0021. From Figure 4.2 we see that the non-relativistic calculation of the speed of a 2-MeV beta particle would be erroneously extreme (well beyond the speed of light), while there would be only a small error in the relativistic calculation of the speed of the massive neutron of the same energy.

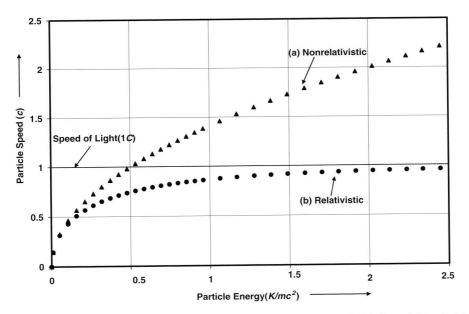

Figure 4.2 A graph comparing particle speeds derived from (a) non-relativistic and (b) relativistic interpretations of the particle kinetic energy. The speeds in units of the speed of light ($c = 2.9979 \times 10^8$ m/sec) are plotted versus particle energy expressed as a ratio of its kinetic energy (K) over its rest energy (mc^2). In the non-relativistic case, the particle kinetic energy is defined as $K = 1/2\,mv^2$ and its speed is calculated according to $v = \sqrt{2K/m}$; whereas in the relativistic case, the particle kinetic energy is the difference between the total energy of the particle and its rest energy ($K = \gamma mc^2 - mc^2$), where $\gamma = 1/\sqrt{1 - (u^2/c^2)}$ and its speed $u = c\sqrt{1 - (K/mc^2 + 1)^{-2}}$ is always less than c.

4.3   SOURCES OF NEUTRONS

The discovery of the neutron had eluded humanity until as late as 1932, because of the particle's neutral charge and high penetrating power when traveling through matter. In 1932 J. Chadwick provided evidence for the existence of the neutron. He placed a source of alpha-particle radiation in close proximity to beryllium. It was known that bombarding beryllium with alpha radiation would produce another source of radiation, which had a penetration power through matter even greater than that of gamma radiation. Chadwick observed that, when a sheet of paraffin (wax) was placed in the path of travel of this unknown radiation, he could detect a high degree of ionization in a gas ionization chamber caused by protons emitted from the paraffin. This phenomenon would not occur when other materials such as metals and even lead were placed in the path of this unknown radiation. On the basis of further measurements of the proton velocities and scattering intensities, it was concluded that the unknown radiation had a mass similar to that of the proton, but with a neutral charge. Only a particle with neutral charge would have a high penetration power through matter. As noted in the previous discussion of beta-particle decay, the neutron is of mass similar to that of the proton and, within the nucleus of an atom, the particle is a close union between a proton and an electron.

4.3.1   Alpha particle-induced nuclear reactions

It is interesting to note that the method used by Chadwick to produce neutrons by alpha particle-induced reactions, described in the previous paragraph, remains an important method of producing a neutron source, particularly when a relatively small or easily transportable neutron source is required. The source may be prepared by compressing an alpha particle-emitting radioisotope substance with beryllium metal. The nuclear reaction, which occurs between the alpha particle and the beryllium nucleus, terminates with the emission of a neutron and the production of stable carbon as follows:

$$^{9}_{4}\text{Be} + ^{4}_{2}\text{He} \rightarrow ^{1}_{0}\text{n} + ^{12}_{6}\text{C} + 5.5\,\text{MeV}_{(average)} \qquad (4.19)$$

Several alpha particle sources are used to produce neutrons via the preceding (α,n) reaction. Among these are the alpha emitters ^{241}Am, ^{242}Cm, ^{210}Po, ^{239}Pu, and ^{226}Ra. The alpha radiation source selected may depend on its half-life as well as its gamma-ray emissions. As noted previously in this chapter, gamma radiation often accompanies alpha decay. The use of an alpha source, which also emits abundant gamma radiation, requires additional protection for the user against penetrating gamma rays. For example, Am–Be sources are preferred over the Ra–Be sources of neutrons used in soil moisture probes (Nielsen and Cassel, 1984; O'Leary and Incerti, 1993), because the latter have a higher output of gamma radiation and require more shielding for operator protection.

The energies of the neutrons emitted from these sources will vary over the broad spectrum of 0–10 MeV. The average neutron energy of 5.5 MeV is shown in eq. (4.19). The neutrons produced by these sources vary in energy as a consequence of several factors, including the sharing of the liberated energy between the neutron and ^{12}C nucleus, the varying directions of emission of neutrons from the nucleus with consequent varying energies and velocities, and the variations in kinetic energies of the bombarding alpha particles.

The neutron activities available from these sources increase up to a maximum as a function of the amounts of alpha emitter and beryllium target material used. For example, as explained by Bacon (1969), the Ra–Be source, prepared by mixing and compressing radium bromide with beryllium powder, increases steadily in neutron activity (neutrons per second) for each gram of radium used as the amount of beryllium is increased to about 10 g; but no significant increase in neutron output is achieved if more beryllium is used. The maximum neutron output achieved is approximately 2×10^7 neutrons per second per gram of radium. Because alpha decay from any alpha particle-emitting source occurs by means of random events, the production of neutrons by (α,n) reactions is also a random event. Therefore, these reactions can be referred to as "not time correlated". This is contrary to the case of neutron sources provided through fission, discussed subsequently.

4.3.2   Spontaneous fission

About 100 radionuclides are known to decay by spontaneous fission (SF) with the emission of neutrons (Karelin et al., 1997) as an alternative to another decay mode, such as alpha decay. Spontaneous fission involves the spontaneous non-induced splitting of the nucleus into two nuclides or fission fragments and the simultaneous emission of more than one neutron on the average. This phenomenon occurs with radionuclides of high mass number, $A \geqslant 230$. The radionuclide ^{252}Cf is a good example of a commercially available spontaneous fission neutron source. It decays with a half-life of 2.65 years primarily by alpha emission (96.91% probability); the remaining of the ^{252}Cf decay processes occur by spontaneous fission with a probability of 3.09% (Martin et al., 2000). Decay of ^{252}Cf by spontaneous fission produces an average number of 3.7 neutrons per fission. Because the sizes of the two fragments resulting from fission are not predictable, average sizes of the two fragments are determined. Consequently, the numbers of neutrons emitted from individual fissions are not the same; and an average number of neutrons produced per fission is determined. The fission rate of ^{252}Cf is 6.2×10^5 SF/μg sec (Isotope Products Laboratories, 1995). The neutron emission from ^{252}Cf in units of neutrons per second per unit mass is reported to be 2.314×10^6 neutrons/μg sec with a specific activity of 0.536 mCi/μg (Martin et al., 2000). If we know the radionuclide specific activity and the % probability of decay by spontaneous fission, we can calculate the fission rate. For example, taking the specific activity and % probability of spontaneous fission reported above for ^{252}Cf, we can calculate the fission rate as the product of decay rate and probability of SF per decay or

$$(0.536 \text{ mCi/μg})(3.7 \times 10^7 \text{ dps/mCi})(0.0309) = 6.13 \times 10^5 \text{ SF/sec μg} \qquad (4.20)$$

which is in close agreement with the value cited above. See Chapter 8 for a discussion of radioactivity units and calculations of radionuclide mass.

The variations in fission fragment sizes and number of neutrons emitted per fission provide variable neutron energies over the range 0–5.5 MeV with an average neutron energy from ^{252}Cf of approximately 2.3 MeV. Small sources of ^{252}Cf are commercially available for a wide range of applications such as prompt gamma neutron activation analysis of coal, cement, minerals, detection of explosives and land mines, neutron radiography, and cancer therapy. These sources are described by Martin *et al.* (1997, 2000) among which include 50-mg sources of ^{252}Cf providing a neutron intensity $>10^{11}$ sec^{-1} and measuring only 5 cm in length $\times$1 cm in diameter. They report also larger sources of mass >100 mg of ^{252}Cf that approach reactor capabilities for neutrons.

Another standard nuclide source of neutrons is ^{248}Cm, which provides spontaneous fission intensity of only 4.12×10^4 SF/mg sec and decays with a half-life of 3.6×10^5 years (Radchenko *et al.*, 2000). The lower neutron flux intensity of this source limits its application, although it has the advantage of a very long half-life providing invariability of sample intensity with time.

Some radionuclides of interest in nuclear energy and safeguards also decay by spontaneous fission. The isotopes of plutonium of even mass number, namely ^{238}Pu, ^{240}Pu, and ^{242}Pu, decay principally by alpha-particle emission but can also undergo spontaneous fission to a lesser extent at rates of 1100, 471, and 800 SF/g sec, respectively. The average number of neutrons emitted per fission is between 2.16 and 2.26 of broad energy spectrum (Canberra Nuclear, 1996). Because the neutrons produced with each fission occurrence are emitted simultaneously, we can refer to these emissions as "time correlated". Other isotopes of uranium and plutonium also undergo spontaneous fission but at a much lower rate.

4.3.3   Neutron-induced fission

When the naturally occurring isotope of uranium, ^{235}U, is exposed to slow neutrons, it can absorb the neutron to form the unstable nuclide ^{236}U (see Radioactivity Hall of Fame, Part IV, Figure IV.4). The newly formed nucleus may decay by alpha particle and gamma ray emission with the long half-life of 2.4×10^7 years. This occurs in approximately 14% of the cases when ^{235}U absorbs a slow neutron. However, in the remaining 86% of the cases, the absorption of a slow neutron by ^{235}U results in the production of the unstable ^{236}U nuclide, which takes on the characteristics of an unstable oscillating droplet. This oscillating nuclear droplet with the opposing forces of two positively charged nuclides splits into two fragments, not necessarily of equal size, with the liberation of an average energy of 193.6 MeV. The general ^{235}U fission reaction may be illustrated by

$$^{235}\text{U} + \text{n} \rightarrow \text{fp} + \nu\text{n} + E \tag{4.21}$$

which represents the fission of one atom of ^{235}U by one thermal neutron n to yield the release of fission products fp of varying masses plus an average yield of $\nu = 2.42$ neutrons and an overall average release of energy $E = 193.6$ MeV (Koch, 1995). Most of this energy (over 160 MeV) appears in the form of kinetic energy of the two fission fragments. The remaining energy is shared among the neutrons emitted, with prompt gamma radiation accompanying fission and beta particles and gamma radiation from decaying fission fragments and neutrinos accompanying beta decay. When a sample of ^{235}U is bombarded with slow neutrons, the fission fragments produced are rarely of equal mass. The mass numbers (A) of the fission fragments vary between 70 and 160, and the most probable values are approximately

96 and 135 (Lilley, 2001). The ^{236}U intermediate nuclide breaks into fragments in as many as 30 different possible ways, producing, therefore, 60 different nuclide fission fragments. In a review Koch (1995) provides a list of the fission fragments and their relative abundances as produced in a typical pressurized water reactor (PWR). The most common fission fragments have a mass difference in the ratio 3:2 (Bacon, 1969). On the average, 2.42 neutrons are emitted per ^{236}U fission (Koch, 1995). Neutrons emitted from this fission process vary in energy over the range 0–10 MeV with an average neutron energy of 2 MeV and are classified as fast neutrons.

Let us take one of many possible examples of ^{236}U fission to calculate the magnitude of energy liberated. One example of many possible fission reactions that may be written conserving mass and charge of the ^{236}U nucleus with the emission of three neutrons is the following:

$$^{236}_{92}U \rightarrow {}^{90}_{38}Sr + {}^{143}_{54}Xe + 3{}^{1}_{0}n \tag{4.22}$$

The energy liberated in the above fission reaction can be calculated by the mass difference (Δm) of the ^{236}U nucleus and the combined masses of the fission products and neutrons, that is,

$$\Delta m = M_U - (M_{Sr} + M_{Xe} + 3M_n) \tag{4.23}$$

The atomic masses of the ^{236}U, ^{90}Sr, ^{143}Xe, and the neutron are obtained from the standard reference tables (e.g., Lide, 1997) and imputed into eq. (4.23) to give

$$\Delta m = 236.045561\,u - [89.907738\,u + 142.9352\,u + 3(1.00866\,u)]$$
$$= 236.045561\,u - 235.86891\,u$$
$$= 0.17665\,u \tag{4.24}$$

Since by definition $1\,u = 1.66053873 \times 10^{-27}$ kg the mass difference can be expressed in kilograms as

$$\Delta m = (0.17665\,u)(1.66053873 \times 10^{-27}\,kg/u) = 2.9333 \times 10^{-28}\,kg \tag{4.25}$$

The mass can be expressed in equivalent units of energy by means of Einstein's equation:

$$E = mc^2 \tag{4.26}$$

where m is the mass in kilograms and c the velocity of light in a vacuum or 2.99792458×10^8 m/sec. If by definition, the mass of $1\,u = 1.66053873 \times 10^{-27}$ kg, the energy equivalent of $1\,u$ is calculated as

$$E = (1.66053873 \times 10^{-27}\,kg)(2.99792458 \times 10^8\,m/sec)^2$$
$$= 14.92417783 \times 10^{-11}\,J \tag{4.27}$$

The energy in units of joules (J) is then converted to units of electron volts, as $1\,eV = 1.602176462 \times 10^{-19}\,J$ or

$$E = \frac{14.92417783 \times 10^{-11}\,J}{1.602176462 \times 10^{-19}\,J/eV}$$

$$= 9.314940 \times 10^8\,eV \tag{4.28}$$

$$= 931.494\,MeV$$

Thus, $1\,u = 931.494\,MeV$ or $931.494\,MeV/c^2$. The latter units of MeV/c^2 are used conveniently when converting atomic mass units to MeV energy in Einstein's equation according to the following equation:

$$E = mc^2 = [(\text{atomic mass units})(931.494\,MeV/c^2)]c^2 \tag{4.29}$$

Thus, from eq. (4.27) the mass difference of the ^{236}U fission reaction illustrated in eq. (4.22) may be quickly calculated as

$$E = [(0.17665\,u)(931.494\,MeV/c^2)]c^2 = 164.5\,MeV \tag{4.30}$$

Most of the energy released appears as kinetic energy of the fission fragments ($\sim$160 MeV). The neutrons carry off about 2 MeV each on the average.

Additional energy is released from fission via gamma radiation and beta decay of the fission products and the neutrino radiation that accompanies beta decay. The overall energy emitted as a result of ^{236}U fission can be approximated from the difference in the binding energies per nucleon of the ^{236}U parent nuclide and the fission product nuclei (see Part 4.6.7, Fission versus fusion, of this section for a discussion of binding energy per nucleon). Let us take two example calculations to estimate the energy released from ^{236}U fission from the difference of binding energies per nucleon of ^{236}U and its fission products.

Example 1:

$$^{236}_{92}U \rightarrow ^{90}_{38}Sr + ^{143}_{54}Xe + 3\,^{1}_{0}n$$

The binding energy (B) of the ^{236}U nucleus, that is, the difference in mass of the ^{236}U nucleus and the combined masses of its nucleons is calculated as follows:

$$B = (M_{92\,\text{protons}} + M_{144\,\text{neutrons}}) - M_{236\,U} \tag{4.31}$$

Using atomic mass units for the masses of ^{236}U nucleus and its constituent nucleons, we calculate the binding energy (B) in atomic mass units (u) as

$$B = [92(1.00782\,u) + 144(1.00866\,u)] - 236.045561\,u$$

$$= 237.96648\,u - 236.045561\,u = 1.92092\,u \tag{4.32}$$

Although the masses of the electrons are included in the atomic masses, the electron masses can be ignored since these are the same for the ^{236}U and the 92 hydrogen atoms. Using

Einstein's equation for the equivalence of mass and energy we can convert the binding energy to MeV or

$$E = mc^2 = [(1.92092 \text{ u})(931.494 \text{ MeV}/c^2)] \, c^2$$
$$= 1789.3254 \text{ MeV} \tag{4.33}$$

Dividing the binding energy by the number of nucleons in the nucleus (mass number, $A = 236$) will provide the binding energy per nucleon for ^{236}U or

$$\frac{B}{A} = \frac{1789.3254 \text{ MeV}}{236} = 7.58 \text{ MeV/nucleon} \tag{4.34}$$

Similar calculations for the binding energies of the fission fragments ^{90}Sr and ^{143}Xe yield $B/A = 8.69$ and 8.19 MeV/nucleon, respectively. Estimated values for the binding energies per nucleon (B/A) as a function of mass number (A) can be obtained from Figure 4.4.

Because the binding energies per nucleon for the two fission products are slightly different, we could average the two values simply to get an estimated figure for the amount of energy released. We can now estimate the energy released from ^{236}U fission, using as an example, the fission reaction of eq. (4.22) as follows:

$$^{236}_{92}\text{U} \rightarrow {}^{90}_{38}\text{Sr} + {}^{143}_{54}\text{Xe} + 3{}^{1}_{0}\text{n}$$

$$E = \left[\left(\frac{B}{A} \right)_{\text{fission products}} - \left(\frac{B}{A} \right)_{{}^{236}\text{U}} \right] \times 236 \text{ nucleons}$$

$$E = \left[\left(\frac{(B/A)_{{}^{90}\text{Sr}} + (B/A)_{{}^{143}\text{Xe}}}{2} \right) - \left(\frac{B}{A} \right)_{{}^{236}\text{U}} \right] \times 236 \text{ nucleons} \tag{4.35}$$

$$E = \left[\left(\frac{8.69 \text{ MeV/nucleon} + 8.19 \text{ MeV/nucleon}}{2} \right) - 7.58 \text{ MeV/nucleon} \right]$$
$$\times 236 \text{ nucleons}$$
$$= (0.86 \text{ MeV/nucleon})(236 \text{ nucleons})$$
$$= 202 \text{ MeV}$$

Example 2: Let us take another example where the fission products are of the same mass number to simplify the calculation.

$$^{236}_{92}\text{U} \rightarrow 2{}^{117}_{46}\text{Pd} + 2{}^{1}_{0}\text{n} \tag{4.36}$$

The binding energy (B) of the ^{117}Pd nucleus is calculated from the difference in mass of the ^{117}Pd nucleus ($M_{117\,Pd}$) and the combined masses of its protons ($M_{46\,protons}$) and neutrons ($M_{71\,neutrons}$)

$$B = (M_{46\,protons} + M_{71\,neutrons}) - M_{117\,Pd}$$

$$= [46(1.00782\,u) + 71(1.00866\,u)] - 116.9178\,u \qquad (4.37)$$

$$= 1.05678\,u$$

Again, atomic mass units (u) can be used because the electron masses included in the atomic mass unit values can be ignored, as the 46 electron masses are included in the mass values of the 46 protons and the ^{117}Pd.

Converting the atomic mass units (u) to energy units gives

$$E = mc^2 = [(1.05678\,u)(931.494\,MeV/c^2)]c^2$$

$$= 984.38422\,MeV$$

And the binding energy per nucleon (B/A) for ^{117}Pd is

$$\frac{B}{A} = \frac{984.38422\,MeV}{117\,nucleons} = 8.41\,MeV/nucleon$$

Quick estimates of binding energies per nucleon can be obtained from Figure 4.4. The energy released from the fission reaction of eq. (4.36) can now be estimated as

$$^{236}_{92}U \rightarrow 2\,^{117}_{46}Pd + 2\,^{1}_{0}n$$

$$E = \left[\left(\frac{B}{A} \right)_{fission\,products} - \left(\frac{B}{A} \right)_{236\,U} \right] \times 236 \text{ nucleons}$$

$$= (8.41\,MeV/nucleon - 7.58\,MeV/nucleon)(236\,nucleons)$$

$$= 196\,MeV$$

Thus, from the two examples of possible fission reactions taken above, we can see that approximately 200 MeV is the amount of energy released from ^{236}U fission, that is, neutron-induced fission of ^{235}U. The magnitude of the energy released from neutron-induced fission of ^{235}U would be an average figure, because the actual amount of energy released would depend on the fission reaction, and there are many possible ways that the ^{236}U could break into two nuclides. Some of the energy arises from fission product radioactivity including gamma-ray emission as fission product nuclides decay to lower energy states, beta-particle emission from fission product decay, and the emission of neutrinos that accompany beta decay. The energy released can be itemized into the various energy categories listed in Table 4.1, which are tallied to provide an approximate average figure for the energy released in neutron-induced fission of ^{235}U fission.

Table 4.1

Approximate energy released from neutron-induced ^{235}U fission

Type	Energy released (MeV)
Prompt energy	
Fission product kinetic energy	165
Neutrons (2.5 neutrons/fission on average, $\sim$2 MeV/neutron on average)	5
Gamma-ray emissions	7
Energy from fission-product decay	
Gamma-ray emissions	7
Beta radiation	6
Neutrinos	10
Total	200

Because more than one neutron is released per fission, a self-sustaining chain reaction is possible with the liberation of considerable energy, forming the basis for the nuclear reactor as a principal source of neutrons and energy. In the case of ^{235}U, slow neutrons are required for neutron capture and fission to occur. The nuclear reactor, therefore, will be equipped with a moderator such as heavy water (D_2O) or graphite, which can reduce the energies of the fast neutrons via elastic scattering of the neutrons with atoms of low atomic weight. The protons of water also serve as a good moderator of fast neutrons, provided the neutrons lost via the capture process ^{1}H(n,γ)^{2}H can be compensated by the use of a suitable enrichment of the ^{235}U in the nuclear reactor fuel (Byrne, 1994). The notation ^{1}H(n, γ)^{2}H is a form of abbreviating a nuclear reaction according to the format:

Target nucleus (projectile, detected particle) product nucleus.

It can be read as follows: the target nucleus of the isotope ^{1}H captures a neutron to form the product isotope ^{2}H with the release of gamma radiation.

The previously described fission of ^{235}U represents the one and only fission of a naturally occurring radionuclide that can be induced by slow neutrons. The radionuclides ^{239}Pu and ^{233}U also undergo slow neutron-induced fission; however, these nuclides are man-made via the neutron irradiation and neutron capture of ^{238}U and ^{232}Th, as illustrated in the following equations (Murray, 1993). The preparation of ^{239}Pu occurs by means of neutron capture by ^{238}U followed by beta decay as follows:

$$^{238}_{92}\text{U} + {}^{1}_{0}\text{n} \rightarrow {}^{239}_{92}\text{U} + \gamma \tag{4.38}$$

$$^{239}_{92}\text{U} \xrightarrow{t_{1/2}\,=\,23.5\,\text{min}} {}^{239}_{93}\text{Np} + \beta^{-} \tag{4.39}$$

$$^{239}_{93}\text{Np} \xrightarrow{t_{1/2}\,=\,2.35\,\text{days}} {}^{239}_{94}\text{Pu} + \beta^{-} \tag{4.40}$$

The preparation of ^{233}U is carried out via neutron capture of ^{232}Th followed by beta decay according to the following equations:

$$^{232}_{90}\text{Th} + ^{1}_{0}\text{n} \rightarrow ^{233}_{90}\text{Th} + \gamma \tag{4.41}$$

$$^{233}_{90}\text{Th} \xrightarrow{t_{1/2} = 22.4\,\text{min}} ^{233}_{91}\text{Pa} + \beta^{-} \tag{4.42}$$

$$^{233}_{91}\text{Pa} \xrightarrow{t_{1/2} = 27.0\,\text{days}} ^{233}_{92}\text{U} + \beta^{-} \tag{4.43}$$

Nuclides that undergo slow neutron-induced fission are referred to as fissile materials. Although ^{235}U is the only naturally occurring fissile radionuclide, it stands to reason that if an excess of neutrons is produced in a thermal reactor, it would be possible to produce fissile ^{239}Pu or ^{233}U fuel in a reactor in excess of the fuel actually consumed in the reactor. This is referred to as "breeding" fissile material, and it forms the basis for the new generation of breeder reactors (Murray, 1993).

Other heavy isotopes, such as ^{232}Th, ^{238}U, and ^{237}Np, undergo fission but require bombardment by fast neutrons of at least 1 MeV energy to provide sufficient energy to the nucleus for fission to occur. These radionuclides are referred to as fissionable isotopes.

4.3.4   Photoneutron (γ,n) sources

Many nuclides emit neutrons upon irradiation with gamma- or x-radiation; however, most elements require high-energy electromagnetic radiation in the range 10–19 MeV. The gamma- or x-ray energy threshold for the production of neutrons varies with target element. Deuterium and beryllium metal are two exceptions, as they can yield appreciable levels of neutron radiation when bombarded by gamma radiation in the energy range of only 1.7–2.7 MeV. The target material of D_2O or beryllium metal is used to enclose a β^{-}-emitting radionuclide, which also emits gamma rays. The gamma radiation bombards the targets deuterium and beryllium to produce neutrons according to the photonuclear reactions $^{2}\text{H}(\gamma,\text{n})^{1}\text{H}$ and $^{9}\text{Be}(\gamma,\text{n})^{8}\text{Be}$, respectively. The photoneutron source ^{124}Sb + Be serves as a good example of a relatively high-yielding combination of gamma emitter with beryllium target. The ^{124}Sb gamma radiation of relevance in photoneutron production is emitted with an energy of 1.69 MeV at 50% abundance (i.e., one-half of the ^{124}Sb radionuclides emit the 1.69-MeV gamma radiation with beta decay). A yield of 5.1 neutrons per 106 beta disintegrations per gram of target material has been reported (Byrne, 1994). The half-life ($t_{1/2}$) of ^{124}Sb is only 60.2 days, which limits the lifetime of the photoneutron generator; nevertheless, this isotope of antimony is easily prepared in the nuclear reactor by neutron irradiation of natural stable ^{123}Sb.

4.3.5   Accelerator sources

The accelerator utilizes electric and magnetic fields to accelerate beams of charged particles such as protons, electrons, and deuterons into target materials. Nuclear reactions are made possible when the charged particles have sufficient kinetic energy to react with target nuclei. Some of the reactions between the accelerated charged particles and target material can be used to generate neutrons.

When electrons are accelerated, they gain kinetic energy as a function of the particle velocity. This kinetic energy is lost as Bremsstrahlung electromagnetic radiation when the accelerated electrons strike the target material. Bremsstrahlung radiation is described in Chapter 3. It is the Bremsstrahlung photons that interact with nuclei to produce neutrons according to the mechanisms described in the previous section under photoneutron (γ,n) sources. The accelerated electron-generated neutrons have been reported to yield in a uranium target as many as 10^{-2} neutrons per accelerated electron at an electron energy of 30 MeV with a total yield of 2×10^{13} neutrons/sec (Byrne, 1994). The accelerator is a good neutron source for the potential generation of nuclear fuels.

Accelerated deuterons can be used to produce high neutron yields when deuterium and tritium are used as target materials according to the reactions ^{2}H(d,n)^{3}He and ^{3}H(d,n)^{4}He, respectively. In the deuterium energy range 100–300 eV it is possible to obtain neutron yields of the order of 10^{10} neutrons/sec from these (d,n) reactions (Byrne, 1994) with relatively small electrostatic laboratory accelerators. Large accelerators can provide charged particle energies > 300 MeV capable of inducing neutron sources, such as accelerated proton-induced charge exchange reactions in ^{3}H and ^{7}Li target nuclei according to the reactions ^{3}H(p,n)^{3}He and ^{7}Li(p,n)^{7}Be as described by Byrne (1994). Practical implications of these neutron sources for the generation of nuclear fuels were noted in the previous paragraph. Murray (1993) pointed out that a yield of as many as 50 neutrons per single 500-MeV deuteron has been predicted and that this source of neutrons could be used to produce new nuclear fuels via neutron capture by ^{238}U and ^{232}Th according to reactions (4.38)–(4.43) described previously.

4.3.6 Nuclear fusion

Nuclear fusion is the process whereby nuclei join together into one nucleus. The fusion of two atomic nuclei into one nucleus is not possible under standard temperature and pressure. This is because the repulsing coulombic forces between the positive charges of atomic nuclei prevent them from coming into the required close proximity of 10^{-15} m before they can coalesce into one. However, as described by Kudo (1995) and the Uranium Information Centre (UIC, 2005) in reviews on nuclear fusion, if temperatures are raised to 100 million degrees Centigrade, nuclei can become plasmas in which nuclei and electrons move independently at a speed of 1000 km/sec, thereby the nuclei have sufficient kinetic energy to overcome their repulsing forces and combine. The process is also referred to as thermonuclear fusion. Nuclear fusion reactors or controlled thermonuclear reactors (CTRs) are under development to achieve nuclear fusion as a practical energy source. The reactors are based on maintaining plasmas through magnetic or inertial confinement as described by Dolan (1982), Kudo (1995), and the UIC (2005). Some fusion reactions produce neutrons.

The energy liberated during nuclear fusion is derived from the fact that the mass of any nucleus is less than the sum of its component protons and neutrons. This is because protons and neutrons in a nucleus are bound together by strong attractive nuclear forces discussed previously in Section 4.3.3. As described by Serway *et al.* (1997), this energy is referred to as the binding energy (B), that is, the energy of work required to pull a bound system apart leaving its component parts free of attractive forces described by

$$Mc^2 + B = \sum_{i=1}^{n} m_i c^2 \tag{4.44}$$

where M is mass of the bound nucleus, the m_is the free component particle masses (e.g., protons and neutrons), and n the number of component particles of the nucleus. From eq. (4.44) we can see that if it is possible to overcome the repulsive forces of protons in nuclei and fuse these into a new nucleus or element of lower mass, energy will be liberated.

Nuclear fusion reactions of two types emit neutrons, and these are of prime interest in man-made CTRs. The first type is fusion between deuterium and tritium nuclei according to

$$^2_1H + ^3_1H \rightarrow ^4_2He + ^1_0n + 17.58 \, \text{MeV} \tag{4.45}$$

We can confirm the energy released in reaction (4.45) by using Einstein's equation for the equivalence of mass and energy. We can write firstly the atomic mass units for each component of the reaction and then calculate the energy equivalence in atomic mass units as follows:

$$^2_1H + ^3_1H \rightarrow ^4_2He + ^1_0n + E \tag{4.46}$$

$$2.014101778 \, \text{u} + 3.01604927 \, \text{u} \rightarrow 4.00260324 \, \text{u} + 1.008664904 \, \text{u} + E$$

$$E = (2.014101778 \, \text{u} + 3.01604927 \, \text{u}) - (4.00260324 \, \text{u} + 1.008664904 \, \text{u})$$
$$= 0.018882904 \, \text{u} \tag{4.47}$$

Converting mass into energy yields

$$E = mc^2$$
$$= [(0.018882904 \, \text{u})(931.494 \, \text{MeV}/c^2)]c^2$$
$$= 17.58 \, \text{MeV}$$

The fusion of deuterium and tritium is illustrated in Figure 4.3.

Two other fusion reactions between two deuterium nuclei may proceed according to either of the following equations, which have approximately equal probabilities of occurring (Kudo, 1995):

$$^2_1H + ^2_1H \rightarrow ^3_2He + ^1_0n + 3.27 \, \text{MeV} \tag{4.48}$$

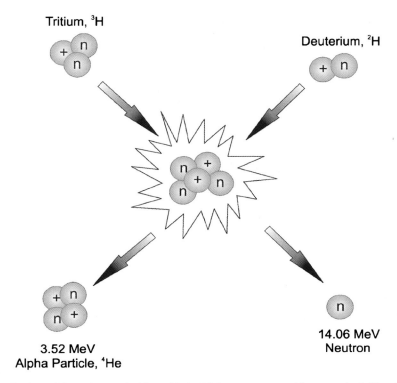

Figure 4.3   Fusion of deuterium and tritium. Under high temperature with magnetic field or inertial confinement, deuterium (^{2}H) and tritium (^{3}H) become plasmas and their electrons, protons (+), and neutrons (n) have sufficient kinetic energy to move about and interact free of their repulsive coulombic barriers. The nucleons coalesce forming a larger nucleus, the alpha particle (^{4}He), and an excess neutron (n) with the emission of energy as kinetic energy of the alpha particle and neutron.

and

$$^2_1H + ^2_1H \rightarrow ^3_1H + ^1_1H + 4.04\,\text{MeV} \tag{4.49}$$

Another possible reaction in deuterium fusion is the formation of an alpha particle with the emission of gamma radiation as described by

$$^2_1H + ^2_1H \rightarrow ^4_2He + \gamma + 23.9\,\text{MeV} \tag{4.50}$$

Although more energy is released in the above reaction, it has a low reaction cross-section, that is, low probability of occurrence, and most of the energy would be carried off by the gamma radiation (Lilley, 2001). The highly penetrating gamma rays would make it extremely difficult to sufficiently trap the energy for heat and electric power generation.

The fusion reaction between deuterium and tritium or D-T reaction (eq. (4.45)) gives rise to a 14.06-MeV neutron and a 3.52-MeV alpha particle. A D-T plasma burning experiment was performed with 0.2 g of tritium fuel with the Joint European Torus (JET) reactor in the UK in November 1991; and a higher power D-T experimental program with 20–30 g of tritium was continued on the Tokamak Fusion Test Reactor (TFTR) at Princeton in December 1993. These are described by JET Team (1994), Strachan *et al.* (1994), Hawryluk *et al.* (1994), Kudo (1995), and the UIC (2005). The word "tokamak" is derived from the Russian *toroidal kamera ee magnetnaya katushka* meaning "torus-shaped magnetic chamber", which was initially designed in 1951 by Soviet physicists Andrei Skaharov and Igor Tamm (UIC, 2005). The International Thermonuclear Experimental Reactor (ITER) project was set up under the auspices of the International Atomic Energy Agency (IAEA) to develop a prototype fusion reactor by the year 2030. The prototype reactor has the purpose of demonstrating that fusion can produce useful and relatively safe energy. Fusion energy production via a commercial reactor is assumed to start around the year 2050 (Sheffield, 2001). The status of the ITER project was summarized by the UIC (2005) as follows:

In 2003 the USA rejoined the project and China also announced it would do so. After deadlocked discussion, the six partners (USA, China, Republic of Korea, European Union, Japan, and the Russian Federation) agreed in mid 2005 to site ITER at Cadarache, in Southern France. The deal involved major concessions to Japan, which had put forward Rokkasho as a preferred site. The European Union (EU) and France will contribute half of the EUR 10 billion total cost, with other partners—Japan, China, Republic of Korea, USA and Russia — putting in 10% each. Japan will provide a lot of the high-tech components, will host a EUR 1 billion materials testing facility and will have the right to host a subsequent demonstration fusion reactor. The total cost of the 500 Megawatt (MWt) ITER comprises about half for the ten-year construction and half for 20 years of operation.

In addition to ITER over 50 countries of the world have active research programs ongoing toward the development of nuclear fusion as a future more efficient and environmentally friendly energy source. The IAEA with headquarters in Vienna, Austria maintains a World Survey of Activities in Controlled Fusion Research (IAEA, 2005a).

Small compact fusion devices have been developed for the purpose of producing neutrons. Under development are compact neutron sources, which utilize either deuterium–deuterium (D-D) or D-T fusion reactions. One instrument described by Miley and Sved (1997) is the inertial electrostatic confinement (IEC) device, which accelerates deuteron ions producing fusion reactions as the ions react with a pure deuterium or deuterium–tritium plasma target. The device is compact measuring 12 cm in diameter and 1 m in length and provides a neutron flux of 10^6–10^7 2.5-MeV D-D neutrons/sec or 10^8–10^9 14-MeV D-T neutrons/sec. Another similar device is described by Tsybin (1997), which utilizes laser irradiation to create a plasma in an ion source. Compact neutron sources of these type can become competitive with other neutron sources previously described such as ^{252}Cf and accelerator solid-target sources, because of advantages including (i) on–off capability, (ii) longer lifetime without diminished neutron flux strength, and (iii) minimum handling of radioactivity.

4.6.7   Nuclear fission versus fusion as an energy source

At this point it is relevant to look at (i) what factors determine whether nuclear fission or fusion is possible, and (ii) which of the two reactions (fission or fusion) should provide the most efficient source of energy.

The potential for either nuclear fission or fusion depends on the stability of the atomic nucleus as measured by the binding energy per nucleon (B/A), which varies for every element in the Periodic Table (Appendix B) and their isotopes. Figure 4.4 illustrates a graph of the binding energy per nucleon (B/A) as a function of the mass number (A) of the nucleus. The binding energy described by eq. (4.44) of Section 4.3.6 is the energy within the nucleus consisting of strong attractive forces that hold the nucleus together, and it would be also the amount of energy required to pull the bound system apart and leave the nucleons (protons and neutrons) free of their attractive forces. It is found by calculating the difference in mass of the bound nucleus and the combined masses of its nucleons. Examples of the calculation of binding energy (B) and binding energy per nucleon (B/A) are given in Section 4.3.3.

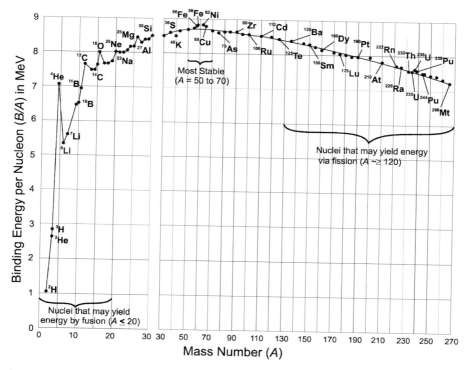

Figure 4.4   Binding energy per nucleon as a function of mass number. The scale between mass number 0 and 30 is broader than that illustrated for mass numbers 30–270. Several nuclides are identified along the curve. The most stable nuclei are ^{62}Ni, ^{58}Fe, and ^{56}Fe with B/A values of 8.790, 8.788, and 8.786 MeV, respectively. Notice the small peak at the top of the curve.

The mass of the bound nucleus is always less than the combined masses of its constituent protons and neutrons. The difference in mass is the binding energy, as mass and energy are equivalent ($E = mc^2$), which is the energy required to hold the nucleus together. The most stable nuclei, that is, those that are most tightly bound, belong to the "iron group" including ^{62}Ni, ^{58}Fe, and ^{56}Fe with binding energies per nucleon (B/A) of approximately 8.8 MeV. A small peak of B/A maxima for the "iron group" can be seen at the top of Figure 4.4. As illustrated in Figure 4.4 there is a broad peak or maximum that encompasses all nuclides with mass number (A) in the range of about 50–70. These are the most tightly bound or stable nuclei. On either side of this maximum, illustrated in Figure 4.4, are (i) the smaller nuclei in the range of $A = 2$–20 with lower binding energies per nucleon that could fuse to yield larger and more stable nuclei or (ii) the very large nuclei in the range of about $A = 120$ and higher, which could break apart by fission to form smaller and more stable nuclei. Consequently the size of the nucleus and its binding energy per nucleon will govern whether nuclear fusion or fission would be feasible.

In Sections 4.3.3 and 4.3.6 we could see that the magnitudes of energy released by neutron-induced fission of ^{235}U and D-T fusion reactions were different by at least 10 orders of magnitude. In review, the reactions and energies released are the following:

Neutron-induced ^{235}U fission:

$$^{235}U + n \rightarrow ^{236}U \rightarrow fp + \nu n + \sim 200\,MeV \qquad (4.51)$$

D-T fusion:

$$^2_1H + ^3_1H \rightarrow ^4_2He + ^1_0n + 17.58\,MeV \qquad (4.52)$$

where in reaction (4.51) fp represents fission products, that is, two nuclides of varying mass, and ν represents, in general, more than one neutron per fission reaction and average number of 2.4 neutrons. Comparing the fission and fusion reactions, we can immediately see that there is more than a 10-fold greater magnitude of energy released by the neutron-induced fission reaction. However, if we evaluate the reactions in terms of fuel mass, we will find that the above D-T fusion reaction is the most efficient for energy production. For example, let us now calculate and compare the energy yields per kilogram of fuel for both nuclear fission and nuclear fusion.

Neutron-induced ^{235}U fission energy yield

$$E\,\text{yield}_{^{235}U\,\text{fission}} = \frac{\text{Reaction energy release}}{\text{Fuel mass}}$$

$$= \frac{200\,MeV}{(M_{^{235}U} + M_{\text{neutron}})(1.660 \times 10^{-27}\,kg/u)} \qquad (4.53)$$

$$= \frac{200\,MeV}{(235.04392\,u + 1.00866\,u)(1.660 \times 10^{-27}\,kg/u)}$$

$$= 0.5104 \times 10^{27}\,MeV/kg$$

Deuterium–tritium fusion

$$E\,\text{yield}_{\text{D-T fusion}} = \frac{\text{Reaction energy release}}{\text{Fuel mass}}$$

$$= \frac{17.6\,\text{MeV}}{(M_{^2\text{H}} + M_{^3\text{H}})(1.660 \times 10^{-27}\,\text{kg/u})} \tag{4.54}$$

$$= \frac{17.6\,\text{MeV}}{(2.01410\,\text{u} + 3.016049\,\text{u})(1.660 \times 10^{-27}\,\text{kg/u})}$$

$$= 2.1077 \times 10^{27}\,\text{MeV/kg}$$

In the above eqs. (4.53) and (4.54) $M_{^{235}\text{U}}$, M_{neutron}, $M_{^2\text{H}}$, and $M_{^3\text{H}}$ refer to the masses of ^{235}U, the neutron, deuterium, and tritium, respectively, in atomic mass units (u). Taking the ratio of energy yields of nuclear fusion over fission yields

$$\frac{E\,\text{yield}_{\text{D-T fusion}}}{E\,\text{yield}_{^{235}\text{U fission}}} = \frac{2.1077 \times 10^{27}\,\text{MeV/kg}}{0.5104 \times 10^{27}\,\text{MeV/kg}} = 4.13 \tag{4.55}$$

Thus, D-T nuclear fusion is more than four times efficient than neutron-induced nuclear fission in terms of energy yield per mass of fuel consumption.

D-T fusion offers great promise as a future energy source, which would be more efficient and more environmentally friendly than neutron-induced fission. The availability and cost of fuel for D-T fusion are also factors that need to be considered when evaluating the advantages of nuclear fusion over the current fission-based nuclear power plants. A recent report by the Uranium Information Centre (UIC, 2005) provides a general assessment of fusion power for the future, which are summarized in the following sections.

Deuterium–tritium fuel

There is an enormous natural supply of deuterium. Deuterium is very abundant in seawater in concentrations of about $30\,\text{g/m}^3$. While tritium is not available naturally, it is currently made in a conventional nuclear power plant. Eventually when D-T fusion is employed in a commercial power plant, tritium can be made as a byproduct in the D-T fusion plant. The fusion reactor would breed tritium from lithium. Because high-energy neutrons are emitted in the D-T fusion reaction (see Figure 4.3), a blanket of lithium would be used to absorb the neutrons, and in so doing, the lithium would be transposed into tritium and helium according to the reactions described by Ongena and van Oost (2004):

$$^{7}_{3}\text{Li} + {}^{1}_{0}\text{n} \rightarrow {}^{3}_{1}\text{H} + {}^{4}_{2}\text{He} + {}^{1}_{0}\text{n} \tag{4.56}$$

$$^{6}_{3}\text{Li} + {}^{1}_{0}\text{n} \rightarrow {}^{3}_{1}\text{H} + {}^{4}_{2}\text{He} \tag{4.57}$$

The blanket of lithium must be very thick (about 1 m) to absorb the high-energy neutrons (14 MeV). The blanket will heat up after absorbing the kinetic energy of the neutrons.

A coolant flowing through the lithium blanket would transfer the heat away to produce steam, which would power turbines to generate electricity in the conventional way. In the long term, tritium could be produced in a D-D fusion reactor (see eq. (4.49)), which would produce even higher energies for power than D-T fusion. Currently, most attention is focused on the development of D-T fusion, because D-D fusion would require higher temperatures.

Environmental impact

D-T fusion, like current nuclear power plants, does not contribute to acid rain or the greenhouse effect. Also, fusion does not generate the radioactive fission products or transuranic elements. The elimination of fission-product waste concerns is a great plus for the development of nuclear fusion. There will be some short-term radioactive waste due to neutron-activation products. Some of the components of the fusion reactor during its lifetime of use would become radioactive upon absorption of neutrons, but the volume of such waste would not be more than any of the activation products produced by the current conventional nuclear power plants. In addition, the radioactive waste from the D-T fusion reactor would be relatively short-lived as compared to the longer-lived transuranic elements produced in the fission reactor.

Reactor safety

D-T fusion presents no danger of any accident due to runaway chain reaction. This is not possible, because the chain reaction occurs only with nuclear fission. Any malfunction of a fusion reactor would result in a sudden shutdown of the plant. Safety of the D-T fusion reactor is nevertheless a concern that is under constant study and review (Cook *et al.*, 2001; IAEA, 2005a,b; UIC, 2005). There is attention drawn to the possibility of a lithium fire, as lithium burns spontaneously when in contact with water or moisture of the atmosphere. Also, tritium release to the atmosphere would have to be monitored and safeguarded against. These concerns are part of the ITER and other international programs for fusion development.

4.4   INTERACTIONS OF NEUTRONS WITH MATTER

If a neutron possesses kinetic energy, it will travel through matter much more easily than other nuclear particles of similar energy, such as alpha particles, negatrons, positrons, protons, or electrons. In great contrast to other nuclear particles, which carry charge, the neutron, because it lacks charge, can pass through the otherwise impenetrable barrier of the atomic electrons and actually collide with nuclei of atoms and be scattered in the process or be captured by the nucleus of an atom. Collision of neutrons with nuclei can result in scattering of the neutrons yielding recoil nuclei with conservation of momentum (elastic scattering) or loss of kinetic energy of the neutron as gamma radiation (inelastic scattering). The capture of a neutron by a nucleus of an atom may result in the emission of other nuclear particles from the nucleus (non-elastic reactions) or the fragmentation of the nucleus into two (nuclear fission). A brief treatment of the various types of neutron interactions, which are based on their scattering or capture of neutrons by atomic nuclei, is provided next.

4.4.1 Elastic scattering

The elastic scattering of a neutron by collision with an atomic nucleus is similar to that of a billiard ball colliding with another billiard ball. A portion of the kinetic energy of one particle is transferred to the other without loss of kinetic energy in the process. In other words, part of the kinetic energy of the neutron can be transferred to a nucleus via collision with the nucleus, and the sum of the kinetic energies of the scattered neutron and recoil nucleus will be equal to the original energy of the colliding neutron. This process of interaction of neutrons with matter results only in scattering of the neutron and recoil nucleus. It does not leave the recoil nucleus in an excited energy state. Elastic scattering is a common mechanism by which fast neutrons lose their energy when they interact with atomic nuclei of low atomic number, such as hydrogen (^{1}H) in light water or paraffin, deuterium (^{2}H) in heavy water, and ^{12}C in graphite, which may be encountered in nuclear reactor moderators. It is easy to conceptualize what would occur when particles of equal or similar mass collide; the event would result in energy transfer and scattering without any other secondary effects, similar to what occurs in billiard ball collisions.

Neutron scattering is the principal mechanism for the slowing of fast neutrons, particularly in media with low atomic number. Let us consider what occurs when a neutron collides with a nucleus and undergoes elastic scattering. Figure 4.5 illustrates the direction of travel of an incident neutron with given kinetic energy (dashed line). The neutron collides with the nucleus. The nucleus is illustrated as undergoing recoil at an angle β, while the neutron is scattered at an angle α to the direction of travel of the incident neutron. The kinetic energy (E_k) lost by the neutron in this collision is defined by the equation:

$$E_k = \frac{4Mm_n}{(M + m_n)^2} \cos^2\beta \tag{4.58}$$

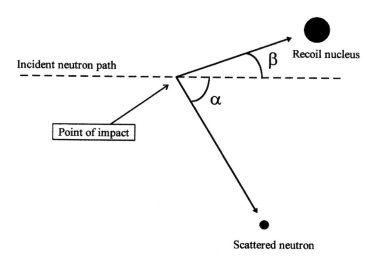

Figure 4.5 Elastic scattering of a neutron by collision of the neutron with an atomic nucleus. The neutron is scattered at an angle α and the nucleus recoils at an angle β to the direction of travel of the incident neutron.

Table 4.2

The maximum fraction of the kinetic energy (E_k) that a neutron can lose upon collision with
the nucleus of various atoms listed in increasing mass in atomic mass units (u)

Nuclide	Nuclide mass (M)	Neutron mass (m_n)	$E_k = (4Mm_n/(M + m_n)^2) \cos^2 \beta$
^{1}H	1.007825	1.008665	4.065566/4.066232 = 0.999 or 99.9%
^{2}H	2.014102	1.008665	8.126217/9.137120 = 0.889 or 88.9%
^{9}Be	9.012182	1.008665	36.36109/100.41737 = 0.362 or 36.2%
^{12}C	12.000000	1.008665	48.41592/169.22536 = 0.286 or 28.6%
^{16}O	15.994915	1.008665	64.53404/289.12173 = 0.223 or 22.3%
^{28}Si	27.976927	1.008665	112.87570/840.16454 = 0.134 or 13.4%
^{55}Mn	54.938047	1.008665	221.65633/3130.0329 = 0.071 or 7.1%
^{197}Au	196.96654	1.008665	787.86616/39194.175 = 0.020 or 2.0%

where M is the mass of the nucleus, m_n the mass of the neutron, and β the recoil angle of
the nucleus. A derivation of eq. (4.58) is provided by Bacon (1969). Let us look at two extreme
examples of elastic collisions between a neutron and a nucleus. In the first example, let us
consider a nuclear recoil angle of $\beta = 90°$. It is intuitively obvious from eq. (4.58) that for
a recoil angle $\beta = 90°$, $\cos^2\beta = 0$ and consequently $E_k = 0$. Under such a circumstance,
the neutron is undeflected by the nucleus and there is no energy transfer to the nucleus. The
neutron continues along its path undeflected until it encounters another nucleus. For the
second case, however, let us consider the other extreme in which the recoil angle, $\beta = 0°$,
where we have a head-on collision of the neutron with the nucleus of an atom. In this case,
the maximum possible energy of the neutron is imparted to the nucleus, where $\cos^2\beta = 1$.
This case is exemplified in Table 4.2, which provides the maximum fraction of the kinetic
energy a neutron can lose upon collision with various atomic nuclei, calculated according
to eq. (4.58). As illustrated in Table 4.2, the neutron can transfer more energy to the nuclei
of atoms, which have a low mass; and the highest fraction of its energy can be transferred to
the nucleus of the proton, which is almost equal in mass to the neutron. Nuclides of low mass
number are, therefore, good moderators for the slowing down of fast neutrons. The substances
often used are light water (H_2O), heavy water (D_2O), paraffin (C_nH_{2n+2}), and graphite (C).

4.4.2   Inelastic scattering

We may picture a fast neutron colliding with a nucleus. The neutron is scattered in another
direction as described in the previous paragraph; however, part of the neutron's kinetic
energy is lost to the recoil nucleus, leaving it in an excited metastable state. Inelastic scattering can occur when fast neutrons collide with nuclei of large atomic number. The recoil
nucleus may lose this energy immediately as gamma radiation or remain for a period of
time in the excited metastable state. In inelastic scattering, therefore, there is no conservation of momentum between the scattered neutron and recoil nucleus. Inelastic scattering
occurs mainly with fast neutron collisions with nuclei of large atomic number.
   Neutron scattering is a common mechanism by which fast and intermediate neutrons are
slowed down to the thermal neutron energy levels. Thermal neutrons have an energy level

at which they are in thermal equilibrium with the surrounding atoms at room temperature. There is an energy range for thermal neutrons as described earlier in this chapter; however, the properties of thermal neutrons are often cited at an energy calculated to be the most probable thermal neutron energy of 0.0253 eV at 20 °C corresponding to a velocity of 2200 m/sec (Gibson and Piesch, 1985). Figure 4.1 may be used to find the velocity of the neutron at energy levels over the range 0.001 eV to 10 MeV. For example, if we select the position 0.025 eV on the X-axis and follow up the graph with a straight line to the upper curve, we find the value 2200 m/sec. At the thermal energy state, the mechanisms of interaction of neutrons with matter change drastically as discussed in the following section.

4.4.3 Neutron capture

Because of the neutral charge on the neutron, it is relatively easy for slow neutrons in spite of their low kinetic energy to "find themselves" in the vicinity of the nucleus without having to hurdle the coulombic forces of atomic electrons. Once in close proximity to nuclei, it is easy for slow neutrons to enter into and be captured by nuclei to cause nuclear reactions. The capture of thermal neutrons, therefore, is possible with most radionuclides, and neutron capture is the main reaction of slow neutrons with matter. The power of a nucleus to capture a neutron depends on the type of nucleus as well as the neutron energy. The neutron absorption cross-section, σ, with units of 10^{-24} cm^2 or "barns", is used to measure the power of nuclides to absorb neutrons. A more detailed treatment of the absorption cross-section and its units and application are given in Section 4.5. However, because capture of thermal neutrons is possible with most radionuclides, references will cite the neutron cross-sections of the nuclides for comparative purposes at the thermal neutron energy of 0.0253 eV equivalent to a neutron velocity of 2200 m/sec. This is also the energy of the neutron, which is in thermal equilibrium with the surrounding atoms at room temperature. For comparative purposes, therefore, Table 4.3 lists the thermal neutron cross-sections for neutron capture reactions in barns (10^{-24} cm^2) for several nuclides. The nuclides selected for Table 4.3 show a broad range of power for thermal neutron capture. Some of the nuclides listed have practical applications, which are referred to in various sections of this book.

The capture of a slow neutron by a nucleus results in a compound nucleus, which finds itself in an excited energy state corresponding to an energy slightly higher than the binding energy of the neutron in the new compound nucleus. This energy of excitation is generally emitted as gamma radiation. Neutron capture reactions of this type are denoted as (n,γ) reactions. Two practical examples of (n,γ) neutron capture reactions were provided earlier in this chapter in the neutron irradiation of ^{238}U and ^{232}Th for the preparation of fissile ^{239}Pu and ^{233}U (eqs. (4.38) and (4.41)), respectively. Another interesting example of a (n,γ) reaction is neutron capture by ^{235}U according to

$$^{235}_{92}\text{U} + {}^{1}_{0}\text{n} \rightarrow {}^{236}_{92}\text{U} + \gamma \tag{4.59}$$

This neutron capture reaction is interesting, because the ^{236}U product nuclide decays by alpha emission in approximately 14% of the cases and decays by nuclear fission with emission of neutrons in the remaining 86% of the cases as discussed previously in Section 4.3.3.

Table 4.3

Cross-sections σ in barns for thermal neutron capture reactions
of selected nuclides in order of increasing magnitude

Nuclide	σ (Barns)
$^{3}_{1}$H	<0.000006
$^{2}_{1}$H	0.00052
$^{16}_{8}$O	0.00019
$^{12}_{6}$C	0.0035
$^{1}_{1}$H	0.332
$^{14}_{7}$N	1.8
$^{238}_{92}$U	2.7
$^{232}_{90}$Th	7.4
$^{55}_{25}$Mn	13.3
$^{233}_{92}$U	530
$^{235}_{92}$U	586
$^{239}_{94}$Pu	752
$^{6}_{3}$Li	940
$^{10}_{5}$B	3,840
$^{3}_{2}$He	5,330
$^{7}_{4}$Be	39,000
$^{155}_{64}$Gd	61,000
$^{157}_{64}$Gd	254,000

Source: Data from Holden (1997b).

The subject of neutron capture is treated in more detail in Section 4.5, which concerns the neutron cross-section and neutron attenuation in matter.

4.4.4   Non-elastic reactions

Neutron capture can occur in nuclei resulting in nuclear reactions that entail the emission of nuclear particles such as protons (n,p), deuterons (n,d), alpha particles (n,α), and even neutrons (n,2n). These reactions may not occur in any specific energy range but may be prevalent at specific resonances, which are energy states of the excited compound nuclei that are specific to relatively narrow energies of the incident neutron. The effect of resonance in neutron capture by nuclei is discussed in more detail subsequently in Section 4.5. The (n,2n) reactions occur at very high incident neutron energies, >10 MeV (Gibson and Piesch, 1985). The (n,p) and (n,α) reactions can occur in the slow neutron capture and reaction with nuclides of low atomic number (low Z), where the Coulomb forces of the electron shells are limited and present less a hurdle for the escape of charged particles from the confines of the atom. Some practical examples of these reactions are the (n,p) reaction used in the synthesis of ^{14}C by slow (thermal) neutron capture by ^{14}N

$$^{14}_{7}N + {}^{1}_{0}n \rightarrow {}^{14}_{6}C + {}^{1}_{1}H \tag{4.60}$$

and the (n,p) and (n,α) reactions used to detect neutrons by the interaction of slow neutrons with ^{3}He and ^{10}B, respectively, according to eqs. (4.60) and (4.61):

$$^{3}_{2}He + {}^{1}_{0}n \rightarrow {}^{1}_{1}H + {}^{3}_{1}H + 0.76\,MeV \tag{4.61}$$

$$^{10}_{5}B + {}^{1}_{0}n \rightarrow {}^{7}_{3}Li + {}^{4}_{2}He + 2.8\,MeV \tag{4.62}$$

Either of these reactions is used to detect neutrons by using gas proportional detectors containing helium or a gaseous form of boron (e.g., boron trifluoride). Slow neutrons that penetrate these detectors produce either radioactive tritium (eq. (4.61)) or alpha particles (eq. (4.62)), which produce ionization in the gas. The ionization events or ion pairs formed can be collected and counted to determine a neutron count rate.

4.4.5   Nuclear fission

The reaction of neutron-induced fission occurs when a neutron interacts with a fissile or fissionable nucleus and the nucleus becomes unstable, taking on the characteristics of an oscillating droplet, which then fragments into two nuclides (fission fragments). At the same time there is the release of more than one neutron (2.4 neutrons on the average for ^{235}U fission) and a relatively high amount of energy (~200 MeV). Fission in natural ^{235}U and man-made ^{233}U and ^{239}Pu is optimal at thermal incident neutron energies; whereas fission in ^{238}U and ^{232}Th requires neutron energies of at least 1 MeV. A more detailed treatment of nuclear fission was provided previously in Section 4.3.3.

4.5   NEUTRON ATTENUATION AND CROSS-SECTIONS

As we have seen in our previous treatment of the neutron, there are several possible interactions of neutrons with nuclei. Among these are elastic scattering, inelastic scattering, neutron capture, non-elastic reactions, and nuclear fission. As we have seen in several examples, probabilities exist for any of these interactions to occur depending on the energy of the incident neutron and the type of nuclide with which the neutron interacts. We can define this probability of interaction by the term cross-section, which is a measure of the capturing power of a particular material for neutrons of a particular energy.

The range of neutrons in matter is a function of the neutron energy and the cross-section or capturing power of the matter or medium through which the neutrons travel. To define cross-section, let us consider an incident beam of neutrons of given intensity or number (I_0), which impinges on a material of unit area (e.g., cm^2) and thickness dx as illustrated in Figure 4.6.

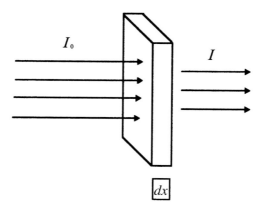

Figure 4.6   Attenuation of a neutron beam of intensity I_0 by an absorber material of unit area (cm²) and thickness dx.

The intensity (I) of the neutron beam traveling beyond the thickness dx will be reduced according to the number of nuclei (n) per unit volume in the material and the "area of obstruction" (e.g., cm²) that the nuclei present to the oncoming beam. This area of obstruction is referred to as the cross-section of the material. On the basis of the description previously given, we can write the equation:

$$\frac{dI}{dx} = -n\sigma I \tag{4.63}$$

which defines the change in beam intensity (dI) with respect to absorber thickness (dx) as proportional to the beam intensity (I) times a proportionality factor, which we may call the absorption coefficient or "obstruction coefficient" that the nuclei pose to the oncoming beam. The coefficient is a function of the number of nuclei (n) in the path of the neutron beam and the stopping power of the nuclei to interact with the neutron beam or, in other words, the neutron cross-section (σ) of the material through which the neutron beam travels. Eq. (4.63) may be written as

$$\frac{dI}{I} = -n\sigma \, dx \tag{4.64}$$

Eq. (4.63) is very similar to eq. (3.35) of Chapter 3 defining the attenuation of gamma radiation in matter with the exception that the absorption coefficients and attenuation coefficients involved for neutron and gamma radiation, respectively, are very different. The negative sign of eqs. (4.63) and (4.64) denotes the diminishing intensity of the neutron beam as a function of absorption coefficient and absorber thickness. The absorption coefficient $n\sigma$ is the combined effect of the number of nuclei (n) in the neutron beam path that might impede the continued travel of neutrons and the power of the nuclei to react with the neutrons.

Eq. (4.64) can be integrated over the limits of beam intensity from I_0 to I and absorber thickness from 0 to x as follows:

$$\int_{I_0}^{I} \frac{dI}{I} = -n\sigma \int_{0}^{x} dx \tag{4.65}$$

to give the equation:

$$\ln \frac{I_0}{I} = n\sigma x \tag{4.66}$$

or

$$I = I_0 e^{-n\sigma x} \tag{4.67}$$

which is the most simplified expression for the calculated beam intensity (I) after passing through an absorber of thickness (x) when the absorber material consists of only one pure nuclide and only one type of reaction between the neutron beam and nuclei is possible. If, however, several types of nuclei and reactions between the neutron beam and nuclei of the absorber material are possible, we must utilize the sum of the neutron cross-sections for all reactions that could take place.

We can use eq. (4.66) to calculate the half-value thickness ($x_{1/2}$) or the thickness of absorber material needed to reduce the incident neutron beam intensity by one-half. If we give the initial beam intensity (I_0) a value of 1 and the transmitted intensity (I) a value of 1/2, we can write

$$\ln \frac{1}{0.5} = n\sigma x_{1/2} \tag{4.68}$$

and

$$\ln 2 = n\sigma x_{1/2} \tag{4.69}$$

or

$$0.693 = n\sigma x_{1/2} \tag{4.70}$$

The half-value thickness for neutron beam attenuation may be written as

$$x_{1/2} = \frac{0.693}{n\sigma} \tag{4.71}$$

where $n\sigma$ is the number of nuclei per unit volume (cm^{-3}) and σ the neutron cross-section in cm^2. The neutron cross-section σ can be defined as the area in cm^2 for which the number of nuclei–neutron reactions taking place is equal to the product of the number of incident neutrons that would pass through the area and the number of target nuclei. The cross-section is defined in units of 10^{-24} cm^2 on the basis of the radius of atomic nuclei being about 10^{-12} cm. It provides a measure of the chances for the nuclei of a material being hit by a neutron of a certain energy. The unit of 10^{-24} cm^2 for nuclear cross-sections is called the barn. Tables in reference sources of nuclear data provide the neutron cross-sections in units of barns for various nuclides and neutron energies. An example is the reference directory produced by McLane et al. (1988), which provides neutron cross-section values in barns and neutron cross-section curves for most nuclides over the neutron energy range 0.01 eV to 200 MeV.

Let us take an example of 10 eV neutrons incident on a water barrier (i.e., neutrons traveling in water). We may use eq. (4.71) to estimate the half-value thickness, if we ignore the less significant interactions with oxygen atoms. This is because the neutron cross-section for hydrogen at 10 eV is about 20 barns (Figure 4.7) and that of oxygen is only 3.7 barns (McLane et al., 1988), and there are twice as many hydrogen atoms as oxygen atoms per given volume of water. The half-value thickness may be calculated as given below.

The value of n for the number of hydrogen nuclei per cm^3 of water may be calculated on the basis of Avogadro's number of molecules per mole. If 1 mol of water is equivalent to 18.0 g and the density of water is 1.0 g/cm^3, we can calculate the number of hydrogen nuclei per cm^3 as

$$6.22 \times 10^{23} \text{ molecules } H_2O/18 \text{ cm}^3 = 3.45 \times 10^{22} \text{ molecules } H_2O/\text{cm}^3$$

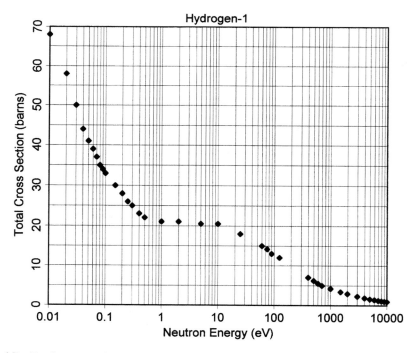

Figure 4.7   Total cross-section curve for hydrogen-1 over the neutron energy range 0.01 eV–10 keV.

$$n = (3.45 \times 10^{22} \text{ molecules } H_2O/cm^3)(2 \text{ proton atoms or } {}^{1}_{1}H/\text{molecule})$$

$$= 6.90 \times 10^{22} \text{ nuclei/cm}^3$$

By definition, 20 barns is equal to $20 \times 10^{-24} \text{ cm}^2$ and the half-value thickness may then be calculated as

$$x_{1/2} = \frac{0.693}{(6.90 \times 10^{22} \text{ cm}^{-3})(20 \times 10^{-24} \text{ cm}^2)}$$

$$= \frac{0.693}{1.38 \text{ cm}^{-1}}$$

$$= 0.502 \text{ cm}$$

If we make the calculation for 1-keV neutrons traversing water and use the value 4.1 barns for the neutron cross-section of hydrogen nuclei at this neutron energy (McLane *et al.*, 1988), we calculate a half-value thickness of

$$x_{1/2} = \frac{0.693}{(6.90 \times 10^{22} \text{ cm}^{-3})(4.1 \times 10^{-24} \text{ cm}^2)}$$

$$= \frac{0.693}{0.283 \text{ cm}^{-1}}$$

$$= 2.45 \text{ cm}$$

As the examples illustrate in the case of the proton, the neutron cross-section (or barns) decreases as the energy or velocity of the neutron increases. That is, the neutron reactions with nuclei obey the general rule of having some proportionality to $1/v$, where v is the velocity of the neutron. This inverse proportionality of cross-section and neutron velocity is particularly pronounced in certain regions of energy as illustrated in the total neutron cross-section curves for protons and elemental boron in Figures 4.7 and 4.8, respectively. However, this is not always the case with many nuclides at certain neutron energies where there exists a resonance between the neutron energy and the nucleus. At sometimes specific or very narrow neutron energy ranges, certain nuclei have a high capacity for interaction with neutrons. The elevated neutron cross-sections at specific neutron energies appear as sharp peaks in plots of neutron cross-section versus energy, such as the cross-section curve illustrated in Figure 4.9 for ${}^{55}_{25}Mn$. These peaks are called resonances and often occur with (n,γ) reactions. The high cross-sections occur when the energy of the incident neutron corresponds exactly to the quantum state of the excited compound nucleus, which is the newly formed nucleus consisting of a compound between the incident neutron and the nucleus. Most nuclides display both the $1/v$ dependence on neutron cross-section and the resonance effects over the entire possible neutron energy spectrum. We should keep in mind that neutron cross-sections can be specific and differ in value for certain reactions, such as proton (σ_p)- and alpha particle (σ_α)-producing reactions, fission reactions (σ_f), or neutron capture cross-sections (σ_c). The total neutron cross-section (σ_{tot}) would be the cross-section representing the sum of all possible neutron reactions at that specific neutron energy.

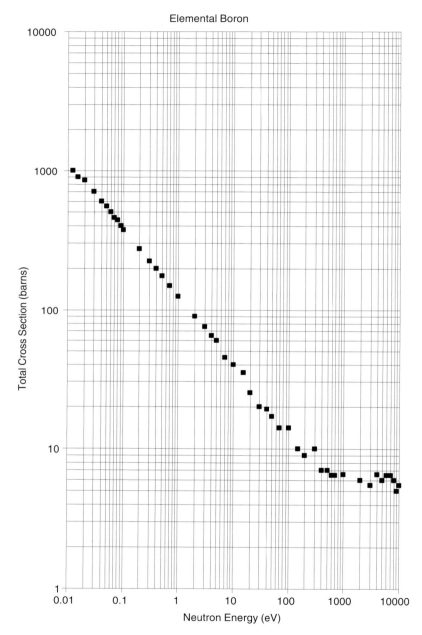

Figure 4.8 Total cross-section curve for elemental boron over the neutron energy range 0.01 eV–10 keV.

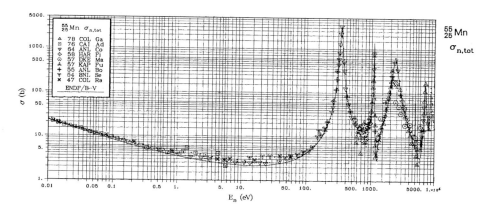

Figure 4.9 Total cross-section curve for manganese-55 over the neutron energy range 0.01 eV–10 keV. The columns in the upper left-hand column provide the number of data points and an abbreviation of the laboratory that provided the data. (From McLane *et al.* (1988), reprinted with permission from Elsevier © 1988.)

For example, the thermal neutron cross section for ^{235}U, which is the neutron cross-section at 0.0253 eV neutron energy corresponding to a neutron velocity of 2200 m/sec at room temperature, can be given as $\sigma_c = 95$ barns for the neutron capture cross-section, $\sigma_f = 586$ barns for the fission cross-section, and $\sigma_\alpha = 0.0001$ barns for the neutron cross-section for the alpha particle-producing reaction. These neutron cross-section values indicate that neutron fission would predominate at the thermal neutron energy of 0.0253 eV, although some neutron capture would also occur. The total neutron cross-section, σ_{tot}, would be the total of the three possible reactions or $\sigma_{tot} = 95$ barns + 586 barns + 0.0001 barns = 681 barns. In our treatment of slow neutron capture by ^{235}U in Radioactivity Hall of Fame, Part IV, illustrated in Figure IV.4, we noted that about 14% of the slow neutron captures by ^{235}U nuclei result in the formation of ^{236}U and gamma radiation, which decays to ^{232}Th and the remaining 86% of the slow neutron captures result in nuclear fission. This is exactly what is predicted by the thermal neutron cross-section values just provided; that is, for ^{235}U

$$\frac{\sigma_c}{\sigma_{tot}} = \frac{95 \text{ barns}}{681 \text{ barns}} = 14\% \text{ neutron capture}$$

and

$$\frac{\sigma_f}{\sigma_{tot}} = \frac{586 \text{ barns}}{681 \text{ barns}} = 86\% \text{ fission}$$

4.6 NEUTRON DECAY

We have seen that fast neutrons may lose their energy through elastic and inelastic collisions with other nuclei, and if these neutrons do not undergo other reactions with nuclei (e.g., fission), they may lose sufficient energy to reach thermal equilibrium with surrounding atoms and possibly be captured by atomic nuclei. The question remains of what would happen to a free neutron that is not absorbed by any atomic nucleus.

In Chapter 2 we discussed the transformation of the neutron within nuclei of radioactive atoms, which have a neutron/proton ratio too high for stability. In these unstable nuclides, the neutron breaks up into a proton, negatron (negative electron), and antineutrino. If the neutron can transform itself in unstable nuclei, it stands to reason that the neutron might be unstable outside the protective boundaries of the stable nucleus. This is just the case, as A.H. Snell and L.C. Miller demonstrated in 1948 followed by further studies by Robson (1950a,b) and Snell *et al.* (1950) that when neutrons were in free flight in a vacuum, they would indeed decay with a lifetime in the range of 9–25 min with a release of 0.782 MeV of energy. More recent and accurate measurements of neutron decay demonstrate the lifetime to be 885.4 ± 0.9 sec (Abele, 2000; Arzumanov *et al.*, 2000; Pichlmaier *et al.*, 2000; Snow *et al.*, 2000). The decay of elementary particles is characterized in terms of lifetime. The lifetime, usually symbolized as τ, is related to the term half-life, $t_{1/2}$, the mean time it takes for one-half of the particles to decay (Sundareson, 2001) according to the relationship

$$t_{1/2} = (\ln 2)\tau = 0.693\tau \qquad (4.72)$$

The free neutron decays according to the scheme

$$n \rightarrow p^+ + e^- + \bar{\nu} + 0.782\,\text{MeV} \qquad (4.73)$$

The 0.782 MeV of energy released in the neutron decay corresponds to the difference in mass of the neutron (1.0086649 u) and the sum of the masses of the products of the neutron decay, the proton (1.0072765 u) plus the electron (0.0005485 u), or 1.0078250 u. Using Einstein's equation of equivalence of mass and energy (Section 4.3.3), this mass difference of 0.0008399 u can be converted to the equivalent of 0.782 MeV of energy. This calculation provides additional evidence for the decay of the neutron into a proton and an electron. The neutron, therefore, outside the protective confines of a stable nucleus, has a very short lifetime.

Radioactivity Hall of Fame — Part V

Niels Bohr (1885–1962), Gustav Hertz (1887–1975) and James Franck (1882–1964), Werner Heisenberg (1901–1976), Erwin Schrödinger (1887–1961), Max Born (1882–1970) and Paul A.H. Dirac (1902–1984), and Clinton Davisson (1881–1958) and George Paget Thomson (1892–1975)

NIELS BOHR (1885–1962)

Niels Bohr was awarded the Nobel Prize in Physics 1922 "for his services in the investigation of the structure of atoms and the radiation emanating from them." He was born in Copenhagen, Denmark on October 7, 1885. His father was Professor of Physiology at Copenhagen University, and his mother also was from a family distinguished in education.

Niels Bohr (1885-1962)

    Bohr earned a Master's degree in Physics in 1909 and the doctorate degree in 1911 at Copenhagen University. His doctoral dissertation was a theoretical explanation of the properties of metals based on the electron theory. In 1911 Niels Bohr had a short stint at the Cavendish Laboratory where he was able to witness the experimental work of J. J. Thomson on the properties of the electron. Subsequently in 1912 he worked in the laboratory of Ernest Rutherford in Manchester where Rutherford was making discoveries on the properties of the atomic nucleus. These experiences with J. J. Thomson and Rutherford helped Bohr formulate his own ideas on the structure of the atom.

    During the time of his doctoral work and following the stint in Rutherford's laboratory, Niels Bohr was confronted with Max Planck's quantum theory of radiation and Rutherford's experimental work on the atomic nucleus. By combining the concepts of Planck's quantum theory and Rutherford's atomic structure, Niels Bohr was able to formulate a quantum theory of atomic structure (Bohr, 1913, 1914), which remains to this day suitable in explaining the chemical and spectroscopic properties of the elements.

Bohr's starting point was Rutherford's theory of the atom based on Rutherford's experiments in alpha-particle scattering (see Radioactivity Hall of Fame—Part I). The Rutherford atom consisted of a positively charged nucleus with electrons revolving in orbits about the nucleus, the number of which were sufficient to neutralize the positive charge on the nucleus (i.e., the number of electrons in each atom equaled the number of protons in the nucleus or its atomic number). Also, the dimensions of the nucleus were very small in comparison to the size of the atom defined by the electron orbits, and almost all of the mass of the atom was concentrated in the nucleus. By taking Rutherford's atom, Max Planck's quantum theory of energy, and the simplest of all atoms hydrogen, Niels Bohr was able to launch his quantum theory of atomic structure, referred to as the Bohr atom. With his quantum theory of atomic structure, Bohr was able to explain the spectra of radiation emitted by atomic electrons and the chemical properties of the elements based on electron groupings in quantum orbits.

Bohr (1913, 1914, 1921a,b) proposed that electrons would exist in orbits about the nucleus at discrete distances and each orbit would represent specific quantized energies. The energy of an electron would be proportional to its orbit, that is, the larger the orbit or the greater its radius, the higher would be the electron energy. Thus, when an atom absorbs energy, from an external form of excitation, an atomic electron would jump from one orbit to another further away from the atomic nucleus. Energy absorbed by the atom could be emitted from the atom as radiation when an electron falls from an outer to an inner orbit, as described by the equation

$$h\nu = E_i - E_f \quad \text{or} \quad h\nu = E_u - E_l \tag{V.1}$$

where h is Planck's constant, ν the frequency of the emitted radiation, E_i and E_f the initial and final electron energies, respectively, and E_u and E_l the upper (higher) and lower electron energies, respectively. Eq. (V.1) and the simplest of all atoms, the atom of hydrogen with its single orbiting electron, are illustrated in the commemorative postage stamp from Denmark shown on the previous page. Bohr explained that energy can be absorbed or emitted by an atom only as a single quantum of energy or light photon ($h\nu$) equal to the energy differences of specific orbitals. He demonstrated that electrons could not possess any orbit, but only specific orbits, which he called "stationary states", and the orbits were defined by a quantum multiple of the angular momentum of the electron round the nucleus. Bohr (1913) stated:

> If we assume that the orbit of the electron in the stationary states is circular, the results of the calculations can be expressed by the simple condition: that the angular momentum of the electron round the nucleus in a stationary state of the system is equal to an entire multiple of a universal value, independent of the charge on the nucleus…The great number of different stationary states we do not observe except by investigation of the emission and absorption of radiation

Bohr demonstrated that the angular momentum (L) of the electron in the various possible orbits was defined by the relation

$$L = n\left(\frac{h}{2\pi}\right) \tag{V.2}$$

where $n = 1, 2, 3, \ldots$ and h is the Planck's constant. Thus, Bohr concluded that the angular momentum of the electron is quantized according to a quantum number n, and that the transition of an electron from one orbit (stationary state) to another as a result of the absorption or emission of energy by an atom could only occur in specific or quantized magnitudes.

With his quantum theory of atomic structure, Bohr derived and explained fully the Rydberg formula, which was a formula devised by the Swedish physicist Janne Rydberg (1854–1919) to calculate and predict the wavelengths of light photons emitted by hydrogen and later applied to the other elements of the periodic table to include, in addition to visible light, other types of electromagnetic radiation emitted by atoms. The Rydberg formula for hydrogen is the following:

$$\frac{1}{\lambda} = R_H \left(\frac{1}{n_l^2} - \frac{1}{n_u^2} \right) \tag{V.3}$$

where λ is the radiation wavelength emitted by the hydrogen atom, R_H the Rydberg constant for hydrogen ($R_H = 1.0967758 \times 10^{-3}\,\text{Å}^{-1}$), and n_l and n_u the lower and upper (higher) integers, respectively, such that

(1) when $n_l = 1$ and $n_u = 2, 3, 4\ldots\infty$, the Lyman series of wavelengths in the UV region are calculated;
(2) when $n_l = 2$ and $n_u = 3, 4, 5\ldots\infty$, the Balmer series of wavelengths in the visible region are calculated;
(3) when $n_l = 3$ and $n_u = 4, 5, 6\ldots\infty$, the Paschen series of wavelengths in the infrared region are calculated; and
(4) when $n_l = 4$ and $n_u = 5, 6, 7\ldots\infty$, the Brackett series of wavelengths in the far infrared region are calculated.

Bohr Atom and Signature of Niels Bohr

While the Rydberg formula was well used by spectroscopists to calculate the lines of atomic absorption or emission, the formula was not fully understood until Niels Bohr derived the Rydberg constant, and explained the origin of the specific lines of radiation emission or absorption on the basis of his quantum theory of atomic structure. Illustrated in the commemorative postage stamp from Sweden is a picture of the Bohr atom with the central nucleus and four of more possible electron quantum orbitals. The Bohr atom is illustrated more clearly in Figure V.1 taken from his Nobel lecture (Bohr, 1922). Using this simplified figure of the atom, Bohr explained the quantized transition that an electron can make in jumping from an inner to outer orbital resulting in energy absorption or the transitions from an outer to an inner electron orbit resulting in the emission of energy as quanta of radiation photons manifested by precise lines of electromagnetic radiation. In his own words, Bohr explained the radiation emission of hydrogen as follows:

We arrive at a manifold of stationary states for which the major axis of the electron orbit takes on a series of discrete values proportional to the square of the whole numbers (The whole numbers that Bohr refers to here are n_l^2 and n_u^2 of the Rydberg formula). The accompanying figure shows

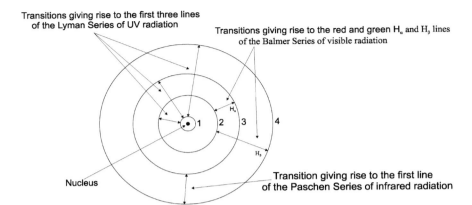

Figure V.1 Bohr's atomic representation of the hydrogen atom with quantized electron transitions according to the quantum numbers 1, 2, 3, and 4 giving rise to specific radiation emissions lines in the Lyman, Balmer, or Paschen series.

the relation diagrammatically. For the sake of simplicity the electron orbits in the stationary states are represented by circles, although in reality the theory places no restriction on the eccentricity of the orbit...The arrows represent the transition processes that correspond to the red and green hydrogen lines, H_α and H_β, the frequency of which is given by means of the Balmer formula (series) when we put $n_l = 2$ and $n_u = 3$ and 4 respectively. The transition processes are also represented which correspond to the first three lines of the series of ultraviolet lines found by Lyman in 1914, of which the frequencies are given by the formula when n_l is put equal to 1, as well as to the first line of the infrared series discovered some years previously by Paschen, which are given by the formula if n_l is put equal to 3. This explanation of the origin of the hydrogen spectrum leads us quite naturally to interpret this spectrum as the manifestation of a process whereby the electron is bound to the nucleus...the state of the atom can only be changed by the addition of energy whereby the electron is transferred to an orbit of larger dimensions corresponding to an earlier stage of the binding process.

Using the Rydberg formula and Bohr's quantum theory of atomic structure, one can calculate the wavelengths (lines) of radiation emission for hydrogen and other non-complex elements and the energy levels of the quantized electron orbits. For example, the electron transition from $n_u = 2$ to $n_l = 1$ will yield a photon in the first line of the Lyman Series with a wavelength calculated according to eq. (V.3) as

$$\frac{1}{\lambda} = R_H \left(\frac{1}{1^2} - \frac{1}{2^2} \right)$$

$$= 1.0967758 \times 10^{-3} \, \text{Å}^{-1} \left(1 - \frac{1}{4} \right) \tag{V.4}$$

$$= 1.0967758 \times 10^{-3} \, \text{Å}^{-1} \, (0.75)$$

$$= 8.2258185 \times 10^{-4} \, \text{Å}^{-1}$$

$$\lambda = 1215 \, \text{Å}$$

Similar calculations for radiation emissions as a result of electron transitions from $n_u = 3$ to $n_l = 1$ and $n_u = 4$ to $n_l = 1$ illustrated in Figure V.1 yield according to eq. (V.3) radiation emissions of 1025 and 972 Å, respectively. Radiation emission for transitions from orbitals further out (e.g., $n_u = 4, 5, 6...\infty$ to the innermost orbital $n_l = 1$) may be calculated in a similar fashion to yield other wavelengths of the Lyman series. If we consider the transition of $n_u = \infty$ to $n_l = 1$, we can calculate the energy level of the lowest or the innermost orbital as

$$E_n = -hcR_H \left(\frac{1}{n_1^2} - \frac{1}{\infty} \right)$$

$$= -(4.136 \times 10^{-15} \, \text{eV sec})(2.9979 \times 10^8 \, \text{m/sec})(1.0967758 \times 10^{-3} \, \mathring{A}^{-1})\left(\frac{1}{n^2} \right)$$

$$= -(12.399 \times 10^{-7} \, \text{eV m})(1.0967758 \times 10^{-3} \, \mathring{A}^{-1})\left(\frac{1}{n^2} \right)$$

$$= -(12.399 \, \text{keV} \, \mathring{A})(1.0967758 \times 10^{-3} \, \mathring{A}^{-1})\left(\frac{1}{n^2} \right)$$

$$= -\frac{0.01359 \, \text{keV}}{n^2}$$

$$= -\frac{13.59 \, \text{eV}}{n^2} \tag{V.5}$$

Thus, the energy of the electron in its lowest orbital closest to the nucleus (quantum number $n = 1$) is -13.59 eV. This energy corresponds exactly to the ionization energy of hydrogen, that is, the energy required to remove completely the electron from the hydrogen atom, that is, move it to an infinite distance from the nucleus to form the hydrogen ion, H^+. The negative sign is used for the energy level because the electron is in a bound state whereby the energy is not free. Higher energy levels of the hydrogen atom would be -3.40 eV for $n = 2$, -1.51 eV for $n = 3$, -0.85 eV for $n = 4$, -0.54 eV for $n = 5$, -0.38 eV for $n = 6$, and -0.28 eV for $n = 7$, etc. Energy levels of the Bohr atomic structure for hydrogen are illustrated in Figure (V.2).

A summary of Bohr's revolutionary quantum theory of atomic structure and the significance of Bohr's hypothesis were highlighted in a Nobel Presentation Speech given by Professor C. W. Oseen, member of the Nobel Committee for Physics of the Royal Swedish Academy of Sciences on December 10, 1926. In Professor Oseen's words:

It was only through a radical break with classical physics that Bohr was able to resolve the spectroscopic puzzles in 1913. Bohr's basic hypothesis can be formulated as follows: Each atom can exist in an unlimited number of different states. Each of these stationary states is characterized by a given energy level. The difference between two such energy levels, divided by Planck's constant h, is the oscillation frequency of a spectral line that can be emitted by the atom [see Eq. V.6]...The extraordinary good agreement with experience obtained in this way, explains why after 1913 almost a whole generation of theoretical and experimental physicists devoted itself to atomic physics and its application in spectroscopy.

$$\nu = \frac{E_2 - E_1}{h} \tag{V.6}$$

The use of electron orbits about a central atomic nucleus as depicted by Bohr (i.e., the Bohr atom) is not considered an accurate depiction of the atom; however, it remains to this day a very didactic method to illustrate the atom and explain radiation emission and absorption of electron origin and the chemical properties of the elements. Because electrons have properties of both particles and waves, that is, a dual nature as described by de Broglie (see Radioactivity Hall of Fame—Part III), Erwin Schrödinger took the dual nature of the electron to describe the electron and its properties with wave equations. Thus, we may also picture atomic electrons as possessing energy levels, such as illustrated on Figure V.2, rather than occupying definite orbits.

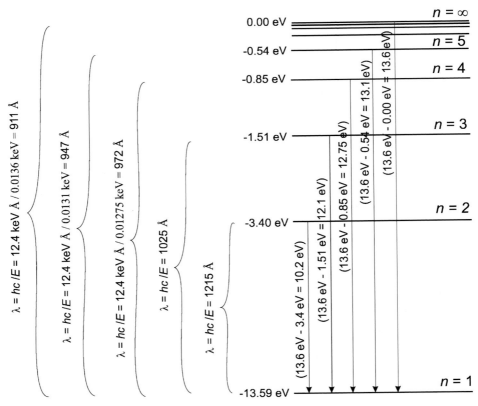

Figure V.2   Energy levels and energy transitions of Bohr's quantized atomic structure for hydrogen that yield the Lyman series (UV) of radiation emissions. The energy levels of hydrogen are represented as plateau and the electron transitions from the outer higher energy orbits to the lower are depicted as arrows pointing downwards. Some energies liberated as radiation quanta in units of electron volts and radiation wavelength in Angstroms are provided.

Bohr's Quantum Characterization of Electron Orbits of the Inert Gases

Helium (2_1)
Neon $(2_1 8_2)$
Argon $(2_1 8_2 8_2)$
Krypton $(2_1 8_2 18_3 8_2)$
Xenon $(2_1 8_2 18_3 18_3 8_2)$
Niton $(2_1 8_2 18_3 32_4 18_3 8_2)$

Figure V.3   Bohr's depiction of the quantum orbits of electrons of the inert gases (Bohr, 1921a). Niton is now known as Radon of atomic number 86. In Bohr's words: "The large numbers denote the number of electrons in groups (shells) starting with the innermost group, and the small numbers (subscripts) denote the total number of quanta characterizing the electron orbits within each group (shell)."

On the basis of his quantum theory of atomic structure, Bohr went further to explain the properties of the elements. In his paper on atomic structure (1921a), Bohr explained how the inert gases were different from the other elements because the atomic orbitals were completely filled providing the atoms with "inherent stability" as it would be difficult (i.e., require excessive energy) to remove any of the electrons from such atoms to form ions. He depicted the electron configurations of the inert gases as illustrated in Figure V.3.

We can compare Bohr's early depiction of the above electron configurations and notice their close similarity to the current configurations taken from a modern Handbook of Chemistry and Physics (Lide, 1997) provided in Table V.1. The letters $K, L, M, \ldots$ are the principle electron energy levels with increasing energy and distance from the nucleus; $s, p, d,$ and f are quantum energy sublevels also with increasing energy and distance from the nucleus; and $n = 1, 2, 3 \ldots$ are the principal quantum numbers. (see Wolfgang Pauli in Radioactivity Hall of Fame—Part II).

During the time of his historic work on the structure of the atom (1913–1914), Niels Bohr was Lecturer in physics at Copenhagen University. He subsequently held a similar position at Victoria University in Manchester during 1914–1916. Bohr returned to his home country in 1916 when he was appointed Professor of Theoretical Physics at Copenhagen University, and in 1920 he became Head of the Institute of Theoretical Physics, which was established for him.

After receiving the Nobel Prize in 1922, Niels Bohr diverted his attention from atomic electrons to the atomic nucleus and the disintegration and transmutation of radioactive atoms. Bohr was the first to conceive of a heavy atomic nucleus (i.e., a nucleus of high atomic number) to display the properties of a liquid drop (Bohr, 1936); and it was Bohr's ideas on the properties of the atomic nucleus that helped Lise Meitner and Otto Frisch (1939) arrive at the discovery of nuclear fission through their interpretation of the work of Otto Hahn and Fritz Strassmann (1939a), who demonstrated the splitting of the uranium-235 nucleus after neutron capture (see Lise Meitner and Otto Hahn in Radioactivity Hall of Fame—Part IV). In his work on the disintegration of heavy nuclei, Bohr (1939) summarized these concepts as follows:

… any nuclear reaction initiated by collisions or radiation involves as an intermediate stage the formation of a compound nucleus in which the excitation energy is distributed among the various degrees of freedom in a way resembling the thermal agitation of a solid or liquid body…Since the effective cross-sections for the fission phenomena seem to be about the same order of magnitude as the cross-sections for ordinary nuclear reactions, we may therefore conclude that for the heaviest nuclei the deformation energy sufficient for the fission is of the same order of magnitude as the energy necessary for the escape of a single nuclear particle…These circumstances find their straightforward explanation in the fact, stressed by Meitner and Frisch that the mutual repulsion between the electric charges in a nucleus will for highly charged nuclei counteract to a large extent the effect of the short-range forces between the nuclear particles in opposing a deformation of the nucleus.

During the Nazi occupation of Denmark in the autumn of 1943, Bohr escaped to Sweden and was invited to England by the British Government. Bohr was then taken into confidence and informed of the secret Manhattan Project for the development of the atomic bomb, which was already at an advanced stage. During the last two years of the war, he spent time in England and America. Niels Bohr and his son Aage Bohr, who later became Nobel Laureate in Physics 1975, joined the secret Manhattan Project at

Table V.1

Electron configurations of neutral atoms in the ground state

Atomic No.	Element	K (n = 1)	L (n = 2)		M (n = 3)			N (n = 4)				O (n = 5)				P (n = 6)		
		s	s	p	s	p	d	s	p	d	f	s	p	d	f	s	p	d
2	He	2																
10	Ne	2	2	6														
18	Ar	2	2	6	2	6												
36	Kr	2	2	6	2	6	10	2	6									
54	Xe	2	2	6	2	6	10	2	6	10		2	6					
86	Ra	2	2	6	2	6	10	2	6	10	14	2	6	10		2	6	

Source: Data from Lide (1997).

the Los Alamos National Laboratories. Bohr was not vital to the development of the atomic bomb, but he was an admired personality and served as a father figure to other scientists in the project. Rhodes (1986) provides much insight into the role of Niels Bohr in the Manhattan Project and provides the following quote from Victor Weisskopf, Austrian émigré theoretician, who worked on the project at Los Alamos, which summarizes Bohr's role in the project:

> In Los Alamos we were working on something which is perhaps the most questionable, the most problematic thing a scientist can be faced with. At that time physics, our beloved science, was pushed into the most cruel part of reality and we had to live it through. We were, most of us at least, young and somewhat inexperienced in human affairs, I would say. But suddenly in the midst of it, Bohr appeared in Los Alamos. It was the first time we became aware of the sense in all these terrible things because Bohr right away participated not only in the work, but in our discussions. Every great and deep difficulty bears in itself its own solution....This we learned from him.

J. Robert Oppenheimer, theoretician and Director of the Manhattan Project at Los Alamos, noted that Bohr was not needed either for technical or theoretical aspects of the projects, but rather his persona had a tremendous positive affect on the project staff. Rhodes (1986) in his book provided insight into the effect of Bohr's presence by quoting Oppenheimer in a postwar lecture as follows:

> Bohr at Los Alamos was marvelous. He took a very lively technical interest...But his real function, I think for almost all of us, was not the technical one."... *In the unedited version of the lecture*: "(Bohr) made the enterprise which looked so macabre seem hopeful"... *In the edited version:* "He made the enterprise seem hopeful, when many were not free of misgiving...Bohr spoke with contempt of Hitler, who with a few hundred tanks and planes had tried to enslave Europe for a millennium. He said nothing like that would ever happen again; and his own high hope that the outcome would be good...

Niels Bohr was a pacifist, and he knew that he was not needed to produce the atomic bomb. He saw his own participation in the Manhattan Project as necessary so that he could bear witness and provide first-hand warning to President Franklin Roosevelt before the bomb was developed and tested and eventually to the world through the United Nations. Bohr feared that Germany, where nuclear fission was discovered in 1939, would eventually develop the atomic bomb, and the USA and England had to beat Germany to the race; however, Bohr had the foresight of the grave menace that nuclear weapons would present to the world and the survival of the human race. In the beginning of 1944, Bohr had the opportunity to bring his views to the attention of the American and British governments, and in a memorandum dated July 3, 1944, Bohr wrote President Roosevelt on his concerns over the project for the nuclear weapon development, the threat that nuclear weapons would present to the human race, and following the development of the weapon, the immediate need for an international nuclear non-proliferation regime to prevent a nuclear weapons race possibly under the auspices of the United Nations. In August of 1944, Niels Bohr was granted a long meeting with the President. The following are some excerpts of Bohr's memorandum to President Roosevelt:

> ...The fact of immediate preponderance is however that a weapon of unparalleled power is being created which will completely change all future conditions of warfare. Quite apart from the question of how soon the weapon will be ready for use and what role it may play in the present war, the situation raises a number of problems which call for most urgent attention. Unless, indeed, some agreement about the control of the use of the new active materials can be obtained in due time, any temporary advantage, however great, may be outweighed by a perpetual menace to human security...especially the terrifying prospect of a future competition between nations about a weapon of such formidable character can only be avoided through a universal agreement in true confidence...The prevention of a competition prepared in secrecy will therefore demand such concessions regarding exchange of information and openness about industrial efforts including military preparations as would hardly be conceivable unless at the same time all partners were assured of a compensating guarantee of common security against dangers of unprecedented

acuteness…Without impeding the importance of the project (Manhattan Project) for immediate military objectives, an initiative, aiming at forestalling a fateful competition about the formidable weapon, should serve to uproot and cause of distrust between the powers on whose harmonious collaboration the fate of coming generations will depend. Indeed, it would appear that only when the question is taken up among the United Nations of what concessions the various powers are prepared to make as their contribution to an adequate control arrangement, it will be possible for any one of the partners to assure themselves of the sincerity of the intentions of the others.

Bohr's fears of an eventual nuclear arms race with the development of weapons of unimaginable destructive power, and the need for the Government to pursue a program for nuclear non-proliferation were elaborated further in a subsequent memorandum to the President dated March 25, 1945, before the weapon was first tested on July 16, 1945. Bohr's memorandum contained the following relevant passages:

Above all, it should be appreciated that we are faced only with the beginning of a development and that, probably within the very near future, means will be found to simplify the methods of production of the active substances and intensify their effects to an extent which may permit any nation possessing great industrial resources to command powers of destruction surpassing all previous imagination.

Humanity will, therefore, be confronted with dangers of unprecedented character unless, in due time, measures can be taken to forestall a disastrous competition in such formidable armaments and to establish an international control of the manufacture and use of the powerful materials.

Any arrangement which can offer safety against secret preparations for the mastery of the new means of destruction would, as stressed in the memorandum, demand extraordinary measures. In fact, not only would universal access to full information about scientific discoveries be necessary, but every major technical enterprise, industrial as well as military, would have to be open to military control.

In this connection it is sufficient that the special character of the efforts which, irrespective of technical refinements, are required for the production of the active materials, and the peculiar conditions which govern their use as dangerous explosives, will greatly facilitate such control and should ensure its efficiency, provided only that the right of supervision is guaranteed.

Detailed proposals for the establishment of an effective control would have to be worked out with the assistance of scientists and technologists appointed by the governments concerned, and a standing expert committee, related to an international security organization, might be charged with keeping account of new scientific and technical developments and with recommending appropriate adjustments of the control measures.

On recommendations from the technical committee the organization would be able to judge the conditions under which industrial exploitation of atomic energy sources could be permitted with adequate safeguards to prevent any assembly of active material in an explosive state.

In an Open Letter to the United Nations dated June 9, 1950 (Bohr, 1950), Niels Bohr provided the above relevant contents of his memoranda to President Roosevelt and implored the United Nations in cooperation with its Member States to take the lead in working toward and achieving nuclear non-proliferation and to advance the peaceful applications of nuclear energy. The following are additional relevant excerpts of his Open Letter to the United Nations:

I address myself to the organization, founded for the purpose to further co-operation between nations on all problems of common concern, with some considerations regarding the adjustment of international relations required by modern development of science and technology… Everyone associated with the atomic energy project (nuclear weapons Manhattan Project) was, of course, conscious of the serious problems which would confront humanity once the enterprise was accomplished (i.e., weapon developed). Quite apart from the role atomic weapons might come to play in the war, it was clear that permanent grave dangers to world security would ensue unless measures to prevent abuse of the new formidable means of destruction could be universally agreed upon and carried out… worldwide political developments have increased the tension between nations and at the same time the perspectives that great countries may compete about the possession of means of annihilating populations of large areas and even making parts of the earth temporarily uninhabitable have caused widespread confusion and alarm…The situation calls for the most unprejudiced attitude towards all questions of international relations. Indeed, proper appreciation of the duties and responsibilities implied in world citizenship is in our time more necessary than ever before…real co-operation between nations on problems of common concern presupposes free access to all information of importance for their relations. Any argument for upholding barriers for information and intercourse, based on concern for national ideals or interests, must be weighed against the beneficial effects of common enlightenment and the relieved tension resulting from openness…The development of technology has now reached a stage where the facilities for communication have provided the means for making all mankind a co-operating unit, and where at the same time fatal consequences to civilization may ensue unless international divergences are considered as issues to be settled by consultation based on free access to all relevant information…I turn to the United Nations with these considerations in the hope that they may contribute to the search for a realistic approach to the grave and urgent problems confronting humanity…

IAEA Emblem

   Niels Bohr in the above statements of his memoranda to President Roosevelt and in his Open Letter to the United Nations displayed tremendous foresight in predicting the need for nuclear arms non-proliferation and the role of the United Nations in this effort. What he wrote more than half a century ago in essence foretold the need for the current nuclear non-proliferation program (NNP) that the International Atomic Energy Agency (IAEA) of the United Nations actively pursues. Through his writings outlined above Niels Bohr in 1950 gave birth to the IAEA's NNP. In recognition of the efforts taken in the direction of nuclear non-proliferation, the IAEA and its Director General Dr. Mohamed ElBaradei were awarded the Nobel Peace Prize 2005. The IAEA was founded in 1957. Its emblem is illustrated on the postage stamp issued in Austria in 1977 commemorating its 20th anniversary. The Headquarters of the IAEA are located in the Vienna International Centre on the banks of the Danube River in Vienna, Austria, which is pictured in the commemorative stamp issued in Hungary in 1980.
   The Nobel award was granted on October 7, 2005 with the following statement:

   The Norwegian Nobel Committee has decided that the Nobel Peace Prize for 2005 is to be shared, in two equal parts, between the International Atomic Energy Agency (IAEA) and its Director

IAEA and UN Headquarters, Vienna

**Mohamed ElBaradei
(1942 -) with permission
© The Nobel Foundation 2005,
Photo: Micheline Pelletier**

General, Mohamed ElBaradei, for their efforts to prevent nuclear energy from being used for military purposes and to ensure that nuclear energy for peaceful purposes is used in the safest possible way.

At the time when the threat of nuclear arms is again increasing, the Norwegian Nobel Committee wishes to underline that this threat must be met through the broadest possible international cooperation. This principle finds its clearest expression today in the work of the IAEA and its Director General. In the nuclear non-proliferation regime, it is the IAEA which controls that nuclear energy is not misused for military purposes, and the Director General has stood out as an unafraid advocate of new measures to strengthen that regime. At a time when disarmament efforts appear deadlocked, when there is a danger that nuclear arms will spread both to states and to terrorist groups, and when nuclear power again appears to be playing an increasingly significant role, IAEA's work is of incalculable importance.

In his will, Alfred Nobel wrote that the Peace Prize should, among other criteria, be awarded to whoever has done most for the "abolition or reduction of standing armies". In its application of this criterion in recent decades, the Norwegian Nobel Committee has concentrated on the struggle to diminish the significance of nuclear arms in international politics, with a view to their abolition.

Niels Bohr retained his post of Head of the Institute for Theoretical Physics at Copenhagen University, created for him in 1920, until his death in Copenhagen in 1962.

GUSTAV HERTZ (1887–1975) AND JAMES FRANCK (1882–1964)

Gustav Hertz and James Franck shared the Nobel Prize in Physics 1925 "for their discovery of the laws governing the impact of the electron upon an atom." Hertz and Franck were awarded the Nobel Prize for their work in providing experimental proof of Bohr's quantum theory of atomic structure. Professor Oseen (1926), member of the Nobel Committee, in his presentation speech on the occasion of the Nobel Award to Hertz and Franck noted the following:

…When all that had been gained in the field of atomic physics seemed to be at stake, there is nobody who would have thought it advisable to proceed from the assumption that the atom can exist in different states, each of which is characterized by a given energy level, and that these energy levels govern the spectral lines emitted by the atoms in the way described [by Bohr's theory]. The fact that Bohr's hypotheses in 1913 have succeeded in establishing this, is because they are no longer mere hypotheses but experimentally proven facts. The methods of verifying these hypotheses are the work of James Franck and Gustav Hertz, for which they have been awarded the Physics Nobel Prize for 1925.

Gustav Ludwig Hertz was born in Hamburg, Germany on July 22, 1887. He started his university studies at the University of Göttingham in 1906 and subsequently continued his studies at the Universities of Munich and Berlin. After graduating in 1911, Hertz was appointed as a Research Assistant at the University of Berlin. James Frank was a professor at the University of Berlin at the time and the two

Gustav Hertz (1887-1975) and his diffusion cascade apparatus

started a collaboration on the interaction of electrons with gas atoms. This work led to their findings that gave experimental proof of Bohr's quantum theory of atomic structure and a joint sharing of the Nobel Prize in Physics 1925, which is described further on in this section. Hertz's research career was interrupted by his conscription into World War I and was wounded in action in 1915. He returned to the University of Berlin in 1917 as Privatdozent and in 1920 entered into the private sector of research with the Philips Incandescent Lamp Factory at Eindhoven where he continued experimental work on electron impact with gas atoms. In 1925 Hertz became Resident Professor of the University of Halle and the following year he became Director of the Physics Institute of the University. At the end of 1927, he returned to Berlin and was given an appointment with the duties of creating a new Physics Institute at the Technische Hochschule (Technological University) at Charlottenburg where he managed to build an institute that paralleled its counterpart at the University of Berlin.

At his new Technische Hochschule in Berlin Gustav Hertz did pioneering work on the separation of isotopes by means of gaseous diffusion cascade. By 1932 he demonstrated the feasibility of isotope separation or isotope enrichment by gaseous diffusion (Hertz, 1933). He successfully separated the isotopes neon-20 from neon-22 as well as deuterium (hydrogen-2), referred to as heavy hydrogen, from its most abundant isotope, which we refer to as normal hydrogen (hydrogen-1). The gaseous diffusion method is based on the principle that, in thermal equilibrium, two isotopes of the same energy will travel or diffuse with different velocities. The lighter isotopes, or the gaseous molecules within which the isotopes are bound, will travel with a higher velocity and therefore travel more rapidly through a porous membrane than the heavier isotopes. The difference in the rates of diffusion is proportional to the square root of the mass ratios or molecular weights of the light isotope over the heavier isotope as described in eq. (V.7) (Wooldridge, 1936; Wooldridge and Smythe, 1936).

$$\mu = \sqrt{\frac{M_L}{M_H}} \tag{V.7}$$

where μ is the velocity of diffusion and M_L and M_H the molecular weights of the gaseous molecules that contain the light and heavier isotopes, respectively. The apparatus used for isotope enrichment as described by Wooldridge (1936) consisted of a length of clay tubing through which the gas containing the isotopes could be led. A region outside the tubing is kept evacuated so that part of the gas will diffuse through the porous walls of the clay tubing as the gas travels along the length of the tubing. As the gas containing the two isotopes travels through the tubing, the gas of lower molecular weight diffuses faster through the porous membrane of the clay tubing. Thus, the gas which travels the entire length of the tubing without passing through the porous walls will be richer in the heavier isotope than the gas that was originally led into the tubing. When only two isotopes are involved in the process, the ratio of the number of heavy to light isotopes is increased by the factor E, the enrichment factor, given by eq. (V.8)

$$E = \sqrt[1-\mu]{\frac{\text{Volume of gas led into tubing}}{\text{Volume led out of tubing without diffusion}}} \tag{V.8}$$

The method devised by Hertz was both simple and ingenious whereby he was able to affect isotope enrichment by leading the gas containing the isotopes automatically through many successive diffusions in a closed system referred to as a diffusion cascade. Hertz led the gas through the successive diffusions by affecting the diffusion process in a chain of numerous chambers, each chamber would consist of the porous clay tubing surrounded by its own evacuated glass jacket. The gas that diffuses through the porous walls was not discarded, but led back into the system at a point in the apparatus where the gas was introduced. The lighter constituent that had diffused through the wall of the tubing in the nth separation chamber (or nth stage) was sent back to the $(n-1)$th separation chamber. The light constituent of the $(n-1)$th chamber is sent back to the $(n-2)$th chamber, etc. This results in a net transfer of light molecules toward one end of the system (cascade) and the heavier molecules toward the other end. A simple illustration of Hertz's gaseous diffusion cascade is illustrated in the commemorative postage stamp seen on the previous page issued by the German Democratic Republic.

Gaseous diffusion was one of the methods used at the Oak Ridge National Laboratories in the Manhattan Project to enrich fissile uranium-235 for the production of the first atomic bomb. Natural uranium contains approximately 0.7% ^{235}U, and to enrich this isotope from ^{238}U the use of a gaseous compound of uranium is required, namely, uranium hexafluoride (UF_6). The production of 99% ^{235}U from natural uranium required approximately 4000 enrichment stages. The enrichment of isotopes for medical applications and biological research is currently in great demand. The enrichment of carbon-13 and nitrogen-15 to percent abundances greater than 99% is a common commercial practice together with the labeling of numerous organic and inorganic compounds with the enriched isotope ^{13}C or ^{15}N for metabolic studies in the biological sciences, which have resulted in many discoveries yielding advances in medical science and increased agricultural production (L'Annunziata, 1984).

In 1934 Hertz could not continue as Director of the Research Institute that he had founded at the Technische Hochschule in Berlin because of the anti-Semitic laws of the German Nazi regime, as his grandfather was of Jewish decent. He left the Technische Hochschule and took on a position in setting up a new research laboratory at Siemens in 1935. About four weeks prior to the testing of the first atomic bomb by the Americans in 1945, Hertz was flown to Moscow and after World War II he helped establish a research institute at Sukhumi on the Black Sea coast of the USSR, which had the principal task of developing the methodology for the separation of uranium isotopes, that is, enrichment of uranium-235, in large quantities. Although there is no evidence of Hertz being involved directly in the Russian nuclear bomb program, he was a Nobel Laureate and respected scientist, who could command the loyalty and following of other prominent scientists in the Russian research effort. In 1954 he took on a position as Professor at the University of Leipzig and became Professor Emeritus of the University in 1961. He died in Berlin on October 30, 1975, where he returned after retiring from the University of Leipzig.

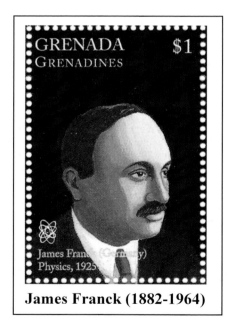

James Franck (1882-1964)

James Franck was born on August 26, 1882 in Hamburg, Germany. He started his university studies in the field of chemistry at the University of Heidelberg, and transferred to the field of physics at the University of Berlin where he received the Ph.D. degree in 1906. After a short stint as Research Assistant at the University of Frankfurt am Main, James Franck returned to the University of Berlin as Research Assistant. In 1911 he became Lecturer at the University of Berlin where he remained until 1918 rising to the position of Associate Professor of Physics. It was at the University of Berlin during 1912–1914 that Franck and Hertz collaborated in research that provided proof of Niels Bohr's quantum theory of atomic structure and for which both were awarded the Nobel Prize in Physics. Their joint work is described further on in this section. During his stint at the University of Berlin, he took time to serve in World War I and was awarded the Iron Cross.

After the war Franck was appointed Head of the Physics Division of the Kaiser Wilhelm Institute for Physical Chemistry at Berlin-Dahlem. In 1920 he was appointed Professor of Experimental Physics at the University of Göttingen. When the Nazi regime ruled Germany, James Franck showed his true character by resigning his post in protest of the Nazi anti-Semitic policies. As explained by Franck's former student and research collaborator (Rosenberg, 2004):

His resignation from his professorship as a public protest against emerging Nazism became a cause celebre in Germany. Although his World War I army service would have spared him from dismissal from his post under the new anti-Semitic laws, although only for the short term as we now know, he refused to accept his orders to dismiss many of his faculty, staff, and students under the new edicts dealing with racial and "politically correct" classifications. Instead, he worked tirelessly as a private citizen from his home in Göttingen to locate scientific employment opportunities throughout the Western world for dozens of his colleagues.

Franck left Germany altogether for Copenhagen, Denmark in 1933 as an academic refugee and took on a position of guest professor for a year in Niels Bohr's laboratory. In 1935 he was appointed Professor at Johns Hopkins University until 1938 when he became Professor of Physical Chemistry at the University of Chicago.

During World War II Franck served as Director of the Chemistry Division of The Metallurgical Laboratory, which was the center established for the secret Manhattan Project for the development of the atomic bomb. Franck and many other scientists of the Manhattan Project worked to develop the horrific bomb for one very specific reason that being the threat of Nazi Germany where nuclear fission was discovered in 1939. Germany led by Hitler represented an evil that could not be tolerated. It had the technological skill needed to develop such a weapon and lacked the moral constraints against its use. But after Germany surrendered, many scientists were confronted with the moral implications of using the bomb and an international nuclear arms race that the use of the bomb could create. They knew that such an arms race would create the potential for a future nuclear war that would cause great worldwide destruction and eventual annihilation of the human race. With this in mind James Franck chaired a Committee on the Political and Social Problems, University of Chicago Manhattan Project "Metallurgical Laboratory". The Committee, which included James Franck as Chairman, Glenn T. Seaborg, Leo Szilard, who was the first to conceive of the atomic bomb, among other scientists of the Manhattan Project drafted "The Franck Report" on June 11, 1945, approximately 5 weeks before the first atomic bomb was tested at Los Alamos, New Mexico. The Franck Report, addressed to the US Government, was by force a secret document asking it to refrain from using the bomb against Japan, but rather, recommending that the Government detonate the weapon of mass destruction in an uninhabited desert area to be witnessed by an international team of United Nations, which would put the United States and the world on a better footing toward the establishment of an international effort and program of safeguards toward nuclear non-proliferation. James Franck personally delivered the Franck Report to the United States Secretary of War, Henry Stimson on June 11, 1945 (Rabinowitch, 1964; Rosenberg, 2004). The following excerpts of The Franck Report, which unfortunately have so far accurately predicted the current state of affairs concerning nuclear proliferation and the threat of nuclear annihilation, are very relevant in reminding the world of the threat of nuclear weapons and the need for a strong United Nations IAEA NNP:

The Franck Report, June 11, 1945 James Franck (Chairman)

I. Preamble

The only reason to treat nuclear power differently from all other developments in the field of physics is its staggering possibilities as a means of political pressure in peace and sudden destruction in war…The scientists on this project do not presume to speak authoritatively on problems of national and international policy. However, we found ourselves, by the force of events, the last five years in the position of a small group of citizens cognizant of a grave danger for the safety of this country as well as for the future of all the other nations, of which the rest of mankind is unaware…the success we have achieved in the development of nuclear power is fraught with infinitely greater dangers than were all of the inventions in the past. All of us, familiar with the present state of nucleonics, live with the vision before our eyes of sudden destruction visited on our own country, of Pearl Harbor disaster, repeated in thousandfold magnification, in every one of our major cities…protection can only come from the political organization of the world. Among all arguments calling for an efficient international organization for peace, the existence of nuclear weapons is the most compelling one. In the absence of an international authority which would make all resort to force in international conflicts impossible, nations could still be diverted from a path which must lead to total mutual destruction, by a specific international agreement barring a nuclear arms race.

II. Prospectives of Armaments Race

…we cannot hope to avoid a nuclear armament race, either by keeping secret from the competing nations the basic scientific facts of nuclear power, or by cornering the raw materials required for such a race…all that these advantages can give us is the accumulation of a larger number of bigger and better atomic bombs…such a quantitative advantage in reserves of bottled destructive power will not make us safe from sudden attack…If no efficient international agreement is achieved, the race of nuclear armaments will be on in earnest not later than the morning after our first demonstration of the existence of nuclear weapons. After this, it might take other nations three to four years to overcome our present head start…

III. Prospectives of Agreement

The prospect of nuclear warfare and the type of measures which have to be taken to protect a country from total destruction by nuclear bombing, must be abhorrent to other nations as to the United States. England, France, and the smaller nations of the European continent, with their congeries of people and industries, are in an entirely hopeless situation in the face of such a threat…There is no doubt that Russia too, will shudder at the possibility of a sudden disintegration of Moscow and Leningrad, almost miraculously preserved in the present war…Therefore, only lack of mutual **trust**, and not lack of **desire** for agreement, can stand in the path of an efficient agreement for the prevention of nuclear warfare. From this point of view, the way in which nuclear weapons, now secretly developed in this country, will be revealed to the world appears of great, perhaps fateful importance…Certain and perhaps important tactical results undoubtedly can be achieved (by the use of the atomic bomb on Japan), but we nevertheless think that the question of the use of the very first available atomic bombs in the Japanese war should be weighed very carefully, not only by military authority, but by the highest political leadership of this country. If we consider international agreement on total prevention of nuclear warfare as the paramount objective, and believe that it can be achieved, this kind of introduction of atomic weapons to the world may easily destroy all our chances of success. Russia, and even allied countries which bear less mistrust of our ways and intentions, as well as neutral countries, will be deeply shocked. It will be very difficult to persuade the world that a nation which was capable of secretly preparing and suddenly releasing a weapon, as indiscriminate as the rocket bomb and a thousand times more destructive, is to be trusted in its proclaimed desire of having such weapons abolished by international agreement…it is not at all certain that the American public opinion, if it could be enlightened as to the effect of atomic explosives, would support the first introduction by our own country of such an indiscriminate method of wholesale destruction of civilian life. Thus from the "optimistic" point of view—looking forward to an international agreement on prevention of nuclear warfare—the military advantages and the saving of American lives, achieved by the sudden use of atomic bombs against Japan, may be outweighed by the ensuing loss of confidence and wave of horror and repulsion, sweeping over the rest of the world, and perhaps dividing the public opinion at home. **From this point of view a demonstration of the new weapon may best be made before the eyes of representatives of all United Nations, on the desert or barren land.** The best possible atmosphere for the achievement of an international agreement could be achieved if

America would be able to say to the world, "You see what weapon we had but did not use. We are ready to renounce its use in the future and to join other nations in working out adequate supervision of the use of this nuclear weapon."...One may point out that the scientists themselves have initiated the development of this "secret weapon" and it is therefore strange that they should be reluctant to try it out on the enemy as soon as it is available. The answer to this question was given above—the compelling reason for creating this weapon with such speed was our fear that Germany had the technical skill necessary to develop such a weapon without any moral constraints regarding its use...

IV. Methods of International Control

We now consider the question of how an effective international control of nuclear armaments can be achieved...Given mutual trust and willingness on all sides to give up a certain part of their sovereign rights, by admitting international control of certain phases of national economy, the control could be exercised ... One thing is clear, any international agreement on prevention of nuclear armaments must be backed by actual and efficient controls. No paper agreement can be sufficient since neither this or any other nation can stake its whole existence on trust into other nations' signatures. Every attempt to impede the international control agencies must be considered equivalent to denunciation of the agreement. It hardly needs stressing that we as scientists believe that any systems of controls envisaged should leave as much freedom for the peace(ful) development of nucleonics as is consistent with the safety of the world."

In 1947 James Franck was named Emeritus Professor at the University of Chicago. He continued research work at the University of Chicago as Head of the Photosynthesis Research Group until 1956. He carried out a lot of pioneering work and theories on photosynthesis that cover a span of almost 30 years. His first paper on photosynthesis was published when he was at Johns Hopkins University (Franck, 1935) and his last with Rosenberg at the University of Chicago (Franck and Rosenberg, 1964). Franck's works in photosynthesis are reviewed by Rosenberg (2004) together with an excellent sketch of his life. Franck wrote on the fate of his theories on photosynthesis in one of his papers (Franck and Herzfeld, 1941) as follows:

> The change in the situation (new methods of observation and new results) is indeed so far-reaching that practically all theories published hitherto are now obsolete...These theories have served the purpose for which they were developed; they have clarified the situation, they have stimulated new experiments, and most of them contained parts which have been used in each subsequent attempt... A theory... by its very nature can contain only a partial truth.

Franck died in Germany when visiting Göttingen on May 21, 1964. In a short tribute to James Franck and his life, Rosenberg (2004) relates a story passed on to him by Franck as follows:

> Another of his anecdotes expressed his pride in sharing the Nobel Prize with Bohr and Planck, because for safekeeping Bohr had dissolved all three of their gold Nobel medals in *aqua regia* and stored them as a solution in an unmarked bottle in his Copenhagen laboratory during World War II, to be reprecipitated and cast into three new medals after the war.

The Franck–Hertz experiment

While at the University of Berlin during 1912–1914, James Franck and Gustav Hertz collaborated in research on the impact of electrons on atoms and the energy loss by electrons upon collision with atoms. In their collaborative work, Franck and Hertz (1914a,b) measured the kinetic energies of electrons before and after collisions with atoms and demonstrated that an electron could lose its kinetic energy upon collision with an atom only when a certain quantum of energy is transferred to the atom. If the electron possessed kinetic energy less than that quantum, no kinetic energy of the electron would be transferred upon collision with an atom. An electron of kinetic energy greater than a given quantum could transfer the quantum of energy to an atom upon collision with the atom, and then continue traveling with any kinetic energy in excess of the transferred energy quantum. The atom with which the electron had collided would increase in energy from E_1 to E_2 corresponding to an absorption of that quantum of energy. Only energy corresponding to fixed quanta could be absorbed by the atom upon collision with electrons. Franck and Hertz were awarded the Nobel Prize in Physics 1925 for this work because it demonstrated Bohr's quantum theory of atomic structure, that is, the atom can exist in only specific quantum energy states whereby the atomic electrons can only absorb and emit energy in discrete quanta. The quantum of energy emitted by the atom would correspond to the quantum of energy absorbed as manifested by spectral lines or radiation frequencies defined by eq. (V.6), that is, $\nu = (E_2 - E_1)/h$.

Gustav Hertz and the Franck-Hertz apparatus and data for Mercury

The experimental arrangement used by Franck and Hertz is illustrated in the postage stamp issued by the GDR in 1987 and in Figure V.4. Their apparatus (Franck and Hertz, 1914a,b; Franck, 1926) consisted of an air-tight glass chamber containing mercury vapor. Within the chamber is located a tungsten wire heated to a bright-red glow by an electric current as a cathode or source of electrons. A few centimeters away from the electron source (G of Figure V.4) is a positively charged wire-screen electrode (N) using the original notation of Franck (1926). The electrons emitted from the cathode will accelerate toward the positive wire screen. The kinetic energy that the electrons gained in the acceleration between G and N without collision with any atoms (i.e., in the absence of any gas in the apparatus) is governed by the voltage potential applied according to the relationship

$$E_e = \frac{1}{2}mv^2 = eV \tag{V.9}$$

where $1/2\,mv^2$ is the kinetic energy of each electron traveling to the anode, e the elementary electrical charge, and V the applied potential difference between G and N. The energy of the electron can therefore, be measured in units of electron volts (eV) according to the accelerating voltage applied. Some of the accelerated electrons are caught by the screen with a specific energy of x eV, the value of which is subject to the voltage potential applied. Some of the electrons fly through the mesh and reach the electrode P to produce a negative current that flows to earth through a galvanometer, provided there is no field difference applied between N and P that would send the electrodes back to N. With the apparatus described, Franck and Hertz introduced an electric field between N and P to determine the energy distribution of the electrons passing through the screen. We can call the electrode P a "collector plate" because it is the electrode that will collect the accelerated electrons passing through the wire-mesh anode N. Frank explains the function of the collector plate as follows:

> By introducing an electric field between N and P the energy distribution of those electrons passing through the screen can be determined. If, for example, we take only 4-volt beams, which pass perpendicularly through the screen, than the electric current measured at the galvanometer as a function of a decelerating potential difference applied between N and P, must be constant, until P becomes 4 volts more negative than N. At this point the current must become suddenly zero since henceforth all electrons will be repelled from P that they return to N.

Franck and Hertz introduced an inert gas, such as mercury vapor, into the chamber at a pressure that would ensure that the electrons accelerated through G and N would make many collisions with atoms of the gas when passing through the space between N and P. They then plotted the energy distribution of the electrons arriving at the collector plate P as a function of the accelerating voltage applied between G and N to determine whether the electrons had lost energy by impact or collision with the atoms of mercury. Franck and Hertz plotted the current collected at electrode P against the accelerating voltage applied across G and N. In this experiment they measured at the collector plate P all electrons with energy greater than the energy of 0.5 V beams. The plot is illustrated in the postage stamp

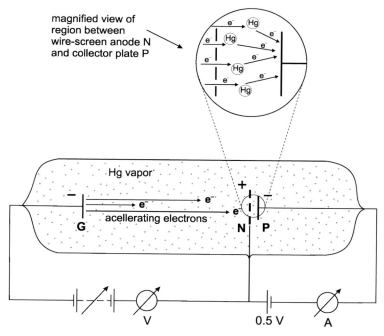

Figure V.4 The Franck–Hertz apparatus with mercury vapor. Electrons accelerated across the potential difference between electrodes G and N with energy greater than 0.5 V were collected at plate P. The electric current (amperes) collected at P was plotted against the accelerating voltage across G and N. The current collected at P would increase as the accelerating voltage was increased until the energy stage of 4.9 eV was reached, which would result in a sudden drop in current collected at P. This would repeat itself at additional increments of 4.9 eV as the accelerating voltage was further increased resulting in current drops at 9.8 and 14.7 V, etc. The current drops in increment of 4.9 eV illustrated that electron energy was transferred upon impact with Hg atoms between plates N and P (see magnified portion of illustration) in quantum increments of 4.9 eV only.

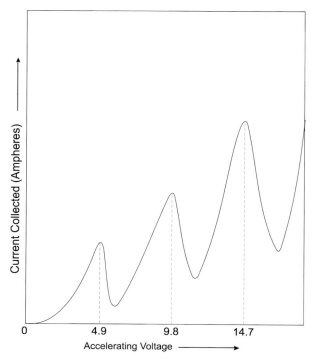

Figure V.5   The Franck-Hertz experiment with mercury vapor. Atoms of Hg absorb increments of 4.9 eV energy upon impact with accelerated electrons (modified from Franck, 1926 and commemorative postage stamp from the former GDR issued in 1987).

issued by the GDR where the current collected (I) is plotted against the accelerating voltage (V) and the graph is illustrated in Figure V.5. The graph illustrates that the electron current at P increases with increasing acceleration until the critical energy stage is reached (4.9 eV) when the current collected falls suddenly. Further increase of the accelerating voltage causes the process to repeat itself, that is, the number of electrons with velocity greater than 0.5 V again increases until the critical value of the accelerating voltage reaches 9.8, a multiple of 4.9, and the current at the collector plate P again falls. The process repeats every time the accelerating voltage overreaches a multiple of the critical voltage of 4.9, that is, 9.8, 14.7 V, etc. as illustrated in the postage stamp and Figure V.5.

The drop in the electron current collected at P at multiples of 4.9 V illustrated that energy was removed from the accelerating electrons upon collision with mercury atoms and transferred to the atoms of mercury only in the quantized energy levels of 4.9 eV. This energy quantum corresponded to the 2537 Å (254 nm) UV line in the emission spectrum of mercury providing evidence to Bohr's theory that an atom can accept only discrete quantities of energy and can subsequently emit the energy only in discrete quantities. Frank (1926) interpreted the results according to Bohr's quantum theory of atomic structure as follows:

> According to Bohr an atom can absorb as internal energy only discrete quantities of energy, namely those quantities which transfer the atom from one stationary state to another stationary state. If following on energy supply an exited energy state results from a transfer to a stationary state of higher energy, then the energy so taken up will be radiated in quanta fashion according to the $h\nu$ relationship. The frequency of the absorption line having the longest wavelength…multiplied by Planck's constant, gives the energy required to reach the first state

of excitation. These basic concepts agree in very particular with our results. The elastic collisions of low electron velocities show that for these impacts no energy is taken up as inner energy, and the first critical energy step results in just the amount of energy required for the excitation of the longest wave absorption line of Hg. Subsequently it appeared to me to be completely incomprehensible that we had failed to recognize the fundamental significance of Bohr's theory...I should like to mention that it later proved successful...to demonstrate also, from the current voltage curves, the stepwise excitation of a great number of quantum transitions, lying before the first excitation level and ionization.

The experimental proof of the quantum atomic absorption and emission of energy provided by James Franck and Gustav Hertz set the stage for future scientific research in spectroscopy an important phenomenon in chemistry and physics. In the words of Oseen (1926): "The extraordinary good agreement with experience obtained in this way (Frank–Hertz experiment), explains why after 1913 almost a whole generation of theoretical and experimental physicists devoted itself to atomic physics and its application in spectroscopy."

WERNER HEISENBERG (1901–1976), ERWIN SCHRÖDINGER (1887–1961), MAX BORN (1882–1970), AND PAUL A. H. DIRAC (1902–1984)

Werner Heisenberg, Erwin Schrödinger, and Paul Dirac were awarded the Nobel Prize in Physics 1932 and 1933 for their work on the "new atomic physics". The awards were presented in a ceremony on December 13, 1933. As expressed by Professor H. Pleijel (1933) in his Nobel Presentation Speech "This year's Nobel Prizes for Physics are dedicated to the new atomic physics...awarded to those men, Heisenberg, Schrödinger, and Dirac, who have created and developed the basic ideas of modern atomic physics." In his Nobel Presentation Speech Pleijel (1933) clearly outlined the historical developments that led to the creation and development of the "new physics", namely, quantum and wave mechanics. Not included by the Nobel Committee at this time was the vital contribution of Max Born, his statistical interpretation of quantum mechanics during 1926–1927, for which the Nobel Prize eluded him until 1954. The historical development described in more detail earlier in this book began with Max Planck, who in 1900, described light as a packet or quantum of energy, that is, a function of the wavelength or frequency and Planck's constant, $E = h\nu$. Then Einstein, through his explanation of the photoelectric effect in 1905, demonstrated the particulate nature of the light photon. Planck demonstrated that light could only be absorbed and emitted by matter in discrete quantities of energy, the light quantum, and, this energy divided by the frequency of the radiation would always yield the universal Planck's constant, h. Then Louis de Broglie in 1923 presented his ideas on the dual nature of matter and light or the particle–wave duality, that is, a particle in motion could be interpreted as having the properties of a particle with mass and as having the properties of light with wavelength and frequency. De Broglie derived the wavelength of the particle as a function of Planck's constant or $\lambda = h/p$ where p is the particle momentum. The dual nature of matter and light presented by Einstein and de Broglie were first received with astonishment, but accepted universally after Davisson and Germer demonstrated experimentally in 1927 the diffraction of electrons by crystals. Thus, atomic particles had to be looked upon not only as having the properties of durable matter but also the properties of light as explained by Pleijel (1933):

De Broglie's theory of matter-waves subsequently received experimental confirmation. If a relatively slowly traveling electron meets a crystal surface, diffraction and reflection phenomena appear in the same way as if an incident beam of waves were concerned. As a result of this theory one is forced to the conclusion to conceive of matter as not being durable, or that it can have definite extension of space. The waves, which form the matter, travel, in fact, with different velocity and must, therefore, sooner or later separate. Matter changes form and extent in space. The picture which has been created, of matter being composed of unchangeable particles, must be modified.

One of the phenomena that needed more clarification was Bohr's theory on the structure of the atom (Bohr, 1913, 1914), which he constructed by taking the heavy positively charged nucleus as the center of the atom around which orbited the atomic electrons in fixed circular paths or stationary states. For the hydrogen atom, Bohr demonstrated that the atom could absorb energy only in fixed quanta resulting in electron orbital transitions from inner to outer orbitals further away from the nucleus, and that the atom would emit this energy as a light photon with a frequency that would be equivalent to the change in energy experienced by the electrons divided by Planck's constant or $\nu = (E_2 - E_1)/h$ where E_2 is the electron energy in an outer orbital after the atom had absorbed energy and E_1 is the initial electron energy in an inner orbital prior to the absorption of energy by the atom. This theory worked well for hydrogen (see Bohr in this section), but full agreement of this theory with atomic spectra did not hold for more complex atoms and molecules, which sparked the creation of the "new physics" as explained by Pleijel (1933):

> The frequencies which Bohr thus obtained held good for a hydrogen atom which has only one electron, but when his method was applied to more complicated atoms and to certain optical phenomena, theory and practice did not agree. The fact that Bohr's hypothesis met the case for the hydrogen atom, however, suggests that Planck's constant was, in one way or another, a determining factor for the light-vibrations of the atoms. On the other hand, one had the feeling that it could not be right to apply the laws of classical mechanics to the rapid movements in the atom. Efforts made from various sides to develop and improve Bohr's theory proved also in vain. New ideas were required to solve the problem of oscillations of atoms and molecules. This solution started in 1925 upon the works of Heisenberg, Schrödinger, and Dirac in which different starting-points and methods were obtained.

The biographies and contributions of Heisenberg, Schrödinger, and Dirac are outlined subsequently:

Werner Heisenberg (1901-1976)

Werner Heisenberg (1901–1976) was born in Würzburg, Germany on December 5, 1901. In 1920 he entered the University of Munich where he pursued studies in theoretical physics obtaining his Ph.D. at the University in 1923. He then joined Max Born in studies of theoretical physics at the University of Göttingen. In 1926 he was appointed Lecturer in Theoretical Physics under Niels Bohr at the University of Copenhagen where he developed the first version of quantum mechanics, a mathematical matrix method of calculating the behavior of electrons and subatomic particles. As noted in his autobiography written at the time of his Nobel Award (Heisenberg, 1933b) his new theory was based on what could be observed, for example, the radiation emitted by an atom. The paths of such radiation had clearly been visualized and photographed ingeniously by C. T. R. Wilson with his cloud chambers (see Radioactivity Hall of Fame—Part II). Heisenberg noted that we cannot always assign to an electron a position in space at a given time, nor follow it in its orbit, so therefore, we cannot assume that the planetary orbits postulated by Niels Bohr actually exist. He represented the electron's position and velocity, not by ordinary numbers, but by abstract mathematical matrices formulating his new theory on the basis of matrix equations (Bohr *et al.*, 1926; Heisenberg, 1926a–c, 1929).

In 1927 Heisenberg discovered and published his Uncertainty Principle, which is a basic concept of quantum theory. Before defining the Uncertainty Principle, it is best to compare the concepts of classical physics with respect to velocity, mass, and position of large bodies of matter as compared to the same properties for the relatively small atomic particles. For example, with classical physics we can observe a baseball as it flies through the air after being hit by a batter and as light reflects off the ball. We can accurately determine the position (x) of the ball at a given time and its momentum (p) according to its mass (m) and velocity (v) where $p = mv$. However, at the small atomic level, when a photon of light hits an atomic particle, it will affect the momentum of the particle due to its low mass as illustrated in Figure V.6. Einstein explained in the photoelectric effect that a light photon is capable of imparting

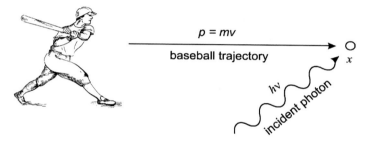

(a) light photon has negligible effect on baseball position (x)

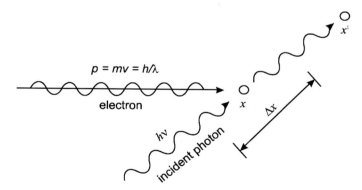

(b) light photon deflects electron matter-wave from position x to x^1

Figure V.6   Artists' depiction of Heisenberg's explanation of The Uncertainty Principle. The momentum (p) and position (x) of a body of relatively large mass, such as a baseball illustrated in (a) above will not be affected to any significant degree by the energy of the incident light photon (Heisenberg used the planets orbiting the sun instead of a baseball in his description; see Heisenberg's explanation taken from his Nobel Lecture in this section); however, the momentum and position of a subatomic particle, such as the electron in (b), will be affected by a light photon. The uncertainty in the position (x) of the electron is depicted by (Δx). Thus, both the location of the electron and its momentum cannot be measured with accuracy at any given time and remain uncertain.

energy to an electron providing it with momentum. Therefore, at the small atomic level, we cannot accurately measure both the momentum (p) and location (x) of the electron, as the particle will have wave as well as mass properties.

Heisenberg (1933a) explained the concept of The Uncertainty Principle in his Nobel Lecture as follows:

> This difference between classical and atomic physics is understandable, of course, since for heavy bodies such as the planets moving around the sun the pressure of the sunlight which is reflected at their surface and which is necessary for them to be observed is negligible; for the smallest building units of matter, however, owing to their low mass, every observation has a decisive effect on their physical behaviour. The perturbation of the system to be observed caused by the observation is also an important factor in determining the limits within which a visual description of atomic phenomena is possible. If there were experiments which permitted accurate measurement of all the characteristics of an atomic system necessary to calculate classical motion, and which, for example, supplied accurate values for the location and velocity of each electron in the system at a particular time, the result of these experiments could not be utilized at all in the formalism, but rather it would directly contradict the formalism. Again, therefore, it is clearly that fundamentally unverifiable part of the perturbation of the system caused by the measurement itself which hampers accurate ascertainment of the classical characteristics and thus permits quantum mechanics to be applied. Closer examination of the formalism shows that between the accuracy with which the location of a particle can be ascertained and the accuracy with which its momentum can simultaneously be known, there is a relation according to which the product of the probable errors in the measurement of the location and momentum is invariably at least as large as Planck's constant divided by 4π. In a very general form, therefore, we should have

$$\Delta p \Delta q \geq \frac{h}{4\pi} \tag{V.10}$$

where p and q are canonically conjugated variables.

The variable p represents the particle momentum and q its location. The postage stamp illustrated on page 310 commemorates Heisenberg's Uncertainty Principle and his original approximation that the product of the uncertainties of p and q are equivalent to the order of magnitude of Planck's constant, h. The more common mathematical term for location is x. Thus, Heisenberg's Uncertainty Principle is most often expressed as

$$\Delta p \Delta x \geq \frac{h}{4\pi} \geq \frac{\hbar}{2} \tag{V.11}$$

where Δp and Δx are the uncertainties of the particle momentum and location, h the Planck's constant ($h = 6.626 \times 10^{-34}$ J sec), and $\hbar = h/2\pi$ or $\hbar = 1.054 \times 10^{-34}$ J sec. Heisenberg's Uncertainty Principle simply states that it is not possible to accurately measure both the atomic particle (e.g., electron) velocity or momentum and its location. Heisenberg's derivation from quantum mechanics showed that it is not possible to determine, in any given instant in time, both the velocity and the position of an atomic particle. It was demonstrated that the more accurately one tries to determine the velocity of the particle, the more uncertain becomes the determination of its position, and vice versa.

Let us take, for example, a hypothetical case where we can measure the velocity of an electron to be 2×10^8 m/sec with an accuracy (e.g., standard deviation) of 0.5%, and we want to know the uncertainty expected in the determination of the position (Δx) of that electron. According to eq. (V.11) we can calculate the uncertainty as follows:

$$\Delta x \geq \frac{\hbar}{2\Delta p} \geq \frac{1.054 \times 10^{-34} \text{ J sec}}{2(0.005)(m_e)(v)} \tag{V.12}$$

where m_e and v are the electron mass and velocity, respectively. The relativistic mass (m_r) of the electron must be used, which is calculated according to eq. (4.6) of Chapter 4 so that

$$m_e = m_r = \frac{m_0}{\sqrt{1 - (v^2/c^2)}} \tag{V.13}$$

where m_0 is the particle rest mass, v the particle velocity, and c the speed of light in a vacuum. The relativistic mass of the electron becomes

$$m_r = \frac{9.109 \times 10^{-31} \text{ kg}}{\sqrt{1 - [(2 \times 10^8 \text{ m/sec})^2/(2.99 \times 10^8 \text{ m/sec})^2]}} = 12.25 \times 10^{-31} \text{ kg}$$

and

$$\Delta x \geq \frac{1.054 \times 10^{-34} \text{ J sec}}{2(0.005)(12.25 \times 10^{-31} \text{ kg})(2 \times 10^8 \text{ m/sec})} \geq 0.43 \times 10^{-10} \text{ m} = 0.43 \text{ Å}$$

If the measurement of the electron velocity were made with a greater accuracy of 0.1%, the expected uncertainty of the electron position would increase to

$$\Delta x \geq \frac{1.054 \times 10^{-34} \text{ J sec}}{2(0.001)(12.25 \times 10^{-31} \text{ kg})(2 \times 10^8 \text{ m/sec})} \geq 2.15 \times 10^{-10} \text{ m} = 2.15 \text{ Å}$$

Thus, increasing the accuracy of the measurement of one of the variables, the electron momentum, results in an increase in the uncertainty of the measurement of the other variable, the electron position.

Let us take another example that can illustrate the use of Heisenberg's Uncertainty Principle to calculate the uncertainty in the measurement of the momentum of an electron in the confines of an atomic nucleus (radius = 0.5×10^{-14} m) and consequently the improbability of finding an electron in the nucleus. If we multiply Heisenberg's eq. (V.11) by the speed of light (c) in both numerator and denominator, we will obtain units of momentum in MeV/c, which are the conventional units for relativistic momentum (see eq. (3.7) of Chapter 3). Thus, eq. (V.11) becomes

$$\Delta p \geq \frac{hc}{4\pi \Delta x c} \tag{V.14}$$

Since $hc = 12.4$ keV Å (see eq. (3.10) of Chapter 3), we can calculate the uncertainty in the measurement of the electron momentum in the atomic nucleus as

$$\Delta p \geq \frac{12.4 \text{ keV Å}}{4(3.14)(5 \times 10^{-5} \text{ Å})c}$$

$$\geq \frac{12.4 \text{ keV Å}}{6.28 \times 10^{-4} \text{ Å } c}$$

$$\geq 1.97 \times 10^4 \text{ keV}/c$$

$$\geq 19.7 \text{ MeV}/c$$

Converting the electron momentum to energy including the rest energy of the electron (m_0c^2) according to eq. (3.4) of Chapter 3 gives

$$
\begin{aligned}
E^2 &= p^2c^2 + (m_0c^2)^2 \\
&= (19.7\,\text{MeV}/c)^2c^2 + [(9.109 \times 10^{-31}\,\text{kg})(2.99 \times 10^8\,\text{m/sec})^2]^2 \\
&= (19.7\,\text{MeV}/c)^2c^2 + (8.143 \times 10^{-14}\,\text{J})^2 \\
&= (19.7\,\text{MeV}/c)^2c^2 + \left(\frac{8.143 \times 10^{-14}\,\text{J}}{1.602 \times 10^{-19}\,\text{J/eV}} \right)^2 \\
&= 388.1\,\text{MeV}^2 + (0.51\,\text{MeV})^2 \\
&= 388.1\,\text{MeV}^2 + 0.26\,\text{MeV}^2 \\
&= 388.4\,\text{MeV}^2 \\
E &= 19.7\,\text{MeV}
\end{aligned}
\tag{V.15}
$$

The energy of the electron would be of magnitude as great as its measured uncertainty of the momentum, that is,

$$E \geq 19.7\,\text{MeV}$$

No electron has ever been detected from isotope decay with an energy as high as 19.7 MeV. The highest beta-particle energy is 4.9 MeV from the decay of the radioactive isotope ^{38}Cl. While an electron (i.e., beta particle) is produced in nuclear beta decay, the Heisenberg Uncertainty Principle demonstrates that the electron from beta decay could not exist in the nucleus, but that the electron is only a product of the nuclear decay.

Heisenberg demonstrated that the uncertainties are not due to any experimental or instrumental error in measurement that may be made on the atomic particle, but rather they are due to the inherent quantum structure of atomic particles, that is, their wave–particle duality. Heisenberg's Uncertainty Principle in terms of energy and time variables is expressed as

$$\Delta E \Delta t \geq \frac{h}{4\pi} \geq \frac{\hbar}{2} \tag{V.16}$$

which states that the precision with which we measure the energy of an atomic particle (ΔE) is limited by the time available (Δt) to measure the particle energy. The energy of the particle can be measured with perfect precision only if an infinite amount of time is used to measure the energy, that is, when $\Delta t = \infty$.

Heisenberg argued from quantum mechanics that the picture of the atom with fixed electron orbitals was not free of ambiguity, that is, we cannot definitely state that such particulate electron orbitals actually exist because of the wave concept of the electron. However, in his Nobel Lecture, Heisenberg (1933a) clarified that the particulate and wave concepts of the atom remain useful for our visual interpretation of atomic structure and spectral emissions when he stated the following:

A visual description for the atomic events is possible only within certain limits of accuracy—but within these limits the laws of classical physics also still apply. Owing to these limits of accuracy as defined by the uncertainty relations, moreover, a visual picture of the atom free from ambiguity has not been determined. On the contrary the corpuscular and the wave concepts are equally serviceable as a basis for visual interpretation.

Niels Bohr (1928) presented his complimentary argument that the clearest analysis that can be derived from quantum mechanics must include both the corpuscular and wave nature of atomic particles, that is, both the particulate and wave nature of the electron are needed to compliment each

other in providing a complete description of the atomic particle. The duality of the wave–particle nature of matter was commemorated in the postage stamp illustrated here issued in Sweden in 1982, which provides an artistic abstract depiction of atoms as corpuscular matter forming a molecule and waves in the background. Heisenberg acknowledged Bohr's concept of complimentarity in his Nobel Lecture, when he stated:

> For the clearest analysis of the conceptual principles of quantum mechanics we are indebted to Bohr who, in particular, applied the concept of complimentarity to interpret the validity of the quantum mechanical laws. The uncertainty relations alone afford an instance of how in quantum mechanics the exact knowledge of one variable can exclude the exact knowledge of the other. This complementary relationship between different aspects of one and the same physical process is indeed characteristic for the whole structure of quantum mechanics.

In 1927 Heisenberg was appointed Chair of Theoretical Physics at the University of Leipzig. He remained in Germany during World War II. In 1941 he was appointed Professor of Physics at the University of Berlin and Director of the Kaiser Wilhelm Institute for Physics. Heisenberg was involved in Germany's effort to develop the atomic bomb and nuclear energy for power together with other notable German scientists; however, the German effort on the bomb development was eventually scuttled in 1942 due to the fact that three to four years would be needed to develop the weapon among other reasons, and Heisenberg (1946) wrote that German physicists "were spared the decision as to whether or not they should aim at producing atomic bombs." (Rhodes, 1986). A great historical debate started with the publication of "Brighter than a Thousand Suns" by Austrian writer and journalist Robert Jungk (1956) first published in German in 1956. After reading Jungk's book in 1956, Heisenberg sent him a letter concerning his role and that of German scientists in the atom bomb project. Jungk then published an excerpt of Heisenberg's letter, out of context, in the Danish edition of the book (Jungk, 1957). The excerpt of Heisenberg's letter in the Danish edition made it appear that Heisenberg was claiming to have purposely derailed the German bomb project. Heisenberg wrote in his letter to Jungk:

> …we were convinced that the manufacture of atomic bombs was possible only with enormous technical resources. We knew that one could produce atom bombs but overestimated the necessary technical expenditure at the time. This situation seemed to us to be a favorable one, as it enabled the physicists to influence further developments. If it were impossible to produce atomic bombs this problem would not have arisen, but if they were easily produced the physicists would have been unable to prevent their manufacture. This situation gave the physicists at that time decisive influence on further developments, since they could argue with the government that atomic bombs would probably not be available during the course of the war…Under these circumstances we

thought a talk with Bohr (Autumn of 1941) would be of value…Being aware that Bohr was under the surveillance of the German political authorities and that his assertions about me would probably be reported to Germany, I tried to conduct this talk in such a way as to preclude my life into immediate danger. This talk probably started with my question as to whether or not it was right for physicists to devote themselves in wartime to the uranium problem—as there was the possibility that progress in this sphere could lead to grave consequences in the technique of the war…He replied as far as I can remember with a counter-question, "Do you really think that uranium fission could be utilized for the construction of weapons?" I may have replied: "I know that this is in principle possible, but it would require a terrific effort, which, one can only hope, cannot be realized in this war." Bohr was shocked by my reply, obviously assuming that I had intended to convey to him that Germany had made great progress in the direction of manufacturing atomic weapons…

Nobel Laureate Hans Bethe (2000) provided evidence that Heisenberg did not work toward or try to build a bomb. Some of the evidence is presented by Bethe as follows:

The best evidence we have suggests that Heisenberg had no interest in building an atomic bomb. In mid-1942, Albert Speer, the weapons minister, asked Heisenberg whether he could produce a weapon in nine months. With a clear conscience he could answer "No."…If he wasn't trying to build a bomb, then why did Heisenberg work on the uranium project? He told me, in 1948, that he wanted to save the lives of a few young German physicists for the post-war period. I believe that this was indeed an important argument for him. An in judging his behavior during the war, we must remember that he was under enormous pressure, pressure virtually unimaginable for Americans or Englishmen. So Heisenberg didn't try to build a bomb, and apparently had no intention of doing so. Why then the visit to Bohr in 1941? What was he trying to communicate?…Perhaps he was trying to get Bohr to be a messenger of conscience, and wanted Bohr to persuade the allied scientists also to refrain from working on a bomb.

Heisenberg and other German scientists were taken prisoners by Allied troops after the war and sent to England. In 1946 he returned to Germany and reorganized the Institute for Physics at Göttingen, which was renamed the Max Planck Institute for Physics in 1948. While remaining Director of the Max Planck Institute Heisenberg carried out extensive lectures in Cambridge in 1948, headed the German delegation to the European Council for Nuclear Research, contemplated the founding of CERN, the European Organization for Nuclear Research, in Geneva, and he delivered the Gifford Lectures on "Physics and Philosophy" at the University of St Andrews, Scotland in the winter of 1955–1956. From 1957 onwards, Heisenberg was interested in plasma physics and a close collaborator of the International Institute of Atomic Physics in Geneva and participated for several years as Chairman of the Scientific Policy Committee of the Institute. He was appointed the first President of The Alexander von Humboldt Foundation until his resignation in October 1975 shortly before his death from cancer at his home in Munich on February 1, 1976.

Erwin Schrödinger (1887–1961) was born in Vienna, Austria on August 12, 1887. He entered the Akademisches Gymnasium in Vienna at the age of eleven in 1898 where he demonstrated to be a gifted student of mathematics and physics. After graduation from the Gymnasium in 1906, he entered the University of Vienna where Fritz Hazenöhrl, the successor to Ludwig Boltzmann, and his lectures in theoretical physics had the greatest influence on him. Schrödinger received the doctorate degree in 1910 and completed his habilitation in 1914. With the outbreak of World War I, he was ordered to serve as artillery officer at the Italian front. While carrying out active military service, he published several papers in theoretical physics (Schrödinger, 1915a,b, 1917a,b) and received commendation for outstanding military service. After the war he continued his work in Vienna where he was appointed Associate Professor at the University where he worked on radioactivity demonstrating the statistical nature of radioactive decay (Schrödinger, 1921). During 1920–1921 in a period of about 18 months, he held three positions as Professor at the University of Stuttgart, University of Breslau, and then Chair of Physics at the University of Zurich.

Erwin Schrödinger (1887-1961)

It was in Zurick where Schrödinger carried out his most fruitful work during 1921–1926. Here Schrödinger studied atomic structure and quantum statistics, and the turning point in his career was his reading of Louis de Broglie's papers on the dual nature of matter and light (de Broglie, 1923a–c). In 1926 Schrödinger published a series of papers revealing his revolutionary wave equation and the new physics of wave mechanics (Schrödinger, 1926a–g, 1927). Schrödinger was not happy with Bohr's quantum theory of atomic structure as it held only for hydrogen and not the more complex atoms. Also, Heisenberg's Uncertainty Principle demonstrated that both the location and momentum of an electron could not be measured accurately and, thus, he considered Bohr's fixed electron orbits about the nucleus as an inadequate representation of the atom. Grasping the theory of matter–wave properties of the electron, Schrödinger created wave mechanics. In Schrödinger's wave mechanics, he explained the movements of an atomic electron as a wave and calculated the coordinates of the electron according to a wavefunction (Ψ) given by his equation

$$-\frac{\hbar^2}{2m}\left\{\frac{d^2\psi}{dx^2}+\frac{d^2\psi}{dy^2}+\frac{d^2\psi}{dz^2}\right\}+U(r)\psi=i\hbar\frac{d\psi}{dt} \qquad (V.17)$$

and described in detail by Levine (1999), McQuarrie (1983), Serway *et al.* (1997), and many others, where $\hbar$ is Planck's constant $h/2\pi$, m the particle mass, Ψ the wavefunction, U the potential energy as a function of the space coordinates $U(r) = U(x, y, z)$, and $i = \sqrt{-1}$. The solution of the Schrödinger equation provides a relation of the wavefunction Ψ to the position or three-dimensional coordinates (x, y, z) of an electron in space whereby Ψ^2 will give the probability of finding an electron within given coordinates of the atom. It was Max Born (1882–1970) who accurately interpreted Schrödinger's wave function Ψ to provide the probability density Ψ^2 of the atomic electron. No longer is the atomic electron envisioned to occupy a fixed orbit, but rather to have a probability of occupying given coordinates. The regions that the electrons occupy in space are defined according to orbitals, which are a region in space where an electron is likely to be found with high probability (e.g., > 90%). Orbitals should not be confused with orbits, as the electron's position in the atom cannot be fixed to specific distances from the nucleus. Because we are dealing with probabilities in defining the locations of the electrons, the orbitals are represented as electron clouds of specific shape representing high densities of the atomic electrons. An electron could be found statistically anywhere from a distance very close to a nucleus (<0.53 Å) to a distance far away from the nucleus; however, high probabilities ($> 90\%$) will define the likelihood of finding an atomic electron at certain coordinates and therefore, define also its orbital and the shape of the orbital.

The assignment of atomic electrons in three-dimensional space, which define atomic orbitals, depends on the following three variables:

(a) *The principle quantum number* (n), which may have values of $n = 1, 2, 3, \ldots$, corresponds to the energy level of the electron and the size or magnitude of the orbital. An orbital with a higher value of n will be larger in size or volume and its electrons of higher energy than an orbital with lower value of n.

(b) *The orbital quantum number* (ℓ), which is assigned values of $\ell = 0, 1, 2, 3, \ldots n - 1$, defines the shape of the orbital and identifies the orbital according to the following orbital types: *s, p, d,* and *f*. For example, spherical orbitals have the quantum number $\ell = 0$ and are identified as *s* orbital types; dumbbell-shaped orbitals have the quantum number $\ell = 1$ and are identified as *p* orbitals; clover-leaf shaped orbitals have the quantum number $\ell = 2$ and are identified as *d* orbitals. The *f* orbitals with quantum number $\ell = 3$ are more complex in shape. The *s, p,* and *d* orbitals are illustrated in Figures V.7–V.9.

(c) *The magnetic quantum number* (m_ℓ) can have values over the range of $m_\ell = -\ell$ to $+ \ell$, and it will define the orientation in space of the orbital. For example, when $\ell = 2$, the magnetic quantum number, m_ℓ, will have values of $-2, -1, 0, +1, +2$, yielding five different spatial orientations of the *d* orbital as illustrated in Figure V.9. When $\ell = 1$, m_ℓ will have values of $-1, 0, +1$ yielding three different spatial orientations of the *p* orbitals as illustrated in Figure V.8. When $\ell = 0$, $m_\ell = 0$ there will be no differing spatial orientations of the *s* orbital as illustrated in Figure V.7. The *s* orbital is a sphere, which cannot differ in shape and differs only in size and energy of the electrons according to the principle quantum number n.

The postage stamp illustrated here was issued by Sweden in 1982 to commemorate Schrödinger's discovery of his wave equation and the awarding of his Nobel Prize. It illustrates two electron orbital types, namely, the *d* and *s* orbitals. The rules applied to the filling of atomic electrons into the various orbitals that constitute the buildup of elements in the Periodic Table are provided in the biographical sketch of Wolfgang Pauli. The simplest of all atoms is, of course, the hydrogen atom of atomic number 1, which

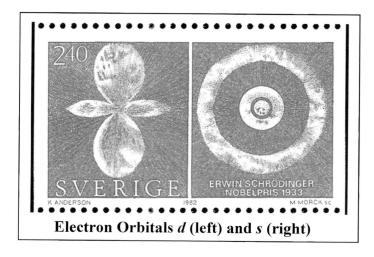

Electron Orbitals *d* (left) and *s* (right)

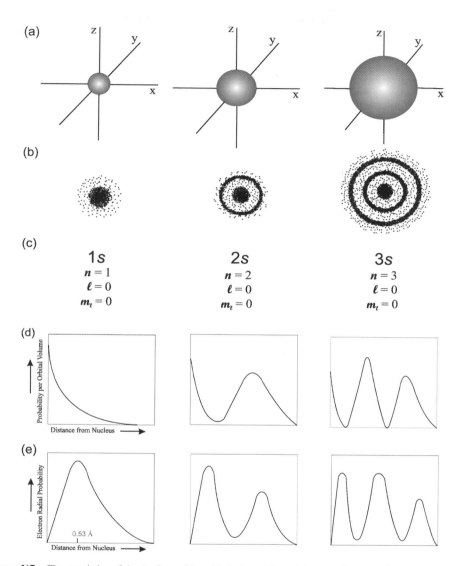

Figure V.7 Characteristics of the 1*s*, 2*s*, and 3*s* orbitals from left to right according to (a) spherical shape of increasing size and electron energy, (b) sketches of the electron density distributions along the cross-sections (radial) of the orbitals, (c) orbital identification with underlying quantum numbers, (d) electron probability positions per orbital volume as a function of distance from the nucleus, and (e) electron radial probability positions as a function of distance from the nucleus.

possesses only one atomic electron. The next atom of atomic number 2 is helium with two electrons. The atomic number of each atom corresponds, of course, to the number of protons in the nucleus, which is also equal to the number of electrons that the neutral atom may possess in its ground state. The buildup of the neutral atoms of some selected elements according to the placement of the atomic electrons into specific orbitals is provided in Table V.2a.

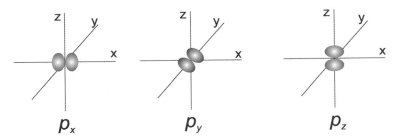

Figure V.8. Shape and orientations of the *p* electron orbitals identified according to their axis of orientation. The quantum numbers that define the *p* orbitals are $\ell = 1$, and m_ℓ will have values of $-1, 0$, or $+1$ yielding three possible orientations in space. The principle quantum number $n = 1, 2, 3\ldots$ are also associated with the *p* orbital to identify energy level. The principle quantum number that may be assigned to the *p* orbital (e.g., $1p$, $2p$, $3p\ldots$, not illustrated) will signify the energy level and size of the orbital.

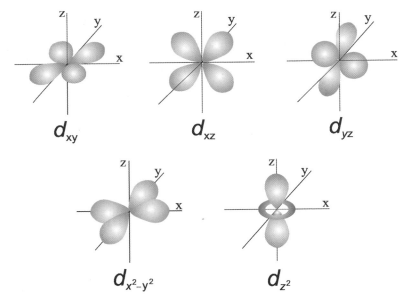

Figure V.9 Shapes and orientations of the *d* orbitals identified according to their axes of orientation. The quantum numbers that define the *d* orbitals are $\ell = 2$, and m_ℓ will have values of $-2, -1, 0, +1$, or $+2$ yielding five possible orientations in space. The d_{xy}, d_{xz}, and d_{yz} orbitals lie in the *xy*, *xz*, and *yz* planes, respectively. The $d_{x^2-y^2}$ orbital lies along the *x* and *y* axes, and the d_{z^2} orbital is oriented along the *z* axis. The principle quantum number that may be assigned to the *d* orbital (e.g., $1d$, $2d$, $3d\ldots$, not illustrated) will signify the energy level and size of the orbital.

The $1s$ orbital is found in all atoms of the elements in the Periodic Table including the hydrogen atom. It is the orbital with electrons of the lowest energy. The hydrogen atom (H) has an atomic number of 1 and only one electron. That electron occupies the $1s$ orbital (see Table V.2a). Its electron configuration is noted as $1s^1$ and it reads as "one s one". The exponent one means "one electron in the *s* orbital". The number 1 before the *s* is the principle quantum number or energy level of the orbital. The next most complex atom is that of helium (He) of atomic number 2 with two electrons (see Tables V.2a and b), and its electron configuration is written as $1s^2$, which reads "one s two" meaning two electrons in the s orbital of lowest energy (principle quantum number or level of 1). The next element in the Periodic Table is lithium (Li) of atomic number 3; and thus, the atom of lithium has three electrons. Its electron configuration is noted as $1s^22s^1$, which reads as "two electrons in the $1s$ orbital and one electron in the $2s$ orbital". Notice here that only two electrons can fill the $1s$ orbital. The third electron in lithium must occupy another orbital that of the next highest level or the $2s$ orbital. The reason for the exclusion of additional electrons to the orbital is explained by Pauli's Exclusion Principle, which permits not more than two electrons of opposite spin to occupy the same quantum or energy state, described in the biographical sketch of Wolfgang Pauli (see Radioactivity Hall of Fame—Part II).

The electrons of the $1s$ orbital are of the lowest energy and closest to the nucleus as illustrated in Figure V.7. According to Schrödinger's wave mechanics, the electrons in the $1s$ orbital can be anywhere from a distance closest to the nucleus to an infinite distance away from the nucleus; however, the electrons in this orbital will be found with highest probability at a distance of 0.53 Å, which is the distance calculated by Niels Bohr of closest approach of the atomic electron to the nucleus. The probability curves for the locations of the electrons in the $1s$, $2s$, and $3s$ orbitals are illustrated in Figure V.7. Two types of probability curves are shown, namely, probability per orbital volume (Figure V.7d) and radial probability (Figure V.7e). The probability per unit volume of the $1s$ orbital shows the electron probability to be most closest to the nucleus because the orbital volume can be relatively large. The radial probability curve provides an electron density probability along the cross-section of the orbital and a clearer picture of the electron density distribution. Notice in Figure V.7e that the electron density or cloud of the $1s$ orbital will be thickest at 0.53 Å distance from the nucleus. Electrons in the $2s$ orbital are found to overlap the $1s$ and extend with highest probability to a distance further away from the nucleus forming a larger sphere with electrons of higher energy than the $1s$ electrons. The electrons in the $3s$ orbital are yet of higher energy and found with highest probability further away from the nucleus.

Another atom we may take as an example is that of sodium (Na), which is more complex, with 11 electrons. Its electron configuration provided in Table V.2b is $1s^22s^22p^63s^1$, which reads "two electrons in the $1s$ orbital, two electrons in the $2s$ orbital, six electrons in the $2p$ orbitals and one electron in the $3s$ orbital". Notice here that there can be as many as six electrons assigned to the p orbitals because there are three types of p orbitals as illustrated in Figure V.8 and only two electrons per orbital of opposite spin are possible. As the p orbitals are filled, the additional electron in the sodium atom is found in the next highest $3s$ orbital. The d orbitals are not formed until the higher quantum level ($n = 3$) is reached as illustrated in Table V.2a. These orbitals can occupy as many as 10 electrons because there are five different d-orbital types, which can be filled with two electrons of opposite spin. The copper (Cu) atom with 29 electrons has all of its d orbitals filled. It has the electron configuration $1s^22s^22p^63s^23p^63d^{10}4s^1$, as given in Table V.2b. The d orbitals are of five types as illustrated in Figure V.9. Each can be filled by two electrons of opposite spin. Electrons will begin to fill the more complex f orbitals (not illustrated) in larger atoms of atomic number 58 (Cerium) and higher.

Schrödinger's wave mechanics was the second formulation of quantum mechanics, the first being the mathematical matrix mechanics of Heisenberg. Schrödinger's wave mechanics provided scientists with a visual picture of the space coordinates occupied by the electron, which Heisenberg's mathematical matrix equations could not provide. The initial reaction of Heisenberg was disagreement with Schrödinger's new wave mechanics; however, Schrödinger demonstrated the equivalence of the matrix and wave versions of quantum mechanics.

In 1927 Schrödinger moved to Berlin to succeed Max Planck, who had retired the previous year, at the University of Berlin. He decided to leave Germany in 1933 with Hitler's coming to power that very year, as he witnessed the persecution of Jews. In late 1933 he left Berlin under a fellowship at Oxford University, the very year that he was awarded the Nobel Prize. Schrödinger was invited to lecture at Princeton in 1934 and offered a permanent position there, which he did not accept. Instead, after returning to Oxford he accepted a position at the University of Graz in 1936 only to encounter again problems with Hitler's regime after Hitler annexed Austria in 1938. As a Nobel Laureate, his preference to live outside the German state was taken as an affront. He fled Austria via Italy and returned to Oxford. While at Oxford, Schrödinger proceeded to the University of Ghent for a 1-year term as visiting professor. In 1939 he took on the position of Director of the newly created School for Theoretical Physics at Dublin where he remained until his retirement in 1955 (Nobel Lectures, Physics, 1922–1941).

Table V.2a

Electron configurations of neutral atoms in the ground state

Atomic no.	Element	K^a ($n^b = 1$)	L ($n = 2$)		M ($n = 3$)			N ($n = 4$)			
		s^c	s	p	s	p	d	s	p	d	f
1	H	1									
2	He	2									
3	Li	2	1								
4	Be	2	2								
5	B	2	2	1							
6	C	2	2	2							
7	N	2	2	3							
8	O	2	2	4							
9	F	2	2	5							
10	Ne	2	2	6							
11	Na	2	2	6	1						
12	Mg	2	2	6	2						
13	Al	2	2	6	2	1					
⋮	⋮										
18	Ar	2	2	6	2	6					
⋮	⋮										
29	Cu	2	2	6	2	6	10	1			
⋮	⋮										
46	Pd	2	2	6	2	6	10	2	6	10	

Source: Data from Lide (1997).
[a]Main quantum level. The electron energy level K is the closest to the nucleus. Other energy levels, namely, L, M, N, and O are of increasing energy and distance from the nucleus.
[b]Principal quantum number.
[c]Orbital type.

Table V.2b

Nomenclature for electron configurations of Table V.2a

Atomic no.	Element	Electron configuration
1	H	$1s^1$
2	He	$1s^2$
3	Li	$1s^2 2s^1$
4	Be	$1s^2 2s^2$
5	B	$1s^2 2s^2 2p^1$
6	C	$1s^2 2s^2 2p^2$
7	N	$1s^2 2s^2 2p^3$
8	O	$1s^2 2s^2 2p^4$
9	F	$1s^2 2s^2 2p^5$
10	Ne	$1s^2 2s^2 2p^6$
11	Na	$1s^2 2s^2 2p^6 3s^1$
12	Mg	$1s^2 2s^2 2p^6 3s^2$
13	Al	$1s^2 2s^2 2p^6 3s^2 3p^1$
⋮	⋮	⋮
18	Ar	$1s^2 2s^2 2p^6 3s^2 3p^6$
⋮	⋮	⋮
29	Cu	$1s^2 2s^2 2p^6 3s^2 3p^6 3d^{10} 4s^1$
⋮	⋮	⋮
46	Pd	$1s^2 2s^2 2p^6\, 3s^2 3p^6 3d^{10} 4s^2 4p^6 4d^{10}$

After his retirement he returned to Vienna, his birthplace, until his death on January 4, 1961. For decades Schrödinger was honored by Austria with his portrait on the coveted Austrian 1000 Schilling monetary bill issued in 1983, which remained legal tender in Austria until the Euro exchange in 2002.

Max Born (1882–1970) was awarded the Nobel Prize in Physics 1954 "especially for his statistical interpretation of the wavefunction". He was born in Breslau, Germany on December 11, 1882, which is now Wroclaw, a capital city of Lower Silesia in southwestern Poland. He attended the Konig Wilhem Gymnasium in Breslau and studied mathematics and physics at the University of Breslau, which is now the University of Wroclaw. Max Born transferred to the University of Göttingen where he studied mathematics, physics, and astronomy graduating in 1907. After a short period of work under J. J. Thomson at Cambridge, Born returned to Breslau in 1908. He accepted an appointment in 1912 to lecture at the University of Chicago in the Department of Physics headed by Albert A. Michelson. In 1915 Born was compelled to join the German armed forces and declined an offer to work under Max Planck at the University of Berlin. After the war he was appointed Professor at the University of Frankfurt am Main where his assistant Otto Stern began his research that led him to the Nobel Prize. In 1921 Max Born was appointed Professor of Physics at the University of Göttingen where he remained until 1933 to carry out his historic statistical interpretations of quantum mechanics collaborating with many great physicists of the time including Heisenberg, Pauli, his student Pascual Jordan, Fermi, Oppenheimer, and Maria Goeppert-Mayer.

Max Born (1882-1970)

Schrödinger had interpreted his wave function Ψ as the electron's matter–wave density distribution, that is, he wanted to dispense with the particulate properties of the electron and consider them to possess a continuous density distribution $|\Psi|^2$ meaning that the electron as a wave could have a high density in some regions of space and lower density in other regions. Max Born, however, interpreted the wave function Ψ as providing a picture of the electron with its coordinates in space, and $|\Psi|^2$ as the probability density of the electron (Born, 1926a–d; Born et al., 1926; Born and Oppenheimer, 1927; Born and Fock, 1928). Born's probability density interpretation is widely accepted today as that which gives the probability of finding the electron particle at particular coordinates in space. Born commented on this interpretation in his Nobel Lecture (Born, 1954) as follows:

> The work, for which I have had the honour to be awarded the Nobel Prize for 1954, contains no discovery of fresh natural phenomenon, but rather the basis for a new mode of thought in regard to natural phenomenon…For a brief period at the beginning of 1926, it looked as though there were, suddenly, two self-contained but quite distinct systems of explanation extant: (Heisenberg's) matrix mechanics and (Schrödinger's) wave mechanics. But Schrödinger himself soon demonstrated their complete equivalence…there came the *dramatic surprise*, the appearance of Schrödinger's famous papers (1926a–e). He took up quite a different line of thought which had originated from Louis de Broglie (1924, 1925). A few years previously the latter had made the bold assertion, supported by brilliant theoretical considerations, that wave-corpuscle duality, familiar to physicists in the case of light, must also be valid for electrons. To each electron moving free of force belongs a plane wave of a definite wavelength, which is determined by Planck's constant and the mass…Wave mechanics enjoyed a very great deal more popularity than the Göttingen or Cambridge version of quantum mechanics. It operates with a wave function Ψ, which in the case of *one* particle at least, can be pictured in space, and it used the mathematical methods of partial differential equations which are in current use by physicists…He (Schrödinger) proposed to dispense with the particle representation entirely, and instead of speaking of electrons as particles, to consider them as a continuous density distribution $|\Psi|^2$ or electric density $e|\Psi|^2$. To us in Göttingen this interpretation seemed unacceptable in face of well established experimental facts. At that time it was already possible to count particles by means of scintillations or with a Geiger counter, and to photograph their tracks with the aid of a Wilson cloud chamber…an idea of Einstein's gave me the lead. He had tried to make the duality of particles—light quanta or photons—and waves comprehensible by interpreting the square of the optical wave amplitudes as probability density for the occurrence of photons. This concept could at once be carried over to the Ψ-function: $|\Psi|^2$ ought to represent the probability density for electrons (or other particles).

Like many German scientists of Jewish decent, Max Born lost his professorship and was forced to leave Germany in 1933. He lectured at Cambridge during 1933–1936 and during the winter of 1935–1936, he spent six months working with C. V. Raman and his students at the Indian Institute of Science at Bangalore. In 1938 Born was appointed Professor of Natural Philosophy at the University of Edinburgh where he remained until his retirement in 1953 [Nobel Lectures, Physics, 1942–62. (1964)]. He died on January 5, 1970 in Göttingen, Germany. For additional reading see his autobiography (Born, 1978).

Paul A. M. Dirac (1902–1984) shared the Nobel Prize in Physics 1933 with Schrödinger "for the discovery of new productive forms of atomic theory". Dirac was born in Bristol, England on August 8, 1902. He started his higher education in the field of electrical engineering at Bristol University. After receiving his Bachelor of Science degree in 1921, Dirac continued his studies at Bristol University until 1923. He then proceeded to St John's College at Cambridge where he received his Ph.D. degree in mathematics in 1926. He did not like the field of electrical engineering *per se*; however, his early studies in this field were useful to his future work, as reflected in a quotation provided in Sir Michael Berry's (1998) biographical sketch of Dirac:

> I owe a lot to my engineering training because it [taught] me to tolerate approximations. Previously to that I thought…one should just concentrate on exact equations all the time. Then I got the idea that in the actual world all our equations are only approximate. We must just tend to greater and greater accuracy. In spite of the equations being approximate, they can be beautiful.

The year following receipt of his doctorate degree Dirac became a Fellow at St. John's College at Cambridge and carried out research in theoretical physics at the Cavendish Laboratory. At the Cavendish Laboratory, Dirac was introduced to Heisenberg's recent papers on quantum mechanics, and the

Paul Dirac (1902-1984)

following year, Dirac (1928a,b) published the quantum theory of the electron, which included new quantum wave mechanics incorporating the theory of relativity for relativistic electrons approaching the speed of light into Schrödinger's equation. Dirac's new quantum theory included a wave equation for the electron, the Dirac equation, which could provide the dynamics of electron waves yielding physical predictions while satisfying the requirements of relativity (Berry, 1998). A result of Dirac's quantum theory of the electron was the prediction of the electron spin and the occurrence of a new electron, one of positive charge, known as the positron, the antiparticle of the negatron or negative electron. In his Nobel Lecture Dirac (1933) described how he came to the conclusion of the existence of the positron as follows:

There is [a] feature of these equations which I should like to discuss, a feature which led to the prediction of the positron....one sees that [the equation] allows the kinetic energy to be either a positive quantity greater than mc^2 or negative quantity less than $-mc^2$...Now, in practice, the kinetic energy of a particle is always positive. We thus see that our equations allow for two kinds of motion for an electron, only one of which corresponds to what we are familiar with. The other corresponds to electrons with a very peculiar motion such that the faster they move, the less energy they have, and one must put energy into them to put them to rest...Thus, in allowing negative energy states, the theory gives something which appears not to correspond to anything known experimentally, but which we cannot simply reject by a new assumption. We must find some meaning for these states...We make use of the exclusion principle of Pauli, according to which there can be only one electron in any state of motion. We now make the assumptions that in the world as we know it, nearly all the states of negative energy for the electrons are occupied, with just one electron in each state, and that a uniform filling of all the negative-energy states is completely unobservable to us. Further, *any unoccupied negative-energy state, being a departure from uniformity, is observable and is just a positron.* An unoccupied negative-energy state, or *hole*, as we may call it for brevity, will have a positive energy, since it is a place where there is a shortage of negative energy. A hole is, in fact, just like an ordinary particle, and its identification with the positron seems the most reasonable way of getting over the difficulty of the appearance of negative energies in our equations. On this view the positron is just a mirror image of the electron, having exactly the same mass and opposite charge. This has already been confirmed by experiment.

Dirac also predicted the phenomenon of annihilation, when a particle and its antiparticle (e.g., negatron and positron) meet and are annihilated by the conversion of their electron masses into electromagnetic radiation. In his Nobel Lecture Dirac explained:

> From our theoretical picture, we should expect an ordinary electron, with positive energy, to be able to drop into a hole and fill up this hole, the energy being liberated in the form of electromagnetic radiation. This would mean a process in which an electron and a positron annihilate one another. The converse process, namely the creation of an electron and a positron from electromagnetic radiation, should also be able to take place. Such processes appear to have been found experimentally, and are at present being more closely investigated by experimenters.

The existence of the positron was experimentally discovered in 1932 by Carl D. Anderson while carrying out research on cosmic rays for which he was awarded the Nobel Prize in Physics 1936. Anderson also demonstrated together with his graduate student Seth Neddermeyer that gamma rays from "ThC" would generate positrons, by pair production, after passing through substances such as lead. Pair production as predicted by Dirac is the creation of an electron and positron pair by the conversion of the energy of electromagnetic gamma radiation into the masses of two electrons of opposite charge as illustrated in Figure 3.12 of Chapter 3. The electron and positron pair is produced when a gamma ray in excess of 1.02 MeV energy passes through the coulomb field of a nucleus. The minimum gamma-ray energy of 1.02 MeV is calculated in Chapter 3 (eqs. (3.28)–(3.31)) as the energy equivalent to twice the electron mass expressed in energy units of electron volts or twice 0.511 MeV. The conversion is written as

$$E_{\text{pair}} = m_{e^-}c^2 + m_{e^+}c^2 = 2mc^2 = 2(0.511 \text{ MeV}) = 1.022 \text{ MeV} \tag{V.18}$$

Energy in excess of 1.02 MeV appears as the kinetic energy of the electron and positron produced and is written as

$$h\nu = 2mc^2 + E_{e^-} + E_{e^+} \tag{V.19}$$

The postage stamp illustrated on the following page was issued by Sweden in 1982 to commemorate Dirac's prediction of the positron, the antiparticle of the electron, which was subsequently discovered by Carl D. Anderson. The stamp shows a copy of an actual picture of the negative and positive electron tracks produced in a bubble chamber following pair production. The tracks are spiral in shape because of a magnetic field running perpendicular to the electron tracks forcing the charged particle paths to bend according to their charge and velocity.

Dirac (1933) also postulated the existence of the antiparticle of the proton, that is, the antiproton or negatively charged proton. In his Nobel Lecture he explains:

> One might perhaps think that the same theory could be applied to protons. This would require the possibility of existence of negatively charged protons forming a mirror-image of the usual positively charged ones…I think it is probable that negative protons can exist, since as far as the theory is yet definite, there is a complete and perfect symmetry between positive and negative electric charge, and if this symmetry is really fundamental in nature, it must be possible to reverse the charge on any kind of particle.

The antiproton was indeed, in due time, discovered in 1955 by Emilio Segrè and Owen Chamberlain for which they were awarded the Nobel Prize in Physics 1955. Dirac even went further

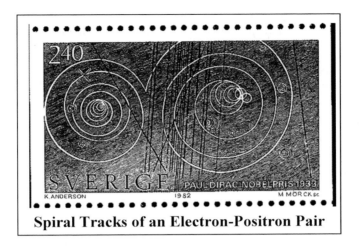

Spiral Tracks of an Electron-Positron Pair

to postulate the possibility of the existence of entire solar systems many light years away that would be built of antiparticles or antimatter, a theory that today still appears plausible. Dirac (1933) explained:

If we accept the view of complete symmetry positive and negative electric charge so far as concerns the fundamental laws of Nature, we must regard it rather as an accident that the Earth (and presumable the whole solar system), contains a preponderance of negative electrons and positive protons. It is quite possible that for some of the stars it is the other way about, these stars being built up mainly of positrons and negative protons. In fact, there may be half the stars of each kind. The two kinds of stars would both show exactly the same spectra, and there would be no way of distinguishing them by present astronomical methods.

Dirac's accomplishments could have provided him with the worldwide fame of Einstein during his lifetime, but he shunned notoriety and remains to this day a figure hardly known outside the scientific community. Chaundy (2002) explains that this is due to the fact that Dirac was an introvert; he was shy and avoided publicity to the extreme. He liked to work alone and had very few graduate students. It is reported that a graduate student once asked Dirac if he could be his thesis advisor; and Dirac replied "Thank you, but I do not need any help at this time." To avoid publicity, Dirac contemplated refusing to accept the Nobel Prize. Chaundy (2002) reports that Dirac accepted the Nobel Prize because he was advised that, by refusing the Prize, he would be given even greater publicity. Sir Michael Berry (1998) provides the following evaluation of Dirac's accomplishments with the following quotation of the mathematician Mark Kac, who divided geniuses into two classes:

There are the ordinary geniuses, whose achievements one imagines other people might emulate, with enormous hard work and a bit of luck. Then there are the magicians, whose inventions are so astounding, so counter to all the institutions of their colleagues, that it is hard to see how any human could have imagined them. Dirac was a magician.

In 1932 Dirac was appointed Lucian Professor of Mathematics at Cambridge, which is the same post once held by Sir Isaac Newton. Dirac was only 31 years of age when he was awarded the Nobel Prize in Physics 1933. He married Margit Wigner, the sister of the Hungarian Nobel Laureate Eugene Wigner. In 1969 Dirac retired from Cambridge and joined Florida State University at Tallahassee. He died in Tallahassee on October 20, 1984. For further reading, one is invited to peruse a biography by Kragh (1990) and his collected works provided by Dalitz (1995).

CLINTON DAVISSON (1881–1958) AND GEORGE PAGET THOMSON (1892–1975)

Clinton Davisson shared the Nobel Prize in Physics 1937 with George Paget Thomson "for their experimental discovery of the diffraction of electrons by crystals". Davisson was born in Bloomington, IL on October 22, 1881. He was the son of Joseph Davisson, a veteran of the Union Army in the Civil War and of Mary Calvert, a schoolteacher from Pennsylvania. In 1902 Clinton Davisson graduated from public High School in Bloomington with a mathematics and physics scholarship to the University of Chicago where he came suddenly under the influence of Professor Robert A. Millikan, who would eventually receive the Nobel Prize in Physics 1923 for his discovery of the elementary charge on the electron. Unable to finance his continued education after his first year at the University Davisson took up employment at a Bloomington telephone company. Upon the recommendation of Professor Millikan he was appointed Assistant in Physics at Purdue University in 1904. The same year he returned to the University of Chicago until 1905. Again under the recommendation of Professor Millikan, Davisson took up an appointment as Instructor in Physics at Princeton University where he remained until 1910 carrying out his duties as instructor as well as student of physics. During his tenure at Princeton, he would study at the University of Chicago during the summer semesters, and eventually graduate with a B.S. degree from the University of Chicago in 1908. Davisson was awarded a Fellowship in Physics during 1910–1911 at Princeton during which time he completed his Ph.D. degree under Professor, and later Nobel Laureate, Owen W. Richardson.

RÉPUBLIQUE DE GUINÉE
OFFICE DE LA POSTE GUINÉENNE
750F
1937
CLINTON DAVISSON

C. Davisson (1881-1985)

Having completed his doctorate degree, Clinton Davisson took on a position of Instructor, Department of Physics, Carnegie Institute of Technology, Pittsburg in 1911 and remained in that position until 1917. During his stint at Carnegie, Davisson spent a summer of leave in 1913 working in the Cavendish Laboratory under J. J. Thomson, who was awarded the Nobel Prize in Physics 1906 for the discovery of the electron. Little did Davisson know during his stint at the Cavendish Laboratory with J. J. Thomson that he would share the Nobel Prize in Physics 24 years later with the son of J. J. Thomson.

Davisson was refused conscription into the army corps in 1917 and, in lieu of military service he accepted wartime employment in the Engineering Department of the Western Electric Company that would later become Bell Telephone Laboratories, New York while on leave-of-absence from Carnegie Tech. After the end of the World War, Davisson remained at the Bell Telephone Laboratories and resigned his position of Assistant Professor of Carnegie Tech.

At the Bell Telephone Laboratories, Davisson carried out a series of experiments, which began in 1919, on thermionic electron emission and the emission and scattering of electrons upon positive ion and electron bombardment (Davisson and Germer, 1920; Davisson and Pidgeon, 1920; Davisson and Kunsman, 1922a,b, 1923). These experiments led to the discovery of electron diffraction by crystalline material in 1927.

Two occurrences led Davisson to the discovery of the diffraction of electrons by crystalline matter and thereby demonstrated the wave nature of the electron. The first was the hypothesis of de Broglie (1923a–d) from work on his doctoral thesis that a particle of momentum, p, would display wave properties, and the wavelength of the matter–wave would be proportional to Planck's constant and inversely proportional to the mass and velocity of the particle or

$$\lambda = \frac{h}{p} = \frac{h}{mv} \tag{V.20}$$

The other occurrence that led to the discovery of electron diffraction was provided by Davisson's previous work with coworkers. This work demonstrated the elastic scattering of electrons. In his Nobel Lecture Davisson noted the following:

> …the work was begun the day after copies of de Broglie's thesis reached America. The work actually began in 1919 with the accidental discovery that the energy spectrum of secondary electron emission has, as its upper limit, the energy of the primary electrons, even for primaries accelerated through hundreds of volts; that there is, in fact, an elastic scattering of electrons by metals…and then chance again intervened; it was discovered purely by accident, that the intensity of elastic scattering varies with the orientations of the scattering crystals. Out of this grew, quite naturally, an investigation of elastic scattering by a single crystal of predetermined orientation.

The breakthrough came in 1927 when Clinton Davisson and Lester Germer were investigating the elastic collisions of electrons as these were reflected off the surface of metallic nickel. A laboratory accident caused the glass housing of the electron gun and target apparatus to break whereby the loss of vacuum formed an oxide layer on the surface of the nickel sample. To remove the oxide layer, they heated the nickel at high temperature in a reducing atmosphere. This caused the surface of the nickel to crystallize. Repeating their experimentation with electron beams on the nickel sample produced different results. They noticed that the intensity of the reflected electron beam was confined to specific angles with respect to the energy and angle of the incident beam. This was characteristic of a diffraction of the electron beam similar to what Sir William and Sir Lawrence Bragg (Nobel Prize, Physics, 1915) observed when x-rays are directed onto a crystalline substance. Their first report of the diffraction of electrons was made in April, 1927 (Davisson and Germer, 1927a).

The apparatus used by Davisson and Germer (1927b) consisted of an elecrtron gun that would direct the electron beam toward the nickel crystal. The intensity of the diffracted electron beam was measured with a Faraday cup rotated from 20° to 90° from the origin of the incident electron beam as illustrated in Figure V.10 The nickel crystal also could be rotated to permit the measurement of refracted electron beam intensities according to various azimuths of the crystal.

Davisson and Germer measured the most intense scattered electron beam at a 50° angle (ϕ) from the primary electron beam. This diffraction was caused by an electron beam created with an acceleration potential of 54 V illustrated in FigureV.11.

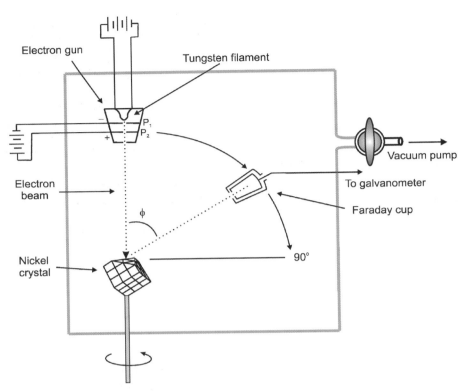

Figure V.10 A schematic diagram illustrating the disposition of the electron beam, nickel crystal, and collector (Faraday cup). The crystal could be rotated to bring a principal azimuth into a plane of observation. The Faraday cup could detect and measure the intensity of the electron matter–wave in varying directions from 20° (closest to the electron gun) to 90° (perpendicular to the plane of electron impact with the crystal (adapted from Davisson, 1937). The electron emission from the tungsten filament was concentrated onto the opening of plate P_1 by giving the plate a potential more negative than that of the filament. The potential of P_2 was given a high positive value, and the difference between the potential of plate P_2 and that of the filament would determine the speed of the emergent electron beam. The distance from the end of the electron gun to the crystal target was only 7 mm, and the nickel crystal measured approximately $8 \times 5 \times 3$ mm. See Davisson and Germer (1927b) for detailed drawings of their apparatus.

They tested the de Broglie relation, $\lambda = h/p$, on the scattered electrons to demonstrate the electron matter–wave. In Davisson's words (1937) the proof of the de Broglie matter wave for the electron was carried out as follows:

> The de Broglie relation was tested by computing wavelengths from the angles of the diffraction beams and the known constant of the crystal [d], and comparing these with corresponding wavelengths computed from the formula $\lambda = h/p$, where p, the momentum of the electrons, was obtained from the potential used to accelerate the beam and the known value of e/m for electrons. If wavelengths computed from the formula agreed with those obtained from the diffraction data, the de Broglie relation would be verified.

For example, the incident electron beam illustrated in Figure V.11 was produced by acceleration across a potential difference of 54 V. The wavelength of this electron beam can be calculated according

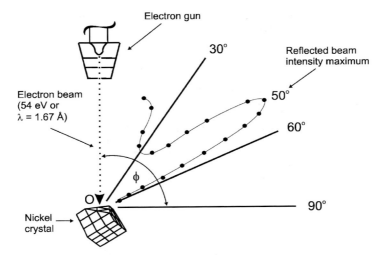

Figure V.11   A schematic diagram of the electron gun, the nickel crystal target, and Davisson and Germer's plot of the electron beam intensity as a function of the angle of reflection, ϕ. The relative intensities of the reflected beam are measured by the distance from the point O to any position on the reflected beam intensity plot. The maximum intensity of the reflected electron beam for an incident beam of 1.67 Å was measured at an angle $\phi = 50°$. A tungsten filament served as the electron source. See Davisson and Germer (1927b) for details.

to the following relation derived by de Broglie according to eqs. (III.6)–(III.9) (see Radioactivity Hall of Fame—Part III):

$$\lambda = \frac{h}{p} = \frac{h}{m_0 v} = \frac{h}{\sqrt{2m_0 eV}} \tag{V.21}$$

where h is Planck's constant, m_0 the rest mass of the electron, e the electron charge, and V the applied voltage. A calculation of the wavelength of the 54 V electron beam according to the de Broglie relation (eq. (V.21)) yields

$$\lambda = \frac{h}{\sqrt{2m_0 eV}} = \frac{6.626 \times 10^{-34} \text{ J sec}}{\sqrt{2(9.11 \times 10^{-31} \text{ kg})(1.6 \times 10^{-19} \text{ C})(54 \text{ V})}} = 1.67 \times 10^{-10} \text{ m} = 1.67 \text{ Å}$$

Taking the electron beam angle of deflection $\phi = 50°$ and the known spacing between the atomic nuclei of the nickel crystal ($d = 2.15$ Å) Davisson and Germer calculated the incident electron beam wavelength according to the refraction eq. (V.22) as follows:

$$\lambda = d \sin \phi = 2.15 \text{ Å} \sin 50° = 1.6469 \text{ Å} = 1.65 \text{ Å} \tag{V.22}$$

which agrees very closely with the 1.67 Å for the wavelength calculated according to the de Brolie matter–wave relation.

Davisson and Germer used low-energy electrons in their diffraction experiments. In such a case, the surface atoms of the crystal acted principally as the diffraction grating because the electrons would not have sufficient energy to penetrate significantly to lattice layers below the outermost ordered layer of the crystal atoms. This was noted by Davisson and Germer (1928) as follows:

> …near grazing first-layer scattering is strongly predominant, and diffraction occurs as if waves were being scattered by a single layer of atoms only. Files of atoms serve as line gratings and ordinary line grating diffraction is observed.

At about the same time as Davisson and Germer performed their diffraction experiments with low-energy electrons George Paget Thomson in Aberdeen, UK independently demonstrated electron diffraction in thin metal foils with high-energy electrons, which could penetrate several layers of crystalline lattice and thereby obey Bragg's law of diffraction $n\lambda = 2d\sin\theta$ as described previously for x-ray diffraction (see eq. (III.14) in Radioactivity Hall of Fame—Part III). A sketch of the life and work of George P. Thomson is provided in the following section.

The significance of the experimental demonstration of the electron matter–wave relation and consequently experimental verification of the de Broglie particle–wave theory was underscored by Davisson in his Nobel Lecture on December 13, 1937, when he stated:

> I will take time only to express my admiration of the beautiful experiments—differing from ours in every respect—by which Thomson in far-away Aberdeen demonstrated electron diffraction and verified de Broglie's formula at the same time as we in New York. And to mention, as closely related to the subject of this discourse, the difficult and beautifully executed experiments by which Stern and Estermann in 1929 (Estermann and Stern, 1930) showed that atomic hydrogen also is diffracted in accordance with the de Broglie-Schrödinger theory.
>
> Important and timely as was the discovery of electron diffraction in inspiring confidence in the physical reality of material waves, our confidence in this regard would hardly be less today, one imagines, were diffraction yet to be discovered, so great has been the success of the mechanics built upon the conception of such waves in clarifying the phenomena of atomic and subatomic particles.

During 1930–1937 Davisson continued his research at the Bell Telephone Laboratories on electron optics and to the applications of electron diffraction in engineering. He subsequently studied the scattering of very slow electrons by metals, and during World War II he researched the development of various electronic devices and crystal physics problems (Nobel Lectures, Physics, 1922–1941, 1965). Davisson retired from Bell Telephone Laboratories in 1946 and was visiting professor at the University of Virginia, Charlottesville, VA during 1947–1949. He died at the age of 76 on February 1, 1958, in Charlottesville.

George Paget Thomson shared the Nobel Prize in Physics 1937 with Clinton Davisson "for their experimental discovery of the diffraction of electrons by crystals".

Thomson was born at Cambridge in 1892, the son of J. J. Thomson, Nobel Laureate in Physics 1906. It has been stated in jest, "J. J. Thomson won the Nobel Prize in Physics 1892 for demonstrating that the electron was an elementary particle, whereas his son, G. P. Thomson, won the Nobel Prize in Physics 1937 for demonstrating that the electron was not a particle." The truth is, however, that G. P. Thomson, through his experimental discovery of electron diffraction, verified the de Broglie formula that describes the dual wave–matter property of the electron, which like other particles with kinetic energy exhibit the properties of a particle and of a wave. The postage stamp issued by the Guinean Republic commemorated G. P. Thomson's scientific discovery; although the reader will notice on the following page that his family name is misprinted on the postage stamp.

George Paget Thomson completed his undergraduate studies in mathematics and physics at Trinity College, Cambridge. He also completed a year of research work under his father at the Cavendish Laboratory, Trinity College, Cambridge. When the First World War broke out, Thomson served in the infantry for a short stint, and then was assigned to numerous scientific research positions with the Royal Flying Corps and several months in the USA as part of the British War Mission. After the war he spent three years as Fellow and Lecturer at Corpus Christi College, Cambridge.

RÉPUBLIQUE DE GUINÉE
OFFICE DE LA POSTE GUINÉENNE
750F
1937
GEORGE THOMPSON

G. Thomson (1892-1975)

In 1922 Thomson became Professor of Natural Philosophy at the University of Aberdeen and remained in that position until 1930. It was at the University of Aberdeen where J. P. Thomson carried out his historic research that led to the discovery of electron diffraction. The experimental demonstration of the diffraction of electrons was a significant step in physics as it provided the evidence to verify the de Broglie equation, $\lambda = h/mv$, defining the wave nature of particles. Thomson arrived at his discovery in Aberdeen at about the same time as Davisson in New York. They announced their discoveries simultaneously in the same volume 119 of the journal *Nature* (Davisson and Germer, 1927a; Thomson and Reid, 1927); however, the experimental approach and the manner in which the diffraction was observed were completely different in the two camps. Davisson's discovery, described previously in this section, was accidental, whereas Thomson, inspired by the theory of de Broglie, set out from the start to prove the wave nature of the electron (Anonymous, 1975). It is not unusual for scientific discoveries to be made by two groups totally independent of each other. When the time is right and the stage is set, the discoveries are made.

Quite opposite to the approach used by Davisson and Germer (1927a), who measured a beam of low-energy electrons (54 eV) as these deflected off the atomic plane of the crystal surface, Thomson and Reid (1927) used a beam of relatively high-energy electrons (20,000–60,000 eV), which would pass directly through the crystalline lattice of thin metal foils. The apparatus used by Thomson and Reid is illustrated in Figure V.12.

Thomson (1927, 1928, 1929) and coworkers calculated the wavelength of the electron beam according to the de Broglie matter–wave theory defined previously in eq. (V.21) corrected for the relativistic mass of the electron or

$$\lambda = \frac{h}{p} = \frac{h}{mv}$$

(V.23)

where m, the relativistic mass of the electron, is defined as

$$m = \frac{m_0}{\sqrt{1 - (v^2/c^2)}}$$

(V.24)

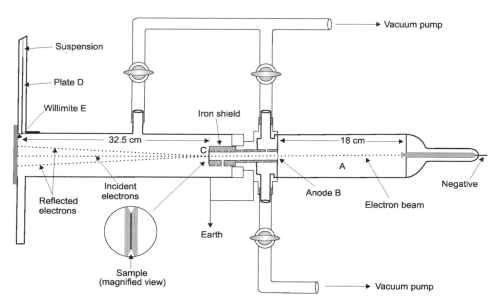

Figure V.12 Experimental arrangement used by Thomson and coworkers to demonstrate the diffraction of an electron beam by the crystalline lattices of thin metal and organic substances (see Thomson, 1928). The chambers of the apparatus were evacuated by a mercury vapor pump and the cathode rays were generated by an induction coil. The electron beam passed from right to left through tube A and through a fine tube of small bore (0.23 mm) and 6 cm length in the anode B. The electron beam exiting the anode would strike the thin foil ($10^{-5} – 10^{-6}$ mm thick) of crystalline substance at C. The electron beam and its diffracted waves would travel on through the foil and onto the willimite screen at E on the far end of the apparatus. The willimite screen would glow green from fluorescence when bombarded by the electron beam. At the willimite screen Thomson and coworkers could observe the image produced by the principal (unreflected) electron beam as a central spot, and the diffracted electron matter–waves as rings of diameter D, which surround the central electron beam. Permanent images of the electron beam and the surrounding rings of the diffracted waves were made by dropping a photographic plate from position D down into the path of the electron beam in front of the willimite screen. Thomson's original apparatus is on display at the Science Museum, Cavendish Laboratory, University of Cambridge. (Adapted from Thomson (1928) reprinted with permission from the Royal Society.)

where m_0 is the electron rest mass, v the speed of the electron, and c the speed of light in a vacuum. Eq. (V.23) can then be written as

$$\lambda = \frac{h}{[m_0/\sqrt{1-(v^2/c^2)}]v} \tag{V.25}$$

and

$$\lambda = \frac{h\sqrt{1-(v^2/c^2)}}{m_0 v} \tag{V.26}$$

Thomson (1928) used eq. (V.26) from the de Broglie matter–wave theory to calculate the wavelength of the electron beam produced in their electron diffraction apparatus. He only needed to know the speed of the electron, v, which was calculated from the electron kinetic energy. The following expression for the electron kinetic energy permits the calculation of the speed of the relativistic electron:

$$K = mc^2 - m_0 c^2 \tag{V.27}$$

where m is the relativistic mass of the electron and m_0 its rest mass. From the expression of m of eq. (V.24), the particle kinetic energy can be written as

$$K = \frac{m_0 c^2}{\sqrt{1 - (v^2/c^2)}} - m_0 c^2 \tag{V.28}$$

The electron rest energy in units of MeV is calculated as

$$E = m_0 c^2 = (9.109 \times 10^{-31}\,\text{kg})(2.9979 \times 10^8\,\text{m/sec})^2 = 8.186 \times 10^{-14}\,\text{J} \tag{V.29}$$

The electron rest energy in joules is converted to units of MeV as

$$\frac{8.186 \times 10^{-14}\,\text{J}}{1.602 \times 10^{-19}}\,\text{J/eV} = 0.511\,\text{MeV} \tag{V.30}$$

The known constant value of the electron rest energy, 0.511 MeV, can be inserted into eq. (V.28) to calculate the electron velocity, v, from the measured value of its kinetic energy, K, as

$$K = \frac{0.511\,\text{MeV}}{\sqrt{1 - (v^2/c^2)}} - 0.511\,\text{MeV} \tag{V.31}$$

Thomson measured the kinetic energy of the electron beam produced in his apparatus in units of eV from an accurate measurement of the potential difference used to accelerate the electron toward the anode. From the measured value of K, he calculated the velocity of the electron according to eq. (V.31).

Thomson used electron beams over the range of 20,000–60,000 V energy, which he reports to have wavelengths varying according to energy over the respective range of 0.8×10^{-9}–0.5×10^{-9} cm (Thomson, 1938). For example, in his early detailed report of electron diffraction Thomson (1928) used an electron beam of 25,000 V energy and calculated the wavelength of the electron beam according to eq. (V.26) to be approximately 0.75×10^{-9} cm. To arrive at the value of the electron wavelength, the velocity of the 25 keV electron is first calculated according to eq. (V.28) as

$$K = \frac{m_0 c^2}{\sqrt{1 - (v^2/c^2)}} - m_0 c^2 \quad \text{or} \quad 0.025\,\text{MeV} = \frac{0.511\,\text{MeV}}{\sqrt{1 - (v^2/c^2)}} - 0.511\,\text{MeV} \tag{V.32}$$

which is transposed to solve for the electron velocity, v, as follows:

$$\sqrt{1 - \left(\frac{v^2}{c^2}\right)} = 0.953 \tag{V.33}$$

and

$$1 - \left(\frac{v^2}{c^2}\right) = (0.953)^2 \tag{V.34}$$

and

$$\frac{v^2}{c^2} = 1 - 0.908 \tag{V.35}$$

and

$$v^2 = 0.092c^2 \tag{V.36}$$

or

$$v = 0.303c \tag{V.37}$$

Thus, the value of $v = 0.303c$ calculated here is the velocity of the electrons of 25,000 V energy, which is 0.303 times the speed of light in a vacuum.

From the calculated value of v, the wavelength λ of the 25 keV electrons is calculated as

$$
\begin{aligned}
\lambda &= \frac{h\sqrt{1-(v^2/c^2)}}{m_0 v} \\
&= \frac{6.626 \times 10^{-34} \text{ J sec}\sqrt{1-([(0.303)(2.9979\times10^8 \text{ m/sec})]^2 /(2.9979\times10^8 \text{ m/sec})^2)}}{(9.109\times10^{-31} \text{ kg})(0.303)(2.9979\times10^8 \text{ m/sec})} \\
&= 0.76 \times 10^{-11} \text{ m} = 0.076 \times 10^{-10} \text{ m} = 0.076 \text{ Å}
\end{aligned}
\tag{V.38}
$$

The above results agree closely with the calculated value of 0.75×10^{-9} cm reported by Thomson (1928).

Because the electrons used by Thomson were of relatively high energy, these would travel directly through the thin (10^{-5} to 10^{-6} mm) foil of crystalline substances such as platinum, aluminum, gold, nickel, and celluloid. By means of the apparatus illustrated in Figure V.12 Thomson could observe that the electron beam would undergo diffraction by the atomic planes in the crystalline substances according to Bragg's law

$$n\lambda = 2d \sin \theta \tag{V.39}$$

which was previously described for x-ray diffraction, where n is an integer 1,2,3..., d the interatomic spacing or atomic planes in the crystalline lattice, and θ the glancing angle at which the electrons are reflected. Only electron beams with wavelengths of 1λ, 2λ, 3λ,... over the path length of travel ($2d\sin\theta$) between each successive crystal atomic plane will produce coherent reflected electron beams (see Figure III.6 in Radioactivity Hall of Fame—Part III).The incident electron beam would deflect off the atomic planes of the crystalline lattice similar to x-ray photons, as illustrated in Figure V.13 and described by Thomson (1938) in his Nobel Lecture as follows:

A narrow beam of cathode rays was transmitted through a thin film of matter. In the earliest experiment of the late Mr. Reid [Thomson's assistant, Alexander Reid] this film was of celluloid, in my own of metal. In both, the thickness was of the order of 10^6 cm. The scattered beam was received on a photographic plate normal to the beam, and when deployed showed a pattern of rings, recalling optical halos and the Debye-Scherrer rings well known in the corresponding experiment with x-rays. An interference phenomenon is at once suggested. This would occur if each atom of the film scattered in phase a wavelet from an advancing wave associated with the electrons forming the cathode rays. Since the atoms of each small crystal of the metal are regularly spaced, the phases of the wavelets scattered in any fixed direction will have a definite relationship to one another. In some directions they will agree in phase and build up a strong scattered wave, in others they will destroy one another by interference. The strong waves are analogous to the beams of light diffracted by an optical grating.

Thomson took crystalline samples of known d spacing, that is, crystalline materials that had interatomic spacing predetermined by x-ray diffraction, and from the diameter of the ring produced by the reflection of an electron beam of known wavelength (λ) off a thin sample of a crystalline material, he could demonstrate that the electron beam satisfied Bragg's law $n\lambda = 2d\sin\theta$.

Thomson could demonstrate that the electron beam behaved as a wave, as it would diffract along the atomic planes of a crystalline substance in accordance with Bragg's law in the same fashion as x-ray photons. For example, an electron beam of 25,000 V corresponding to a wavelength of 0.076 Å according to the de Broglie matter–wave theory (see calculations according to eqs. (V.27)–(V.38)) was directed toward a thin foil sample of crystalline platinum. The d spacing of the platinum was previously measured by x-ray diffraction to be 3.92 Å (Thomson, 1927). If the distance of travel (L) of the diffracted beam from its origin in the platinum foil to the end of its path at the photographic plate was 32 cm (distance from the sample to the photographic plate as illustrated in Figure (V.13)), Thomson could calculate the first-order (1λ) angle of deflection that should correspond to Bragg's law (eq. (V.39)) as

$$\sin\theta = \frac{\lambda}{2d} = \frac{0.076\,\text{Å}}{(2)(3.92\,\text{Å})} = 0.00969$$

(V.40)

$$\theta = \sin^{-1} 0.00969 = 0.555°$$

(V.41)

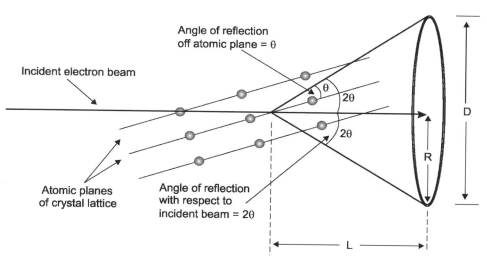

Figure V.13 The diffraction of an electron beam off an atomic plane of a crystalline lattice as the beam passes through the lattice. The beam diffracts at an angle θ to the atomic plane and an angle 2θ with respect to the path of the incident electron beam. A ring of diffracted electrons is formed around the central electron beam, which was observed in real time by the willimite screen or by the image formed after exposure to a photographic plate placed normal to the electron beam as illustrated in Figure V.14. The diameter of the ring is a function of the angle of reflection (θ) and the distance or length (L) of travel of the diffracted beam.

If the diameter (D) of a first-order diffraction ring on the photographic plate measured 1.24 cm the angle of deflection of the electron beam should agree with the following relationship derived by Ironside (1928), an assistant to G. P. Thomson:

$$\sin \theta = \frac{(1/4)(D)}{L} = \frac{D}{4L} \tag{V.42}$$

$$\sin \theta = \frac{1.24 \, \text{cm}}{(4)(32 \, \text{cm})} = 0.009687 \tag{V.43}$$

$$\theta = \sin^{-1} 0.009687 = 0.555° \tag{V.44}$$

Also, the angle of diffraction of the electron beam calculated above according to Bragg's law (eq. (V.40)) should satisfy the following geometric relation to the ring size:

$$\tan 2\theta = \frac{R}{L} \tag{V.45}$$

where R is the radius of the ring as illustrated in Figure V.13, and

$$\tan 2(0.555°) = \frac{R}{32 \, \text{cm}} \tag{V.46}$$

$$0.019375 = \frac{R}{32 \, \text{cm}} \tag{V.47}$$

$$R = 0.62 \, \text{cm} \tag{V.48}$$

Thomson thus demonstrated the de Broglie electron matter–wave. The materials tested by Thomson were polycrystalline meaning that the substance is built up of many crystals having different orientations, such as a powdered sample of single crystal grains. The effect of x-ray or electron diffraction of poly-crystalline materials is the production of a diffraction pattern consisting of several concentric rings of various diameters about the central spot produced by the electron beam. An example of a typical electron ring diffraction pattern produced by Thomson (1935) on a photographic plate is shown in Figure V.14

During the winter of 1929–1930 Thomson was a non-resident Lecturer at Cornell University, and afterwards appointed Professor at Imperial College, London. After completing his major work on electron diffraction, George Thomson became interested in neutron physics in the 1930s and took up the study of nuclear energy once fission was discovered in 1939. When it became clear that fission had military possibilities during the Second World War, he became Chairman of the confidential MAUD Committee, the British Committee set up in 1940 to study the possibilities of the utilization of nuclear fission for atomic weapons (Nobel Lectures, Physics, 1965; Anonymous, 1975.). The MAUD Committee was established following a memorandum by O. R. Frisch and R. E. Peirls, both at Birmingham University, in which they predicted that a small mass of uranium-235 could sustain a chain reaction that would serve as the basis for the making of an atomic weapon. MAUD was the code name for the secret British Committee commissioned to research the military potential of nuclear fission. The origin of the code name MAUD is related in an historical account by Fakley (1983). When Denmark was occupied by the Germans, Niels Bohr sent a telegram to Otto Frisch, who had worked in Bohr's Copenhagen Laboratory, and at the end of the telegram asked Frisch to "Tell Cockroft [Prof. J. D. Cockroft] and Maud Ray Kent". Fakley (1983) explains that Maud Ray Kent was a cryptic reference to possible uranium disintegration, and after the war the code name Maud Ray was identified to be a

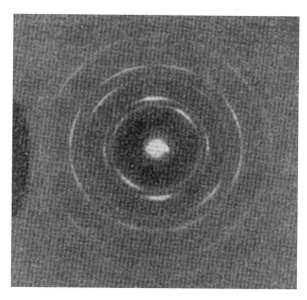

Figure V.14   Diffraction patter of gold foil irradiated with an 85 keV electron beam. (From Thomson (1935) reprinted with permission from Macmillan Publishers Ltd., © 1935.)

former governess to Bohr's children, who was living in the county of Kent. From this cable the code name MAUD was chosen for the secret British Committee to study the feasibility of utilizing uranium fission for weapons development. The MAUD Committee concluded in 1941 that the development of a nuclear weapon was feasible, and George Thomson was authorized to hand over the conclusions of the Committee to American scientists Vaneveer Bush and James Conant (Nobel Lectures, Physics, 1965).

George Thomson held his post as Professor at Imperial College, London until 1952. That same year he became Master of Corpus Christi College, Cambridge where he had carried out his historic electron diffraction experiments and remained in that post until his retirement in 1962. As a hobby Thomson enjoyed building model ships. He died on September 17, 1975.

– 5 –

Atomic Electron Radiation

5.1 INTRODUCTION

Atomic electrons may be emitted from atoms as a result of a nuclear decay process referred to as internal conversion. In such a process an unstable nucleus decaying to a lower more stable state transmits its decay energy to an atomic electron, which is emitted from the atom with an energy corresponding to the nuclear decay energy less the binding energy of the electron to the atom. Another source of atomic electron radiation is the emission of Auger (pronounced OH-ZHAY) electrons. Auger electrons may be emitted by radioactive atoms, which have electron shell vacancies and undergo electron energy-level transitions. Radioactive atoms that decay by electron capture (EC) or internal-conversion (IC) decay processes leave vacancies in electron shells. When a vacancy occurs in an electron shell an outer electron will move to fill that vacancy. In such an electron energy-level transition, energy may be released in the form of either x-radiation or the energy of transition is transferred to an atomic electron emitted from the atom as an Auger electron. The origins and properties of IC and Auger electrons are described in this chapter.

5.2 INTERNAL-CONVERSION ELECTRONS

Decay by IC results in the emission of an atomic electron. This electron, called the IC electron, is emitted from an atom after absorbing the excited energy of a nucleus. This mode of decay accompanies and even competes with gamma-ray emission as a deexcitation process of unstable nuclei.

The kinetic energy of the electron emitted is equivalent to the energy lost by the nucleus (energy of transition of the excited nucleus to its ground or lower energy state) less the binding energy of the electron. This is illustrated by the following equation:

$$E_e = (E_i - E_f) - E_b \tag{5.1}$$

where E_e is the kinetic energy of the IC electron, $(E_i - E_f)$ the energy of transition between the initial, E_i, and the final, E_f, nuclear energies normally associated with gamma-ray emission, and E_b the binding energy of the atomic electron.

An example of radionuclide decay by IC is found in Figure 5.1, which illustrates the decay of the parent–daughter nuclides ^{109}Cd ($^{109\,m}$Ag). Note that the $^{109\,m}$Ag daughter decays

341

by IC with a 96% probability (i.e., 45% for IC from the *K*-shell +48% from the *L*-shell +3% from higher electron shells – the latter is not illustrated in Figure 5.1) and decay occurs via gamma emission with the remaining 4% probability (Rachinhas *et al.*, 2000).

Because the emission of IC electrons competes with gamma-ray emission as an alternative mode of nuclear deexcitation, many radioactive nuclei that emit gamma radiation will also emit IC electrons. The degree to which this competition occurs is expressed as the IC

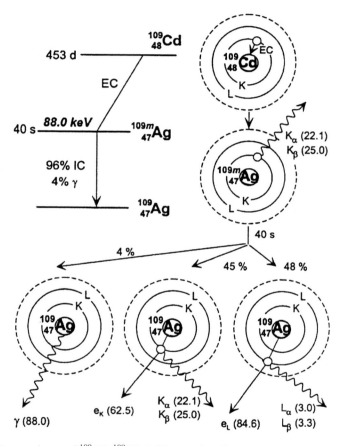

Figure 5.1   Decay scheme of ^{109}Cd ($^{109\,\mathrm{m}}$Ag). The numbers in parenthesis indicate energy values in keV. The electron-capture (EC) process occurs from *K*, *L*, and outer shells with probabilities of 79, 17, and 4%, respectively, but only *K* capture is represented above. The $^{190\,\mathrm{m}}$Ag daughter decays by emission of 88.0 keV gamma-rays with a 4% probability or by internal conversion (IC) with the probabilities of 45 and 48% for *K*- and *L*-shells. IC from shells higher than *L* contribute the remaining 3% (not illustrated). The *K* and *L* IC decay illustrated involve the ejection of a conversion electron with energy $e_K = 62.5$ keV or $e_L = 84.6$ keV, accompanied by the emission of a Ag *K*- or *L*-fluorescence x-ray photon ($K_\alpha = 22.1$, $K_\beta = 25.0$ keV or $L_\alpha = 3.0$, $L_\beta = 3.3$ keV) or by the emission of Auger electrons (not represented) and x-ray photons following Auger electron emissions. (From Rachinhas *et al.*, 2000, reprinted with permission from Elsevier © 2000.)

coefficient, which is the ratio of the rate of emission of IC electrons to the rate of emission of gamma-rays of equivalent energy. In other words, the IC coefficient is a quantitative measure of the number of IC electrons divided by the number of gamma-rays emitted from a radionuclide sample. The IC coefficient is denoted by α or e/γ.

IC electrons may be emitted from specific electron shells of atoms and may be expressed in terms of IC electrons and gamma-rays of the same energy less the energy difference resulting from the binding energy of the electron. When expressed in terms of electrons emitted from specific shells, the IC coefficient is written with a subscript denoting the electron shell of origin, for example, α_K or e_K/γ, α_L or e_L/γ, and α_i or e_i/γ, where $i = K, L, M$, and so on electron shells.

Values of IC coefficients are provided in many reference tables on isotope decay. In general, IC coefficients are small for gamma-ray emitting nuclides of low Z and high-energy transitions and larger for nuclides of high Z and low-energy transitions. This relationship is illustrated in Table 5.1, which lists a few radionuclides selected at random as examples in order of increasing Z number. As can be seen, large IC coefficients occur when IC electrons are emitted with low-energy nuclear transitions as indicated by the large values of α associated

Table 5.1

Relationship between gamma radiation and IC electron radiation, e^-, associated with several nuclides listed in the order of increasing Z number

Nuclide $_Z^A X$	Gamma radiation (MeV)[a]	e^- (MeV)	$\alpha = e/\gamma$	x-rays[a]
$_4^7$Be	0.477 (10%)		7.0×10^{-7}	
$_{11}^{22}$Na	1.275 (100%)		6.7×10^{-6}	
$_{22}^{44}$Ti	0.068 (90%)	0.065	0.12	Sc K
	0.078 (98%)	0.073	0.03	
$_{27}^{57}$Co	0.014 (9%)	0.013	8.2	Fe K (55%)
	0.122 (87%)	0.115	0.02	
	0.136 (11%)	0.129	0.15	
$_{29}^{64}$Cu	1.34 (0.6%)	1.33	1.3×10^{-4}	Ni K (14%)
$_{38}^{87m}$Sr[b]	0.388 (80%)	0.386	0.21	Sr K (9.4%)
$_{50}^{119m}$Sn[b]	0.024 (16)	0.020	5.13	Sn K (28%)
$_{53}^{125}$I	0.035 (7%)	0.030	13.6	Te K (138%)
$_{53}^{129}$I	0.040 (9%)	0.034	22	Xe K (69%)
$_{68}^{169}$Er	0.008 (0.3%)	0.006	220	Tm M
$_{74}^{181}$W	0.006 (1%)	0.004	46	Ta K (65%)
$_{80}^{203}$Hg	0.279 (82%)	0.275	0.23	Tl K (13%)
$_{94}^{239}$Pu	0.039 (0.01%)	0.033	461	U K (0.012%)
	0.052 (0.02%)	0.047	269	

[a]Values in percent are radiation intensities or abundances.
[b]m denotes a metastable state.

with low gamma-ray energies. It should also be pointed out that the IC electron (e^-) energies are slightly lower than the gamma-ray energies. This is because the energy of the IC electron is equal to the energy absorbed from the decaying nucleus (transition energy) less the binding energy of the atomic electron described previously in eq. (5.1). On the other hand, gamma-ray energies serve as a measure of the exact quanta of energies lost by a nucleus.

The loss of atomic electrons through the emission of IC electrons leaves vacancies in atomic electron shells. The vacancies are filled by electrons from outer higher energy shells, whereby there is a concomitant loss of electron energy as internal Bremsstrahlung or x-radiation. Emission of x-radiation resulting from electron filling of vacancies in electron shells (K, L, M...) is also listed in Table 5.1. This is a process that occurs in the daughter atoms; the x-rays are a characteristic of the daughter rather than of the parent.

IC electrons are identical in their properties to beta particles. They differ, however, in their origin. Beta particles originate from the nucleus of an atom, whereas IC electrons originate from atomic electron shells. A characteristic difference between these two types of electron is their energy spectra. Beta particles, as discussed in Chapter 2, are emitted as a product of nuclear decay with a broad spectrum or smear of energies ranging from near zero to E_{max}. However, IC electrons are emitted from the atoms of decaying nuclei with discrete lines of energy of a magnitude equivalent to that of the energy lost by the nuclei less the electron-binding energy. The energy of an IC electron can be used to estimate the energy lost by a nucleus.

Like beta particles, IC electrons dissipate their energy by ionization they cause in matter. The abundance of IC electrons emitted from some nuclide samples can be significant and should not be ignored. In certain cases it can play a significant role in radionuclide detection and measurement. IC electron energies are slightly lower than the true gamma-decay energy because of the energy consumed in the ejection of the bound atomic electron (E_b in eq. (5.1)).

An IC coefficient of large magnitude does not, however, necessarily signify the emission of a high abundance of IC electrons. For example, ^{239}Pu with a high IC coefficient ($\alpha = 461$) corresponding to a 0.039 MeV gamma-decay process emits only a trace of IC electrons because of the low abundance of gamma-decay (0.01%, see Table 5.1).

5.3 AUGER ELECTRONS

An Auger (pronounced OH-ZHAY) electron can be considered as the atomic analogue of the IC electron. In the EC decay processes, vacancies are left in electron shells (K, L, M...) that can be filled by atomic electrons from higher energy levels. In the process of falling to a lower energy shell to fill a vacancy, electron energy may be lost as a photon of x-radiation (see Chapter 3, Section 3.6). Alternatively the energy liberated in the shift of an electron from its higher energy state to a lower one can be transferred to an electron in an outer shell resulting in the emission of the electron as an Auger electron. See Figure VI.24 in the biographical sketch of Pierre Auger. The figure provides an illustration of the Auger effect, that is, the process that gives rise to Auger electrons. Pierre Auger (Auger, 1923, 1925a,b) and Lise Meitner (1923) independently discovered Auger electrons.

Whenever an atomic electron is lost from an inner shell, such as, by EC emission of an IC electron, emission of an Auger electron, ionization by external x-ray or electron beams, or by other means, another electron from an outer shell can fall to a lower one to fill the vacancy left behind by the electron emitted. There can be a cascading effect of electrons falling from yet more distant shells to fill vacancies left behind until the atom reaches the ground or stable state. The downward transitions of electrons in this fashion produce x-ray photons. The production of x-ray photons in this fashion is referred to as x-ray fluorescence.

The energy of an Auger electron is low, because it is equivalent to the energy of the x-ray photon that would result in the transition of an electron from an outer to inner shell less the binding energy of the electron emitted as an Auger electron. For example, if an electron from the K-shell is lost from an atom as a result of EC, an x-ray photon may result from an electron transition from the L-shell to the K-shell. The energy of the x-ray photon would be equivalent to the energy difference between the L and K-shells or

$$E_{\text{x-ray}} = hv = E_L - E_K \tag{5.2}$$

In lieu of the emission of an x-ray photon, the energy of the electron transition described in eq. (5.2) may be transferred to an electron in an outer shell resulting in the emission of the electron from the atom as an Auger electron. The energy of the Auger electron would be equivalent to the energy of the x-ray photon less the binding energy of the electron prior to its emission as an Auger electron, or

$$E_{\text{Auger}} = (E_L - E_K) - E_b \tag{5.3}$$

where E_L and E_K are the electron energies in the L and K-shells, respectively, and E_b is the binding energy of the electron prior to its ejection as an Auger electron. Other transitions may be described such as $E_M - E_L$ for M and L electron shells. Eq. (5.3) may be written also as

$$E_{\text{Auger}} = hv - E_b \tag{5.4}$$

where hv is the x-ray photon energy expressed as a product of Planck's constant, $h(h = 6.62 \times 10^{-27} \text{erg sec} = 4.14 \times 10^{-15} \text{eV sec} = 6.62 \times 10^{-34} \text{J sec})$, and the photon frequency, v, in units of sec^{-1}. A more detailed treatment of Auger electron emission is given in the biographical sketch of Pierre Auger in Radioactivity Hall of Fame—Part VI.

Auger electron emission competes with x-ray emission, and it can accompany any decay process that results in the production of x-rays. Like IC electron emission described previously, the EC decay process (see Chapter 3, Section 3.6) also results in the emission of appreciable quantities of x-radiation. Thus, Auger electron emissions also accompany EC decay. Auger electron emission can reduce appreciably the abundance of x-ray emission normally expected to accompany radionuclide decay processes. The two competing processes of Auger electron emission and x-ray emission are important to consider in the detection and measurement of nuclides that decay by EC. This is measured by both the fluorescence yield and Auger yield.

The fluorescence yield is the fraction of vacancies in a given electron shell that is filled with accompanying x-ray emission, and Auger yield is the fraction of vacancies that are filled resulting in the emission of Auger electrons (Friedlander et al., 1964). The fluorescence yield is important in the measurement of nuclides that decay by EC, as it is the x-ray fluorescence photons that are usually detected (Mann, 1978). Figure 5.2 illustrates the

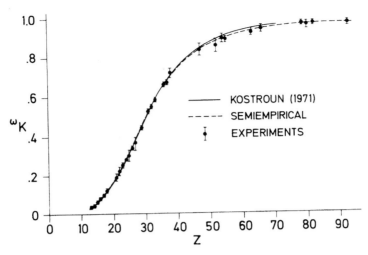

Figure 5.2   Fluorescence K-shell yields, ω_K, as a function of atomic number, Z. (a) according to Kostroun *et al.* (1971); (b) a best fit to selected experimental data; and (c) critically evaluated experimental results. (From Bambynek *et al.*, 1972, reprinted with permission from The American Physical Society.).

K-shell fluorescence yield as a function of nuclide atomic number. The L-shell fluorescence yield also varies similarly with atomic number as the K-shell fluorescence yield, but is several times lower in magnitude (Friedlander *et al.*, 1964).

Radioactivity Hall of Fame — Part VI

Victor F. Hess (1883–1964), Carl D. Anderson (1905–1991), Patrick M.S. Blackett (1897–1974), Hideki Yukawa (1907–1981), Cecil F. Powell (1903–1969), Donald A. Glaser (1926–), and Pierre Victor Auger (1899–1993)

VICTOR F. HESS (1883–1964)

Victor Franz Hess was born on June 14, 1883 in Waldstein Castle near Peggau approximately 75 km northwest of the capitol city of Graz, Steiermark, Austria. He received the Nobel Prize in Physics 1936 "for his discovery of cosmic radiation".

Victor Hess received his entire academic education in Graz including Graz Gymnasium during 1893–1901 followed by a higher education at the University of Graz where he received the doctoral degree in 1910. Hess then went to the Physics Institute of the University of Vienna located on the Turkenstrasse where he studied briefly the science of radioactivity under the prominent physicist Professor Egon von Schweidler, who was among the first to discover toward the end of 1899 that radiation emissions from radium and polonium could be deflected by a magnetic field. Others who made this observation independently about the same time were Friedrich Giesel of Braunschweig, Germany and Stefan Meyer of Vienna (Becquerel, 1903). Also during 1899–1900 Egon von Schweidler, Friedrich Giesel, Stefan Meyer, Henri Becquerel, and Pierre and Marie Curie had also determined that the beta rays from radioactive sources were fast electrons.

Victor F. Hess (1883-1964)

During 1910–1920, Victor Hess was an assistant under Stefan Meyer at the Institute of Radium Research during which time Hess did his pioneering observations that lead to his discovery of cosmic radiation. The original building of the Institute of Radium Research can be visited today. It is now occupied by the Institute of Physical Chemistry located at the intersection of Währinger Strasse and Boltzmanngasse in the 9th District of Vienna. The original entrance with the sign "Radium Institut", set in concrete, can be seen today above the door of the Boltzmanngasse entrance.

Around 1910, radioactivity was commonly measured with electroscopes. A schematic illustration of the basic components of a gold-leaf electroscope of the period is illustrated in Figure VI.1. Electroscopes contained a metal rod to which was attached a thin metal foil such as a gold leaf at one end of the rod. The metal rod and gold leaf were enclosed in an air-tight chamber often made of metal or glass through which one could observe the position of the gold leaf against a transparent scale placed in front or behind the gold leaf in the line of sight of the viewer. The metal rod, insulated from the chamber walls, was first charged by connecting the rod to a few hundred volts with a battery, and the charged rod and gold leaf would cause the gold leaf to be repelled away from the rod, since like charges repel each other, as illustrated in the figure. When radiation enters the chamber, such as high-energy beta- and gamma-radiation or cosmic rays, ionization of the air within the electroscope would occur and the ion pairs produced would collect on the gold leaf and rod causing their charge to become reduced whereby the gold leaf would move back toward the direction of the metal rod. The rate of fall of the gold leaf back toward the metal rod would provide a measure of the radiation intensity. The instrument could not provide the number of rays entering the chamber, but it was a very sensitive instrument that could be calibrated with known sources to provide a measure of the number of ions

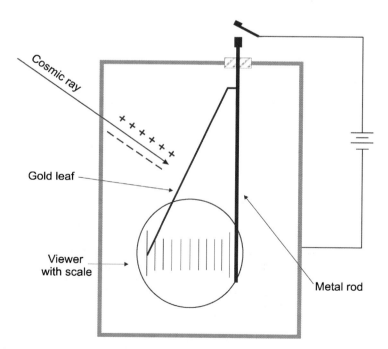

Figure VI.1   Schematic diagram of a gold-leaf electroscope. After charging with a battery, the gold leaf moves to the left away from the metal support rod due to charge repulsion with the metal rod. When cosmic radiation enters the chamber ion pairs are formed, and ions will discharge the gold leaf and rod whereby the flexible gold leaf returns toward the rod at a rate dependent on the rate of discharge. The rate of fall of the gold leaf toward the rod provides a measure of the radiation intensity.

collected in the chamber per unit of time. Hess calibrated his electrometers to measure the number of ions per cubic centimeter per second (Hess, 1912).

During the early 1900s, high penetrating radioactivity was detected with the electroscope in all parts of the globe, on the ground in valleys and mountain tops, in buildings, in the air, in lake bottoms, on top of the Eiffel Tower, and even when protected with lead to shield the electroscope from known mineral sources of radioactivity. It was thought by some that the radioactivity was emanating from minerals on the earth's surface, but the radioactivity persisted at altitudes of the tallest building or structure such as the Eiffel Tower. Some thought that the radioactivity originated from sources in the air. Some scientists made balloon ascents during 1909–1911 to measure the ionization caused by this mysterious radioactivity, but their instrumentation developed defects that could not provide irrefutable evidence as to the source of the radioactivity (Moritz, 1963).

During 1911–1913, Victor Hess made seven balloon ascents into the atmosphere up to heights of 5300 m. He carefully calibrated his instruments and took along with him three electroscopes to measure the ionization caused by the external radiation. To eliminate doubt and reduce error he thought it best to make simultaneous measurements of ionization at various heights with more than one instrument as one instrument could fail or demonstrate erratic readings. If the data from three instruments coincided then the results would provide the confidence needed for an interpretation. He found that at an altitude of 500 m the ionization dropped to about half that obtained from the soil surface. However, the readings would increase proportionally with altitude as he would ascend further. He found the ionization at 1000 m and upwards the radioactivity would noticeable increase, and further ascents to altitudes of 5000 m would provide ionization readings at several times that at the ground level (Hess, 1912, 1936, 1940). It was obvious to Hess that gamma rays, from radium in air, could not cause any increase in ionization with altitude, and he interpreted his results in the following words:

> The only possible way to interpret my experimental findings was to conclude to the existence of a hitherto unknown and very penetrating radiation, coming mainly from above and being most probably of extra-terrestrial origin...

The discovery of Victor Hess opened the door to a new field of science as underscored by Professor H. Pleijel, Chairman of the Nobel committee for Physics, when he stated in his Presentation Speech on December 10, 1936

> The mystery of the origin of this radiation remained unsolved until Prof. Hess made it his problem...With superb experimental skill Hess perfected the instrumental equipment used and eliminated its sources of error. With these preparations completed, Hess made a number of balloon ascents to heights up to 5,300 m, in 1911 and 1912. His systematic measurements showed that a decrease in ionization did occur up to 1,000 m, but that it increased considerably thereafter... Later ascents and investigations made by successors of Hess in free balloons equipped with recording instruments showed that, at a height of 9,300 m, radiation is about 40 times as intensive as on the earth's surface...From these investigations Hess drew the conclusion that there exists an extremely penetrating radiation coming from space, which enters the earth's atmosphere. This radiation, which has been found to come from all sides in space, has been called cosmic radiation...The new rays surpass in intensity and penetrating power everything previously known. They are capable of penetrating lead plates one meter thick and they have been detected on the floor of lakes with a depth of 500 m.

In the autumn of 1931, Victor Hess set up an observatory on a 2300 m-high mountain, the Hafelekar at Innsbruck, Austria for the continuous recording of the fluctuations in intensity of cosmic rays. Higher mountains were available, but not accessible during all seasons of the year. Hess chose the Hafelekar, as the summit could be reached by suspended cable car from Innsbruck in 40 min during the whole year (Hess, 1932). He recorded a great number of results during continuous studies throughout the various seasons including the measurement of small, regular, daily fluctuations of radiation attributed to atmospheric influences, particularly electrical and magnetic effects in the highest layers of the atmosphere (Hess and Demmelmair, 1937; Hess et al., 1938a,b; Hess, 1939).

Following the historic discovery of cosmic radiation by Victor Hess in 1912 other researchers studied the physical properties of this penetrating radiation. One of the methods used was to observe and study the tracks of the cosmic rays by means of the Wilson Cloud Chamber. This led to the discovery of antimatter by Carl D. Anderson, the first observation of the positron, the positive electron, and antiparticle of the negatively charged electron. Carl Anderson shared in the Nobel Prize Physics 1936 and a biographical sketch of Anderson and a brief description of his work is provided in the following section.

In 1919, Victor Hess was awarded the Ignaz L. Lieben Prize by the Austrian Academy of Sciences. It was the "Nobel Prize equivalent" established by the former Danube Monarchy of Old Austria, but discontinued by the National Socialists in 1938. The Prize was revived in 2004 with a foundation from a U.S. patron, Alfred Bader, chemist and founder of the chemicals company Sigma-Aldrich, who had to escape Vienna in 1939 in a children's transport to England. This award was first granted in 1862 and one of the most important science prizes in the history of Old Austria offered originally by the bankers' family for the "continuing promotion of scientific research". The new Lieben Prize is to be granted to young scientists from Bosnia-Herzegovina, Croatia, Slovakia, Slovenia, the Czech Republic, Hungary, and Austria for "outstanding work in the field of molecular biology, chemistry and physics".

In 1919, Hess was named Extraordinary Professor of Experimental Physics at Graz University. During 1921–1923, Hess took leave of absence from his post in Austria to work in the USA as Director of Research at the U.S. Radium Corporation and consultant to the U.S. Bureau of Mines. He returned to the University of Graz in 1923, and in 1931 Hess was appointed Professor at Innsbruck University and Director of its new Institute of Radiology. It was in 1931 that Hess established his Cosmic Ray Observatory at the summit of Hafelekar mountain at Innsbruck.

In 1937, Hess returned to the University of Graz as Professor of Physics and Director of the Physics Institute. However, 2 months after the "Anschlus" on 13 March, 1938, Victor Hess was dismissed from his post because he had a Jewish wife and also because he was a representative of the sciences in the government of Chancellor Kurt von Schuschnigg (Moritz, 1963). Chancellor Schuschnigg opposed the German absorption of Austria, which was successful until he lost support of Benito Mussolini, and he was a Nazi prisoner until 1945. Fortunately a sympathetic Gestapo officer warned the Hess family that they would be taken to a concentration camp if they remained in Austria. Even a Nobel Laureate could not escape such a fate (see the biographical sketch of James Franck and his story of the pact with Niels Bohr and Max Planck to hide their gold Nobel Medals from the Nazi Regime in a solution of Aqua Regia). The Hess family escaped to Switzerland 4 weeks before the order came for their arrest (Moritz, 1963).

Victor Hess was appointed to a full professorship in 1938 at Fordham University in New York and became an American citizen in 1944. At Fordham, he carried out research on the measurement of environmental radioactivity including the measurement of radium, artificial radioactivity in the atmosphere from nuclear testing, and the effects of radioactivity on human health. Hess retired from Fordham University in 1958. He died on December 17, 1964. The Victor F. Hess Prize was established in 1987 in Austria in his honor. The prize is awarded to Ph.D. graduates, who have produced an outstanding Ph.D. dissertation at an Austrian university in the field of nuclear and particle physics. An excellent summary of the work and life of Victor Hess can be found in a thesis by Federmann (2003).

CARL D. ANDERSON (1905–1991)

Carl David Anderson was born in New York City on September 3, 1905. He shared the Nobel Prize in Physics 1936 with Victor Hess. Anderson's share of the prize was awarded "for his discovery of the positron", the first experimental evidence of antimatter. He was only 31 years of age at the time of the award and then the youngest person to have received the Nobel Prize. A yet younger Nobel Laureate would emerge, Tsung-Dao Lee from China, who was awarded the Nobel Prize Physics in 1957.

**Carl D. Anderson
(1905-1991)**

Carl Anderson's parents moved to Los Angeles from New York when he was 7 years of age. He got all of his academic education in California. After High School in Los Angeles he completed his undergraduate and graduate degrees at the California Institute of Technology (Caltech) in Pasadena. Even his entire professional career and scientific accomplishments were fulfilled at Caltech. He received his B.Sc. degree in Physics and Engineering in 1927. Nobel Laureate Robert Millikan became his graduate advisor for his doctoral degree, receiving the Ph.D. in 1930 after research on photoelectrons produced by x-rays (Anderson, 1930). In his autobiography, Anderson wrote concerning his Ph.D. thesis and Millikan's involvement

"For this I thanked him, but not once during the three years of my graduate thesis work did he visit my laboratory or discuss the work with me". Then came postdoctoral work [that led to Anderson's improvement of the Wilson Cloud Chamber and discovery of the positron from cosmic-ray showers and the Nobel Prize], again, loosely supervised by Millikan, during which Anderson built and ran the Caltech Magnet Cloud Chamber (Weiss, 1999; Brown, 2000).

Although such statements might seem unusual and cruel toward one's graduate advisor and a person as famous as Millikan, graduate students who are self driven and gifted have had similar experiences at other institutions. Concerning this William H. Pickering (1991), a former colleague of Anderson wrote

Millikan became Carl's graduate advisor. In those days graduate students had a great deal of freedom in their research. Consequently, Carl received very little direction from Millikan... In retrospect Anderson's achievement was due in part to R.A. Millikan's intuition that the study of cosmic rays was important and that Anderson had the experimental ability to build a superior cloud chamber. The intellectual climate at Caltech encouraged the young physicist. This was still the period when physics was being done with "love and string and sealing wax". Brilliant scientists with very little money or other support were pushing back the frontiers of physics and in the process giving us new concepts of the world. Anderson's "anti-matter" was the first step that led to the understanding of the nucleus.

After receiving his doctoral degree at the age of 25 Millikan suggested to Anderson that he get a National Research Council Fellowship and work at another institution to broaden his experience. He agreed and was accepted to work with A. H. Compton at Chicago when Millikan changed his mind and convinced Anderson to stay at Caltech to work on cosmic radiation. Thus Anderson remained at Caltech as Research Fellow during 1930–1933 during which time he made his historic discovery of antimatter.

To observe cosmic-ray interactions with matter Anderson firstly made improvements on the Wilson Cloud Chamber detector (See C.T.R. Wilson in Radioactivity Hall of Fame—Part II) to provide clearer images of the tracks of elementary particles that would be produced when high-energy cosmic radiation interacted with the supersaturated vapor of the cloud chamber and with various metals such as lead and aluminum inserted into the path of the particles. For his cloud chamber, Anderson developed a vacuum-controlled piston that would expand rapidly a mixture of water and alcohol. The supersaturation produced in the cloud chamber by the water–alcohol mixture provided clearer images of the particle tracks than pure water vapor originally used by Wilson. When a charged particle passes through the supersaturated vapor, very minute water droplets condense and become visible along the path of the ionizing particle producing tracks like those of a jet traveling at high speed in the upper atmosphere. The tracks are photographed with proper illumination and timing of the camera.

Anderson added a magnetic field around his cloud chamber. The magnetic field perpendicular to the path of the charged particle would cause the particle path of travel to bend. The degree of bend in the particle track produced by the magnetic field would provide information as to the charge, mass, and energy of the particle. For example, when an applied magnetic field was directed normal to the path of a charged particle, the positively charged particle would travel in an upward direction and a negatively charged particle in the downward direction. The lower the mass of the particle and the lower its energy, the greater would be the bend or sharper the curvature of the particle path under a given magnetic field. With sufficient magnetic field strength the charged particles would travel in circular paths when traveling across the magnetic field. By counting the number of droplets per unit length along the particle tracks Anderson could make measurements of the specific ionization of both positive and negative particles. The specific ionization and number of droplets formed would be proportional to the particle mass and charge. Anderson was also able to determine the direction of travel of the charged particles by photographing their tracks in the cloud chamber containing lead or aluminum plates. Particles that traveled through the plates would lose energy and thus the track left by the particle after passing through the metal plate would have a sharper bend under the effect of an applied magnetic field. In his Nobel Lecture, Anderson explained his technique as follows:

...In the spring of 1932, a preliminary paper, on the energies of cosmic-ray particles was published [Anderson, 1932, 1933a] in which energies over 1 billion electron-volts were reported. It was here shown that particles of positive charge occurred about as abundantly as did those of negative charge, and in many cases several positive and negative particles were found to be projected simultaneously from a single center. The presence of positively charged particles and the occurrence of "showers" of several particles showed clearly that the absorption of cosmic rays in material substances is due primarily to a nuclear phenomenon of a new type.

Measurements of the specific ionization of both the positive and negative particles, by counting the number of droplets per unit length along the tracks, showed the great majority of both the positive and negative particles to possess unit electric charge. The particles of negative charge were readily interpreted as electrons, and those of positive charge were at first tentatively interpreted as protons, at that time the only known particle of unit positive charge.

If the particles of positive charge were to be ascribed to protons than those of low energy and sharp curvature in the magnetic field...should be expected to exhibit an appreciably greater ionization than the negatively charged electrons. In general, however, the positive particles seemed to differ in specific ionization only inappreciably from the negative ones...it seemed inadequate to account for the large number of particle tracks which showed a specific ionization anomalously small if they were to be ascribed to protons.

To differentiate with certainty between the particles of positive and negative charge it was necessary only to determine without ambiguity their direction of motion. To accomplish this purpose a plate of lead was inserted across a horizontal diameter of the chamber. The direction of motion of the particle could then be readily ascertained due to the lower energy and therefore the smaller radius of curvature of the particles in the magnetic field after they have traversed the plate and suffered a loss of energy.

Results were then obtained which could logically be interpreted only in terms of particles of a positive charge and a mass of the same order of magnitude as that normally possessed by the free negative electron. In particular one photograph (see Figure VI.2) shows a particle of positive charge traversing a 6 mm plate of lead. If electronic mass is assigned to this particle, its energy before it traverses the plate is 63 MeV and after it emerges its energy is 23 MeV. The possibility that this particle of positive charge could represent a proton is ruled out on the basis of range and curvature. A proton of the curvature shown after it emerges from the plate would have an energy of 200 keV, and according to previously well-established experimental data would have a range of only 5 mm whereas the observed range was greater than 50 mm. The only possible conclusion seemed to be that this track, was the track of a positively charged electron.

Anderson found many examples of positron tracks similar to that described above and published his findings in 1932 and 1933 (Anderson 1932, 1933b). In addition, he determined the relative specific ionizations of the positive and negative electrons by counting the droplets of water vapor photographed in the particle tracks. He demonstrated that the mass and charge of the positive electron did not differ significantly from the magnitude of the mass and charge of the negative electron.

Another interesting finding first reported by Anderson and his first graduate student Seth Neddermeyer was that gamma rays from Thorium C (ThC or ^{212}Bi) would give rise to positrons (Anderson, 1933c; Anderson and Neddermeyer, 1933). Such an observation was also observed independently by Irène Joliot-Curie and her husband Frédéric Joliot (1933) and Lise Meitner and K. Philipp (1933). An interesting example is that illustrated in Figure VI.3, which shows the track of a positron emerging from a lead plate bombarded by ThC gamma rays and the passage of the positron through an aluminum plate. The charge and mass of the particle was determined by its direction of travel in a magnetic field and its specific ionization. The positron would travel in the opposite direction to that of the electron in a magnetic field normal to the path of travel and its specific ionization would not differ greatly from that of the electron.

Occasionally Anderson and Neddermeyer would observe the simultaneous appearance of two particles, both a positron and negatron, either upon bombarding lead with ThC gamma rays or from the

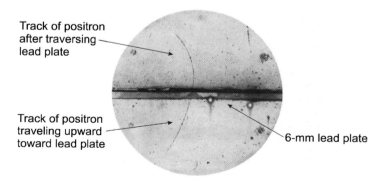

Figure VI.2 Historic cloud chamber photograph taken by Anderson of a track made by a 63 MeV positron that had traveled in the upward direction under the influence of a 15,000 G magnetic field in a cloud chamber. The positron emerges after traversing a lead plate with a lower energy of 23 MeV. Notice the sharper curvature of the particle track after it had lost energy in the lead plate. From the length of travel of the positively charged particle and its curvature Anderson could conclude that the particle was not a proton, but a positively charged electron. (From Anderson, 1933b, 1936, reprinted with permission from The American Physical Society.)

absorption of cosmic gamma radiation. Their photographs of the cloud chamber tracks were the first evidence of the conversion of energy into mass by the process now known as "pair-production". Figure VI.4 illustrates two examples of their cloud chamber photographs of the tracks produced by positron and negatron pairs after the absorption of cosmic gamma rays in a lead barrier.

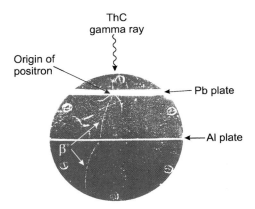

Figure VI.3   The cloud chamber track of a positron (β^+) originating from a 2-mm lead plate bombarded with ThC gamma rays from above. The track illustrates the path of the positron as it traveled toward a 0.5-mm aluminum plate and passed through the aluminum. Notice the slight curvature of the positron track before entering the aluminum and the sharper curvature after passing through the aluminum. The sharper curvature of the positron path in the magnetic field was due to its reduction in energy from 0.8 MeV before entering the aluminum to 0.5 MeV after passing through the aluminum. (From Anderson and Neddermeyer, 1933 reprinted with permission from The American Physical Society.)

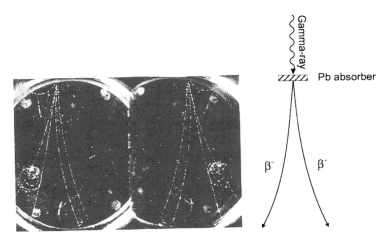

Figure IV.4   Photograph of cloud chamber tracks of positrons (curved tracks on the right) and negatrons (curved tracks on the left) originating from a common point of origin where gamma-ray photons of cosmic origin were absorbed by a lead barrier. The magnetic field was applied in the direction of the page of this book. The photograph was presented as a stereoscopic pair, and the direct image is at the left (From Anderson and Neddermeyer, 1936 reprinted with permission from The American Physical Society). A drawing to the right illustrates pair-production, that is, the production of a negatron–positron pair from the energy of a gamma ray photon in the region of the nucleus of an absorber material such as lead. The curved tracks illustrate the direction of the negatron–positron pair by the magnetic field.

In a biographical sketch by William Pickering (1991), a colleague and coworker of Anderson in cloud-chamber measurements of cosmic-ray showers (Anderson et al., 1934), he noted that following his observation of the positive electron tracks in the cloud chamber photographs, he consulted with Oppenheimer and Richard B. Feynman for their view on the phenomena of the origin and fate of the positive electron. Pickering relates the following:

> Anderson said that he discussed the problem of the formation and disappearance of the positron with J. Robert Oppenheimer and, in retrospect, was surprised that Oppenheimer did not come up with this pair-production mechanism. He also commented that it was very difficult to understand Oppenheimer's answers to his questions. Later, when he talked with Richard P. Feynman, just the opposite was true. Feynman was clear and precise. Anderson's positron and J. Chadwick's neutron, discovered and reported in the *Proceedings of the Royal Society* in 1932, were the first new fundamental particles.

In the report of their findings Anderson and Neddermeyer (1933) stated their observation as follows:

> One of the most striking phenomena, which have been observed in this experiment is the occasional simultaneous appearance of paired tracks consisting of one positive particle and one negative with a common point of origin…The process, which gives rise to the positrons is at present not known, but so far as these data go they are in accord with the view expressed by Blackett and Occhialini (1933) that the two particles may be formed by a process in which the energy of approximately 1 MeV required for the formation of a positive and negative electron is supplied by the impinging radiation.

It was Nobel Laureate Patrick Blackett and his coworker G.P.S. Occhialini (1933) who recognized that the two particles, positron and negatron, are formed by the process known as "pair production" whereby the energy of 1.02 MeV required for the production of the two particles is provided by the gamma radiation. In his Nobel Lecture, Anderson (1936) explained the pair-production phenomena as follows:

> …that the appearance of pairs of positive and negative electrons could be understood in terms of this theory [Dirac's theory] as the "creation" of a positive-negative electron pair in the neighborhood of an atomic nucleus. The energy corresponding to the proper mass of both of the particles, as well as to their kinetic energies is supplied, according to this view, by the incident radiation. Since the energy corresponding to the proper mass of a pair of electrons is approximately 1 MeV, one should expect gamma rays of energy greater than this amount to produce positrons in their passage through matter, and further that the sum of the kinetic energies of the positive and negative electrons should be equal to the energy of the radiation producing them diminished by approximately 1 MeV.

In the above statement, Anderson described the process of pair-production, first theorized by Dirac (1928a,b), observed through cloud-chamber track photographs by Anderson and Neddermeyer (1933), and interpreted by Blackett and Occhialini (1933). The gamma-ray photon energy required for pair production is, as noted above by Anderson, equivalent to the sum of the masses of the electron and positron described by the equation

$$E_{pair} = m_{e^-}c^2 + m_{e^+}c^2 = 2\,mc^2 = 2(0.511 \text{ MeV}) = 1.022 \text{ MeV} \qquad \text{(VI.1)}$$

The gamma-ray photon energy in excess of 1.022 MeV appears as the kinetic energy of the electron and positron created, and the overall pair-production phenomena can be expressed as

$$h\nu = 2mc^2 + E_{e^-} + E_{e^+} \qquad \text{(VI.2)}$$

where $h\nu$ is the gamma-ray photon energy, $2mc^2$ the energy of the mass of two electrons, and E_{e^-} and E_{e^+} the kinetic energies of the electron and positron, respectively. By the time of his Nobel Lecture in

1936, three sources of positrons were known, and Anderson took the opportunity of his lecture to describe these which are the following: (1) cosmic-ray showers of electron–positron pairs *via* pair production from cosmic gamma radiation, (2) pair production *via* the absorption of gamma radiation in excess of 1 MeV from radionuclide sources, and (3) positron emission *via* beta decay of radioactive elements, such as ^{30}P, ^{13}N, and ^{27}Si, artificially produced by Joliot-Curie and Joliot (1934a–c).

Subsequent studies reported as joint papers, Anderson and Neddermeyer (1936) and Neddermeyer and Anderson (1937, 1938) provided evidence for the muon in the cosmic-ray showers, a particle of "intermediate" mass, that is, of mass greater than the electron but less than the proton or neutron. They used a cloud chamber with a Geiger counter placed within the chamber and coupled by means of a coincidence circuit to another Geiger counter placed above the cloud chamber. With such an arrangement, the counters could trigger illumination and camera thereby increasing the probability of photographing cosmic-ray particle tracks and include images of tracks as the particles neared the end of their ranges, that is, up to the location where the particles came to rest. Such images could provide Anderson and Nedermeyer with the mass and stability properties of the particle. An historic cloud-chamber photograph illustrated in Figure VI.5 is the first demonstrating the existence of the muon in cosmic-ray showers.

Neddermeyer and Anderson (1938) were able to calculate that the specific ionization of the particle before it traversed the glass and copper barrier was greater than that of a fast electron, and the very high specific ionization after traversing the barrier provided evidence for its intermediate mass. Data on the specific ionization together with that provided on the curvature of the track before and after traversing the glass and copper barrier provided them with sufficient information to report an initial estimate of the particle mass at 240 electron masses. Their findings were subsequently confirmed by Street and Stevenson (1937). Subsequent studies by Leighton *et al.* (1949) determined the muon mass to be 217 ± 4 electron masses. The muon is now classified to have a mass of 106 MeV/c^2, which is equivalent to 207 electron masses, and a lifetime of 2.2×10^{-6} sec. Neddermeyer and Anderson

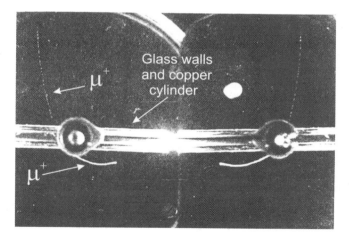

Figure VI.5 A track of a positive muon (μ^+) of about 10 MeV energy that had entered from above the cloud chamber and passed trough the glass walls and copper cylinder of a Geiger counter in the center of the cloud chamber. The muon track was curved toward the right under the effect of a magnetic field of 7900 G. Notice the high droplet density (thickness) of the muon track after the muon emerged from the glass and copper barrier with an energy of only 0.21 MeV. The reduction in energy of the muon increased its specific ionization in the chamber. Its residual range after emerging from the glass and copper barrier was 2.9 cm in the gas of the cloud chamber containing 2/3 He, 1/3 Ar, and alcohol vapor at a pressure of 1 atmosphere equivalent to 1.5 cm in air. The photograph was presented as a stereoscopic pair, and the direct image is at the left. (From Neddermeyer and Anderson, 1938 reprinted with permission from The American Physical Society.)

(1938) were also able to provide initial information concerning the decay scheme of the muon from its track illustrated in Figure VI.5. In their paper they noted the following:

An interesting feature of the photograph is the fact that the particle is actually observed to come to rest in the gas of the chamber. No completely certain evidence of a subsequent disintegration can be found on the photograph. There are, however, three droplets which appear on the left-hand image, which is the direct view, and also on the right-hand mirror image. Stereoscopic observation shows that these droplets line up so as to indicate a short segment of an electron track emanating from the point in the gas at which the particle came to rest and directed toward the counter...These droplets may therefore indicate that the particle, after coming to rest, disintegrated by the emission of a positive electron.

Their deductions from the early observations made in this historic photograph have proven to be true. The positive muon is known to decay to a positive electron and a muon neutrino, and the negative muon decays to a negative electron and antimuon neutrino illustrated by the following equation:

$$\mu^{+/-} \rightarrow e^{+/-} + \nu_\mu + \bar{\nu}_\mu \tag{VI.3}$$

Anderson became Assistant Professor at Caltech in 1933 the very year he made his discovery of antimatter providing him the Nobel Prize 3 years later at the young age of 31. He did not become full professor until 1939. It is difficult to understand why Caltech would keep Anderson from the title of Professor of Physics until 3 years after his award of the Nobel Prize. Possibly he was simply too young with insufficient experience as Assistant Professor. To have made a major discovery from work carried out during 1930–1933 as a Research Fellow at the young age of 25–28 years with the Nobel prize awarded only a couple of months after his 31st birthday was astounding. One can only surmise that the administration at Caltech would not award him the Professorship until 1936, because of his young age.

In his autobiography edited by Weiss (1999) Anderson revealed that in May of 1942, Arthur H. Compton had asked him "to head a project to design and build an atomic bomb", but Anderson turned down the offer simply "on economic grounds". General Leslie R. Groves offered the job to J. Robert Oppenheimer 5 months later on October 15, 1942. In his autobiography Anderson wrote

I believe my greatest contribution to the World War II effort was my inability to take part in the development of the atomic bomb. Thinking so brings me peace of mind.

However, Anderson did contribute significantly to the war effort. During World War II, he took part in the research effort at Caltech for the development of rocket launchers. Anderson's main responsibility was to help develop rocket launchers from aircraft that could be used against submarines detected by magnetic sensors. Rockets and launchers were also developed to be used from ships to facilitate Pacific island landings. Pickering (1991) noted that Anderson traveled to the Normandy beachhead in 1944 to observe the Caltech rockets under battlefield condition. His student Seth Neddermeyer received his Ph.D. at Caltech in 1935, then joined the staff at the University of Washington. However, he was later recruited into the Manhattan Project at Los Alamos and one of the originators of the implosion technology for the atomic bomb. Rhodes (1986) wrote about Neddermeyer's involvement in the bomb development "His idea of using explosives to squeeze a nuclear core to criticality saved the plutonium bomb when impurities threatened its design".

High-energy cosmic-ray nucleons collide with nuclei of the atmosphere smashing the nuclei into subatomic particles creating the cosmic-ray showers described in Chapter 6. After the war, Anderson continued his research on cosmic-ray showers at Caltech. He flew in B-29 bomber airplanes equipped with cloud chambers to study the cosmic-ray showers with the support of the Office of Naval Research. On one of these flights at 9300 m of altitude he obtained a cloud chamber photograph of a muon or μ-meson and its disintegration (Anderson et al., 1947), which belong to the group of particles that are strongly interacting and bind protons and neutrons together in the nucleus. He and cowork-ers reported the decay of the positive muon to a positron. The muon neutrino obviously could not be detected in the cloud chamber.

$$\mu^+ \rightarrow e^+ + \nu_\mu \tag{VI.4}$$

The muon came to rest in the cloud chamber, and the positron made a track at a 90° angle with respect to the disintegrating muon. The mass of the muon can be determined from the track left by the parti-cle in the cloud-chamber image based on the following parameters: particle range, specific ioniza-tion, and track curvature in the magnetic field. On the basis of the data obtained they measured the muon mass at 200 electron masses, which is very close to the currently known mass of the muon at 207 electron masses. A subsequent study carried out by Anderson and coworkers (Leighton *et al.*, 1949) at Caltech involved the analysis of 75 cloud-chamber tracks of decaying cosmic-ray muons at sea level. Again the cloud chamber was subjected to a magnetic field (7250 G) to permit the charge determination of the particles and measurements of the track curvatures. From the numerous tracks analyzed, they observed a cosmic-ray muon spectrum over the range of 9–55 MeV with a continuous distribution of intermediate energy values and a mean energy of 34 MeV.

Further studies of Anderson and coworkers focused on the cloud-chamber measurements of *V*-particle decay in cosmic-ray showers (Leighton *et al.*, 1953). *V*-particles got their name from the V-shaped tracks that they formed from a given point in cloud chambers. These were produced by the decay of neutral kaons (K^0) or charged kaons ($K^\pm$) also called K mesons, which are strongly interacting particles holding protons and neutrons together in the nucleus. The neutral kaon exists in a short-lived (K_S^0) or long-lived (K_L^0) state with relative lifetimes of 0.9×10^{-10} and 5×10^{-8} sec; and the mass of the neutral kaon is 498 MeV/c^2. The charged kaons ($K^\pm$) have a lifetime of 1.2×10^{-8} sec and mass of 494 MeV/c^2. The predominant decay modes of the charged kaons are the following (Sundaresan, 2001):

$$K^\pm \rightarrow \mu^\pm + \nu_\mu (63\%) \tag{VI.5}$$

$$K^\pm \rightarrow \pi^\pm + \pi^0 (21\%) \tag{VI.6}$$

$$K^\pm \rightarrow \pi^\pm + \pi^+ + \pi^- (5.5\%) \tag{VI.7}$$

$$K^\pm \rightarrow \pi^0 + e^\pm + \nu_e (5\%) \tag{VI.8}$$

$$K^\pm \rightarrow \pi^0 + \mu^\pm + \nu_\mu (3\%) \tag{VI.9}$$

$$K^\pm \rightarrow \pi^\pm + \pi^0 + \pi^0 (2\%) \tag{VI.10}$$

Other modes of decay of the charged kaon exist with abundances of less than 1%. The first-order neu-tral kaon modes of decay are

$$K^0 \rightarrow \pi^+ + \pi^- (88\%) \tag{VI.11}$$

and

$$K^0 \rightarrow \pi^0 + \pi^0 (12\%) \tag{VI.12}$$

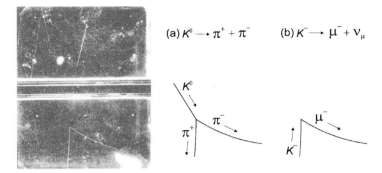

Figure VI.6 A cloud-chamber photograph of the tracks of a probable V^0 particle in a magnetic field of 5000 G. The cloud chamber consists of upper and lower chambers separated by the glass walls and copper tubing of a central Geiger counter. The V-shaped track appears left center in the lower-chamber. The V-shaped track was the result of one of two possible modes of kaon decay, namely, (a) a neutral kaon (K^0) decaying into positive and negative pions where the neutral kaon does not leave a track because of its lack of charge, or (b) a negative kaon (K^-) entering the cloud chamber from below and decaying into a negative muon and neutrino. The neutrino obviously does not leave a track (Cloud chamber photograph from Anderson and coworkers (Leighton *et al.*, 1953) reprinted with permission from The American Physical Society.)

The *V*-particles or particles that yield V-shaped tracks, were discovered by Rochester and Butler (1947). These result from the decay of the neutral kaon, which has a mass of about 490 MeV/c^2. The neutral kaon cannot produce a track in the cloud chamber due to its lack of charge; however, its decay into positive and negative π-mesons (pions) according to eq. (VI.11) yields two tracks in a cloud chamber subjected to a magnetic field. The two tracks take the shape of a V due to the opposing pion paths and lack of track from an incoming neutral kaon. Figure VI.6 illustrates a cloud-chamber photograph of the tracks left by a neutral V^0 particle (i.e., neutral kaon, K^0). Notice that the V-shaped tracks appear in the left-center of the lower chamber of the apparatus. No track is left by the incoming particle, because it has no charge. Alternatively, Anderson and coworkers (Leighton *et al.*, 1953) proposed that the V-shaped track illustrated in Figure VI.6 could also be due to a negative kaon decaying to a negative muon according to eq. (VI.5).

Carl Anderson devoted his entire professional career at Caltech. He served as Professor there until 1976 at the age of 71. He died on January 11, 1991.

PATRICK M. S. BLACKETT (1897–1974)

Patrick Maynard Stuart Blackett was born in London, England on November 18, 1897. He was awarded the Nobel Prize in Physics 1948 "for his development of the Wilson cloud chamber method, and his discoveries therewith in the fields of nuclear physics and cosmic radiation".

Blackett started his career as a naval cadet in 1914 taking part in the World War I naval battles of Falkland Islands in December of 1914 and Jutland in May of 1916. He graduated from the Osborne Naval College in 1917. After the war, he took up physics at Cambridge University, and after graduation in 1921 began research on cloud chambers under Lord Rutherford at his Cambridge Laboratory. Blackett first improved the Wilson Cloud Chamber by making it fully automatic in the sense that a photograph of the illuminated tracks left by the elementary particles would be taken every 15 sec (Blackett, 1922, 1927, 1929). His first challenge under Rutherford would be to use the cloud chamber to photograph the process of transmutation of nitrogen into oxygen following the collision of an alpha particle with nitrogen. In 1919, Rutherford had made the discovery that nuclei of certain light elements, such as nitrogen, could be changed to another element by the impact of very fast alpha particles from radioactive

RÉPUBLIQUE DE GUINÉE
OFFICE DE LA POSTE GUINÉENNE
750F

1948
PATRICK BLACKETT

**Patrick Blackett
(1897-1974)**

sources, and in the process fast protons were emitted. Rutherford could not provide experimental evidence for the process of transmutation at the time of his discovery, and it was the next step to use the Wilson Cloud Chamber to see and measure, through photographic images of the particle tracks, what nuclear processes were occurring.

Patrick Blackett was meticulous in his approach to the use of the cloud chamber to observe and study nuclear transmutation. He first used the cloud chamber to photograph and measure the collisions of alpha particles with nuclei of oxygen, hydrogen, and helium atoms to verify that normal collisions, where transmutations did not occur, were truly elastic collisions, that is, collisions where no energy was lost in the process. From the photographs of the nuclear tracks Blackett measured the angles between the track of the α particle before collision and the tracks of both the deflected α particle and recoil atom after the collision denoted as ϕ and θ, respectively, as illustrated in Figure VI.7. Notice that in Figure VI.7b the α particle undergoes only a slight deflection (small ϕ), that is, the α particle continues to travel in almost a straight path after collision with hydrogen atoms because of the large difference in mass between the He and H nuclei.

Blackett (1923, 1948a) and coworkers (Blackett and Hudson, 1927; Blackett and Champion, 1931; Blackett and Lees, 1932b,c) based their study on the fact that, for truly elastic collisions, energy and momentum are conserved, and the angles of atom recoil and α-particle deflection would be a function of the mass of the α particle (M) and the mass of the recoil atom (m) according to the following equation:

$$\frac{M}{m} = \frac{\sin(2\theta + \phi)}{\sin \phi} \tag{VI.13}$$

By measuring the angles of α-particle deflection and atomic recoil he calculated the mass (m) of the recoil atom for each of the three types of collisions illustrated in Figure VI.7 and compared the calculated values with the known atomic masses. Table VI.1 gives the measured angles of α-particle deflection and atomic recoil and calculated and known values of atomic mass. The very close agreement between the calculated and known masses of the recoil atoms demonstrated to Blackett that the

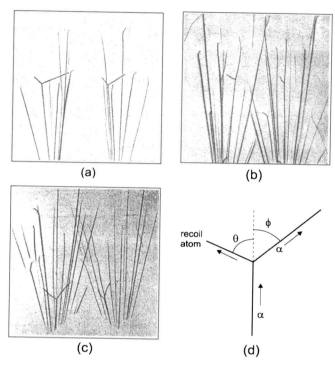

Figure VI.7   Cloud chamber tracks illustrating elastic collisions of alpha particle with (a) oxygen nucleus, (b) hydrogen nucleus, (c) helium nucleus, and (d) a drawing of the geometry of the collisions. Only a few collisions are observed in the photographic images. (From Blackett, 1923, 1948a, reproduced with permission from The Royal Society.)

Table VI.1

Measurements of collisions of α-particles with nuclei of oxygen, hydrogen, and helium and the calculated atomic mass (m) of the recoil atoms

Recoil atom	α-particle deflection (ϕ)	Angle of recoil (θ)	m (calculated)	m (known)
Oxygen	76°6′	45°12′	16.72	16.00
Hydrogen	9°21′	65°39′	1.024	1.008
Helium	45°49′	43°56′	4.032	4.00

Source: Blackett (1948) reproduced with permission from The Royal Society.

collisions were not nuclear transmutations and were truly elastic collisions. This study on the forked tracks was one of the first quantitative measurements ever conducted on the dynamics of single collisions of subatomic particles.

As the next step in his quest to identify and measure nuclear transmutation via α-particle collision Blackett (1922, 1923, 1925 and Blackett and Lees 1932a,b) conducted detailed studies of the relationship between the range of a recoil nucleus and its velocity. He determined the velocities of the recoil nuclei from the angles of the collision and the initial and final velocity of the α particle on the basis that the collisions were elastic. Studies on the bombardment of hydrogen, helium, nitrogen, and argon with α particles provided Blackett with the following formula for the range in air of a nucleus of mass m and atomic number z:

$$R \approx mz^{-1/2} f(u) \tag{VI.14}$$

where $f(u)$ was roughly proportional to $v^{3/2}$. He needed the range measurements to facilitate the identification of the recoil particles emerging from collision that were "abnormal" or inelastic.

In 1924, Blackett took 23,000 photographs of α-particle collisions with nitrogen during a period of a few months. Each photograph had an average of 18 tracks yielding in total over 400,000 tracks, which had to be scrutinized for anomalous behavior, which might show a nuclear transmutation. He found eight forked tracks, which were different than the usual tracks of normal elastic collisions. These photographs were of the interaction of the α-particle with the nitrogen nucleus resulting in the transmutation of the nucleus to that of oxygen after the emission of a proton from the nucleus. One of these photographs is seen in Figure VI.8. Blackett (1925) and Blackett and Lees (1932a) provided evidence that the transmutation included the capture of the α particle by a nucleus of nitrogen with

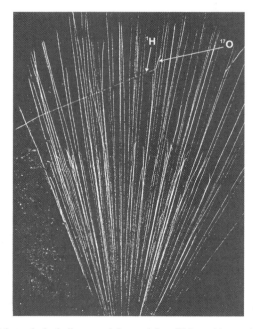

Figure VI.8   Alpha-particle tracks including one alpha-particle collision with a nucleus of nitrogen resulting in the transmutation of the nucleus of a nitrogen atom into a nucleus of an oxygen atom. The photograph illustrates the capture of an alpha particle by a nitrogen nucleus with the emission of a proton (^{1}H). The thin beaded track toward the left and slightly backward is that of the ejected proton. The oxygen nucleus (^{17}O) created from a nucleus of nitrogen (^{14}N) made a short thick track upward and slightly to the right of the alpha-particle trajectory. Blackett (1925) noted that the beaded appearance of the proton tracks was evidence of the small ionization along them, since the ionization due to any particle is proportional to the square of its charge and the reciprocal of its velocity. The source of alpha particles was a mixture of Thorium B + C, more commonly referred to as ^{212}Pb(^{212}Bi). The alpha-particle tracks ending in the middle of the photograph were from the lower-energy alpha particles of the ^{212}Bi alpha-particle source. Notations and arrows are inserted into the original figure to facilitate identification of the tracks produced by the products of the nuclear transmutation. (From Blackett and Lees, 1932a and Blackett, 1948a reproduced with permission from The Royal Society.)

the concomitant emission of a proton from the new and larger nucleus. In his Nobel Lecture, Blackett (1948a) described the evidence from the photographs of the particle tracks and his reasoning as follows:

> Rutherford's original experiments, using the scintillation technique, were only capable of proving that when an alpha particle struck a nitrogen nucleus a fast proton occasionally was ejected, but they were not able to reveal what happened to the alpha particle after the collision. There were two possibilities. The alpha particle might leave the nucleus again as a free particle, or it might be captured, so forming a heavier nucleus. In the former case, one would expect to find a forked track showing the track of the incident alpha particle, with three emergent tracks due to the alpha particle, the ejected proton, and the recoil nucleus. The eight anomalous tracks all showed only two emergent particles, so proving that the assumed "disintegration" of nitrogen by alpha particles was in reality an "integration" process. Applying the principle of conservation of charge and mass, it was immediately deduced that the new nucleus formed must be a heavy isotope of oxygen $^{17}_{8}O$; the nuclear reaction being

$$^{4}_{2}He + ^{14}_{7}N \rightarrow ^{1}_{1}H + ^{17}_{8}O \qquad\qquad (VI.15)$$

> …These experiments gave for the first time detailed knowledge of what is now known to be a typical nuclear transformation process. Owing to the laborious nature of the task of photographing the collisions of natural alpha particles with nuclei, not very much subsequent work has been carried out with this method.

During 1924–1925, Patrick Blackett worked at the University of Göttingen with Nobel Laureate James Franck and then returned to Cambridge. In the autumn of 1931, he began a collaboration with Giuseppe Occhialini to study the high-energy particles in cosmic radiation with the cloud chamber as the tool for the identification and measurement of the cosmic-ray particle interactions with atomic nuclei. Guiseppe "Beppo" Occhialini (nickname was Beppo) was a renowned physicist, born in Pesaro, Italy in 1907. He collaborated successfully with Patrick Blackett from 1931 to 1934 and with Nobel Laureate Cecil Powell from 1944 to 1947. Italian Physicist Bruno Pontecorvo (b.1913–d.1993) once praised Occhialini when he toasted "I drink not to Beppo, but to us all: may we collaborate with him, it is a practically sure way of winning a Nobel Prize".

Early work with the use of the cloud chamber to photograph tracks of cosmic-ray particles was very tedious, as the instant of expanding the chamber and taking the photograph, it was not known whether or not one or more cosmic-ray particles or interactions would be occurring in the chamber. Hundreds sometimes thousands of photographs had to be taken in the hope of photographing tracks of cosmic-ray particles and the cosmic-ray showers, that is, the particles resulting from the collision of high-energy cosmic-ray particles with atomic nuclei in the cloud chamber.

Blackett and Occhialini (1932, 1933) made a great leap in the study of cosmic radiation by developing the counter-controlled cloud chamber. Anderson et al. (1934) also reported the construction and use of a similar apparatus for the study of cosmic-ray showers. The apparatus utilized the coincidence circuitry developed by Nobel Laureate Walther Bothe in collaboration with Hans Geiger and Bruno Rossi at the Physikalisch-Technische Reichsanstalt in Berlin-Charlottenburg (Bothe, 1926, 1954). A coincidence circuit is an electronic device that analyzes the outputs of two or more radiation detectors, and when the electronic signals from the detectors are received by the coincidence circuit within a specified period of time (generally within a few microseconds), known at the coincidence time or coincidence window, the signal is accepted. If, for example, a cosmic-ray particle passes through two detectors, for example, two Geiger–Müller counters, each of the detectors will produce simultaneously an output signal. The signal from each detector will pass the coincidence circuit, and the output signal from the coincidence circuit can be used to trigger any device such as a radiation counter or a cloud chamber and its camera. Any signal received from only one of the detectors generally would be rejected by the coincidence circuit, because two signals from the two detectors would need to occur in coincidence within the very short time window of a few microseconds. A coincidence circuit was employed to trigger the detection of a cosmic-ray particle and the simultaneous cloud-chamber expansion and photography of its track. An example of a coincidence circuit linked to a cloud chamber and camera is illustrated in Figure VI.9.

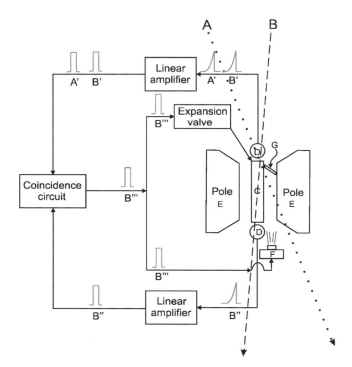

Figure VI.9   A schematic illustration of a counter-controlled cloud chamber apparatus. Two cosmic rays (A and B) are illustrated as traveling through the apparatus from top to bottom. The heart of the apparatus is the cloud chamber (C), which is in the vertical position. A typical cloud chamber would measure approximately 17 cm in diameter and 3 cm in depth fully expanded. Above and below the cloud chamber are Geiger–Müller counters (D) illustrated as circular tubes. The poles of an electromagnet (E) are facing the cloud chamber. Cosmic ray A is illustrated as passing through the Geiger–Müller counter above the cloud chamber and through the right-hand corner of the cloud chamber. Cosmic ray A, produces ionization in the Geiger–Müller counter resulting in an electric pulse A′ shown as a tail pulse traveling in the direction of the linear amplifier. Cosmic ray B travels through both Geiger–Müller counters and directly through the cloud chamber. The cosmic ray is detected by both Geiger–Müller counters producing electric tail pulses B′ and B″ illustrated traveling simultaneously toward the two linear amplifiers. The pulses are shaped after passing the linear amplifier. Pulse A′ is rejected by the coincidence circuit, because it was not received as two pulses arriving simultaneously at the circuit with a preset resolving time (window) of generally a few microseconds. Pulses B′ and B″, which arrive simultaneously at the coincidence circuit, are accepted; and the output of the coincidence circuit is a single pulse B‴. This pulse, resulting from cosmic ray B, travels simultaneously to an expansion valve to trigger the cloud chamber expansion and to a camera (F) to trigger its shutter and the illumination needed to photograph the tracks in the chamber. A mirror (G) located between the horizontal face of the cloud chamber and a pole of the magnet reflects the image of the cosmic-ray tracks in the direction of the camera lens.

The significance of the development of the counter controlled cloud chamber was underscored by Blackett in his Nobel Lecture (1948a) when he stated

> The method used, that of making an expansion of a cloud chamber at a random time and taking the chance that one of the rare cosmic rays would cross the chamber during the time of sensitivity—generally less than ¼ second—was much consuming of time and photographic film, since in a small chamber only some 2% to 5% of photographs showed cosmic-ray tracks.
> Occhialini and I set about, therefore, the devising of a method of making cosmic rays take their own photographs, using the recently developed 'Geiger–Müller counters' as detectors of

the rays. [Walter] Bothe and [Bruno] Rossi had shown that two Geiger counters placed near each other gave a considerable number of simultaneous discharges, called coincidences, which indicated in general the passage of a single cosmic ray through both counters. Rossi devised a neat valve circuit by which such coincidences could easily be recorded.

Occhialini and I decided to place Geiger counters above and below a vertical cloud chamber, so that any ray passing through the two counters would also pass through the chamber. By a relay mechanism, the electric impulse from the coincident discharge of the counter was made to actuate the expansion of the cloud chamber, which was made so rapid that the ions produced by the ray had no time to diffuse much before the expansion was complete. Having made the apparatus ready, one waited for a cosmic ray to arrive and take its own photograph. Instead of a small fraction of photographs showing a cosmic ray track, as when using the method of random expansion, the counter-controlled chamber yielded a cosmic-ray track on 80% of the photographs.

A detailed description of the design and technique of the counter-controlled cloud chamber is given by Blackett (1934). The expansion of the cloud chamber would take only approximately 0.005 sec to travel its full movement of 1 cm. Expansion was required to get the super saturation of the water or alcohol in air to yield visible tracks by the collection of droplets along the paths left by the cosmic-ray particles. About 1/100 sec after the cloud-chamber expansion, the camera and illumination flash would be triggered. The flash would last 1/30 sec. Initial magnetic field strengths of 3000 G were used; however, high-energy cosmic rays would maintain straight paths. The magnetic field strengths were increased to 14,000 G to produce a curvature and direction to the paths of the charged particles facilitating the identification of their mass, charge, and energy (Blackett, 1936; Blackett and Brode, 1936).

With the newly developed counter-controlled cloud chamber Blackett confirmed Anderson's discovery of the positive electron, and he demonstrated the existence of the cascading effect of the collisions of cosmic rays with atomic nuclei of the atmosphere producing a multitude of charged particles, which were named "the cosmic-ray showers". In his words during his Nobel Lecture, Blackett explained

During the late autumn of 1932, Occhialini and I, using our new counter-controlled cloud method, accumulated some 700 photographs of cosmic rays, among which grouped or associated rays were so striking a feature as to constitute a new phenomenon and to deserve a name. From their appearance they came to be known as "showers" of cosmic-ray particles.

Often half of the rays in the cosmic-ray showers were positively charged and the other half negatively charged particularly when the cosmic radiation passes through a high atomic number absorber such as a barrier of lead placed in the outer perimeter of the cloud chamber. He also observed that the mass of the positively charged particles were close to the mass of the electron as determined by their range and specific ionization in the cloud chamber. Patrick Blackett was first to conclude that these positive and negative electrons were due to the phenomenon of "pair production" theorized by Dirac. In his Nobel Lecture, Blackett (1948a) reflected on his finding as follows:

So not only was Anderson's discovery of the positive electron further confirmed by a wealth of evidence, but it was proved that the newly discovered particles occurred mainly in showers along with approximately an equal number of negative electrons. This fact of the rough equality of numbers of positive and negative electrons, and the certainty that the former do not exist as a normal constituent of matter on earth, led us inevitably to conclude that the positive electrons were born together in collision processes initiated by high-energy cosmic rays. The energy required to produce such a pair is found from Einstein's famous equation to be $2mc^2 \cong 1$ MeV. So was demonstrated experimentally for the first time the transformation of radiation into matter.

Blackett thus explained that the abundant cosmic gamma radiation is converted in large part to positron–electron pairs, in the cosmic-ray showers of our atmosphere *via* pair production, and that finally the positron or antimatter of the electron, which was only previously theorized by Dirac to exist, was now a known entity. The equations describing pair production and the derivation of the energy required for it to occur (a minimum gamma-ray energy of 1.02 MeV) are given in eqs. (VI.1) and (VI.2) of the previous biographical sketch on Carl Anderson and in eqs. (3.28)–(3.32) of Chapter 3. In his Nobel Lecture, Patrick Blackett went further by explaining also his observation, for the first time, of the phenomenon called "annihilation", where matter is changed into electromagnetic gamma radiation. He explained his historic observations as follows:

> Dirac's theory predicted that a positive electron would disappear by uniting with a negative electron to form one or more quanta. Occhialini and I suggested that the anomalous absorption of hard gamma rays by nuclei might be the result of the process of pair production, and that the observed re-emission of softer radiation might represent the emission of two 0.5 MeV quanta resulting from the annihilation of a positive and negative electron. Subsequent work has confirmed this suggestion.

From Einstein's equation $E = mc^2$, Brackett knew that the energy equivalent of the mass of an electron was 0.51 MeV (see eqs. (3.28)–(3.30) of Chapter 3 for the calculation of the conversion of the electron rest mass of 9.109×10^{-31} kg to its energy equivalent of 0.51 MeV). Since the positron and electron are of the same mass, but opposite in charge, they each would possess the energy equivalence of 0.51 MeV, and when the positive electron comes to rest, it encounters an electron and becomes annihilated with the liberation of two quanta of 0.51 MeV gamma radiation or

$$e^+ + e^- = 2h\nu = 2(0.51\,\text{MeV}) \tag{VI.17}$$

The two 0.51 MeV gamma rays from positron–electron annihilation are emitted in opposite directions.

By 1933, Blackett had made the major discoveries cited above for which he was awarded the Nobel Prize. He became Professor at Birkback College, London in 1933 where he continued his research on cosmic radiation. In 1937, he succeeded Sir Lawrence Bragg at Manchester University at the post previously held by Ernest Rutherford, where Brackett established his school of cosmic-ray research.

During the Second World War, Blackett helped establish a research effort to combat U-boats and became Director of Naval Operational Research at the Admiralty. He also worked on antiaircraft defense of England during the blitz. Mary Jo Nye (2004) writes in a biography on Blackett "Outside physics, Blackett was well known for his role in developing the scientific field of operational research, and he is regarded as the 'father' of operational research following his critical role on its diffusion throughout the British military command structure in World War II (Kirby, 2003). Although he had been a youthful naval officer in the First World War, he became a hero in the Second World War, credited with determining operational strategies for the antisubmarine campaign that turned the tide of the European war by late 1943".

After the war in 1945, Blackett resumed his research on cosmic radiation at the University of Manchester. In his Manchester laboratory, Rochester and Butler (1947) discovered the neutral kaon in the cosmic-ray showers. The particle has a mass of about 490 MeV/c^2. It is a product of collision of a high-energy cosmic-ray particle with an atomic nucleus in the atmosphere. Cosmic radiation was the current natural atom-smasher and creator of subatomic particles in the cosmic-ray showers. The kaon is discussed in more detail in the previous biographical sketch of Carl Anderson.

It is not unusual in scientific research for discoveries to be made independently and almost simultaneously, that is, within weeks of each other, by two research groups at opposite sides of the globe. When the time is right in science and the stage is set with previous discoveries, new findings will be made. Such was the case of Clinton Davisson in the USA and George Paget Thomson in England. Both had discovered experimentally the diffraction of electrons. Both had published their findings in the same volume of the journal *Nature* in 1927 and both shared the Nobel Prize for their discovery. Fate is not always so rewarding. In the case of the discovery of the positron, Carl Anderson was first to report the discovery in late 1932 and was thus rewarded the Nobel Prize for the discovery of this elementary particle. Some controversy arose in the British press over this, as Blackett and Occhialini waited until February 1933 to publish the discovery. In an essay McLean (2002) writes about Blackett

In late 1932, Carl D. Anderson at Caltech published an observation of a positively charged particle, with mass smaller than a proton, in cosmic radiation…he initially characterized its production as a rare event. By February 1933, Blackett and Occhialini had completed a paper summarizing their analysis of some 500 tracks of cosmic-ray particles, of which 14 tracks were evidence of the antielectron or positive electron, which they explicitly linked to Dirac's relativistic electron theory. Some physicists thought it unfortunate that Blackett and Occhialini appeared to have delayed publication, in order to get firmer data on the positive electron within the framework of Dirac's theory, so that it was Anderson who received the Nobel Prize in Physics in 1936 for the "discovery of the positron".

Concerning the credit for the discovery of the positron Nye (2004) writes in her authoritative biography of Blackett

One of the great controversies over the awarding the Nobel Prize in Physics entered on Blackett's work, not in 1948, when he received the Prize, but in 1936 when he did not. The issue was that Carl Anderson alone received recognition in 1936 for discovery of the positron or positively charged electron, and British physicists continued to discuss the failure to recognize Blackett even after Blackett singly received the 1948 Prize…Blackett emphasized that Anderson at Pasadena had previously found tracks that seemed to indicate the existence of a positive electron. Only in one interview did Blackett add that his and Occhialini's work had been in progress for some months before they published their results from the Cavendish Laboratory in early 1933…What everyone acknowledged and admired in this matter was that Blackett behaved impeccably and did not himself engage in disputes about recognition for the positron.

Blackett was very outspoken in science and politics. Nye (2004) writes that during the 1940s and 1950s Blackett was increasingly committed to the development of science and technology in India and the Third World, which proved controversial. He also pleaded for setting up a special foreign-aid fund for former colonies in the Third World, which evoked anger and ridicule from many of his former colleagues. She explains, however, that "by 1964 the British political climate changed, and Blackett, who had been meeting with a scientific group advising the Labour Party's leadership, became scientific advisor in Harold Wilson's Ministry of Technology". Blackett had foresight in foreign affairs and the necessary ingredients to world peace by his push for massive foreign aid to Third World countries, and the following quotation, attributed to Blackett (Lovell, 1975) reflects current bilateral and international foreign aid programs:

The uneven division of power and wealth, the wide differences of health and comfort among the nations of mankind, are the sources of discord in the modern world, its major challenge and, unrelieved, its moral doom.

Patrick Blackett became Head of the Department of Physics at the Imperial College of Science and Technology, London in 1953. During the last decade of his scientific career he and a group under his direction studied magnetism in rocks with the objective of determining the history of the earth's magnetic field. The work at the Imperial College under his direction and in other laboratories provided evidence for the continental drift that has occurred over the course of geological history (Nobel Lectures, 1964).

Patrick Blackett was the author of several books including *Military and Political Consequences of Atomic Energy* (1948b), *Lectures on Rock Magnetism* (1954), *Studies of War* (1962), and *Science, Technology and Aid in Developing Countries* (1971). He retired in 1963 and died in London on July 13, 1974.

HIDEKI YUKAWA (1907–1981)

Hideki Yukawa was born in Tokyo, Japan on January 23, 1907. He was awarded the Nobel Prize in Physics 1949 "for his prediction of the existence of mesons on the basis of theoretical work on nuclear forces". The mesons were first theorized in 1935 by Yukawa as a particle, which could account for the strong interactions that hold protons and neutrons together in the atomic nucleus. The mesons were discovered in 1947 in the cosmic radiation showers by Cecil F. Powell and Guiseppe P. S. Occhialini.

Hideki Yukawa (1907-1981)

Hideki Yukawa grew up in Kyoto and graduated from Kyoto University in 1929. Since graduation he had continued his studies and research in theoretical physics with emphasis on elementary particles. During 1932–1939, he was Assistant Professor and Lecturer at Osaka University as well as Lecturer at Kyoto University.

At the very young age of 27 from Osaka University he took a bold step to publish (Yukawa, 1935) a revolutionary theory of nuclear forces that predicted the existence of the meson, an elementary particle or gluon that serves as an exchange particle of strong interaction holding protons and neutrons together in the nucleus. The underlying question was what type of force strongly binds the protons and neutrons together in the atomic nucleus? The major forces, which physicists were already well familiar with, were the gravitational and electromagnetic forces. The reasoning used by Yukawa in his conception of the meson was provided in his Nobel Lecture of December 12, 1949 when he stated the following:

The meson theory started from the extension of the concept of the field of force so as to include the nuclear forces in addition to the gravitational and electromagnetic forces. The necessity of introduction of specific nuclear forces, which could not be introduced to electromagnetic interactions between charged particles, was realized soon after the discovery of the neutron, which was to be bound strongly to the protons and other neutrons. As pointed out by Wigner in 1933 [Eugene Wigner, Nobel Laureate in Physics 1963], specific nuclear forces between two nucleons, each of which can be either in the neutron state or the proton state, must have a very short range of the order of 10^{-15} m, in order to account for the rapid increase of the binding energy from the deuteron [nucleus of one proton and one neutron] to the alpha-particle [nucleus of two protons and two neutrons]. The binding energies of nuclei heavier than the alpha-particle do not increase as rapidly as if they were proportional to the square of the mass number A, *i.e.*, the number of nucleons in each nucleus, but they are in fact approximately proportional to A [See Fig. 4.4 of Chapter 4]. This indicates that nuclear forces are saturated for some reason.

In his Nobel Lecture, Yukawa (1949) explained further that certain concepts of Heisenberg and Fermi lead him to the meson theory. Heisenberg (1932a,b, 1933c) had suggested that a force could be assumed between a neutron and proton due to an exchange of an electron or more generally an electric charge, as in the case of a chemical bond between a hydrogen atom and a proton [e.g., H · H$^+$, the chemical bond, depicted by the one dot between the hydrogen atom and proton, consists of a sharing of the electron between two hydrogen atoms or, in the words of Dunne (2002) "in this molecule the binding forces are associated with the continuous migration of the single electron from one atom to the other"]. Yukawa added that shortly after Heisenberg's theory was published Fermi (1934a) developed a theory of beta-decay based on the hypothesis of Pauli, which a neutron, as an example, could decay into a proton, an electron, and a neutrino, a neutral particle of near zero mass. Yukawa explained that he searched for the particle that would represent the charge and binding force shared between the proton and neutron in the atomic nucleus.

In his Nobel Lecture, Yukawa explained that the above-mentioned concepts of Heisenberg, Pauli and Fermi…"gave rise, in turn, to the expectation that nuclear forces could be reduced to the exchange of a pair of an electron and a neutrino between two nucleons, just as electromagnetic forces were regarded as due to the exchange of photons between charged particles. It turned out, however, that the nuclear forces thus obtained were much too small (Tamm and Ivanenko, 1934), because the beta decay was a very slow process compared with the supposed rapid exchange of the electric charge responsible for the actual nuclear forces. The idea of the meson field was introduced in 1935 in order to make up this gap".

Hideki Yukawa proposed his concept of the strong force holding protons and neutrons together as a particle (the name meson for this particle came later) with positive or negative charge that would be continuously emitted and absorbed by the proton and neutron, respectively. In other words, the meson would be an exchange particle that would hold the protons and neutrons as a unit in the atomic nucleus. In his words Yukawa explained

In order to obtain exchange forces, we must assume that these mesons have the electric charge, and that a positive (negative) meson is emitted (absorbed) when the nucleon jumps from the proton state to the neutron state, whereas a negative (positive) meson is emitted (absorbed) when the nucleon jumps from the neutron to the proton. Thus a neutron and a proton can interact with each other by exchanging mesons just as two charged particles interact by exchanging photons.

Yukawa calculated the mass of the meson to be equivalent to approximately 200 electron masses. He knew that the nuclear forces between the proton and neutron states were of short range and of the order of 10^{-15} m, which would be the distance that the meson would travel when jumping from proton to neutron or in the reverse direction. In harmony with the Heisenberg uncertainty principle, the creation and disappearance of an exchange particle such as the meson would be possible. The Heisenberg uncertainty principle in terms of energy and time according to eq. (V.16) can be expressed as

$$\Delta E \Delta t \geq \frac{\hbar}{2} \tag{VI.18}$$

The existence of the meson as an exchange particle, because of its additional mass, would break the laws of energy conservation ($E = mc^2$), if the meson existed for a time that would exceed

$$\Delta t \approx \frac{\hbar}{2\Delta E} \tag{VI.19}$$

The meson could travel from a proton to another proton or as an exchange particle between proton and neutron if its lifetime (Δt) during that exchange was short enough to satisfy Heisenberg's uncertainty principle. Since the meson would travel at a speed less than that of light, the distance of travel of the meson would be defined by the term $c\Delta t$. Also the energy of the meson would be defined by Einstein's $\Delta E = mc^2$. Then eq. (VI.19) can be written as

$$d \approx c\Delta t = \frac{\hbar c}{2\Delta E} = \frac{\hbar c}{2mc^2} \tag{VI.20}$$

and

$$mc^2 \approx \frac{\hbar c}{2d} \approx \frac{(1.054 \times 10^{-34} \text{ J sec})(2.99 \times 10^8 \text{ m /sec})}{(2)(1 \times 10^{-15} \text{ m})} \approx 1.57 \times 10^{-11} \text{ J} \tag{VI.21}$$

or

$$mc^2 \approx \frac{1.57 \times 10^{-11} \text{ J}}{1.60 \times 10^{-19} \text{ J eV}^{-1}} \approx 100 \text{ MeV} \tag{VI.22}$$

and

$$m \approx 100 \text{ MeV}/c^2 \tag{VI.23}$$

which is the estimated mass of the meson in energy units. Since the mass of the electron $m_e = 0.51$ MeV/c^2, the mass of the meson, compared to the electron mass, would be

$$\frac{m}{m_e} = \frac{100 \text{ MeV}/c^2}{0.51 \text{ MeV}/c^2} \approx 200 \times \text{electron mass} \tag{VI.24}$$

Yukawa estimated the mass of his theorized meson to be approximately 200 times that of the electron. His revolutionary theoretical work on the origin of the strong nuclear forces is commemorated in the postage stamp illustrated on page 368, which was issued in Japan in 1985. Yukawa's meson was discovered finally in 1947 by Cecil Powell and Guiseppe Occhialini (Occhialini and Powell, 1947; Lattes *et al.*, 1947) at the University of Bristol in the cosmic radiation showers. Much interest in Yukawa's discovery ensued after the Second World War, which opened up an entire new field in the study of nuclear forces. The words of Professor I. Waller, member of the Nobel Committee on Physics, in the Nobel Award Presentation Speech on December 10, 1949, underscored the importance of Yukawa's work when he stated

...According to Yukawa's theory, the nuclear forces can be traced back to the exchange of mesons between the nucleons. These are continually emitting and absorbing mesons. Yukawa also studied the important question of whether the mesons can appear outside the nuclei. He found that the mesons can be created during the interaction of nucleons if these can deliver a sufficient amount of energy. Therefore, mesons cannot be created in ordinary nuclear reactions. Yukawa emphasized, however, that they can be expected to appear in the cosmic radiation, in which particles of great energy are found...Prof. Hideki Yukawa, in 1934, when you were only 27 years old, you boldly predicted the existence of new particles, now called "mesons" which you anticipated to be of fundamental importance for the understanding of the forces acting in the atomic nucleus. Recent experiments have provided brilliant support for your essential ideas. These ideas have been exceedingly fruitful and are a guiding star in present day theoretical and experimental work on atomic nuclei and on cosmic rays.

The name meson, attributed to Yukawa's particle, is derived from the Greek word for "middle", as the meson acts as a mediator or go-between permitting one particle to change into another thereby acting as a strong force holding the particles together in the nucleus. The meson predicted by Yukawa and discovered in the cosmic radiation showers is called the π-meson or pion. Yukawa predicted that the π-meson would come in two charged forms as noted earlier. The positive and negative forms (i.e., the π^+ and π^-) were discovered in the cosmic radiation showers, which are products of high-energy collision of cosmic-ray particles with nuclei of the earth's atmosphere. The mass of the charged pions are 139.6 MeV/c^2 and they have a short lifetime of 2.6×10^{-6} sec. More information on the abundance and properties of the pions are provided in Chapter 6 on cosmic radiation. First evidence for the neutral pion (π^0) was provided by Bjorklund et al. (1950) at the Lawrence Berkeley Laboratory from the gamma-ray photon emissions following the bombardment of various targets with protons accelerated to energies of 180 MeV. The neutral pion decays to two gamma-ray photons. It has a mass of 135 MeV/c^2 and has a lifetime of 8.3×10^{-17} sec (see Chapter 6). Confirmation of the neutral pion was provided by Carlson et al. (1950) by observing the tracks in photographic emulsions of scattered pairs of fast electrons produced by interactions with gamma-ray photon pairs, which are products of π^0 decay. These observations were made at an altitude of 70,000 ft, which provided evidence that the π^0 was produced in the cosmic radiation showers. Additional evidence for the π^0 was provided the following year by Panofsky et al. (1951) at the Lawrence Berkeley laboratories by measuring the gamma-ray photon emission following the bombardment of targets with protons accelerated to energies of 330 MeV.

Prior to receipt of the Nobel Prize and only about 3 years after the surrender of Japan in World War II Yukawa was Visiting Professor at the Institute of Advanced Study at Princeton University during 1948–1949. In July of 1949 he became Visiting Professor at Columbia University in New York until 1950. The importance of Yukawa's work and his international collaboration with the USA shortly after the war was a great example for peace and international cooperation. This was underscored in the following words of Carl Skottsberg, President of the Royal Academy of Sciences, on December 10, 1949 on the occasion of the Banquet Speech following the awarding of the Nobel Prize to Professor Hideki Yukawa

Your fellow-workers in atomic research recognize the high intrinsic value of what you have done, and that's the reason why you are here today, the first Japanese to receive a Nobel prize. You have told us already what this means, not only to you personally, but also to scientific life in Japan, an enlivening stimulation in these days of great distress. We can understand this. And are you not the very best example of the importance of science in bringing nations and races together? Not many years ago Japan and the United States stood against each other, armed to the teeth, bent upon destruction. Today you work happily in the midst of American colleagues.

In 1932, Hideki Yukawa became the first Chairman of the Yukawa Institute for Theoretical Physics at Kyoto University. He was among the ten leading scientists and intellectuals that signed the Russell–Einstein Manifesto in 1955 discussed previously in this book. Hideki Yukawa died on September 8, 1981 in Kyoto, Japan.

CECIL F. POWELL (1903–1969)

Cecil Frank Powell was born in Tonbridge, Kent Southeast of Greater London on December 5, 1903. He was awarded the Nobel Prize in Physics 1950 "for his development of the photographic method of studying nuclear processes and his discoveries regarding mesons made with this method".

Cecil Powell grew up in humble surroundings. His grandfather was a successful and crafted gunsmith and his father had inherited the company and craft. However, the business went bankrupt after his grandfather spent considerable funds fighting a lawsuit over a hunting accident and the eventual loss of business with the advent of the industrial manufacture of guns making the handicraft production of guns less economic. His father had to work as a clerk in his brother's electrical business, and they had to take up lodging in their residence to make ends meet. In his autobiography Powell (1987) relates

> My parents' life was hard and my mother was very anxious that her children should escape from the drudgery, which had been her lot. Her brother…had succeeded in getting into Trinity College, Cambridge, and had become an electrical engineer. My mother determined that I should secure a similar emancipation.

Cecil Powell (1903-1969)

We can never underestimate the important role of our primary and secondary school teachers. Often it is our school teacher in our formative youth, who will impress us to the extent, that he or she helps us find the vocation where we have the greatest potential to make a contribution to society. This is clear in the following statement from Powell's autobiography:

> I was fortunate that at Judd School I came under the influence of a physics master who was devoted to his profession and an able teacher.…the experiments in his class were well designed and gave scope for originality and skill. I used to be allowed to spend a lot of time there on my own. And he was a man of some taste and originality.

Powell did well in secondary school and, after two attempts, won a scholarship in 1921 to Sidney Sussex College, Cambridge. C.T.R. Wilson, who would become Nobel Laureate in Physics 1927, was a Fellow at the same college when Powell began his studies there. It was difficult for Powell to adjust to the transition from a small town to a big university, but he adjusted and graduated with honors. He thought of becoming a school teacher and was accepted to a teaching post at Uppingham School; however, after serious thought about his future calling and ambitions, he approached Ernest Rutherford in 1925 about the possibility of becoming a research student at the Cavendish. Rutherford agreed and arranged for C.T.R. Wilson to be his supervisor. The Cavendish was bustling with new scientific developments at the time Powell entered as a research student. Charles Eryl Wynn-Williams (1931, 1932) was developing automatic electronic scalers for the counting of radiation emissions, Patrick Blackett (1934) was building his automatic counter-controlled cloud chamber described previously, and John Cockcroft and Ernest Walton (1932a,b), who would share the Nobel Prize in Physics 1951, were developing methods for the acceleration of protons to high energies for experimentation on proton collisions with atomic nuclei. The environment at the Cavendish was perfect for any young budding scientist like Powell. His initial research under C.T.R. Wilson was the development of an improved all-glass cloud chamber for the study of the nature of condensation by different degrees of expansion of dust-free air. In his autobiography, Powell relates how impressed he was that Wilson would ever have attempted to make successfully a cloud chamber 20 years earlier that would be capable of observing the paths of charged subatomic particles. Such work taught Powell the importance of perseverance, as Powell relates in his autobiography

> I always found it remarkable that Wilson should ever have attempted to make such a device [cloud camber] but he succeeded 20 years before: Rutherford used to tell a story that on one occasion just before he went on a long vacation, he had left Wilson in the attempt, sitting glass-blowing at the blow-pipe and pedaling the foot-bellows. When he returned Wilson was in the same position and seemed to be engaged in exactly the same task.

Powell's work on the cloud chamber led him to the discovery of the nature of steam discharge through nozzles, which had some impact on the design of steam turbines. He received his doctorate degree in 1927 and, in 1928, moved to the University of Bristol as Research Assistant to A. M. Tyndall to study the mobility and nature of ions in gases. In 1935, he joined an expedition to Montserrat in the West Indies to study seismic activity on the island. That same year he became very interested in nuclear physics and decided to return to the Cavendish, which was bustling with new discoveries. It was an opportune time at the Cavendish, as Blackett and Occhialini had discovered pair production in 1932, Chadwick discovered the neutron the same year, and Cockcroft and Walton successfully caused the disintegration of lithium by accelerated protons also in 1932. Powell wanted to study particle scattering in a Wilson Cloud Chamber following the bombardment of light elements, such as lithium, beryllium, boron, carbon, and oxygen, by protons and deuterons with a Cockcroft and Walton accelerator. However, research in Vienna by Marietta Blau (1894–1970) and her assistant Hertha Wambacher (1903–1950) at the Radium Institute in Vienna had demonstrated that photographic emulsions could be used to detect subatomic particles, which could be more versatile than the cloud chamber. Powell relates in his autobiography

> The original intention was to study the scattering of fast neutrons by protons using a Wilson Chamber filled with hydrogen …But about this time W. Heitler, who had been in Bristol for some years, pointed out that Blau and Wambacher had successfully used 'halftone' photographic emulsion to detect particles in the cosmic radiation and, since the method had the advantage of extreme simplicity, he thought we might begin by sending similar plates on to a mountain to see if we could simulate the Viennese results.

The author would like to divert from the life and work of Powell to briefly touch on the life and contributions of Marietta Blau, whose work on photographic emulsions helped pave the way for some of Powell's discoveries. She was born in Vienna in 1894 to a Jewish family and, after Lise Meitner, was among the first women to study physics at the University of Vienna where she was awarded the doctorate degree in 1919. She did research without pay at the Radium Institute, University of Vienna from 1923 to 1938. She was forced to flee Austria in 1938 after the annexation of Austria by Germany. Marietta Blau was the supervisor for the doctorate studies of Hertha Wambacher at the University of Vienna. After completion of her doctorate, Wambacher joined Blau in research, and they had a fruitful collaboration for about 6 years mostly in the development of the emulsion technique for the detection and measurement of subatomic particles originating from cosmic-ray collisions with atomic nuclei in the earth's atmosphere (Blau and Wambacher, 1935, 1937; Blau, 1938). The work by Blau and Wambacher in 1937 yielded the detection of numerous images on photographic emulsions called "stars", as these images showed the track of a high-energy cosmic-ray particle colliding with an atomic nucleus in the emulsion resulting in numerous tracks leading from the point of impact. As many as 12 tracks would be observed as straight lines stretching out in many directions from the point of impact of the cosmic-ray particle with the atomic nucleus. These tracks illustrated the shattering of an atomic nucleus into as many as 12 particles. Nobel Laureate Max Born credited Blau and Wambacher with the discovery of the photographic emulsion technique for the detection and measurement of tracks left by subatomic particles when he noted the following in his book published in 1969:

> Another great advance was made by two Viennese ladies. Misses Blau and Wambacher (1937), who discovered a photographic method of recording tracks of particles. The grains of emulsion [i.e., silver bromide grains suspended in an emulsion of gelatin] are sensitive not only to light but also to fast particles; if a plate exposed to a beam of particles is developed and fixed, the tracks are seen under the microscope as chains of black spots.

Erwin Schrödinger nominated Blau two times for the Nobel Prize (Moore, 1989; Perlmutter, 2000). Blau escaped from Austria with the assistance of Albert Einstein, and went to Mexico where she worked as Professor at the National Polytechnic University in Mexico City from 1939 to 1944. She held positions in industry during 1944–1948 in New York and was a Research Physicist at Columbia University from 1948 to 1950. Her career continued as Associate Physicist at the Brookhaven National Laboratory from 1950 to 1955 and finally Associate Professor at the University of Miami from 1955 to 1960 (Byers, 2000). While at Columbia and Brookhaven Blau did pioneering work on the study of high-energy particle interactions with atomic nuclei (Blau and De Felice, 1948; Blau et al., 1953; Blau and Caulton, 1954; Blau, 1956; Blau and Oliver, 1956). Like Lise Meitner she was born in an era, which required great fortitude and perseverance for a woman to overcome many obstacles in scientific research. Her Jewish heritage added to the hurdles and tortuous obstacles she would have to overcome during her productive years, which kept her from achieving her maximum potential. Marietta Blau returned to her native Austria in 1960 at the age of 66 and died of cancer in Vienna in 1970. For additional reading on the life and scientific accomplishments of Marietta Blau the reader is invited to peruse a biography by Rosner and Strohmaier (2003), an extensive treatise by Perlmutter (2000), a book chapter by Halpern (1993), and a book chapter and article by Galison (1997a,b).

Cecil Powell grasped the photographic emulsion technique (or nuclear emulsion technique) pioneered by Blau amd Wambacher and improved its application to study nuclear disintegrations produced by artificially accelerated beams of protons and deuterons and later to nuclear disintegrations produced by collisions of cosmic-ray particles. The purpose of the technique is to detect and measure atomic particles (i.e., measure their mass, charge, and energy) by the tracks they leave behind in nuclear emulsions. The emulsion was a suspension of silver bromide crystals in gelatin supported on a transparent glass plate, currently transparent plastic plates are used. Charged particles produce ionization in the emulsion as the particle travels through the emulsion or until it comes to a stop within the emulsion. The ionization that occurs within the silver bromide crystals yields electrons that are free to travel from one ion to another in the silver bromide until they are trapped at sites, which consist of defects within the silver bromide crystal structure. At a structural defect a silver ion absorbs the electron to become a silver atom. The crystals affected in this way by nuclear radiation become visible only after the application of a chemical "developer", which reduces all of the silver ions of the affected crystals to metallic silver (observed as black grains or spots). The crystals, which do not contain reduced silver, are dissolved out of the emulsion by the chemical "fixer" leaving a black image produced by the nuclear radiation. A series of dark grains in the film emulsion denotes the track left by the nuclear particle. The intensity of the dark grains or the thickness of the track is a function of the specific ionization caused by the particle. Thus, the thickness of a particle track in a nuclear emulsion is proportional to the mass and charge of the particle, as particles of higher mass and charge produce higher ionization in matter. However, the thickness of a particle track will be inversely proportional to the particle velocity, because a particle produces more ion pairs as it slows down and before it comes to rest (see Figure 1.4 of Chapter 1). To observe the particle tracks in nuclear emulsions it is necessary that the photographic film be wrapped in opaque paper to protect it from light, which would expose and darken the film. After exposure to nuclear radiation the protective wrapping is removed in a darkroom, and the film treated with chemical developer and fixer to yield dark particle tracks on a clear background. For additional reading on the theory and practice of nuclear emulsions see Gurney and Mott (1938), Barkas (1963), Rogers (1979), Mundy *et al.* (1983), and L'Annunziata (1987).

In his autobiography, Powell (1987) explained clearly the advantages of nuclear emulsions over the cloud chamber to detect and measure atomic particles. These advantages were: (1) nuclear emulsions are continuously sensitive, that is, as a detector the emulsion is sensitive throughout the duration of its exposure to the nuclear particles until the time of chemical development of the emulsion, unlike the cloud chamber, which is sensitive only for a few milliseconds at the instant of expansion of the chamber; (2) nuclear emulsions can record with precision the point of entry of a particle and its direction of motion; (3) the nature of the particle (e.g., mass and charge) and its range and energy can be determined by observations on the character of the tracks (e.g., some nuclear particles can come to a full stop in the emulsion and their range measured, and the thickness and shape of the track can characterize the speed, mass, and charge of the particle). The versatility of nuclear emulsion was demonstrated by Powell and his coworkers, and this was underscored by Professor A. E. Lindh, member of the Nobel Committee, in his Nobel Presentation Speech given on December 11, 1950. The following is an excerpt:

> The photographic method used by Prof. Powell is based on the fact that after an electrically charged particle has passed through a photographic emulsion, the silver bromide grains of the emulsion can be developed, making the path of the particle appear as a dark line which is, actually, a series of blackened grains with longer or shorter intervals between. The distance between the grains is proportional to the speed of the particle; the greater the speed of the particle, the greater the distance, which circumstance is connected with the fact that a swift particle has less power of ionizing than a slow one.

To demonstrate how clearly, the ionizing power of a particle, governed by its speed, mass, and charge can be distinguished by the nuclear emulsion technique Powell, in his Nobel Lecture on December 11, 1950, showed seven particle tracks, illustrated in Figure VI.10, of primary cosmic-ray particles recorded in various emulsions carried by high-flying balloons at altitudes of approximately 95,000 ft. From Figure VI.10 it can be seen that, if the particle speeds are similar, the thickness of the particle track in the emulsion is a function of the particle mass. Powell (1950) noted that experimental data including data of Freier *et al.* (1948) and Bradt and Peters (1948) taken on the frequency of the nuclear tracks made by primary cosmic-ray particles at high altitudes provided much information on the relative abundances and speed of the nuclei of different chemical elements in primary cosmic radiation. He noted the following:

> By observations at great altitudes we now know that the primary cosmic radiation is made up of atomic nuclei moving at speeds closely approaching that of light. It is possible to record the tracks of the incoming particles and to determine their charge; and thence the relative abundance of the different chemical elements. Recent experiments prove that hydrogen and helium occur most frequently, and the distribution in mass of the heavier nuclei appears to be similar to that of the matter of the universe. Thus elements more massive than iron or nickel occur, if at all, very infrequently.

After the Second World War, Cecil Powell joined forces with Italian Physicist Giuseppe Occhialini, who had previously worked with Nobel Laureate Patrick Blackett at the Cavendish Laboratory on cosmic radiation using the Wilson Cloud Chamber that lead to the interpretation of "pair production" and the discovery of the nuclear process of "annihilation" in the cosmic-ray showers. Occhialini left the Cavendish in 1938 to return to Florence, Italy and a few years subsequently went on to the University of São Paulo in Brazil to conduct research on cosmic radiation. During the war he took refuge in the Itatiaya mountain between São Paulo and Rio de Janeiro. Occhialini joined forces with Powell at the University of Bristol in 1945. Together with other researchers including Brazilian physicist Césare Lattes they employed nuclear emulsions to study the collisions of cosmic-ray particles with atomic nuclei.

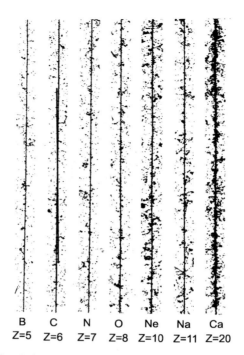

| B | C | N | O | Ne | Na | Ca |
| Z=5 | Z=6 | Z=7 | Z=8 | Z=10 | Z=11 | Z=20 |

Figure VI.10   Examples of tracks in photographic emulsions of primary nuclei of the cosmic radiation moving at relativistic velocities. The tracks produced by particles of higher mass and charge have a higher density of black grains. (From Powell, 1950 with permission from The Nobel Foundation © 1950.)

The experimental setup involved stacking several nuclear emulsions supported on glass plates in a vertical position to improve chances that an incident cosmic radiation would enter one emulsion only and travel through it or even come to a stop in that single emulsion. Only a cosmic-ray incident to the vertical orientation of a plate of nuclear emulsion would have the opportunity to travel through the one emulsion and produce a track in the emulsion. Cosmic radiation that would enter the plates at an acute angle such as scattered cosmic radiation showers would not produce a track. The experimental arrangement is illustrated in Figure VI.11.

Powell and coworkers took two approaches to the study of cosmic radiation with nuclear emulsions, namely: (1) to expose the emulsions at very high altitudes of up to 95,000 ft (about 29 km) by placing the plates in a gondola aboard a large (37 m) hydrogen-filled balloon, or (2) by placing the plates of nuclear emulsion on mountain tops of various altitudes, such as 5500 m in the Bolivian Andes (Lattes *et al.*, 1947), atop the Jungfraujoch in central Switzerland at 3000 m, the Kilimanjaro in Tanzania at 5500 m, and Chacaltaya in Bolivia at 5500 m (Occhialini and Powell, 1948). Powell's objective in exposing nuclear emulsions at very high altitudes was to measure the abundances of the atomic nuclei in the primary cosmic radiation, which cannot be observed at lower altitudes because they undergo collisions and disintegrate with atomic nuclei of the atmosphere. Also, the primary cosmic-ray nuclei arriving from intergalactic space, would interact with atomic nuclei of the nuclear emulsions to produce "stars", which are tracks produced by the disintegrations of the colliding nuclei and the elementary nuclear particles (e.g., alpha particles, neutrons, protons, electrons, and mesons) emitted in many directions from the point of impact (see Figure VI.12). The tracks of these particles could then be studied for the determination of their mass, charge, and energy. On the other hand, at the lower altitudes of mountain peaks, the atmosphere would be free of the primary cosmic nuclei of high

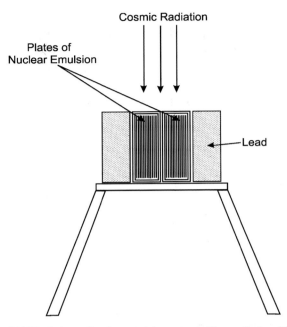

Figure VI.11 Stacks of 24 Ilford plates of nuclear emulsion supported in a vertical position between two blocks of lead shielding. (The Ilford plates came from Ilford Imaging, which is now a worldwide company. The Ilford Company was founded by Alfred Harmon, who made gelatine dry photographic plates in the basement of his house on Cranbook Road in Ilford, Essex, UK in 1879) The plates each measured approximately 2 × 1 cm in area and each contained a layer of silver bromide emulsion approximately 100–250 microns thick. The plates were kept in their original light-tight packing. Stacks of about 12 Ilford plates were housed in 0.5 mm-thick tinned iron layered together in a vertical sandwich position between two lead shields. The lead shielding reduces background radiation from scattered cosmic-ray showers that could enter the plates at angles other than from directly above the plates. (From Camerini *et al.*, 1948, by permission from Macmillan Publishers Ltd., © 1948.)

mass, and the nuclear emulsion exposures would therefore show the interactions of the nuclear particles of low mass and secondary cosmic radiation (i.e., the products of the nuclear collisions of primary cosmic rays with nuclei of the atmosphere) with nuclei of the emulsion. In his Nobel Lecture, Powell (1950) elucidated on his experimental approach based on the differences of cosmic radiation as a function of altitude as follows:

> The detailed study of the 'mass spectrum' of the incoming nuclei [Examples of the tracks of such incoming nuclei are provided in Fig. VI.10] has an important bearing on the problem of the origin of the primary particles; but it is complicated by the fact that, because of their large charge, the particles rapidly lose energy in the atmosphere by making atomic and nuclear collisions. They therefore rarely penetrate to altitudes less than 70,000 ft (~21 km). It is for this reason that exposures at high altitudes are of particular interest.
>
> A second reason for making experiments at extreme altitudes is that the primary nuclei commonly suffer fragmentation in making nuclear collisions. A primary nucleus of magnesium, aluminum [or nitrogen, See Figure VI.12], for example, may decompose into lighter nuclei such as lithium, α particles and protons. The mass spectrum at a given depth is therefore different from that of the primary radiation, and such effects are appreciable at 90,000 ft [27 km, where the air density is low, ~0.05 kg/m^3]. They would be much reduced at 12,000 ft [3.6 km, where the air density is higher, ~0.9 kg/m^3].

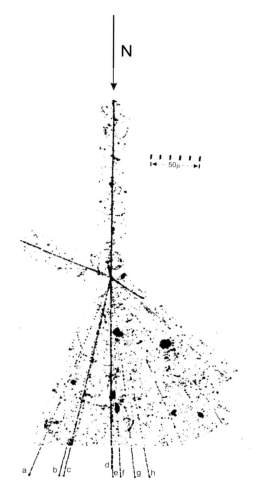

Figure VI.12   A cosmic radiation "star" produced in nuclear emulsion. The name "star" was attributed to the patterns of tracks produced when a high-energy particle collided with an atomic nucleus and the nuclei disintegrated in nuclear emulsions yielding a starlike pattern of particle tracks in many directions. Here an incoming nitrogen nucleus of the primary cosmic radiation collides with a nucleus of the nuclear emulsion, and splits up into a lithium nucleus, an alpha particle, a deuteron, and a number of protons. Some π-mesons are created in the collision. The event was reported by Powell (1950) to be very exceptional, because, by chance, the tracks of many of the particles in the emulsion are long so that they can be identified and their energy determined. Powell (1950) identified the particles that produced the tracks and the energies of the particles. These were at the top of the "star": N, 28,000 MeV; at the bottom: (a) α-particle, 220 MeV; (b) π-meson, 800 MeV; (c) proton, that is, $^1_1$H, 81 MeV; (d) Li, ~10,000 MeV; (e) $^1_1$H, 2000 MeV, (f) deuteron, that is, $^2_1$H, 3800 MeV; (g) π-meson, 1000 MeV, and (h) π-meson, 1200 MeV. (From Camerini *et al.*, 1948, by permission from Macmillan Publishers Ltd., © 1948.)

The primary protons and α-particles, because of their smaller charge, penetrate to much lower altitudes. In collisions they disintegrate the nuclei which they strike [see Figure VI.12] and, in the process, lead to the creation of new forms of matter, the π-mesons of mass 274 m_e (Lattes *et al.*, 1947, Piccioni, 1950, Fowler, 1950). These particles are usually ejected with great speed and proceed downward towards the earth.

In February of 1947, Giuseppi Occhialini and Cecil Powell (Occhialini and Powell, 1947) reported the existence of the π-meson. They observed six "star" formations in nuclear emulsions, exposed to cosmic radiation, which demonstrated the existence of a previously unknown particle of small mass, but of mass greater than the electron mass. In each of the six star-like nuclear disintegrations out of 800 "stars" studied they observed that the disintegrations appeared to be caused by the entry of a slow charged particle, the meson, into the nucleus. They provided evidence by grain density measurements that the meson slowed down and entered a nucleus causing the nucleus to disintegrate. In their words Occhialini and Powell (1947) reported

> The conclusion receives additional support from the observation that the number of grains per unit length of the [meson] track, which can be taken as a measure of the ionization produced by the particle, is greatest in the immediate neighborhood of the disintegrating nucleus and becomes less as we recede from it. [It is a known fact that the ionization per unit length of travel by a charged particle is greatest when the particle is approaching the end of its range as illustrated in Figure 1.4 of Chapter 1]. Powell went on to comment "We have now observed six of these events among a total of eight hundred stars. The probability, in any one case, that a charged particle, unrelated to the star, has, by chance, come to the end of its range within 1 micron of the disintegrating nucleus, is less than 1 in 10^5. We must therefore conclude that the particle entered the nucleus and produced a disintegration with the emission of heavy particles".

About the same time that Powell's group reported tracks produced by this new particle Don Perkins (1947) of the Imperial College in London reported in the same volume of the journal *Nature* the observation of one track of the meson in nuclear emulsion. This track was obtained by exposing the emulsion to cosmic radiation for several hours aboard an aircraft at an altitude of 30,000 ft (9100 m). From the meson track he observed that the meson was absorbed by a nucleus causing the nucleus to explode, as a star-like image was produced by numerous tracks of subatomic particles emitted in various directions from the point of impact. In his Nobel Lecture, Powell (1950)

> The observation of two 'events' [Powell's and Perkin's research] of a similar nature seemed to exclude completely the possibility of a juxtaposition of unrelated tracks and it appeared certain that we were observing the consequences of the capture of a negative meson by a nucleus of an atom in the emulsion and its resulting disintegration.

Shortly after reporting the observation of the π-meson tracks Césare Lattes, Occhialini and Powell reported in October of the same year (Lattes *et al.*, 1947) the observation of pion decay to a muon after examining plates of nuclear emulsion exposed in the Bolivian Andes at an altitude of 5500 m. They found 40 examples of nuclear disintegrations leading to the production of the muon or μ-meson. In their paper they noted

> That many of the observed mesons [π- and μ-mesons] are locally generated in the 'explosive' disintegration of nuclei.

In his Nobel Lecture, Powell (1950) described the origins and fate of the π- and μ-mesons in the cosmic radiation showers, as knowledge derived from the tracks of these nuclear particles in nuclear emulsions exposed at high altitudes to cosmic radiation. Nuclear emulsion exposure times at mountain tops were often of the order of 35 days, after which the emulsions were treated with chemical "developer", chemical fixer, and the tracks studied under the microscope. The origins of the π-mesons already discussed were: (a) the collisions of primary cosmic-ray nuclei with atomic nuclei of the upper atmosphere (above 70,000 ft) resulting in nuclear disintegrations into smaller nuclear particles as illustrated in Figure VI.12, and (b) the absorption of a negatively charged π-meson by an atomic nucleus in the atmosphere, which also results in the explosion of the nucleus into smaller particles, (c) other sources of mesons described by Powell (1950) including high-energy protons and α particles that collide with atomic nuclei in the earth's atmosphere that break up or "smash" the nuclei into numerous particles including protons, neutrons, α particles and mesons. Thus, the forces in inter-galactic space have accelerated primary high-energy cosmic particles to relativistic speeds, which through collisions with atomic nuclei in the earth's atmosphere, smash atoms to produce high-energy secondary particles (e.g., high-energy protons and α particles). These secondaries, in turn, also serve as natural "atom smashers", producing a cascade or shower of protons, neutrons, electrons, pions, muons, and neutrinos in the earths atmosphere (see Figure 6.3 of Chapter 6).

The pions and muons were the new particles observed by Powell and his coworkers by their tracks in nuclear emulsions. In his Nobel Lecture, Powell (1950) discussed the origin and fate of the pions and muons of which the following is an excerpt:

When brought to rest in a photographic emulsion, the positive π-particles decay with the emission of a μ-meson of mass 212 m_e [See Fig. VI.13]. This particle commonly emerges with a constant velocity, so that its range varies only within the narrow limits due to straggling [Notice in Fig. VI.13, that the μ-meson tracks are of relatively constant length, which indicates a relative constant velocity among the particles]. It follows that in the transmutation of the π-meson, the μ-meson is accompanied by the emission of a single neutral particle. It has been shown that this neutral particle is of small rest-mass and that it is not a photon. It is therefore reasonable to assume, tentatively, that it is a neutrino, the particle emitted in the process of nuclear β-decay (see Eqs. 6.19 and 6.20 of Chapter 6 for the equations for charged pion decay to muon and neutrino).

When a negative π-meson is arrested in a solid material, it is captured by an atom, interacts with a nucleus, and disintegrates it (Occhialini and Powell, 1947; Perkins, 1947). It follows that the particle has a strong interaction with nucleons, and in this respect its properties are similar to those predicted for the 'heavy quanta' of Yukawa. [Powell here points out that they have discovered the π-meson as the transitory particle of very short lifetime predicted by Yukawa in 1935. Thus, the π-meson was identified here by Powell as the particle within the nucleus that

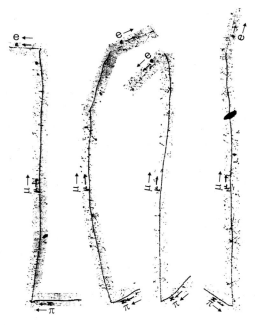

Figure VI.13 Photo-micrographs of four examples of the successive decay π–μ–e as recorded in photographic emulsions. (From Powell, 1950, with permission from The Nobel Foundation 1950.)

acts as a go-between permitting one particle to change into another, such as a proton into a neutron and vice versa acting as a strong force holding the particles in the nucleus together].

When π-mesons are created in nuclear collisions occurring in the atmosphere, they commonly transform, whilst in flight, into μ-mesons and neutrinos. It is these μ-particles, which form the "hard" or "penetrating" component of the cosmic radiation and they are responsible for most of the residual ionization in air at sea level [see Fig. 6.3 of Chapter 6]. The μ-mesons are penetrating because, unlike the π-mesons, they are able to traverse the nuclei with which they collide without interacting with them, and some of them reach great depths underground...

In addition to producing the charged π-mesons, the primary protons in making nuclear collisions also produce neutral π-particles (Bjorklund et al., 1950). The neutral π-mesons are very short-lived and each transforms spontaneously into two quanta of radiation [see eq. 6.15 of Chapter 6]. Such a quantum, when it happens to pass near an atomic nucleus, can in turn transform into a pair of electrons, one positive and one negative; and the electrons can generate new photons in further collisions. A succession of such processes results in the production of the well-known cascades of electrons and photons which form the "soft" or easily absorbed component of the cosmic radiation (Carlson et al., 1950). See Fig. 6.3 of Chapter 6 for an illustration of the cascade of particles in the cosmic radiation showers described by Powell.

One of Powell's assistants, H. Muirhead (see Occhialini and Powell, 1947), made a quantitative study of the grain densities along the tracks of the mesons (later identified as π-mesons) observed in six "star" formations to get an initial estimate of the particle mass. He counted the black grains in the nuclear emulsion to quantify the variation of the grain density along the tracks of protons so that he could extrapolate or predict the distribution of grain density expected for particles of the same charge as the proton but of different mass. His initial estimate of the mass of the new particle was in the range of $100\,m_e$–$230\,m_e$, where m_e is the mass of the electron. This was relatively close to the current figure for the mass of the charged pion, which is $139.6\,MeV/c^2$ or $274\,m_e$, since the ratio of the masses of the pion and electron in energy units is $139.6\,MeV/c^2/0.51\,MeV/c^2$).

Powell and coworkers (Lattes et al., 1947) were able to deduce the masses of the π- and μ-mesons by determining their ranges in the nuclear emulsions and comparing them to the ranges of the proton at various energies. Under the magnification of a strong optical microscope they could count the grain densities in a particle track, that is, the number of grains per unit length of particle track. The method is described by Lattes et al. (1947) as follows:

In determining the grain density in a track, we count the number of individual grains in successive intervals of length $50\,\mu$ along the trajectory, the observation being made with optical equipment giving large magnification (2,000 $\times$), and the highest available resolving power. Typical results for protons and mesons are shown in Fig. VI.14,...and it will be seen that there is satisfactory resolution between the curves for particles of different types...We can therefore conclude that there is a significant difference in the grain-density in the tracks of the primary and secondary mesons [i.e., π- and μ-mesons], and therefore a difference in the mass of the particles. The conclusion depends, of course, on the assumption that the π- and μ-particles carry equal charges.

Lattes et al. (1947) concluded that the charges on the π- and μ-mesons were equal and unitary equivalent to the magnitude of the electric charge (e). Their conclusion was based on the observation that the grain-densities at the ends of the tracks for both particle types were consistent with particles of unitary electronic charge. For particles of double charge the grain densities would increase at the end of the particle track, that is, at the end of the particle range. See Figure 1.4 of chapter 1, which illustrates how the α particle, which is of double charge, causes more ionization as it slows down and the highest ionization just before it reaches the end of its range. This is due to the fact that, as the α particle slows down and comes close to a stop the Coulomb attractive forces between the particle and atomic electrons have more time to act forcing the atomic electrons to latch on to the alpha particle and increased ionization occurs. Lattes et al. (1947) made a more precise comparison of the masses

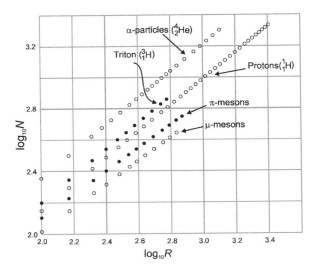

Figure VI.14   An approximate depiction of plots by Lattes *et al.*, (1947) of the total number N of grains in a particle track of residual range R (scale divisions) where 1 scale division = 0.85 microns of nuclear emulsion. Notice that the greater the mass and charge of the cosmic-ray particle, the greater is the developed grain densities of the particle track.

of the π- and μ-mesons from the track lengths. The π-meson had a track length of about 600 μ, whereas, the length of the tracks of the μ-mesons in the nuclear emulsion exceeded 1000 μ. They estimated the ratio of the masses of the π- and μ-mesons (m_π/m_μ) to be about 1.5. This is quite close to the currently known value of 1.3, as the masses of the π- and μ-mesons are 139.6 and 106 MeV/c^2, respectively.

Nuclear emulsions served as excellent media for the recording and measurement of particle tracks and considered as an improved method over the Wilson Cloud Chamber. The major improvements offered by the nuclear emulsions were: (a) the possibility of measuring particle tracks over a continuous period of time, that is, film exposures to cosmic radiation over extended periods beyond 30 days were common, while the Wilson Cloud Chamber was sensitive for only a few milliseconds during it expansion; and (b) the higher density of nuclear emulsions over the gaseous cloud chamber permitted the measurement of the ranges and consequently the energies of cosmic-ray particles that would come to a full stop in the more dense emulsions. The disadvantage of nuclear emulsions was the practical impossibility of causing magnetic deflections of the charged particles as these passed through the emulsions. Very high magnetic fields of the order of 10^6 G would be required, and the angle of Coulomb scattering would be very small in the denser nuclear emulsion complicating the interpretations of deflections. This disadvantage was clearly underscored by Powell and Rosenblum (1948) when that noted

> It is well known that it is not possible, to make magnetic deflection experiments with photographic emulsions—analogous to those which have been so successful with the Wilson expansion chamber in experiments on cosmic radiation—by observations on the curvature, due to the magnetic field, of the trajectory of a particle in an emulsion.

To circumvent this shortcoming Powell and Rosenblum (1948) designed the experimental arrangement illustrated in Figure VI.15, which would permit the use of two face-to-face nuclear emulsions for the measurement of particle deflections in a magnetic field as the particle passes through a vacuum or gas. They arranged two photographic plates with their emulsions face-to-face in a suitable plate holder, which would keep the emulsions separated by a distance of 3 mm. The assembly was placed between

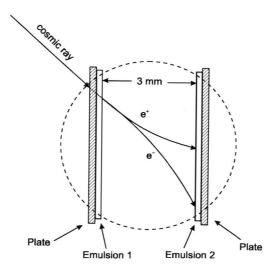

Figure VI.15 Experimental arrangement of Powell and Rosenblum (1948) designed for the use of nuclear emulsions in conjunction with applied magnetic fields for the measurement of the deflections of cosmic-ray particles and the calculations of their charge and mass. Two photographic plates are placed 3 mm apart, and their nuclear emulsions are oriented face-to-face. The distance between the emulsions is exaggerated in the figure to illustrate the deflections produced by the magnetic field. The dashed circle represents the positions of the poles of a magnet in the front and rear of the photographic plates. The poles of the magnet provide a 8600-G magnetic field normal to the nuclear emulsions. Positive and negative particles (e.g., electrons) originating from a cosmic-ray interaction in Emulsion 1 on the left-hand side of the apparatus exit from Emulsion 1 and are deflected in the magnetic field while traveling in the 3-mm space between the two emulsions. The positive electron curves upward and the negative electron bends downwards into Emulsion 2. The electrons may travel through the second emulsion or come to a full stop within the emulsion. The position of exit of a particle from Emulsion 1 and its point of entry in Emulsion 2 will define the arc or curvature of deflection.

the poles of an electromagnet providing a field of 8600 G and the planes of the emulsions arranged normal to the magnetic field. After traveling through Emulsion 1 and while traversing the space between the two emulsions, positive particles will be deflected upwards and negative particles downwards toward Emulsion 2. The points of exit from Emulsion 1 and point of entry into Emulsion 2 of a particle permitted Powell and Rosenblum (1948) to calculate ρ, the radius of the curvature of the particle path. The radius of curvature of the particle path is a function of the magnetic field strength (H), and the particle's mass and velocity or its momentum. In the words of Powell and Rosenblum (1948)

…in the space between the plates, it [the particle] will have moved in a spiral path. If we can identify the track of the same particle in the two emulsions, then the observed change in the direction of motion in traversing the air gap, and the distance between the points of exit and entry of the particle in the emulsions, allows us to calculate $H\rho$ for the particle; the magnitude of its momentum in the plane perpendicular to the field is thus determined. Further, if the particle comes to rest in the emulsion, its observed range gives a measure of its velocity as a function of its mass. The two observations taken together define the mass of the particle.

Blackett and Brode (1936) demonstrated that the relation between the energy of an electron in electron volts (eV) and the magnetic field strength (H) and the radius of curvature of the particle path (ρ) was defined as

$$E = 300\,H\rho$$

(VI.25)

For example, they calculated that the radius of curvature of the track of an electron of 10^{10} eV energy in a field of 14,000 G would be 24 m, or

$$\rho = \frac{1 \times 10^{10} \text{ eV}}{(300)(14000 \text{ G})} = 0.24 \times 10^4 \text{ cm} = 24 \text{ m} \tag{VI.26}$$

The nuclear emulsion technique developed by Powell and his coworkers was a vital tool used by many researchers to determine the properties of high-energy cosmic radiation and the physics of elementary particles in general. Among these, a very important example can be taken from the work of Emilio Segrè, Owen Chamberlain, and others, who used Powell's nuclear emulsion technique to study antinucleon annihilation and measure the resulting particle energies. They identified particle tracks and measured the energy released by the explosive effect of antiproton interactions with atomic nuclei, which yielded conclusive evidence for proton and antiproton annihilation (Chamberlain et al., 1956). An example is the annihilation star illustrated in Figure VI.16. Antiproton annihilation gives rise to mostly π mesons, on the average about 4.8 pions per annihilation together with other nucleons that result from the disintegration of a nucleus.

In the case of nucleon–antinucleon (e.g., proton–antiproton or neutron–antineutron) interactions the resulting annihilation is more complex than the electron–positron annihilation. The electron–positron annihilation yields the practically immediate conversion of the two electron masses to electromagnetic energy (see eq. (V.18)). However, in the nucleus of the atom there exists also the meson predicted by Yukawa, which is an exchange particle between the nucleons, representing the strong nuclear forces binding the nucleons. Thus, the nucleon–antinucleon annihilation process involves the emission of π mesons, which eventually in a matter of microseconds decay to photons or particles of zero rest mass.

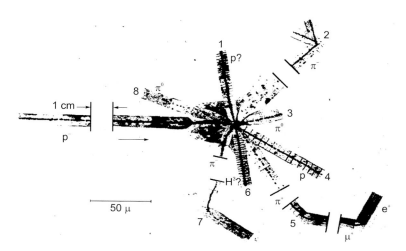

Figure VI.16 An annihilation "star" in nuclear emulsion resulting from an antiproton–proton interaction. The track of the antiproton indicates where it traveled from left to right. It comes to rest in the vicinity of a proton of a nucleus in the photographic emulsion resulting in the antiproton–proton annihilation and destruction of the nucleus. Tracks of particles resulting from the interaction show their various directions of travel away from the spot where the collision occurred. The particles are identified as (1) proton?, (2) π^--meson, (3) π^0-meson, (4) proton, (5) π^+-meson, (6) H^3?, (7) π^--meson, and (8) π^0-meson. The total visible energy released = 1300 MeV. (From Segrè (1959) taken from the work of Gösta Ekspong at Berkeley with permission from The Nobel Foundation © 1959.)

Segrè (1959), in his Nobel Lecture on December 11, 1959 explained the nucleon–antinucleon annihilation as follows:

> By now we know a great deal about [nucleon] annihilation. It gives rise prevalently to π mesons. These, in a time of the order of 10^{-8} seconds, decay into μ mesons and neutrinos. The μ mesons, in a time of the order of microseconds decay into electrons or positrons and neutrinos, and the electrons and positrons finally recombine to give gamma rays. In a few microseconds the total rest mass of the nucleon-antinucleon pair degrades to particles with rest mass zero, traveling away from the spot of the annihilation with the velocity of light. Direct annihilation into photons may occur, but is expected to be rare and thus far has never been observed with certainty. The reason for this difference between the behavior of electron-positron and nucleon-antinucleon pairs is, of course, that the latter can annihilate not only through the electromagnetic interaction giving rise to light quanta, but also through the specific nuclear interaction whose quanta are the pions [i.e., π mesons]. This last interaction is much stronger than the electromagnetic one and when both are simultaneously present its effects overwhelm those of the electromagnetic interaction, which is the only [type of interaction] available to the electron-positron pair. The most significant result of the [nucleon-antinucleon] annihilation is that the annihilation process gives rise to an average of 4.8 pions per annihilation, about equally divided among positive, negative, and neutral pions [See Fig. VI.16]. These pions escape with a continuous energy distribution, the average kinetic energy being about 200 MeV. In about 4 percent of the annihilation cases at rest strange particles, K mesons, are emitted. The escaping pions give rise in complex nuclei to secondary processes and thus a number of nucleons or light nuclei are also found among the particles emitted on annihilation. Sometimes the relatively rare K mesons interact producing a Λ hyperon and even more complicated hyperfragments have been observed.

The Λ hyperon is a particle that interacts with the nuclear proton and neutron through the strong force. It is an electrically neutral and unstable particle with a mass higher than the neutron ($1115.6\,MeV/c^2$) and short lifetime (2.63×10^{-10} sec). The Λ hyperon has the following two principle modes of decay (Sundaresan, 2001):

$$\Lambda^0 \rightarrow p^+ + \pi^- \ (63.9\%) \tag{VI.27}$$

$$\Lambda^0 \rightarrow n + \pi^0 \ (35.8\%) \tag{VI.28}$$

The predominant mode of decay depicted by eq. (VI.27) above was observed in cloud chamber images as V-shaped tracks such as the track illustrated previously in Figure VI.6.

As previously mentioned, Powell's nuclear emulsion technique permitted a measurement of particle energies from the particle ranges and grain densities of their tracks. Thus Segrè, Chamberlain, and their coworkers were able to measure the energy liberated from the nucleon–antinucleon annihilation. To demonstrate unequivocally proton–antiproton annihilation they needed to provide evidence that the energy released in the initial encounter of the proton and antiproton, according to Einstein's equation, would be greater than mc^2 and possibly near the order of magnitude of $2mc^2$, that is, the energy, E, released in the annihilation would be in the range

$$mc^2 < E < 2mc^2 \tag{VI.29}$$

where m is the mass of the proton or antiproton and c the velocity of light in a vacuum. $2mc^2$ is derived from the following. The maximum energy released from the annihilation would be calculated as

$$
\begin{aligned}
E &= m_p c^2 + m_{\bar{p}} c^2 = 2mc^2 \\
&= 2(1.672 \times 10^{-27}\ \text{kg})(2.9979 \times 10^8\ \text{m/sec})^2 \\
&= 30.053 \times 10^{-11}\ \text{J}
\end{aligned}
\tag{VI.30}
$$

Converting the energy in Joules to electron volts gives

$$E = \frac{30.053 \times 10^{-11} \text{ J}}{1.602 \times 10^{-19} \text{ J /eV}}$$

$$= 18.76 \times 10^8 \text{ eV} = 1876 \text{ MeV}$$

The 1876 MeV would be the maximum energy released in the annihilation, as it would represent the conversion of the two proton masses to energy. However, the energy measured with a nuclear emulsion would be expected to fall short of the maximum energy released, because of the various neutrino and gamma-ray energies following pion and muon decays, the tracks of which would not be visible in the nuclear emulsion. In his Nobel Lecture, Segrè commented on the energy released from the annihilation as follows:

> Initially the effort was mainly directed toward establishing the fact that the energy released was $2mc^2$ (where m is the mass of the proton, c the velocity of light), thus furnishing a final proof of the annihilation. In the early investigations with photographic emulsions carried out in my group by Gerson Goldhaber and by a group in Rome led by Amaldi, we soon found stars showing a visible energy larger than mc^2, giving conclusive evidence of the annihilation in pairs of proton and antiproton (Chamberlain *et al.*, 1956).

Cecil Powell staunchly apposed nuclear weapons and was one of the original proponents for total nuclear disarmament. He was cosignatory of the Russell–Einstein Manifesto of 1955 (Bone, 2005) calling for nuclear disarmament and world peace (see Leo Szilard—Radioactivity Hall of Fame—Part IV). Powell also chaired the first Pugwash Conference in 1957, which focused on the then and yet most difficult and vital challenge facing mankind, nuclear disarmament and world peace. Cecil Powell died on August 9, 1969 in Milan, Italy.

DONALD A. GLASER (1926–)

Donald Arthur Glaser was born in Cleveland, Ohio on September 21, 1926. He was awarded the Nobel Prize in Physics 1960 "for the invention of the bubble chamber".

Donald Glaser completed his primary and secondary education in the public schools of Cleveland Heights, OH. He received a B.Sc. degree in physics and mathematics in 1946 at the Case Institute of Technology, which is now the Case Western Reserve University. The new university was formed in 1967 by the federation of the Case Institute of Technology, founded in 1880 by philanthropist Leonard Case, Jr., and the Western Reserve University founded in 1826. He taught mathematics during the 1946 spring semester at the Case Institute of Technology and the following autumn began his graduate studies at the California Institute of Technology (Caltech). He completed his Ph.D. work under the supervision of Nobel Laureate Carl Anderson in the autumn of 1949 and was awarded the degree the following year. His doctoral research was a study on the momentum spectra of high-energy cosmic radiation and mesons of cosmic-ray showers at sea level.

The same year he completed his doctoral work Glaser began a teaching and research position in the Physics Department of the University of Michigan, and was promoted to the rank of Professor in 1957. It was at the University of Michigan where Donald Glaser carried out his pioneering research that led to his invention of the bubble chamber for the identification and measurement of elementary atomic particles by the tracks of bubbles these leave behind in various liquids. He was aware of the need of a detector that could record many particle tracks on a real-time basis to permit the collection of a lot of data, which would improve our knowledge of the lifetimes, energies, and interactions of the elementary nuclear particles. Particularly there was the need for a detector of large volume (several liters) to speed the collection of rare events in cosmic-ray interactions unlike the thin photographic emulsions. With the development of a large-volume detector, sufficient data could be collected on rare events that would

RÉPUBLIQUE DE GUINÉE
OFFICE DE LA POSTE GUINÉENNE
750F

1960
DONALD A. GLASER

Donald Glaser (1926-   )

improve the accuracy of the information that could be acquired from the particle tracks. One example of a rare event was the occurrence of the "*V*-particles" or "pothooks" as they were called, because of the unusual V shape that would appear in some rare tracks in nuclear emulsion (see Figure VI.6) such as those produced by kaons of the cosmic radiation showers. Tracks of this type were observed only infrequently, about one a day with nuclear emulsions or cloud chambers placed on mountain tops. A large-volume detector as noted by Glaser (1960) could provide more experimental information concerning the production, decay, and interactions of the new and yet rarely observed particles.

Glaser based his invention of the bubble chamber on the basis that nucleation of bubble formation could result from the passage of an ionizing (charged) particle through a superheated liquid. In other words, the ionization that would occur along the path of travel of a charged particle would create microscopic bubbles or a vapor trail of the liquid analogous to a vapor trail produced by a high-speed aircraft traveling in the upper atmosphere. The particles would be very small (10^{-6} cm) one or two milliseconds after they are created and would grow larger as each millisecond of time would transpire. Such an effect occurs when a high-flying aircraft produces a vapor trail. The trail starts very fine and then grows thicker with time. If the liquid selected was transparent, the particle path through the liquid could be photographed only a few milliseconds after a charged particle traverses the liquid. Glaser selected diethyl ether as the liquid medium for his initial experiment, because it had a relatively low boiling temperature (34.5°C), critical temperature (192.6°C), and critical pressure (35.6 atm.), and it was relatively inexpensive. Also, he found fortunately a relatively old publication by Kenrick *et al.* (1924) where it was reported that diethyl ether could be maintained quiescent at 130°C and 1 atmosphere pressure for hours under very clean conditions; but the authors had complained that at 140°C the liquid would erupt at erratic time intervals after being brought rapidly to 140°C. In disgust over this unexpected occurrence and prior to abandoning their experiment, Kenrick *et al.* (1924) reported a typical series of 30 consecutive times that this random boiling would erupt. Glaser (1960, 1994) relates that he made a histogram of the time intervals and got a Poisson distribution that corresponded to a random event with a mean time interval of 60 sec. Glaser estimated that the combination of cosmic radiation and terrestrial background radiation would produce an ionizing event, on average, approximately once every 60 sec in the hot diethyl ether of Kenrick *et al.* (1924). Glaser then proceeded to demonstrate that ionizing radiation could trigger boiling in the superheated diethyl ether. His experiment is illustrated in Figure VI.17.

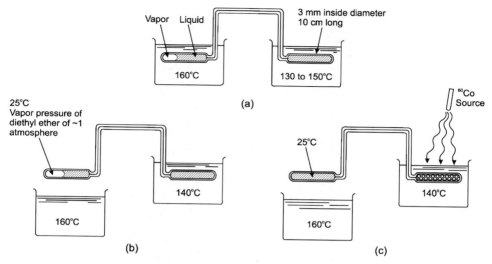

Figure VI.17 Glaser's experiment demonstrating that superheated diethyl ether could be made to boil by ioniz-ing radiation. The experimental apparatus consisted of two glass bulbs containing diethyl ether connected with narrow-bore glass tubing. The diethyl ether in the bulbs was maintained at different temperatures to control the pressure on the enclosed liquid. The experiment proceeded as follows: (a) The pressure on the enclosed diethyl ether was initially high (~20 atm) due to the vapor pressure of the liquid at 160°C. The temperatures were main-tained by oil baths. (b) The oil bath of 160°C was removed quickly whereby the temperature in the bulb on the left would drop to 25°C, and the pressure on the overall system would drop to ~1 atmosphere. The diethyl ether in the bulb on the right at 140°C then became highly superheated. (c) Boiling is triggered in the superheated diethyl ether by radiation from the ^{60}Co source. (Adapted from Glaser, 1960, 1994, with permission from The Nobel Foundation © 1960.)

   His first report on the formation of bubbles in liquids by ionizing radiation was the result of initial tests with diethyl ether in a glass tube kept at 140°C under a pressure of 20 atmospheres (Glaser, 1952). He showed that the diethyl ether would erupt to a boil as soon as the pressure was released if exposed to a 12.6 mCi ^{60}Co source. When the diethyl ether was not exposed to the radiation source eruptive boil-ing would not occur until about 68 sec after release of the pressure keeping in mind that a hard cosmic-ray particle could traverse the liquid on the average of one in every 34 sec. His initial tests clearly showed that external radiation could cause bubble formation in superheated liquids and the potential that the tracks of the very first minute bubbles formed by ionizing radiation might be observed if photographed almost immediately (few milliseconds) after the particle traverses the liquid. Figure VI.17 illustrates Glaser's initial experiment the results of which were first reported in 1952 and subsequently in more detail in his Nobel Lecture (Glaser, 1960) and recollections later in life (Glaser, 1994).
   After demonstrating that ionizing radiation from ^{60}Co could initiate boiling in a superheated liq-uid Glaser needed to determine whether or not a minimum ionizing particle could initiate boiling and if such a particle would form very minute bubbles along its path of travel in the superheated liquid in the few milliseconds prior to the eventual growth of the bubbles into a more eruptive effect such as boiling. If minute bubbles along the particle path of a minimum ionizing particle were not observable, the device would have no application in cosmic-ray or high-energy particle physics. By minimum ionizing particle is meant any charged particle of low mass and high (relativistic) speed, such as cosmic-ray mesons.
   To demonstrate the effect of the passage of a minimum ionizing particle in the superheated liquid Glaser constructed numerous bubble chambers of various shapes (round, oval, etc.) of Pyrex glass and of small volume (few cm^3). He used a hot oil bath to keep the diethyl ether at the appropriate

Figure VI.18   One of the first penetrating cosmic-ray tracks photographed by Glaser in a 3 cm^3 diethyl ether bubble chamber at 140°C with random expansion and counter-controlled flashlamp. The photograph was taken with a 10 μsec flash delay and 20 μsec flash duration. (From Glaser, 1953 reprinted with permission from The American Physical Society.)

temperature; and a mechanical piston operated by a hand crank controlled the pressure above the diethyl ether. With the liquid at its appropriate temperature Glaser would reduce the pressure with a quick turn of the hand crank to achieve the superheated state. A few seconds later when the diethyl ether was in its quiescent superheated state a high-speed movie camera would record the formation of bubbles and the onset of boiling. Glaser (1994) reported his recollections of the historic findings in the following words:

> From these movies taken at 3000 pictures/second, one sees that the bubbles grow to more than a millimeter in diameter in 300 microseconds. Finer tracks were obtained by constructing an automatic device for expanding and recompressing the chamber every 10 seconds. Photographs of the chamber were taken with a xenon flashlamp whenever a vertical Geiger counter telescope indicated the passage of a penetrating cosmic ray particle during the few seconds of sensitive time following each expansion. These photographs [one is illustrated in Fig. VI.18] proved that bubble chambers could yield precision measurements of events involving minimum ionizing tracks.

Diethyl ether was the first liquid tested by Glaser, because it was conveniently available in his laboratory. After his initial demonstration of the first particle tracks by penetrating cosmic radiation in superheated diethyl ether Glaser theorized that bubble nucleation in superheated propane and other liquids would also occur and proceeded to research other liquids. He thought that liquid hydrogen would be a good candidate particularly since physicists would be able to observe cosmic-ray or other high-energy particle interactions with protons. He did not have a cryogenics laboratory at his University of Michigan, Ann Arbor facilities; so in 1953 he joined with Roger Hildebrand and Darragh Nagle at the University of Chicago to work on liquid-hydrogen bubble chambers. Glaser (1994) related the momentum of work on the development of his invention at the time as follows:

> I was pleased to learn that Roger Hildebrand and Darragh Nagle at the University of Chicago wanted to make a liquid-hydrogen chamber and invited me to join them. Liquid hydrogen was soon shown to be radiation sensitive under the predicted conditions [Hildebrand and Nagle, 1953] and the race was on to make a track-forming chamber. John Wood at Berkeley was the first to succeed using liquid hydrogen with a chamber fabricated of glass and metal (Wood, 1954), a so-called dirty chamber, because boiling always began at the metal walls. The same technique works for other liquids but shortens the sensitive time making the chamber very inefficient for cosmic-ray experiments, but not interfering with pulse-beam accelerator experiments at all. Construction of large hydrogen bubble chambers became feasible because of this critical development.

Subsequent research by Glaser and coworkers demonstrated that other liquids such as isopentane (Glaser and Rahm, 1955), liquid xenon (Brown et al., 1956), and propane (Meyer et al., 1957) in addition to diethyl ether and hydrogen were suitable for track formation in bubble chambers. As noted by Glaser in his Nobel Lecture on December 12, 1960

> Many other liquids were tested in our laboratory and in other places. No liquid that has been tested seriously has failed to work as a bubble chamber liquid. The choice of liquids depends only on the physical objectives of the experiment and on engineering and economic considerations…Water, of course, turns out to be the wrong substance to use in a bubble chamber, because it has a large surface tension and a high critical pressure.

Glaser et al. (1956) demonstrated that the particle velocities could be determined by the measurement of the relative ionization produced by the charged particles as they traveled through the bubble chamber. The relative ionization was measured by a determination of the bubble density along the particle track, that is, the number of bubbles formed per particle path length. The bubble densities produced by particles of known velocity and mass would be used as standards to determine the velocities and mass of the unknown particles. They could count bubble densities up to approximately 150 bubbles per centimeter on a photograph negative without error due to fusion of neighboring bubble images along the particle track. Thus particle velocity and mass measurements were possible with the new bubble chamber.

Following the demonstration of Wood (1954), that bubble chambers could be constructed with metal walls, larger bubble chambers were made of metal frames and glass windows. One of the first of these was a circular bubble chamber with 5-cm diameter glass windows and aluminum frame, followed by the construction of a 30-cm, 21 liter propane chamber and Luis Alvarez's (1911–1988) 180-cm, 500 liter liquid hydrogen chamber at Berkeley (Glaser, 1960). Magnetic fields could be applied to the new bubble chambers by external magnets to facilitate the measurement of charge, momentum, and mass of a particle from its direction of deflection and radius of curvature as described previously. As charged particles slowed they would take on a very small radius of curvature in the magnetic field and even display a winding circular path. The new bubble chambers thus provided much information on the masses, lifetimes, decay schemes, and branching ratios, etc. of the elementary particles. Some illustrative examples are provided in Figures VI.19–VI.23.

The K^+ meson has several modes of decay as described by eqs. (VI.5)–(VI.12). Some of the decay schemes yield one or more π^0 mesons. The neutral pion will not produce a track in any bubble chamber; however, its conversion to positron–negatron (e^+e^-) pairs, which are formed in the proximity of the vertex of the K^+-meson decay, provide evidence for the π^0 mesons such as provided by the example illustrated in Figure VI.21. Kalmus and Kernan (1967) took 240,000 photographs of 3×10^6 tracks

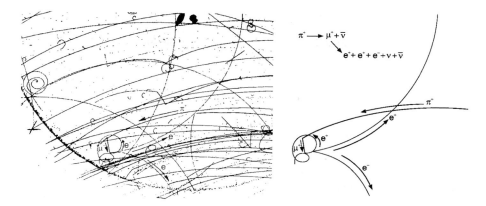

Figure VI.19   Bubble chamber tracks of the decay of a μ^+ meson. (From Lee and Samios, 1959 reprinted with permission from The American Physical Society.). On the left is the bubble chamber image with its many tracks, and on the right is a drawing of the tracks produced by the meson decay and the identified decay scheme. The slow meson track was stopped in a liquid hydrogen bubble chamber 30.5 cm in diameter and 15.25 cm in depth placed in a magnetic field of 8800 G. Out of 22,000 positive muons observed by Lee and Samios (1959) they found three events of μ^+ decay with three charged prongs (two positrons and one electron) such as the event illustrated here. The authors concluded that a positron and electron pair arise from the internal conversion of a gamma ray (internal pair production) arising from the decay scheme $\mu^+ \rightarrow e^+ + \nu + \bar{\nu} + \gamma$.

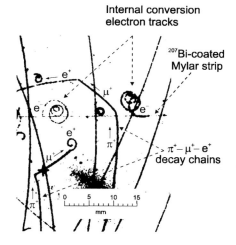

Figure VI.20   Particle tracks formed in a 10-l hydrogen bubble chamber exposed to a beam of π^+. Curvatures to particle paths were produced by a magnetic field of 20.85 kG whereby positive and negative charged particles bend in opposite directions and the radius of curvature is a function of the particle momentum. The picture illustrates two 1.4 MeV/c internal conversion electron tracks. The internal conversion electrons originated from a ^{207}Bi source coated on the surface of a Mylar® (DuPont trademark) strip stretched across the bubble chamber. The tracks of the conversion electrons were used to calibrate the μ^+ momentum measurements. The tracks of two π^+-μ^+-e^+ decay chains are seen here. (From Derenzo (1969) reprinted with permission from The American Physical Society.)

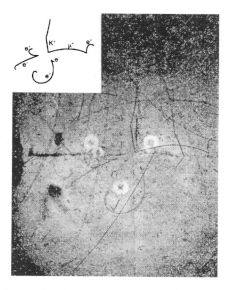

Figure VI.21   An enhanced photograph of the tracks produced by a K^+ meson that was stopped in a 76 cm Freon (C_3F_8) bubble chamber. A drawing of the relevant tracks is provided in the inset in the upper left-hand corner of the photograph. The positive kaon decayed according to the path $K^+ \rightarrow \pi^0 + \mu^+ + \nu$. The μ^+ decays to a positron and neutrino. The neutrinos, of course, do not form tracks because of their lack of charge and near-zero mass. The π^0 decays according to the pathway $\pi^0 \rightarrow 2\gamma$ whereby the two gamma-ray photons convert to positron–negatron (e^+e^-) pairs. Notice the gap between the end of the K^+-meson track and the start of the two tracks formed by the (e^+e^-) pairs. The absence of tracks here is due to the neutral charge of the π^0 meson. (From Callahan *et al.*, 1966, reprinted with permission from The American Physical Society.)

$$K^+ \longrightarrow \pi^0 + e^+ + \nu$$
$$2\gamma \longrightarrow 2(e^+ + e^-)$$

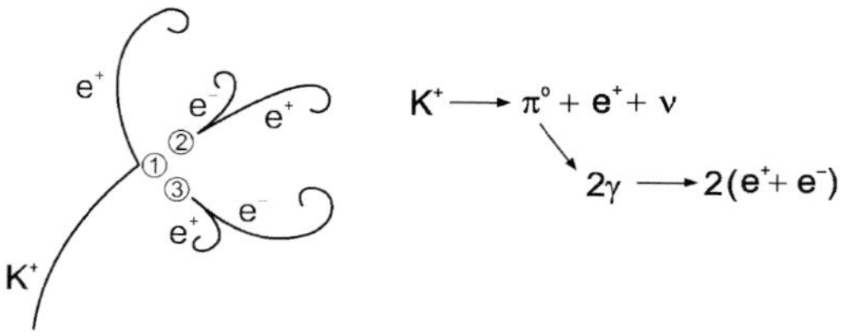

Figure VI.22   Schematic diagram of the bubble-chamber particle tracks produced by the decay sequence $K^+ \rightarrow \pi^0 + e^+ + \nu$ at vertex (1), followed by $\gamma \rightarrow e^+ + e^-$ at vertices (2) and (3). (From Kalmus and Kernan, 1967 reprinted with permission from The American Physical Society.)

of stopped K^+ mesons in a 76-cm bubble chamber filled with Freon. They were able to identify the K^+-meson decay schemes according to the characteristic patterns of particle tracks formed by π^0-meson conversion to e^+e^- pairs as illustrated in Figure VI.22.

In the Nobel Prize Presentation Speech Professor Kai Siegbahn, member of the Swedish Academy of Sciences and later a recipient of the Nobel Prize in Physics 1981 for his work toward the development of high-resolution electron microscopy, spoke the following words on December 12, 1960 concerning the significance of Glaser's invention of the bubble chamber:

> The Wilson [cloud] chamber has certainly played a tremendously important role, especially during the 1930's, which is referred to as 'the golden age of nuclear physics', and there is no doubt

that it was the Wilson chamber which made possible the greatest nuclear physical discoveries during that decade...This situation is completely different in the nuclear physics of today, where one now has at one's disposal particle accelerators with energies as high as 25 billion volts, for example the accelerator which has recently been built in the European Nuclear Research Center in Geneva. In other words, [particle] energies which are more than 1,000 times larger than those which were earlier obtainable...Its is obviously necessary to use a medium other than gas in order to be able to bring such particles to rest. Donald Glaser has succeeded in solving this problem, and his so-called 'bubble chamber' is the high-energy nuclear physics counterpart to the low-energy nuclear physics Wilson chamber...One could say that Glaser's bubble chamber is an anti-Wilson chamber. Particle tracks in Glaser's chamber are composed of small gas bubbles in a liquid...The largest bubble chamber is surrounded by a powerful electromagnet which is capable of bending the paths of the particles so that the faint bubble tracks become slightly curved. In this way one is able to identify the unknown atomic particles when they, traveling very close to the speed of light, pass through the chamber...Already a great amount of information has been obtained in this way and many important discoveries will no doubt follow in the near future by means of your method. It is unusual for a development in modern nuclear physics to be due to such a large extent to one single man [person].

Among the prominent findings facilitated by Donald Glaser's bubble chamber was evidence for the mechanisms of antiproton and antineutron annihilation. In 1955, Emilio Segrè and Owen Chamberlain together with coworkers discovered the antiproton (Chamberlain et al., 1955); and the following year Cork et al., (1956) discovered the antineutron. For their discovery, Segrè and Chamberlain were awarded the Nobel Prize in Physics 1959. The discovery of the antinucleons was long awaited since Dirac had predicted the positron or antiparticle of the electron in 1928, and the positron discovered by Anderson in 1933. Dirac (1933) went further to predict the antiparticles of the nucleons. It is now known that the antiproton is a component of cosmic radiation, but its numbers are very small compared to the cosmic-ray protons (see Table 6.1). Thus, it was not until the Bevatron, the high-energy accelerator capable of accelerating particles to energies of the order of several billion electron volts was constructed at Berkeley, CA, could sufficiently high proton accelerations and collisions be possible to create a sufficient number of antiprotons to demonstrate their existence (Segrè, 1959; Chamberlain, 1959). It was expected that the newly discovered antiproton would undergo immediate annihilation upon direct contact with a proton (see Figure VI.16). Another interaction of the antiproton ($\bar{p}$) would be a charge-exchange reaction with a proton (p), as described by eq. (VI.31), resulting in the production of a neutron (n) and antineutron ($\bar{n}$), which on annihilation would produce an explosive star of particles consisting mostly of π mesons (Segrè, 1959).

$$p + \bar{p} \rightarrow n + \bar{n} \qquad\qquad (VI.31)$$

The charge-exchange process and subsequent antineutron annihilation, photographed in Glaser's bubble chamber, is illustrated in Figure VI.23.

Donald Glaser continued his work as Professor of Physics at the University of California at Berkeley during 1959–1964. In 1962, Donald Glaser decided to change course in scientific research and entered the field of molecular biology. At the University of California, Berkeley he and his students have carried out significant research in the identification of human genes where defects lead to a form of human skin cancer (Nobel Lectures, 1964). In 1970, he and two associates founded the first biotechnology company, which started an industry that now provides the findings of research in molecular biology to advances in the fields of agriculture and medicine. Donald Glaser is currently Professor of the Graduate School, Department of Molecular and Cell Biology at the University of California, Berkeley. He and his students currently research the field of neurobiology and the human visual system, which is the best-known part of the brain. Their work currently includes the construction of computational models of the human visual system that explain its performance in terms of its physiology and anatomy (Chaudhuri and Glaser, 1991; Kumar and Glaser, 1993, 1995; Glaser and Barch, 1999).

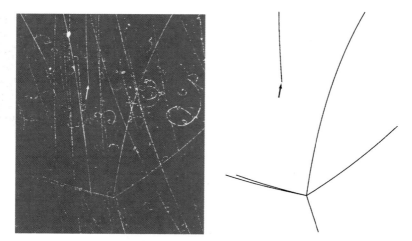

Figure VI.23   The photograph on the left shows tracks in a propane bubble chamber. The track of an antiproton is seen ending at the point marked by the arrow where it underwent charge exchange with a proton producing a neutron and antineutron neither of which produce tracks due to their lack of charge. However, the antineutron undergoes annihilation by causing the explosive star directly below. The distance between the charge exchange and star was 0.95 cm. (From Agnew *et al.*, 1958 and Segrè, 1959 reprinted with permission from The American Physical Society.)

PIERRE VICTOR AUGER (1899–1993)

Pierre Auger (pronounced OH-ZHAY) was born in Paris, France on May 14, 1899. He did not win the Nobel Prize, but he made great contributions to the science of radioactivity including the discovery of the atomic Auger effect, Auger electrons, and the showers of the cosmic radiation.

Pierre Auger studied at the Ecole Normale Supérieure de Paris during 1919–1922 and devoted his professional career to atomic and cosmic-ray physics. At the young age of 24 years, he discovered the emission of electrons from atoms excited by x-rays (Auger, 1923, 1925a,b). Pierre Auger irradiated krypton and argon gas with x-rays and discovered the emission of electrons. He reasoned that an x-ray photon causes the ejection of an atomic electron from a given shell or quantum level leaving a vacancy behind. He further reasoned that an electron from the next outer shell could fill the vacancy causing the emission of electromagnetic radiation that would correspond to the difference in the energy levels of the electrons in the outer and inner shells. For example, Auger noted that, if an electron in the K-shell was ejected by an artificially produced x-ray photon, an electron from the next outer L-shell could fill the vacancy resulting in the emission of energy, E, as electromagnetic radiation, which he described by the following equation:

$$E_{\text{x-ray}} = h\nu = E_L - E_K \tag{VI.32}$$

where E_L and E_K are the energy levels of the electrons in the L and K shells, respectively. The transition energy $h\nu$ may be emitted as an x-ray photon characteristic of the atom or, alternatively, the transition energy may be absorbed by an atomic electron resulting in its emission from the atom. The electron emitted in this fashion is identified as an Auger electron in the name of Pierre Auger for his discovery and interpretation of this phenomenon. Lise Meitner (1923) also independently discovered Auger electrons the same year as Pierre Auger; however, the Auger electrons and the phenomenon that gives rise to these electrons, the Auger Effect, are named after Pierre Auger.

Auger electrons are defined as atomic electrons that are emitted from atoms after acquiring energy from an atomic electron transition within the atom. Electron transitions will occur when an atom

**Pierre Auger (left) profile with Niels Bohr (right)
with pipe [UNESCO photograph, courtesy AIP
Emilio Segre Visual Archives]**

becomes ionized by the loss of an electron from an inner shell. For example, an atom may become ionized by the ejection of an electron from an inner shell by one of several mechanisms such as: (i) irradiation with artificial external x-rays or irradiation with external electron beams, (ii) the emission of an internal conversion electron such as K-shell internal conversion, or (iii) by electron capture (EC) also known as K-capture (see Chapter 5). The vacancy left in the K-shell can be filled by an electron from an outer L-shell. In turn, the vacancy left in the L-shell could be filled by another electron from the outer M-shell, etc. The process of non-radiative rearrangement of atomic electrons as a result of the ionization of the atom in one of its inner shells is defined as the Auger effect (Borisenko and Ossicini, 2004). In the process of falling to a lower energy shell to fill a vacancy, electron energy may be lost as a photon of x-radiation. The energy of the photon radiation is equivalent to the differences in the energy levels of the electrons in the outer and inner shells as described by the above eq. (VI.32). The x-ray photon may be emitted from the atom, or alternatively the energy released in the electron transition will be transferred to an electron of an outer shell and cause its emission from the atom as an Auger electron. The energy of the Auger electron would be that defined by the above eq. (VI.32) less the binding energy, E_b, of the electron or

$$E_{\text{Auger}} = (E_L - E_K) - E_b \qquad (VI.33)$$

The energy of the Auger electron is thus equivalent to the energy of the x-ray photon less the binding energy of the electron or

$$E_{\text{Auger}} = E_{\text{x-ray}} - E_b \qquad (VI.34)$$

Values of the binding energies of electrons in various shells ($K, L, M,\dots$) are found in references texts such as Lide (1997). Either of two processes, the emission of an x-ray photon or the emission of an Auger electron, can occur as a result of electron energy-level transitions from higher to lower energies. Auger-electron emission competes with x-ray emission. An example of the Auger effect and the resultant emission of an Auger electron as compared to x-ray emission are illustrated in Figure VI.24.

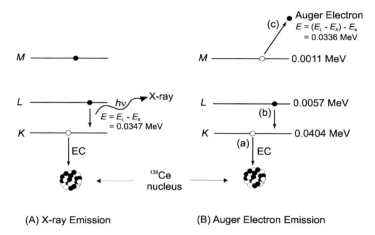

(A) X-ray Emission   (B) Auger Electron Emission

Figure VI.24   Decay of the radionuclide ^{139}Ce by electron capture (EC) resulting in (A) x-ray emission or (B) Auger-electron emission. An electron vacancy in a shell is illustrated as an empty circle and an electron as a filled circle. Specific binding energies for Cerium atomic electrons in K, L, and M shells were obtained from Lide (1997) and are provided along the right-hand side of lines illustrating the various electron energy levels. The Auger effect or process by which an Auger electron is emitted from an atom occurs according to the following sequence as noted in the figure: (a) an atom losses an atomic electron from an inner shell leaving a hole or vacancy, (b) an electron from a higher energy level fills the vacancy, and (c) the energy emitted in the electron transition in step (b) is transferred to an outer atomic electron causing the electron to be emitted from the atom. The energy of the K x-ray illustrated above is equal to the electron transition ($E_{\text{x-ray}} = h\nu = E_L - E_K = -0.0057 - (-0.0404) = 0.0347\,$MeV. The energy of the Auger Electron is equal to the electron transition energy ($E_L - E_K$) less the binding energy of the electron (E_b) or, in other words, the x-ray energy less the binding energy of the electron ($E_{\text{Auger}} = 0.0347 - 0.0011 = 0.0336\,$MeV. The daughter nucleus of the ^{139}Ce remains in an excited state after EC and emits a gamma-ray (not illustrated). Also the loss of an Auger electron will leave an electron vacancy, which can be filled by an electron of an outer shell. This will produce another x-ray, which is not illustrated here. The electron filling the vacancy will, in itself leave another vacancy, which could be filled by yet another electron in an outer shell. The production of x-rays by this cascading effect of electron transitions from outer to inner shells is known as x-ray fluorescence.

The emission of an Auger electron from an atom will occur as a consequence of the following transitions: (i) an atom losses an atomic electron from an inner shell leaving a hole or vacancy in that shell, (ii) an electron from a higher energy level fills the vacancy, and (iii) the energy emitted in the transition of the electron from the higher to lower energy levels is transferred to an outer atomic electron causing the electron to be emitted from the atom.

From this discovery of Meitner and Auger has developed the very important analytical technique of Auger-electron spectroscopy (AES). The technique of AES involves irradiating the surface of a sample with an electron beam of energy sufficient to ionize the inner orbits of atoms thereby inducing the concomitant emission of Auger electrons. The Auger-electron energy spectra drawn from the induced emissions serve as fingerprints of different atoms to enable their identification and quantitative analysis (Chourasia and Chopra, 1997; Mehlhorn, 1998). In 1978, Pierre Auger was recognized with the Gaede-Langmuir Award "For establishing the fundamental principle of Auger spectroscopy which has led to the most widely used surface analysis technique of importance to all aspects of vacuum science and technology".

Pierre Auger together with French physicist Roland Maze (Auger and Maze, 1938, 1939; Auger *et al.*, 1938, 1939) discovered the extensive air showers, that is, the all-encompassing cascade of high-energy particles in the earth's atmosphere produced by the impact of relativistic galactic cosmic-ray particles with nuclei in the earth's atmosphere. In a comprehensive review on cosmic radiation

Cronin (1999) relates the work of Auger and Maze (1938), who demonstrated in Jean Perrin's laboratory in Paris that cosmic-ray particles separated by distances as large as 20 m arrived in time coincidence, which indicated that the particles were secondary particles, or part of a cascade of particles, from a primary or common source. Cronin (1999) relates that subsequent experiments by Auger, Maze, and coworkers (Auger et al., 1939) in the Alps demonstrated that particle coincidences were observed even at distances of 200 m apart. They observed cosmic-ray particles of very high energy up to 10^{15} eV. In their paper Auger et al. (1939) reported

> One of the consequences of the extension of the energy spectrum of cosmic rays up to 10^{15} eV is that it is actually impossible to imagine a single process able to give to a particle such an energy. It seems much more likely that the charged particles, which constitute the primary cosmic radiation, acquire their energy along electric fields of a very great extension.

Yet cosmic-ray particles in excess of 10^{20} eV have been discovered since the first particle of about 10^{20} eV was discovered by Linsley (1963). Thus, in honor of Pierre Auger in an effort to study further properties of these ultra-high-energy particles the Pierre Auger Cosmic Ray Observatory is being established (Argirò, 2004). The observatory is an international effort designed to measure cosmic-ray particles of energy in excess of 10^{19} eV. These are rare, as the expected particle flux at these energies is approximately $1/km^2/yr$. Cosmic-ray particles of energy greater then 10^{20} eV are even more uncommon, as these strike the earth at about $1/km^2/century$. The Pierre Auger Observatory will constitute the largest cosmic-ray detector ever built, encompassing two sites, one in the northern hemisphere in Utah and the other in the Southern hemisphere in Mendoza, Argentina. As described by Argirò (2004) the observatories will be equipped with 1600 water Cherenkov detectors and four fluorescence detectors, each covering $3000 km^2$. The large size of the detectors is required to achieve consistent statistics, and the combination of the two detection techniques will be used to reconstruct the air showers to the greatest precision yet achievable. The overall objective is to measure the arrival direction, energy, and mass composition of 60 particle events per year with energies above 10^{20} eV and 6000 events per year above 10^{19} eV. The project is supported by UNESCO, the United Nations Educational, Scientific and Cultural Organization with headquarters in Paris.

During 1948–1956, Auger was Director of the Department of Sciences for UNESCO during which time he helped promote the creation of international research programs and the founding of CERN, the European Organization for Nuclear Research in 1954 with headquarters in Geneva. During 1967 until his retirement in 1970 Auger was Director of the Centre des Faibles Radioactivités (CFR), a joint laboratory of the Centre National de la Recherche Scientifique (CNRS) and the Commissariat à l'Energie Atomique (CEA), France. Pierre Auger died in Paris on December 25, 1993.

– 6 –

Cosmic Radiation

6.1   INTRODUCTION

Stable charged particles and nuclei with lifetimes of 10^6 years or longer originating from space, which strike the top of the atmosphere (TOA) from all directions, constitute what is known as the cosmic radiation. Air showers of the cosmic radiation consist of cascades of subatomic particles and electromagnetic radiation resulting from nucleon–nucleon collisions of high-energy cosmic-ray particles from space with atomic nuclei of the earth's atmosphere. Cosmic rays are classified according to their origin in space, and the air showers of the cosmic radiation are characterized according to the products of collisions of high-energy nuclei with the earth's atmosphere.

The discovery of cosmic radiation is attributed to Victor Hess, who made seven balloon ascents during 1911–1913 into the atmosphere during the daytime, during the evening darkness, and even during a solar eclipse. On these ascents he took along three electroscopes that would measure the ionization caused by external radiation. He thought it best to make simultaneous measurements of ionization with more than one instrument to eliminate doubt in the event one instrument might give erroneous readings. He found that at an altitude of 500 m, the ionization dropped to about half that obtained from the soil surface. However, the readings would increase proportionally with altitude as he would ascend beyond 1000 m. He found the ionization at 1500 m would increase to be approximately equal to that at the soil surface, and further ascents to altitudes of 5000 m would provide ionization readings of several times that at the ground level (Hess, 1912, 1936, 1940). It was obvious to Hess that gamma rays, from radium in air, could not cause any increase in ionization with altitude where air got thinner, and he interpreted his results in the following words:

> The only possible way to interpret my experimental findings was to conclude to the existence of a hitherto unknown and very penetrating radiation, coming mainly from above and being most probably of extra-terrestrial (cosmic) origin …

Victor Hess was awarded the Nobel Prize in 1936 for his discovery of cosmic radiation.

Cosmic radiation is of most concern to astronauts, airline pilots, and persons living at high altitude, as our atmosphere attenuates considerably cosmic radiation and the dose to the human body consequently increases with altitude. When a scientist analyzes low levels of radioactive nuclides in the earth's environment, cosmic radiation is of concern as background radiation that causes interference in the measurement of the levels of radioactivity in soil, plants, and waters. The accurate measurement of radionuclides in our environment requires the suppression and accurate measurement of the background interference from cosmic rays.

6.2   CLASSIFICATION AND PROPERTIES

Cosmic radiation incident on the earth's atmosphere is classified into "primary" or "secondary" cosmic rays. As defined by Gaisser and Stanev (2002), primary cosmic rays are stable charged particles and nuclei accelerated at astrophysical sources, and secondary rays are particles produced via the interaction of the primaries with interstellar gas. The nuclear interactions of the accelerated primary cosmic-ray particles with interstellar medium produce stable nuclei of the light elements (e.g., Li, Be, and B), as well as many fundamental particles including pions or π mesons of zero charge (π^0) that decay into gamma rays (see eqs. (6.15) and (6.16)). Galactic gamma radiation has been mapped recently by the Compton Gamma Ray Observatory satellite. The satellite has provided an image of the galaxy produced by gamma rays of approximately 100 MeV (Simpson, 2001). The gamma-ray image compares closely to the visible-light image of the galaxy of which we are most familiar. Galactic gamma rays can cover the full energy range from <100 MeV to >10 TeV.

Primary cosmic-ray particles include protons, helium nuclei, electrons, and nuclei of most elements of the Periodic Table (e.g., carbon, iron, oxygen, etc.) of stellar origin. Nuclei of the light elements lithium, beryllium, and boron ($Z = 3–5$) are classified as secondary radiation. The abundance of these light elements in cosmic radiation is highly enriched over the abundance of these elements in the universe, which is evidence that these are secondary nuclei created via collision (nuclear spallation interactions) with heavier primary particles during the interstellar propagation of primary nuclei. The radionuclide ^{10}Be ($t_{1/2} = 1.6 \times 10^6$ years) is found among other isotopes of the light elements in cosmic rays. From the measured abundance analysis of ^{7}Be, ^{9}Be, and the radioactive ^{10}Be in cosmic radiation, it is found that cosmic-ray particles remain contained in galactic magnetic fields for approximately 10^7 years before escaping into our atmosphere or intergalactic space. The mean interstellar density for propagation of the cosmic radiation is $\sim$0.2 atom/cm^3 (Simpson, 2001).

Nucleons of practically all elements of the Periodic Table will be found in cosmic radiation. Free protons account for about 80% of the primary nucleons and approximately 15% are nucleons bound in helium nuclei (equivalent to alpha particles). Electrons constitute about 2% of the primaries. Nuclei of the elements of the Periodic Table other than the previously mentioned (H and He) make up the remaining components of cosmic radiation. Nuclei of the light elements (Li, Be, and B), which constitute secondary cosmic-ray particles, account for a small fraction ($\sim$0.3%) of cosmic-ray nuclei.

The composition of cosmic radiation will vary according to the 11-year solar cycle and the earth's magnetic latitude. The earth's magnetic field, which extends well into space, affects the composition of the charged particles of cosmic radiation. The geomagnetic latitude effect was first reported by Clay (1928) and confirmed by Compton (1932, 1933), who measured cosmic-ray showers with ionization detectors while traversing latitudes onboard ocean vessels. Cosmic-ray particles will concentrate in the earth's Van Allen radiation belts, which extend from about 1000 to 60,000 km from the earth. The sea-level latitude effect of the electromagnetic and meson components of cosmic-ray interactions in the earth's atmosphere measured by Compton corresponded to an approximately 12–15% increase between the geometric equator and high latitudes. However, no information was obtained then on the latitude effect on the nucleonic component of cosmic radiation at sea level. Subsequent studies (Simpson, 1948, 1951, 2001) demonstrated a dramatic increase of 300–400% in the nucleonic component compared with a 10–15% increase in the meson component of cosmic radiation at sea level as one traveled from 0° to 70° latitude. The latitude effect on cosmic radiation intensity is greater at higher altitudes as illustrated in Figure 6.10. This is particularly relevant to international airline pilots who are concerned about the cosmic radiation dose they accumulate when they travel frequently at high altitudes (12 km) and high latitudes (40–60°) where the radiation intensities are higher. This is discussed in more detail in Section 6.7.

Positrons and antiprotons are components of secondary cosmic radiation. Positrons are much lower in abundance than electrons. The differential flux of negatrons and positrons incident at the TOA from galactic space is a function of particle energy. The positron fraction (e^+/e^-) decreases from ~0.2 below 1 GeV, to ~0.1 around 2 GeV, and to ~0.05 at higher energies of 5–20 GeV (Alcaraz et al., 2000; Boezio et al., 2000a,b; DuVernois et al., 2000; Sundaresan, 2001; Clem et al., 2002; Gaisser and Stanev, 2002). Antiprotons in the cosmic radiation were discovered by Golden et al. (1979, 1984). These are classified as secondaries, as antiprotons are expected to be products of interactions of primary cosmic radiation, principally protons with the interstellar medium (Gaisser and Maurer, 1973; Webber and Potgieter, 1989; Gaisser and Schaefer, 1992). The proportions of antiprotons to protons ($\bar{p}/p$) is a function of energy with ratios varying from ~3 × 10^{-5} at the energy range of approximately 0.2–1 GeV to ~2 × 10^{-4} at approximately 10–20 GeV (Hof et al., 1996; Mitchell et al., 1996; Basini et al., 1999; Beach et al., 2001). The antiproton lifetime is of the order of magnitude of the storage lifetime of cosmic rays in the galaxy (10^7 years) estimated from the abundance of ^{10}Be (Steigman, 1977; Simpson and Garcia-Munoz, 1988). In summary the approximate composition of galactic cosmic-ray particles is provided in Table 6.1.

Table 6.1

Approximate composition of cosmic radiation of galactic origin incident on TOA[a]

Radiation type	Approximate fraction (%)
H nuclei or protons (p^+ and p^-)[b]	80
He nuclei (alpha particles)	15
Electrons (e^- and e^+)[c]	2
Heavier nuclei (e.g., C, O, Mg, Fe, Si, etc.)[d]	1
Lighter nuclei ($Z = 3$–5, i.e., Li, Be, and B)[e]	0.2

[a] Cosmic radiation incident at the top of the terrestrial atmosphere includes all stable charged particles and nuclei with lifetimes of the order of 10^6 years or longer (Gaisser and Stanev, 2002).
[b] The ratio of antiprotons to protons ($\bar{p}/p$) is about 2 × 10^{-4} in the particle energy range of 10–20 GeV.
[c] The ratio of positrons to negatrons (e^+/e^-) varies from ~0.2 below 1 GeV, to ~0.1 around 2 GeV, and to ~0.05 at higher electron energies of 5–20 GeV.
[d] Nucleons of all of the elements of the Periodic Table stripped of their atomic electrons with the exception of the lighter nucleons listed above.
[e] The nuclei of Li, Be, and B are classified as secondary cosmic radiation, that is, particles produced by the interaction of the primaries with interstellar gas.

Most galactic cosmic-ray particles possess energies from about 100 MeV to 10 GeV as illustrated in Figure 6.1, which provides the energy distributions of some of the major components of primary cosmic radiation for a particular period of the solar cycle. A series of the most precise measurements of primary protons and helium nucleon intensities have been made by the AMS Collaboration (2000) and Bellotti *et al.* (1999), Boezio *et al.* (1999), Menn *et al.* (2000), and Sanuki *et al.* (2000). Particle energies up to 10^{15} eV were first observed by Auger and coworkers (1938, 1939), and a cosmic-ray particle of about 10^{20} eV energy was observed by Linsley (1963).

Protons in the kinetic energy range of 100 MeV to 10 GeV would have relativistic speeds ranging from 42.8 to 99.6% the speed of light (i.e., 0.428c to 0.996c) calculated according to eq. (4.12) of Chapter 4. For example, the relativistic speed of a proton with 10 GeV kinetic energy would be calculated according to the following equation:

$$u = c\sqrt{1 - \left(\frac{K}{mc^2} + 1\right)^{-2}}$$ (6.1)

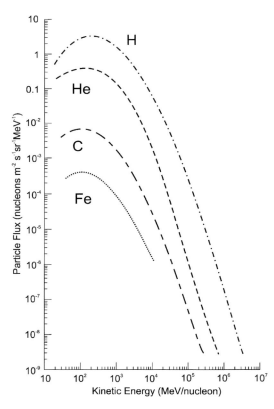

Figure 6.1 Approximate flux distributions of major components of primary cosmic radiation. See Simpson (1983) and Gaisser and Stanev (2002) for more precise data on nucleon flux as a function of particle energy.

The rest energy of the proton (mc^2) is calculated as

$$mc^2 = (1.6726 \times 10^{-27}\,\text{kg})(2.9979 \times 10^8\,\text{m/sec})^2 = 1.503 \times 10^{-10}\,\text{J} \qquad (6.2)$$

which can be expressed in units of electron volts as follows:

$$\frac{1.503 \times 10^{-10}\,\text{J}}{1.602 \times 10^{-19}\,\text{J/eV}} = 0.9382 \times 10^9\,\text{eV} = 938.2\,\text{MeV} \qquad (6.3)$$

The relativistic speed of the 10 GeV proton can then be calculated as

$$u = c\sqrt{1 - \frac{1}{(10^4\,\text{MeV}/938.2\,\text{MeV} + 1)^2}} = 0.996c \qquad (6.4)$$

A higher kinetic energy 100-GeV (10^5 MeV) cosmic-ray proton would have a relativistic speed of

$$u = c\sqrt{1 - \frac{1}{(10^5\,\text{MeV}/938.2\,\text{MeV} + 1)^2}} = 0.999956c \qquad (6.5)$$

The speed of the 100 GeV proton may also be calculated on the basis of the definition of the kinetic energy of the particle being the difference between the *total energy* of the particle and its rest energy or

$$K = \gamma mc^2 - mc^2 = (\gamma m - m)c^2 \qquad (6.6)$$

or

$$10^5\,\text{MeV} = (\gamma m - m)c^2$$

Converting the units of electron volts in the above equation to joules and imputing the speed of light provides

$$(10^{11}\,\text{eV})(1.602 \times 10^{-19}\,\text{J/eV}) = (\gamma m - m)(2.9979 \times 10^8\,\text{m/sec})^2 \qquad (6.7)$$

from which the differences of the relativistic and rest masses of the proton can be calculated as

$$\gamma m - m = \frac{1.602 \times 10^{-8}\,\text{J}}{8.9874 \times 10^{16}\,\text{m}^2/\text{sec}^2} = 0.1782495 \times 10^{-24}\,\text{kg} \qquad (6.8)$$

Inputting the known rest mass m of the proton (1.6726×10^{-27} kg) permits the calculation of the relativistic mass of the 100 GeV proton as

$$\gamma m = 178.2495 \times 10^{-27}\,\text{kg} + 1.6726 \times 10^{-27}\,\text{kg} = 179.9221 \times 10^{-27}\,\text{kg} \qquad (6.9)$$

Since by definition, $\gamma m = m/\sqrt{1-(u^2/c^2)}$, we can write

$$\frac{\gamma m}{m} = \frac{1}{\sqrt{1-(u^2/c^2)}} \tag{6.10}$$

which transforms to read

$$\left(\frac{m}{\gamma m}\right)^2 = 1 - \frac{u^2}{c^2} \tag{6.11}$$

or

$$\frac{u^2}{c^2} = 1 - \left(\frac{m}{\gamma m}\right)^2 \tag{6.12}$$

Inputting the values for the relativistic and rest masses of the proton yields

$$\frac{u^2}{c^2} = 1 - \left(\frac{1.6726 \times 10^{-27}\,\text{kg}}{179.9221 \times 10^{-27}\,\text{kg}}\right)^2$$

$$\frac{u^2}{c^2} = 1 - (0.00929624)^2 \tag{6.13}$$

$$\frac{u^2}{c^2} = 1 - 8.64200 \times 10^{-5}$$

$$u^2 = (0.9999135)c^2$$

and

$$u = 0.999956c$$

where u is the relativistic speed of the 100 GeV proton exactly as calculated previously by eq. (6.5).

High-energy cosmic-ray protons, as demonstrated above, approach the speed of light. Primary cosmic-ray nucleon intensity drops rapidly according to energy beyond several GeV as illustrated in Figure 6.1. The primary nucleon intensity from a few GeV to 100 TeV is described according to the power law

$$I_N(E) \approx 1.8E^{-\alpha}\,\text{nucleons/cm}^2\,\text{sec sr GeV} \tag{6.14}$$

where E is the energy per nucleon (including rest mass energy) and α is approximately 2.7.

As illustrated in Figure 6.1 primary nucleon energies in excess of 1 TeV (i.e., 10^{12} eV or 10^6 MeV) are relatively few in number; and primary cosmic-ray particles with energies in excess of 10^{20} eV or 10^{14} MeV are very rare, but have been reported (Linsley, 1963; Hayashida *et al.*, 1994; Bird *et al.*, 1995; Nagano and Watson, 2000). The rarity of such extremely high-energy primary cosmic-ray particles can be visualized with the graph of primary cosmic-ray flux as a function of nucleon energy illustrated in Figure 6.2.

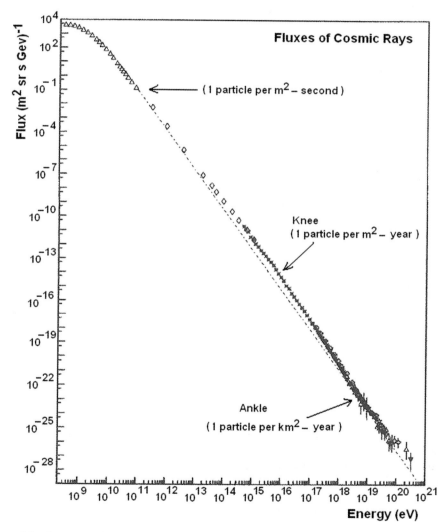

Figure 6.2 Spectrum of cosmic rays greater than 100 MeV. This figure was produced by S. Swordy, University of Chicago and published in Cronin (1999). The cosmic rays consist predominantly of nucleons ranging in species from protons to iron nuclei with traces of nuclei of the heavier elements. Not included in the figures are notations for estimated cosmic-rays fluxes of about 1 particle/cm^2 sec at the low-energy end of 10^8 eV and only of the order of 1 particle/km^2 century at the high-energy end of 10^{20} eV of the spectrum. (From Cronin (1999), reprinted with permission from The American Physical Society.)

The cosmic-ray scientific community is studying the knee and ankle regions of the cosmic-ray spectrum illustrated in Figure 6.2. The knee region at particle energies of 10^{15}–10^{16} eV shows a decline in the particle flux, while the ankle region of the spectrum occurring between 10^{18} and 10^{19} eV shows a rise in particle flux. Sundaresan (2001) summarized the current thinking for the production of the knee and ankle regions of the spectrum. It appears that the ankle is caused by a higher-energy cosmic-ray population mixed in with a lower-energy population at energies a couple of orders of magnitude below the ankle. The spectrum below 10^{18} eV is considered to be of galactic origin, and the higher energy population in excess of 5×10^{18} eV is considered to be of extragalactic origin (Cronin, 1999; Sundaresan, 2001).

Much attention in the cosmic-ray community is also focused on the mechanisms by which the cosmic-ray nucleons are accelerated in space to energies in excess of 10^{20} eV. Enrico Fermi (1949) proposed and calculated the acceleration of cosmic rays resulting from a series of collisions with magnetic fields or magnetic clouds traversing the universe whereby the cosmic-ray particles gain energy each time they bounce off the magnetic fields. This is now referred to as Fermi acceleration, and Cronin (1999) explains that subsequent work has shown that multiple "bounces" off turbulent magnetic fields associated with supernova shock waves is the more efficient acceleration process (Drury, 1983).

6.3   SHOWERS OF THE COSMIC RADIATION

Cosmic radiation consisting of high-energy nucleons striking the TOA collide with atoms of the air to produce a cascade of secondary subatomic particles and electromagnetic radiation referred to as showers of the cosmic radiation. During the years that preceded the development of man-made high-energy particle accelerators, cosmic rays provided much information to the field of high-energy particle physics. The collisions of high-energy nucleons with atomic nuclides of the atmosphere provided natural "atom-smashers" that led to the early discovery of subatomic particles.

When a cosmic-ray particle strikes an atomic nucleus of a gaseous molecule of the atmosphere (e.g., N_2, O_2, etc.), a nuclear disintegration follows producing high-energy secondary nucleons and charged and neutral π mesons (pions). These secondary nucleons collide with additional atomic nuclei of the atmosphere producing a cascade of particles including mesons, nucleons of various masses, neutrons, and products of meson–nuclear interactions and meson decay, namely, electromagnetic radiation, positrons, negatrons, and neutrinos. Figure 6.3 provides a good schematic of the various possible nucleon interactions that produce the cascade of secondary cosmic radiation in the atmosphere.

Mesons are strongly interacting particles of mass intermediate between that of the electron ($0.511 \, \mathrm{MeV}/c^2$) and the proton ($938 \, \mathrm{MeV}/c^2$). Among the mesons produced are (a) kaons (K-mesons), which have a mass of about $490 \, \mathrm{MeV}/c^2$, lifetimes of approximately 10^{-10} and 10^{-8} sec for the neutral (K^0) and charged ($K^{\pm}$) kaon, respectively, decaying to pions and/or muons (see eqs. (VI.5)–(VI.12) of Radioactivity Hall of Fame—Part VI), and (b) the pions (π-mesons) that can decay into muons (μ-mesons). A neutral pion has a mass of $135 \, \mathrm{MeV}/c^2$ and a lifetime of 8.3×10^{-17} sec. It decays into two gamma-ray photons, as illustrated in Figure 6.3, or into a positron–negatron pair and gamma ray with branching

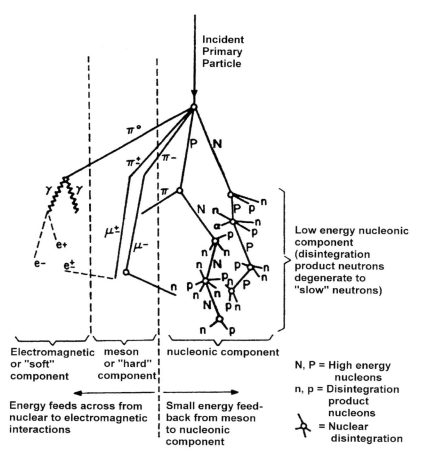

Figure 6.3 Schematic representation of the typical development of the secondary cosmic radiations within the atmosphere arising from an incident primary particle. (From Simpson *et al.* (1953), reprinted with permission from The American Physical Society.)

ratios of approximately 98.8 and 1.2%, respectively, for the two decay modes (Glasser *et al.*, 1961; Sundaresan, 2001) as follows:

$$\pi^0 \rightarrow \gamma + \gamma \; (98.798 \pm 0.032\%) \tag{6.15}$$

$$\pi^0 \rightarrow e^+ + e^- + \gamma \; (1.198 \pm 0.032\%) \tag{6.16}$$

The positron and negatron pair (e^+e^-) as products of the neutral pion decay, illustrated by eq. (6.16), is a result of internal conversion of a gamma-ray photon (i.e., internal pair production). The neutral pion decay may also manifest itself in yet a more rare form of dual internal conversion whereby two gamma-ray photons are converted to two positron–negatron pairs. The decay scheme is illustrated by eq. (6.17), and the tracks from such a decay scheme produced in a bubble chamber are illustrated in Figure 6.4.

$$\pi^0 \rightarrow e^+ + e^- + e^+ + e^- \tag{6.17}$$

Out of eight million π^0 decays in 836,000 bubble chamber pictures Samios *et al.* (1962) were able to find only 206 neutral pion decays with double internal conversion. The numerous neutral pions were produced by Samios *et al.* (1962) by the interactions of π^- mesons from a cyclotron, with protons of the bubble chamber liquid according to the reaction

$$\pi^- + p^+ \rightarrow n + \pi^0 \tag{6.18}$$

The π^- mesons were slowed down with a polyethylene absorber and then allowed to stop in the hydrogen bubble chamber. The numerous bubble chamber photographs could be analyzed for the identification of specific tracks by a digitized scanning machine.

Any of the gamma rays can, in turn, produce positron–negatron pairs via pair production, and the positron can undergo annihilation to gamma radiation. The positive and negative pions are antiparticles. Charged pions have a mass of 139.6 MeV/c^2 and a lifetime of 2.6×10^{-6} sec, much longer than that of the neutral pion. The positive pion decays to a positive muon and muon neutrino, whereas the negative pion decays into a negative muon and muon antineutrino as illustrated by the following equations:

$$\pi^+ \rightarrow \mu^+ + \nu_\mu \tag{6.19}$$

$$\pi^- \rightarrow \mu^- + \overline{\nu_\mu} \tag{6.20}$$

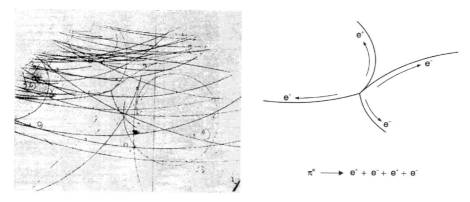

Figure 6.4 Photograph of a typical double internal conversion following π^0 decay on the left with a drawing of the specific tracks provided on the right. The tracks were produced by Samios *et al.* (1962) in a hydrogen bubble chamber 30.5 cm in diameter and 15.25 cm in depth. A magnetic field of 5.5 G produced the curved deflection of the oppositely charged electrons. (From Samios *et al.* (1962), reprinted with permission from The American Physical Society.)

The muon has a mass of $106\,\text{MeV}/c^2$ and lifetime of $2.2 \times 10^{-6}\,\text{sec}$. The positive muon decays to a positron and muon neutrino, and the negative muon to a negatron and antimuon neutrino as illustrated in Figure 6.3 and by the following equation:

$$\mu^{\pm} \rightarrow e^{\pm} + \nu_{\mu} + \overline{\nu_{\mu}} \tag{6.21}$$

Slow negative muons undergo nuclear absorption for nucleons with sufficiently high atomic number ($Z \approx 10$ or greater); however, for low Z nuclides such as carbon ($Z = 6$) slow negative muon decay occurs rather than absorption by the nucleus (Sundaresan, 2001).

At sea level muons are the most numerous of the charged particles resulting from cosmic-ray interactions with atomic nuclei of the atmosphere. These have been measured with an intensity of $100\,\text{m}^{-2}\text{sec}^{-1}$ at sea level. The relative abundances of the cosmic radiation shower components in the atmosphere at different altitudes with energies in excess of 1 GeV are illustrated in Figure 6.5. As illustrated in the figure, most muons are produced high in the atmosphere at an altitude of $\sim 15\,\text{km}$ where high-energy cosmic-ray nucleons encounter and undergo collision with nuclei of the atmospheric gases. Pions likewise are produced high in the atmosphere; however, their number at sea level is highly diminished due to the

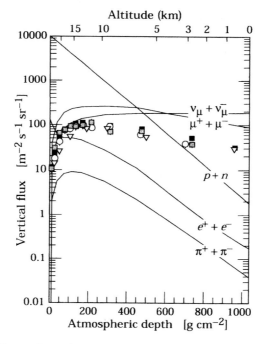

Figure 6.5 Vertical fluxes of cosmic rays in the atmosphere with $E > 1\,\text{GeV}$ estimated from the nucleon flux of eq. (6.14). The points show measurements of negative muons with $E_{\mu} > 1\,\text{GeV}$. (From Bellotti *et al.* (1996, 1999); Boezio *et al.* (2000b); Coutu *et al.* (2000); Gaisser and Stanev (2002), reprinted with permission from The American Physical Society.)

prompt decay of the neutral pion to gamma radiation and the decay of the charged pions to muons. The muon produced high in the atmosphere loses about 2 GeV of energy to ionization before reaching the ground and the mean energy of the muon at the ground is ≈ 4 GeV (Gaisser and Stanev, 2002), and these travel at speeds close to the speed of light ($> 0.99c$).

One might ask at this point the following question: If muons of the cosmic radiation showers are produced high in the earth's atmosphere (at altitudes of ($\sim$15,000 m or higher), travel at speeds close to the speed of light, and have an average lifetime of only 2.2×10^{-6} sec, how could such muons be detected to reach the earth's surface? The question is an obvious one, as it would appear with classical mathematics that, with such a short lifetime, muons would never be expected to reach the earth. Let us say, for example, a muon is produced at the altitude of 15,000 m and it travels at the speed v of $0.9993c$, which would correspond to a speed in units of m/sec as

$$0.9993c = (0.9993)(2.9979 \times 10^8 \, \text{m/sec}) = 2.9958 \times 10^8 \, \text{m/sec} \qquad (6.22)$$

The classical calculation of its average distance d of travel during its lifetime would be

$$d = (v)(t) = (2.9958 \times 10^8 \, \text{m/sec})(2.2 \times 10^{-6} \, \text{sec}) = 659.1 \, \text{m} \qquad (6.23)$$

Thus, the muon would be expected to decay, according to classical transformations, after traveling only 659 m on the average and, therefore, never be detected an earth. However, for such a muon viewed by an observer on earth, he or she would view the two events: (A) the creation of the muon and (B) the decay of the muon, to occur in two locations in space–time coordinates. The Lorentz time transformations demonstrate that the two events, A and B, separated by the time interval Δt, will be dilated to the time interval Δt_{earth}, when viewed by an observer on earth. The time dilation is calculated according to the following Lorentz transformation derived by Gautreau and Savin (1999) and Serway *et al.* (1997):

$$\Delta t_{\text{earth}} = \frac{\Delta t}{\sqrt{1 - (v^2/c^2)}} \qquad (6.24)$$

where v is the velocity of the particle, and c the invariable speed of light in a vacuum. For the muon traveling at $0.9993c$, we can calculate Δt_{earth} or the dilated lifetime of the muon as follows:

$$\Delta t_{\text{earth}} = \frac{\Delta t}{\sqrt{1 - (v^2/c^2)}} = \frac{2.2 \times 10^{-6} \, \text{sec}}{\sqrt{1 - (0.9993)^2}} = \frac{2.2 \times 10^{-6} \, \text{sec}}{0.0374} = 58.8 \times 10^{-6} \, \text{sec} \qquad (6.25)$$

According to the dilated time of 58.8×10^{-6} sec, the observer on earth will determine that the muon would travel, before it decays, a distance of

$$d = (v)(\Delta t_{\text{earth}}) = (2.9958 \times 10^8 \, \text{m/sec})(58.8 \times 10^{-6} \, \text{sec}) = 17615 \, \text{m} \qquad (6.26)$$

which is well beyond the distance of 15,000 m elevation when it was created, permitting its detection on earth. The distance is only an average figure, because the muon lifetime of

2.2×10^{-6} sec is an average lifetime. For thorough treatments of relativistic space–time measurements and the applications of Lorentz transformations to these measurements see books by Gautreau and Savin (1999) and Serway *et al.* (1997).

The example provided here is a realistic one, and the relativistic time dilation has been demonstrated by Bailey *et al.* (1977, 1979) based on the historical tests of Rossi and Hall (1941).

The historical work of Rossi and Hall entailed the measurement of the muon flux on Mt. Washington in New Hampshire at the altitude of 2000 m and at the mountain base. They found that the muon flux at both altitudes did not differ greatly, only by the ratio of 1.4, that is, the muon flux was 1.4 times as high at the altitude of 2000 m compared to the ground level. Calculations based on a muon half-life of 1.5×10^{-6} sec, the muon flux ratio for the two altitudes, should have been approximately 22 even when attributing the highest unrealistic speed of light to the muons. The actual low ratio of 1.4 could only be explained by applying the time dilation calculations for muons traveling at the speed of $0.994c$.

More recent experiments at CERN by Bailey *et al.* (1977, 1979) involved accelerating muons in a storage ring to speeds of up to $0.9994c$. The rates of decay of the accelerated muons were measured by detecting their electron decay-product emissions. The accelerated muons were found to have a lifetime of 29.3 times higher than a muon at rest. This agrees perfectly with the calculated dilated lifetime of the muon according to the Lorentz transformation

$$\Delta t = \frac{\Delta t_0}{\sqrt{1 - (v^2/c^2)}} \tag{6.27}$$

where Δt_0 is the lifetime of the muon at rest, v the speed of the muon, and c the speed of light. The dilated lifetime of the muon at $0.9994c$ is calculated to be

$$\Delta t = \frac{2.2 \times 10^{-6} \sec}{\sqrt{1 - (0.9994)^2}} = 63.58 \times 10^{-6} \sec \tag{6.28}$$

The increase in the observed lifetime of the accelerated muons over the muons at rest is

$$\frac{\Delta t}{\Delta t_0} = \frac{63.58 \times 10^{-6} \sec}{2.2 \times 10^{-6} \sec} = 28.9 \tag{6.29}$$

6.4 COSMIC RAYS UNDERGROUND

The only significant cosmic radiation with energy sufficient to penetrate considerable depths of earth are charged muons and muon neutrinos. High-energy muons will traverse rock and lose energy via either direct ionization or radiative processes. The radiative processes include Bremsstrahlung production, which lead to positron–electron pair production, and photonuclear interactions. The muon cosmic-ray intensities diminish with depth as expected. This is a subject of concern to scientists who must measure very low levels of radioactivity such as those encountered in ^{14}C-dating measurements. Consequently, underground laboratories have been built to achieve lowest possible background radiation interference in the counting of very low levels of natural radioactivity. Muon ranges in rock are measured in units of km-water-equivalent (km.w.e) where 1 km.w.e. $= 10^5$ g/cm^2 of standard rock. Average ranges in standard rock ($A = 22$, $Z = 11$, $\rho = 2.65$ g/cm^3) for muons of 10, 100, 1000, and 10,000 GeV are 0.05, 0.41, 2.45, and 6.09 km.w.e. The vertical muon intensity versus underground depth of standard rock has been determined. The muon vertical intensity drops from 10^{-6} to 10^{-12} cm^{-2} sr^{-1} sec^{-1} over the range of rock depths from 1 to 10 km.w.e., respectively (Gaisser and Stanev, 2002).

6.5 ORIGINS OF COSMIC RADIATION

Cosmic rays can originate from (i) energetic particles associated with mass ejections from solar flares and similar energetic solar events, (ii) anomalous cosmic rays, which are particles of interstellar origin accelerated at the edge of the heliopause and accelerated at the termination shock in the solar wind, (iii) high-energy particles of galactic origin far outside the heliosphere or our solar system, and (iv) extragalactic sources.

The sun has been identified as one of the sources of high-energy cosmic-ray particles that hit the TOA, and these particles can be accelerated during solar flares. The mass ejections associated with solar flares drive shocks into the interplanetary medium where particles undergo acceleration via shocks in the heliosphere. Linked to these are the anomalous cosmic rays, which are classified as a subset of particles accelerated in the heliosphere; these are particles of interstellar origin accelerated at the termination shock in the solar wind (Gaisser, 2000). Neutron monitors are used to detect the secondary neutrons produced by the cosmic-ray nucleon collisions with nuclides of the earth's atmospheric molecules. The gamma-ray images of the galaxy taken in space provide evidence for the gamma radiation produced by interactions of the accelerated particles in the solar atmosphere in addition to gamma-radiation originating from neutral meson decay described previously.

Galactic cosmic rays can arise from a supernova, which is the final stage in a star's evolution. During the death of a star it explodes as a supernova producing heavy particles and accelerating them in a shock wave into interstellar gas. Supernova are sources of galactic cosmic rays with particles undergoing acceleration to $\sim 10^{15}$ eV or higher. The interactions of these accelerating particles also serve as sources of gamma radiation that we detect throughout the galaxy.

Accelerated particles in excess of approximately 10^{18} eV or higher are considered to be possibly of extragalactic origin. Most galactic cosmic rays are confined by the magnetic field lines of the galaxy. Extragalactic cosmic-ray particles with sufficient energy to invade our galaxy (10^{18} to beyond 10^{20} eV) are relatively few in number as discussed previously. Several theories and models exist for the origins of extragalactic cosmic rays and these are reviewed by Gaisser (2000).

6.6 COSMIC BACKGROUND RADIATION

The hot big bang or "explosive" nucleosynthesis theory of the origin of the universe predicts the existence of a relic cosmic microwave background (CMB) radiation left over by

Georges Lemaître
(1894-1966)

the big bang. The first to describe a very hot and explosive beginning of the universe on a scientific basis was Georges Lemaître, a Belgian mathematician and Catholic Priest. Georges-Henri Lemaître obtained a doctorate in mathematics at the University of Leuran in 1920 and ordained a Catholic Priest in 1923. That same year he studied at the University of Cambridge under astronomer Arthur Eddington and went on to study the following year at Harvard College Observatory in Cambridge, MA and also registered that year at the Massachusetts Institute of Technology (MIT) for a doctoral degree in the sciences. By 1927 he obtained his Ph.D. degree at MIT and returned to Belgium as Professor at the University of Leuven. The very year he received his doctorate, Lemaître (1927) published his conclusions that the universe was expanding based on Einstein's general theory of relativity and the red-shift surrounding extragalactal bodies. In a biographical account Midbon (2000) relates that following Edwin Hubble's systematic observations confirming the red shift, Lemaître sent a copy of his 1927 paper to British astronomer Arthur Eddington, who then realized that Lemaître had bridged the gap between observation and theory. Eddington therefore requested that an English translation of Lemaitre's paper be published in the March 1931 *Monthly Notices of the Royal Astronomical Society* (Lemaître, 1931a). Georges Lemaitre went further to propose in a letter published in the journal *Nature* (Lemaître, 1931b) his interpretations that would lead to the big bang theory. In a book Mather and Boslough (1996) summarized Lemaître's theory as follows:

> A letter Lemaître wrote to *Nature* magazine in 1931 was effectively the charter of what was to become the Big Bang theory. He theorized that this primordial explosion, occurring on 'a day without yesterday,' had burst forth from an extremely dense point of space and time. He began calling this the 'primeval atom'. By now Lemaître had

become a celebrity in his own right for his revolutionary ideas. At an immense gathering of the British Association for the Advancement of Science in London the same year, he speculated before an audience of several thousand scientists that the cosmic rays may have originated in the primordial explosion. Eventually, he thought, they might prove to be material evidence of the universe's natural beginning.

In January of 1933 Georges Lemaître traveled with Albert Einstein for a series of seminars in California including a lecture at the California Institute of Technology at Pasadena. At one of these seminars after Lemaître detailed his theory on the explosive beginning of the universe, Einstein stood up, applauded and acclaimed, "This is the most beautiful and satisfactory explanation of creation to which I have ever listened." (Midbon, 2000). There was, of course, some opposition to Lemaître's theory on the creation of the universe as explained by Midbon (2000), particularly at Cambridge University, and that it was Fred Hoyle, an astronomer at Cambridge, who had coined the term "Big Bang" in sarcasm to Lemaître's theory. Nevertheless, before Lemaître's death he was able to get satisfaction from the discovery by Nobel Laureates Arno Penzias and Robert Wilson in 1965 of the cosmic microwave radiation, the primordial radiation, that would be key evidence for the big bang theory. These findings are subsequently discussed.

A primordial relic radiation from a hot early beginning of the universe was first predicted by George Gamow in 1948 and by his students Ralph Alpher and Robert Herman in 1949 (see Alpher and Herman, 1988). Gamow (1948) made calculations of the conditions in the early stage of the universe and understood that the universe had to have a very hot beginning to prevent the coalescence of all hydrogen nuclei into heavier elements. In a story related by Hawking (1988), the often humorous Gamow convinced Hans Bethe to coauthor a paper in 1948 with Gamow's student Ralph Alpher whereby the authors would be cited as Alpher, Bethe and Gamow so that the first letters of each author would coincide with the familiar names of the "alpha, beta and gamma" radiations, which would be most appropriate for a paper on the beginnings of the universe. In their theory they predicted a very hot and dense beginning of the universe, which today with the expanding universe would exhibit a residual photon radiation corresponding to a temperature that would have reduced to only a few degrees above absolute zero of the thermodynamic temperature Kelvin (i.e., few degrees $>0\,K$). Alpher and Herman (1949) followed the temperature evolution of the early beginning of the universe and predicted a cooling down to a current temperature of $5\,K$. Alpher et al. (1953) went even further to carry out what was considered by Nobel Laureate Robert W. Wilson (1978) as the first thorough analysis of the early history of the universe, but failed to calculate or even mention their earlier prediction of 1949 that the universe would have cooled down to about $5\,K$.

It was not until 1965 did Arno Penzias and Robert Wilson at Bell Telephone Laboratories in New Jersey accidentally discover a CMB radiation. They were building and testing a new sensitive radio receiver, when they observed an excess noise that was of equal intensity regardless of the direction that the detector or antenna was pointing in the atmosphere. They concluded that the microwave radiation had originated from deep space even beyond our solar system and even our galaxy, because as the earth orbited the sun and rotated on its axis the antenna would be pointing to a different direction in space. At the same time a team of physicists led by Robert Dicke at Princeton University was studying microwaves to devise a system that would demonstrate the CMB predicted by Gamow and Alpher. When Penzias and Wilson learned about this research work in search of the CMB, they realized that they had discovered it. Penzias and Wilson (1965) reported their discovery in the *Astrophysics Journal* and Dicke and coworkers (1965) at Princeton reported the interpretation of the discovery in the same journal as back-to-back papers in the same issue. Penzias and Wilson shared the Nobel Prize in Physics 1978 for this discovery.

In the Nobel Presentation Speech, Professor Lamek Hulthén of the Royal Academy of Sciences on December 8, 1978, in Stockholm, Sweden, underscored the importance of the work and discovery of Penzias and Wilson. An excerpt of his speech is the following:

> In the early 1960's a station was set up in Holmdel [Bell Laboratories, Holmdel, New Jersey] to communicate with the satellites Echo and Telstar. The equipment, including a steerable horn antenna, made it a very sensitive receiver for microwaves, i.e. radio waves of a few cm wave length. Later radio astronomers Arno Penzias and Robert Wilson got the chance to adapt the instrument for observing radio noise, e.g. from the Milky Way … The task of eliminating various sources of errors and noise turned out to be very difficult and time-consuming, but by and by it became clear that they had found a background radiation, equally strong in all directions, independent of the time of the day and the year, so it could not come from the sun or our Galaxy … Continued investigations have confirmed that this background radiation varies with wavelength in the way prescribed by well known laws for a space, kept at the temperature [of] 3 K. Our Italian colleagues call it "la luce fredda" – the cold light … But where does the cold light come from? A possible explanation was given by Princeton physicists Dicke, Peebles, Roll and Wilkinson (1965) and published together with the report of Penzias and Wilson (1965). It leans on a cosmology theory, developed about 30 years ago by the Russian born physicist George Gamow and his collaborators Alpher and Herman. Starting from the fact that the universe is now expanding uniformly, they concluded that it must have been very compact about 15 billion years ago and ventured to assume that the universe was born in a huge explosion, the 'Big Bang'. The temperature must then have been fabulous: 10 billion degrees [K], perhaps more. At such temperatures lighter chemical elements can be formed from existing elementary particles, and a tremendous amount of radiation of all wavelengths is released. In the ensuing expansion of the universe, the temperature of the radiation rapidly goes down. Alpher and Herman estimated that this radiation would still be left with a temperature around 5 K … Have Penzias and Wilson discovered 'the cold light from the birth of the universe'? It is possible – this much is certain that their exceptional perseverance and skill in the experiments led them to a discovery, after which cosmology is a science, open to verification by experiment and observation.

The significance of the discovery of the CMB radiation corresponding to a temperature of approximately 3 K by Arno Penzias and Robert Wilson was commemorated in a postage stamp issued in Sweden. The stamp illustrated on the following page shows a temperature of 3 K. A more precise figure has been determined to be a cosmic radiation background corresponding to a 2.73 K blackbody radiation spectrum. This is discussed in more detail subsequently.

The precise spectrum of the CMB has been determined over three decades of frequency via many satellite, rocket, balloon, and ground measurements, and the cumulated data of radiant flux intensity as a function of radiation frequency is plotted in Figure 6.6. The CMB spectrum fits exactly to that predicted by a 2.73 K blackbody radiation spectrum whereby the brightness (B_ν) or intensity (I_ν) per unit of radiation frequency or wavelength is a function of the absolute temperature of the blackbody. The term "brightness" is used when measuring the radiant flux from a projected area of the surface of a blackbody, while the term "intensity" is used when reception of the radiant flux is measured (Woan, 2000). The Planck functions for the brightness or intensity are given by the following equations:

$$I_\nu(T) = B_\nu(T) = \frac{2h\nu^3}{c^2}\left[\exp\left(\frac{h\nu}{kT}\right) - 1\right]^{-1} \tag{6.30}$$

and

$$I_\lambda(T) = B_\lambda(T) = B_\nu(T)\frac{d\nu}{d\lambda} = \frac{2hc^2}{\lambda^5}\left(\frac{hc}{\lambda kT} - 1\right)^{-1} \tag{6.31}$$

where I_ν and B_ν are the intensity or brightness, respectively, of blackbody radiation per unit frequency (W/m^2 Hz sr) and I_λ and B_λ are the intensity and brightness, respectively, of blackbody radiation per unit wavelength (W/m^2 m sr), h is Planck's constant (6.626×10^{-34} J sec), c is the speed of light (2.99×10^8 m/sec), k is the Boltzmann constant (1.38×10^{-23} J/K), and T is the temperature in Kelvin (Woan, 2000).

The blackbody radiation spectra plotted according to eq. (6.30) over a broad range of radiation frequencies for three temperature levels, namely, 5.46, 2.73, and 1.36 K are illustrated in Figure 6.7. Compare the measured spectrum of the CMB of Figure 6.6 of radiation intensity per unit frequency with the calculated spectrum for a 2.73 K blackbody of Figure 6.7 (curve b). The measured CMB spectrum of Figure 6.6 is identical to the calculated radiation spectrum for a 2.73 K blackbody. The current precise calculated temperature corresponding to the measured cosmic radiation background is 2.73 K (Smoot and Scott, 2002).

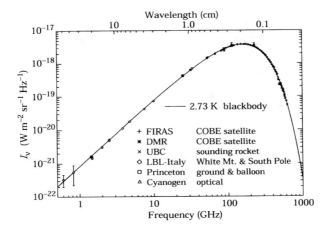

Figure 6.6 Precise measurement of the CMB radiation spectrum. The line represents a 2.73 K blackbody, which describes the spectrum very well, especially around the peak of intensity. The spectrum is less well constrained at 10 cm and longer wavelengths. FIRAS, DMR, UBC, LBL-Italy, Princeton, and Cyanogen are sources of the experimental data points. These sources are cited in the references. (From Smoot and Scott (2002), reprinted with permission from The American Physical Society.)

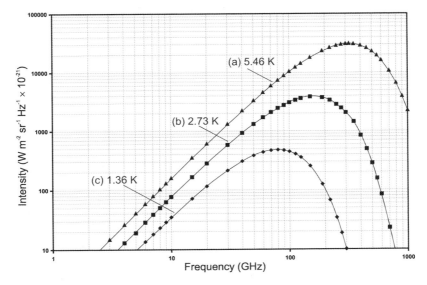

Figure 6.7 Logarithmic plots of the spectra of blackbody radiation depicting the intensity (I_ν) of the CMB radiation as a function of radiation frequency (Hz) when the expanding universe was (a) hotter at 5.46 K and one-half its current size, (b) its current temperature of 2.73 K at its current size, and (c) its predicted one-half the current temperature when the universe expands to twice its size and cools to 1.36 K in approximately 10 billion years.

A quick calculation of the temperature that corresponds to the CMB spectrum can be made directly from Figure 6.6 and Wien's displacement law that describes the inverse proportionality between the wavelength of maximum intensity of emission, λ_{max}, and temperature of a blackbody, that is, as the temperature of the blackbody decreases the wavelength of maximum brightness (emission intensity) increases. This inverse proportionality was defined by the German Nobel Laureate Wilhelm Wien in 1893 as

$$\lambda_{max}T = 5.1 \times 10^{-3}\,\text{m K} \qquad \text{for } B_\nu \tag{6.32}$$

or

$$\lambda_{max}T = 2.9 \times 10^{-3}\,\text{m K} \qquad \text{for } B_\lambda \tag{6.33}$$

Eq. (6.32) or (6.33) can be applied to calculate the temperature as a function of the wavelength of maximum intensity or brightness when the brightness is expressed as a function of radiation frequency, B_ν, or wavelength B_λ, respectively (Woan, 2000). For example, if we take the value of 0.185 cm from the top axis of Figure 6.6 as the wavelength of maximum intensity or brightness ($I_\nu = B_\nu$) we can calculate the temperature corresponding to the microwave background radiation according to eq. (6.32) as

$$T = \frac{5.1 \times 10^{-3}\,\text{m K}}{\lambda_{max}} = \frac{5.1 \times 10^{-3}\,\text{m K}}{1.85 \times 10^{-3}\,\text{m}} = 2.7\,\text{K} \tag{6.34}$$

The temperature calculated in eq. (6.34) is very close to the more precisely determined temperature of 2.73 K (Smoot and Scott, 2002). The residual temperature of only a few degrees (2.73 K), the CMB radiation spectrum that precisely corresponds to the blackbody radiation spectrum for 2.73 K (Figure 6.6), and the isotropy (uniformity in all directions) of the radiation among other data (see Smoot and Scott, 2002) are compelling evidence for the hot big bang hypothesis for the origin or creation of the universe.

Wilhelm Wien (1864–1928) received the Nobel Prize in Physics 1911 for his discoveries concerning the laws of thermal radiation and, in particular, for his displacement law described above. The importance of Wien's work, reflected in the most recent findings on cosmic radiation, was underscored by E. W. Dahlgren, President of the Royal Swedish Academy of Sciences, on December 10, 1911, in the following excerpt of his Presentation Speech of the Nobel Prize to Wilhelm Wien:

> The importance of Wien's displacement law extends in various directions… it provides one of the conditions which are required for the determination of the relationships between energy radiation, wavelength and temperature for black bodies, and thus represents one of the most important laws in the theory of heat radiation… Thus it became possible to determine the temperature of bodies, within fairly narrow limits, simply by seeking the wavelength at which radiation is greatest. The method has successfully been applied to the determination of the temperature of our light sources, of the sun, and of some of the fixed stars…

Wilhelm Wien
(1864-1928)

What could not be foreseen at the time of Wilhelm Wien was that his law would be instrumental also in determining the temperature of the universe from the wavelength of the CMB radiation.

The hot big bang hypothesis describes the universe as having a very hot ($\sim 10^{10}$ K a second after the big bang) and small, dense beginning with an expansion and cooling of the universe ever since the big bang. The measured CMB radiation illustrated in Figure 6.6 provides evidence for the prediction of the big bang creation and the ever expansion and cooling of the universe. When the universe expands to twice its size, its temperature falls by half (Hawking, 1988). Consequently, when the universe was one-half its present size, its temperature would have been twice as hot (5.4 K) and the CMB radiation would have had a brightness peaking at a higher photon frequency as illustrated in Figure 6.7. Likewise, if we go further back in time when the universe was 1/100th its current size, the universe would have been a hundred-fold hotter than now or 273 K equivalent to 0° C. The CMB radiation would have had a spectrum shifted to yet higher frequencies and shorter wavelengths.

As we go further and further back in time to the big bang, the universe was proportionally hotter, smaller and more dense. The postage stamp and souvenir sheet issued in Macao, China and pictured on the following page illustrates the changes in the temperature of the universe from about 10^{32} K some 10^{-42} sec after the big bang to the current temperature of about 3 K some 15 billion years after the big bang. The temperature at about a second after the big bang is estimated to have been about 10 billion degrees Kelvin or 10^{10} K, and about 1 million years later the universe consisting then of only neutral hydrogen and smaller amounts of light elements would have expanded and cooled to about a few thousand degrees Kelvin or 10^3–10^4 K (Hawking, 1988). Because the microwave background photons interact very

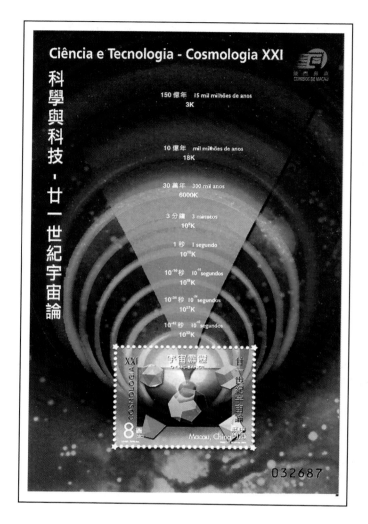

weakly with neutral hydrogen, the constitution of the universe at that time could be considered as the "surface of last scattering", that is, when the CMB radiation was scattered and continued to travel in all directions through the expanding universe with time. The CMB radiation shifts to longer wavelengths as the universe expands and cools. The universe is presently expanding by 5–10% every 10^9 years as determined by the Doppler effect of the velocity of travel of galaxies away from earth (Hawking, 1988). We could extrapolate, therefore, that the universe would expand to twice its current size and cool to one-half its current temperature to 1.36 K in approximately 10^{10} years. The CMB radiation at that time in the future would be expected to be that corresponding to the spectrum of blackbody radiation at about 1.36 K as illustrated in Figure 6.7c.

The temperatures depicted by the blackbody radiation spectra of Figure 6.7 can be confirmed with Wien's law given above in eq. (6.34). The temperatures for the three spectra (a), (b), and (c) of Figure 6.7 are calculated from the wavelength of maximum brightness (λ_{max}) as follows:

Spectra (a): From spectra (a), the frequency, ν, at maximum radiation intensity is found to be approximately 320 GHz, which is converted to wavelength in meters as

$$\lambda = \frac{c}{\nu} = \frac{2.99 \times 10^8 \, \text{m/sec}}{320 \times 10^9 \, \text{sec}^{-1}} = 9.34 \times 10^{-4} \, \text{m} \tag{6.35}$$

and the temperature according to Wien's law is calculated as

$$T = \frac{5.1 \times 10^{-3} \, \text{m K}}{\lambda_{max}} = \frac{5.1 \times 10^{-3} \, \text{m K}}{9.34 \times 10^{-4} \, \text{m}} = 5.46 \, \text{K} \tag{6.36}$$

Spectra (b): The frequency, ν, at maximum intensity is found to be approximately 160 GHz, and the temperature is calculated as

$$T = \frac{5.1 \times 10^{-3} \, \text{m K}}{\lambda_{max}} = \frac{5.1 \times 10^{-3} \, \text{m K}}{2.99 \times 10^8 \, \text{m/sec}/160 \times 10^9 \, \text{sec}^{-1}} = \frac{5.1 \times 10^{-3} \, \text{m K}}{1.87 \times 10^{-3} \, \text{m}} = 2.73 \, \text{K}$$

$$\tag{6.37}$$

Spectra (c): The frequency, ν, at maximum intensity is found to be approximately 80 GHz, and the temperature is calculated as

$$T = \frac{5.1 \times 10^{-3} \, \text{m K}}{\lambda_{max}} = \frac{5.1 \times 10^{-3} \, \text{m K}}{(2.99 \times 10^8 \, \text{m/sec})/80 \times 10^9 \, \text{sec}} = \frac{5.1 \times 10^{-3} \, \text{m K}}{3.74 \times 10^{-3} \, \text{m}} = 1.36 \, \text{K} \tag{6.38}$$

The differences between the current microwave background radiation intensity of the universe compared to that predicted for the universe when it expands to twice its current size are most easily recognized when plotting the blackbody radiation spectra at the current temperature of $T = 2.73$ K and the predicted temperature of 1.36 K on linear scales as illustrated in Figure 6.8.

6.7 DOSE FROM COSMIC RADIATION AND OTHER SOURCES

Radiation exposure was historically measured by the roentgen (R), which is a measure of the quantity of radiation deposited in air from the amount of charge or ionization produced by the radiation in air. By definition 1 R = 2.58×10^{-4} C/kg of air at STP, that is, one roentgen will produce 2.58×10^{-4} coulombs of ion pairs in 1 kg of air. The roentgen is a unit of exposure that is mostly historical and seldom used; it still occasionally appears on some dosimeter readings. Of more significance is the measure of absorbed dose, that is, the energy of radiation absorbed per unit mass of absorber. The original unit of absorbed dose is the rad, which is derived from the term "radiation absorbed dose". The rad has been

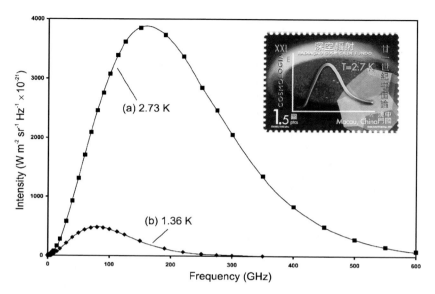

Figure 6.8   Linear plots of the spectra of blackbody radiation depicting the intensity of the cosmic background radiation as a function of radiation frequency at (a) the current temperature ($T = 2.7$ K) and size of the universe and (b) the predicted temperature ($T = 1.3$ K) when the universe expands to twice its current size. The inset in the upper right corner of the graph is a picture of a postage stamp issued in Macao, China in 2004 commemorating the finding of the current CMB radiation spectra corresponding to the linear spectra of blackbody radiation at $T = 2.7$ K.

replaced with the gray (Gy), which is the SI unit of absorbed dose. The use of SI units is recommended by the International Commission on Radiation Units and Measurements (ICRU). The rad and gray have the following equivalents:

$$100 \, \text{rad} = 10^4 \, \text{erg/g} = 1 \, \text{Gy} = 1 \, \text{J/kg} \tag{6.39}$$

$$1 \, \text{rad} = 10 \, \text{mGy} = 100 \, \text{erg/g} \tag{6.40}$$

$$1 \, \text{mrad} = 10 \, \mu\text{Gy} \tag{6.41}$$

As $1 \, \text{eV} = 1.602 \times 10^{-19}$ J we can convert the gray to units of electron volt energy deposited in a kilogram of absorber or

$$1 \, \text{Gy} = 6.24 \times 10^{12} \, \text{MeV/kg} \tag{6.42}$$

Another important unit of radiation dose is the rem. The rem is a measure of absorbed dose in biological tissue. The unit of measure is derived from the term "roentgen equivalent for man".

The rem was created as a measure of dose of ionizing radiation to body tissue in terms of its estimated biological effect; its SI unit is the sievert (Sv) and

$$100\,\text{rem} = 1\,\text{Sv} = 1\,\text{J/kg} \tag{6.43}$$

$$1\,\text{rem} = 10\,\text{mSv} \tag{6.44}$$

The rem or Sv (sievert) are referred to as units of equivalent dose, because the dose is measured on the basis of a weighting factor (w_R), which defines the relative hazard of radiation on the basis of the types and energies of the radiations by placing all radiation classes on the same dose level or equivalent (L'Annunziata, 1987). The weighting factor was formerly known as the quality factor (QF), which is defined as the ratio of the gamma-ray or x-ray dose to the dose required to produce the same biological effect by the radiation in question. Table 6.2 lists the radiation weighting factors according to radiation type and energy. The weighting factor of a given radiation is a function of the radiation linear energy transfer (LET), that is, the radiation energy loss per path length of travel. As discussed in Chapter 2, Section 2.5, radiations of high mass and charge, such as alpha particles or fission products will have a higher LET than electrons or beta particles, muons, or x- or gamma-ray photons. Because photons will produce electrons as secondary particles, x- or gamma-ray photons are classified to have the same radiation weighting factor as electrons. According to the weighting factors provided in Table 6.2, alpha particles are considered to be 20 times more damaging to cells of the human body than beta particles, muons, and x- or gamma-radiation. However, alpha particles and other heavy nuclei have short ranges in matter (see Chapter 1) compared to other radiation types. Thus distance provides the best protection from alpha particles and other heavy nuclei. Also, the weighting factors for neutrons in Table 6.2 classify the neutrally charged neutrons as more hazardous to biological tissue than the charged electrons or muons. This is because neutrons will produce recoil protons and ions of higher mass via neutron collisions in body tissue.

Table 6.2

Radiation weighting factors

Radiation	w_R
x- and gamma-rays, all energies	1
Electrons and muons, all energies	1
Neutrons	
<10 keV	5
10–100 keV	10
>100 keV to 2 MeV	20
2–20 MeV	10
>20 MeV	5
Protons (other than recoils) >2 MeV	5
Alphas, fission fragments, and heavy nuclei	20

Source: ICRP (1991) and Donahue and Fassó (2002).

The equivalent dose in Sv can be calculated from the weighting factors as

$$Sv = \text{absorbed dose in grays} \times w_R \tag{6.45}$$

where the weighting factor expresses potential damage to cells, and long-term risk (primarily cancer and leukemia) from low-level chronic exposure, which is dependent on radiation type and other factors (ICRP, 1991; Donahue and Fassó, 2002). Using the notation of Lilley (2001) for average absorbed dose in tissue eq. (6.45) can be written as

$$Sv = D_{T,R} \times w_R \tag{6.46}$$

where Sv is the equivalent dose in sievert and $D_{T,R}$ is the average absorbed dose in tissue T from a given type of radiation R. For example, from the weighting factors of Table 6.2 and eq. (6.46) we can calculate that an average absorbed dose of 0.01 Gy from alpha particles ($w_R = 20$) will produce the same biological effect as a 0.2 Gy dose from x- or gamma-radiation ($w_R = 1$). Likewise, a 0.1 Gy dose from 50-keV neutrons ($w_R = 10$) will produce the same biological effect as a 1 Gy dose of electrons ($w_R = 1$). When more than one radiation type contributes to the absorbed dose, the equivalent dose is calculated according to the weighted sum of the contributions from each radiation type.

The ICRP (1991) recommendation for limits of exposure for radiation workers for whole body dose is 20 mSv/year (2 rem/year) averaged over 5 years, with the dose in any one year ⩽50 mSv (5 rem/year). It is estimated by the United States National Safety Council that the average person in the United States receives only 3.6 mSv (0.36 rem or 360 mrem) of accumulated radiation dose per year from all sources including cosmic radiation, medical x-rays, natural radioactivity, etc. In most world areas the whole-body equivalent dose rate is ≈0.4–4 mSv/year or 40–400 mrem/year, although in certain areas of the world the dose rate can range up to 50 mSv/year or 5 rem/year (Donahue and Fassó, 2002).

The radiation sources and their approximate average relative contribution to the dose received by the human body are radon (~55%), medical x-rays (~11%), internal radiation (~11%), terrestrial radiation (~8%), cosmic radiation (~8%), nuclear medicine (~4%), consumer products (~3%), and other sources (~1%). Cosmic radiation contributes only a relatively small (~8% or ~30 mrem) portion of the total annual radiation dose the average person may receive from both man-made and natural radioactivity. Cosmic radiation, therefore, contributes only a small part to the total radiation that the human body receives by the average person during his or her lifetime. Natural radon is the largest source of radioactivity that may affect our health. This is followed with x-rays that we may receive from medical and dental diagnostic procedures and internal radiation arising from radionuclides within our own bodies. Radioactive sources used in nuclear medicine, such as in cancer treatment, and radioactivity from consumer products, such as smoke detectors, lawn fertilizer, cigarettes, stone or brick masonry, television sets, etc., contribute yet smaller amounts of radiation to the overall dose our bodies receive during our lifetimes. Finally, the vague term of other sources (~1%) cited earlier in this paragraph refers to radioactivity that might reach our bodies from radioactive waste associated with man-made radionuclides from nuclear power plants, nuclear weapons production and testing, and peaceful applications of radionuclides used in scientific research.

As we travel above sea level to higher altitudes in the earth's atmosphere, the dose rate from cosmic radiation will increase obviously since the atmosphere serves as a shield to cosmic radiation. Persons who live several thousand feet above sea level should expect to accumulate a higher dose from cosmic radiation over a period of a year than another person living at sea level in the same general region of the country. For example, Figure 6.9 illustrates approximate cosmic radiation dose rates for persons in the State of Idaho. A person living in the city of Idaho Falls, ID (altitude = 4740 ft) could expect to receive more than twice the cosmic radiation dose over the period of a year than a person living in Sandpoint, ID (altitude = 2150 ft). A person living in Leadville, CO, which is the highest city (10,150 ft) in the United States should expect to receive several times the annual dose rate than a person at or near sea level in the same State. Nevertheless, the accumulated dose received by the human body at inhabitable altitudes of the world are very low and contribute only a small percentage (~8%) to the total body dose received from all sources.

If we travel to altitudes accessible only by aircraft (7–12 km above the earth), the dose from cosmic radiation increases, and the amount of increase will be a function of latitude as illustrated in Figure 6.10. The effect of altitude on cosmic radiation was discussed in Section 6.3. At these high altitudes, the cosmic radiation of most concern are neutrons, charged particles including muons and protons (see Figure 6.5) and gamma radiation. Although cosmic radiation is incident from above the earth's atmosphere, exposure to cosmic radiation showers in the atmosphere aboard airplanes is considered isotropic, that is, it should be incident from all directions. This is the result of scattering of radiation in all directions as a result of nucleon collisions in the atmosphere.

Exposure to cosmic radiation is mostly a concern to intercontinental airline pilots, who fly at high altitudes and high degrees of latitude and who compile large numbers of flight hours per year over long careers. Hammer *et al.* (2000) report a mean cosmic radiation dose estimate of 35 mSv for 509 pilots, who averaged 26.6 years of employment at an average of 481 flight hours per year. The accumulated cosmic radiation dose of 35 mSv over 26.6 years will yield an average dose rate of 1.3 mSv/year. The dose to aircrew is a function of altitude, latitude, and flight time (flight hours) at various altitudes and latitudes of flight,

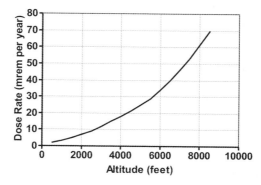

Figure 6.9 Approximate cosmic radiation dose rates as a function of inhabitable altitudes above sea level. (Data obtained from State of Idaho, Idaho National Engineering and Environmental Laboratory, Idaho Falls (http://www.oversight.state.id.us).)

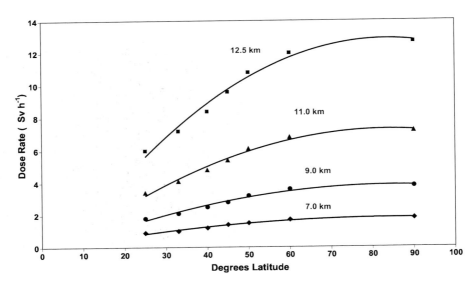

Figure 6.10 Cosmic radiation dose rate as a function of altitude above sea level and degrees latitude. The latitudes plotted are not precise, rather the seven data points are plotted at the upper ends of the general latitude ranges of 0–25°, 25–33°, 33–40°, 40–45°, 45–50°, 50–60°, and 60–90°, respectively. (Data from NRL (1998), National Radiation Laboratory, New Zealand.)

Table 6.3

Cosmic radiation dose to aircrew

Flight altitudes (km)	Flight latitudes (degrees)	Flight hours (h/year)	Dose rate (mSv/year)
7.0	–	500	<0.2
9.3–8.2	–	500–700	<0.5
11 (70%)[a]	Middle latitudes	1000	>3.5
12 (90%)[b]	0–40 (50%)[c]	1000	>6.5
	40–60 (50%)[c]		

Source: Taken from data provided by NRL (1998).
[a] Percentage of flight time at 11 km altitude.
[b] Percentage of flight time at 12 km altitude.
[c] Flight time equally divided (50%) between low (0–40°) and high (40–60°) latitudes.

among other factors. The National Radiation Laboratory in New Zealand (NRL, 1998) provide statistics for national aircrew. They separated aircrew into four categories according to their altitudes, latitude, and yearly flying hours as provided in Table 6.3. Their conclusions are that flight crew exposure is of limited health significance. The dose rates are well within the ICRP (1991) recommendations for limits of exposure for radiation workers for whole body dose set at 20 mSv/year. For pregnant women, however, the ICRP limit of dose is 2 mSv for the duration of pregnancy. Consequently, the study by NRL (1998) concludes that airline companies may apply a limit of 2 mSv for a pregnant member of the aircrew from the time of conception.

Radioactivity Hall of Fame — Part VII

Sergei Ivanovich Vavilov (1891–1951), Pavel Alekseyevich Cherenkov (1904–1990), Il'ja Mikhailovich Frank (1908–1990), and Igor Yevgenyevich Tamm (1895–1971)

Sergei Ivanovich Vavilov (1891–1951), Pavel Alekseyevich Cherenkov (1904–1990), Il'ja Mikhailovich Frank (1908–1990), and Igor Yevgenyevich Tamm (1895–-1971), while at the P. N. Lebedev Institute of Physics, Moscow, shared a role in the discovery and interpretation of Cherenkov radiation. Pavel Cherenkov, Il'ja Frank, and Igor Tamm shared the Nobel Prize in Physics 1958 "for the discovery and interpretation of the "Cherenkov Effect". Sergei Vavilov would have shared in the prize had he not passed away several years before.

SERGEI IVANOVICH VAVILOV (1891–1951)

Sergei Vavilov was born in Moscow in 1891. Sergei and his elder brother Nikolai both became prominent scientists. Their father was a successful merchant and educated his sons with the hope that they would inherit his business. However, Sergei studied physics and mathematics at Moscow University and Nikolai studied biology and became a prominent plant breeder. Biodata on Sergei Vavilov presented here was taken in part from comprehensive and authoritative reviews of Vavilov's life, contributions and biography by Bolotovski *et al.* (1998, 2004) of the Lebedev Physics Institute, Moscow, and information provided in Moscow News (2006). Such information is yet very limited and made possible only after the transformations that have made Russian society more open and free of censorship during the past decade. Bolotovski *et al.* (1998) provide the most thorough and authoritative historical review of Vavilov's contribution to the development of Russian physics in the face of unconceivable political repression and persecution during the leadership of Joseph Stalin from the mid-1920s until Stalin's death in 1953 and unyielding power of Stalin's henchman, Lavrenty Beria during 1935–1953. Vavilov's biography by those who knew him (Bolotovski *et al.*, 1998) provide insight into life during the Great Purge of the Soviet Union and the personal sacrifices made by Sergei Vavilov to assure the development of physics in the Soviet Union during those repressive years. The postage stamp illustrated on the following page was issued in the former USSR in 1961 to commemorate 10 years following the death of Sergei Vavilov.

During his second year at Moscow University Sergei Vavilov started research work in the laboratory of Pyotr Nikolaevich Lebedev (1866–1912), who was the first physicist to demonstrate the pressure exerted by light photons on solids and gases, which confirmed James Clerk Maxwell's theory that light has momentum and could thus exert pressure on solids. Vavilov graduated with honors from Moscow University in 1914 and shortly thereafter began service in the military during the First World War. He was taken prisoner by the German forces in 1917. Bolotovski *et al.* (2004) related the following account of his capture and escape:

He was interrogated by a German officer who happened to be a physicist, and they spent all night discussing physics, especially Max Planck's new theory of light. By morning the officer had helped Vavilov escape, and in February 1918 he appeared in Moscow.

Sergei I. Vavilov (1891-1951)

His father had lost all of his property and emigrated from Russia following the Russian Revolution of 1917 culminating in the defeat of the counterrevolutionaries of the civil war of 1918–1922. Sergei and his brother Nikolai, as budding scientists, remained to work in their homeland. Bolotovski *et al.* (2004) explains

> …they realized that they had to coexist with the Soviet government as many other Russian scientists did. The government had decided that use should be made of scientific and technical specialists 'in spite of the fact that they have been nourished upon capitalist ideology'. The 1920's were thus a period of great liberty for Russian scientists, despite civil war, widespread famine and economic collapse.

Sergei Vavilov began his career in 1918 working in the Physics and Biophysics Institute, Moscow, headed by Pyotr Petrovich Lazarev, a disciple of Lebedev. Here Vavilov started a research career in physical optics and specifically research in the field of photoluminescence of which little was known at the time. Vavilov also lectured at Moscow University. He and his coworkers defined the laws of luminescence and mechanisms of luminescence quenching. They were the first to introduce the term "luminescent yield" as the ratio of luminescent energy to the energy of light creating the luminescence. In 1926 Vavilov and Vadim Levshin discovered a violation of Bouguer's Law in uranium glass. The law defines light attenuation by optically homogeneous or transparent media, variations of which are also known as Beer's Law or the Beer–Lambert Law. They found a reduction in the light absorption by the glass as the intensity of the incident light was increased. Bolotovski *et al.* (2004) explain that this is known as the photorefractive effect, which is described as resulting from the depopulation of the ground state by the incident beam. Vavilov thus introduced the term "non-linear optics", which is now a branch of physics.

Sergei Vavilov devoted his entire scientific career in Russia with the exception of a 6-month research stint at the Physikalisches Institut der Universität Berlin with Peter Pringsheim, the then authoritative figure in the science of fluorescence (Pringsheim, 1943, 1951, 1963). Bolotovski *et al.* (2004) report, that in 1930 when the political climate in the USSR changed abruptly for the worse as Joseph Stalin consolidated his power, many scientists were persecuted. Pyotr Lazarev, Director of the Physics Institute was suddenly arrested in March 1931 and exiled in the Urals and many colleagues at the institute lost their jobs including Vavilov. Although Vavilov still kept his position as university professor, and in an attempt to preserve the Lebedev-Lazarev school of physics, colleagues nominated Vavilov to the USSR Academy of Science, and he was elected in 1932. Around this time Vavilov was invited by the director of the Optical Institute of Leningrad (now Saint Petersburg) to bolster the activities of the institute and save its programmes from being stopped by the government that wanted to make way for the production of optical equipment particularly for the military. Vavilov then became head of research of the State Optical Institute in 1932 and also head of the Academy of Science's Physics Department in Leningrad. Here Vavilov organized teams of young scientists to carry out research in neutron physics and luminescence.

In 1933 Sergei Vavilov had a doctoral student whose name was Pavel Alekseyevich Cherenkov. For his doctoral research Vavilov proposed that Pavel Cherenkov work on the topic of *The luminescence of uranyl salt solutions under the influence of hard gamma radiation*. High-energy gamma radiation would induce the production of a weak blue light from solutions of uranium salts, and it was thought at the time that this might be due to the occurrence of luminescence. The mysterious weak blue light was historically documented as seen by Marie Curie in 1910 with bottles of concentrated radium solutions aglow with a pale blue light, as related in a biography by her daughter Eve Curie (1941). Lucien Mallet (1926, 1928, 1929) also had studied the properties of the light and found that it would be emitted by a variety of transparent bodies when exposed to radioactive sources. However, Mallet did not determine the origin of the blue light. Vavilov put Cherenkov to investigate the origins and properties of the mysterious light, which he thought to be fluorescence. Vavilov and Cherenkov apparently were not aware of the earlier observations of Curie or the previous studies by Mallet and, in any case, they started the research to investigate what they thought was fluorescence. Sophisticated photomultiplier detectors were not available in 1933; however, Vavilov had developed a visual method of measuring low light intensities very close to the thresholds of human vision. To observe and measure the intensities of this weak light, Cherenkov had to adapt his eyes by remaining in a dark room for an hour or more before taking any measurements.

Cherenkov found that the weak light was emitted not only from the uranium salt in sulfuric acid solution when irradiated with strong gamma rays, but that there was a background glow from the solvent itself, that is, when sulfuric acid solution without the uranium solute was irradiated with the hard gamma. Bolotovski *et al.* (2004) give an account that Vavilov asked Cherenkov to check other solvents that were highly purified to see if these also emitted light when irradiated. Cherenkov investigated 16 different pure solvents and observed about the same intensity of blue light in all solvents when these were irradiated with hard gamma radiation. Various attempts to quench the effect were unsuccessful, which provided evidence that the phenomenon was not fluorescence. What was occurring was obvious to them at this point. This was not fluorescence, but that the light emission was a new phenomenon of origin yet unknown to mankind. They had discovered a new radiation, which they reported in two papers back-to back in the 1934 Proceedings of the USSR Academy of Sciences (Cherenkov, 1934a,b; Vavilov, 1934). In his paper Vavilov provided a preliminary theory on the origin of the radiation. He proposed that Compton electrons produced by the hard gamma radiation would yield a type of Bremmstrahlung radiation. He was correct in the conclusion that the mysterious light was being produced by the Compton electrons; however, the mysterious light was not the result of Bremmstrahlung, it was yet of unknown origin. It was not till a few years later (1935–1938), with more research by Cherenkov (1936, 1937a–d, 1938a–c,) under the directorship of Vavilov and theoretical calculations by Frank and Tamm (1937), was it understood finally that the new radiation was produced by the Compton electrons when they had sufficient energy to travel at a speed in excess of the speed of light in the transparent medium. The new radiation soon came to be known as "Vavilov–Cherenkov radiation" or "Cherenkov radiation".

Vavilov–Cherenkov radiation eventually became a very important tool in the studies of high-energy physics (see Chapter 7). Cherenkov counting, that is, the instrumental technique of detecting and counting the individual Cherenkov radiation wave fronts or emissions, has become a common means of analyzing the levels of activity or disintegration rates (disintegrations per minute) of radionuclides that emit high-energy beta particles (L'Annunziata, 2003b). As a consequence of the practical importance of the Vavilov–Cherenkov Effect, the Nobel Prize in Physics 1958 for the discovery and interpretation of the Cherenkov Effect was awarded to Pavel Cherenkov, Il'ja Frank, and Igor Tamm. Sergei Vavilov would also have shared the award had he not passed away in 1951, as the Nobel Prize is not awarded posthumously. Those who have had the opportunity to view the core of an operating nuclear reactor at the bottom of a pool of water may have witnessed the weak blue Cherenkov light surrounding the core. This Cherenkov radiation in the water surrounding the reactor core is produced mostly from Compton electrons from high-energy gamma-ray interactions with atomic electrons of the water or high-energy beta particles emitted as a result of fission-product decay. The postage stamp illustrated here was issued in Germany to commemorate the 25th anniversary of the discovery of nuclear fission, and it is an abstract illustration of the cross-section of a reactor core with a surrounding blue light.

In addition to his elected membership to the USSR Academy of Sciences in 1932 Vavilov was appointed Head of the State Optics Institute (currently the Vavilov State Optical Institute or Gosudarstvennyy Opticheskiy Institute or GOI), and Head of Institute of the Physics Department of the Institute of Physics and Mathematics of the USSR Academy of Sciences in Leningrad. From this time on for the rest of his life Sergei Vavilov devoted his professional career to helping and supervising young scientists and performing administrative duties required to build up research groups that would put Russian physics research to the forefront and at a par with the leading scientific physics research groups of the world. In 1934 the USSR Academy of Sciences moved from Leningrad to Moscow, and Sergei Vavilov and his Physics Department moved as well to Moscow where they occupied the same building formerly used by the exiled Pyotr Lazarev. After the move to Moscow, Vavilov's Physics Department was transformed into the Physics Institute of the USSR Academy of Sciences (Fizicheskii Institut Akademya Nauk, USSR or FIAN), and Sergei Vavilov was appointed its director. Following Vavilov's proposal the new institute was named the P. N. Lebedev Physics Institute after Pyotr Nikolaevich Lebedev (1866–1912), the pioneering physicist in who's laboratory Vavilov had the

**Reactor Core and Blue Cherenkov
Light**

opportunity to work when he was a young student at Moscow University. Vavilov remained Director of FIAN his entire life. Under Vavilov's direction, it developed from no more than about a couple dozen staff members and graduate students in 1932 to one of the largest and leading physics research institutes of the world. This was not easy for Vavilov to accomplish as related by Bolotovski *et al.* (1998):

> Though the thirties were a difficult time of political hysteria in Russia, FIAN surprisingly managed to continue working productively while preserving a positive internal ambience. That was largely due to the efforts of Vavilov. It is well illustrated by G. E. Gorelink (1995) in his essay *Moscow Physics 1937* where he cites the transcript of a conference of the top FIAN researchers held soon after the Plenary Session of the Central Committee of the Communist Party in February and March of 1937 which had purged the former top communist leaders Bukarin and Rykov who had then been arrested by secret police as Japanese and German spies. At this conference Vavilov's behavior was both bold and decent as he fielded political accusations. He spoke of B. M. Gesen, who had been jailed as 'an enemy of the people' in August 1936, as a normal human being and admitted responsibility for inviting him to FIAN to the position of deputy director. He shielded [future pioneer] Grigory Landsberg and [future Nobel Laureate] Igor Tamm from attacks by Communist zealots (the latter's brother, the chief engineer of the Berezniki Chemical Factory, had been arrested for political reasons) and thus prevented a hate campaign that raged in the country from taking over the institute.

To survive and to be able to build FIAN as its director into an institute packed with the most qualified physicists who could work in relative tranquility at the forefront of science, and as President of the USSR Academy of Sciences (1945–1951), Vavilov had to bow to the Communist Party and make ceremonial statements during symposia and conferences on many occasions. Bolotovski *et al.* (1998) noted

> for instance, he [Vavilov] started his address to the conference with quotes from Stalin's speech, but such were the rules of the game and as an administrator holding high rank he was helpless in this respect…It is not surprising, therefore, that Vavilov managed to attract to FIAN outstanding physicists…and to maintain an atmosphere at FIAN that differed strikingly from the political frenzy ruling over the nation at the time.

V.I. Veksler (1907-1966) and Phase Stability Principle

Sergei Vavilov would attract the best scientists to FIAN. As an example, in 1936 Vavilov had invited the young physicist Vladomir Iosifovich Veksler to join FIAN. Veksler would later be the first to develop the principle of phase stability (Veksler, 1944), which would permit the acceleration of particles to relativistic speeds in modern particle accelerators. This was independently discovered by E. M. McMillan (1945) at the University of California, Berkeley. For their discovery Veksler and McMillan shared the Atoms for Peace Award in 1963. The postage stamp illustrated on the previous page commemorates V. I. Veksler and his discovery of the principle of phase stability. The stamp illustrates a "star" consisting of numerous particle tracks emitted in various directions following the collision of a charged particle at relativistic speeds with an atomic nucleus in a synchrotron, developed as consequence of Veksler's phase stability principle. Bolotovski et al. (2004) relate the development of particle accelerators under Vavilov's direction. They detail the account of how Vavilov believed that experiments in nuclear physics required particle accelerators that were larger than the 10 MeV cyclotron already available at the Radium Institute. Vavilov consequently called together a "cyclotron group" in the Lebedev Physics Institute to evaluate the possibility of constructing a larger more powerful cyclotron. The general feeling of the group was that it would be impossible to jump over the relativistic barrier. However, after Veksler wrote two papers in 1944 on his new principle of phase stability (Veksler, 1944, 1945), Vavilov immediately recommended the papers for publication in the national journals and initiation of construction of a 30 MeV synchrotron at the Lebedev Institute. Veksler's new concept of phase stability was subsequently published in the international journal *Physical Review* the following year. Veksler spoke the following of Vavilov at the time of the award presentation (Bolotovski et al., 1998):

> I was lucky in that as a young scientist I was invited to join the staff of the Lebedev Physics Institute [FIAN], which included such exceptional scientists as Vavilov, [Leonid] Mandel'shtam, [Igor]Tamm and many others…an exciting atmosphere of complete commitment to science prevailed in the institute. I had opportunities for regular live contacts with these outstanding scientists…

It was in the 1940s that Sergei Vavilov was struck with one of the hardest blows of his lifetime, the arrest and death of his brother Nikolai Ivanovich Vavilov (1887–1943). Bolotovski et al. (1998, 2004), Vavilov (2002), and the Moscow News (2006) provide detailed accounts of the tragic history of Nikolai's arrest and demise at the peak of his career. Nikolai Vavilov was the elder brother of Sergei and a famous plant breeder and geneticist. By 1920 Nikolai had already made many scientific advances in the field of plant breeding and genetics. He had discovered the law of biological variability essential to plant breeding for crop improvement, and he subsequently during 1922–1933 organized numerous expeditions to 60 countries of the world and collected about a quarter-million specimens of grain. This study of the world's plant resources enabled Nikolai to work out foundations for selection and breeding of improved varieties in Russia. By 1929 Nikolai was elected President of the Agricultural Academy and elected member of the Academy of Sciences. Nikolai was head of the research institutes of plant breeding in Leningrad and genetics in Moscow. He set up many experimental stations in various parts of the USSR and directed their research programs. Nikolai's pioneering work was recognized worldwide, and the field of plant genetics and breeding, which Nikolai helped establish, would eventually become universally recognized as a vital field for agricultural production. The importance of plant breeding in the improvement of crop production and the reduction of famine in the world was underscored by the Nobel Committee when it awarded the Nobel Prize for Peace 1970 to Norman Borlaug for his work in plant breeding and the green revolution. Unfortunately in early 1935, a campaign was launched in the Soviet Union against genetics, which was branded the "handmaiden of the bourgeoisie" (Moscow News, 2006), and Nikolai Vavilov as a world leader of genetics had to defend his field of work, particularly against the sinister figure of Trofim Lysenko, his arch enemy. Lysenko promised quick crop improvements, against Nikolai Vavilov's slow process of systematic hybridization and selection, and Nikolai had to bear constant accusations of holding "idealist" Mendelian theories (Bolotovski et al., 2004). Stalin was particularly receptive to Lysenko's claims. As related in Moscow News (2006)

> together with millions of other victims of the Great Terror, Nikolai Vavilov was arrested and jailed. There was talk of allowing him to work in one of the research institutes in the Gulag

which employed camp inmates to work on scientific projects. But apparently Vavilov was too important a figure. After the war broke out in 1941, Vavilov was thrown into prison in Saratov.... On January 26, 1943, the scientist who had probably done more in his life than anyone else in the country to prevent humanity from suffering famine, died of hunger in a prison cell...the Soviet regime declared [Nikolai] Vavilov an enemy of the people, his name was erased from all printed sources, and his works were banned.... Nikolai Vavilov's name was cleared, and his works began to be published again in the country in the 1960s. But the harm to Russian agriculture had been done and could not be reversed.

The postage stamp illustrated here was issued in the Russian Federation in the year 2000 to commemorate the pioneering work of Nicolai Vavilov and his discovery of the law of biological variability in 1920.

   Bolotovski *et al.* (1998) analyze in detail the historical data including quotations from those who knew Sergei Vavilov to ascertain how Sergei could manage to continue to work diligently toward the creation of a national physics research institute of world renown for a government that, in the short period of only three years, had unjustly imprisoned and taken the life of his dear brother. They quoted Nobel Laureate Andrei Sakharov (1921–1989) who commented

Nikolai I. Vavilov (1887-1943)

I had an appointment with the FIAN director, a prominent optical physicist, academician Sergei Ivanovich Vavilov, who was a brother of another, even better known academician, Nikolai Ivanovich Vavilov, a biologist who had been arrested and died in prison a few years before. It was one of the most terrible pages in the tragic history of Soviet biology. [Sergei] Vavilov was soon appointed (or had already been appointed) President of the USSR Academy of Sciences. In this capacity he had to meet regularly (at least once a week) with T. D. Lysenko, an Academy Presidium member who had been a principle prosecutor of his brother, responsible for his death. I just can't imagine how he took these meetings...[Sergei] Vavilov was a nice person to meet, good hearted and gentle.... He had placed a number of envelopes with cash in a drawer of his desk (his personal money) and he pressed on the destitute visitors these envelopes as in most cases he was unable to give them the support they really needed. The authorities learned of this and attempted to prohibit it...he carried out his duties without sparing himself.... As Academy President he often had to deliver official addresses. In one such speech he referred to Stalin as the 'Coryphaeus of Science' and this expression later became a part of Stalin's official title (apparently he liked it). The fates of two brothers (one dying of starvation while cleaning latrines in a Saratov prison while the other was a President enjoying all official honors) was a rare paradox even for that time though it was highly typical in some ways, too.

Bolotovski *et al.* (1998) comment on Sakharov's observations as follows:

Vavilov's official position made it imperative to glorify Stalin in public speeches and in written contributions. Vavilov could not avoid it. He published articles entitled *Stalin's Scientific Genius, Science of Stalin's Era* and so on…. Vavilov became President of the USSR Academy of Sciences in July of 1945. People often ask why Vavilov agreed to take the position (after Stalin had given his approval, election to office was just a formality). One should also ask why Vavilov was chosen for the position…In our opinion one of the main reasons was that after the Second World War…the country could hardly retain its place as a leading world power unless it possessed a well developed science (including the fundamental science that provides a basis for developing most applications in military technologies and engineering)…Stalin could have had appointed one of his trusted henchmen as the President (for instance A. Vyshinski or T. Lysenko)…one of the reasons why Vavilov agreed to take the position was that, if he had refused, it would have been given to one of those people. Under the circumstances, the person to occupy the position had to be a reputable scholar, preferably a scientist, possessing effective administrative skills, rather than just a tough executive (the main objectives of the Academy were in sciences and nuclear weapons development was becoming an immediate concern).

Sergei Vavilov was concerned with saving Soviet physics and the lives and welfare of those he could help. The state of affairs in physics was a shambles after the Second World War and Stalin ordered assessments of army officers, executives of the defense ministries, and others, who might be helpful to the state and were imprisoned as enemies of the people so that they may be released and offered positions. Bolotovski *et al.* (1998) elucidate that Stalin instructed his secret police chief (Lavrenty Pavolich) Beria to process the assessments. They noted the fact that Sergei Vavilov's brother Nikolei had been purged was not an impediment to his appointment as President of the Academy of Sciences, and that Sergei's assessment states as follows: "Brother N. I. Vavilov, geneticist arrested in 1940 for sabotage in agriculture, sentenced to a 15-year term of imprisonment, died in Saratov prison…". They noted also that Stalin "approved" of the arrangement under which close relatives of top government and public figures were purged.

To further explore the reason why Sergei Vavilov continued to work as Director of FIAN and President of the Academy of Sciences Bolotovski *et al.* (1998) provide the following statement of Nobel Laureate Il'ja Frank, who considered Vavilov's appointment as President of the Academy as something Vavilov could not avoid considering the political repression and persecutions of the time:

"Many now are asking how Vavilov could agree to become President even though his beloved brother had perished in prison. A pertinent question to those asking is what would have happened if he had refused? I am not sure that he would have been allowed to stay alive as had happened with [Pyotr] Kapitsa, who had shown obstinacy…. Even if Stalin had not destroyed him he would definitely have been stripped of his rank and removed from all his positions, in particular, from running his favorite creation, the institute of Physics of the USSR Academy of Sciences. I am absolutely convinced that the least of Vavilov's concerns was his own destiny. He felt deeply his personal responsibility for the destiny of science and culture. I am sure that if Vavilov had been purged the Institute of Physics would have been branded an asylum for enemies of the people. It is a well-known fact that we, his absolutely loyal disciplines in science and life, were greatly indebted to him. He would not have been able to protect us from inevitable persecution…. The entire system of overall control in the country was such that the slightest motion of Stalin's hand, anything he uttered, was supreme law. I know not a single instance when somebody refused to follow Stalin's instructions…. When he was already Academy President he said to me 'Each time I go for an appointment at the Kremlin I am not sure whether I shall return home or they will take me to Lubyanka (the headquarters of the secret police).' Later [Nikita] Kruschev said the same thing in public and it became general knowledge.

Pyotr L. Kapitsa (1894-1984)

The case of Pyotr Kapitsa (1894–1984) cited by Il'ja Frank in the above quote is an interesting example. Kapitsa was awarded the Nobel Prize in Physics 1978 "for his basic inventions and discoveries in low-temperature physics". The postage stamp illustrated here was issued in 1994 by the Russian Federation to commemorate the 10th anniversary of the death of Pyotr Kapitsa. He was born in Kronstadt near Leningrad, USSR in 1894 and graduated in 1918 from the Petrograd Polytechnical Institute. His wife and two small children perished of illness during the chaos of the civil war that followed the Russian Revolution. After their deaths Kapitsa left USSR in 1921 to study at Cambridge University and eventually worked at the Cavendish Laboratory at Cambridge, England with Ernest Rutherford. He became a renowned physicist in England starting as a Clerk Maxwell Student at Cambridge University from 1923 to 1926, Fellow of Trinity College in 1925, Assistant Director of Magnetic Research at the Cavendish Laboratory from 1924 to 1932, Fellow of the Royal Society in 1929, Messel Research Professor of the Royal Society from 1930 to 1934, and Director of the Royal Society Mond Laboratory from 1930 to 1934. In 1934 Kapitsa went to a scientific meeting in the USSR. His passport was seized and he was detained on Stalin's orders and not permitted to leave the country. He was required to form the new Institute for Physical Problems of the USSR Academy of Sciences in Moscow, and he thus arranged through the help of Rutherford to have his equipment at the Mond Laboratory shipped to Moscow. As head of the new institute Kapitsa made many discoveries in the field of low-temperature physics and discovered superfluidity in helium. He was of great assistance to the success of the USSR in the Second World War assisting particularly in the production of large quantities of liquid oxygen for the steel industry, and after the USSR victory was awarded the title of Hero of Socialist Labor in 1945, the highest award granted then to a civil servant. In 1946 he refused to work on the development of the hydrogen bomb, and thus fell out of favor with Stalin. He was punished and removed of his post as Head of the Institute of Physical Problems in August of 1946 by a decree of the USSR Council of Ministers signed by Stalin and exiled to his dacha near Moscow. He was forced to remain in exile until Stalin's death in 1953. Sergei Vavilov secretly helped Kapitsa at this time of need. Bolotovski *et al.* (1998) provide the following quotation of

A. A. Kapitsa, the widow of Pyoty Kapitsa, which illustrates one of many examples of how Vavilov helped his colleagues whenever possible:

> Vavilov rendered him great assistance. He always helped Kapitsa in any way he could but he never made his assistance known. I believe that Kapitsa was not even aware of the many things Vavilov had done for us in those years and this has come to light only now.

Sergei Vavilov used his position as President of the USSR Academy of Sciences to try to attain exoneration for his brother Nikolei. Bolotovski *et al.* (1998) note that in 1955, 4 years after Sergei died, V. F. Sennikov, a staff member of FIAN, examined documents in the archives of the Ministry of State Security with the purpose of rehabilitating the innocent victims of Stalin's purges. They note that Sennikov was "particularly impressed (by Sergei) Vavilov's letter of 1949 addressed personally to Stalin in which he appealed for the exoneration of his brother. The letter included a detailed account of the life and work of N. Vavilov, described his openness and honest and straightforward manner of speaking. Vavilov firmly rejected all subversive actions attributed to his brother claiming that the accusations against him were slanderous. Vavilov ended his letter saying "if my brother N. Vavilov is not exonerated, I cannot remain President of the USSR Academy of Sciences." They note that the letter bears an inscription by Beria "Not authorized", and that there were no marks indicating that the letter was shown to Stalin.

As Director of the P. N. Lebedev Physics Institute and President of the USSR Academy of Sciences, Sergei Vavilov did all he could to help and protect staff and colleagues from Stalin's purges. There are numerous examples and the reader is invited to peruse the detailed accounts given by Bolotovski *et al.* (1998). The degree to which Vavilov played in the establishment of physics in the USSR is evidenced by the relatively numerous Nobel Laureates that emerged from the P. N. Lebedev Physics Institute, which he established from 1934 to the time of his death in 1951 and from the USSR Academy of Sciences over which Vavilov presided. Among these Noble Laureates are Pavel A. Cherenkov (1904–1990), Il'ja M. Frank (1908–1990), Igor Y. Tamm (1895–1971), Lev D. Landau (1908–1968), Aleksandr M. Prokhorov (1922–), Nicolay G. Basov (1922–), Andrei D. Sakharov (1921–1989) was awarded the Nobel Prize for Peace, Pyotr L. Kapitsa (1894–1984), and Vitaly L. Ginzburg (1916–).

A final example of the numerous cases where Sergei Ivanovich Vavilov tried to help fellow scientists within the stressful atmosphere of Stalin's oppression within which they had to live and work is taken from the life of Vitaly Ginzburg, who shared the Nobel Prize in Physics 2003 "for pioneering contributions to the theory of superconductors and superfluids." Vitaly Ginzburg did pioneering work on the interpretation of the Vavilov–Cherenkov Effect in collaboration with Il'ja Frank during 1939–1947. His contributions to the interpretation of the Vavilov–Cherenkov radiation did not begin to appear in print until 1940 in the *Journal of Physics* of the former USSR and in the *Proceedings of the USSR Academy of Sciences* some three years after the famous paper of Il'ja Frank and Igor Tamm (1937), which explained the radiation discovered by Vavilov and Cherenkov. Consequently Ginzburg did not share in the Nobel Prize for Physics 1958 with Cherenkov, Frank and Tamm. However, he was awarded the Nobel Prize for Physics 45 years later as noted above on a totally different subject matter. Vitaly Ginzburg provides the following account of Sergei Ivanovich Vavilov in his book *The Physics of a Lifetime* (Ginzburg, 2001) with kind permission of Springer Science and Business Media © Springer:

> I have been working at FIAN since 1940 and, hence, for ten years I was at the Institute with S. I. Vavilov as its director.... The point is that my wife Nina Ivanova Ginzburg (born Ermakova) was arrested in 1944, and after nine months in prison was 'sentenced' by the notorious KGB Special Consultation to a three-year confinement in a camp according to the ill-famed 'counter-revolutionary' 58th article of the Criminal Code...the 1945 amnesty for those sentenced for a term below three years was extended to the 58th article. Nina was freed in September 1945, but with a limited number of cities in which she was allowed to reside (Moscow was certainly excluded).... I began writing applications to the NKVD (I believe that this was the name of the VChK-OGPU-KGB at that time with a request to register my wife in Moscow. But someone made me listen to reason and not write directly to the 'organs', but to do it through the Institute instead, because millions of people were imprisoned and exiled...I must say at once that I never

received a positive answer and my wife could only return to Moscow in 1953 after a new amnesty which followed the death of the 'coryphaeus of all sciences' [i.e., Josef Stalin].... When I went to Sergei Ivanovich [Vavilov] in this connection for the first time (it was at the end of 1947, or rather at the beginning of 1948), he agreed to 'back up' my application. The same was repeated the next year. When I went [to him] for the third time, he said approximately the following: 'I shall of course support you, but you know, my 'belle soeur'—a sister not only of my wife, but also of the wife of Vensin, the President of the Academy of Architecture—is also exiled. And we— two presidents of Academies (S. I. was then already President of the USSR Academy of Sciences) are applying for the registration of our belle soeur in Moscow, but they refuse. Well the poor woman sometimes comes illegally to Moscow for a short time, but she may not live here.'...For residing without registration one could be put in prison for three years. For the eight years of exile, my wife also came several times to Moscow.

Sergei Ivanovich [Vavilov] died in 1951 when he was not yet sixty. He had an infarction and looked bad; I met him in the hall of the Institute not long before his death. He was very sad, and that is how I remember him. Dimitrii Vladimirovich Skobeltsyn became director of FIAN [after the death of Vavilov], and the last two times I asked for my application to be seconded I addressed him. When I went to him (for the second time, I believe), he said to me, 'My brother is in exile in Tsarevo-Kokshaisk (before 1919 this was the name of the present Yoshkar-Ola), and my application to register him at my place in Moscow has been declined.'

This is, properly speaking, the story of two directors. Of course, in light of what we know today about life in Stalin's times, nobody will be surprised at such information. But all the same S. I. Vavilov was President of the USSR Academy of Sciences and D. V. Skobeltsyn was not only an academician, but also the chief expert or even the Head of the Soviet Delegation at UN negotiations on banning nuclear weapons, and, from 1950, Chairman of the Committee for the International Lenin Prize "for Consolidation of Peace Among Nations". And even such [important] people were refused permission to lodge their ostensibly guilty relatives in the places where they lived. In truth, nowhere could a man *breathe freely*.

Sergei Vavilov, codiscover of Cherenkov radiation, did not live a very long life. Nevertheless, he did a lot for physics, his colleagues, and his nation. His life should serve as an example for all. In a seminar at CERN on *The Work and Life of S. I. Vavilov* given by Youri N. Vavilov (Sergei's nephew and the son of Nikolai Vavilov and also scientist at the P. N. Lebedev Physics Institute) on November 21, 2002 the following is taken:

Vavilov was concerned with the general issues of national science administration, the development of higher education, practical applications of the knowledge gained from fundamental research, and the broad cultural progress of the nation. Unfortunately the heavy administrative responsibilities did not leave him enough time to pursue his own research objectives. Vavilov could not behave otherwise, though: he was a man of integrity, and his own interests were never his primary concern. Acting as President of the Academy under the brutal dictatorial regime of Stalin was the source of an appalling stress. With his eldest brother (the biologist Nikolai Vavilov) having been murdered by this regime, Vavilov's health was seriously damaged, and he died two months before his sixtieth birthday.

PAVEL ALEKSEYEVICH CHERENKOV (1904–1990)

Pavel Alekseyevich Cherenkov was born on July 28, 1904 in the village of Novaya Chigla in the fertile agricultural region of Voronezh, USSR. He shared the Nobel Prize in Physics 1958 "for the discovery and interpretation of the Cherenkov Effect." The unique radiation that is produced as a result of the Cherenkov Effect would subsequently become referred to as Cherenkov radiation or Vavilov–Cherenkov radiation in honor of Sergei Vavilov and Pavel Cherenkov, who were the first to discover the origin of the radiation.

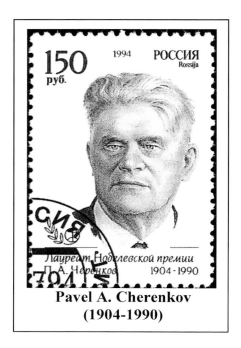

Pavel A. Cherenkov
(1904-1990)

Detailed information on the personal aspects of Cherenkov's life in the English language is limited. The biography provided by Pavel Cherenkov at the time of his Nobel Prize Award in Stockholm in 1958 is one of the briefest biographies among all of the Nobel Laureates (see Nobel Lectures, Physics, 1942–1962), a sign of his modesty. A paper by his daughter Elena Cherenkov (2005) recently filled the gap with information on his life and experiences as a person and physicist. Also, the same year one of his former colleagues, Boris Borisovich Govorkov (2005) published a paper on Cherenkov's equipment and work in fundamental physics. Personal information on Cherenkov's life and work is taken from these limited sources.

Pavel Cherenkov had a simple and humble upbringing as a young boy. He skated in winter on home-made ice skates. Pavel was a passionate reader, and would prefer reading at the village library in lieu of dancing or other amusements of young boys at the time. Pavel's father was a skilled farmer and shop owner. His mother died when Pavel was only two years of age. In 1928 Pavel Cherenkov graduated from Voronezh State University with major studies in physics and mathematics. After graduation from the university Pavel Cherenkov worked as a school teacher in the Michurinsk county center. It was in Michurtinsk in 1930 where he met his wife Maria Alekseevna Putintseva, a graduate also of Voronezh State University in the field of philology. Maria's parents were well educated. Her father Aleksey Mihaylovich Putintsev was a professor of philology at Voronezh University and her mother Maria Mihaylovna taught foreign languages at the same university. This was a time of great turmoil for both Pavel and Maria and for many in the USSR as related by the following taken from Elena Cherenkov (2005):

My parents—beautiful, intellectual, industrious, well-educated, believed in bright horizons opened for the country and its young people. But the event took a very unexpected turn. In the late 1920s, the communist government launched an expropriation of farmer's property and forced collective farms. Aleksey Egorovich, Pavel's father, was exiled to the northern regions. In autumn of 1930. Maria's father Aleksey Mihaylovich Putsinsev was arrested in connection with the so-called 'case of regions specialists'. More than 90 people were persecuted in this high-scale trial.

Similar processes were launched against both all independently-mined groups and all not independently-minded groups all over the country. Aleksey Mihaylovich was sentenced to five years of concentration camps. Pavel's and Maria's freedom was hanging in the balance too; they could be convicted by association.

In 1930 Pavel Cherenkov went to Leningrad to carry out postgraduate studies in the School of the USSR Academy of Sciences. He and Maria were married in 1931 and Pavel was able to get a single room in a student's dormitory where they had to accommodate five persons. The hardship and personal anguish with which Pavel and Maria had to cope is related by Cherenkov (2005) as follows:

> They [Pavel and Maria] settled in the postgraduate student's dormitory. Pavel managed to get a separate room! They had to house five people there—including Tatiana, Pavel's youngest sister (the life in the country was very hard in those years). Also Maria's mother joined them after her husband was imprisoned. In 1932, their son Aleksey was born. During the first years of his life, he was a sickly child. Food rationing was enforced in the country at that time. Maria was often unemployed. Her friends routinely shared their bread-cards. This bread was dried and sent to grandfather to the concentration camp. Pavel's only winter coat was also sent to her father. In 1932, Maria's father was released from the camp. He lost his strength and health. In spring of 1937 he died. At the end of 1937 Aleksey Egorovich, Pavel's father, was arrested again. By the sentence of 'the tribunal of three', without the investigation or trial, he was executed. Such was the family environment in which the young physicist Pavel Cherenkov had to live while undertaking his graduate studies and research.

One can only imagine the determination and discipline that Pavel Cherenkov had to have to concentrate on his studies and research. It was in 1932 when Sergei Ivanovich Vavilov became Pavel Cherenkov's research adviser. Sergei Vavilov was renowned for his research studies on luminescence and fluorescence, which are processes that result in the emission of light by certain compounds as a consequence of atomic electron energy state transitions. Certain organic or inorganic compounds may absorb energy and become elevated to excited energy states via electron transitions whereby the compounds may emit the excitation energy as light photons as the electrons fall back to lower energy states. For his Ph.D. studies Vavilov assigned Cherenkov the task of studying the luminescence of uranyl salt solutions under the effect of hard gamma radiation. In his recollections published when he was approximately 81 years of age Cherenkov (1986) explained

> The main purpose of these investigations was to make clear the mechanisms for the light emission and to compare it with the properties of ordinary photoluminescence radiation. Therefore, all measurements were carried out in parallel: exciting the luminescence by both radioactive sources and by visible light. As the luminescent materials, liquids and in particular solutions of uranium salts were, for reason of methodology chosen…at the same time, control experiments with solutions of other lumifors such as esculinium or fluorescein were carried out. But the uranium salts had become the main object of the investigations. It was planned to determine the laws of their light emission under radioactive rays, and to ascertain whether, in this case, there was something in common with photoexcitation luminescence…Making use of liquid lumifors gave one the possibility of choosing, over wide limits, the characteristics such as concentration, viscosity, temperature, which affect one of the general signs of luminescence: the finite lifetime of the excited states. This was approximately the theme of my dissertation work. One can see that there was no task to discover a new effect here.

Thus Cherenkov clearly explained that his initial task was to study luminescence by the uranium salts, and that the outcome of the work, the discovery of a new phenomenon (Cherenkov Effect) and new radiation (Cherenkov radiation) was a fortuitous outcome of the work. Although it was not Cherenkov's intention to find a new type of radiation, his discovery, like most in science, are not accidental. In the words of Cherenkov (1986) in his recollections of the work

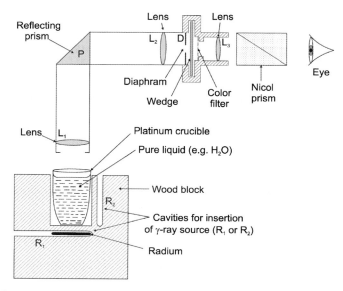

Figure VII.1   Cross section of the experimental apparatus used by Cherenkov that resulted in his discovery of Cherenkov radiation (Cherenkov, 1934a). The setup was made to observe and measure the intensity of luminescence in liquids in a dark room with the naked eye, and resulted in the discovery of Cherenkov radiation. A crucible containing a pure liquid, such as water, is located within a hole bored into the center of a wood block. Cavities labeled R_1 and R_2 are drilled in the wood to permit the insertion of a capsule containing 104 mg of radium as a source of gamma rays. Light emitted from the open top of the crucible would pass through the collimating lens L_1, refracting prism P, and telescope lenses L_2 and L_3. The diaphragm D would define the field of view. A transparent graded wedge controlled by the observer was used to measure the intensities of the radiation. Filters of different colors were used to provide a rough estimate of the radiation wavelength. The Nicol prism provided information on the polarization of the radiation.

> …it would be perhaps incorrect to consider the discovery as an accidental one, since the conditions for it were, in fact, prepared by preceding successes in different spheres of physics,

Cherenkov did not have a spectrophotometer at his disposal to measure light intensities; however, Vavilov had developed a technique of dark-adapting the human eye and a graded wedge to enable a visual measurement of relative intensities of light. Cherenkov used the experimental arrangement illustrated in Figure VII.1

The light emitted by the radiation, which he was about to discover, was extremely weak and therefore difficult to measure with the apparatus illustrated in Figure VII.1. Each day before beginning work he had to remain in a dark room for an hour to dark-adapt his eyes before starting the observations. Cherenkov's (1986) recollection on the crude apparatus, which he used in 1933, is the following:

> The main difficulty of the work was in the necessity to detect this extremely weak light with a new method (using the threshold of vision), which had a distinctly subjective character. The accuracy of the method depended upon the conditions of the observer and was determined by many factors (tiredness, lack of sleep, and even mood) and to take strict account of these was, of course, complicated…as the detector the eye was used. This has a rather high sensitivity after adaptation to total darkness.

In his Nobel Lecture in 1958 he commented

> This method makes use of the human eye instead of the use of a light-measuring device. Since the sensitivity of an eye adapted to darkness is some tens of thousands of times greater than its sensitivity by daylight, this method was superior to others by virtue of its high sensitivity. Notwithstanding its subjectivity and the comparatively large errors in the measurements, this method was at the time the only one that could be used which permitted a quantitative determination of those extremely low light intensities.

Pavel Cherenkov (1934a) measured the intensity of light emitted from the crucible when various substances were irradiated with the gamma rays from the radium source. The 140 mg of radium yields high doses of radiation, which under today's standards of radiation safety, would not be permitted without shielding. High levels of gamma rays were needed, because of the weak nature of the light emitted. He measured light intensity while irradiating various substances dissolved in solvents, and he would also irradiate the pure solvents to serve as blanks as no luminescence would be expected from solvents such as water, sulfuric acid, propyl alcohol, acetone, carbon tetrachloride, etc. However, Cherenkov noticed a very weak light from the solvents when irradiated with the high intensity gamma rays of the radium. The light emitted by the solvents was weaker in intensity than that emitted by the luminescent solutes. He could have simply subtracted this weak light from the blank solvents, as the objective of his research was to study luminescence. Fortunately he did not ignore this anomaly. Cherenkov discussed his observation with Sergei Vavilov, and they agreed that Cherenkov should investigate this weak light from the solvents by further purifying the solvents to ascertain whether the light was a result of luminescent impurities or a property of the solvent when irradiated by the gamma radiation. To his amazement Cherenkov observed no change in light intensity from the solvents even for the purest samples. Cherenkov (1934a) tested as many as 16 pure solvents and would observe the same light when irradiated with the radium source. Yet he had to make one more test to ascertain whether he was dealing with luminescence in these solvents or a totally new phenomenon. This test was based on the property of quenching of luminescent substances by the addition of chemical additives, that is, chemical quenchers, or by heating. Thus when luminescence occurs, the addition of chemical quenchers or heat will reduce the light intensity. Also, Cherenkov (1958) noted in his Nobel Lecture that heating would also alter the polarization of the luminescence photons as this would alter the mobility of the particles. He made the following comments concerning the results of his test for quenching:

> …the intensity of the light of fluids [pure solvents] cannot be influenced either by heating or by dissolving in them such active fluorescence quenchers as potassium iodide, silver nitrate, and many others. It has also been shown that the perceptible polarization which appears in this light cannot be altered either…Confirmation [of the discovery a new phenomenon] was forthcoming also in the unusual character of the polarization of this light. The main direction of the vector of the electric vibrations did not run perpendicularly to the excited beam, as is the case with polarized fluorescence, but parallel to it. Taken altogether, the results collected even during the first stage consequently gave rise to the statement that the light produced in fluids by the action of gamma rays is not a trivial phenomenon.

Pavel Cherenkov (1934a,b) published his initial findings (i.e., discovery paper) in the *Proceedings of the USSR Academy of Sciences*, and his supervisor Sergei Vavilov (1934) offered an interpretation for the origin of this unusual light in a short paper back-to-back with Cherenkov's. Vavilov thought the light might be due to Compton electrons that were undergoing some type of Bremmstrahlung in the liquids irradiated by the gamma rays of the radium source. A subsequent interpretation by Frank and Tamm (1937) would show that Vavilov was correct concerning the Compton electrons as the origin of this new light, but that a Bremmstrahlung Effect was not the process generating the light.

Only following Cherenkov's subsequent discovery in 1936 of the asymmetric properties of this light would it be confirmed that he was dealing with a phenomenon and type of radiation so far unknown. Cherenkov (1936) was able to demonstrate the space asymmetry of the light by providing experimental data to confirm that when solvents were irradiated with the gamma rays the light would be emitted forward in one direction with respect to the gamma-ray beam. Also Cherenkov demonstrated that an applied magnetic field would affect the direction of the light emitted thus providing evidence that the Compton electrons deflected in the magnetic field were involved in the production of the light. To provide experimental evidence of this asymmetric property of the radiation and the effect of a magnetic field on the direction of photon emissions Cherenkov devised the experimental arrangement illustrated in Figure VII.2.

The table inserted in Figure VII.2 provides the results obtained by Cherenkov. In the absence of a magnetic field, Cherenkov would set the measured intensity of the light produced by the gamma rays as unity, and any deviation of the measured light intensity as a result of the applied magnetic field as relative to unity. When the gamma-ray source was placed at positions R_1 or R_2 the observed light intensity would either double or be reduced to one-half depending on the direction of the magnetic field. Charged particles are deflected either upwards or downwards depending on their charge and the direction of the magnetic field. For example, if the magnetic field is directed into the plane of the page of this book, particles of positive charge are directed upwards and those of negative charge (e.g., negative electrons) are directed downward. In the reverse, if the magnetic field is directed out of the plane of the page, negative electrons are deflected upwards and positively charged particles deflected downwards. The light intensity measured from sources R_1 and R_2 were a function of the direction of the magnetic field indicating that electrons deflected upwards in the direction of the photometer and produced a measured light intensity approximately twice that produced in the absence of the magnetic field.

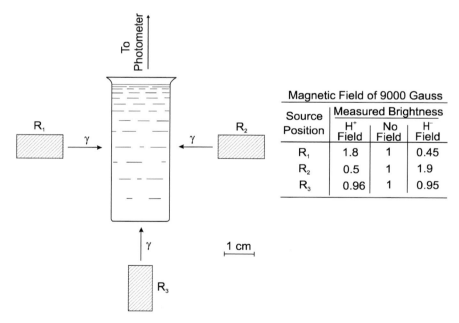

Source Position	Magnetic Field of 9000 Gauss		
	Measured Brightness		
	H^+ Field	No Field	H^- Field
R_1	1.8	1	0.45
R_2	0.5	1	1.9
R_3	0.96	1	0.95

Figure VII.2 Cherenkov's experimental arrangement that provided evidence of the asymmetric property of the mysterious light and the involvement of Compton electrons in the production of the light (Cherenkov, 1936). The sample vial located in the center of the figure was made of platinum or glass and contained either water or pentane. The inside walls of the vial were coated with a black layer to reduce reflection of the photons off the surfaces. A magnetic field of 9000 G was applied perpendicular to the plane of the figure, and Cherenkov could reverse the field to be directed into (H^+ field) or out of (H^- field) the plane of the figure. The source of gamma rays, equivalent to 0.5–1.0 g of radium, was placed at position R_1, R_2, or R_3 relative to the sample vial. The radium source was shielded with 1 mm of lead to absorb beta particles. The photometer consisted of the visual periscope illustrated in Figure VII.1.

Electrons deflected downwards in the magnetic field produced a measured light intensity of about one-half of that measured in the absence of a magnetic field. In the case of source R_3 located at the bottom of the sample vial, the Compton electrons are produced by the gamma rays in the upward direction toward the photometer, and the magnetic field would deflect the electrons either to the right or left (but not downwards) thereby reducing only slightly the intensity of the measured light relative to the intensity produced in the absence of the magnetic field. Cherenkov now had firm evidence that electrons were the particles producing the mysterious light and that the light was asymmetric, that is, the photons were emitted in a direction similar to that of the Compton electrons. Cherenkov (1937a) subsequently demonstrated that high-energy beta particles produced the same light, thus confirming that high-energy electrons in transparent media were the origin of the light.

   With the firm evidence at hand of the space asymmetry of the mysterious light, Cherenkov and his supervisor Sergei Vavilov knew that they had a new phenomenon at hand. Thus, in 1936 two other physicists at the same institute as Cherenkov took interest in this new radiation. Il'ja Frank and Igor Tamm of the Theoretical Division of the P. N. Lebedev Institute of Physics of the USSR Academy of Sciences began their theoretical study on the origin and properties of the radiation. The theory established by Frank and Tamm (1937) following the initial discovery of the radiation and its properties by Cherenkov states that when an electron traveling in a transparent medium of refractive index n with a velocity exceeding the velocity of light in that medium,

$$\beta > \frac{1}{n} \tag{VII.1}$$

where β is the relative phase velocity of the particle, that is, the velocity of the particle in a transparent medium divided by the speed of light in a vacuum, and the index of refraction n, is the ratio of the velocity of light in a vacuum to its velocity in the medium. The above condition is met when the particle travels in a transparent medium with a velocity exceeding that of light in the same medium. Under such circumstances the particle will produce an electromagnetic "shock" wave analogous to the acoustical shock wave or sonic boom produced by supersonic aircraft. Light will be propagated in the direction of the electron forming an angle θ with the path of the electron as a cone of light. The angle of emission of the photons would be defined by the equation

$$\cos \theta = \frac{1}{\beta n} \tag{VII.2}$$

As the electron energy increases, the value of β approaches one. If water ($n = 1.332$) is the medium within which the electron travels, and we assign β its maximum value of unity, the value of θ would be calculated according to eq. (VII.2) to be 41.3°. The angle of emission of the photons relative to the direction of travel of the electron in water will vary according to the energy of the electron, and it will not exceed 41.3°. For other substances with different indexes of refraction (n), the maximum angle of emission will therefore differ. Cherenkov (1937b) calculated the values of θ for three substances having distinctly different indexes of refraction and for electrons of different energies traveling in water, that is, different values of β. His calculated values for θ are provided in Table VII.1.

Table VII.I

Dependence of the direction of the radiation produced by electrons traveling in a medium as function of the electron velocity and index of refraction of the medium (Cherenkov, 1937b)

n ($\lambda = 4861.5$ Å)	β				
	1	0.95	0.90	0.85	0.80
1.3371 (eater)...	41°40′	38°10′	34°0′	28°30′	21°0′
1.51327 (benzene)...	48°40′	46°0′	42°40′	39°0′	34°20′
1.65439 (CS$_2$)...	52°50′	50°30′	47°50′	44°40′	41°0′

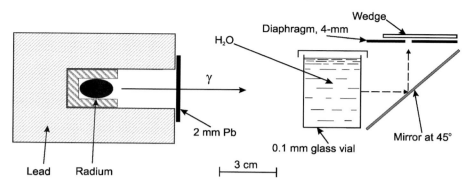

Figure VII.3   Experimental arrangement used by Cherenkov to measure angles of emission (θ) of light produced by the Compton electrons in water from a radium source (Cherenkov, 1937b).

From Cherenkov's calculations it could be seen that, from a determination of the angle of emission in a given transparent medium of known index of refraction (n), it would be possible to determine the energy or velocity of the electron from a measured value of θ. This concept forms the basis of measuring particle velocity with the modern ring imaging Cherenkov (*RICH*) counters described in Chapter 7. Cherenkov (1937b) made an approximate measurement of the angles of emission (θ) for radiation produced by the Compton electrons in water from a radium source employing the experimental arrangement illustrated in Figure VII.3.

The radium source used by Cherenkov had a very high activity (200–350 mCi) to provide sufficient light intensity to be measurable with the ocular system periscope arrangement illustrated in Figure VII.1. As in his previous experiments the level of radiation was extremely high yielding excessive radiation doses to the experimenter. Present day radiation safety levels would require adequate shielding and means for remote readings of the light intensity. A 2 mm lead plate at the orifice of the lead housing of the radium source was used to absorb any hard beta particles providing a pure gamma-ray source. Horizontal movement of the diaphragm permitted readings of the light intensity at various angles of reflection relative to the direction of the gamma rays. The wedge, which is moved horizontally along the diaphragm slit, was calibrated to provide an approximate measure of the light intensity. The direction of the Compton electrons was considered to be that of the gamma radiation. Thus, readings of the light intensities at various angles of reflection provided an approximation of the angles of emission of the Cherenkov radiation relative to the direction of travel of the incident gamma rays.

Cherenkov measured the angles of emission with respect to the direction of the incident gamma rays, and found the highest intensities of light between 0 and 30° as illustrated in Figure VII.4. He used the very simple and imprecise visual apparatus illustrated in Figure VII.3. As illustrated in the graph of Figure VII.4, the highest intensities of the light were emitted in the range of 0 and 30°, which would obey the relationship $\cos \theta = 1/\beta n$. In other words, as Cherenkov had calculated previously (Table VII.1) when β is assigned the value of 1 for the highest-energy electrons possible, the value of θ is calculated to be 41°40′. Thus according to the relation $\cos \theta = 1/\beta n$ the angles of emission of Cherenkov photons in water could range anywhere between 0 and 41°40′ and cannot exceed 41°40′ in water. However, the data illustrated in Figure VII.4 indicated emissions of photons beyond 41°40′ of diminishing intensity to 90° degrees and beyond. This can be explained by the fact that, Compton electrons produced by the incident gamma rays can be emitted at a large angle relative to the incident gamma ray. When high-energy beta particles or electrons, in lieu of gamma rays, are incident on water, the angle of emission of Cherenkov photons does not exceed 41°40′, and the angle of emission will vary according to the energy or phase velocity β of the electrons.

The data provided the first experimental evidence of the relationship $\cos \theta = 1/\beta n$. In the case of water, which was assigned an index of refraction $n = 1.3371$, and taking the value of $\theta = 30°$, that is, the angle of highest intensity of light emission, Cherenkov calculated the phase velocity of the electrons

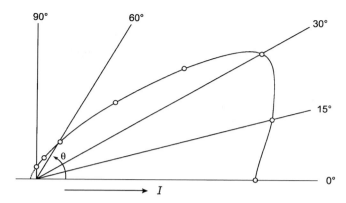

Figure VII.4 Cherenkov's plot of light intensity (I) as a function of angle of emission (θ) with respect to the direction of incident gamma radiation (Cherenkov, 1937b).

to be $\beta = 0.86$. This was the first estimate of the use of the Cherenkov Effect to measure the velocity of an elementary particle.

Cherenkov continued his experimentation and obtained recorded evidence of the unique asymmetric nature of the mysterious light using photographic plates. He was now able to demonstrate by means of photographs the direction of emission of the light photons with respect to the direction of the incident gamma rays on a sample of water. In June of 1937 Cherenkov submitted his historic findings concerning the evidence for the space symmetry of the radiation to the journal *Nature* for publication. However, Cherenkov's paper was rejected, and the letter of rejection is reproduced in Figure VII.5.

In an article on his recollections of his historic work Cherenkov (1986) related his reaction to the rejection to publish his work by the editors of *Nature* and the reactions of the scientific community. An excerpt of his comments is the following:

> …I can only say it was not particularly a pleasure for me to be aware that my experiments were now and again classified as "spiritualism" and often paralleled with an erroneous effect, the notorious N-rays of Blondlot, which were finally settled by Wood. [Cherenkov refers here to the history of René R. Blondlot (1849–1930), who in 1903 claimed to have discovered N-rays, and many scientists subsequently published articles on the origins and characteristics of the N-radiation. R. W. Wood of John Hopkins University later exposed the experimental results of Blondlot to be erroneous, misconceived, and the N-rays to be imaginary. The reader is invited to read an interesting historical account of the Blondlot N-rays by Lagemann, 1977].… Of course, sometimes there were pleasant exceptions. I still recall with great pleasure the exclamation by the greatest scientist of that epoch, Niels Bohr—"Wunderbar, Wunderschön", repeated several times after I demonstrated to him one of the most essential properties of the radiation, namely its space symmetry. Nevertheless, the atmosphere of distrust of the new effect from scientific opinion remained. The most open and sharp manifestation of this distrust was the refusal by the journal *Nature* to publish a short paper submitted to me, summarizing the essence of the phenomena and its main properties. Evidently, it was a mistake of the editorial staff of *Nature* who did not take the article sufficiently seriously. Receiving a negative answer from them and, fortunately, also the text of the article, I consulted S. I. Vavilov and right then transferred it, to another envelope and sent it to the editors of the American journal *Physical Review*. They were more attentive to it and, rather soon, published it in the next issue of the journal. This was the beginning of an overall recognition of the experimental discovery of a hitherto unknown effect: the radiation of charged particles moving faster than light in a material medium.

One Shilling Weekly

Telegraphic Address   Publishing and Editorial Offices
    PHUSIS LESQUARE LONDON   MACMILLAN & CO LTD
Telephone Number: ST MARTIN'S STREET,
    WHITEHALL 8871   LONDON W C 2

        RAG.AH/N.   29. 6. 37.

        The Editor of "NATURE" presents his

    compliments to Mr. P. A. Cherencov

    and regrets he is unable to make use of

    the communication, returned herewith,

    entitled "VISIBLE RADIATION PRODUCED

        BY ELECTRONS MOVING IN A MEDIUM

        WITH VELOCITIES EXCEEDING THAT OF

        LIGHT".

Figure VII.5   Letter to Pavel Cherenkov from the editors of the journal *Nature* rejecting his historic findings for publication. (From Cherenkov (1986) reprinted with permission from Elsevier © 1986.)

In only six weeks after *Nature* had rejected Cherenkov's article did Cherenkov's paper appear in the August 15, 1937 issue of *Physical Review* (see Cherenkov, 1937c). In his paper Cherenkov described the experimental arrangement, illustrated in Figure VII.6, that he used to demonstrate unequivocally the asymmetric character of the light. A liquid contained in a thin-walled glass vessel would be bombarded with gamma rays from one direction only, and light emitted from the liquid would be reflected off the cylindrical mirror completely surrounding the vessel. The light intensity would be recorded onto a photographic plate.

With the experimental setup illustrated in Figure VII.6 Cherenkov used a high-intensity gamma ray source with an activity equivalent to 794 mg of radium and a long exposure time of 72 h to produce a suitable photographic image of the light emitted by the vessel. He demonstrated that water and benzene would emit photons only in the onward direction, that parallel to the gamma-ray beam. The photographic prints are reproduced in Figure VII.7. To illustrate the distinction of the asymmetric character of the newly discovered radiation Cherenkov compared the images of the radiation produced by the water and benzene with that produced by a luminescent compound esculin.

Any doubts about the validity of the mysterious radiation discovered by Cherenkov were finally put to rest after Cherenkov's radiation and its properties were investigated in detail and confirmed by George Collins and Victor Reiling (1938) at the University of Notre Dame. They published their experimental results in the journal *Physical Review*, the same journal that accepted Cherenkov's paper the year before. They carried out a detailed study confirming Cherenkov's findings on the

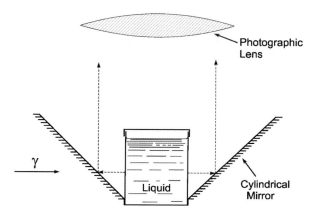

Figure VII.6   Experimental arrangement of Cherenkov used to demonstrate the unique direction of emission of the Cherenkov light photons with respect to the direction of the incident gamma rays on a sample of water. A glass vessel covered at the top was used to contain a pure liquid. The vessel was surrounded with a cylindrical mirror. Above the vessel was located a photographic lens and photographic plate. Gamma rays from thorium C (ThC) or radium (Ra) were incident on the sample from one side only, and the light emitted from the liquid reflected off the cylindrical mirror up toward the photographic lens and onto the photographic plate. (From Cherenkov (1937c, 1958) with permission © The Nobel Foundation 1958.)

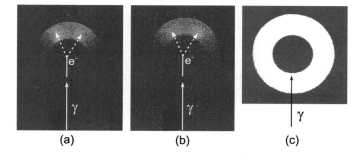

Figure VII.7   Photographs showing asymmetry of the radiation emitted by (a) water, $n = 1.337$, and (b) benzene, $n = 1.513$, and the symmetric light emission in all directions by (c) a solution of the luminescent compound esculin (Cherenkov, 1958). The light produced a black image on the photographic negative after chemical development of the plate. The black image on the photographic negative is seen as a white image above on the positive photographic prints. The arrows show the direction of the incident gamma rays. The white solid lines in (a) and (b) were drawn over the prints by the writer to depict the direction of travel of the gamma rays and photoelectrons (e^-) once the gamma rays penetrated the liquid. The dashed lines were drawn to depict the direction of travel of the Cherenkov photons. In the case of (c) the luminescence in the center vessel resulted in the symmetrical emission of light from the vessel in all directions, thus the white circle from the more intense and symmetrical luminescence light reflected off the circular mirror that surrounded the vessel (see Figure VII.6).(From Cherenkov (1937c) reprinted with permission © The American Physical Society.)

direction of emission, spectroscopic properties, and intensity of the unique radiation and reported the following:

Electrons of two million volts energy from an electrostatic generator were used to investigate the properties of the asymmetric radiation discovered by Cherenkov. This radiation is produced when electrons traverse a material medium with a velocity greater than the velocity of light in that medium. It was found for several solids and liquids [water, mica, glass, cellophane, etc.] that the direction of the emission of the radiation is accurately expressed by the relation, $\cos \theta = 1/\beta n$, and that the intensity maximum is quite sharp. The nature of the radiation from all solids and liquids investigated was found to be continuous and identical in appearance. The radiation apparently extends with increasing intensity from the infra-red to the ultraviolet absorption limit of the medium in which it is produced. Rough quantitative measurements of its intensity indicate that one 1.9 million volt electron in being brought to rest in water produces 40 quanta in a wavelength range 4000 Å to 6700 Å. These results are in good agreement with the classical explanation of the phenomenon given by Frank and Tamm [1937].

Next Cherenkov (1938a–c) carried out a series of thorough experiments to demonstrate the theory presented by Frank and Tamm (1937) on his newly discovered radiation. Cherenkov made more meticulous measurements of the spatial distributions of Cherenkov photons which, according to the theory of Frank and Tamm (1937), obeyed the relationship $\cos \theta = 1/\beta n$. He measured the spatial distributions of the photons induced by high-energy electrons traveling in four liquids having different indexes of refraction (Cherenkov, 1938a). These are in order of increasing index of refraction n, water ($n = 1.3371$), cyclohexane (C_6H_{12}, $n = 1.4367$), benzene (C_6H_6, $n = 1.5133$), and ethyl cinnamate ($C_{11}H_{12}O_2$, $n = 1.58043$). At the same time he tested two sources of gamma rays with differing energies, which would produce photo- and Compton-electrons of different energies when a liquid was irradiated. The two gamma-ray sources were (i) radium, which would produce electrons with a phase velocity $\beta = 0.847$, and (ii) thorium C'' or ThC'' (i.e., ^{208}Tl), which emits a high-energy, 2.16 MeV gamma ray, yielding electrons with a phase velocity $\beta = 0.896$. To guarantee only the hardest components of the gamma-ray emissions from the two sources Cherenkov filtered the lower-energy gamma rays of radium with 2 mm of lead shielding and those of the higher energy gamma rays of ThC'' with 12 mm lead shielding. He accurately measured the angles of emissions (θ) of the Cherenkov photons in the four liquids when irradiated with the gamma-ray sources Ra or ThC''. By measuring θ in liquids of differing index of refraction (n) and sources of two electron energies (β) Cherenkov set out to demonstrate that the angle θ, which is the direction of emission of the Cherenkov photons relative to the line of flight by the electrons traveling in a medium at a velocity exceeding that of light in the same medium, satisfies the relation $\cos \theta = 1/\beta n$. The graphical representation of his measurements of θ in the four liquids when irradiated by either Ra or ThC'' is illustrated in Figure VII.8.

From the experimental data provided in Figure VII.8 Cherenkov was able to demonstrate that the angle θ for the light intensity maxima in the four liquids would increase as with increasing index of refraction n as well as with increasing electron velocity or β. The index of refraction increases with each liquid tested from top to bottom of Figure VII.8 and the electrons produced by the ThC'' are faster than those produced by the Ra source. In the four liquids the light intensity maxima, measured as θ with respect to the direction of electron travel, enlarged with increasing n and increasing β. Cherenkov compared the experimentally measured light intensity maxima (θ) obtained from Figure VII.8 with the values of θ calculated according to the relation $\cos \theta = 1/\beta n$. Cherenkov's calculated and observed values of θ for the four liquid media tested and two sources of fast electrons are provided in Table VII.2. The experimentally observed and those calculated according to the Frank and Tamm (1938) theoretical relation $\cos \theta = 1/\beta n$ are very close demonstrating that the theory holds with experiment.

Pavel Cherenkov continued his thorough investigation into the properties of the newly discovered radiation. In another work published the same year, as that previously described, Cherenkov (1938b) investigated the spectrum of the visible radiation produced by charged particles that traveled in a transparent medium at speeds faster than the speed of light in the medium. To measure the light intensity as a function of the wavelength of the visible light Cherenkov used the experimental arrangement illustrated in Figure VII.9. A source of up to 280 mCi of Ra(B+C) or ^{214}Pb (^{214}Bi) in a 2 mm glass tube was suspended in benzene by means of a thin wire in the center of a thin-walled (0.1 mm) glass vessel measuring 3 cm in diameter and 4.5 cm in height. The outside of the inner glass tube containing the radium was coated with silver to eliminate any Cherenkov photons that might be produced in

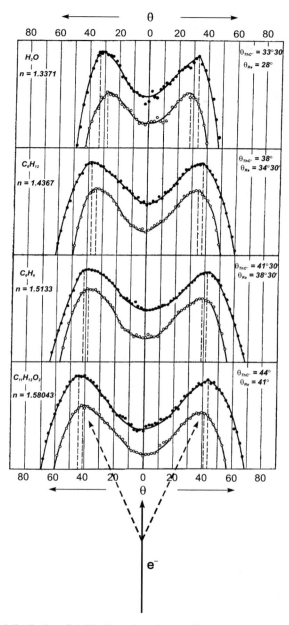

Figure VII.8 Spatial distribution of visible Cherenkov photon radiation produced by fast electrons in four liquid media and two sources of fast electrons. The liquid media tested were (from top to bottom) water, cyclohexane, benzene, and ethyl cinnamate. The sources of fast electrons used were ThC″ producing electrons in the liquid media with a phase velocity $\beta = 0.896$ and radium producing electrons with $\beta = 0.847$. In the lower part of the figure the author has inserted an arrow with solid line to indicate the direction of travel of the electrons at $\theta = 0$ and two arrows with dashed lines toward the light intensity maxima indicating the direction of travel of the Cherenkov photons at an angle θ with respect to direction of the electrons. (From Cherenkov (1938a, 1958) with permission © The Nobel Foundation 1958.)

Table VII.2

Observed and calculated values of θ for four liquid media and two sources of fast electrons
(ThC″ and Ra gamma rays)

Liquid	Chemical formula	n ($\lambda = 4861.5$ Å)	ThC″ gamma rays		Ra gamma rays	
			θ (experimental)	θ (calculated)[a]	θ (experimental)	θ (calculated)[b]
Water	H_2O	1.3371	33°30′	33°30′	28°	28°
Cyclohexane	C_6H_{12}	1.4367	38°	39°	34°30′	34°40′
Benzene	C_6H_6	1.5133	41°30′	41°35′	38°30′	38°40′
Ethyl cinnamate	$C_{11}H_{12}O_2$	1.58043	44°	45°10′	41°	41°50′

Note: The calculated values of θ were obtained by the relation $\cos\theta = 1/\beta n$ (Cherenkov, 1938a).
[a]$\beta = 0.896$.
[b]$\beta = 0.847$.

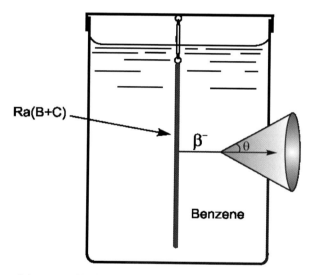

Figure VII.9   Source of photons used by Cherenkov to study the spectrum of visible radiation created by the high-energy beta particles from a source of radium hanging in the center of a glass vessel containing benzene (Cherenkov, 1938b). The radium was contained in a narrow glass tube in the center of benzene liquid contained in a 3 × 4.5 cm glass vessel. A high-energy beta particle is illustrated traveling through the benzene and creating a cone of Cherenkov radiation at an angle θ to the particle direction of travel.

the glass walls of the tube. The source provided high-energy beta particles to create the Cherenkov photons in benzene while a minor fraction of the photons would arise from gamma-ray interactions with electrons in the liquid.

Cherenkov placed the radiation source illustrated in Figure VII.9 at a distance of 45 cm from the entrance slit S_1 of the dual monochromator and lens arrangement in the experimental setup illustrated in Figure VII.10. With this arrangement he was able to observe and measure the intensities of light at various wavelengths of the visible part of the spectrum. The luminous volume of the glass vessel was projected onto the entrance slit S_1 of the monochromator by means of a lens. The part of the spectrum needed for measurement was defined by slit S_2. Cherenkov measured the intensity of the light at specific

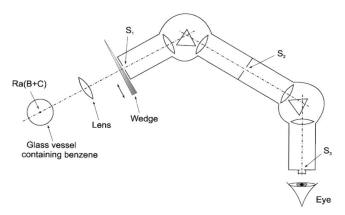

Figure VII.10   The experimental arrangement used by Cherenkov (1938b) to measure the visible spectrum of Cherenkov radiation. The source of high-energy beta particles suspended in benzene, as illustrated in Figure VII.9, is located to the left. Cherenkov measured the radiation intensities at selected wavelengths by moving the wedge manually in the direction of the double arrow to the point where the light intensity as seen at slit S_3 would be diminished to the threshold of the visual extinction of the eye.

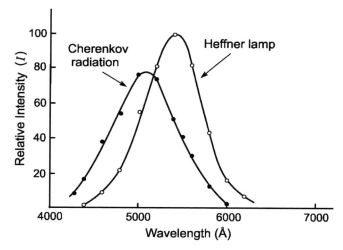

Figure VII.11   The visible spectrum of Cherenkov radiation compared with that from a Heffner lamp (Cherenkov, 1938b).

wavelengths visually by means of a wedge as described previously and illustrated in Figure VII.1. He would have to accustom his eyes to the dark for about an hour before making the measurements. Cherenkov would move the wedge manually until it diminished the light intensity down to the threshold of visual extinction. The wedge thickness and the degree of light transmission by the wedge would depend on its position. The light intensities at various wavelengths with this arrangement were calibrated by Cherenkov with a Heffner lamp, which was located in place of the glass vessel of radium and benzene. Instead of directing the light from the Heffner lamp directly onto slit S_1, he reflected the light of the lamp off a plate coated with magnesium oxide. The light intensity (I) at various wavelengths of the visible spectrum of the Cherenkov photons and of the Heffner lamp is illustrated in Figure VII.11.

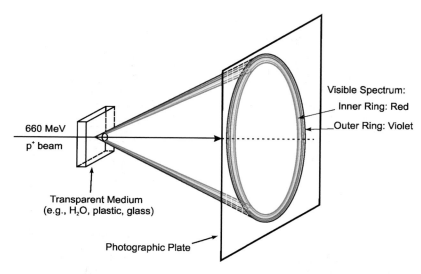

Figure VII.12 Diagram of an experimental arrangement for obtaining photographs of the cone section in the plane of a photographic plate placed perpendicular to the path of high-speed (660 MeV) protons in an accelerator beam. (From Cherenkov's Nobel Lecture (1958) with permission © The Nobel Foundation 1958.)

Later in life, approximately 20 years after his historic research on the visible spectrum of the newly discovered radiation, Cherenkov commented on the spatial distribution and the visible spectrum of the radiation in his Nobel Lecture given on December 11, 1958 as follows:

> If we consider the picture [of the radiation] not in the [2-dimensional] plane but spatially, then the radiation must spread out along the surface of a cone whose axis forms the path of an electrically charged particle while the surface line forms with this axis the angle [θ].
>
> If we place [a] photographic plate perpendicular to the beam of high-speed particles [See Figure VII.12], we shall obtain, in addition to an image of the track of the beam, also a photograph of the radiation in the form of a ring. This photograph [see Figure VII.13] was obtained with the aid of a fine beam of protons in the accelerator of the United Institute for Nuclear Research at Dubna.
>
> …we have in our considerations assumed some fixed frequency. In reality, however, the radiation spectrum is continuous. Since the medium exhibits dispersion, i.e., the refractive index is dependent on the frequency, this means that the light of different wavelengths is propagated at angles which, even with strictly constant velocity of the particles, differ somewhat from one another.
>
> Thus the radiation is broken up as in spectral analysis. The radiation cone will consequently show a definite intensity, and in the case of a medium with normal dispersion the spectral red will lie in the inner part of the cone while the violet is on the outside. That this is actually so was shown by a photograph showing part of the ring with a colour plate.

Finally Cherenkov (1938c) measured the energy output of the radiation produced by electrons traveling in a medium at superlight velocity. This study was important to determine what portion of the total energy of the electron is dissipated in the generation of the Cherenkov radiation. According to the theory of Frank and Tamm (1937) the energy W emitted by an electron traveling in a medium at superlight velocity is calculated according to the equation

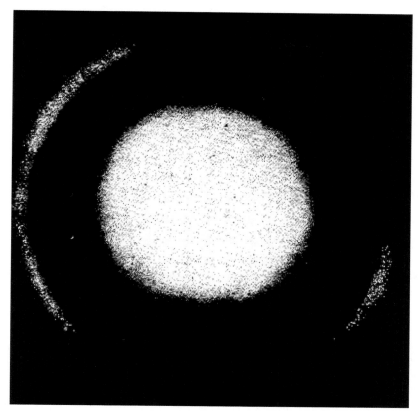

Figure VII.13 Photograph of a section of the Cherenkov radiation cone (outer circle) obtained by the experimental arrangement illustrated in Figure VII.12. The central patch is that produced by the high-energy proton beam. (From Cherenkov (1958) with permission © The Nobel Foundation 1958.)

$$W = \frac{e^2 l}{c^2} \int \omega \, d\omega \left(1 - \frac{1}{\beta^2 n^2} \right) \qquad \text{(VII.3)}$$

where l is the length of the electron path, c the speed of light in a vacuum, ω the cyclic frequency of the radiation, β the electron phase velocity, that is, v/c as previously defined, and n is the refractive index of the medium. Cherenkov (1938c) explained that the velocity of the electron must exceed the phase velocity of light in the medium, where $\beta n > 1$. Also Cherenkov added that the motion of the electron in the medium involves a considerable change in velocity and, accordingly the magnitude of its phase velocity β will vary greatly for different sections of the electron path. Therefore, he concluded that, on the assumption that a relatively small region of the radiation spectrum is examined, within which n may be considered a constant, eq. (VII.3) with reference to the true motion of the electron would be written as follows:

$$W = \frac{e^2}{c^2} \int \omega \, d\omega \int_{\beta n > 1} \left(1 - \frac{1}{\beta^2 n^2} \right) dl \qquad \text{(VII.4)}$$

where β depends on l. To facilitate the integration along l, Cherenkov substituted β and dl with V and dV where V is the kinetic energy of the electron in keV and dV is the energy lost by the electron in the section of path dl.

Cherenkov next expressed the kinetic energy of the particle as the difference between its total energy and rest energy. As described in eqs. (4.3–4.5) of Chapter 4 the total energy (E) of a particle is the sum of its kinetic (V) and rest energies (mc^2) or

$$E = V + mc^2 = \gamma mc^2 \tag{VII.5}$$

The kinetic energy V of the particle is then expressed as

$$V = \gamma mc^2 - mc^2 \tag{VII.6}$$

The term γmc^2 is dependent on the particle speed where

$$\gamma = \frac{1}{\sqrt{1 - (v^2/c^2)}} = \frac{1}{\sqrt{1 - \beta^2}} \tag{VII.7}$$

Thus, the kinetic energy (V) of the electron can be expressed as

$$V = \frac{mc^2}{\sqrt{1 - \beta^2}} - mc^2 \tag{VII.8}$$

Cherenkov considered the rate of electron energy lost per path length in a given medium to be constant or $-dV/dl = k$, and that the value of k would differ for each medium and be proportional to the density of the medium. From eq. (VII.8) Cherenkov expressed β^2 as

$$\beta^2 = 1 - \frac{1}{[(V/mc^2) + 1]^2} \quad \text{or} \quad \beta^2 = 1 - \left[\frac{1}{(V/mc^2) + 1}\right]^2 \tag{VII.9}$$

Substituting the value of β^2 from eq. (VII.9) and the value $-dV/k$ for dl into eq. (VII.4) Cherenkov (1938c) expressed the energy emitted by an electron at relativistic speed in a medium as

$$W = \frac{e^2}{c^2} \int \omega \, d\omega \frac{1}{k} \int \frac{dV}{n^2[1 - (1/[(V/mc^2) + 1]^2)]} \tag{VII.10}$$

He carried out the integration for W the total energy emitted by the electron in water within a range of frequencies from 536 to 556 mμ (5360–5560 Å) and electron energy range from 260–3000 keV. The electron energy spectrum would be that corresponding to a source of Ra(B+C), as Cherenkov would subsequently use this source of beta particles to determine experimentally the energy output of the Cherenkov photons. The value of k, a constant for each medium, was given the value of 1710 keV/cm for water. Beta-particle energies below 260 keV were ignored, because only electron energies above 260 keV satisfy the threshold condition of $\beta > 1/n$ for the production of Cherenkov photons in water medium where $n = 1.334$ (see Chapter 7 for the calculation of the threshold electron energy for the production of Cherenkov photons in water). Cherenkov thus calculated W, the theoretical output of the radiation energy to be 3.5×10^{-4} ergs/sec.

To compare the above theoretically calculated value of the energy output W with an experimentally determined value Cherenkov (1938c) would measure the energy output for the same Cherenkov

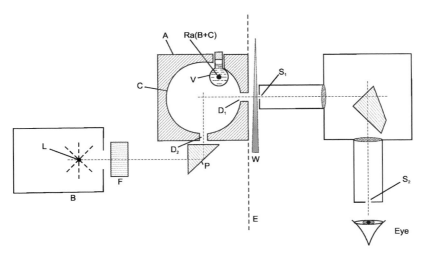

Figure VII.14 Experimental arrangement used to measure the radiation intensity of Cherenkov photons produced by beta particles of a Ra(B+C) source (Cherenkov, 1938c).

photon spectral region of 5360–5560 Å using the experimental arrangement illustrated in Figure VII.14. The Cherenkov photons were produced in a lead block (A) with a 13 cm diameter spherical cavity coated with reflecting magnesium oxide, an Ulbricht sphere. The liquid medium under investigation (H_2O was the first liquid to be used and other liquids were tested subsequently.) was placed inside the cavity in a round thin-walled glass vessel (V), 3.5 cm in diameter. The Cherenkov photons were produced by beta particles from a high-activity (135–193 mCi) source of Ra(B+C) located in a 8 mm diameter spherical thin-walled (0.12 mm) glass ampoule inside the liquid in the center of the vessel. The ampoule was coated with silver on its outer surface to reflect and eliminate Cherenkov photons produced in the glass ampoule. The lead block protected the experimenter from the high levels of gamma radiation.

The Ulbricht sphere provided homogeneous Cherenkov photons in all directions even though the emission of beta particle and consequent Cherenkov photon was asymmetrical. The size of the vessel (V) was carefully selected to provide a thickness layer of liquid sufficient to completely absorb the beta particles emitted by the Ra source, while at the same time insufficient to absorb to an appreciable extent the hard gamma rays from the source. In this way Cherenkov assured that any contribution by gamma rays to the production of Cherenkov photons would be insignificant.

The visible radiation would exit the cavity via aperture D_1 and pass through a 0.2 mm entrance slit S_1 of the monochromator and finally exiting at slit S_2. Here the observer would select the region of the spectrum ranging from 5360 to 5560 Å with the mean wavelength at 5460 Å corresponding to the green line of the mercury arc. The wedge (W) was controlled by the observer to determine radiation intensities using the criterion of the threshold of the eye conditioned to the dark. A screen (E) shielded the eye from external light. Cherenkov made comparisons with known light intensities observed from a mercury lamp (L) situated in an external reflector chamber (B). The green line of the mercury arc was selected by directing the light through a preliminary liquid filter (F) and prism (P) before passing through aperture D_2 and reaching the spherical cavity (C). The intensity of the beam was adjusted to permit measurement with a thermopile placed inside the spherical cavity (C) in front of aperture D_2 in a manner that the working surface of the thermopile would completely overlap the aperture D_2. His readings were calibrated with the light output of a standard Heffner lamp.

Cherenkov carried out three series of experiments with Ra(B+C) sources of 193, 162, and 135 mCi activity. The results of the work are provided in Table VII.3. The experimental mean value provided

Table VII.3

Absolute energy output of Cherenkov photons produced by beta particles from
Ra(B+C) in water in the spectral region from 5360 to 5560 Å

Experiment series	Absolute output (per 1 mCi of emanation in ergs/sec)
1st series, 193 mCi Ra(B+C)	4.28×10^{-4}
2nd series, 162 mCi Ra(B+C)	3.91×10^{-4}
3rd series, 135 mCi Ra(B+C)	4.11×10^{-4}
Mean	4.1×10^{-4}
Calculated (theoretical)	3.5×10^{-4}

in Table VII.3 was very close to the theoretically calculated value and Cherenkov (1938c) commented on the results as follows:

> As seen from this table [Table VII.3] the experimentally obtained value of the radiation output of water produced by β-particles of Ra(B+C) in the spectrum region extending from 536 to 556 mμ constitutes 4.1×10^{-4} ergs/sec. per mCi…this figure has to be somewhat reduced, since, under the conditions of the experiments above described, part of the energy of the β-particles is expended on the passage through the walls of the ampoule containing the emanation. The loss in energy occasioned by the circumstance equals for a separate β-particle 60–70 keV, which constitutes with respect to the mean energy of only those β-particles which satisfy the condition $\beta n > 1$ a magnitude of the order of 12%. Since the passing through of the walls of the ampoule takes place in the early part of the path of β-particles, it will be possible to assume approximately that the lowering of the output caused by this passage also constitutes a magnitude of the same order. Accordingly, the expected magnitude of absolute output should be 3.5×10^{-4} ergs per sec. instead of 4.1×10^{-4}…. Thus, we are free to assume that the experimentally obtained and theoretically calculated values of absolute output are in good agreement.

Cherenkov (1938c) repeated the above experiment of theoretically calculated and experimentally measured energy outputs in five additional liquid media, namely benzene, cyclohexane, carbon disulfide, isobutyl alcohol, and carbon tetrachloride. The theoretical values were calculated according to eq. (VII.10) varying only n and k for each medium. The absolute outputs differed for each medium because of the differences in index of refraction (n) and density (thus different k) of each medium; however, the experimental and calculated values for each medium were in close agreement.

One of the major outcomes of this work was Cherenkov's finding that the energy expended by the beta particles in the production of Cherenkov photons was small, of the order of 0.1%. His conclusions were the following:

> From a comparison of the figures…it will be evident that for all liquids subjected to investigation, as well as for water, the values of the output experimentally obtained agree sufficiently well with those expected from theory. It seems of interest to note the fact that the energy expended by β-particles of Ra(B+C) on the excitation of the radiation under consideration [Cherenkov photons] constitutes a value of the order of only 0.1% of the total energy of these particles.

In his Nobel Lecture given on December 11, 1958 Cherenkov summarized the outcome of his discovery in 1934, his experimental research on the newly discovered radiation during 1934–1938 and the theory presented by Frank and Tamm in 1937 as follows:

> The development of this theory [Frank and Tamm, 1937] completes a great cycle of research work which covers discovery, all-round experimental investigation, and the development of

the theoretical basis of this phenomenon which established a new branch of physics—the optics of rays moving faster than light.

Although, as noted previously, the production of Cherenkov radiation is not a principle means by which charged particles may dissipate their energy in matter, the production of Cherenkov photons and their detection in matter later developed to be a vital tool in the biological, environmental, and physical sciences. In the biological sciences the counting of Cherenkov photons is used commonly to analyze quantitatively the activity in disintegrations per minute (DPM) of the beta-particle emitting nuclide ^{32}P, an important radioisotope used as a tracer for phosphorus. This technique is discussed in detail in a previous book (L'Annunziata, 2003b). In the environmental sciences, Cherenkov counting of the nuclide ^{90}Y, which emits high-energy beta particles, is performed to analyze for its parent nuclide ^{90}Sr (L'Annunziata, 1971; L'Annunziata and Passo, 2002). In the physical sciences the Cherenkov Effect has become a useful tool for the identification of elementary particles and information on their mass and velocity particularly following the development of Ring Imaging Cherenkov (RICH) detectors reviewed by Sequinot and Ypsilantis (1994), Ypsilantis and Sequinot (1994), and L'Annunziata (2003b). Cherenkov (1958) in his Nobel Lecture summarized the unique properties of Cherenkov radiation that would open a new field of physics proving to be of immense value in the detection and measurement of radionuclides and the identification of elementary particles and their mass and velocity. An excerpt of his comments listed according to application, is the following:

1. The presence of an energy threshold makes this type of counter [Cherenkov counter or detector] insensitive to the slow particles whose energy lies below the threshold. [In this statement Cherenkov clearly points out that radionuclides which emit low-energy β-particles, for example ^{3}H or ^{14}C, will cause no interference in the detection and counting of Cherenkov photons produced by β-particle emissions of ^{32}P in a liquid medium such as water. This is so because the threshold energy of for the production of Cherenkov photons in water is 260 keV and the energy maxima of the β-particles emitted by ^{3}H and ^{14}C are 18.6 and 155 keV, respectively, well under the threshold energy of 260 keV].

2. …The asymmetry of the radiation facilitates with this counter the measurement of those particles which are moving in the radiator in the direction of the cathode of the photomultiplier. Particles moving in the opposite direction are not measured by the counter. In other words, this type of counter has the distinction of focusing on a definite directional effect.

3. …one of the most important parameters of a particle—its mass—can be determined on the basis of measurements of its momentum and speed…It is evident without more ado that where the speed of a particle is within a certain range where β [particle phase velocity], that fulfills the condition $\beta n > 1$, is still sufficiently different from 1, the speed of the particle, on the basis of the equation $\cos \theta = 1/\beta n$ and starting from the measured quantity θ and the known refractive index, can easily be calculated. When the type of the particle is known, speed measurements enable us to determine their energy also.

4. …We have already noted the fact of the [Cherenkov] radiation having an energy threshold [that is, the particle must possess an energy in excess of the threshold to produce Cherenkov radiation], which renders the counter insensitive to low-energy particles. We thus have the possibility of altering the threshold energy E_0 when we choose a radiator with a suitable quantity n. It is clear that two counters set at degrees of threshold energy E'_0 and E''_0, and switched on in the proper sequence according to the scheme of anticoincidence will only measure those particles whose velocity is in the region of E'_0 and E''_0. A procedure of this kind was successfully employed by [Emilio] Segrè and his collaborators in their outstanding work which led to the discovery of the antiproton.

An interesting paper on Cherenkov's professional work and his equipment is provided by one of his former colleagues Boris Borisovich Govorkov (2005) at the Lebedev Physical Institute (LPI), Moscow from which some of the following notes on Cherenkov's professional career are provided here. In 1949 Pavel Cherenkov collaborated with Vladimir Iosifovich Veksler in the startup of the first

European electron synchrotone, which accelerated electrons to energies of up to 250 MeV. Cherenkov became chief investigator in particle physics after the accelerator was put into operation. In 1959 he became the leader of the Photomesonic Laboratory in the LPI of the Russian Academy of Sciences, which he managed until 1986. Pavel Cherenkov then was appointed director of the High Energy Physics Department of the LPI, which staffed approximately 150 physicists, engineers, and technicians carrying out research in photonuclear processes and accelerator physics.

Pavel Alekseyevich Cherenkov died on January 6, 1990. In August of the following year John Hubbel (1991) wrote a short article entitled *Faster Than a Speeding Photon* commemorating Cherenkov's paper published in 1934 announcing his discovery of the light created by charged particles that traveled at speeds faster than light in a transparent medium. In his article Hubbel underscored the significance of Cherenkov's work by the following statement:

> Pavel Alekseyevich Cherenkov casts a long shadow (or light actually) across both the history and future of radiation physics.

IL'JA MIKHAILOVICH FRANK (1908–1990) AND IGOR YEVGENYEVICH TAMM (1895–1971)

Il'ja Mikhailovich Frank and Igor Yevgenyevich Tamm shared the Nobel Prize in Physics 1958 with Pavel Alekseyevich Cherenkov "for the discovery and interpretation of the Cherenkov Effect". It was Pavel Cherenkov, who discovered the effect in 1934 and measured the physical properties of Cherenkov radiation; however, it was Il'ja Frank and Igor Tamm, who in a joint publication in 1937, provided the interpretation of the origin and properties of the radiation. Their biographical sketches and work are presented jointly here since they worked together in providing the theoretical concepts needed to interpret the Cherenkov Effect.

Il'ja Frank was born in Leningrad (now Saint Petersburg) Russia on October 23, 1908. He was a pupil of Sergei I. Vavilov at Moscow State University graduating in 1930. The following year he was appointed scientific officer of the State Optical Institute where S. I. Vavilov would become Head of

**Il'ja M. Frank (1908-1990)
with permission
© The Nobel Foundation 1958**

Research the following year. In 1934 Frank joined the P. N. Lebedev Institute of Physics in Moscow, which was newly created under the directorship of S. I. Vavilov. It was that same year that Cherenkov (1934a,b) reported his discovery of a new radiation produced by high-energy electrons in transparent media, and Vavilov (1934) presented his initial understanding of the origin of this unique radiation, which is currently referred to as Cherenkov radiation or Vavilov–Cherenkov radiation. Frank was awarded the doctorate in physical and mathematical sciences in 1935.

Идея фононов. И.Е. ТАММ. 1929 г.

РОССИЯ ROSSIJA·2000   1.75

Igor Y. Tamm (1895-1971)

Igor Tamm was born in Vladivostok, Russian Empire on July 8, 1895. When he was five years of age he moved with his family to Elizavetgrad (now Kiroovograd, Ukraine). He graduated from the Elizavetgrad Gymnasium in 1913 and then studied at the University of Edinburgh during 1913–1914. Like his future colleague Il'ja Frank, Igor Tamm studied physics and math at Moscow State University graduating from the university in 1918. Upon graduation he continued his studies and teaching of physics and mathematics in several institutions including the Crimean and Moscow State Universities, the Polytechnical and Engineering Physical Institutes, and the J. M. Sverdiov Communist University (Nobel Lectures, Physics, 1942–1962). He eventually achieved the rank of full professor and was awarded the doctorate in physics and mathematical sciences. Again like Il'ja Frank, in 1934 Igor Tamm joined the newly created P. N. Lebedev Institute of Physics in Moscow, which was established under the directorship of Sergei I. Vavilov.

Once Cherenkov had discovered a unique visible radiation produced by high-energy Compton electrons or high-energy (fast) beta particles and once Cherenkov found that this light was polarized and that the light emission was asymmetric, that is, it was emitted only in the direction of the fast electrons (Cherenkov, 1934a,b, 1936). Only then did Cherenkov and his doctoral dissertation advisor and Director of the P. N. Lebedev Instutite, Sergei I. Vavilov, know that they were dealing with a yet unknown phenomenon. Il'ja Frank and Igor Tamm of the Theoretical Physics Division of the same institute took notice and went to work on the interpretation of this new phenomenon. Their findings were published in a joint paper entitled *Coherent Visible Radiation of Fast Electrons Passing through Matter* (Frank and Tamm, 1937). They explained the phenomenon as caused by a charged particle traveling in a medium at a speed in excess of the speed of light in that medium. It was known from Einstein's theory of relativity that matter could not travel in excess of the speed of light in a vacuum ($c = 2.99 \times 10^8$ m/sec); however, in other gaseous, liquid, or solid media the velocity of light will be less than its velocity in a vacuum, and an elementary charged particle with sufficient energy could travel in such media at speeds exceeding that of light. The charged particle in passing through the electron clouds of a medium would create an electromagnetic shock wave analogous to that of a "sonic boom" created by a jet airplane or projectile traveling in the atmosphere at a speed exceeding that of sound. In the words of Frank and Tamm (1937):

> In 1934 P. A. Cherenkov has discovered a peculiar phenomenon, which he has since investigated in detail. All liquids and solids [later discovered that gases as well] if bombarded by fast electrons, such as β-electrons or Compton electrons produced by γ-rays, do emit a peculiar visible radiation…This radiation is particularly polarized, the electric oscillation vector being parallel to the electron beam, and its intensity can be reduced neither by temperature nor by addition of quenching substances to the liquid…the most peculiar characteristic of the phenomenon was discovered, namely, its highly pronounced asymmetry, the intensity of light emitted in the direction of the motion of electrons…the phenomenon can be explained both qualitatively and quantitatively if one takes in account the fact that an electron moving in a medium does radiate light even if it is moving uniformly provided that its velocity is greater than the velocity of light in the medium…
>
> We shall consider an electron moving with constant velocity v along [an] axis through a medium characterized by its index of refraction n [The index of refraction, n, is by definition the ratio of the speed of light in a vacuum to its speed in a particular medium, and the value of n will vary from one medium to another]. The field of the electron may be considered as the result of superposition of spherical waves of retarded potential, which are being continually emitted by the moving electron and are propagated with the velocity (c/n). [see Figure VII.15] It is easy to see that all these consecutive waves emitted will be in phase along the directions making the angle θ with the axis of motion [of the particle], if only v, n, and θ do satisfy the condition

$$\cos \theta = \frac{1}{\beta n} \tag{VII.11}$$

where $\beta = v/c$. Thus there will be a radiation emitted in the direction θ, whereas the interference of waves will prevent radiation in any other direction. Now the condition [of eq. (VII.11)] can be fulfilled only if

$$\beta n > 1, \tag{VII.12}$$

that is, only in the case of fast electrons in a medium, whose index of refraction n for frequencies in question is markedly larger than 1.

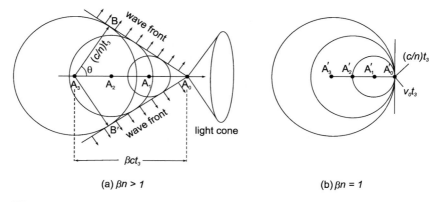

(a) $\beta n > 1$ (b) $\beta n = 1$

Figure VII.15 A Huygens construction of the spherical waves of retarded potential produced by a charged particle traveling along an axis in a refractive medium from points A_3 to A_0 at two velocities, namely (a) $\beta n > 1$ and (b) $\beta n = 1$. (From Nobel Lecture of Il'ja Frank (1958) with permission © The Nobel Foundation 1958.)

The term "fast electrons" used by Frank and Tamm (1937) above refers to electrons of such energy that they travel at a speed in excess of the speed of light in a particular medium. Also the term β is called the relative phase velocity of the particle ($\beta = v/c$) or the velocity of the particle in a medium divided by the speed of light in a vacuum.

Eq. (VII.11) derived by Frank and Tamm can be found from the distance of travel of the spherical wave front relative to the distance traveled by the electron as seen in Figure VII.15. In the words of Frank (1958) in his Nobel Lecture:

> ...We have to consider each point of the particle trajectory as a source of waves. In this case the wave phase is determined by the instant of passage of the particle through a given point. Let us assume that at moment $t = -t_3$ the emitter was at point A_3, at moment $t = -t_2$ at point A_2, at moment $t = -t_1$ at A_1, and, finally, at the moment of observation $t = 0$ at A_0.
>
> ...the velocity of the waves is equal to the phase velocity of light c/n...The surfaces of the rays are simply spheres whose radii for points A_3, A_2, A_1, and A_0 are $(c/n)t_3$, $(c/n)t_2$, $(c/n)t_1$, and 0. respectively [see Fig. VII.15a]. The envelope of these spheres evidently represents a cone of circular cross section with the apex at A_0. Its generatrices in the plane of the drawing are A_0B and A_0B'.
>
> According to the Huygens' principle, the directions of the rays are defined by the radius vectors drawn from some centre of the waves to the point of tangency with the envelope. For example, in Figure VII.15 it is A_3B and A_3B' coinciding with the generatrices of the wave normal cone...
>
> From Figure VII.15 it is not difficult to determine the magnitude of the threshold velocity. When the velocity diminishes, the distances between points A decrease. The threshold case arises when point A, occupies the position A'_0 on the surface of the sphere. (This case is depicted separately in Figure VII.15b.) At lower velocities, one of the spheres lies completely within the other and they do not have a common envelope. In the threshold case [$\beta n = 1$], they have only a common point of tangency, A'_0. Thus evidently $(c/n)t_3 = v_0t_3$, i.e. $v_0 = c/n$. The cone of wave normals is compressed in the direction of velocity v, and the wave cone transforms into a plane perpendicular to the axis of motion at point A'_0 [Figure. VII.15b]."

From Figure VII.15a the equation defining the angle θ as a function of the particle phase velocity β and index of refraction n, that is, $\cos\theta = 1/\beta n$, is derived from the distance of travel of the wave front with respect to that traveled by the charged particle. The velocity v of travel of the wave front in a particular medium is a function of the index of refraction n of the medium, since by definition $n = c/v$. The distance traveled by the wave front from A_3 to B of Figure VII.15a is the product of velocity and time t or $A_3B = vt_3$, and the distance traveled by the charged particle in the same time span from A_3 to A_0 is the product of the particle velocity in the medium v_p and time t_3 or $A_3A_0 = v_pt_3$, which may be written as $A_3A_0 = (v_p/c)ct_3$. Since the particle phase velocity β is defined as v_p/c, the distance A_3A_0 traveled by the charged particle becomes $A_3A_0 = \beta ct_3$. The angle of emission θ of the Cherenkov radiation with respect to the direction of travel of the charged particle is defined as

$$\cos\theta = \frac{AB}{A_3A_0} = \frac{(c/n)t_3}{\beta ct_3} \quad \frac{1}{\beta n} \qquad \text{(VII.13)}$$

Frank and Tamm (1937) calculated the total energy, W, radiated by an electron via the Cherenkov Effect through a cylindrical region of medium of length, l, where the axis of the cylinder coincided with the line of motion of the electron. Under the assumption that β is constant over a short length, dl, of the electron track, they derived the equation

$$W = \frac{e^2l}{c^2}\int_{\beta n > 1}\omega\,d\omega\left(1 - \frac{1}{\beta^2 n^2}\right) \qquad \text{(VII.14)}$$

where ω is the frequency of the Cherenkov radiation. The lower limit of the integral, $\beta n > 1$, is the threshold condition for the production of Cherenkov photons in any medium. Frank and Tamm (1937) established the equation to be valid provided that l should be large in comparison with the wave length λ of the Cherenkov radiation emitted. Cherenkov (1938c) took eq. (VII.14) of Frank and Tamm and evaluated the energy, W, radiated by an electron taking under consideration that the magnitude of β will vary greatly over the entire path length of the electron. Cherenkov stated (1938c) the following:

> In reality, the motion of an electron in the medium involves a considerable change in velocity and, accordingly, magnitude β will vary greatly in different sections of the path of the electron. Hence on the assumption that a but relatively small region of the spectrum is examined, within which n may be considered as constant, formula VII.14 with reference to the true motion of the electron should be written as follows:

$$W = \frac{e^2}{c^2} \int \omega \, d\omega \int\limits_{\beta n > 1} \left(1 - \frac{1}{\beta^2 n^2}\right) dl \qquad \text{(VII.15)}$$

where β depends on l.

Frank and Tamm (1937) thus calculated the rate of radiated energy per path length of travel, dW/dl, to be of the order of only several kilovolts per centimeter. Thus the electron energy loss via the emission of Cherenkov radiation would be a negligible quantity of energy in comparison to the energy lost by the electron by other mechanisms such as ionization. Cherenkov (1938c) confirmed this through calculations and experimentation and concluded that the radiation expended by beta particles from Ra(B+C) via the emission of Cherenkov radiation amounts to only a value of the order of only 0.1% of the total energy of the particles.

From eq. (VII.14) Frank and Tamm (1937) deduced the following equation for the number of photons, N, emitted by an electron within the limits of a given spectral region confined by the wavelengths λ_1 and λ_2 for a constant β, that is, for a short length of the electron track where the phase velocity of the electron remains relatively constant to be

$$N = 2\pi\alpha \left(\frac{1}{\lambda_2} - \frac{1}{\lambda_1}\right)\left(1 - \frac{1}{\beta^2 n^2}\right) \qquad \text{(VII.16)}$$

where α is the fine structure constant ($\alpha \sim 1/137$). From eq. (VII.16) Frank and Tamm (1937) calculated that a 500 keV electron, which would correspond to $\beta^2 = 0.75$, traveling in water, where $n = 1.33$ would produce approximately 10 photons per 0.1 cm within the visible region between $\lambda_1 = 4000$ Å and $\lambda_2 = 6000$ Å as follows:

$$N = 2(3.14)\frac{1}{137}\left(\frac{1}{4 \times 10^{-5} \text{ cm}} - \frac{1}{6 \times 10^{-5} \text{ cm}}\right)\left(1 - \frac{1}{(0.75)(1.33)^2}\right)$$

$$= 94.1 \text{ photons/cm} \approx 10 \text{ photons/0.1 cm}$$

The above approximation by Frank and Tamm was confirmed with more precise calculations by Belcher (1953), who calculated the integral number of Cherenkov photons produced by an electron as a function of its varying phase velocity over the entire effective range of travel of the electron in water ($n = 1.33$) and Perspex plastic medium ($n = 1.50$). The effective range of travel of the electron is that of its highest energy or highest phase velocity ($\beta = \beta_{\text{max}}$) to a lower energy as a result of reduced velocity in the medium corresponding to the threshold phase velocity or $\beta = 1/n$ at which point Cherenkov photons are not produced. The calculations of Belcher are treated in more detail in Chapter 7.

Following the publication of their joint work on the Cherenkov Effect in 1937 described above, for which they shared with Cherenkov the Nobel Prize in Physics 1958, Frank and Tamm concluded their collaboration. However, Il'ja Frank continued to study the Cherenkov Effect publishing a work on the possibility of interference phenomena in Cherenkov radiation emission (Frank, 1944). In this paper Frank summarized the conclusions of this and previous studies with Tamm as follows:

> The radiation discovered by Cherenkov in 1934 is of essentially different nature from the usual sources of light. The radiator here are electrons [or other charged particles] moving in a liquid at a speed higher than the phase velocity of light. Now, in relation to Cherenkov's radiation the question of interference phenomena has never been studied in detail. This type of radiation is known to be one of pronouncedly directional character. Practically the entire energy of the radiation is concentrated in a very thin conic layer, and the angle formed by the cone's generatrix with the axis is θ ($\cos \theta = 1/\beta n$ where n is the refractive index of the medium and $\beta = v/c$ is the ratio of the electron velocity to that of light). From this it may be inferred that the cases where the radiation of an electron will proceed at once in two directions [that is, two electrons traveling simultaneously in two directions] forming a given angle with each other, are likely to happen very rarely even if that angle is not above 2θ. It was supposed, therefore, that the interference visibility of Cherenkov's radiation must be very low in the case of divergent rays, and if the angle between the rays is above 2θ, no interference would take place at all.

Il'ja Frank continued to study Cherenkov radiation undertaking numerous theoretical works with Vitaly Ginzburg on various extensions of the Cherenkov Effect as produced by dipoles, multipoles, and oscillators. These studies are reviewed in a comprehensive treatise on Cherenkov radiation and its applications by Jelley (1958). Frank became Head of the Physics Department of Moscow State University in 1944 and was appointed Head of the Neutron Laboratory of the Joint Institute of Nuclear Research in 1957. He composed an authoritative text entitled *Vavilov-Cherenkov Radiation, Theoretical Aspects*, which was published in 1988 in the Russian language two years before his death. Ginzburg (2001) also a Nobel Laureate and former collaborator of Frank commented on Frank's theoretical treatise as follows:

> …I naturally looked through Frank's book once again. What hard work was done, how many results were obtained! This [the book] is a memorial…. This memorial and the name of Il'ja Mikhailovich will remain in the history of physics forever.

In an historical account Bolotovski *et al.* (2004) wrote about Il'ja Frank's final work as follows:

> Frank preserved his respect and affection for his beloved teacher [Sergei I. Vavilov] until the end of his days. He compiled and edited a collection of reminiscences about Vavilov, which was published in three volumes [see Frank, 1979, 1981, 1991]. Frank was very ill when he was working on the third volume in 1990 and afraid he would die without finishing the work. When he completed the manuscript he emerged from his study at home and joyfully informed his family that the book had been finished, adding: "Now at last I can die," He died a few days later.

Il'ja Mikhailovich Frank died on June 22, 1990.

Igor Tamm, after his collaboration on the theoretical interpretation of the Cherenkov Effect with Il'ja Frank, continued to work on other concepts of theoretical nuclear physics and turned his attention to plasma physics. From 1934 and for the remainder of his life, Igor Tamm was head of the Department of Theoretical Physics of the P. N. Lebedev Physical Institute in Moscow. In 1948 Igor Tamm was commissioned by Igor Kurchatov, Head of the Soviet atomic bomb project, to work on the development of the hydrogen (fusion) bomb. Tamm and some of his students including Nobel Laureate for Peace 1975,

Andrei Sakharov, became part of a specialized group at FIAN that worked on the fusion bomb. As an outcome of their collaboration Tamm and Sakharov in 1952 proposed a Tokamak system as a means of achieving a controlled thermonuclear fusion (CTF) as a peaceful source of energy. The Tokamak is a word of Russian origin that is an acronym for a Russian phrase meaning "toroidal-chamber-magnetic coils", and the first Tokamak was successfully developed in the USSR around 1960. The Tokamak utilizes current-carrying coils that create a toroidal magnetic field confinement of heated plasma within a doughnut-shaped reactor, which is designed to achieve and maintain the high temperatures needed for the fusion of hydrogen nuclei. A current transmitted through the plasma creates magnetic field lines around the current to improve plasma confinement by preventing ion migration toward the walls of the torus. The Tokamak design remains the most promising for the future utilization of nuclear fusion as a peaceful source of energy. An excellent review of the advances and various components of the Tokamak is available from Murray (2001). Fusion energy production by means of a commercial-scale reactor is assumed to be developed around the year 2050 (Sheffield, 2001).

In a recent book Nobel Laureate Vitaly Ginzburg (2001) provided his personal reminiscences of Igor Tamm from which a small excerpt is quoted as follows (with kind permission of Springer Science and Business Media © Springer):

> The theoretical Department of FIAN, established by Tamm in 1934, now bears his name. Originally it had five or six staff members and now it is one of the largest departments of its kind in the world. It has 60 researchers on its staff and hundreds of former postgraduate students and other researchers from the department now work successfully in laboratories all over the world. In more than 50 years there was not a single instance of serious interpersonal friction, let alone conflict. This is a rare situation, indeed, and it was, of course, Tamm who made it happen. It was unthinkable for him to disregard junior staff, to allow senior personnel or supervisors to add their names as authors to papers written by their subordinates, or to take precipitate or coercive administrative steps. On the contrary, he was ready to provide to his subordinates support, freedom of action, sympathy, and friendly comments.... That was a very simple 'secret' of Tamm as a leader and mentor. Mature people in a team require sympathy and lack of any hindrance from the leader, while young people need freedom and guidance, when indicated, and that is all that is needed for the team to be successful. Its sounds simple but, unfortunately, it is not always practiced by administrators in science.... He was typically fascinated with genuine mysteries, the problems of a fundamental character. Problems of this kind are especially hard to work on because years may pass without one getting and significant results. Such concerns, though, never worried Tamm; he never was guided by such desires as publishing a paper or, in general, 'to maximize the production rate.'...A person is called ambitious not only when he strives to attain a high position and control other people, but also when he strives to fulfill himself by producing good results and hoping to see them acclaimed. The latter kind of ambition is typically a necessary condition for successful research. So many gifted people...have failed to fulfill themselves because of their indifference and laziness, that is, essentially, lack of ambition. Tamm was ambitious in that sense of the term.

Ginzburg (2001) described the last three years of Tamm's life, which were tragic. In 1967 he was struck with lateral amyotrophic sclerosis, and in February 1968 he became partially paralyzed by the illness and could survive only with a breathing machine, and other machines were provided him to enable him to sit and work at his desk. In his book Ginzburg (2001) wrote the following recollections of Tamm's final years:

> With a bitter smile, he used to say, "I am like a beetle on a pin." The first two years of his illness he managed to work a lot, he played chess, and was glad to see visitors. The fatal illness did not make him angry; indeed, he became gentler. When healthy he tended to conceal his warm feelings, apparently thinking they were unseemly, and illness freed him of that inhibition.

Igor Tamm died on April 12, 1971.

– 7 –

Cherenkov Radiation

7.1   INTRODUCTION

Photons of light are produced when a charged particle travels through a transparent medium at a speed greater than the speed of light in that medium. The medium may be a gas, liquid, or solid. These photons are referred to as Cherenkov radiation in honor of the Russian physicist Pavel A. Čerenkov for his discovery of this unique radiation (Cherenkov, 1934a,b) and his basic research to measure the properties of the radiation and the phenomenon upon which the radiation finds its origin (Cherenkov, 1936, 1937a–d, 1938a–c), now referred to as the Cherenkov effect. In the scientific literature, three variations of the contemporary spelling of Cherenkov radiation can be found, namely (i) Čerenkov after the Russian, (ii) Cherenkov, which provides the English pronunciation from the Russian Č, and (iii) Cerenkov. For consistency throughout this text, the writer will adopt the spelling Cherenkov, which conforms with that used by *Chemical Abstracts*. The reader may also find in the scientific literature reference to Vavilov–Cherenkov radiation. This is another name given to the same radiation in honor of Sergei I. Vavilov, who was Cherenkov's research director and doctoral dissertation advisor and hence involved in the research that led to Cherenkov's discovery. See the biographical sketches on Sergei Vavilov and Pavel Cherenkov provided earlier in this book.

The Cherenkov effect and the resulting radiation produced by a charged particle traveling in a transparent medium at a velocity exceeding the speed of light in that medium were predicted theoretically as early as 1888 by Oliver Heaviside (1888, 1889, 1892, 1899, 1912a,b), but long forgotten until the radiation was discovered finally by Cherenkov (1934a,b) and interpreted by Frank and Tamm (1937). Govorkov (2005) reviewed the historical significance of Heaviside's prediction, namely, that Heaviside first predicted theoretically that a point charge (electron) would produce a conical wave front whenever it were to travel at a speed greater than the speed of light in a medium. Heaviside's prediction was before Einstein's theory of relativity, which holds that matter cannot travel faster than light in a vacuum but that matter can travel faster than light in a medium. At the time of Heaviside, even space was not considered to be a vacuum but to consist of an "ether", and therefore, according to Heaviside, the conical wave front would be produced in the "ether" as well as any other medium provided the electron or point charge travels at a speed exceeding that of light. Of course, electrons as well as other charged particles with sufficient energy can indeed travel faster than light in media where light travels at speeds less than its speed in a vacuum.

The speed of light in a transparent medium is defined by the index of refraction, n, of the medium where $n = c/v$ and c is the velocity of light in a vacuum and v the velocity of light in the medium. Heaviside (1912a) had even calculated the geometry and angle of emission of the electromagnetic wave front relative to the axis of travel of the charged particle. Also, Afanasiev et al. (1999) explains that the early theoretical predictions of Heaviside, namely, that a conical radiation wave front would arise when the point charge velocity exceeds the light velocity in a medium, had been forgotten by the scientific community until 1974 when they were revived by Tyapkin (1974) and Kaiser (1974).

In the words of Govorkov (2005):

> A genius, Heaviside was half a century in advance of his time in his calculations and predictions, but unfortunately he was not understood by his contemporaries. His works were forgotten, and scientists had to start from the beginning, this time from experiment [the experiments and discovery by Cherenkov (1934a,b, 1936, 1937a–d, 1938a–c)].

Cherenkov radiation was first observed by Marie Curie in 1910 as reported by E. Curie (1937), and it was thought then to be some type of luminescence. The radiation was researched by Lucien Mallet (1926, 1928, 1929) who photographed the spectrum of the radiation but did not determine the origin of the radiation. It was Cherenkov (1934a,b, 1936, 1937a–d), who demonstrated the origin of the radiation and its unique properties after whose work the radiation is now known. Il'ja Frank and Igor Tamm (1937), who shared the Nobel Prize in physics with Cherenkov (see the biographical sketches on Frank and Tamm in this book), are responsible for much of the theoretical work that went into the understanding of Cherenkov radiation. Comprehensive treatments on the theory and properties of Cherenkov radiation are available from Marshall (1952), Belcher (1953), Jelley (1958, 1962), and Ritson (1961).

Cherenkov radiation consists of a continuous spectrum of wavelengths extending from the near UV region to the visible part of the spectrum peaking at approximately 420 nm (Kulcsar et al., 1982; Claus et al., 1987). Only a negligible amount of photon emissions is found in the infrared or microwave regions.

Cherenkov photon emission is the result of local polarization along the path of travel of the charged particle with the emission of electromagnetic radiation when the polarized molecules return to their original states (Gruhn and Ogle, 1980). This has been described by Marshall (1952) and Siegbahn (1958) and others as the electromagnetic "shock" wave that is analogous to the acoustical shock wave or sonic boom created by supersonic aircraft. In the Presentation Speech of the Nobel Prize in Physics 1958 awarded to Pavel Cherenkov, Il'ja Frank, and Igor Tamm for "the discovery and interpretation of the Cherenkov effect" on December 11, 1958, Professor K. Siegbahn described Cherenkov radiation in the following words:

> The phenomenon can be compared to the bow wave of a vessel that moves through the water with a velocity exceeding that of the waves. In air, an analogous phenomenon occurs when a jet plane penetrates the so-called sound barrier at about 1,000 km/hr, i.e., when the jet velocity exceeds the propagation velocity of the sound waves [See Figure 7.1a] ... The condition that is required to form the corresponding Cherenkov bow wave of ordinary light when a charged particle, e.g. an electron, traverses a medium is, analogously, that the particle moves with a velocity greater than that of light in the medium. [Figure 7.1b]

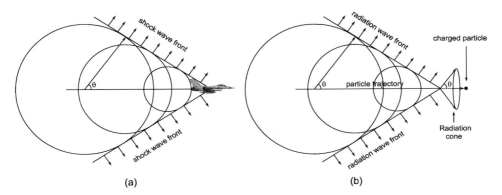

Figure 7.1 Conical wave fronts produced by (a) a jet traveling at supersonic speed and (b) a charged particle traveling at superlight speed in a transparent medium, that is, a speed exceeding the speed of light in the medium. The supersonic jet produces a conical shock wave front, while the charged particle produces a conical wave front of Cherenkov radiation. The Cherenkov photons are emitted at an angle θ to the particle trajectory.

The production of Cherenkov photons by a charged particle traveling in a transparent medium at a speed exceeding that of light in the medium is depicted as a conical wave front in a two-dimensional diagram in Figure 7.1b. The photons are asymmetric and emitted in the direction of the particle trajectory at an angle θ relative to the particle track. The emission angle θ is a function of the particle phase velocity and the index of refraction of the medium as described previously in the biographical sketches of Pavel Cherenkov, Il'ja Frank, and Igor Tamm. The conical Cherenkov radiation wave front is illustrated in three dimensions in Figure 7.2. A cross-section of the wave front yields rings of photon emissions in the direction of the particle trajectory. As will be discussed further on in this chapter, the photon ring can be imaged and its radius measured. The ring radius may be used to calculate the phase velocity of the charged particle when the index of refraction of the medium is known.

7.2   THEORY AND PROPERTIES

The theory explaining the production of Cherenkov radiation in various types of transparent media was provided by Frank and Tamm (1937), and many of the properties of Cherenkov radiation, upon which the theory was based, were demonstrated experimentally by Cherenkov (1934a,b, 1936, 1937a–d, 1938a–c). These studies are described in detail in the biographical sketches on Cherenkov, Frank, and Tamm in the section on Radioactivity Hall of Fame—Part VII in this book. A brief summary of the theory will be provided here.

7.2.1   Threshold condition

The threshold condition for the production of Cherenkov radiation in a transparent medium is given by

$$\beta n = 1 \tag{7.1}$$

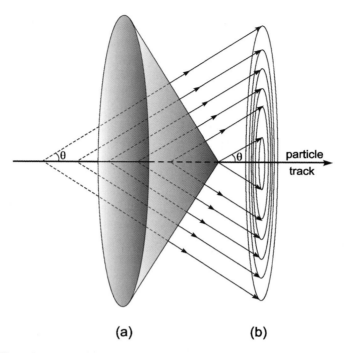

(a)   **(b)**

Figure 7.2 Cherenkov conical wave front in (a) three dimensions and (b) photon trajectories at angle θ (indicated by arrows) in two dimensions illustrating Cherenkov cones and the rings of Cherenkov photons produced along the particle track. The shaded region (a) is not the Cherenkov cone, but rather a complementary wave front.

where β is the relative phase velocity of the particle, that is, the velocity of the particle divided by the speed of light in a vacuum and n the refractive index of the medium (i.e., the ratio of the velocity of light in a vacuum to its velocity in the medium). Only charged particles that possess

$$\beta > \frac{1}{n} \tag{7.2}$$

produce Cherenkov photons in transparent media (Frank and Tamm, 1937). The value of β of the charged particle is dependent on its kinetic energy, as reported by Jelley (1962).

$$E = m_0 c^2 \left[\frac{1}{(1 - \beta^2)^{1/2}} - 1 \right] \tag{7.3}$$

where m_0 is the particle rest mass in grams, c the velocity of light ($= 2.99 \times 10^{10}$ cm/sec), and E the kinetic energy in ergs (where 1.602×10^{-12} ergs $= 1$ eV). Eq. (7.3) can be solved for

β to give the expression for the phase velocity of the particle as a function of its energy or

$$\beta = \left[1 - \left(\frac{1}{(E/m_0 c^2) + 1} \right)^2 \right]^{1/2} \tag{7.4}$$

The value of the rest energy for an electron is $m_0 c^2 = (9.11 \times 10^{-28}\,g)$ $(2.99 \times 10^{10}\,cm/sec)^2 = 8.14 \times 10^{-7}$ ergs. The energy in ergs may be converted to electron volts as follows: $(8.14 \times 10^{-7}\,ergs)/(1.602 \times 10^{-12}\,ergs/eV) = 511\,keV$. Substituting the value of $511\,keV$ for $m_0 c^2$ in eq. (7.4) and solving for β gives

$$\beta = \left[1 - \left(\frac{1}{(E/511) + 1} \right)^2 \right]^{1/2} \tag{7.5}$$

Thus, the value of β, where electrons or beta particles are concerned, is dependent on the electron energy, E, in keV according to eq. (7.5).

7.2.2   Threshold energies

As noted above, the particle phase velocity, β, is a critical factor in the production of Cherenkov photons in a medium of given index of refraction, n, that is, Cherenkov radiation will occur only when $\beta > 1/n$. Consequently the particle energy, upon which the particle phase velocity is dependent, is a critical factor. Thus, a minimum or threshold energy of the particle must be reached when traveling in a medium of given index of refraction before Cherenkov photons are produced. If we apply the threshold value of β for the production of Cherenkov photons from eq. (7.1), that is, $\beta = 1/n$, to eq. (7.5), we will obtain the threshold energy that electrons or beta particles must possess for the production of Cherenkov radiation as a function of the index of refraction of the medium as

$$E_{th} = 511\,keV \left[\left(1 - \frac{1}{n^2} \right)^{-1/2} - 1 \right] \tag{7.6}$$

When water is the transparent medium, where $n = 1.332$, the threshold energy for the production of Cherenkov photons by electrons or beta particles is calculated according to eq. (7.6) to be $263\,keV$ (or $0.263\,MeV$). Thus, only electrons or beta particles that possess energy in excess of $263\,keV$ produce Cherenkov photons in water.

We can write eq. (7.6) in general for the calculation of the threshold energy of charged particles of any mass as follows

$$E_{th} = m_0 c^2 \left[\left(1 - \frac{1}{n^2} \right)^{-1/2} - 1 \right] \tag{7.7}$$

where m_0 is the particle rest mass and c the velocity of light in a vacuum ($2.99 \times 10^{10}\,cm/sec$).

The threshold energy for Cherenkov production will vary according to the particle rest mass and index of refraction of the medium. The energy required for the production of Cherenkov photons, E_{th}, increases with particle mass, and it will be lower for media of higher index of refraction. For example, electrons or beta particles traveling through a transparent ceramic of refractive index $n = 2.1$, such as that studied by Takiue *et al.* (2004), will have the following calculated threshold energy of 0.070 MeV, which is much lower than the 0.263 MeV required for the production Cherenkov photons in water:

$$E_{th} = (9.10938 \times 10^{-28}\,\text{g})(2.99792 \times 10^{10}\,\text{cm/sec})^2 \left[\frac{1}{\sqrt{1 - \frac{1}{(2.1)^2}}} - 1 \right]$$

$$= 8.18707 \times 10^{-7}\,\text{ergs} \left[\frac{1}{\sqrt{0.77324}} - 1 \right]$$

$$= \left(\frac{8.18707 \times 10^{-7}\,\text{ergs}}{1.602 \times 10^{-12}\,\text{ergs/eV}} \right) \left[\frac{1}{0.87934} - 1 \right] \tag{7.8}$$

$$= (0.5110 \times 10^6\,\text{eV})(0.1372) = 0.070\,\text{MeV}$$

Table 7.1 lists the calculated threshold energies of particles of differing mass in media of differing index of refraction. A medium within which Cherenkov photons are produced is referred to as a Cherenkov radiator.

Table 7.1

Particle threshold energies in MeV for the production of Cherenkov photons in media (Cherenkov radiators) of different index of refraction

Medium	Index of refraction (n)	Particle threshold energy in MeV[a]		
		Electron[b]	Muon[c]	Proton[d]
Air	1.00027712[e]	21.2	4380.9	38925.9
Silica aerogel[f]	1.05	1.16	240.7	2139.0
Water	1.332	0.263	54.3	482.1
Glass[g]	1.47	0.186	38.5	341.9
Plastic[h]	1.52	0.167	34.6	307.6
Ceramic[i]	2.1	0.070	14.5	128.7
Diamond	2.4	0.051	10.6	93.9

[a]Calculated according to eq. (7.7).
[b]Energy equivalent mass of electron (m_0c^2) = 0.511 MeV.
[c]Energy equivalent mass of muon (m_0c^2) = 105.6 MeV.
[d]Energy equivalent mass of proton (m_0c^2) = 938.3 MeV.
[e]Index of refraction of air measured at the sodium D line at STP (Lide, 2001).
[f]Silica aerogel index of refraction (n) will vary according to mode of manufacture. The value of $n = 1.05$ is taken from experiments by Brajnik *et al.* (1994, 1995) and Pestotnik *et al.* (2002).
[g]Borosilicate glass tested for electrons or beta particles by L'Annunziata and Passo (2002).
[h]Polyethylene plastic tested for electrons or beta particles by L'Annunziata and Passo (2002).
[i]Commercial ceramic tested for electrons and beta particles by Takiue *et al.* (2004).

Data in Table 7.1 illustrate that the threshold energy for a particular particle decreases in media of increasing index of refraction. Also, the threshold energy required for the production of Cherenkov photons in any particular medium increases proportionally with the mass of the particle. Thus, by the selection of a medium of given index of refraction it is possible to discriminate between particles of different energies or mass. Some practical applications of the effect of refractive index on the threshold energy are discussed in Section 7.4.1.

7.2.3   Photon spatial asymmetry

An important property of Cherenkov radiation is its asymmetry, that is, the directional emission of the Cherenkov photons. Cherenkov radiation is not emitted in all directions. When Cherenkov (1936) discovered the asymmetric properties of this radiation, he knew he had discovered a type of radiation yet unknown. Frank and Tamm (1937) explained the spatial asymmetry of the radiation as a result of the charged particle traveling in a medium at a velocity exceeding the velocity of light in the medium. Frank and Tamm (1937) theorized that when the phase velocity of the charged particle, β, exceeded the reciprocal of the index of refraction, n, of the medium, that is, $\beta > 1/n$, Cherenkov photons are emitted as a cone at an angle, θ, to the direction of the charged particle as depicted in Figure VII.15a in the biographical sketch of Il'ja Frank and in Figures 7.1b and 7.2. According to the theory of Frank and Tamm (1937), the angle, θ, of photon emission would be defined by the relationship

$$\cos\theta = \frac{1}{\beta n} \tag{7.9}$$

Cherenkov (1937b) tested the theory of Frank and Tamm (1937) using gamma rays directed at a vessel of water to create high-energy electrons in the water via the Compton effect (see Figure VII.3). He then measured the angles of emission of the photons relative to the direction of the incident gamma radiation. Cherenkov found a broad range of photon emission angles in water between 0° and 90° with a maximum around 30°. The maximum angle of emission would depend on the phase velocity of the Compton electron, β, and thus on the energy of the electron. Although Cherenkov definitely was able to demonstrate the spatial asymmetry of the newly discovered photons, the Compton electrons unfortunately are not unidirectional and can display wide angles of deflection relative to the direction of the gamma rays that created them. Consequently, Cherenkov's results were not precise to demonstrate irrefutably Frank and Tamm's theory that $\cos\theta = 1/\beta n$. It was not until monoenergetic electrons were used by researchers such as Collins and Reiling (1938) and Wyckoff and Henderson (1943) could the angle of photon emission, θ, be accurately measured. They directed electrons of uniform energy into thin sections of transparent media such as mica, glass, cellophane, and water. A thin section of Cherenkov radiator assured insignificant loss of electron energy in traversing the medium.

According to the theory of Frank and Tamm (1937), the angle, θ, illustrated in Figure 7.3 is defined by the equation:

$$\cos\theta = \frac{\text{distance of travel of radiation front in the medium}}{\text{distance of travel of charged particle in the medium}} \tag{7.10}$$

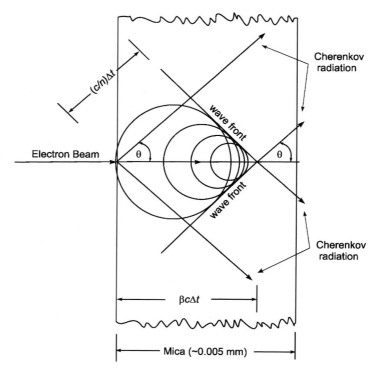

Figure 7.3   A thin section of mica used by Collins and Reiling (1938) and Wyckoff and Henderson (1943) as the transparent medium for the production of Cherenkov photons by a beam of high-energy electrons of uniform energy. The thin section of Cherenkov radiator would assure minimal loss of electron energy in traversing the mica.

or

$$\cos \theta = \frac{(\text{speed of light in the medium})(\Delta t)}{(\text{speed of particle in the medium})(\Delta t)} \tag{7.11}$$

where Δt is a time interval. Using the notation provided in Figures VII.15 and 7.3 we can then write

$$\cos \theta = \frac{(c/n)t}{\beta c t} \tag{7.12}$$

The term $(c/n)t$ represents the distance of travel of the radiation wave front measured by the product of the velocity of the light in the medium (c/n) and time (t), while the term $\beta c t$ measures the distance of particle travel according to $\beta c t = (v/c)ct = vt$ where v is the velocity of the particle in the medium, c the speed of light in a vacuum, and t the time.

From eq. (7.12) we obtain the expression

$$\cos \theta = \frac{1}{\beta n} \qquad (7.13)$$

When $\beta = 1/n$, which is the threshold condition for the production of Cherenkov photons (eq. (7.1)), the charged particle possesses a threshold energy where Cherenkov radiation is yet to be produced, that is, $\cos \theta = 1$ and $\theta = 0$. When $\beta > 1/n$, the charged particle possesses sufficient energy and velocity to produce Cherenkov photons, which are emitted in the exact forward direction of the charged particle at an angle θ, as defined according to eq. (7.13).

As the particle energy, E, of eq. (7.4) increases to relativistic speeds, the value of β approaches unity and $\cos \theta$ approaches $1/n$ of eq. (7.13). Thus, the maximum angle of emission of a Cherenkov photon produced by a charged particle in any transparent medium would occur when the charged particle possesses a high energy to the point where the value of β of eq. (7.4) approaches unity, or

$$\cos \theta = \frac{1}{n} \quad \text{when } \beta \text{ approaches 1} \qquad (7.14)$$

If we consider the case of mica as the transparent medium with index of refraction $n = 1.59$, such as that tested by Collins and Reiling (1938) and Wyckoff and Henderson (1943), the maximum possible angle of emission of Cherenkov photons in the mica is calculated to be

$$\cos \theta = \frac{1}{1.59} = 0.6289$$

and

$$\theta = 51.0°$$

According to the theory of Frank and Tamm (1937), Cherenkov photons are emitted at more acute angles for charged particles just above their threshold energy and cannot exceed an emission angle of 51.0° in mica, even for the most energetic particles.

Collins and Reiling (1938) and Wyckoff and Henderson (1943) tested the theory of Frank and Tamm by shooting electrons of various energies at mica, and they measured accurately the angles of emission of the Cherenkov photons as a function of electron energy. They could then compare the experimentally obtained emission angles, θ, with those calculated according to the relation $\cos \theta = 1/\beta n$. The key component of their experimental arrangement illustrated in Figure 7.4 included a conical mirror to reflect the Cherenkov photons, such as that designed by Cherenkov (1937b) in his initial experiments with Compton electrons.

The theoretical angles of emission (θ) of Cherenkov photons produced by electrons in mica with an index of refraction, $n = 1.59$, are plotted in Figure 7.5. According to the Frank and Tamm (1937) theory, where $\cos \theta = 1/\beta n$, the theoretical angles of emission will vary from zero at the threshold electron energy to the highest emission angle for the relativistic electrons of highest energy where β approaches 1, its highest possible value. Collins and Reiling (1938)

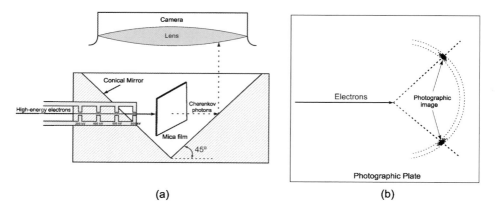

(a)   (b)

Figure 7.4   Basic experimental arrangement used by Collins and Reiling (1938) reprinted with permission from The American Physical Society and Wyckoff and Henderson to demonstrate the relation $\cos \theta = 1/\beta n$. In (a) an electron gun provides electrons of uniform energy by accelerating electrons across potentials varying from 200 to 800 kV. Wyckoff and Henderson (1943) produced electrons with energies over the range of 240–815 keV and Collins and Reiling (1938) accelerated electrons up to 2000 keV. The electrons shown in (a) traverse the mica film and produce cones of Cherenkov radiation. The Cherenkov photons are produced only in the direction of the electrons, and the photons are reflected off the conical mirror toward the camera. The photographic plate shown in (b) provides two spots produced by the Cherenkov radiation cone reflected in two dimensions from which the angle θ can be calculated.

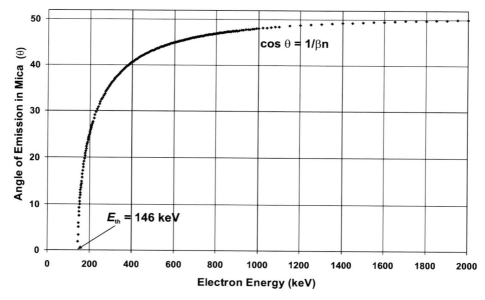

Figure 7.5   Cherenkov photon emission angles (θ) in mica of index of refraction, $n = 1.59$, as a function of electron energy calculated according to $\cos \theta = 1/\beta n$. The threshold energy E_{th} of electrons (146 keV) for the production of Cherenkov photons in the mica was calculated according to eq. (7.7).

and Wycoff and Henderson (1943) found the experimentally measured angles of emission to agree very closely to the theoretical values over the electron energy range of 240–1900 keV within approximately 1°. In addition to mica thin films of other transparent media of different index of refraction, including thin films of glass and cellophane, were tested by Wyckoff and Henderson (1943); and the measured and theoretically calculated values of θ agreed to be within 1°. The slight discrepancy between the measured and calculated values was considered due to slight unavoidable bending or imperfections of the thin films used.

We can conclude from above that for a particle of given mass and charge the angle of emission of the Cherenkov photons will depend on the particle velocity (β), which is a function of its energy as well as the index of refraction of the medium. The maximum angle of emission of Cherenkov photons will, of course, differ for media of different index of refraction. In the fields of particle and nuclear physics, the Cherenkov effect has been exploited over the last decade for the identification of particles based on their velocity, which in any given medium of known index of refraction, will govern the angle of emission of the Cherenkov photons. The concepts of particle identification (PID) by measurement of Cherenkov emissions are discussed briefly further on in this chapter and reviewed in detail by Va'vra (2000), Križan (2001), and Joram (2002).

7.2.4 Photon spectrum and radiation intensity

As noted earlier, Cherenkov radiation consists of a continuous spectrum of wavelengths extending from the near UV region to the visible part of the spectrum peaking at approximately 420 nm (Kulcsar et al., 1982; Claus et al., 1987). Only a negligible amount of photon emissions is found in the infrared or microwave regions. In Cherenkov's early work (see Figure VII.12), he found the Cherenkov cone of radiation to comprise the visible spectrum. He observed specifically that the outer ring of the radiation cone was violet and its inner ring was red. Jelley (1958) explains that a medium is always dispersive, and consequently, the radiation is restricted to those frequency bands for which $n(\omega) > 1/\beta$ where $n(\omega)$ is the index of refraction at a particular frequency band. Jelley adds that the absorption bands in media, which are transparent at visible wavelengths, limit the radiation to the near UV and longer wavelengths.

In their classic interpretation of the Cherenkov effect, Frank and Tamm (1937) calculated the total energy, W, radiated by an electron via the Cherenkov effect through a cylindrical region of medium of length, l, where the axis of the cylinder coincided with the line of motion of the electron. Under the assumption that β is constant over a short length, dl, of the electron track, they derived the equation

$$W = \frac{e^2 l}{c^2} \int_{\beta n > 1} \omega \, d\omega \left(1 - \frac{1}{\beta^2 n^2} \right) \tag{7.15}$$

where ω is the frequency of the Cherenkov radiation. The lower limit of the integral, $\beta n > 1$, is the threshold condition for the production of Cherenkov photons in any medium. Frank and Tamm (1937) established the equation to be valid provided that l should be large in comparison with the wavelength λ of the Cherenkov radiation emitted. Cherenkov (1938c) took eq. (7.15) of Frank and Tamm and evaluated the energy, W, radiated by an electron taking into consideration that the magnitude of the electron velocity, that is, the magnitude of β, will vary greatly over the entire path length of travel of the electron. Cherenkov (1938c) thus expressed eq. (7.15) with β as a variable depending on its length of travel l or

$$W = \frac{e^2}{c^2} \int \omega \, d\omega \int_{\beta n > 1} \left(1 - \frac{1}{\beta^2 n^2} \right) dl \tag{7.16}$$

Frank and Tamm (1937) thus calculated the rate of radiated energy per path length of travel, dW/dl, to be of the order of only several kilovolts per centimeter. Thus, it was found that the electron energy loss via the emission of Cherenkov radiation would be a negligible quantity of energy in comparison to the energy lost by the electron by other mechanisms such as ionization. Cherenkov (1938c) confirmed this through calculations and experimentation and concluded that the radiation expended by beta particles from Ra(B+C) via the emission of Cherenkov radiation amounts to a value of the order of only 0.1% of the total energy of the particles.

From eq. (7.15), Frank and Tamm (1937) deduced the equation for the Cherenkov photon radiation intensity, that is, the number of photons emitted by an electron per path length of travel, as follows:

$$dN = 2\pi\alpha \left(\frac{1}{\lambda_2} - \frac{1}{\lambda_1} \right) \left(1 - \frac{1}{\beta^2 n^2} \right) dl \tag{7.17}$$

where dN is the number of photons emitted in path length dl within the spectral region defined by the wavelengths λ_1 and λ_2, α the fine structure constant approximately 1/137, and β and n the particle phase velocity and the index of refraction of the medium, respectively. From eq. (7.17), Belcher (1953) assumed the index of refraction to be constant over the spectral range studied and expressed the total number, N_E, of photons radiation along the track of a particle of initial energy E and brought to rest within the medium as

$$N_E = 2\pi\alpha \left(\frac{1}{\lambda_2} - \frac{1}{\lambda_1} \right) \int_{\beta=\beta_{\max}}^{\beta=1/n} \left(1 - \frac{1}{\beta^2 n^2} \right) dl \tag{7.18}$$

where $\beta_{\max}$ is the maximum phase velocity, that is, the initial velocity of the particle, and $1/n$ the limiting phase velocity of the particle, that is, the velocity at which Cherenkov photons cease to be produced. Belcher (1953) reduced eq. (7.18) to read

$$N_E = A \int_{\beta=\beta_{\max}}^{\beta=1/n} \left(1 - \frac{1}{\beta^2 n^2} \right) dl \tag{7.19}$$

where $A = 2\pi\alpha[(1/\lambda_2) - (1/\lambda_1)]$. For the spectral range of $\lambda_2 = 3000$ to $\lambda_1 = 7000\,\text{Å}$, the value of A is calculated to be

$$A = 2\pi\alpha\left(\frac{1}{\lambda_2} - \frac{1}{\lambda_1}\right) = 2(3.14)(0.007297)\left(\frac{1}{3\times10^{-5}\,\text{cm}} - \frac{1}{7\times10^{-5}\,\text{cm}}\right) = 874\,\text{cm}^{-1}$$

(7.20)

Over the defined spectral range eq. (7.19) is reduced to

$$N_E = 874\int_{\beta=\beta_{\max}}^{\beta=1/n}\left(1 - \frac{1}{\beta^2 n^2}\right)dl$$

(7.21)

Belcher (1953) evaluated the integral defined by eq. (7.21) by first determining the range-energy function for an electron in water ($n = 1.33$ and $\rho = 1.00\,\text{g/cm}^3$) and Perspex plastic ($n = 1.50$ and $\rho = 1.18\,\text{g/cm}^3$). He plotted the values of $(1 - 1/\beta^2 n^2)$ against the range of electrons in water and Perspex plastic as illustrated in Figure 7.6. Belcher then evaluated eq. (7.21) by graphical integration to obtain the expected values of N_E as a function of electron energy in water or Perspex plastic, which is illustrated in Figure 7.7. As can be seen in Figure 7.7, photons are not produced by electrons below 0.175 MeV in the Perspex plastic or 0.260 MeV in water, which are the respective threshold energies for the Cherenkov effect in these two materials.

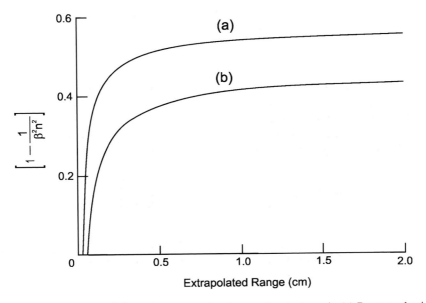

Figure 7.6 Plot of $(1 - 1/\beta^2 n^2)$ against extrapolated range for electrons in (a) Perspex plastic with density $1.18\,\text{g/cm}^3$ and (b) water with density of $1.00\,\text{g/cm}^3$. (From Belcher (1953) reprinted with permission from The Royal Society.)

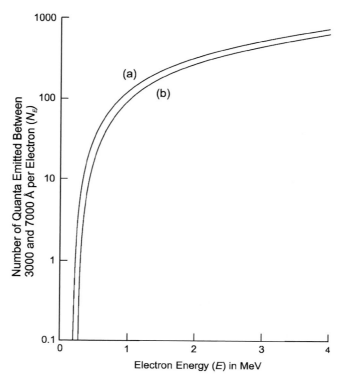

Figure 7.7 A plot of the number of Cherenkov photons in the spectral range of 3000–7000 Å produced by an electron in (a) Perspex plastic ($n = 1.50$) and (b) water ($n = 1.33$) over its entire range of travel as a function of electron energy. (From Belcher (1953) reprinted with permission from The Royal Society.)

Frank and Tamm (1937) made an approximate calculation that a 500 keV electron, which would correspond to $\beta^2 = 0.75$, traveling in water, where $n = 1.33$, would produce approximately 10 photons per 0.1 cm within the visible region between $\lambda_1 = 4000$ Å and $\lambda_2 = 6000$ Å (see eqs. (VII.16) and (VII.17) in the previous section on Radioactivity Hall of Fame—Part VII). Their approximation came very close to the above more precise calculations of Belcher (1953), which indicate that a 0.500 MeV electron would produce approximately 20 photons in water (Figure 7.7). With a range of 0.2 cm in water, the rate of photons production by a 0.5 MeV electron would be 10 photons per 0.1 cm.

The above calculations for the number of photons emitted as a function of a single electron of known energy in a transparent medium are more complex when dealing with beta particles. This is so because beta particles are emitted from radioactive isotopes with a broad spectrum of energies between 0 and E_{max} as depicted graphically in Figure 2.1 of Chapter 2. In such a case, Belcher (1953) noted that it is necessary to combine the function N_E of eq. (7.18) or (7.19) with the energy distribution function of the electrons released within the medium. He therefore elaborated that if p_E is the probability of release of an electron possessing energy between the limits E and $(E + dE)$ then the average total intensity I of Cherenkov photon emission per electron is given by the following:

$$I = \int_0^{E_{max}} p_E N_E \, dE \tag{7.22}$$

The property of photon intensity is of particular relevance in studies on the measurement of Cherenkov radiation for PID and radionuclide analysis, a subject to be dealt with further on in this chapter. The photon intensity is also expressed in terms of the number of Cherenkov photons created per unit photon energy. As noted by Križan (2001) for particles above the threshold, that is, when $\beta > 1/n$, the number of Cherenkov photons emitted per unit photon energy in a medium of particle path length L is

$$\frac{dN}{dE} = L \frac{\alpha}{\hbar c} \sin^2 \theta \tag{7.23}$$

where α is the fine structure constant $(1/137)$, $\hbar$ the Planck's constant divided by 2π (i.e., $h/2\pi = 6.582 \times 10^{-16}$ eV sec, c the speed of light in a vacuum or 2.998×10^{10} cm/sec, and θ is, as previously defined, the angle of emission of the Cherenkov photons relative to the direction of travel of the charged particle. Taking the values of the aforementioned constants, the term $\alpha/\hbar c$ of eq. (7.23) becomes 370 eV^{-1}cm^{-1}. From eq. (7.23), Križan (2001) calculated that in 1 cm of water a particle track where $\beta = 1$ (ultra-relativistic or most energetic particle possible) emits $N = 320$ photons in the spectral range of visible light ($\Delta E \approx 2$ eV). Since the high-energy electron (e.g., 1000 GeV) does not lose significant energy in 1 cm of water, otherwise the angle of emission would diminish as the electron loses energy and its velocity reduces, the above number of 320 photons is calculated as follows:

Since the angle of emission θ of the Cherenkov photons for a particle of given energy and velocity is defined by

$$\cos \theta = \frac{1}{\beta n} \tag{7.24}$$

and, because the charged particle at ultra-relativistic speeds approaches the speed of light in a vacuum or β approaches 1, we can write

$$\cos \theta = \frac{1}{n} \tag{7.25}$$

If water is the medium

$$\cos \theta = \frac{1}{1.33} = 0.7518 \tag{7.26}$$

and then

$$\cos^{-1} 0.7518 = 41.3° \tag{7.27}$$

Thus, the number of photons emitted (ΔN) over the energy range of the visible spectrum ($\Delta E \approx 2\,\text{eV}$) in the particle path length of 1 cm in water is calculated according to eq. (7.23) as

$$\Delta N = L\,\frac{\alpha}{\hbar c}\,\sin^2\theta\,\Delta E$$

$$= (1\,\text{cm})(370\,\text{eV}^{-1}\text{cm}^{-1})(\sin^2 41.3°)(2\,\text{eV}) \tag{7.28}$$

$$= 320\,\text{photons}$$

In the instrumental detection of individual Cherenkov photons there is an efficiency factor, which defines the efficiency of an instrument to detect the photons. With an average instrumental detection efficiency of $\epsilon = 0.1$ over the spectral interval in the example given above, only $N = 32$ photons would be measured. The detection efficiency for Cherenkov photons should not be confused with the term counting efficiency of each particle that interacts with the medium, which is employed in radionuclide analysis, as it will be seen in this chapter that Cherenkov counting efficiencies can be $>70\%$ or >0.70 (see Table 7.2 in Section 7.5 at the end of this chapter).

The number of photons emitted per path length of electron travel in the wavelength interval between λ_1 and λ_2 is calculated by Sowerby (1971) according to

$$\frac{dN}{dl} = 2\pi\alpha z^2\left(\frac{1}{\lambda_2} - \frac{1}{\lambda_1}\right)\sin^2\theta \tag{7.29}$$

where α is the fine structure constant ($e^2/\hbar c = 1/137$), z the particle charge ($z = 1$ for electrons or beta particles), and the refractive index of the medium and particle velocity appear in the $\sin^2\theta$ term. Over the visible range of wavelengths from $\lambda_2 = 400$ to $\lambda_1 = 700\,\text{nm}$, eq. (7.29) becomes

$$\frac{dN}{dl} = 2(3.14)(.007297)(1)^2\left(\frac{1}{4\times10^{-5}\,\text{cm}} - \frac{1}{7\times10^{-5}\,\text{cm}}\right)\sin^2\theta \tag{7.30}$$

$$= 490\sin^2\theta\,\text{cm}^{-1}$$

Sundaresan (2001) estimated the number of photons N per path length L to be

$$\frac{N}{L} = 490\sin^2\theta\,\text{cm}^{-1} \tag{7.31}$$

The emission angle θ would diminish as the particle loses energy and velocity. It can be considered constant over particle path lengths where the particle does not undergo significant energy loss.

7.2.5 Duration of Cherenkov light flash

The Cherenkov effect is the result of a physical disturbance caused by the high-energy charged particle along its path of travel resulting in a directional anisotropic emission of light. Therefore, there is no chemical fluorescence nor the relatively long excitation decay times associated with fluorescence. Jelley (1958) and Burden and Hieftje (1998) calculated the width of a typical Cherenkov-generated pulse or, in other words, the duration of the photon flash, Δt, depicted in Figure 7.8, which is a function of the spread of the Cherenkov wave front and the position of observation with respect to the particle trajectory. They approximated the duration of the light pulse, Δt, over the range of Cherenkov photon wavelengths λ_2 and λ_1 observed at a distance r parallel to the particle path, as illustrated in Figure 7.8, according to the following equation:

$$\Delta t = \frac{r}{\beta c}\left(\sqrt{\beta^2 n^2(\lambda_2)-1} - \sqrt{\beta^2 n^2(\lambda_1)-1}\right) \qquad (7.32)$$

and

$$\Delta t = \frac{r}{\beta c}(\tan\theta_2 - \tan\theta_1) \qquad (7.33)$$

Cherenkov radiation would encompass the entire spectrum of visible radiation as noted by Cherenkov (1958) whereby the inner ring of the Cherenkov cone would be red (longer

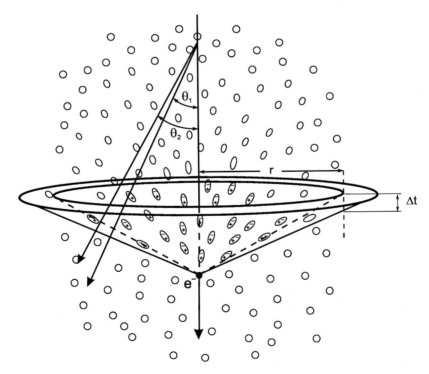

Figure 7.8 Depiction of the production of Cherenkov radiation in a dispersive medium and the resulting wave front expansion. The wave front spreading lengthens the excitation pulse on a time scale that is small in comparison to fluorescence decay. Δt is the duration of the light pulse along a line parallel to the axis of the particle at a distance r from the axis. (From Burden and Hieftje (1998) reprinted with permission from American Chemical Society © 1998.)

wavelength) and the outer limit of the cone would be violet (shorter wavelength). Jelley (1958) calculated a flash duration $\Delta t = 5 \times 10^{-12}$ sec for a light flash observed between the wavelength limits of 4000 Å (violet) and 6000 Å (orange) at a distance $r = 10$ cm to the side of the track of a fast electron traversing water where $\beta = 1$, average value of index of refraction $\bar{n} = 1.33$, the angle of emission at its maximum value of 41°. Burden and Hieftje (1998) calculated a light pulse duration $\Delta t = 326$ fsec or 0.326×10^{-12} sec for a 1-MeV electron traversing water where the light pulse is detected at a distance $r = 1$ cm parallel to the particle path of travel between the wavelengths of 300 and 350 nm or 3000 and 3500 Å.

Conventional liquid scintillation counters equipped with photomultiplier tubes (PMTs) are most often used to detect and count the individual Cherenkov light flashes that may be emitted from samples containing beta-particle-emitting radioisotopes in solution within transparent liquid media or deposited onto the surface of other types of media such as glass or plastic (L'Annunziata and Passo, 2002, and L'Annunziata, 2003b). The directional emission of Cherenkov photons is a disadvantage when conventional liquid scintillation spectrometers are used for counting Cherenkov photons. The photocathodes of most liquid scintillation counters consist of two PMTs positioned at 180° relative to each other. This is not an optimum arrangement for the detection of Cherenkov photons; however, reflector material on the surface of the counting chamber walls facilitates the detection of Cherenkov photons, which otherwise would not reach the PMTs in coincidence. Thus, when conventional liquid scintillation counters are employed for Cherenkov counting, the counting efficiencies are inferior to the theoretical maximum efficiencies.

Extensive treatments of the origin and interpretation of Cherenkov radiation are given by Marshall (1952), Belcher (1953), Cherenkov (1958), Jelley (1958, 1962), Gruhn and Ogle (1980), and Kulcsar et al. (1982). Practical reviews on the application of Cherenkov counting to the measurement of radionuclides are available from L'Annunziata (2003b), Takiue et al. (1993, 1996), and Al-Masri (1996) and a comprehensive theoretical and practical treatment of Cherenkov radiation and its application to radionuclide measurement is available in a book by Grau Carles and Grau Malonda (1996).

7.3 CHERENKOV PHOTONS FROM GAMMA-RAY INTERACTIONS

Cherenkov (1934a,b) discovered the radiation named after him by directing gamma rays from a radium source into pure transparent media such as water and other pure solvents. Gamma radiation produces Cherenkov photons indirectly through gamma-ray photon–electron interactions, namely, Compton interactions, as the gamma radiation travels through the transparent medium.

The transfer of gamma-ray photon energy to an atomic electron via a Compton interaction produces a Compton electron with energy, E_e, within the range between zero and a maximum defined by

$$0 < E_e \leq E_\gamma - \frac{E_\gamma}{1 + 2E_\gamma/0.511} \tag{7.34}$$

where E_γ is the gamma-ray photon energy in MeV and the term $E_\gamma - [E_\gamma/(1 + 2E_\gamma/0.511)]$ defines the Compton electron energy at 180° Compton scatter according to eqs. (3.22)–(3.27) previously defined in Chapter 3. To produce Cherenkov photons, the Compton electron must possess energy in excess of the threshold energy, E_{th}, defined by eq. (7.6) previously in this chapter. For example, the threshold energy for electrons in water ($n = 1.332$) according to eqs. (7.6) or (7.7) is calculated to be 263 keV (see also Table 7.1). A Compton electron must possess, therefore, energy in excess of 263 keV to produce Cherenkov photons in water. In this case, however, the gamma-ray photon must possess an energy in excess of 422 keV calculated according to the inverse of eq. (3.25) of Chapter 3 or

$$E_\gamma = E_e + E'_\gamma + \phi \qquad (7.35)$$

where E_e is the Compton electron energy, E'_γ the energy of the Compton scattered photon defined by eq. (3.21) of Chapter 3 as

$$E'_\gamma = \frac{E_\gamma}{1 + (E_\gamma/0.511\,\text{MeV})(1 - \cos\theta)} \qquad (7.36)$$

and ϕ is the electron binding energy. The electron binding energy is negligible and can be ignored. Thus, eq. (7.35) at $\theta = 180°$ Compton scatter or complete backscatter of the gamma-ray photon becomes

$$E_\gamma = E_e + \frac{E_\gamma}{1 + 2E_\gamma/0.511} \qquad (7.37)$$

The value of E'_γ at 180° Compton scatter is used in eq. (7.37) because maximum energy is imparted to the Compton electron when the gamma-ray photon is completely backscattered at 180°.

For example, if we take E_e to be 0.263 MeV, the threshold electron energy for Cherenkov photon production in water, and E'_γ as the scattered-photon energy at 180° Compton scatter, eq. (7.37) becomes

$$E_\gamma = 0.263\,\text{MeV} + \frac{E_\gamma}{1 + 2E_\gamma/0.511} \qquad (7.38)$$

where $E_\gamma = 0.422$ MeV is the threshold gamma-ray energy for the production of Cherenkov photons in water. Threshold energies will vary according to the index of refraction of the medium, and these are provided graphically in Figure 7.9 for gamma radiation and electrons or beta particles.

The number of photons emitted by a Cherenkov detector (medium) is generally only approximately 1% of the number emitted by a good scintillator for the same gamma-ray energy loss (Sowerby, 1971). Although Cherenkov detection efficiencies for gamma radiation are low, the phenomenon is applied to create threshold detectors. A variety of media, which vary significantly in refractive index, can be selected to discriminate against gamma radiation of specific energy. For example, silica aerogels of low refractive index ($n = 1.026$) can be

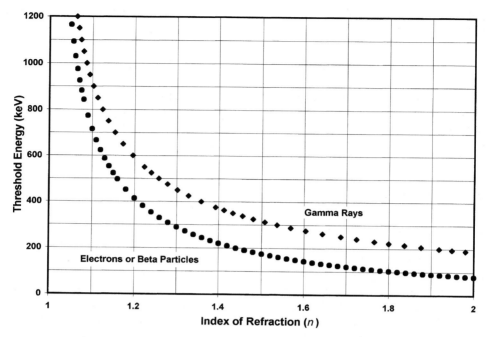

Figure 7.9 Threshold energies for the production of Cherenkov radiation as a function of the index of refraction of the medium for gamma rays and electrons or beta particles. The threshold energies for electrons or beta particles are calculated according to eq. (7.6), and the gamma-ray threshold energies are calculated according to eq. (7.37), as the gamma rays that yield Compton electrons of the threshold energy via 180° Compton scatter.

used to discriminate against gamma rays of relatively high energy (2.0 MeV), while a transparent medium of high refractive index such as flint glass ($n = 1.72$) can serve to discriminate against relatively low-energy gamma radiation (0.25 MeV). Figure 7.9 illustrates the potential for gamma-ray energy discrimination according to refractive index of the detector medium.

Another application of gamma-ray analysis with Cherenkov photons is the Cherenkov verification technique used in nuclear safeguards to verify the authenticity of irradiated nuclear fuel, which is one of the important tasks performed by the International Atomic Energy Agency (IAEA). The IAEA nuclear safeguards program audits the national declarations of fuel inventories to ensure that no illicit diversion of nuclear material has occurred. High levels of gamma radiation are emitted by fission products in irradiated nuclear fuel. The irradiated fuel stored under water will produce Cherenkov light as a result of Compton scattering in the water surrounding the fuel. A Cherenkov viewing device containing a UV-transmitting lens coupled to a UV-sensitive charge-coupled device (CCD) and image monitor enables the real-time imaging of the UV light portion of the Cherenkov radiation in the presence of normal room lighting (Attas *et al.*, 1990, 1992, and Attas and Abushady, 1997; Kuribara, 1994; Kuribara and Nemeto, 1994; Lindsey *et al.*, 1999). The presence of fission products and the nature of their distribution, as indicated by the Cherenkov glow, are used as evidence of fuel verification.

7.4   PARTICLE IDENTIFICATION (PID)

Cherenkov counters are applied in particle physics research for the determination of particle mass (m), velocity (β), and PID. Cherenkov detectors of various designs are applied to the discrimination and identification of high-energy particles, among which are threshold Cherenkov counters, ring-imaging Cherenkov (RICH) detectors, as well as time of flight (TOF) and time of propagation (TOP) measurements.

7.4.1   Threshold Cherenkov counters

Threshold Cherenkov counters consist of Cherenkov detectors of differing refractive index employed to discriminate particles of different mass based on the differing Cherenkov threshold energies that the particles have in the detectors. For example, if we consider a beam of two types of particles of different mass (m), such as pions ($\pi^\pm$, $m = 139.6 \, \text{MeV}/c^2$) and kaons $K^\pm$, $m = 493.7 \, \text{MeV}/c^2$), a Cherenkov detector of a given refractive index (n) may be selected such that the particle of higher mass does not produce Cherenkov radiation. This would be the case if the threshold condition for the production of Cherenkov radiation is not met by the particle of higher mass, that is, $\beta < 1/n$. Sundaresan (2001) describes another example of the application of two Cherenkov detectors of different refractive index, namely, silica aerogel ($n = 1.01$–1.03) and pentane ($n = 1.357$) whereby particles of lower mass such as 10 GeV kaons ($m = 493.7 \, \text{MeV}/c^2$) produce Cherenkov photons in the two Cherenkov detector media, whereas particles of higher mass such as protons ($m = 939.3 \, \text{MeV}/c^2$) produce Cherenkov photons only in the silica aerogel. The difference in count rates from Cherenkov photons produced in the two detectors yield the relative numbers of the heavier and lighter particles. Adachi *et al.* (1995) describe a threshold Cherenkov counter for the identification of $\pi^\pm$ and $K^\pm$ in a particle beam with momentum in the region of 1.0–2.5 GeV/c. Silica aerogel with refractive index of 1.0127 with 14 cm thickness was used as the detector coupled to PMTs for the measurement of Cherenkov photons. The threshold momentum for the detection of $\pi^\pm$ and $K^\pm$ was determined to be 0.863 and 3.05 GeV/c, respectively. A Cherenkov detector arrangement reported by Perrino *et al.* (2001) provides an example of two Cherenkov detectors operated in tandem together with TOF measurements to discriminate between pions (π^+), positrons (e^+), and protons (p^+). The experimental setup consisted of CO_2 gas detectors ($n = 1.00041$) providing excellent detection for positrons, and a silica aerogel detector with a refractive index $n = 1.025$ providing (π, p) discrimination in the 1–4 GeV/c range with Cherenkov thresholds of 0.62 and 4.2 GeV/c for pions and protons, respectively. Pions and protons at 1 and 2 GeV/c are below the Cherenkov threshold in the CO_2 gas. Complementary data provided by TOF measurements between two BC408 scintillation detectors separated at a 23 m distance along the particle beam enabled the tagging of protons. TOF measurements are determined by signals between detectors permitting the determination of the speed of a particle, and with its total energy signal, the mass of the ion can be identified (Lilley, 2001).

7.4.2 Ring Imaging Cherenkov (RICH) counters

The RICH detector is designed principally for PID, as it can provide information on the velocity, β, and the charge of the particle, z, and complementary information provided by rigidity measurements using a magnetic tracker can provide the identity of the particle according to its mass (Pinto da Cunha et al., 2000). The detector is designed to accept particles that originate from any 4π direction. Several detector geometries and designs are reviewed by Glässel (1999), and the classical RICH geometry is illustrated in Figure 7.10. The distance ($d = 2f$) from the source of the charged particles or interaction vertex (Figure. 7.10) defines the radius of a spherical mirror, R_s. The Cherenkov photon detector has a concentric spherical surface of smaller radius. The space between the outer surface of the photon detector and inner spherical mirror is filled with a transparent medium of a given refractive index [e.g., gas, C_4F_{10} ($n = 1.0015$), liquid, C_6F_{14} ($n = 1.276$), crystalline NaF ($n = 1.33$), or silica aerogel ($n = 1.01–1.02$)] to serve as the Cherenkov radiator (Pinto da Cunha et al., 2000). The Cherenkov radiator is chosen according to the mass and momentum of the particle to be identified, as the emission of Cherenkov radiation at an angle θ must satisfy the threshold condition $\beta > 1/n$. At the moment the particle penetrates the Cherenkov radiator, the Cherenkov radiation is emitted at an angle θ according to the particle velocity, β, and the refractive index (n) of the Cherenkov radiator as defined by eq. (7.9), that is, $\cos \theta = 1/\beta n$. The radiator dimensions used by Pinto da Cunha et al. (2000) were 2 cm thickness and 50 cm radius. The Cherenkov photons are reflected off the inner surface of the outer spherical

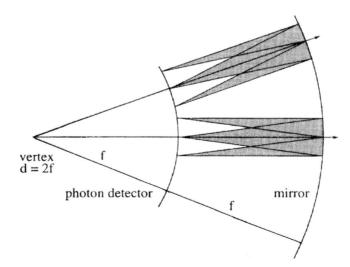

Figure 7.10 Classical RICH detector geometry. A spherical mirror surrounds a spherical photon detector. The two arrows illustrate two charged-particle trajectories. The shaded areas surrounding the particle trajectories illustrate the emitted conical Cherenkov light and cone image (ring) reflected onto the detector surface. The radius (r) of the light cone (not indicated in the figure) is that distance on the detector surface from the line of particle trajectory to the focal point (ring) of the reflected light. (From Glässel (1999) reprinted with permission from Elsevier © 1999.)

mirror as a ring of light onto the conical detector surface. The ring has a radius, r, which is measured to determine the particle velocity. An imaging detector is used to provide an image of the ring of Cherenkov light and its coordinates relative to the vertex. According to the derivations of Sundaresan (2001), the focal length, f, of the mirror is defined as $f = R_s/2$; and if $r = f\theta$, we can write $r = R_s\theta/2$ and

$$\cos\theta = \cos\frac{r}{f} = \cos\frac{2r}{R_s} \tag{7.39}$$

When the threshold condition for the emission of Cherenkov photons is met, that is, $\beta n > 1$, the Cherenkov photons are emitted at an angle θ to the particle trajectory according to eq. (7.9), namely, $\cos\theta = 1/\beta n$ and

$$\beta n = \frac{1}{\cos(2r/R_s)} \tag{7.40}$$

and

$$\beta = \frac{1}{n\cos(2r/R_s)} \tag{7.41}$$

Thus, the particle velocity, β, in units of speed of light can be obtained from the radius r of the Cherenkov ring image, and its coordinates from which the emission angle θ can be derived. The particle charge can be derived from the Cherenkov photon intensity.

The medium employed for the Cherenkov radiator is critical for the discrimination of charged particles with certain energy ranges as was noted earlier. Gaseous radiators, which have low indexes of refraction relative to liquid and solid media, may be used to detect and discriminate between very high-energy relativistic particles ($\beta > 0.99$). For example, Križan et al. (2001) used perfluorobutane (C_4F_{10}, $n = 1.00135$) in a 2.8 m long Cherenkov radiator to discriminate between particles of different mass, namely, protons ($m = 938.27\,MeV/c^2$), kaons ($m = 493.67\,MeV/c^2$), pions ($m = 139.56\,MeV/c^2$), and electrons ($m = 0.511\,MeV/c^2$), and it was specifically capable of discriminating between kaons and pions of momentum of approximately $50\,GeV/c$. Early work of Giese et al. (1970) used gaseous radiators of 9–12 m in length for proton and kaon discrimination at energies of 200 GeV. The radiators are placed in the tracks of high-energy particle beams. Such gaseous radiators are much larger (e.g., several meters in length) than their liquid or solid counterparts (centimeters in length), as the number of photons produced in these detectors is proportional to their length, which is limited by transmission loss as length increases. The proportionality of radiator length (L) to detected photons on a Cherenkov ring (N) is parameterized by Križan et al. (2001) as

$$N = N_0 L \sin^2\theta_c \tag{7.42}$$

where N_0 is the detector response parameter and θ_c the Cherenkov angle of emission relative to the particle track defined by eq. (7.25) or $\cos\theta_c = 1/n$ when $\beta = 1$. The Cherenkov

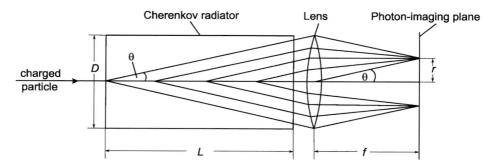

Figure 7.11 Principle of Cherenkov light focusing. (From Giese *et al.* (1970) reprinted with permission from Elsevier © 1970.)

light radiated from these long detectors often requires optical lenses and or mirrors to focus the Cherenkov ring onto a photosensitive imaging device such as position-sensitive flat-panel multianode PMTs. Once imaged, the dimensions of the Cherenkov ring can be measured to determine the particle velocity and momentum (Roberts, 1960; Giese *et al.*, 1970; Križan *et al.*, 2001; Matsumoto *et al.*, 2004). A schematic diagram of the Cherenkov radiation emitted at angle θ relative to the particle track in a gaseous detector of length L and the focusing of the Cherenkov ring onto a photosensitive layer is illustrated in Figure 7.11.

The figure as described by Giese *et al.* (1970) illustrates the focusing of the Cherenkov light by a lens or mirror of focal length f to a ring image of radius r, which is related to the Cherenkov emission angle θ and the phase velocity β of the charged particle according to the relation

$$r = f \tan \theta \tag{7.43}$$

and the aperture of the focusing lens is

$$\frac{D}{f} = \frac{2\theta^2 L}{r} \tag{7.44}$$

The particle velocity is obtained from

$$\cos \theta = \frac{1}{\beta n} \tag{7.45}$$

In the early days of RICH detectors photographic film plates were used to image the Cherenkov ring (Roberts, 1960; Giese *et al.*, 1970). Modern methods of ring imaging include the use of planar position-sensitive PMTs such as in the experimental arrangement illustrated in Figure 7.12.

The RICH detector of Matsumoto *et al.* (2004) illustrated in Figure 7.12 employs two $5 \times 5\,\text{cm}^2$ multiwire proportional chambers (MWPCs) to provide precise measurement of the particle track. The MWPCs are constructed of 20 µm diameter, gold-plated tungsten anode wires with 2 mm pitch equipped with 90% Ar + 10% CH$_4$ gas flow providing a position

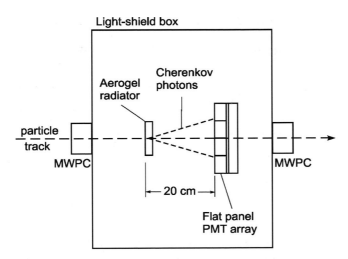

Figure 7.12   Experimental arrangement of a RICH detector with a 4 × 4 array of flat-panel position-sensitive PMTs. (From Matsumoto *et al.* (2004) reprinted with permission from Elsevier © 2004.)

readout by delay lines on the *x*- and *y*-cathode strips. MWPCs are gas ionization detectors, first developed by Nobel Laureate Georges Charpak and coworkers (Charpak *et al.*, 1968, 1970; Charpak, 1969, 1970; Charpak and Sauli, 1978, 1979), contain an array or grid of uniformly spaced anode wires sandwiched between two other grids of uniformly spaced cathode wires. The two outer cathode planes are oriented at right angles to one another, and the sandwich of these grids is contained in a chamber of proportional counting gas such as an Ar + CH$_4$ gas mixture. Output signals provide the *x*- and *y*-coordinates of the ionizing particles entering the grid. See L'Annunziata (1987) and Buchtela (2003) for additional references and information on MWPCs.

As illustrated in Figure 7.12 after passing through the first MWPC, the high-energy particles continue to penetrate a silica aerogel Cherenkov radiator (10–25 mm thick). The refractive index ($n = 1.05$) of the aerogel (see Table 7.1) provided discrimination of pions and kaons in the momenta range of 0.5–4 GeV/c. The Cherenkov photons produced in the aerogel are detected by a 4 × 4 array of flat-panel multianode PMTs, that is, 16 PMTs. The surface of each PMT is divided into 64 (8 × 8) channels with a 6 × 6 mm^2 pixel size. The multichannel PMTs are connected to an analog memory board providing a readout of the PMT signals. The position sensitivity of the readout yielded a Cherenkov angle resolution per photon of 10 mrad (Matsumoto *et al.*, 2004).

Križan *et al.* (2001) provide another excellent example of the use of position-sensitive PMTs to provide high-resolution images of Cherenkov rings (Figure 7.13a). They used 2300 position-sensitive PMTs with a 2.8 m long C$_4$F$_{10}$ gas radiator ($n = 1.00135$) yielding an average number of 32 Cherenkov photons per ring produced by 50 GeV/c kaons or pions and approximately twice that number of photons per ring when produced by overlapping e^+e^- pairs. Cherenkov photon resolutions of approximately 1 mrad provided discrimination of kaons and pions of momentum as high as 50 Gev/c. Examples of Cherenkov rings imaged by the signal outputs of position-sensitive PMTs is illustrated in Figure 7.13.

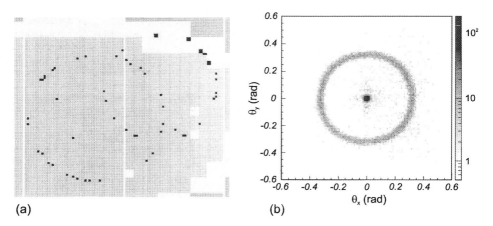

(a)   (b)

Figure 7.13 Cherenkov rings imaged by multichannel position-sensitive PMTs. (a) an event with two Cherenkov rings with single photon resolutions of 0.7 mrad produced by two different particle trajectories in a 2.8 m perfluorobutane gaseous medium. (From Križan *et al.* (2001) reprinted with permission from Elsevier © 2001.) (b) a broader Cherenkov ring produced by a distribution of numerous PMT hits in the Cherenkov *x,y* space by 3 Gev/*c* pions traveling through a 2 cm thick silica gel Cherenkov radiator. The central dark spot was produced by the particle beam. Background spots outside of the ring are produced by Cherenkov photons that are Rayleigh scattered in the radiator. (From Matsumoto *et al.* (2004) reprinted with permission from Elsevier © 2004.)

The relation between Cherenkov angle θ_c and particle momentum was established by Križan *et al.* (2001) to be

$$\theta_c^2 = \frac{\theta_0^2 - m^2}{p^2} \tag{7.46}$$

where $\theta_0^2 = 2(n-1)$, m the particle mass, and p the particle momentum. The relation between the variables θ_c^2 and $1/p^2$ as defined by eq. (7.46) is linear, and thus the plot of θ_c^2 against $1/p^2$ for a particle of given mass would be a straight line. Figure 7.14 illustrates the linear relationship between θ_c^2 and $1/p^2$ for particles of different mass, namely, protons ($m = 938.27$ MeV/c^2), kaons ($m = 493.67$ MeV/c^2), pions ($m = 139.57$ MeV/c^2), and electrons ($m = 0.511$ MeV/c^2). Thus, measurement of the Cherenkov angle of emission in radians (rad) can be used to determine the velocity of a particle from its momentum or identify the particle from its known momentum.

An historical review and a thorough treatment of the theory of RICH counters are available from Sequinot and Ypsilantis (1994) and Ypsilantis and Sequinot (1994).

7.4.3   Time-of-Propagation (TOP) Cherenkov counters

A relatively new concept in the application of Cherenkov detectors for PID is via the measurement of the TOP and horizontal emission angle, Φ, of Cherenkov photons described by

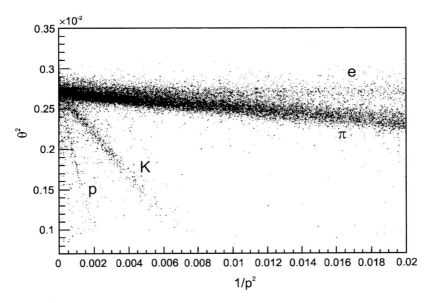

Figure 7.14 θ_c^2 versus $1/p^2$ plots in which particles of given mass lie on a straight line. Plots illustrated are those of the proton (p), kaon (K), pion (π), and electron (e). (From Križan *et al.* (2001) reprinted with permission from Elsevier © 2001.)

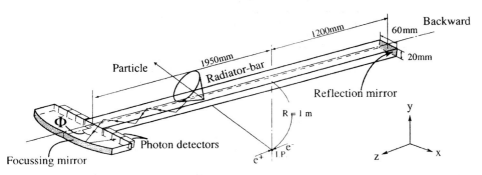

Figure 7.15 Structure of the TOP Cherenkov counter. Basic parameters are indicated in the figure. The bar (Cherenkov radiator) and mirrors made of synthetic optical quartz ($n = 1.47$ at $\lambda = 390$ nm) is configured z-asymmetric to the interaction point (IP) of an asymmetric collider. The counters are placed 1 m radially distant from the IP to form a cylindrical structure. (From Ohshima (2000) reprinted with permission from Elsevier © 2000.)

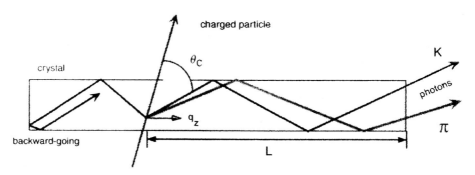

Figure 7.16   Side view of propagating photons. TOP is inversely proportional to z-component q_z of the light velocity. TOP $= (L \times n(\lambda))/cq_z = 4.90$ nsec $\times Lm/q_z$. θ_c is the Cherenkov angle and L is the particle injection position from the bar end in meters. At the opposite end, a mirror reflects the backward-going photons. (From Ohshima (2000) reprinted with permission from Elsevier © 2000.)

Akatsu *et al.* (2000) and Ohshima (2000). The basic structure of the TOP Cherenkov counter is illustrated in Figure 7.15. The TOP detector consists of a quartz Cherenkov radiator bar (20 mm thick, 60 mm wide, 3150 mm long). Two mirrors are located at both ends for focusing the Cherenkov photons. The mirror is flat at the backward end to reflect the Cherenkov light toward the forward end where butterfly-shaped mirrors are located. The Cherenkov photons are focused horizontally onto the photon detector plane, and the TOP and angle Φ are measured by position-sensitive multichannel phototubes. The method is based on the following principles: (1) the Cherenkov photon emission angle (θ_c) illustrated in Figure 7.16 is a function of the particle velocity (β) according to the relation cos $\theta_c = 1/\beta n$ where n is the refractive index of the Cherenkov radiator, (2) the TOP of photons in a light guide with internal-total-reflection characteristics can be calculated as a function of the photon emission angle, and (3) a correlation between TOP and photon emission angle would provide information on PID. Notice from the illustration provided in Figure 7.16, the Cherenkov photon emission angle created by the pion (π) is less acute than that created by the kaon (K) and the TOP of the Cherenkov photons derived from the pion is shorter than that derived by the kaon. TOP differences of 100 psec or more are found for normal incident 4 GeV/c K and π at 2 m long propagation (Ohshima, 2000).

7.5   APPLICATIONS IN RADIONUCLIDE ANALYSIS

Cherenkov radiation, when produced at significant levels, can be employed for the efficient measurement of radioactivity. This was first demonstrated by Belcher (1953), who used a liquid-nitrogen-cooled single photomultiplier to measure Cherenkov radiation intensity in terms of count rates per mCi of various radionuclides in aqueous solution. However, it was not until dual photomultiplier liquid scintillation analyzers became available commercially and in widespread use in the early 1960s, namely, the Packard Tri-Carb 314 liquid scintillation analyzer, did practical research into Cherenkov counting for the measurement of radioactivity begin. (Modern Tri-Carb liquid scintillation counters are now produced by

PerkinElmer Life and Analytical Sciences, Boston, MA, and other manufacturers are found worldwide). Although the production of Cherenkov radiation does not involve the scintillation phenomenon, a conventional liquid scintillation counter can detect and count Cherenkov photons emitted from a given sample in a standard counting vial. The beta-particle emissions of certain radionuclides may be counted in water or in other suitable transparent medium without the use of scintillation fluor or any other chemical reagents. The radionuclide beta-particle emissions may even be counted in the dry state, albeit at a diminished counting efficiency, where the glass or plastic vial wall containing the sample serves as the transparent medium or Cherenkov radiator. The percent counting efficiencies are calculated according to the equation

$$\%E = \frac{\text{cpm}}{\text{dpm}} \times 100 \tag{7.47}$$

where E is the counting efficiency defined by the ratio of the radioactive sample count rate in counts per minute (cpm) to the sample disintegration rate or radioactivity in units of disintegrations per minute (dpm). When the sample radioactivity (dpm) is unknown and the sample radiation count rate is determined experimentally, the radioactivity is calculated from the detection efficiency according to the equation

$$\text{dpm} = \frac{\text{cpm}}{E} \tag{7.48}$$

The application of Cherenkov counting to the activity analysis of radionuclides is popular in those cases where the Cherenkov counting efficiency of the radionuclide of interest is adequate to meet particular detection limit requirements. Table 7.2 provides approximate Cherenkov counting efficiencies of radionuclides measured and listed according to the E_{max} of the beta particles emitted by each radionuclide. The E_{max} is listed because the Cherenkov detection efficiency of radionuclides is a function of the threshold energy (E_{th}) and the refractive index (n) of the medium calculated according to eq. (7.6). For example, if water is the medium ($n = 1.332$), the Cherenkov counting efficiency would be a function of the number of beta particles of $E > 263\,\text{keV}$ relative to the total number of beta particles emitted by the radionuclide. Cherenkov counting is popular, when counting efficiencies are adequate, because of the ease of sample preparation and low expense incurred in the preparation and disposal of samples. Because water is generally the medium of counting, samples are often left in a state suitable for subsequent tests such as chemical analysis, spectrometric analysis, chromatographic tests, or even chemical compound extraction and isolation. The application of Cherenkov counting to the analysis of specific radionuclides is reviewed in a previous book (L'Annunziata, 2003d).

Table 7.2

Experimentally determined Cherenkov counting efficiencies

Nuclide	E_{max} (keV)[a]	Counting efficiency[b] (%)	References
^{14}C	155 (100%)	1.5[c]	Takiue et al. (2004)
^{45}Ca	258 (100%)	2.3[c]	Takiue et al. (2004)
^{99}Tc	292 (100%)	1.0[d]	Scarpitta and Fisenne (1996)
^{59}Fe	273 (48.5%)	5.8[d]	Scarpitta and Fisenne (1996)
	475 (51.2%)		
^{90}Sr	546 (100%)	1.0	Rucker (1991); Chang et al. (1996)
^{60}Co	Compton electrons[e]	5.6	Grau Carles and Grau Malonda (1995)
^{36}Cl	710 (98%)	6.6	Grau Carles and Grau Malonda (1995)
^{204}Tl	763 (98%)	4.0	Grau Carles and Grau Malonda (1995)
^{137}Cs	510 (92%)	4.8	Grau Carles and Grau Malonda (1995)
	1170 (8%)		
^{198}Au	960 (99%)	5.4	Parker and Elrick (1970)
^{47}Ca	660 (83%)	7.5	Parker and Elrick (1970)
	1940 (17%)		
^{210}Pb(^{210}Bi)	1160 (from ^{210}Bi >99%)	18	Blais and Marshall (1988)
^{115m}Cd	680 (3%)		Bem et al. (1978); Ramesh and Subramanian (1997)
	1620 (97%)	35	
^{89}Sr	1490 (100%)	42	Rucker (1991); Chang et al. (1996)
^{228}Th	580 (from ^{212}Pb)[f]	53	Al-Masri and Blackburn (1994)
	1790 (from ^{208}Tl)[f]		
	2250 (from ^{212}Bi)[f]		
^{86}Rb	680 (8.5%)	53	L'Annunziata and coworkers (see Noor et al., 1996)
	1770 (91.5%)		
^{40}K	1310 (89%)	55	Pullen (1986)
^{32}P	1710 (100%)	57	Takiue et al. (1993)
^{234}Th	(from ^{234m}Pa)[f]	60	Navarro et al. (1997), Nour et al. (2002)
	2290 (99.8%)		
^{238}U	(from ^{234m}Pa)[f]	62	Blackburn and Al-Masri (1994)
^{90}Y	2280 (100%)	72	L'Annunziata and Passo (2002)
^{118}Re	2120 (79%)	53	Kushita and Du (1998)
	1970 (20%)		
	<1900 (1%)		
^{106}Ru(^{106}Rh)	2000 (from ^{106}Rh, 3%)	62	Carmon and Dyer (1987)
	2440 (from ^{106}Rh, 12%)		
	3100 (from ^{106}Rh, 11%)		
	3530 (from ^{106}Rh, 68%)		
^{234m}Pa	2290 (99.8%)	54	Grau Malonda and Grau Carles (1998a,b, 2000)
^{42}K	1970 (18.4%)	75	Buchtela and Tschurlovits (1975)
	3560 (81.3%)		
^{226}Ra/^{222}Rn	1505 (from ^{214}Bi, 16%)[f]	77	Blackburn and Al-Masri (1993) Al-Masri and Blackburn (1995a,b, 1999)
	1540 (from ^{214}Bi, 16%)[f]		
	3270 (from ^{214}Bi, 24%)[f]		
	650 (from ^{214}Pb)[f]		

(*Continued*)

Table 7.2   (*Continued*)

Nuclide	E_{max} (keV)[a]	Counting efficiency[b] (%)	References
^{34m}Cl	Several beta-decay branches up to a energy maximum of 4500	57	Wiebe *et al.* (1980)
^{38}Cl	4913 (57.6%) 2770 (11.1%) 1111 (31.3%)	66	Wiebe *et al.* (1980)
^{18}F	635 (97%)	50[g]	Wiebe *et al.* (1978)
^{214}Bi(^{214}Pb)	(see ^{226}Ra in this table)		

[a]Where more than one beta-decay modes occur for a given radionuclide, the percentage abundance or relative intensities of each decay process is provided in parenthesis.

[b]Counting efficiencies will vary from instrument to instrument and vial type (e.g., glass or plastic). Also, measurements made with older less efficient PMTs provided lower counting efficiencies than can be expected from contemporary more efficiency phototubes. No attempt is made in this table to categorize the counting efficiencies according to the aforementioned governing factors that will affect the counting efficiency values reported. However, the counting efficiencies listed are intended to provide a general idea of their orders of magnitude relative to the energies of the beta-particle emissions. The counting efficiencies are values reported of nuclides in water without wavelength shifter unless otherwise noted.

[c]Measured in the dry state deposited onto the surface of high refractive index ceramic ($n = 2.1$).

[d]Measured in 10 ml aqueous 25 mM 7-amino-1,3-naphthalene disulfonic acid (ANSA) wavelength shifter.

[e]The Cherenkov photons from ^{60}Co are due to Compton electrons produced by its 1.33 and 1.17 MeV gamma-ray emissions.

[f]Daughter nuclide.

[g]A combination of Cherenkov and scintillation counting in high refractive index methyl salicylate ($n = 1.5369$).

Radioactivity Hall of Fame — Part VIII

Ernest O. Lawrence (1901–1958), John Douglas Cockcroft (1897–1967) and Ernest Thomas Sinton Walton (1903–1995), Hans A. Bethe (1906–2005), and Willard F. Libby (1908–1980)

ERNEST LAWRENCE (1901–1958)

Ernest Orlando Lawrence was born in Canton, South Dakota on August 8, 1901. He was awarded the Nobel Prize in Physics 1939 "for the invention and development of the cyclotron and for results obtained with regard to artificial radioactive elements". Ernest Lawrence received the Nobel Prize at a ceremony on the campus of the University of California, Berkeley on February 29, 1940. He was unable to travel to Stockholm for the traditional award presentation and ceremony due to the precarious travel conditions in the early stages of World War II. At the Berkeley ceremony, Professor R. T. Birge (1940), Chairman of the Department of Physics at Berkeley delivered a detailed account of Lawrence's life and work, which was published in *Nobel Lectures, Physics 1922–1941*. Some details of Lawrence's biography are taken from Professor Birge's speech.

**Ernest Lawrence
(1901-1958)**

Lawrence attended elementary and secondary schools in Canton and Pierre, South Dakota and undertook undergraduate studies at St. Olaf College and the University of South Dakota where he received the B.A. degree in Chemistry in 1922. He shunned the college fraternities to devote his time to studies. Initially he had planned to enter medicine, but it became evident to him and his teachers that his interest was physics. He carried out graduate studies at the University of Minnesota where he received his Masters degree and spent a year at the University of Chicago before going on to Yale University where he was awarded the Ph.D. in Physics in 1925. Often there is a secondary school teacher or college professor that has a tremendous influence on a successful student. In Lawrence's case it was Professor W. F. G. Swann who was his teacher for advanced study in physics at the University of Minnesota. Lawrence followed Swann, his thesis advisor, to the University of Chicago and to Yale where he completed his doctoral research in physics on the photoelectric effect. Lawrence became a very practical and relentlessly hard-working physicist, who was able to put theory into practice in the development of the cyclotron, an instrument that became vital for the study of the nucleus and the production of new radioactive isotopes useful in medicine, biology, and other fields. Lawrence learned from Swann, who is quoted as saying "Cut-and-try sometimes is more important than too prolonged theorizing". This proved to be the practice of Ernest Lawrence throughout his career. He would grasp the theory and put it into the development of the cyclotron that appeared possible only in theory.

Upon graduation at Yale, Lawrence remained there as a National Research Fellow and then Assistant Professor of Physics. Swann left Yale and the University of California saw the opportunity to attract Lawrence to the Berkeley campus where he was appointed Associate Professor of Physics in 1928 and then Professor in 1930.

As a new member of the faculty at the University of California, Berkeley, Lawrence decided to research the atomic nucleus. As a young scientist he recalled the pioneering work of Ernest Rutherford, who bombarded the nucleus of atoms with alpha particles emitted by radium and discovered that almost the entire mass and energy of the atom is in the nucleus. It was in 1919, 10 years earlier, that Rutherford discovered that bombarding atoms of nitrogen with alpha particles resulted in the capture of the alpha particles by the nitrogen nucleus with the emission of a proton thus converting the nitrogen into an isotope of oxygen as follows:

$$\ce{^4_2He} + \ce{^{14}_7N} \rightarrow \ce{^{17}_8O} + \ce{^1_1H} \tag{VIII.1}$$

He related in his Nobel Lecture on December 11, 1951 in Stockholm, 12 years after receiving the Nobel Prize, that he clearly saw, following the work of Rutherford and his school, that "the next great frontier for the experimental physicist was surely the atomic nucleus". Unfortunately, alpha particles emitted by radionuclides were limited in their energies to approximately 4–8 MeV, and their intensities were very much limited to the amount or activity of the radioactive alpha-particle source. Also, protons that were much lighter particles would have a better chance of penetrating the atomic electron coulombic barrier to collide with the nucleus. Lawrence thus contemplated the need to accelerate protons and other charged particles to energies sufficient to collide with atomic nuclei and explore the energy of the nucleus.

It was during one evening in 1929 that Lawrence was reviewing the literature in the University of California, Berkeley Library when he came across a paper published in the German language by a Norwegian engineer Rolf Wideröe (1928) concerning the multiple acceleration of positive ions. Lawrence was not able to read German easily, but he was able to understand the principle approach to the problem from the various drawings in the article. In his article, Wideröe had described a glass linear acceleration apparatus where he used a 25,000 V potential to impart to positive potassium ions a 50,000 V potential. He achieved 50,000 V potassium ions with only 25,000 V, a twofold voltage amplification. Wideröe's demonstration of positive ion acceleration was based on the original concept of Swedish Professor Gustaf Ising (1924) for a linear accelerator of positive ions. Ising's principle for the multiple acceleration of positive ions involved the application of radiofrequency oscillating voltages to a series of cylindrical electrodes positioned in a linear fashion as illustrated in Figure VIII.1. The principle involved lining up hollow cylinders on the same axis through which the positive ions would be accelerated through a series of successive pushes across the same voltage potential.

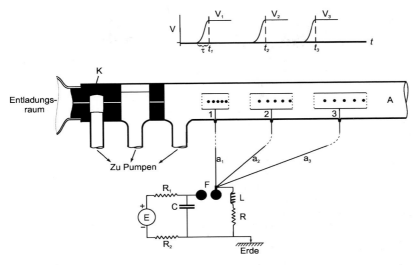

Figure VIII.1 The pioneering diagram of Ising illustrating his concept of the multiple acceleration of ions. Positive potassium ions produced at K are accelerated from left to right through hollow cylindrical electrodes 1, 2, and 3. The voltage difference applied to each electrode is identical and illustrated graphically above the apparatus. (Adapted from Ising, 1924 as illustrated by Ernest Lawrence in the text of his Nobel Lecture reprinted with permission © The Nobel Foundation 1939.)

The accelerating electrodes, through which the ions pass, are successively longer in length due to the increasing velocity of the ions. Three such accelerating electrodes were illustrated in Ising's drawing of his original concept. On the basis of this idea Ising is considered the father of the development of multiple acceleration.

Upon seeing Widerøe's paper employing the concept of multiple linear acceleration, Lawrence began to make some calculations. He concluded that, to effectively penetrate the Coulomb barrier of atoms, he would have to accelerate positive ions to energies in excess of 1 MeV, which would require an accelerator too long for practical laboratory purposes. Lawrence then began to visualize an accelerator not consisting of a large number of cylindrical electrodes in line, but one that had only two electrodes, which would be used over and over again by sending the positive ions back and forth through the electrodes in a circular fashion in a magnetic field. In relating his reaction after seeing Widerøe's paper Lawrence (1951) stated

...a little analysis of the problem showed that a uniform magnetic field had just the right properties—that the angular velocity of the ions circulating in the field would be independent of their energy so that they would circulate back and forth between suitable hollow electrodes in resonance with an oscillating electrical field of a certain frequency...

Thus, Lawrence concluded two hollow electrodes would be needed with a relatively low voltage (e.g., 4000 V) and oscillating positive and negative potential so that, each instant the ions exit one electrode and enter the other, the charges on the electrodes would change in accord with the proper frequency providing a "push" to the ions as they exit an electrode and thereby increase their energy and velocity. In the Nobel Award Ceremony at Berkeley Professor R. T. Birge (1940) described Lawrence's idea eloquently as follows:

Widerøe had used two hollow cylinders, lined up on the same axis. Lawrence sketched a series of such cylinders, but in the case of atoms of small mass, which are most effective in nuclear disintegration, the necessary length of the apparatus would be too great. He next thought of the possibility of using a curved path. Now an electrically charged particle, entering into a magnetic field directed at right angles to the motion of the particle, proceeds to move in a circle at constant speed. Moreover, the time to move through a half circle depends only on the charge and mass of the particle and on the strength of the magnetic field. It does not depend on the speed of the particle. The greater the speed, the greater the radius of the circle in which the particle moves [See Fig. VIII.2]. This important fact, which Lawrence immediately noted by writing down a very simple mathematical relation, gave him the idea of the present essential features of the cyclotron. All this happened within a few minutes of the time he had seen Wideröe's paper.

On that day in 1929 after reading Wideröe's paper Lawrence invented the cyclotron or the principle of the cyclotron. Conceiving the principle of a new concept is one matter, but making it happen may be altogether a seemingly impossible hurdle. Birge (1940) related

The next morning Dr. Lawrence told his friends that he had found a method for obtaining particles of very high energy, without the use of any high voltage. The idea was surprisingly simple and in principle quite correct—everyone admitted that. Yet everyone said, in effect, "Don't forget that having an idea and making it work are two very different things".

The above statement is a lesson to us all in all fields of work and whatever the pursuit. Many good ideas that conform with known laws of nature may come to us, but to bring these ideas or inventions into reality can be to most an insurmountable burden. Often it is difficult to see the toil, almost endless hours, frustration, and sometimes border despair that one encounters coupled with the needed stubborn determination to find solutions to make an idea become a reality. We can be certain that Lawrence and many persons, who have achieved greatness, have battled these tremendous hurdles. Birge (1940) quoted the remarks of Dr W. D. Coolidge, then Director of the Research Laboratory of the General Electric Company in 1947 when he presented the Comstock Prize of the National Academy of Sciences to Ernest Lawrence. The prize is awarded only once in 5 years, and it is considered one of the greatest honors to be bestowed by the Academy. Dr Coolidge stated the following:

Dr. Lawrence envisioned a radically different course—one which did not have those difficulties attendant upon the use of potential differences of millions of volts. At the start, however, it presented other difficulties and many uncertainties, and it is interesting to speculate on whether an older man, having had the same vision, would have ever attained its actual embodiment and successful conclusion. It called for boldness and faith and persistence to a degree rarely matched.

The experimental work on the development of the cyclotron began in 1930 and, in the Spring of that year, Lawrence's first graduate student at Berkeley, Nels Edlefsen, constructed a crude model, which gave evidence of working. In September of 1930, they gave a preliminary report before a meeting of the National Academy of Sciences (Lawrence and Edlefsen, 1930). In the fall of the same year another graduate student, M. Stanley Livingston improved on the model, which was only 4.5 in. (11.43 cm) in diameter. The diameter of the accelerator refers to the size of the chamber in which the ions move in a circular path of increasing radius. The first working cyclotron produced 80,000 V protons with less than 1000 V applied to the semicircular accelerating electrode. This was reported before the American Physical Society (Lawrence and Livingston, 1931a) demonstrating the cyclotron to be a practical apparatus and principle for the acceleration of ions. The same year Lawrence and Livingston (1931b) reported the acceleration of protons to 0.5 MeV with only 5000 V potential applied to the electrodes.

Lawrence and Livingston (1932) described the basic design and principle of the cyclotron, with a simple diagram such as that illustrated in Figure VIII.2. The cyclotron utilizes the principle of repeated acceleration of ions by means of resonance with an oscillating field. The ions circulate from the interior of one electrode to the interior of another back and forth with increasing speed in a circular fashion. The hollow electrodes A and B of Figure VIII.2 are placed between the poles of a magnet, and the magnetic

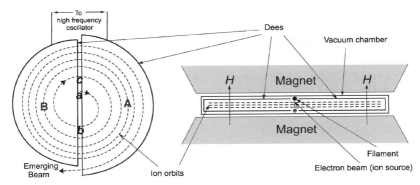

Figure VIII.2   Diagrams illustrating the cyclotron method of multiple acceleration of ions. On the left is a drawing of the top or bottom of the cyclotron electrodes and on the right is a drawing of a view from the side with the electrodes between magnet poles. Two electrodes A and B in the form of semicircular hollow plates (called Dees' because they have the shape of the letter D) are mounted in a vacuum chamber with their diametral edges adjacent. The electrodes of the early cyclotron of Lawrence and Livingston measured only 24 cm in diameter; and their hollow semicircular chambers, through which the ions traveled, had a width of only 1 cm. The system is placed between the poles of a magnet whereby the magnetic field H is introduced normal to the plane of the plates. High-frequency electric oscillations are applied to the plates so that there results an oscillating electric field in the diametral region between them. A filament placed above the diametral region creates a stream of electrons that produce the needed positive ions, which are pulled out sideways by the oscillating electric field. The electrons are not drawn out into the cyclotron electrodes because of their very small radii of curvature in the magnetic field. The positive ions (e.g., protons, deuterons, or helium nuclei) are accelerated by the oscillating electric field and travel with increasing speed and energy in a circular fashion of increasing radius in the magnetic field as described in the text. The beam of positive ions emerges from the cyclotron to collide with specific targets creating new elements, stable isotopes, and radioisotopes, etc. (Adapted from Lawrence and Livingston, 1932.)

field normal to the plane of the hollow plates cause the ions to deflect in a circular fashion with a radius of curvature dependent on the mass, charge, and velocity of the ions.

   The oscillating electric field between the electrodes is vital to the operation of the cyclotron. For example, as Lawrence and Livingston (1932) explained, if at one instant there is an ion in the region between the electrodes, and if electrode A is negative with respect to electrode B, the ion will be "pushed" and accelerated to the interior of electrode A. Within the hollow electrode the ion will travel along a circular path under the effect of the magnetic field. After completing a semicircular path in electrode A, as illustrated by the arc $a...b$ in Figure VIII.2, the ion emerges again between the electrodes. If the time consumed in the travel of this semicircular path (e.g., from point a to point b of Figure VIII.2) is equal to the half-period of the electric oscillations, the electric field will have reversed and the ion will receive a second "push" and be accelerated into the interior of electrode B with higher velocity. The ion continues to travel in a circular path in electrode B and completes another semicircular path with greater velocity and greater radius of curvature. Upon completion of the semicircular path in electrode B at point c, the ion again emerges between the electrodes, and the ion is again accelerated by the oscillating electric field to travel at greater speed into the interior of the opposing electrode with a path of greater radius of curvature. The process of acceleration repeats itself over and over until the ion achieves a radius of curvature whereby it emerges from the cyclotron.

   A vital characteristic of the cyclotron principle is that the time of travel of an ion to complete a semicircular path is independent of the ion's velocity and thus independent of the radius of the ion path. This is demonstrated by calculations provided below and the following elucidation provided by Lawrence and Livingston (1932) following their demonstration of a 11-in. (28-cm) diameter cyclotron:

   ...the radius of the path is proportional to the velocity, so that the time required for the travel of a semi-circular path is independent of the ion's velocity. Therefore, if the ion travels its first half-circle in a half-cycle of the oscillations, it will do likewise on all succeeding paths. Hence it will circulate around on ever widening semi-circles from the interior of one electrode to the interior of the other, gaining an increment of energy on each crossing of the diametral region that corresponds to the momentary potential difference between the electrodes. Thus, if, as was done in the present experiments, high frequency oscillations having peak values of 4000 volts are applied to the electrodes, and protons are caused to spiral around in this way 150 times, they will receive 300 increments of energy, acquiring thereby a speed corresponding to 1,200,000 volts.

The independence of the time of travel t of an ion along a semicircular path and its radius r and velocity v was demonstrated quantitatively by Lawrence and Livingston (1932). They equated the centrifugal force of an ion traveling along a circular path within an electrode with the magnetic force on the ion as follows:

$$\frac{mv^2}{r} = \frac{Hev}{c} \tag{VIII.2}$$

where m is the mass of the particle, H the magnetic field, e the ion charge, and c the velocity of light. Eq. (VIII.2) may be inverted to read

$$\frac{r}{mv^2} = \frac{c}{Hev} \tag{VIII.3}$$

or

$$\frac{r}{v} = \frac{mc}{He} \tag{VIII.4}$$

Since the circumference of a circle is defined as $2\pi r$, the distance d of a semicircle would be πr. The time of travel of the ion along the semicircular path of an electrode, would be defined as

$$t = \frac{d}{v} = \frac{\pi r}{v} = \frac{\pi mc}{He} \tag{VIII.5}$$

Thus, the time of ion travel along a semicircular path is independent of the radius r and the velocity v. The particle of mass m and charge e was made to travel in phase with the oscillating electric field in the radiofrequency range by adjustment of the magnetic field H of the order of magnitude of 10,000 G.

The 1.2 million-V hydrogen ions produced by Lawrence and Livingston in their early 11-in. (28-cm) diameter accelerator were of sufficient energy to disintegrate the light lithium nucleus; but much higher energies would be needed to penetrate larger atomic coulombic barriers and disintegrate the heavier elements. Larger diameter cyclotrons would require larger magnets, because the spiraling path taken by the ions in the cyclotron are produced by a uniform magnetic field normal to their plane of travel. Thus the pole faces of the magnets had to be at least as large as the diameter of the cyclotron vacuum chamber.

The largest magnets in 1932 that would be found in any scientific laboratory was 11-in. (28 cm). Fortunately, however, Lawrence found a larger magnet through the assistance of Professor L. F. Fuller of the University of California, Berkeley, who was previously Vice-President of the Federal Telegraph Company. The magnet was once intended for shipment to China, but considered obsolete for its intended purpose and never sent. Consequently, it was of no use to the Telegraph Company and a godsend for Lawrence. The diameter of the pole faces of the magnet measured 37 in. (94 cm). In addition, deuterium or heavy hydrogen ($^2_1$H) was discovered by Harold C. Urey, which opened the door to the acceleration of deuterons, a particle with twice the mass of the proton. Such particles could have twice the disintegrating power of protons of the same energy. With the larger magnet at hand Lawrence proceeded to build a 27-in. (69 cm.) diameter cyclotron with the electrodes ("dees") measuring 20 in. (51 cm) having a width of 1.12 in. (2.8 cm). With the new and larger cyclotron, Lawrence and Livingston (1934) were able to accelerate protons ($^1_1$H) to 5 MeV energy. With yet further modification by essentially increasing the size of the electrodes "dees" to a diameter of 24.5 in. (62 cm) and width of 1.75 in. (4.4 cm) Lawrence and Cooksey (1936) produced deuterons (heavy hydrogen, $^2_1$H) ions and helium ions (alpha particles, $^4_2$He) of 5- and 10-MeV energy, respectively. The helium ions, would be accelerated to twice the energy of the deuterons, explained by Lawrence and Cooksey (1936) as follows:

> …The helium ions having twice the charge and mass relative to deuterons, receive twice the increments in energy each time they pass between the dees and arrive at the periphery with twice as much energy. Thus, apart from difficulties of producing doubly charged helium ions at the center, the cyclotron produces 10 MeV alpha particles as readily as 5 MeV deuterons.

The most significant outcome of the new and improved cyclotron demonstrated by Lawrence and Cooksey (1936) was the production of neutrons and artificial radioactive isotopes. Neutron fluxes higher than ever achieved were reported. For example, the neutron flux by the deuteron bombardment of various elements including beryllium was 10^5 times greater than that produced by a mixture of 1 Curie of radon and beryllium. Lawrence began to produce many new radioisotopes as noted by Lawrence and Cooksey (1936)

> With 5 MeV deuterons, it has been found possible (Lawrence, 1935) to produce radioactive isotopes of many of the elements throughout the periodic table. In many cases, the yields of the radioactive substances are quite large; as for example, a day's bombardment of sodium metal with 20 microamperes of 5 MeV deuterons produces more than 200 milligrams—equivalent of radio sodium.

By 1939, the year Lawrence received the Nobel Prize, he and coworkers (Lawrence *et al.*, 1939) at the Berkeley lab completed a larger 60-in. (1.5 m) cyclotron that produced 16 MeV deuterons. It had a magnet weighing 200 tons. In his Nobel Lecture at Stockholm Lawrence commented about the larger cyclotron

> …the cyclotron for the first time began to look like a well-engineered machine. It was with this machine that the discoveries of the transuranium elements were made, which have been rewarded this year by the award of the Nobel Prize in Chemistry [1951] to Edwin M. McMillan and Glenn T. Seaborg.

Lawrence and his coworkers wanted to engineer even bigger and more powerful cyclotrons that could produce protons and deuterons at yet higher energies. In his Nobel Lecture, Lawrence (1951) explained that Hans Bethe predicted that they would encounter difficulties at this point, because of the relativistic increase in mass of the particles as they increase in energy in the course of acceleration. This would cause the ions to get out of resonance with the oscillating electric field. Lawrence (1951) added

> …the war prevented the building of this new machine and immediately afterwards Edwin McMillan (1945) and Vladomir Veksler (1944, 1945) independently came forward with the principle of phase stability, which transformed the conventional cyclotron to a much more powerful instrument for higher energies—the synchrocyclotron.

Particle accelerators of various types have been developed over the years from the 1930s to the present. Modern particle accelerators now produce ion beams with energies beyond 10,000 TeV where

1 TeV $= 10^{12}$ eV. For a modern review of the history and state-of-the-art of accelerators the reader is invited to peruse review works by Bryant (1992), Kullander (2005), Steere (2005), and a book by Wilson (2001).

Lawrence received the Nobel Prize not only for the invention and development of the cyclotron but equally, if not more so, for the results obtained therein in the production of new radioactive isotopes. Many of these new isotopes became, soon after their creation from the cyclotron, very useful tools for peaceful applications in nuclear medicine, radiation biology, and agricultural research among other fields of science that in short term improved our lives and wellbeing. The significance of this development was underscored by Birge (1940) only 10 years after Lawrence and Edlefsen (1930) first reported the feasibility of the cyclotron, when he stated

> With the cyclotron one can, as stated, transform every stable element into other forms. Some of the final products are themselves stable, but most of them are radioactive. The cyclotron is by far the best device for producing new radioactive substances. ...there are about 335 artificially produced radioactive substances, of which 223 have been discovered by means of the cyclotron... Many of the artificially produced radioactive substances are proving of extraordinary value in medicine and biology.

Marie Curie was one of the first to treat cancer patients with radium radiation. The cyclotron production of many new radioactive sources differing in their modes of decay and consequently in the radiation types and energies emitted as well as differing half-lives opened the door to a new frontier of nuclear medicine. Birge (1940) emphasized the importance of this by highlighting recent developments in medicine as follows:

> The great importance of radioactive elements in medicine and biology results chiefly from their use as so-called 'tracer atoms'...They are now being used also in the direct treatment of various diseases...and shown very promising in the treatment of cancer, and other medical uses are constantly being found.

Dr Ernest Lawrence recognized the possible future application of the newly created isotopes shortly after these were discovered. At the Radiation Laboratory at Berkeley, Ernest Lawrence collaborated closely with his brother and medical scientist Dr John Lawrence in the research for medical applications of radioactive isotopes. His brother became Director of the University of California's Medical Physics Laboratory and Ernest Lawrence became consultant to the Institute of Cancer Research at Columbia. The development of nuclear medicine continues to this day such as in the diagnosis and treatment of cancer with the use of isotopes and radiation in such techniques as radiotherapy, computerized tomography (CT), single photon emission computerized tomography (SPECT), and positron emission tomography (PET). These applications of radioactive isotopes in modern medicine are discussed in the Introduction to this book.

Ernest Lawrence died on August 27, 1958 at Palo Alto, CA.

JOHN D. COCKCROFT (1897–1967) AND ERNEST T. S. WALTON (1903–1995)

John Douglas Cockcroft and Ernest Thomas Sinton Walton shared the Nobel Prize in Physics 1951 "for their pioneer work on the transmutation of atomic nuclei by artificially accelerated atomic particles". They joined forces in 1928 at the Cavendish Laboratory of Cambridge University under the directorship of Ernest Rutherford to work on the high-voltage acceleration of protons. In 1932 by means of their newly developed accelerator, Cockcroft and Walton became the first to artificially split the atom or produce a nuclear disintegration by artificial means. They were, in turn, able to demonstrate Einstein's equation of equivalence of mass and energy ($E = mc^2$) by measuring the total energy of the alpha particles liberated from a nucleus of Li-7 upon impact with a proton and measurement of the differences in mass of the interacting proton and Li-7 nucleus and masses of the alpha particles. Some of the biographical data provided below is obtained from the autobiographies of Cockcroft and Walton written at the time of their Nobel Award and available in *Nobel Lectures, Physics 1942–1962*. Details concerning their collaborative research are taken from their scientific research papers published in *Nature* and the *Proceedings of the Royal Society of London*.

John Cockcroft (1897-1967) with permission from The Nobel Foundation © 1951

Ernest Walton (1903-1995)

   John Cockcroft was born at Todmorden, England on May 27, 1897 where he acquired his secondary education. During 1914–1915, he studied mathematics at Manchester University. Cockcroft served during the First World War in the Royal Field Artillery returning afterwards to study electrical engineering at the College of Technology. He served 2 years apprenticeship with the Metropolitan Vickers Electrical Company. His education and experience in electrical engineering would serve him well in the collaborative work with Ernest Walton on the development of a high-voltage proton accelerator. Following his apprenticeship he entered St. John's College and took the Mathematical Tripos in 1924. Afterwards he went on to work under Lord Rutherford at the Cavendish Laboratory. At first he collaborated on the production of intense magnetic fields and low temperatures with Piotr Kapitsa, who would be awarded eventually the Nobel Prize for studies in low-temperature physics. In 1928, he joined with Ernest Walton at the Cavendish to work on the development of a high-voltage accelerator.

   Ernest Walton was born at Dungarvan, County Waterford located on the south coast of Ireland on October 6, 1903. He became a boarder at the Methodist College, Belfast in 1915 where he pursued studies in mathematics and science. In 1922 he entered Trinity College, Dublin on a scholarship and completed honors courses in mathematics and experimental science with a major in physics. Walton graduated in 1926 with top honors in mathematics and physics, and was awarded the M.Sc. degree in 1927. The same year Walton received a Research Scholarship and went on to Cambridge University to work under Lord Rutherford at the Cavendish Laboratory where he would soon team up with John Cockcroft to develop a high-voltage accelerator with which the two partners would become the first to produce the first nuclear disintegration by artificial means.

   In 1919 Rutherford was the first to demonstrate the disintegration of light elements by bombarding them with alpha particles from radioactive sources, In particular, the most celebrated reaction was that whereby he demonstrated the transmutation of nitrogen into oxygen accompanied by the emission of a proton after bombarding the nitrogen with alpha particles from a radium source, illustrated by the following:

$$^{14}_{7}\text{N} + ^{4}_{2}\text{He} \rightarrow ^{17}_{8}\text{O} + ^{1}_{1}\text{H} \qquad\qquad (\text{VIII.6})$$

Alpha particles from radioactive sources were limited in energy. Rutherford was aware that particles including protons, the newly discovered deuteron, and even higher-energy alpha particles were needed at intensities and energies much higher than any particles that could be provided or produced from any natural sources were needed particularly to split large atoms, which would possess greater binding energies than nitrogen. Steere (2005) in his review of particle accelerators noted "In a speech before the Royal Society of London in 1927, Rutherford expressed publicly the desire of the scientific community to accelerate charged particles to energies greater than those of natural α-decay in order to disintegrate nuclei with higher binding energies than nitrogen" (Rutherford, 1929). The energies and intensities of these particles had to be controlled in the laboratory for thorough studies into the constitution of the atomic nucleus and possible nuclear transmutations into various elements. It was 1927 at the Cavendish and Rutherford was aware that there was a race on to be the first and at the cutting edge in this field. Gustaf Ising (1924) from Sweden had set the stage with his concept for particle acceleration and Rolf Wideröe (1928) from Norway would soon thereafter demonstrate the feasibility of Ising's principle of particle acceleration. Ernest Lawrence at Berkeley and Robert Van de Graaff at Princeton would in a few years produce high-energy protons with their newly developed cyclotron and electrostatic generator, respectively. Rutherford wanted the Cavendish to be the first to penetrate the atomic nucleus with accelerated particles, and an excellent historical account on "how a group of Cambridge scientists won the international race to split the atom" can be read in Cathcart's book (2004) *The Fly in the Cathedral*.

John Cockcroft began his research on the atomic nucleus under the directorship of Lord Rutherford at the Cavendish in 1928. That same year George Gamow, a Ukrainian born physicist was visiting the Cavendish Laboratory and he discussed the theory of tunneling where a relatively low-energy proton of only 100 keV, with its de Broglie wave property, had a high probability of penetrating the nucleus of boron and even greater probability of penetrating the smaller lithium nucleus (Gamow and Houtermans, 1929). Cockcroft discussed the theory with Rutherford in 1928, who agreed he begin work on the problem. Cockcroft was soon joined by Walton, who had already done some previous work on the development of particle accelerators.

They started their work by accelerating protons in a canal ray tube with voltages up to 280 kV and bombarded lithium, beryllium, and other targets and searched for gamma radiation from the targets using a gold-leaf electroscope detector (Cockcroft and Walton, 1930). They did not detect any gamma rays from their targets, and then decided to proceed to develop an apparatus that would accelerate protons to higher energies and to search for alpha particles from the disintegration of lithium. With this objective in mind they needed a system that could provide a steady high voltage potential, and they devised a system of rectifiers, which could provide a steady potential of 800 kV. Such high-voltage sources were not available for general laboratory use, but they devised a voltage multiplier circuit that would produce the needed high voltage by charging capacitors in parallel and release the charge in series (Cockcroft and Walton, 1932c; Walton, 1951).

With the capacity of achieving a voltage potential of 800 kV, Cockcroft and Walton successfully constructed an accelerator following tedious experimentation. The basic components of the accelerator are illustrated in Figure VIII.3. Details in the design and description of the apparatus can be found in Cockcroft and Walton (1932c). In his Nobel Lecture, Cockcroft (1951) commented

...the new apparatus was completed in early 1932. We were soon able to bring a narrow beam of 500 kilovolt protons out through a thin mica window in the base of the experimental tube, and to measure their range as a function of energy.

The first working accelerator constructed by Cockcroft and Walton, illustrated in Figure VIII.3, was constructed of two glass tubes, A, which measured 3 feet (91 cm) in height and 14 in. (35.5 cm)

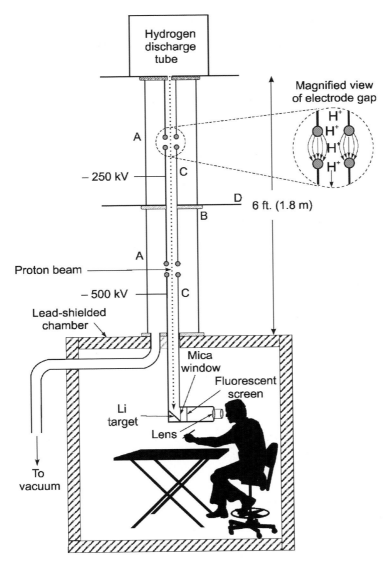

Figure VIII.3 Basic design (cross section) of Cockcroft and Walton accelerator of 1932. In a magnified view of a gap between electrodes are drawn electric field lines illustrating the acceleration of protons across the gap.

in diameter. Between the glass tubes was placed a steel plate, B, to form support for the electrodes, C. To the steel plate, B, was attached externally a 3 ft^2 (8 m^2) piece of sheet metal, D, which acted as a stress distributor and which was maintained at one-half of the potential applied to the electrodes. The electrodes, C, were made of thin-walled steel tubes to which the ends were fitted with thick rings made from ¾-in. (2 cm) diameter steel to prevent automatic electronic emission. Protons (H$^+$) were produced by electron-impact ionization of hydrogen in the hydrogen discharge tube at the top of the accelerator; and the protons were accelerated at two gaps between electrodes, C, held at increasing potential difference descending to the bottom of the accelerator. The protons traveled from top to bottom through the center of the electrodes to hit specific targets, such as lithium, illustrated in Figure VIII.3. A thin mica window was placed after and close to the target. The mica window would absorb any scattered protons of 600 kV energy and, at the same time, thin enough to enable alpha particles from the target to escape and then interact with a zinc sulfide fluorescent screen to produce scintillations and emission of visible light. An observer in a shielded chamber below the accelerator tower could count the scintillations of light produced by the alpha particles by observing these through the magnifying lens of a microscope.

Cockcroft and Walton resumed their experimentation by bombarding the accelerated protons against a round 5 cm diameter lithium target and searching for alpha particles as illustrated in Figure VIII.4a. In his Nobel Lecture, Cockcroft (1951) describes their experience as follows:

> Almost at once, at an energy of 125 kV, Dr. Walton saw the bright scintillations characteristic of α-particles, and a first primitive absorption experiment showed that they had a range of 8.4 cm. We then confirmed by a primitive coincidence experiment, carried out with two zinc sulfide (ZnS) screens and two observers tapping keys [Fig. VIII.4b], that the α-particles were emitted in pairs…More refined experiments showed that the energy of the α-particles was 8.6 million volts [MeV]. It was obvious then that lithium was being disintegrated into two α-particles with a total energy release of 17.2 million volts [MeV].

Their finding was reported in *Nature* (Cockcroft and Walton, 1932b) and in the *Proceedings of the Royal Society of London* (Cockcroft and Walton, 1932d). They achieved the first nuclear disintegration by purely artificial means.

The detection of two alpha particles of measured energy of 8.6 MeV each emitted in coincidence was sufficient evidence for Cockcroft and Walton (1932d) that a proton ($_1^1$H) penetrated the nucleus of lithium ($_3^7$Li) and the resulting nucleus would split into two alpha particles ($_2^4$He) according to the reaction

$$_3^7\text{Li} + _1^1\text{H} \rightarrow _2^4\text{He} + _2^4\text{He} + 17.2 \text{ MeV} \qquad \text{(VIII.7)}$$

In his Nobel Lecture, Cockcroft (1951) related that Kenneth Bainbridge (b.1904–d.1996) with whom he had established a friendship during Bainbridge's Guggenheim Fellowship at Rutherford's Cavendish Laboratory in 1933, had measured the mass of $_3^7$Li by mass spectrometry to be 7.0130 atomic mass units (u). This was the most precise measurement at the time for the mass of ^{7}Li.

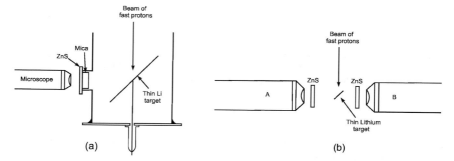

Figure VIII.4 Experimental arrangements of Cockcroft and Walton for (a) the detection of alpha particles by a single observer or (b) recording the emissions of alpha particles in coincidence by two observers. The alpha particles were observed and counted during the bombardment of lithium with a beam of protons in the proton energy range of 125–500 kV. (From Cockcroft, 1951 with permission from The Nobel Foundation © 1951.)

Cockcroft and Walton used the mass of the lithium nucleus, the known masses of the proton and alpha particle and the measured energies of the alpha particles to confirm Einstein's equation of mass–energy equivalence ($E = mc^2$). Walton (1951) calculated the mass balance for the above reaction (VIII.7) as follows:

Sum of masses of ^{7}Li and ^{1}H less the sum of masses of the two alpha particles gives a mass decrease of

$$
\begin{array}{rl}
^7\text{Li} & 7.0130\,\text{u} \\
^1\text{H} & 1.0072\,\text{u} \\
\hline
& 8.0202\,\text{u} \\
\text{Less 2 }^4\text{He} & 8.0022\,\text{u} \\
\hline
\text{Mass decrease} & 0.0180\,\text{u}
\end{array}
\tag{VIII.8}
$$

The mass decrease or difference in mass of the combined lithium and proton and the two alpha particles should be manifested in the reaction as energy, that is, the combined energies of the two alpha particles measured to be 17.2 MeV. The mass difference of 0.0180 u converted to energy according to $E = mc^2$ gives

$$
\begin{aligned}
E &= mc^2 \\
&= (0.0180\,\text{u})(931.494\,\text{MeV}/c^2)c^2 \\
&= 16.8\,\text{MeV},
\end{aligned}
\tag{VIII.9}
$$

since by definition 1 u = 931.494 MeV/c^2. The calculated energy of 16.8 MeV for the mass difference was very close to their measured combined alpha particle energy of 17.2 MeV. Thus, the energy lost in mass was equivalent to the energy emitted by the alpha particles.

In addition to splitting the lithium atom with protons Cockcroft and Walton (1932d) measured the minimum proton energy required to penetrate the lithium nucleus. They varied the energy of the proton beam by controlling the potential difference between the accelerator electrodes and determined the rates by which the alpha particles were emitted by a lithium target as a function of proton energy. This was accomplished by counting the number of alpha particles emitted for a given time period of proton bombardment and converting the count/time to the unit counts/minute, that is, alpha particle counts per minute. The first sign of alpha particles appeared at the proton-beam energy of approximately 100 kV. The alpha particles count rate increased exponentially as the proton-beam energy was increased to 500 kV. Their experiment demonstrated clearly that Gamow's prediction was correct, that the proton, being of relatively low mass would penetrate the small atomic nucleus of lithium at the low energy of approximately 100 kV.

Cockcroft and Walton (1933) proceeded to study the disintegration of other light elements by the accelerated protons. Their next target was boron. They observed that, in most interactions the proton would enter the boron nucleus in the proton-energy range of 115 kV and higher followed by its disintegration into three alpha particles, that is

$$
^{11}_{5}\text{B} + ^{1}_{1}\text{H} \rightarrow 3\,^{4}_{2}\text{He}
\tag{VIII.10}
$$

Additional evidence for reactions (VIII.7) and (VIII.10) for the disintegration of lithium and boron was provided by photographs of particle tracks produced in a Wilson Cloud Chamber. P. I. Dee was at the Cavendish Laboratory as a Stokes Student working under C. T. R. Wilson in research with Wilson's Cloud Chamber. Rutherford diverted Dee to work with Cockcroft and Walton. A second accelerator was constructed by Dee and Walton with a Wilson Cloud Chamber positioned directly below the proton beam. A shutter was constructed to absorb the beam when in the closed position and to open at the appropriate times that the cloud chamber would expand timed also with the shutter of a camera and illuminating lamp. Dee and Walton (1933) produced photographs of the tracks produced by two alpha particles when lithium was the target, and three alpha particle tracks were photographed when boron was the target.

Harold Urey had discovered heavy hydrogen (i.e., deuterium) ($^2_1$H) in 1931, an isotope of hydrogen with twice the mass of hydrogen. A small sample was made available to Cockcroft and Walton in 1933. They were then able to accelerate the deuterons, and study the effect of collisions with various targets. They observed the following reactions with accelerated deuterons with lithium and boron:

$$^6_3\text{Li} + {}^2_1\text{H} \rightarrow {}^4_2\text{He} + {}^4_2\text{He} \tag{VIII.11}$$

$$^6_3\text{Li} + {}^2_1\text{H} \rightarrow {}^7_3\text{Li} + {}^1_1\text{H} \tag{VIII.12}$$

$$^{10}_5\text{B} + {}^2_1\text{H} \rightarrow {}^{11}_5\text{B} + {}^1_1\text{H} \tag{VIII.13}$$

Reactions (VIII.12) and (VIII.13) demonstrated the transmutation of one isotope into another. Cloud chamber photographs of particle tracks provided additional evidence of their findings.

Proceeding on to targets of greater mass, Cockcroft, Gilbert, and Walton irradiated carbon in the form of graphite ($^{12}_6$C) with accelerated protons or deuterons. They looked for alpha particles as products of a nuclear reaction, but could not find any after bombarding a graphite target with either protons or deuterons. In early 1934 they heard of the discoveries of Joliot Curie and Joliot (1934a–c), who had produced radioactive positron-emitting elements by bombarding targets of aluminum, boron, and magnesium with alpha particles from polonium. They borrowed a Geiger counter from Kenneth Bainbridge and discovered that delayed positrons were emitted after graphite was bombarded with a beam of 400–500 kV protons. The same effect occurred when the graphite was irradiated with a beam of deuterons. The radioisotope produced had a measured half-life of 10 min corresponding to that of the radioisotope $^{13}_7$N (Cockcroft et al., 1934a,b), and the radiation emissions were demonstrated to be positrons by their direction of deflection in a magnetic field and their absorption properties characteristic of electrons. The reactions were therefore described to occur as follows:

$$^{12}_6\text{C} + {}^1_1\text{H} \rightarrow {}^{13}_7\text{N} \tag{VIII.14}$$

$$^{12}_6\text{C} + {}^2_1\text{H} \rightarrow {}^{13}_7\text{N} + {}^1_0\text{n} \tag{VIII.15}$$

The product ^{13}N decays by positron emission at a half-life ($t_{1/2}$) of 10 min as follows:

$$^{13}_7\text{N} \xrightarrow{t_{1/2}=10\,\text{min}} {}^{13}_6\text{C} + \beta^+ + \nu \tag{VIII.16}$$

Cockcroft and Walton had now demonstrated the transmutation of carbon into the radioactive isotope ^{13}N. The same year Joliot Curie and Joliot (1934a–c) had produced positron-emitting radionuclides from stable elements using alpha particles from polonium, as noted previously. These transmutations were not performed by totally artificial means, because the alpha particles were emitted naturally by polonium and consequently their energies are not controlled by man. The production of ^{13}N by Cockcroft, Gilbert, and Walton (Cockcroft et al., 1934a,b) was carried out by artificially produced protons or deuterons of controlled energies. All variables were entirely under human control. This fact was emphasized by Professor I. Waller, member of the Nobel Committee for Physics in Stockholm,

Sweden on the occasion of the presentation of the Nobel Prize to Cockcroft and Walton on December 11, 1951 when he stated

> The work of Cockcroft and Walton was a bold thrust forward into a new domain of research. Great difficulties had to be overcome before they were able to achieve their first successful experiments at the beginning of 1932. By then they had constructed an apparatus which, by multiplication and rectification of the voltage from a transformer, could produce a nearly constant voltage of about six hundred thousand volts. They had also constructed a discharge tube in which hydrogen nuclei were accelerated. Causing these particles to strike a lithium layer, Cockcroft and Walton observed that helium nuclei were emitted from the lithium. Their interpretation of this phenomenon was that a lithium nucleus into which a hydrogen nucleus has penetrated breaks up into two helium nuclei, which are emitted with high energy, in nearly opposite directions. This interpretation was fully confirmed. Thus, for the first time, a nuclear transmutation was produced by means entirely under human control

Professor Waller further underscored the significance of their work when he stated

> The analysis made by Cockcroft and Walton of the energy relations in a transmutation is of particular interest, because a verification was provided by this analysis for Einstein's law concerning the equivalence of mass and energy [$E = mc^2$]. Energy is liberated in the transmutation of lithium, because the total energy of the helium nuclei produced is greater than that of the original nuclei. According to Einstein's law, this gain in energy must be paid for by a corresponding loss in the mass of the atomic nuclei…In subsequent work, Cockcroft and Walton investigated the transmutations of many other atomic nuclei. Their techniques and results remain a model for nuclear research.

Cockcroft and Walton (1934) and Cockcroft and Lewis (1936a,b) studied other reactions resulting from the bombardment of the light elements including Li, Be, B, C, N, and O with protons and deuterons. The Cockcroft and Walton accelerator is still used today often as an ion injector for more sophisticated and powerful accelerators. For a review of the history and state-of-the-art of particle accelerators the reader is invited to peruse reviews by Bryant (1992), Kullander (2005), and Steere (2005) and a book by Wilson (2001).

After concluding his collaboration with Walton in 1934, Cockcroft became Head of the Royal Society Mond Laboratory in Cambridge. In 1939 he took up a war-time assignment as Assistant Director of Scientific Research in the Ministry of Supply and worked on the application of radar to coastal and air defense. In 1944, Cockcroft headed the Canadian Atomic Energy Program and directed the Montreal and Chalk River Laboratories. In 1946, he returned to England to become Director of the Atomic Energy Research Establishment at Harwell. During 1954–1959 Cockcroft was a member of the scientific research staff of the U.K. Atomic Energy Authority. He was elected Master of Churchill College, Cambridge in 1959. He was also Chancellor of the Australian National University, Canberra and during 1960–1962 President of the Institute of Physics of the Physical Society and the British Association for the Advancement of Science during 1961–1963. John Cockcroft died on September 18, 1967.

When Ernest Walton began his collaboration with Cockcroft on a Research Scholarship at the Cavendish in 1928 he was only 25 years of age. He received his Ph.D. 3 years later after completing the first publication on the preliminary development of their accelerator (Cockcroft and Walton, 1930). Before he reached his 29th birthday he had developed with Cockcroft the working accelerator (Cockcroft and Walton, 1932a) and split the atom of lithium with protons accelerated to energies of 125–400 keV (Cockcroft and Walton, 1932b). He was a Clerk Maxwell Scholar at the Cavendish Laboratory during his continued collaboration with Douglas Cockcroft from 1932 to 1934. Afterwards Ernest Walton returned as a Fellow to Trinity College, Dublin. In 1946 he was appointed Erasmus Smith's Professor of Natural and Experimental Philosophy at Trinity College and elected Senior Fellow there in 1960. During his remaining life Walton served on numerous scientific, governmental, and church committees in his home country of Ireland. Ernest Walton died in Belfast on June 25, 1995.

HANS A. BETHE (1906–2005)

Hans Albrecht Bethe was born on July 2, 1906 in Strasbourg, Alsace-Lorraine, which was then a part of Imperial Germany. He was awarded the Nobel Prize in Physics 1967 "for his contributions to the theory of nuclear reactions, especially the discoveries concerning the energy production of stars". Bethe wrote a brief autobiography at the time of his Nobel Prize Award, which was published in *Nobel Lectures, Physics 1963–1970*. Specific dates and places of his study and work were obtained from this source.

Hans Bethe (1906-1995)

After completing his studies in the Gymnasium in Frankfurt in 1924 Hans Bethe attended the University of Frankfurt for 2 years and completed his Ph.D. degree in theoretical physics at the University of Munich in July 1928 under the guidance of Professor Arnold Sommerfeld. After completing his doctorate degree Bethe taught at the Universities of Frankfurt and Stuttgart for only a semester at each institution. In the autumn of 1929, he transferred to the University of Munich to become Privatdozent in 1930 where he remained until the fall of 1933. The following winter semester of 1932–1933 Hans Bethe was appointed to the faculty of the University of Tübingen. However, not long after Hitler came to power in 1933 a law was passed that forbade appointments in civil service to anyone with Jewish ancestry. Bethe lost his faculty position, because his mother was Jewish. He personally did not consider himself Jewish. Although his mother was Jewish, she had converted to the Lutheran Church and his father was also Christian. At the time Hans Bethe's expectations to pursue physics were low. He thus emigrated to England in October of 1933 and was appointed to a temporary position of Lecturer at the University of Manchester during 1933–1934. A fellowship at the University of Bristol followed in the fall of 1934.

In February of 1935 Hans Bethe was offered an assistant professorship position at Cornell University in Ithaca, New York. He emigrated to America, and after a little more than two years Bethe was promoted to full Professor at Cornell in 1937. He remained a member of the physics faculty at

Cornell for the rest of his life, and would be absent only temporarily for sabbatical leaves and work in the Manhattan project during Second World War.

During the 1930s, Hans Bethe made many of his major contributions to nuclear physics. The first was a series of three lengthy papers in the journal *Reviews of Modern Physics* in 1936 and 1937 comprising almost 500 pages of a comprehensive review of experimental and theoretical nuclear physics, which he composed alone and with Robert F. Bacher and M. Stanley Livingston, who had collaborated previously with Ernest Lawrence in the development of the cyclotron during 1931–1934. The three papers (Bethe and Bacher, 1936; Bethe, 1937; Livingston and Bethe, 1937) were so comprehensive that together they became known as the Bethe Bible and 50 years later republished as a book by the American Institute of Physics (Bethe *et al.*, 1986). The significance of his work in theoretical physics was underscored by Professor O. Klein, member of the Swedish Academy of Sciences, when he stated at the presentation of the Nobel Prize to Hans Bethe on December 11, 1967

> Bethe belonged to the small group of young theoretical physicists who, through skill and knowledge, were particularly qualified for tackling the many theoretical problems turning up in close connection with the rapidly appearing experimental discoveries. The centre of these problems was to find the properties of the force that keeps the protons and neutrons together in the nucleus, the counterpart of the electric force that binds the atomic electrons to the nucleus. Bethe's contributions to the solution to these problems have been numerous and are still continuing... Moreover, about the middle of the thirties he wrote, partly alone, partly together with some colleagues, what nuclear physicists at the time used to call the Beth Bible, a penetrating review of about all that was known of atomic nuclei, experimental as well as theoretical.

In April of 1938, George Gamow convened a small conference of physicists and astrophysicists in Washington, D.C. sponsored by the Department of Terrestrial Magnetism of the Carnegie Institution. At the conference, the astrophysicists had the opportunity to discuss with the nuclear physicists all that they know about the internal elemental constitution and properties of the stars. The conference highlighted a key problem that the astrophysicists needed to solve, which was the origin of the energy of the sun. How could the sun continue for thousands of millennia during the evolution of the human race and before to emit light and heat without exhausting its source of energy? At one time it was thought that radioactivity might be the source of energy of the sun and even gravity forcing particles to reach high velocities and kinetic energies, but the energy yield from such sources would not be sufficient or long lasting and the lack of sufficient amount of radioactive substances in the sun's matter obviated radioactivity. Only months after the conference, taking into account the properties of the sun including the temperature, pressure, and elemental constitution in sun matter, did Hans Bethe provide evidence for nuclear fusion and nuclear reactions as the key processes that would provide the long-lasting energy of the sun. This work was summarized in a lengthy paper on "Energy Production in Stars" submitted to the journal *Physical Review* in September of 1938 and published in that journal in March of the following year (Bethe, 1939). Hans Bethe grasped the current facts from the astrophysicists presented at the conference and, in only 3 to 4 months provided the answers. The facts and reasoning used by Bethe were eloquently described in Professor Klein's Nobel Presentation Speech when he stated

> During that conference and afterwards he [Bethe] seems also to have acquired the necessary astrophysical knowledge. This knowledge depended mainly on a pioneer work by [Arthur S.] Eddington from the year 1926, according to which the innermost part of the sun is a hot gas mainly consisting of hydrogen and helium. Owing to the high temperature, about 20 million degrees,—these atoms being dissolved into electrons and nuclei—the mixture, despite the high density—about 80 times that of water—really behaves like a gas. The amount of energy generation necessary to maintain this state was known from measurements of the radiation falling on the earth. Taken as a whole it is enormous, but very slow as compared to the size of the sun...This very slow burning together with the very high energy release from a given weight of fuel gives this source the high durability required by geology and the long existence of life on the earth.

Hans Bethe explained that all nuclei in a normal star are positively charged, and for them to react the nuclei would need to penetrate each others Coulomb potential barrier. He demonstrated that for a given temperature and equal conditions for all nuclei the reactions that could occur most easily would be reactions between nuclei of the smallest value of W defined by the following:

$$W = AZ_0^2 Z_1^2 \qquad \text{(VIII.17)}$$

where

$$A = \frac{A_0 A_1}{(A_0 + A_1)} \qquad \text{(VIII.18)}$$

and A_0 and A_1 are the atomic weights of the two colliding nuclei and Z_0 and Z_1 are their atomic numbers or the charges on the nuclei. Thus, as explained by Bethe, the simplest of all possible reactions would be that between two proton nuclei or

$$_1^1\text{H} + _1^1\text{H} \rightarrow _1^2\text{H} + \beta^+ + \nu \qquad \text{(VIII.19)}$$

The reaction was first suggested by von Weizsäcker (1937) and calculated by Bethe and Critchfield (1938), who demonstrated that the reaction is relatively very slow, because it involves the beta disintegration, and that there would be no chance to observe such a slow reaction on earth. However, Bethe explains that the unlimited time and very large supply of protons of very high energy in the stars would yield a rate of energy production that would correspond to the observed energy production in the sun. The deuterons formed by the above reaction would quickly react with protons to yield ^{3}He according to the reaction

$$_1^2\text{H} + _1^1\text{H} \rightarrow _2^3\text{He} + \gamma \qquad \text{(VIII.20)}$$

which would, in turn with the buildup on ^{3}He and a temperature of 15 million degrees, coalesce to form ^{4}He according to the following

$$_2^3\text{He} + _2^3\text{He} \rightarrow _2^4\text{He} + _1^1\text{H} + _1^1\text{H} \qquad \text{(VIII.21)}$$

Bethe demonstrated that at higher temperatures the following reaction would compete with the above reaction (VIII.21):

$$_2^4\text{He} + _2^3\text{He} \rightarrow _4^7\text{Be} + \gamma \qquad \text{(VIII.22)}$$

Bethe (1968) showed that the ^{7}Be formed above could further react with a proton to form ^{8}B

$$_4^7\text{Be} + _1^1\text{H} \rightarrow _5^8\text{B} + \gamma \qquad \text{(VIII.23)}$$

which would yield two helium nuclei according to the reaction

$$_5^8\text{B} \rightarrow _2^4\text{He} + _2^4\text{He} + \beta^+ + \nu \qquad \text{(VIII.24)}$$

yielding neutrinos of very high energy. Ahmad et al. (2002) have detected solar neutrinos and found that the total number of neutrinos coming from the sun do match predictions of nuclear fusion in the sun.

In his Nobel Lecture given at Stockholm, Sweden on December 11, 1967 Hans Bethe provided arguments in favor of certain reactions as follows:

I examined (Bethe, 1939) the reactions between protons and other nuclei, going up in the periodic system. Reactions between ^{1}H and ^{4}He lead nowhere, there being no stable nucleus of mass 5. Reactions of ^{1}H with Li, Be, and B, as well as with deuterons, are all very fast at the central temperatures of the sun, but just this speed of the reaction rules them out...In fact, and just because of this reason, all of the elements mentioned from deuterium to boron, are extremely rare on earth and in the stars, and can therefore not be important sources of energy.

However, Bethe found that the next element in the periodic table, carbon, reacts very differently. Firstly, it is a relatively abundant element in the sun, and, in the gaseous plasma of stellar temperature, the nucleus of ^{12}C would undergo the following cycle of reactions, referred to as the carbon–nitrogen (CN) cycle:

$$^{12}_{6}\text{C} + {}^{1}_{1}\text{H} \rightarrow {}^{13}_{7}\text{N} + \gamma \tag{VIII.25}$$

$$^{13}_{7}\text{N} \rightarrow {}^{13}_{6}\text{C} + \beta^+ + \nu \tag{VIII.26}$$

$$^{13}_{6}\text{C} + {}^{1}_{1}\text{H} \rightarrow {}^{14}_{7}\text{N} + \gamma \tag{VIII.27}$$

$$^{14}_{7}\text{N} + {}^{1}_{1}\text{H} \rightarrow {}^{15}_{8}\text{O} + \gamma \tag{VIII.28}$$

$$^{15}_{8}\text{O} \rightarrow {}^{15}_{7}\text{N} + \beta^+ + \nu \tag{VIII.29}$$

$$^{15}_{7}\text{N} + {}^{1}_{1}\text{H} \rightarrow {}^{12}_{6}\text{C} + {}^{4}_{2}\text{He} \tag{VIII.30}$$

Bethe (1967) explained that reactions (VIII.25), (VIII.27), and (VIII.28) are radiative captures whereby the proton is captured by the nucleus and the energy of the reaction is emitted in the form of gamma radiation, which is converted into thermal radiation of the sun. The remaining reactions (VIII.26) and (VIII.29) are spontaneous beta-decays with lifetimes of only 10 and 3 min, respectively, which are negligible in comparison to stellar lifetimes. The last reaction (VIII.30) is a nuclear reaction resulting in two nuclei as products of the collision. It is the most interesting, because the reaction closes the CN cycle, as the ^{12}C nucleus is reproduced from the same nucleus from which the reaction started. Thus Bethe (1967) classified the ^{12}C as a catalyst of a reaction resulting in the production of one ^{4}He nucleus in which process two neutrinos are emitted that take away approximately 2 MeV energy together. The remaining energy, approximately 25 MeV per cycle, is released to keep the sun warm. Bethe demonstrated that at higher stellar temperatures, when the CN cycle prevails, there is a high probability for the reaction chain

$$^{16}_{8}\text{O} + {}^{1}_{1}\text{H} \rightarrow {}^{17}_{9}\text{F} + \gamma \tag{VIII.31}$$

$$^{17}_{9}\text{F} \rightarrow {}^{17}_{8}\text{O} + \beta^+ + \nu \tag{VIII.32}$$

$$^{17}_{8}\text{O} + {}^{1}_{1}\text{H} \rightarrow {}^{14}_{7}\text{N} + {}^{4}_{2}\text{He} \tag{VIII.33}$$

The above reaction chain is not cyclic, but Bethe indicated that it feeds into the CN cycle whereby the entire set of reactions is referred to as the CNO bi-cycle.

Bethe (1942) summarized his findings and the rationale that he used to investigate nuclear transmutations as the source of stellar energy. He pointed out the following:

> We know, however, from radioactive measurements [radioisotope dating of rocks] that the age of the earth is at least 1,500 million years. The sun is likely to be at least as old, and there is no reason to assume that its radiation 1,500 million years ago was much less than it is now. In fact, it is certain that the radiation did not change appreciably during the last 500 million years, because during all this time life existed on earth. [because life can tolerate only a narrow range of temperature].

Bethe explained that the solution to the problem of energy production in the sun was unlocked when Rutherford discovered in 1919 the transmutation of atomic nuclei. He noted that Atkinson and Houtermans (1929) first suggested the possibility of nuclear transmutations as the source of stellar energy. Bethe proceeded then to demonstrate that nuclear reactions, and specifically which nuclear reactions, would give sufficient energy and reaction rates to provide the energy supplied by the sun over hundreds of millions of years. By using Einstein's equation of equivalence of energy and mass, $E = mc^2$, Bethe (1942) gave an illustrative example of the energy released when a nucleus of oxygen, namely ^{16}O was created by the fusion of its constituent nucleons, that is, the fusion of 8 protons and 8 neutrons. Bethe showed that the mass of the ^{16}O nucleus was smaller than the sum of the masses of its 8 protons and 8 neutrons. He then provided the following explanation to the mass difference:

> Where did the remaining mass go to? Here the special relativity theory gives the answer: The smaller mass of ^{16}O indicates that this nucleus contains less energy than the 16 separate protons and neutrons. Energy is set free when the protons and neutrons combine to form a nucleus; in fact, that is the reason why a nucleus stays together at all. The amount of energy set free, the binding energy, can be calculated from the decrease in mass by m corresponding to a decrease of energy by mc^2 where c is the velocity of light.

Bethe (1942) calculated firstly the sum of the masses of the constituent protons and neutrons of ^{16}O. He then took the difference of the sum of the masses of the constituent nucleons and the mass of the lighter ^{16}O nucleus. From the mass difference he calculated the energy released as follows (The reader should note that the proton and neutron masses used by Bethe in his calculations were based on the most accurately measured values at the time. These differ slightly from current more accurately measured values.):

Mass of 8 neutrons	8.0714 u (atomic mass units)
Mass of 8 hydrogen atoms	8.0650
Total	16.1364
Minus Mass of ^{16}O atom	16.0000
Mass change	0.1364 u

Corresponding energy change = 0.000205 ergs = 205 microergs, which was based on the conversion factor of 0.0015 ergs of energy per atomic mass unit (u).

On the above calculation, Bethe (1942) commented the following:

> This energy may seem very small, but it must be remembered that it is for a single oxygen nucleus! If we could produce a *gram* of oxygen nuclei by bringing half a gram of protons and half a gram of neutrons together, the energy set free would be 7.7×10^{18} ergs, or 210,000 kilowatt-hours. The ordinary combustion of a gram of coal gives about 1/100 of a kilowatt-hour. Thus it is seen what tremendous energies are liberated when nuclei are formed from protons and neutrons: a very promising result when we are trying to explain the energy production in stars.

In the above calculation Bethe did not show details, for reasons of brevity, on how he arrived at the energy equivalence of 210,000 kWh. For the benefit of the reader, the writer provides the following calculations with all of the details:

From the mass difference of 0.1364 amu calculated above by Bethe, we can convert this to energy using Einstein's equation of equivalence of mass and energy as follows:

$$E = mc^2$$

$$= (0.1364\,u)(1.66 \times 10^{-27}\,\text{kg/u})(3 \times 10^8\,\text{m/sec})^2 \qquad \text{(VIII.34)}$$

$$= 2.04 \times 10^{-11}\,\text{J}$$

The energy in joules (J) converted to energy in ergs becomes

$$\frac{2.04 \times 10^{-11}\,\text{J}}{1 \times 10^{-7}\,\text{J/erg}} = 2.04 \times 10^{-4}\,\text{ergs} = 204\,\text{microergs}$$

The above calculations provide the energy released in the fusion of only 8 protons and 8 neutrons to form one nucleus of oxygen-16. To calculate the energy released by combining 0.5 g of protons and 0.5 g of neutrons to form 1.0 g of oxygen, we may first calculate the number of protons and neutrons contained in 0.5 g of each as follows:

$$1\,\text{proton} = 1.6726 \times 10^{-24}\,\text{g} \quad \text{or} \quad \frac{1\,\text{proton}}{1.6726 \times 10^{-24}\,\text{g}} = 5.9787 \times 10^{23}\,\text{protons/g}$$

$$1\,\text{neutron} = 1.6749 \times 10^{-24}\,\text{g} \quad \text{or} \quad \frac{1\,\text{neutron}}{1.6749 \times 10^{-24}\,\text{g}} = 5.9705 \times 10^{23}\,\text{neutrons/g}$$

The number of protons or neutrons per gram of each nucleon $\approx 6 \times 10^{23}$ and 0.5 g of either nucleon would number $\sim 3 \times 10^{23}$. Since $8p + 8n = 1\ ^{16}O = 1$ fusion reaction, we should divide the number of protons (p) or number of neutrons (n) contained in 0.5 g by a factor of 8 to arrive at the number of fusion reactions achieved with 0.5 g of p + 0.5 g of n as follows:

$$\frac{3 \times 10^{23}\,\text{protons/0.5 g}}{8\,\text{protons/fusion reaction}} = 3.75 \times 10^{22} \frac{\text{fusion reactions}}{0.5\,\text{g}\,\text{p} + \text{n}}$$

If one fusion reaction as calculated above yields 204 microergs of energy, then 3.75×10^{22} fusion reactions would yield

$$\left(\frac{204 \times 10^{-6}\,\text{ergs}}{\text{fusion reaction}} \right)(3.75 \times 10^{22}\,\text{fusion reactions}) = 7.7 \times 10^{18}\,\text{ergs}$$

The energy liberated in ergs is converted to energy units in joules (J) and kWh as follows:

$$(7.7 \times 10^{18}\,\text{ergs})(1 \times 10^{-7}\,\text{J/erg}) = 7.7 \times 10^{11}\,\text{J}$$

and

$$\frac{7.7 \times 10^{11}\,\text{J}}{3.6 \times 10^3\,\text{J/Wh}} = 2.1 \times 10^8\,\text{Wh} = 2.1 \times 10^5\,\text{kWh} = 210{,}000\,\text{kWh}$$

Bethe (1942) thus demonstrated that fusion and other nuclear reactions are immense sources of energy that the sun generates. He also proceeded to demonstrate the reaction rates would correspond to the long life of the sun based on laboratory data for nuclear reaction probabilities and on values for the temperature of the sun as calculated by astrophysicists on the basis of calculations of Eddington (1926). Taking the nuclear reactions of the CN cycle described by eqs. (VIII.25) to (VIII.30) as examples, Bethe calculated how long any nucleus at the center of the sun will live, on the average, before it is attacked by a proton and undergoes a nuclear transmutation. Thus for a temperature of $20,000,000\,°C$, Bethe found the average lifetime for the following nuclei to be

^{12}C	2,500,000 years
^{13}N	10 min (spontaneous decay)
^{13}C	50,000 years
^{14}N	4,000,000 years
^{15}O	2 min (spontaneous decay)
^{15}N	20 years

Thus, a complete CN cycle of the six reactions, whereby ^{12}C in the first reaction, acts as a catalyst for the formation of one ^4He nucleus consuming four protons (see reactions (VIII.25) to (VIII.30)), takes about 6,000,000 years. Bethe (1942) provided the following conclusions based on the reaction rates and nuclear fuel converted in the sun:

We have found that about 40 microergs are set free for each helium nucleus produced in the sun, or 10 for each proton destroyed. One gram of the sun's material has been calculated to contain about 2×10^{23} protons. Therefore if all the protons can be converted into helium, the available energy supply is 2×10^{18} ergs per gram. At present the sun radiates about 2 ergs per gram of its mass. At this rate the energy supply will last for about 30 billion years. It is evident that our nuclear reactions represent a sufficiently plentiful source of energy.

Hans Bethe was recruited by Oppenheimer to head the theoretical division of the secret nuclear weapons laboratory at the Los Alamos Nuclear Laboratories during World War II to work on the fission bomb. After the success of the Manhattan Project he returned to Cornell University. Bethe had decided to work on the bomb as he told his biographer "Rose [his wife] asked me to consider carefully whether I really wanted to continue to work on this. Finally, I decided to do it. The fission bomb had to be done, because the Germans were presumably doing it….The main reason was, I felt in this way I could make the greatest contribution to the war effort. After all, nuclear physics was my field" (Hall, 1995). Edward Teller tried to convince Bethe to later work on the hydrogen fusion bomb, but Bethe declined. Bethe became deeply involved with disarmament in the 1950s and later. He became a member of the US president's Science Advisory Committee in the 1950s and remained involved in advisory capacity nationally and internationally for decades later. Although not active in the Pugwash movement, he participated in the Pugwash meeting that took place at Stowe in the US in 1961; and he also shared the Albert Einstein Peace Award with Nobel Peace Laureate Joseph Rotblot in 1992 (Tucker, 2005).

Hans Bethe died at the age of 98 on March 8, 2005.

WILLARD F. LIBBY (1908–1980)

Willard Frank Libby was born in Grand Valley, Colorado on December 17, 1908. He was awarded the Nobel Prize in Chemistry 1960 "for his method to use carbon-14 for age determination in archeology, geology, geophysics, and other branches of science".

Libby got his primary and secondary education at schools near Sabastopol, California during 1913–1926. He entered the University of California at Berkeley in 1927 and was awarded the Ph.D. degree there in 1933. The same year Libby was appointed Instructor in the Department of Chemistry

Willard Libby (1908-1980)

at the University of California, Berkeley, and by 1943 he rose in the ranks at Berkeley to become Associate Professor of Chemistry. In 1941, he was awarded a Guggenheim Fellowship and chose to work at Princeton University. His fellowship was interrupted in December of that year to work on America's war effort at Columbia University in the Manhattan District Project until 1945. All this time Libby remained on leave from his post at the Department of Chemistry of the University of California, Berkeley.

After the World War, Libby was appointed in 1945 to the post of Professor of Chemistry at the Institute of Nuclear Studies of the University of Chicago. In 1954, President Eisenhower appointed him a Member of the U.S. Atomic Energy Commission in which capacity he headed Eisenhower's international "Atoms for Peace" program, which helped promote peaceful applications of nuclear energy in various fields of science including agriculture, medicine, hydrology, and industry. He resigned from this position in 1959 to take up a new post as professor of Chemistry at the University of California, Los Angeles and subsequently appointed as Director of the Institute of Geophysics and Planetary Physics in 1962. He died in Los Angeles on September 8, 1980 (Nobel Lectures, Chemistry, 1964).

In his Nobel Lecture given in Stockholm on December 12, 1960 and at an IAEA International Symposium on Radioactive Dating, Libby (1967) related the history of the carbon-14 dating technique. The sequence of events related here are taken in part from his historical account.

The origins of natural atoms of carbon-14, upon which Libby's dating technique is based, are neutrons produced in the earth's atmosphere. In 1939, Korff and Danforth developed a boron-trifluoride counter for the detection of neutrons. He and coworkers flew the counter on a balloon and discovered that neutrons were produced in the atmosphere by the collision of cosmic rays at the top of the atmosphere (Korff, 1940). The intensity of the neutrons was found to be two neutrons per square cm per second. In this report, Korff indicated that the absorption of neutrons by nitrogen in the atmosphere would likely make carbon-14 according to the following (n, p) reaction first demonstrated artificially in a Wilson Cloud Chamber by Kurie (1934), Bonner and Brubaker (1936a,b), and Burcham and Goldhaber (1936):

$$^{14}_{7}\text{N} + ^{1}_{0}\text{n} \rightarrow ^{14}_{6}\text{C} + ^{1}_{1}\text{H} \qquad \text{(VIII.35)}$$

The likely production of ^{14}C in the Earth's atmosphere was vital to the conception of the radiocarbon dating technique as stated by Libby (1960) in his Nobel lecture

Professor Korff noted in one of his papers announcing the discovery of neutrons [in the upper atmosphere] that the principal way in which the neutrons would disappear would be to form radiocarbon…knowing that there are about 2 neutrons formed per square centimeter per second, each of which forms a carbon-14 atom, and assuming that the cosmic rays have been bombarding the atmosphere for a very long time in terms of the lifetime of carbon-14 [half-life is 5730 years]…we can see that a steady-state condition should have been established, in which the rate of formation of carbon-14 would be equal to the rate at which it disappears to reform nitrogen-14 [Radiocarbon decays by the emission of a beta-particle to form stable nitrogen-14 according to Eq. VIII.36]. This allows us to calculate quantitatively how much carbon-14 would exist on earth [See Fig. VIII.5]

$$^{14}_{6}C \rightarrow {}^{14}_{7}N \text{ (stable)} + \beta^- + \bar{\nu} \tag{VIII.36}$$

Anderson and Libby estimated the expected level of natural radiocarbon activity on the earth from its expected rate of disintegration and the estimated amounts of carbon on earth (Table VIII.1). Their estimation of carbon on earth was obtained from a summation of the carbon content of the various carbon reservoirs on earth reported by Libby (1960) in his Nobel Lecture:

Table VIII.1

Carbon reservoir make-up

Reservoir	Carbon content (g/cm^2)
Ocean, "carbonate"	7.25
Ocean, dissolved organic	0.59
Biosphere	0.33
Humus	0.20
Atmosphere	0.12
Total	8.5

Source: From Libby (1960) with permission from The Nobel Foundation © 1960.

As illustrated in Figure VIII.5 neutrons are formed in the upper atmosphere at the rate of 2 neutrons per second per square centimeter forming 2 ^{14}C atoms at the same rate. The newly formed ^{14}C exchanges with carbon on the earth constituting 8.5 g per square centimeter. A steady-state condition is reached over a period of many thousands of years with the disintegration rate of the ^{14}C = 2 disintegrations per second (2 dps) equaling its rate of formation. Converting the units of disintegrations per second to disintegrations per minute (dpm) and dividing by the amount of exchangeable carbon on earth yields the estimated level of activity of ^{14}C on earth as calculated by Libby and coworkers as follows:

$$^{14}C \text{ Activity on Earth} \approx \frac{2 \text{ dps}}{8.5 \text{ g C}} \approx \frac{(2 \text{ dps})(60 \text{ m/sec})}{8.5 \text{ g C}} \approx 14 \text{ dpm/g C} \tag{VIII.37}$$

The production of carbon-14 in the atmosphere leading to the continuous labeling of all forms of carbon on earth, as illustrated in Figure VIII.5, was the basis for the conception of the carbon-14 dating technique as Libby (1967) explained

The real beginning of the history of radiocarbon dating as such was in the realization that the cosmic-ray production of radiocarbon in the high atmosphere leads to a continuous labeling of

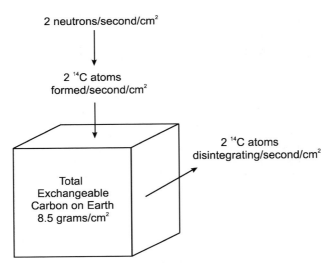

Figure VIII.5 Radiocarbon genesis and mixing. (From Libby, 1960 with permission from The Nobel Foundation © 1960.).

the biosphere and living matter, which is terminated at death. It is difficult to know when this idea was born, but it was very soon after the author's plan was fixed to detect the effects of cosmic rays on the Earth's atmosphere. It was necessary to know how a test could be made for the radiocarbon that must be present on Earth, according to Korff's measurements [Natural radiocarbon was yet undiscovered.]; in considering where to look for this natural radiocarbon, and how to detect it, it was realized that the supply of radiocarbon from the well-mixed system, consisting of the carbon dioxide in the air, and the dissolved salts in the ocean and in the biosphere would be cut off to any living being at the instant of death. For this, of course, the principle of radiocarbon dating was obvious—namely, that the decay of the radiocarbon normally present during life would measure the time elapsed since death—half of it being lost every 5730 years.

In his Nobel Lecture, Libby (1960) elaborated further the historical account of the conception of carbon-14 dating as follows:

...the agreement between the predicted and observed assays is encouraging evidence that the cosmic rays have indeed remained [relatively] constant in intensity over many thousands of years and that the mixing time, volume, and composition of the oceans have not changed either. We are in the radiocarbon dating business as soon as this has been said, for it is clear from the set of assumptions that have been given that organic matter, while it is alive, is in equilibrium with the cosmic radiation; that is, all the radiocarbon atoms which disintegrate in our bodies are replaced by the carbon-14 contained in the food we eat, so that while we are alive we are part of a great pool which contains the cosmic-ray produced radiocarbon [See Fig. VIII.6]. The specific activity is maintained at the level of about 14 disintegrations per minute per gram by the mixing action of the biosphere and hydrosphere. We assimilate cosmic-ray produced carbon-14 atoms at just the rate that the carbon-14 atoms in our bodies disappear to form nitrogen-14. At the time of death, however, the assimilation process stops abruptly. There is no longer any process by which the carbon-14 from the atmosphere can enter our bodies. Therefore, at the time of death the radioactive disintegration process takes over in an uncompensated manner and, according to the law of radioactive decay, after 5,600 years [The most precise figure now is

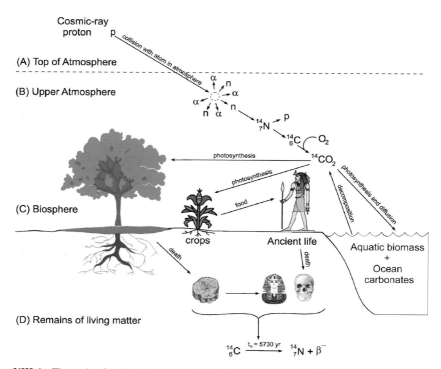

Figure VIII.6 The cycle of radiocarbon on earth. Cosmic-ray protons enter (A) the top of the atmosphere and collide with atoms in (B) the earth's upper atmosphere. The collisions smash the atoms into their various nuclear components including alpha-particles, protons, and neutrons. The neutrons are captured by nuclei of nitrogen and converted to nuclei of radioactive carbon-14. The newly created carbon-14 is immediately oxidized to radioactive $^{14}CO_2$, which together with stable carbon dioxide is absorbed into the earth's reservoirs of organic and inorganic carbon including all forms of living matter in its biosphere (C). At the moment of death the remains of living matter (D) are no longer able to assimilate new radioactive ^{14}C whereby the level of the radiocarbon begins to decay to lower levels of radioactivity losing one-half of its activity after 5730 years from the time of death. This would occur in all forms of former living matter such as human and animal remains, wood, artifacts made from wood such as the mask of the mummy illustrated in (D), and pottery containing embedded charcoal.

5730 years.] the carbon that is in our bodies while we are alive will show half the specific carbon-14 radioactivity that it shows now. Since we have evidence that this has been true for tens of thousands of years, we should expect to find that a body 5,600 [now known to be 5730] years old would be one-half as radioactive as a present-day living organism

The production of ^{14}C by cosmic radiation in the earth's atmosphere, its thorough mixing in the carbon pool on earth, and the decay of ^{14}C after the death of living systems is illustrated in Figure VIII.6.

Using the value of 14 dpm/g C as the estimated uniform specific activity of radiocarbon in the earth's carbon reservoirs Libby (1960) explained that the specific activity of the ^{14}C would diminish according to the radioactivity decay law given by eq. (VIII.38)

$$N = N_0 e^{-\lambda t}$$

(VIII.38)

where N is the number of ^{14}C atoms remaining after time t, N_0 the number of atoms of ^{14}C at the time of initiating observations or time $t = 0$, e the base to the natural logarithm, and λ the decay constant defined as $\ln 2/t_{1/2}$ where $t_{1/2}$ is the half-life of the radioactive isotope. A derivation of eq. (VIII.38) and some of its practical applications are provided in Chapter 8. The terms N and N_0 may be replaced by A and A_0 as given in eq. (VIII.39) for the activity or specific activity of the radioactive isotope at times t and t_0, respectively, because the number of atoms of a radioactive isotope in a given sample of material and the radioactivity of the isotope in the sample expressed in units of disintegration rate (e.g., dpm) are proportional.

$$A = A_0 e^{-\lambda t} \tag{VIII.39}$$

Thus, Libby and his coworkers based their conception of the ^{14}C-dating technique on the unalterable half-life of ^{14}C being 5730 years and the fact that assimilation of ^{14}C by a living organism would cease upon death. They reasoned that the level of ^{14}C in the sample would diminish with time after death according to the half-life of ^{14}C and the decay eq. (VIII.38) or (VIII.39). The decay of carbon-14 and the consequent reduction in its activity with time in the carbon remains of an organism (e.g., mummy) from its time of death is illustrated in Figure VIII.7.

The graph was constructed using eq. (VIII.39) substituting the value of 1 or unity for A_0 and then calculating for A, the fraction of ^{14}C activity remaining as a function of time t. For example, the fraction of the ^{14}C activity in the remains of a mummy 3000 years after the Pharaoh's death would be calculated according to eq. (VIII.39) as

$$
\begin{aligned}
A &= A_0 e^{-\lambda t} \\
&= (1)\left(e^{-(\ln 2/5730 \text{ years})(3000 \text{ years})}\right) \\
&= 0.696 \text{ or } 69.6\%
\end{aligned}
\tag{VIII.40}
$$

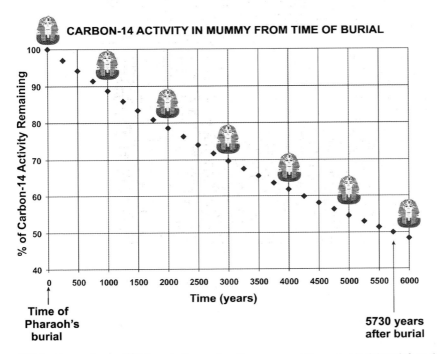

Figure VIII.7   Expected rate of ^{14}C decay in the remains of a mummy and its wooden burial mask from the time of the Pharaoh's death.

In turn, Libby reasoned that the radioactivity decay law would permit the calculation of t, the time in years from death or age of a mummy, if the specific activity of ^{14}C in the reservoirs of carbon on earth (A_0) could be determined accurately, and if the ^{14}C specific activity, A, in the carbon remains of the mummy could be experimentally determined. Thus, the time that had transpired from the time of death of the remains could be calculated from the following eq. (VIII.41), which is derived from eq. (VIII.40):

$$t = \frac{1}{\lambda} \ln \frac{A_0}{A} \qquad\qquad (VIII.41)$$

As it was confirmed that radiocarbon was produced in the earth's atmosphere, it was expected to exist in living material; however, Willard Libby and his collaborator E. C. Anderson had to demonstrate that it did indeed exist. The Chart of the Nuclides at that time had a vacant space where the isotope ^{14}C would be located. The radioactive measurement devices, such as gas ionization counters, were not very sensitive at that time, so they proceeded to concentrate the carbon-14 isotope in methane gas obtained from the sewage of the City of Baltimore with the assistance of A. V. Grosse of Temple University, who was developing thermal diffusion columns for the concentration of the carbon-13 isotope for medical applications (Anderson et al., 1947a,b). Methane from the city sewage was used, because it contained carbon from living organic matter whereby the radioactive carbon-14 would not have time to decay to any appreciable level. In contrast, petroleum methane would be absent of ^{14}C, because it would have all decayed in the oil deposits of the earth. Anderson, Libby, and coworkers (Anderson et al., 1947a,b) did indeed find carbon-14 in the methane of living organic matter at about the specific activity predicted.

The next step in the process was to develop an instrument, which could measure very low levels of radioactivity in a sample of carbon. Libby and coworkers could not use the isotope enrichment technique to routinely analyze samples for ^{14}C, because such methods were expensive and time consuming. An isotope dating technique would not be useful, if it would cost several thousands of dollars to analyze only one sample as would an isotope enrichment method. It was necessary to find a routine method of measuring the radioactivity of carbon-14 and from as small a sample as possible, which would permit the analysis of numerous antiquities without sacrificing a large sample of the antiquity for the analysis.

Gas ionization detectors, such as proportional counters or Geiger–Müller counters, were popular instruments during the 1940s and 1950s, and they remain today reliable instruments for the detection and measurement of radioactivity. In his Nobel Lecture (1960), Libby related that such detectors then yielded an unshielded background count rate of about 500 counts per minute (cpm). The background radioactivity was due to high-energy radiation from natural radionuclides of terrestrial origins such as uranium, thorium, and potassium in cement used in the building foundation and other structural materials as well as cosmic radiation. Shielding the detector with 8 in. of iron would absorb terrestrial sources of radioactivity and reduce the background count rate to about 100 cpm from principally cosmic-ray μ mesons. The cosmic-ray muons could penetrate readily the metal shielding, and consequently this background could not be eliminated by brute passive shielding. The background was yet too high to detect carbon-14 in antiquities that would contain less than 15 dpm/g of carbon. Libby and his coworkers needed to achieve a 50- to 100-fold reduction in background to provide a high signal-to-noise ratio of sample-to-background radioactivity particularly if old artifacts would produce carbon-14 count rates of only 1 or 2 counts per minute. They devised a counter assembly illustrated in Figure VIII.8 known as the anticoincidence counter described by Libby as follows:

The 100 remaining counts due in main part to μ mesons had to be removed. In order to do this we surrounded the counter with the carbon-dating sample in it with a complete layer of Geiger counters in tangential contact with one another and wire them so that when any one of these counters counts, the central counter with the dating sample is turned off for about 1/1000th of a second. In this way the μ mesons are eliminated from the record, so the background comes down to something between 1 and 6 counts per minute, depending on detailed counter and shield design. This is for a counter of about 1 liter volume capable of holding up to 5 grams of carbon with counting rates of 75 counts per minute for living carbon, 37.5 counts for 5,600 year-old carbon, 18.7 counts for 11,200 years, and 0.07 counts for 56,000 years.

Libby's statement above refers to the decay of radiocarbon in dead matter based on a half-life of 5600 years. The half-life of carbon-14 is now more accurately measured to be 5730 years. Samples were initially counted as carbon black deposited onto the inner detector wall with the detector filled with a suitable counting gas such as argon. Also, Libby converted samples to CO_2 or acetylene, which served as the counting gas, thereby, providing higher sample count rates. The beta particles emitted by the decaying carbon-14 atoms in the sample detector or cosmic-ray muons that may penetrate the guard and sample detectors produced ion pairs in the gaseous media of the detectors. The ion pairs become separated by an applied potential and collected at a central anode and outer cathode wall of the detector producing an output pulse, which is registered by the scaler and counted or rejected by the anticoincidence circuit illustrated in Figure VIII.8.

With a counter providing low background count rates, Libby and his graduate student Ernest Carl Anderson proceeded to analyze carbon-14 in living matter from various parts of the globe. The variable cosmic-ray intensity with latitude discussed in detail in Chapter 6 was a concern. It was estimated that about 8000 years would be required for the carbon-14 to be mixed in the atmosphere and ocean bodies, and thereby provide a relatively uniform carbon-14 specific activity around the globe. They needed to confirm that the known variation of cosmic-ray intensity with latitude and expected long time of mixing would not produce a wide variation in carbon-14 deposits in the various carbon reservoirs throughout the globe. With a counting method at hand, E. C. Anderson devoted much of his Ph.D. doctoral thesis on this problem (Anderson, 1949). He analyzed the ^{14}C specific activity (dpm/g C)

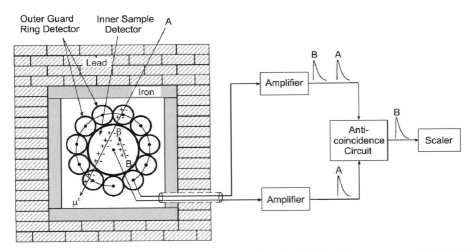

Figure VIII.8 Low-background anticoincidence counter and passive shielding. The counter assembled by Libby and coworkers consisted of 11 gas proportional or Geiger–Müller tubes surrounding a larger 1-l gas ionization sample detector located in the center. The outer tubes are coupled electronically so that any ionizing radiation that might penetrate one or more tubes within a time window of 0.001 sec would produce an output pulse. A cosmic-ray muon (A) is illustrated passing through the guard ring detector and central sample detector producing ionization in the guard ring and sample detectors and coincident output pulses from each detector. An atom of carbon-14 (B) inside the sample detector emits a beta particle, which cannot penetrate outside the central detector wall producing only one output pulse B. The coincident pulses produced by the cosmic-ray muon are rejected by the antico-incidence circuit, and only the pulse produced by the carbon-14 passes on to be counted by the scaler. The counts collected over a preset period of time are summed by the scaler. Sample counting times were as long as 48 h.

in 18 samples from almost as many countries from geomagnetic latitudes starting north from 60° north latitude to the equator and as far south as 65° south latitude. The carbon-14 specific activities of the samples were relatively uniform, varying only from 14.5 to 16.3 and providing an average value of 15.3 ± 0.1 dpm ^{14}C/g C. Libby (1960) commented on the ^{14}C specific activities found in the earth's reservoirs as follows:

> They [data on natural abundance or specific activity of radiocarbon in the earth] show no appreciable differences even though they come from places varying in latitude from near the South Pole to near the North Pole. At the present time, ten years later, no evidence for variation has been found except in areas of extensive carbonate deposits where the surface waters may carry a considerable amount of old carbon dissolved, and thus reduce the carbon-14 level below the world-wide average for the biosphere-atmosphere-ocean pool as a whole. Fortunately, such conditions are relatively rare and generally easily recognized.

Now equipped with the knowledge that living matter had a relatively uniform carbon-14 specific activity Willard Libby and his coworkers were able to plot a curve, similar to that illustrated in Figure VIII.9, which would predict the content of carbon-14 in ancient organic matter. Their dating curve was plotted using the universal radioisotope decay eq. (VIII.40) and the relatively uniform content of carbon-14 in living matter or 15.3 dpm/g C. Libby (1955) expressed his dating equation as follows (Libby, 1955):

$$I = I_0 e^{-\lambda t} \tag{VIII.42}$$

$$= 15.3 e^{-(\ln 2/5568 \text{ years})(t)} \tag{VIII.43}$$

where I is the content of carbon-14 in units of dpm/g C in any dead matter after time t in years, I_0 is the carbon-14 content in living matter, and λ is the decay constant for carbon-14 or the ln 2/half-life. In 1955, the half-life of carbon-14 was measured to be 5568 years. The more precise figure for the half-life of carbon-14 accepted today is 5730 years. Libby reasoned that, if the content of carbon-14 in ancient dead organic matter is determined experimentally in units of dpm/g C, then the age of the organic matter or time t that had transpired from the time of death would be calculated from eqs. (VIII.42) and (VIII.43) as

$$t = \frac{1}{\lambda} \ln \frac{I_0}{I}$$
$$t = \frac{1}{\ln 2/5568 \text{ years}} \ln \frac{15.3 \text{ dpm/g C}}{I} \tag{VIII.44}$$

For example, if the ^{14}C activity in a sample of a mummy was determined to be 8.2 dpm/g C, and using Libby's initial estimate of the specific activity of the earth's carbon reservoirs to be 15.3 dpm/g C, the age of the mummy would be calculated as

$$t = \frac{1}{\ln 2/5568 \text{ years}} \ln \frac{15.3 \text{ dpm/g C}}{8.2 \text{ dpm/g C}}$$

$$= \frac{1}{1.245 \times 10^{-4} \text{ years}^{-1}} \ln 1.86$$

$$= 4985 \text{ years}$$

With the accepted level of carbon-14 in living matter expressed as unity or given the value of 1, Libby (1955) calculated the fraction of carbon-14 that would be remaining in ancient dead matter after t years according to the equation

$$I = 1e^{-(0.693/5568 \text{ years})(t)} \qquad\qquad\qquad\text{(VIII.45)}$$

In the above equation, he used the value of 5568 years for the half-life of carbon-14, which was 2.8% lower than the current more precisely determined value of 5730 years. Eq. (VIII.45) was used by the writer to plot the carbon-14 decay with time in years with the more precise half-life value of 5730 years whereby the carbon-14 radioactivity content in a given sample would relate to the age in years of that sample from its time of death. An example of such a curve is plotted in Figure VIII.9.

With a decay curve such as the one illustrated in Figure VIII.9 Libby and coworkers had yet only to prove that antiquities of dead matter could be dated accurately. They needed to analyze antiquities of historically known age and compare the age determined by the carbon-14 dating method. Remains and artifacts of historically known age could be found only as far back as 5000 years. The youngest would be tree rings, remains from Pompeii, Dead Sea Scrolls, and remains from identified Egyptian tombs, etc. Their known ages correlated quite closely with ages determined by the carbon-14 dating method (see Figure VIII.9). Thus, Libby and coworkers had developed a technique that could be used

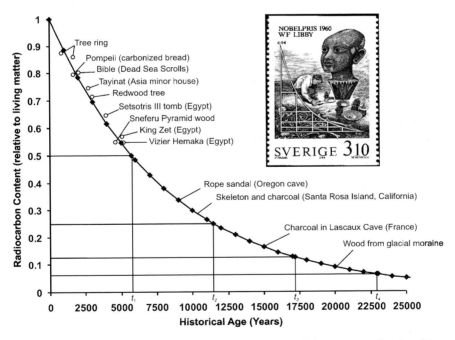

Figure VIII.9   The level of carbon-14 in dead organic matter relative to living matter as a function of the age of the organic matter from time of death. The curve is plotted according to the 5730 year half-life of carbon-14. The hollow circles along the curve from 1000 to 5000 years represent carbon samples of known age and measured carbon-14 content reported by Arnold and Libby (1949) on the basis of the then measured ^{14}C half-life of 5568 years. Samples identified on the curve from 9000 to 20,000 years have historical ages determined only by their carbon-14 content. Four carbon-14 half-lives are identified on the curve, namely, t_1, t_2, t_3, and t_4 at 5730, 11,460, 17,190, and 22,920 years, respectively. The insert is a copy of a postage stamp issued by Sweden in 1988 to commemorate the discovery of the carbon-14 dating technique, which illustrates an archeologist with mummy, pottery, and carbon-14 decay curve.

to determine the ages of remains containing dead carbon as far back as approximately nine half-lives of carbon-14 or approximately 50,000 years. Beyond 50,000 years of age, the level of carbon-14 would be too low and too close to background to determine accurately. Figure VIII.9 illustrates the ages of remains determined by their carbon-14 content up to 20,000 years reported by Libby (1960). In his Nobel Lecture given on December 12, 1960 Libby summarized the state-of-the art

> The dating technique itself is one which requires care, but which can be carried out by adequately trained personnel who are sufficiently serious-minded about it. It is something like, the discipline of surgery—cleanliness, care, seriousness, and practice. With these things it is possible to obtain radiocarbon dates which are consistent and which may indeed help role back the pages of history and reveal to mankind something more about his ancestors, and in this way perhaps about his future.

An important finding, which affects the carbon-14 dating technique, is the Suess Effect named after its discoverer (Suess, 1955). During the late 19th century and early 20th century the use of fossil fuels for industry and domestic heating injected relatively large amounts of CO_2 free of carbon-14 into the atmosphere. Fossils fuels do not contain any carbon-14, as fossil fuels such as coal and oil are much older than the lifetime of the radioisotope. The infusion of ^{14}C-depleted CO_2 into the atmosphere lowered the overall specific activity of radiocarbon in the atmosphere by about 2%, referred to as the Suess Effect. In the 1950s the Suess Effect was abated when atmospheric tests of nuclear weapons, which infused carbon-14 into the atmosphere from nuclear weapon thermal-neutron capture reactions with atoms of nitrogen. Thus, artificial carbon-14 was infused into the atmosphere by neutrons from nuclear weapons testing according to the same reaction that occurs naturally when thermal neutrons from cosmic-ray collisions in the upper atmosphere are captured by nitrogen (eq. (VIII.35)). This produced the opposite effect of increasing the carbon-14 level in the atmosphere. Thus, for carbon-14 dating a reference standard had to be made, which was uncontaminated with fossil-fuel CO_2 or the Suess Effect. The Primary Modern Reference Standard for radiocarbon measurements, as described by Cook et al. (2003), is wood from the AD 1890 tree ring, formed before the Suess Effect and consequently uncontaminated by fossil-fuel CO_2. The carbon-14 activity of the AD 1890 wood was measured in the 1950s and corrected for ^{14}C decay to the year of growth to yield the absolute carbon-14 activity found to be 226 Bq/kg C equivalent to 13.56 dpm/g C. Thus, radiocarbon ages are expressed in years BP (before present) where present is the year 1950. For example, AD 0 would be expressed as 1950 BP, and BC 3000 would be expressed as 4950 BP.

Laboratories involved in radiocarbon dating use a Secondary Modern Reference Standard, namely Oxalic Acid SRM 4990C provided by NIST, against which their sample activities are measured. The following equation described by Cook et al. (2003) is used in the calculation of radiocarbon age:

$$t = \frac{1}{\lambda} \ln \frac{A_{ON}}{A_{SN}}$$ (VIII.46)

where t is time or age, λ the decay constant $= \ln 2/5568$ years or $\ln 2$/Libby half-life of ^{14}C, A_{ON} is 74.59% of the measured net Oxalic Acid Secondary Standard SRM 4990C activity normalized for ^{13}C fractionation, and A_{SN} is the sample activity normalized for sample fractionation to $-25‰$. Notice the similarity of the above equation and that used by Libby described by eqs. (VIII.41) and (VIII.44). The activities are normalized for ^{13}C fractionation, because the isotopes of carbon (^{12}C, ^{13}C, and ^{14}C) undergo fractionation or fluctuation in their ratios as a result of isotope effects in natural biochemical processes, because of their mass differences. For in-depth treatments of carbon-14 aging measurements readers are invited to peruse articles by Stuiver and Polach (1977), Mook and van der Plicht (1999), Taylor (2000), and Currie (2004). State-of-the-art techniques for the measurement of ^{14}C are reviewed by Buchtela (2003), Huber et al. (2003), and Cook et al. (2003).

– 8 –

Radionuclide Decay, Mass, and Radioactivity Units

8.1   INTRODUCTION

The activity of a radioactive source or radionuclide sample is, by definition, its strength or intensity, or in other words, the number of nuclei decaying per unit time (e.g., dpm or dps). The activity of a given radionuclide in a sample is proportional to the number of radioactive atoms present in that sample. Consequently the activity of a radionuclide is also a measure of its mass. The activity decreases with time as a result of radionuclide decay. A unit of time in which there is an observable change in the activity of a given quantity of radionuclide may be very short, of the order of seconds, or very long, of the order of many years. The rate of decay of some nuclides is so slow that it is impossible to measure any change in radioactivity during our lifetime.

8.2   HALF-LIFE

Rates of radionuclide decay are usually expressed in terms of half-life. This is the time, t, required for a given amount of radionuclide to lose 50% of its activity. In other words, it is the time required for one-half of a certain number of nuclei to decay. The decay curve of ^{32}P (Figure 8.1) illustrates the concept of half-life. In Figure 8.1, the activity of the ^{32}P is plotted against time in days. It can be seen that, after every interval of 14.3 days, the radioactivity of the ^{32}P is reduced by half. Thus, the half-life, $t_{1/2}$, of ^{32}P is 14.3 days. It is not possible to predict when one particular atom of ^{32}P will decay; however, it is possible to predict statistically for a large number of ^{32}P radionuclides that one-half of the atoms would decay in 14.3 days.

The phenomenon of radioactivity decay and its measurement in terms of half-life was first observed by Ernest Rutherford (1900). He published his observations on the rate of decay of a certain isotope of thorium in the *Philosophical Magazine* in January 1900 less than 4 years after Henri Becquerel first reported the existence of radioactivity to the French Academy of Sciences in February and March 1896. Rutherford (1900) measured the rate of

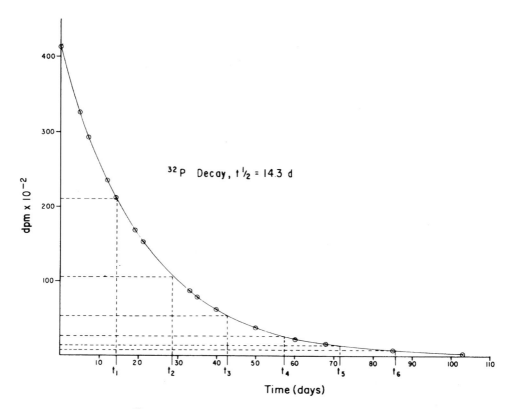

Figure 8.1 Decay of ^{32}P represented as a linear plot of activity in dpm against time in days. Horizontal and vertical lines between the ordinate and abscissa delineate ^{32}P activities (dpm) after six half-lives identified by the symbols $t_1, t_2, t_3,...t_6$. (From L'Annunziata, 1965, unpublished work.)

radioisotope decay by the level of ionization the radiation would produce in air as a function of time, and he noted the following:

> ...the intensity of the radiation has fallen to one-half its value after an interval of about one minute. The rate of leak [meaning rate of decay or rate of ionization caused by the radiation] due to the emanation was too small for measurement after an interval of 10 minutes (i.e., 10 half-lives]...The current [produced by the ionization] reaches half its value in about one minute—a result which agrees with the equation given, for $e^{-\lambda t} = 1/2$ when $t = 60$ seconds.

In his original paper Rutherford (1900) did not use the term "half-life"; but he defined mathematically the concept of half-life. As we shall see subsequently in eq. (8.6) that the ratio of the number of radionuclides, N, in a sample at a given time, t, to the original number of nuclides, N_0, in that sample at time $t = 0$, is equal to $e^{-\lambda t}$, where e is the base to the natural logarithm, λ a decay constant of that radionuclide, and t the interval of time.

When radionuclide decay can be recorded within a reasonable period of time, the half-life of a nuclide can be determined by means of a semilogarithmic plot of activity versus time (as shown in Figure 8.2). Radionuclide decay is a logarithmic relation, and the straight line obtained on the semilogarithmic plot permits a more accurate determination of the half-life.

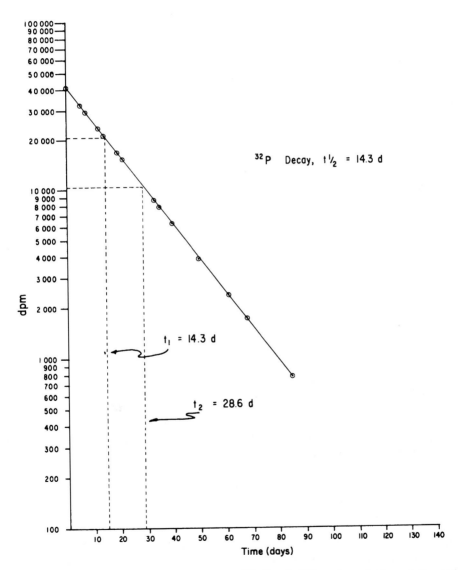

Figure 8.2 Semilogarithmic plot of the decay of ^{32}P. Two half-lives (t_1 and t_2) are delineated by horizontal and vertical lines between the ordinate and abscissa. (From L'Annunziata, 1965, unpublished work.)

Ernest Rutherford and Frederick Soddy (1902) in the *Philosophical Magazine* summarized their findings on radioactivity and radionuclide decay as follows:

> Turning from the experimental results to their theoretical interpretation, it is necessary to first consider the generally accepted view of the nature of radioactivity. It is well established that this property is the function of the atom and not of the molecule. Uranium and thorium, to take the most definite cases [examples], possess the property in whatever molecular condition they occur...So far as the radioactivity of different compounds of different density and states of division can be compared together, *the intensity of the radiation appears to depend only on the quantity of active element [radioisotope] present.*

The above statement by Rutherford and Soddy in 1902 that "the intensity of the radiation appears to depend only on the quantity of active element present" is defined subsequently in mathematical terms.

The number, ΔN, of atoms disintegrating in a given time, Δt, is proportional to the number, N, of radioactive atoms present. This relationship may be written as

$$\frac{\Delta N}{\Delta t} = \lambda N \tag{8.1}$$

or

$$\frac{dN}{dt} = -\lambda N \tag{8.2}$$

where λ is a proportionality constant, commonly referred to as the decay constant, and the negative sign signifies a decreasing number of radionuclides with time.

One condition must be fulfilled for eq. (8.1) to be rigorously applicable: the total number of radioactive atoms being considered must be large enough to make statistical methods valid. For example, in the case of a single isolated atom of ^{32}P there is no way to predict when the atom will decay. In fact, the atom might decay in the first second after $t = 0$ (the moment observations are initiated) or it might decay days or weeks later. The concept of half-life is a statistical one, which, when applied to a large number of atoms, as is usually the case, allows an accurate calculation of the activity of radionuclides after a given time interval.

For radionuclide decay calculations, eq. (8.2) must be transformed into a more suitable form and may be expressed as

$$\frac{dN}{N} = -\lambda \, dt \tag{8.3}$$

which can be integrated between the limits N_0 and N and between t_0 and t, where t_0 is 0 (the moment observations are initiated), N_0 the number of atoms originally present at time t_0, and N the number of atoms remaining after time t:

$$\int_{N_0}^{N} \frac{dN}{N} = -\lambda \int_{t_0}^{t} dt \tag{8.4}$$

to give

$$\ln \frac{N}{N_0} = -\lambda t \tag{8.5}$$

Eq. (8.5) may be written in exponential form as

$$N = N_0 e^{-\lambda t} \tag{8.6}$$

where e is the base of the natural logarithm, λ the decay constant, and t the interval of time. Eq. (8.6) is the form used to determine the decay of a radionuclide sample after a given time interval. To use eq. (8.6), the value of the decay constant λ, must be known, and this is different for each radionuclide. To determine λ for a particular radionuclide, a relationship between the decay constant and the half-life may be derived from the decay equation (eq. (8.5)), which may be transposed to

$$\ln \frac{N_0}{N} = \lambda t \tag{8.7}$$

By definition, we know that, after an interval of time corresponding to the half-life, half of the original activity remains. Therefore, we may assign the original activity N_0 as unity whereby after one half-life the remaining activity N would be one-half of unity, and eq. (8.7) would become

$$\ln \frac{1}{1/2} = \lambda t \tag{8.8}$$

or

$$\ln 2 = \lambda t_{1/2} \tag{8.9}$$

where $t_{\frac{1}{2}}$ is the half-life of the radionuclide in units of time, namely, seconds, minutes, hours, days, or years,
and

$$0.693 = \lambda t_{1/2} \tag{8.10}$$

The decay constant can then be defined as

$$\lambda = \frac{0.693}{t_{1/2}} \tag{8.11}$$

The value of λ can be calculated easily from the half-life of an isotope with eq. (8.11). The units used for λ are expressed in reciprocal time, $\sec^{-1}$, $\min^{-1}$, h^{-1}, $days^{-1}$, or $years^{-1}$, depending on the half-life of the radionuclide and also on the time interval t used in eq. (8.6). For example, if ^{32}P, which has a half-life of 14.3-days, is used in an experiment, λ may be expressed in $days^{-1}$. The unit of the decay constant must agree with the time interval t of eq. (8.6).

The following example illustrates the use of eq. (8.6) to calculate the decay of a radionuclide sample within any time interval.

If a sample contained 3.7 MBq of ^{32}P on a given date and an investigator wished to determine the amount remaining after a 30-day period, he or she would first determine the decay constant for ^{32}P according to eq. (8.11) and then calculate the activity after the specified time period using the decay eq. (8.6) as illustrated below. The decay constant in units of days^{-1} is determined by

$$\lambda = \frac{0.693}{t_{1/2}} = \frac{0.693}{14.3 \text{ days}} = 4.85 \times 10^{-2} \text{day}^{-1}$$

With the calculated value of λ and the time interval t equal to 30 days, the activity of the remaining ^{32}P is determined by eq. (8.6) expressed in terms of activity or

$$A = A_0 e^{-\lambda t} \tag{8.12}$$

and

$$A = 3.7 \times 10^6 \text{dps} \cdot e^{-[(4.85 \times 10^{-2} \text{day}^{-1})30 \text{ days}]}$$
$$= 3.7 \times 10^6 \text{dps} \cdot e^{-1.455}$$
$$= 3.7 \times 10^6 \text{dps} \cdot 0.2334$$
$$= 8.64 \times 10^5 \text{ dps} = 0.864 \text{ MBq}$$

where A is the activity of the radionuclide in units of decay rate (e.g., dps) after time t, and A_0 the initial activity at time t_0 or $A_0 = 3.7 \text{ MBq} = 3.7 \times 10^6 \text{dps}$ since by definition 1 MBq $= 1 \times 10^6 \text{dps}$. Thus, after the 30-day period, the activity of ^{32}P decayed from an initial activity of 3.7 to 0.864 MBq.

The decay equation has many practical applications, as it can be used as well to calculate the time required for a given radionuclide sample to decay to a certain level of activity. Let us consider the following example:

A patient was administered intravenously 600 MBq of [^{99m}Tc]-methylene diphosphate, which is a radiopharmaceutical administered for the purposes of carrying out a diagnostic bone scan. The doctor then wanted to know how much time would be required for the ^{99m}Tc radioactivity in the patient's body to be reduced to 0.6 MBq (0.1% of the original activity) from radionuclide decay alone ignoring any losses from bodily excretion. The half-life, $t_{1/2}$, of ^{99m}Tc is 6.0 h. To calculate the time required we can transpose eq. (8.12) to

$$\ln \frac{A_0}{A} = \lambda t \tag{8.13}$$

or

$$t = \frac{1}{\lambda} \ln \frac{A_0}{A} \tag{8.14}$$

By definition (eq. (8.11)) the decay constant λ of ^{99m}Tc is $0.693/t_{1/2}$ or $0.693/6$ h. Solving eq. (8.14) above after inserting the value of λ and the relevant activities of ^{99m}Tc gives

$$t = \left(\frac{6.00 \, \text{h}}{0.693} \right) \ln \left(\frac{600 \, \text{MBq}}{0.6 \, \text{MBq}} \right) = 59.8 \, \text{h} = 2.5 \, \text{days}$$

In the case of a mixture of independently decaying radionuclides, the rate of decay of each nuclide species does not change. However, the rate of decay of the overall sample is equal to the sum of the decay rates of the individual nuclide species. The cumulative decay of a mixture of independently decaying nuclides from the simplest case of a mixture of two nuclides to a more complex case of n number of nuclides is described by

$$N = N_1^0 e^{-\lambda_1 t} + N_2^0 e^{-\lambda_2 t} + \ldots N_n^0 e^{-\lambda_n t} \tag{8.15}$$

where N is the number of atoms remaining after time t, and N_1^0, N_2^0, and N_n^0 are the numbers of atoms originally present at time t_0 of 1, 2, and, n number of nuclide species, and λ_1, λ_2, and λ_n are the decay constants of 1, 2, and n number of nuclide species, respectively.

The semilogarithmic decay plot of a mixture of two independently decaying nuclides is not a straight line, contrary to pure radionuclide samples, but is a composite plot, as in the case of a mixture of ^{32}P and ^{45}Ca (see Figure 8.3). If the half-lives of the two nuclides are significantly different, the composite curve may be analyzed so that these may be determined. If the decay of the composite mixture can be observed over a reasonable period of time, the composite curve will eventually yield a straight line representing the decay of the longer-lived nuclide after the disappearance of the shorter-lived nuclide (depicted in Figure 8.3). This straight line may be extrapolated to time $t = 0$ so that the activity (dpm) of this nuclide at $t = 0$ can be found. The difference between the activity at $t = 0$ of the longer-lived nuclide and the total activity of the sample at $t = 0$ gives the activity at $t = 0$ of the shorter-lived nuclide. Likewise, further subtraction of points of the extrapolated decay curve from the composite curve yields the decay curve of the shorter-lived nuclide.

The half-lives of the two radionuclides are determined from the slopes of the two decay curves isolated from the composite curve. Eq. (8.7), which is expressed in natural logarithms, may be transformed to logarithms to the base 10 by

$$2.30 \log \left(\frac{N_1}{N_2} \right) = \lambda (t_2 - t_1) \tag{8.16}$$

or

$$\log \left(\frac{N_1}{N_2} \right) = \frac{\lambda}{2.30} (t_2 - t_1) \tag{8.17}$$

where N_1 and N_2 are the numbers of atoms or activity of the sample at times t_1 and t_2, respectively. As $\lambda/2.30$ of eq. (8.17) is equal to the slope of the straight-line decay curve,

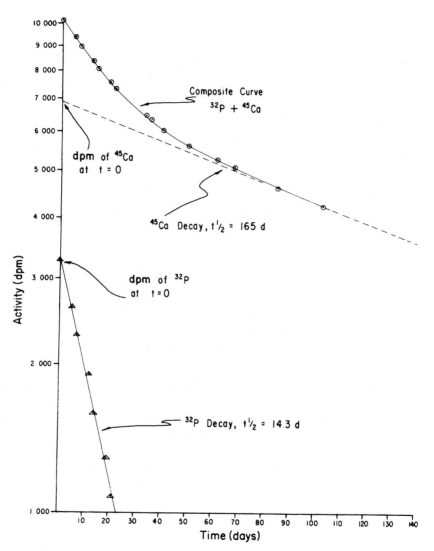

Figure 8.3   Semilogarithmic decay curves of ^{32}P and ^{45}Ca isolated from a composite decay curve of a mixture of ^{32}P + ^{45}Ca. (From L'Annunziata, 1965, unpublished work.)

the decay constant, λ, may be calculated from a graphical determination of the slope. With a calculated value of λ, the half-life of the nuclide is then calculated from eq. (8.11).

Many radionuclides have very long half-lives, which make the graphic determination of their decay impossible. Some examples are ^{3}H ($t_{1/2} = 12.3$ years), ^{14}C ($t_{1/2} = 5.73 \times 10^{3}$ years), ^{40}K ($t_{1/2} = 1.3 \times 10^{9}$ years), and ^{174}Hf ($t_{1/2} = 2 \times 10^{15}$ years). In such cases the half-lives can be calculated from eqs. (8.2) and (8.11). The decay rate or activity, A, in disintegrations per year (DPY) of a given nuclide sample, defined by dN/dt of eq. (8.2), is measured experimentally. The number of atoms of the radioassayed sample, defined by N of eq. (8.2), must be known or determined. This is simple for pure samples. For example, the number of atoms of ^{40}K in a pure sample of KCI is easily calculated from Avogadro's number (6.022×10^{23} molecules/mol) and the percentage natural abundance of ^{40}K (0.012%). Samples of unknown purity and isotopic abundance require a quantitative analysis of the element such as that provided by a mass spectral analysis of the isotopic abundance. The value of λ in year^{-1} is calculated as

$$\lambda = \frac{dN/dt}{N} = \frac{A}{N} = \frac{CPM/E}{N} (5.25 \times 10^5 \, m/year) \qquad (8.18)$$

where A is the sample nuclide activity in DPY, N the number of atoms of the nuclide in the sample, CPM the sample count rate (counts per minute) provided by the instrument radioactivity detector, E the instrument counting efficiency, and 5.25×10^5 min/year the factor used to convert CPM to counts per year (CPY). The half-life can then be calculated according to eqs. (8.11) and (8.18) both of which define the value of λ.

Let us look at a practical example of the use of the above equations to determine the half-life of ^{40}K taken from the work of Grau Malonda and Grau Carles (2002). The accurate determination of the half-life of ^{40}K has very practical implications, as it is currently used by geologists to date a rock's formation based on the measurement of the quantity of the stable daughter nuclide ^{40}Ar. Grau Malonda and Grau Carles (2002) report the accurate determination of the half-life of ^{40}K by measuring accurately the activity of ^{40}K in a sample of pure KNO$_3$ and applying the relationships of half-life to λ according to eqs. (8.11) and (8.18). They measured the ^{40}K specific activity in KNO$_3$ to be 12.24 ± 0.014 Bq/g using a very accurate CIEMAT/NIST efficiency tracing liquid scintillation radionuclide standardization method first conceived by Agustín Grau Malonda in 1978 and developed by A. Grau Malonda in collaboration with Bert M. Coursey as described in detail by Grau Malonda (1999) and L'Annunziata (2003a). Also, applying the known isotopic concentration of ^{40}K in KNO$_3$ of 0.01167% and the value of Avogadro's number 6.022×10^{23} atoms/mole, they could calculate the number of atoms of ^{40}K in 1 g of KNO$_3$ as follows: (6.022×10^{23} molecules/101.103 g KNO$_3$)(0.0001167) = 6.951×10^{17} atoms ^{40}K per gram of KNO$_3$. From eqs. (8.11) and (8.18) we can write

$$\frac{1}{\lambda} = \frac{t_{1/2}}{0.693} = \frac{N}{A} \qquad (8.19)$$

or

$$t_{1/2} = 0.693 \left(\frac{N}{A} \right) \qquad (8.20)$$

From the determined specific radioactivity of ^{40}K in KNO$_3$ and the number of atoms of ^{40}K per gram of KNO$_3$, Grau Malonda and Grau Carles (2002) calculated the half-life of ^{40}K as

$$t_{1/2} = 0.693 \left(\frac{6.951 \times 10^{17} \, atoms \, ^{40}K/g \, KNO_3}{(12.24 \, dps \, ^{40}K/g \, KNO_3)(60 \, sec/min)(5.25 \times 10^5 \, min/year)} \right)$$

and

$$t_{1/2} = 1.248 \times 10^9 \text{ years}$$

From the mean of nine determinations, Grau Malonda and Grau Carles (2002) were able to assign the value of the half-life ($t_{1/2}$) of ^{40}K to be $(1.248 \pm 0.004) \times 10^9$ years at a 95% confidence level.

Other radionuclides have very short half-lives such as ^{209}Ra ($t_{1/2} = 4.6\,\text{sec}$), ^{215}At ($t_{1/2} = 1.0 \times 10^{-4}\,\text{sec}$) and ^{212}Po ($t_{1/2} = 2.98 \times 10^{-7}\,\text{sec}$). The methods of determination of half-lives of such short duration can be determined by delayed coincidence methods (Schwarzschild, 1963; Ohm et al., 1990; Morozov et al., 1998), which involve the use of scintillation detectors with detector response times as short as 10^{-11} sec. These methods are applicable when a parent nuclide of normally perceptible or long half-life produces a daughter of very short half-life. Radiation detectors with resolving times of fractions of a microsecond are set electronically so that a delay circuit will detect a radiation-induced pulse from the parent in coincidence with a radiation pulse produced from the daughter. Varying the delay time of the coincidence circuit results in a delay of the coincidence pulse rate from which a decay curve of the very short-lived daughter nuclide can be plotted and the half-life determined.

8.3   GENERAL DECAY EQUATIONS

The simplest decay relationship between parent and daughter nuclides that can be considered is that of a parent nuclide which decays to form a stable daughter nuclide.

The decay of the radionuclide ^{33}P serves as an example. The parent nuclide ^{33}P decays with a half-life of 25 days with the production of the stable daughter ^{33}S, as indicated by

$$^{33}_{15}\text{P} \rightarrow {}^{33}_{16}\text{S(stable)} + \beta^- + \bar{\nu} \tag{8.21}$$

Numerous radionuclides, such as ^{3}H, ^{14}C, ^{32}P, ^{35}S, ^{36}Cl, ^{45}Ca, and ^{131}I, decay by this simple parent–daughter relationship.

However, numerous other radionuclides produce unstable daughter nuclides. The simplest case would be that in which the parent nuclide A decays to a daughter nuclide B, which in turn decays to a stable nuclide C:

$$\text{A} \rightarrow \text{B} \rightarrow \text{C(stable)} \tag{8.22}$$

In such decay chains, the rate of decay and production of the daughter must be considered as well as the rate of decay of the parent. The decay of the parent is described by the simple rate equation

$$\frac{-dN_\text{A}}{dt} = \lambda_\text{A} N_\text{A} \tag{8.23}$$

which is integrated to the form

$$N_A = N_A^0 e^{-\lambda_A t} \tag{8.24}$$

where N_A^0 is the number of atoms of the parent at the time $t = 0$ and N_A the number of atoms after a given period of time $t = t_1$.

The decay rate of the daughter is dependent on its own decay rate as well as the rate at which it is formed by the parent. It is written as

$$-\frac{dN_B}{dt} = \lambda_B N_B - \lambda_A N_A \tag{8.25}$$

where $\lambda_B N_B$ is the rate of decay of the daughter alone and $\lambda_A N_A$ is the rate of decay of the parent or rate of formation of the daughter. Eqs. (8.24) and (8.25) may be transposed into the linear differential equation

$$\frac{dN_B}{dt} + \lambda_B N_B - \lambda_A N_A^0 e^{-\lambda_A t} = 0 \tag{8.26}$$

which is solved for the number of atoms of daughter, N_B, as a function of time to give

$$N_B = \frac{\lambda_A}{\lambda_B - \lambda_A} N_A^0 (e^{-\lambda_A t} - e^{-\lambda_B t}) + N_B^0 e^{-\lambda_B t} \tag{8.27}$$

Although unnecessary in this treatment, the solution to eq. (8.26) is given by Friedlander *et al.* (1964).

In decay schemes of this type, the following three conditions may predominate: (1) secular equilibrium, (2) transient equilibrium, and (3) the state of no equilibrium. Each of these cases will now be considered in detail.

8.4   SECULAR EQUILIBRIUM

The phenomenon of radioisotope decay equilibrium was first observed by Ernest Rutherford and Frederick Soddy in 1902, which they reported in their classic paper on *The Cause and Nature of Radioactivity*. They reported their observations as follows:

Radioactivity is shown to be accompanied by chemical changes in which new types of matter are being continuously produced. These reaction products are at first radioactive, the activity diminishing regularly from the moment of formation. Their continuous production maintains the radioactivity of the matter producing them at a definite equilibrium value.

Secular equilibrium is a steady-state condition of equal activities between a long-lived parent radionuclide and its short-lived daughter. The important criteria upon which secular equilibrium depends are:

1. The parent must be long-lived; that is, negligible decay of the parent occurs during the period of observation, and
2. The daughter must have a relatively short half-life. The relative difference in half–life in this latter criterion is further clarified by

$$\frac{\lambda_A}{\lambda_B} \leq {\sim}10^{-4} \tag{8.28}$$

that is,

$$\lambda_A \ll \lambda_B \tag{8.29}$$

where λ_A and λ_B are the respective decay constants of the parent and daughter nuclides. The importance of these two requirements can be clearly seen if the ^{90}Sr(^{90}Y) equilibrium is taken as an example.

The nuclide ^{90}Sr is the parent in the decay scheme

$$^{90}_{38}\text{Sr} \xrightarrow{\ t_{1/2}\,=\,28.8\,\text{years}\ } \,^{90}_{39}\text{Y} \xrightarrow{\ t_{1/2}\,=\,2.7\,\text{days}\ } \,^{90}_{40}\text{Zr (stable)} \tag{8.30}$$

The long half-life of ^{90}Sr definitely satisfies the first requirement for secular equilibrium, because over a quarter of a century is needed for it to lose 50% of its original activity. As will be seen, less than 3 weeks are required for secular equilibrium to be attained and, in this interim period, negligible decay of ^{90}Sr occurs.

To satisfy the second requirement the decay constants for ^{90}Sr and ^{90}Y, λ_A and λ_B, respectively, must be compared. The decay constants for ^{90}Sr and ^{90}Y are easily calculated from their half-lives and eq. (8.11), and the values are 6.60×10^{-5} and 2.57×10^{-1} day^{-1}, respectively. Consequently, in the comparison $\lambda_A/\lambda_B = 2.57 \times 10^{-4}$, and this is in agreement with the order of magnitude required for secular equilibrium.

An equation for the growth of daughter atoms from the parent can be obtained from eq. (8.27) by consideration of the limiting requirements for secular equilibrium. Since $\lambda_A \approx 0$ and $\lambda_A \ll \lambda_B$, $e^{-\lambda_A t} = 1$ and λ_A falls out of the denominator in the first term. If the daughter nuclide is separated physically from the parent (L'Annunziata, 1971), $N_B^0 = 0$ at time $t = 0$ (time of parent–daughter separation) and the last term would fall out of eq. (8.27). Thus, in the case of secular equilibrium, the expression of the ingrowth of daughter atoms with parent can be written as

$$N_B = \frac{\lambda_A N_A^0}{\lambda_B}(1 - e^{-\lambda_B t}) \tag{8.31}$$

If the observation of the ingrowth of the daughter is made over many half-lives of the daughter, it is seen that the number of atoms of daughter approaches a maximum value

$\lambda_A N_A^0 / \lambda_B$, which is the rate of production of daughter divided by its decay constant. The final form of eq. (8.31) to be used for the calculation of the ingrowth of daughter can be expressed as

$$N_B = (N_B)_{max}(1 - e^{-\lambda_B t}) \tag{8.32}$$

Since the activity of the daughter atoms, A_B, is proportional to the number of daughter atoms, or $A_B = k\lambda_B N_B$, where k is the coefficient of detection of the daughter atoms, eq. (8.32) may also be written as

$$A_B = (A_B)_{max}(1 - e^{-\lambda_B t}) \tag{8.33}$$

Rutherford and Soddy (1902) were the first to write and interpret eqs. (8.32) and (8.33) when they studied the equilibrium existing between radioactive thorium and a daughter radionuclide. They noted the following:

The radioactivity of thorium at any time is the resultant of two opposing processes: 1. The production of fresh radioactive material at a constant rate by the thorium compound, and 2. The decay of the radiating power of the active material with time. The normal or constant radioactivity possessed by thorium is an equilibrium value, where the rate of increase of radioactivity due to the production of fresh active material [daughter nuclide] is balanced by the rate of decay of radioactivity of that already formed...The experimental curve obtained with the hydroxide [this was the chemical form they used to separate the parent nuclide from the daughter] for the rate of rise of its activity from a minimum to a maximum value will therefore be approximately expressed by the equation $I_t/I_0 = 1 - e^{-\lambda t}$, where I_0 represents the amount of activity recovered when the maximum is reached, and I_t the activity recovered after time t, λ being the same constant as before.

(Notice the similarity of Rutherford's and Soddy's equations, which may be transposed to read $I_t = (I_0)(1 - e^{-\lambda t})$ and eqs. (8.32) and (8.33)).

Let us take an arbitrary example of equal activities of 100 dpm of parent ^{90}Sr and 100 dpm of daughter ^{90}Y in secular equilibrium and calculate and graphically represent the ingrowth of ^{90}Y with its parent and also the decay of ^{90}Y subsequent to the separation of parent and daughter nuclides (L'Annunziata, 1971). Identical activities of ^{90}Sr and ^{90}Y are arbitrarily chosen, because their activities are equal while in secular equilibrium prior to their separation. Figure 8.4 illustrates the calculated growth of ^{90}Y (curve B) as produced by ^{90}Sr using eq. (8.33) with $(A_B)_{max} = 100$. The decay of separated ^{90}Y (curve A) is plotted by simple half-life decay ($t_{1/2} = 2.7$ days). The dashed line (line C) represents the decay of ^{90}Sr, which is negligible during the period of observation ($t_{1/2} = 28.8$ years). The total activity (curve D) is the result of both ^{90}Sr decay and the ingrowth of ^{90}Y after the separation of the latter and is obtained by the addition of curve B to line C. It may be noted from Figure 8.4 that after approximately six half-lives of ^{90}Y ($\sim$18 days) the growth of ^{90}Y has reached the activity of ^{90}Sr, after which both nuclides decay with the same half-life, that of the parent ^{90}Sr (28.8 years).

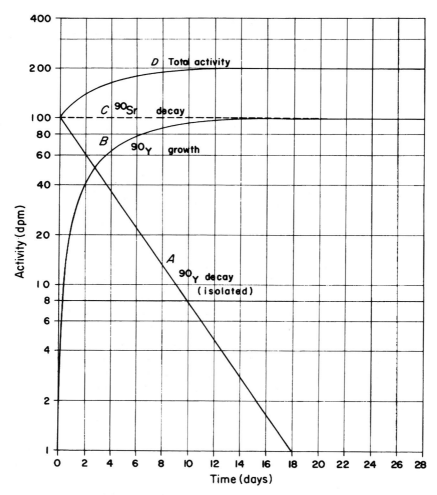

Figure 8.4 Growth and decay curves following the separation of ^{90}Sr (^{90}Y) in secular equilibrium. (A) Decay of isolated ^{90}Y. (B) Ingrowth of ^{90}Y with ^{90}Sr. (C) Decay of isolated ^{90}Sr. (D) Total activity from isolated ^{90}Sr, representing both ^{90}Sr decay and ^{90}Y growth until secular equilibrium is attained. (From L'Annunziata (1971) reprinted with permission © 1971, Division of Chemical Education, Inc., American Chemical Society.)

Ernest Rutherford and Frederick Soddy in their classic paper on *The Cause and Nature of Radioactivity* in 1902 in the *Philosophical Magazine* made the first published observation of what is now known as secular equilibrium. They were studying the radioactivity of an isotope of thorium and a daughter isotope, which they called ThX. The term "daughter isotope" for an isotope decay product was not then established, because knowledge of radioactivity was only then at its infancy. They summarized their findings with the following:

The foregoing experimental results may be briefly summarized. The major part of the radioactivity of thorium—ordinarily about 54 percent—is due to a non-thorium type of matter, ThX, possessing distinct chemical properties, which is temporarily radioactive, its activity falling to half value in about four days. The constant radioactivity of thorium is maintained by the production of this material at a constant rate. Both the rate of production of the new material and the rate of decay of its activity appear to be independent of the physical and chemical condition of the system.

(We now know that the activity of parent and daughter nuclides are equal in secular equilibrium and the value of "about 54%" reported by Rutherford and Soddy was precisely 50%.)

As an example of the practical utility of this phenomenon, the application of secular equilibrium theory to the analysis of ^{90}Sr in biological systems is presented here.

One method reported by the Los Alamos National Laboratory (see Gautier, 1995) entails the initial chelation (complex formation) of the sample strontium with the sodium salt of ethylenediaminetetraacetic acid (EDTA). The complexed strontium is then isolated by elution on an ion-exchange column. The eluted strontium is then precipitated as a carbonate. The activity of radioactive strontium, which will include ^{89}Sr + ^{90}Sr in the sample, is determined by low-background counting. Low-background liquid scintillation counting is most often used for the total ^{89}Sr + ^{90}Sr analysis as described by Passo and Cook (1994) and Cook et al. (2003). The isolated radiostrontium is then allowed to remain in the sample without further treatment for a period of about 2 weeks to allow ingrowth of ^{90}Y. About 2 weeks are needed to ensure that the parent and daughter radionuclides are in secular equilibrium before the chemical separation of yttrium from strontium. From eq. (8.33) it is calculated that after 2 weeks the activity of ^{90}Y grows to 97.4% of its original level. Carrier yttrium is then added to the dissolved radiostrontium, and the yttrium is precipitated as the hydroxide, redissolved, and reprecipitated as an oxalate. The step involving the precipitation of yttrium from the sample results in the separation of ^{90}Y from the radiostrontium. The separated ^{90}Y can then be assayed by suitable low-background counting using liquid scintillation or Cherenkov counting (Passo and Cook, 1994; L'Annunziata and Passo, 2002). The ^{90}Sr activity in the sample is determined from the activity of ^{90}Y by calculating the ^{90}Y decay from the time of separation (precipitation) of yttrium from strontium. This is possible because the parent and daughter radionuclides were at secular equilibrium (i.e., ^{90}Sr dpm = ^{90}Y dpm) at time $t = t_0$ when the precipitation and separation of yttrium from strontium were carried out. The ^{89}Sr activity in the sample is determined from the difference between the total radiostrontium activity (^{89}Sr + ^{90}Sr) and the measured activity of ^{90}Sr.

Certain chemical processes in natural and biological systems can preferentially select either the parent or daughter nuclide and, in this manner, separate the two. For example, a research investigator could administer nuclides in secular equilibrium to a soil and plant system. At the time of administration, the nuclides are in secular equilibrium; that is, both the parent and daughter activities are equal. However, if in the course of the experiment the investigator obtains a plant sample for radioassay, which had preferably absorbed either the parent or daughter radionuclide, problems ensue if the equilibrium phenomenon is not considered. Radioassay of plant tissue that had selectively concentrated the parent could show an initial progressive rise in radioactivity due to ingrowth of daughter, whereas a selective concentration of daughter would result in a sample showing an initial decrease in radioactivity. In cases such as these, it is necessary to isolate the parent radionuclide chemically and wait for a period of time sufficient to permit secular equilibrium to be reached ($\sim$2 weeks for the ^{90}Sr(^{90}Y) example) before analyzing the sample.

8.5 TRANSIENT EQUILIBRIUM

Like secular equilibrium, transient equilibrium is a steady-state condition between the parent and daughter nuclides. However, in transient equilibrium the parent–daughter nuclides do not possess the same activities, but rather they decay at the same half-life, that of the parent nuclide.

The criterion upon which transient equilibrium rests is that the parent nuclide must be longer lived than its daughter, but not of the order of magnitude described by eq. (8.28); that is, it is necessary that $\lambda_A < \lambda_B$. However, the ratio λ_A/λ_B should fall within the limits $10^{-4} < \lambda_A/\lambda_B < 1$.

The decay chain of ^{100}Pd serves as an example of parent–daughter nuclides that may attain transient equilibrium. ^{100}Pd decays by electron capture to ^{100}Rh with a half-life of 96 h. The daughter nuclide ^{100}Rh decays by electron capture and positron emission to the stable nuclide ^{100}Ru. The half-life of the daughter nuclide is 21 h. The decay scheme may be represented as

$$^{100}_{46}\text{Pd} \xrightarrow{t_{1/2}\,=\,96\,\text{h}} {}^{100}_{45}\text{Rh} \xrightarrow{t_{1/2}\,=\,21\,\text{h}} {}^{100}_{44}\text{Ru (stable)} \tag{8.34}$$

The first criterion for transient equilibrium is satisfied in this case; the half-life of the parent nuclide is greater than that of the daughter. If the decay constants λ_A and λ_B are now calculated, we can determine whether or not the second criterion ($10^{-4} < \lambda_A/\lambda_B < 1$) is satisfied.

The value of λ_A, given by 0.693/96 h, is $7.2 \times 10^{-3}\text{h}^{-1}$, and that of λ_B, given by 0.693/21 h, is $3.3 \times 10^{-2}\text{h}^{-1}$. Consequently, the ratio $\lambda_A/\lambda_B = 2.2 \times 10^{-1}$ and lies within the limits of the second criterion.

If the general decay eq. (8.27) of the daughter nuclide is considered, the term $e^{-\lambda_B t}$ is negligible compared with $e^{-\lambda_A t}$ for sufficiently large values of t. Thus, the terms $e^{-\lambda_B t}$ and $N_B^0 e^{-\lambda_B t}$ may be dropped from eq. (8.27) to give

$$N_B = \frac{\lambda_A}{\lambda_B - \lambda_A}(N_A^0 e^{-\lambda_A t}) \tag{8.35}$$

for the decay of the daughter nuclide as a function of time. Because $N_A = N_A^0 e^{-\lambda_A t}$, eq. (8.35) may be written as

$$\frac{N_B}{N_A} = \frac{\lambda_A}{\lambda_B - \lambda_A} \tag{8.36}$$

From eq. (8.36), it can be seen that the ratio of the number of atoms or the ratio of the activities of the parent and daughter nuclides is a constant in the case of transient equilibrium.

Since $A_A = k_A \lambda_A N_A$ and $A_B = k_B \lambda_B N_B$, where A_A and A_B are the activities of the parent and daughter nuclides, respectively, and k_A and k_B are the detection coefficients of these nuclides, eq. (8.36) may be written in terms of activities as

$$\frac{A_B}{k_B \lambda_B}(\lambda_B - \lambda_A) = \frac{A_A}{k_A \lambda_A} \lambda_A \qquad (8.37)$$

or

$$\frac{A_B}{A_A} = \frac{k_B \lambda_B}{k_A(\lambda_B - \lambda_A)} \qquad (8.38)$$

If equal detection coefficients are assumed for the parent and daughter nuclides, eq. (8.38) may be written as

$$\frac{A_B}{A_A} = \frac{\lambda_B}{(\lambda_B - \lambda_A)} \qquad (8.39)$$

Thus, for transient equilibrium eq. (8.39) indicates that the activity of the daughter is always greater than that of the parent by the factor $\lambda_B/(\lambda_B - \lambda_A)$. Eq. (8.39) may likewise be written as

$$\frac{A_A}{A_B} = 1 - \frac{\lambda_A}{\lambda_B} \qquad (8.40)$$

whereby the ratio A_A/A_B falls within the limits $0 < A_A/A_B < 1$ in transient equilibrium.

If an activity of 100 dpm is arbitrarily chosen for the daughter nuclide ^{100}Rh in transient equilibrium with its parent ^{100}Pd, the activity of ^{100}Pd can be found using either eq. (8.39) or (8.40). Eq. (8.39) gives

$$\frac{100\,\text{dpm}}{A_A} = \frac{3.3 \times 10^{-2}\,\text{h}^{-1}}{3.3 \times 10^{-2}\,\text{h}^{-1} - 7.2 \times 10^{-3}\,\text{h}^{-1}}$$

or

$$A_A = 78\,\text{dpm}$$

With the use of eq. (8.39) or (8.40), the decay of the daughter nuclide may be calculated as a function of parent decay in transient equilibrium. The ^{100}Pd–^{100}Rh parent–daughter decay in transient equilibrium is illustrated by curves A and B, respectively, of Figure 8.5. The parent and daughter nuclides are shown to have respective activities of 78 and 100 dpm at time $t = 0$. As curves A and B show, the parent and daughter nuclides in transient equilibrium decay with the same half-life, that corresponding to the half-life of the parent.

If the parent and daughter nuclides were to be separated, the daughter nuclide would decay according to its half-life as indicated by curve C. The isolated parent nuclide would, however, show an increase in activity with time owing to the ingrowth of daughter until

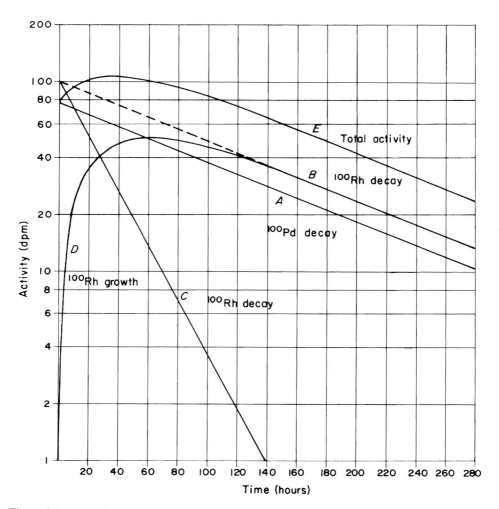

Figure 8.5   Growth and decay curves following the separation of ^{100}Pd (^{100}Rh) in transient equilibrium. (A) Decay of isolated parent nuclide ^{100}Pd. (B) Decay of ^{100}Rh daughter nuclide in transient equilibrium. The dashed portion of this curve represents ^{100}Rh decay if parent and daughter nuclides were not separated. (C) Decay of ^{100}Rh after separation from its parent. (D) The ingrowth of ^{100}Rh with the isolated parent ^{100}Pd. (E) Total activity from the isolated ^{100}Pd representing both ^{100}Pd decay and ^{100}Rh growth until transient equilibrium is attained.

transient equilibrium is attained. Curve D of Figure 8.5 shows the ingrowth of daughter nuclide from a freshly isolated parent. Because $N_B^0 = 0$ at time $t = 0$ (time of separation of parent and daughter), the last term of eq. (8.27) falls out to give

$$N_B = \frac{\lambda_A N_A^0}{\lambda_B - \lambda_A}(e^{-\lambda_A t} - e^{-\lambda_B t})$$

(8.41)

The term $\lambda_A N_A^0/(\lambda_B - \lambda_A)$ describes the rate of production of the daughter divided by the difference between the daughter and parent decay constants, which will approach a maximum value that may be written as

$$N_B = (N_B)_{max}(e^{-\lambda_A t} - e^{-\lambda_B t}) \qquad (8.42)$$

similar to the case of eq. (8.32). Since the activity, A_B, of the daughter atoms is proportional to the number of daughter atoms, or $A_B = k_B \lambda_B N_B$, where k is as defined previously, eq. (8.42) may also be written as

$$A_B = (A_B)_{max}(e^{-\lambda_A t} - e^{-\lambda_B t}) \qquad (8.43)$$

Because the maximum daughter activity in this example is 100 dpm, eq. (8.43) may be used to calculate the ingrowth of daughter nuclide with $(A_B)_{max} = 100$.

Curve E of Figure 8.5 illustrates the total activity, that is, the activity of the isolated parent nuclide plus the ingrowth of daughter nuclide. It is found by summing curves A and D and consequently accounts for the simultaneous decay of the parent nuclide and the ingrowth of the daughter. Notice that the slopes of curves A, B, and E are identical when transient equilibrium is attained, that is, the rates of decay of both the parent and daughter are identical.

8.6 NO EQUILIBRIUM

The cases of secular equilibrium and transient equilibrium, which involve decay schemes whereby the parent nuclide is longer lived than its daughter, were just considered. In other cases in which the daughter nuclide is longer lived than its parent, $\lambda_A > \lambda_B$, no equilibrium is attained. Instead, the parent nuclide of shorter half-life eventually decays to a negligible extent, leaving only the daughter nuclide, which decays by its own half-life. The following decay scheme of ^{56}Ni serves as an example:

$$^{56}_{28}\text{Ni} \xrightarrow{t_{1/2} = 6.4 \text{ days}} {}^{56}_{27}\text{Co} \xrightarrow{t_{1/2} = 77.3 \text{ days}} {}^{56}_{26}\text{Fe (stable)} \qquad (8.44)$$

The parent nuclide ^{56}Ni decays by electron capture with a half-life of 6.4 days, whereas its daughter ^{56}Co decays with the longer half-life of 77.3 days by electron capture and β^+ emission. Curve A of Figure 8.6 illustrates the decay of initially pure ^{56}Ni parent nuclide. The decay of ^{56}Ni is followed by the ingrowth (production) of the ^{56}Co daughter nuclide, shown by curve B. The ingrowth of daughter is calculated from eq. (8.27), of which the last term, $N_B^0 e^{-\lambda_B t}$, falls out because $N_B^0 = 0$ at time $t = 0$. The number of daughter atoms N_B of eq. (8.27) may be converted to activity, A_B, by the term $A_B = k_B \lambda_B N_B$ as discussed previously. The total activity illustrated by curve C of Figure 8.6 depicts both the simultaneous decay of parent nuclide and the growth and decay of daughter determined by summing curves A and B. Notice from Figure 8.6 that the parent nuclide activity in this example becomes negligible after around 55 days, after which the total activity, curve C, has a slope corresponding to the decay rate of the daughter nuclide.

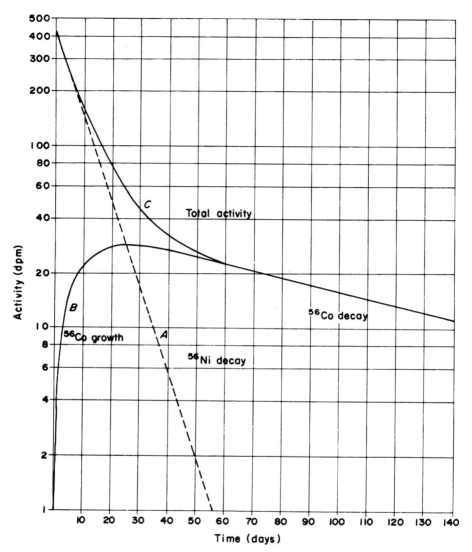

Figure 8.6 Growth and decay curves of the ^{56}Ni (^{56}Co) parent–daughter nuclides following the isolation or fresh preparation of the parent nuclide ^{56}Ni. (A) Decay of pure parent nuclide ^{56}Ni. (B) Ingrowth of daughter nuclide ^{56}Co. (C) Total activity representing both ^{56}Ni decay and the simultaneous growth and decay of ^{56}Co daughter.

8.7   MORE COMPLEX DECAY SCHEMES

Other decay schemes exist that involve a chain of numerous nuclides such as

$$A \rightarrow B \rightarrow C \rightarrow \ldots N \tag{8.45}$$

where nuclides A, B, and C are followed by a chain of a number N of decaying nuclides. A long decay chain of this type may be observed in the complex decay schemes of high atomic number natural radionuclides such as ^{235}U, ^{238}U, and ^{232}Th. The complex decay scheme of ^{232}Th is illustrated in Figure 8.7. The decay sequence of $^{232}_{90}Th$ to $^{212}_{83}Bi$ is described by the general eq. (8.45). However, the continuation of this decay scheme with $^{212}_{83}Bi$ involves a branching decay of the type

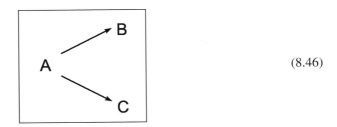

$$(8.46)$$

In this example $^{212}_{83}Bi$ is the parent of the two daughter nuclides $^{212}_{84}Po$ and $^{208}_{81}Ti$. The half-life of ^{212}Bi is written under the nuclide symbol rather than along the arrows of Figure 8.7, because the ^{212}Bi half-life is a function of the two decay processes and may be written as

$$t_{1/2} = \frac{0.693}{(\lambda_A + \lambda_B)}$$

$$(8.47)$$

where λ_A and λ_B are the decay constants of the two separate decay processes.

8.8 RADIOACTIVITY UNITS AND RADIONUCLIDE MASS

8.8.1 Units of radioactivity

The units used to define radioactivity or, in other words, the activity of a radioactive sample are written in terms of the number of atoms, N, disintegrating per unit of time, t. We can use eq. (8.1) previously discussed in this chapter to calculate the activity of any given mass of radionuclide. The equation, namely $\Delta N/\Delta t = \lambda N$, defines the proportionality between the rate of decay of a radionuclide and the number of atoms of the radionuclide in a sample. As an example, we may use eq. (8.1) to calculate the activity of 1 g of ^{226}Ra as follows:

$$\frac{\Delta N}{\Delta t} = \lambda N$$

$$(8.48)$$

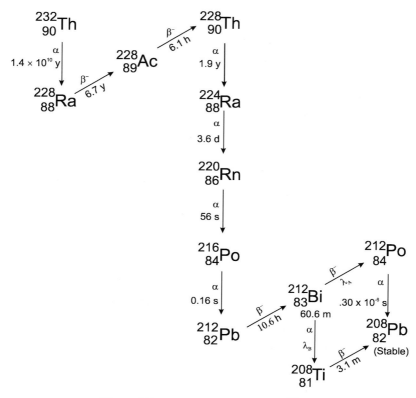

Figure 8.7   Decay scheme of natural ^{232}Th.

$$\frac{\Delta N}{\Delta t} = \left(\frac{0.693}{t_{1/2}}\right)(N) \tag{8.49}$$

where $\lambda = 0.693/t_{1/2}$ as derived previously (eq. (8.11)). If we take the half-life, $t_{1/2}$, of ^{226}Ra to be 1599 years and substitute for N, in the preceding equation, the number of atoms per mol of ^{226}Ra, we can write

$$\frac{\Delta N}{\Delta t} = \left(\frac{0.693}{1599 \text{ years}}\right)\left(\frac{6.022 \times 10^{23} \text{ atoms}}{226 \text{ g}}\right) \tag{8.50}$$

where, according to Avogadro's number, there are 6.022×10^{23} atoms per gram mole of substance. If we now convert the half-life of ^{226}Ra from units of years to minutes, we can calculate the number of atoms of ^{226}Ra dpm per gram according to

$$\frac{\Delta N}{\Delta t} = \left[\frac{0.693}{(1599 \text{ years})(365 \text{ days/year})(24 \text{ h/day})(60 \text{ min/h})}\right]\left(\frac{6.022 \times 10^{23} \text{ atoms}}{226 \text{ g}}\right)$$

$$\tag{8.51}$$

$$\frac{\Delta N}{\Delta t} = \left(\frac{0.693}{8.404 \times 10^8 \, min}\right)(2.665 \times 10^{21} \, atoms/g)$$

$$= 2.19 \times 10^{12} \, atoms/min/g$$

The activity of 1 g of ^{226}Ra is the basis of the unit of radioactivity known as the curie (Ci). One curie is almost equal to the activity of 1 g of ^{226}Ra or, by definition,

$$1Ci = 2.22 \times 10^{12} \, dpm = 3.7 \times 10^{10} \, dps \qquad (8.52)$$

Therefore, one curie of activity or any multiple of the curie of any radionuclide defines the number of atoms disintegrating per unit of time in minutes or seconds.

The rate of decay in terms of time in seconds gives rise to a more internationally adopted Système International d'Unités (SI) unit of activity, which is the becquerel (Bq), where by definition

$$1 \, Bq = 1 \, dps \qquad (8.53)$$

Therefore, we can interrelate the curie and becquerel as follows:

$$1 \, Ci = 2.22 \times 10^{12} \, dpm = 3.7 \times 10^{10} \, dps = 37 \, GBq \qquad (8.54)$$

Likewise, smaller units of the curie, namely the millicurie (mCi) and microcurie (μCi), may be interrelated with the becquerel as follows:

$$1 \, mCi = 2.22 \times 10^9 \, dpm = 3.7 \times 10^7 \, dps = 37 \, MBq \qquad (8.55)$$

and

$$1\mu Ci = 2.22 \times 10^6 \, dpm = 3.7 \times 10^4 \, dps = 37 \, kBq \qquad (8.56)$$

Another unit of activity recommended in the early 1960s by the International Union of Pure and Applied Physics, but less frequently used, is the rutherford, where 1 rutherford $= 10^6$ dps and 1 microrutherford would be equivalent to 1 dps or 1 Bq (Buttlar, 1968; Das and Ferbel, 1994).

8.8.2   Correlation of radioactivity and radionuclide mass

From eq. (8.48) and calculations made in the previous Section 8.8.1, we can see that, for samples of a given level of activity, radionuclides of shorter half-life will contain a smaller number of radioactive atoms than radionuclides of longer half-life.

We can use eq. (8.48) again to compare two radionuclides of relatively short and long half-lives to see the magnitude of the differences in radionuclide masses we would encounter for any given level of radioactivity. For example, we may take the radionuclide ^{32}P of 14.3-day half-life and the radionuclide ^{14}C of 5730-year half-life and calculate the activity per gram (e.g., dpm/g) and grams per activity (e.g., g/Ci) of each radionuclide for comparative purposes. These calculations are as follows.

1. ^{32}P, half-life = 14.3 days:

$$\frac{\Delta N}{\Delta t} = \left(\frac{0.693}{t_{1/2}}\right)(N)$$

(8.57)

$$\frac{\Delta N}{\Delta t} = \left[\frac{0.693}{(14.3 \text{ days})(24 \text{ h/day})(60 \text{ min/h})}\right]\left(\frac{6.022 \times 10^{23}}{32 \text{ g}}\right)$$

$$= 6.32 \times 10^{17} \text{ dpm/g } ^{32}\text{P}$$

If, by definition, 1 curie = 2.22×10^{12} dpm, we can convert this activity per gram of ^{32}P to grams ^{32}P per curie as follows:

$$\frac{2.22 \times 10^{12} \text{ dpm/Ci}}{6.32 \times 10^{17} \text{ dpm/g } ^{32}\text{P}} = 3.51 \times 10^{-6} \text{ g } ^{32}\text{P per Ci}$$

(8.58)

$$= 3.51 \times 10^{-6} \text{ mg} ^{32}\text{P per mCi}$$

2. ^{14}C, half-life = 5730 years:

$$\frac{\Delta N}{\Delta t} = \left[\frac{0.693}{(5730 \text{ years})(365 \text{ days/year})(24 \text{ h/day})(60 \text{ min/h})}\right]\left(\frac{6.022 \times 10^{23}}{14 \text{ g}}\right)$$

(8.59)

$$= 9.90 \times 10^{12} \text{ dpm/g } ^{14}\text{C}$$

This activity per gram of ^{14}C is converted to grams ^{14}C per curie as follows:

$$\frac{2.22 \times 10^{12} \text{ dpm/Ci}}{9.90 \times 10^{12} \text{ dpm/g } ^{14}\text{C}} = 0.224 \text{ g} ^{14}\text{C per Ci}$$

(8.60)

$$= 0.224 \text{ mg} ^{14}\text{C per mCi}$$

The calculated mass of ^{32}P in 1 curie of activity is almost a million fold less than the calculated mass of ^{14}C in 1 curie of activity. In general, research with radionuclides involves the handling and analysis of lower levels of radioactivity in millicuries, microcuries, and picocuries, and so on. The masses of radioactive atoms in the milli-, micro-, and picocurie levels of radioactivity are obviously much smaller than encountered at the curie level.

Appendix A

Particle Range–Energy Correlations

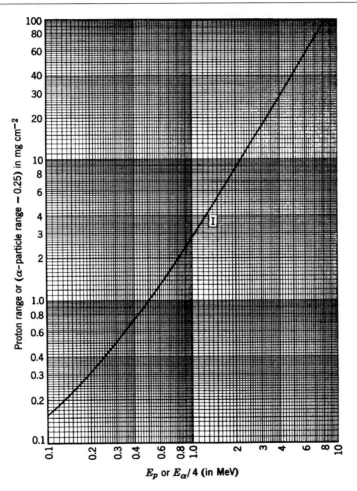

Figure A.1 Range–energy relations for protons and alpha particles in air. When applying the graph to alpha particles, the alpha particle energy is divided firstly by 4 and the value of $0.25 \, \text{mg cm}^{-2}$ added to the range read off the graph. For example, to obtain the range for a 5.5 MeV alpha particle we read $5.5/4 = 1.375$ MeV off the graph and find the range $4.7 \, \text{mg cm}^{-2} + 0.25 = 4.95 \, \text{mg cm}^{-2}$. The range in cm can be obtained by dividing the above range by the density of air ($\sigma = 1.226 \, \text{mg cm}^{-3}$ at STP) or $4.95 \, \text{mg cm}^2/1.226 \, \text{mg cm}^{-3} = 4.0 \, \text{cm}$. (From Friedlander *et al.*, 1964, Copyright © John Wiley and Sons, Inc. This material is used by permission of John Wiley & Sons, Inc.)

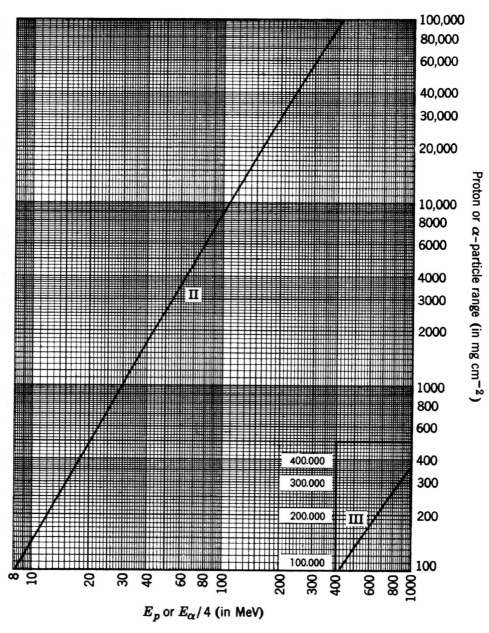

Figure A.2 Range–energy relations for protons and alpha particles in air. (From Friedlander *et al.*, 1964, Copyright © John Wiley and Sons, Inc. This material is used by permission of John Wiley & Sons, Inc.)

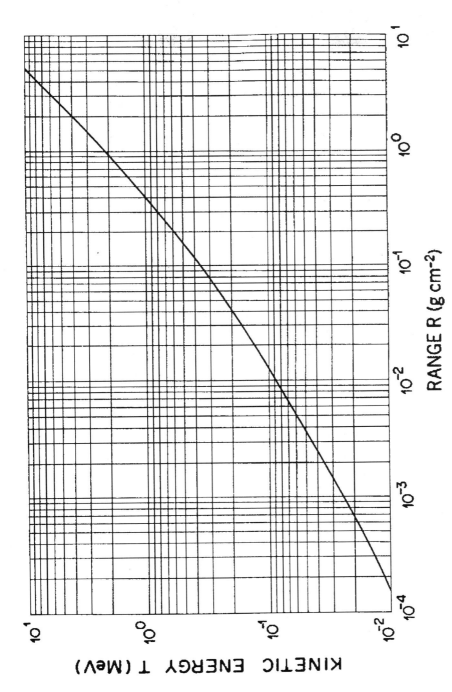

RANGE R (g cm⁻²)

Figure A.3 Beta-particle range–energy curve for absorbers of low atomic number. The curve is described by the formulas $R = 0.412T^{1.27-0.0954\ln T}$ for $0.01 \leq T \leq 2.5$ MeV, and $R = 0.530T - 0.106$ for $T > 2.5$ MeV, where T is the beta-particle energy in MeV. (From U.S. Public Health Service. (1970). Radiological Health Handbook. Publication No. 2016, Bureau of Radiological Health, Rockville, MD.)

Appendix B

Periodic Table of the Elements

Key to Chart

Atomic Number → 79
Symbol → Au
Atomic Weight → 196.97
Name → Gold

Electron configuration from top to bottom in K, L, M, N, O, P to Q energy levels

(Periodic table of the elements, showing groups IA–VIIIA, IIIB–IIB, Lanthanides, and Actinides with atomic numbers, symbols, atomic weights, names, and electron configurations.)

*Lanthanides (57–71)
**Actinides (89–103)

[a]A value in parenthesis denotes the mass number of the longest lived isotope of the element.
[b]Atomic weight of the most commonly available long-lived isotope of the element.
[c]The names of elements 112–118 are the Latin equivalents of the atomic numbers. Element 112 was created in 1996 at the GSI, Darmstadt, Germany. Element 116 was created in 1996. Elements 113–116 are reported created during 1999 to 2004 at the Joint Institute for Nuclear Research, Dubna, Russia and Lawrence Livermore National Laboratories. Elements 117 and 118 are yet unknown.

References

Aaserud, F. (2003). Hilde Levi: 1909–2003, Obituary, August 2003. Niels Bohr Institute Archive, Copenhagen, Denmark.

Abbe, J. C., and Marques-Netto, A. (1975). Szilard-Chalmers effects in halfnium complexes. *J. Inorg. Nucl. Chem.* **37**, 2239–2242.

Abele, H. (2000). The standard model and the neutron β-decay. *Nucl. Instrum. Methods Phys. Res. A* **440**, 499–510.

Adachi, I., Sumiyoshi, T., Hayashi, K., Iida, N., Enomoto, R., Tsukada, K., Suda, R., Matsumoto, S., Natori, K., Yokoyama, M., and Yokogawa, H. (1995). Study of a threshold Cherenkov counter based on silica aerogels with low refractive indices. *Nucl. Instrum. Methods Phys. Res. A* **355**, 390–398.

Adkisson, P., and Tumlinson, J. (2003). Edward F. Knipling 1909–2000: A biographical memoir. *Biograph. Mem.* **83**, 3–15.

Afanasiev, G. N., Kartavenko, V. G., and Stepanovsky, Yu. P. (1999). On Tamm's problem in the Vavilov-Cherenkov radiation theory. *J. Phys. D Appl. Phys.* **32**, 2029–2043.

Argirò, S. (2004). Status and perspectives of the Pierre Auger Cosmic Ray Observatory. *Eur. J. Phys. C* **33**, s947–s949.

Agnew, L. E., Elioff, T., Fowler, W. B., Gilly, L., Lander, R., Oswald, L., Powell, W., Segrè, E., Steiner, H., White, H., Wiegand, C., and Ypsilantis, T. (1958). $\bar{p}$–p elastic and charge exchange scattering at about 120 MeV. *Phys. Rev.* **110**, 994–995.

Ahmad, Q. R., *et al.* (SNO Collaboration). (2002). Direct evidence for neutrino flavor transformation from neutral–current interactions in the Sudbury Neutrino Observatory. *Phys. Rev. Lett.* **89** (011301), 1–6.

Akatsu, M., Aoki, M., Fujimoto, K., Higashino, Y., Horose, M., Inami, K., Ishikawa, A., Matsimoto, T., Misono, K., Nagai, I., Ohshima, T., Sugi, A., Sugiyama, A., Suzuki, S., Tomoto, M., and Okuno, H. (2000). Time-of-propagation Cherenkov counter for particle identification. *Nucl. Instrum. Methods Phys. Res. A* **440**, 124–135.

Alcaraz, J., Aplat, B., and Ambrosi, G., *et al.* (2000). Leptons in near earth orbit. *Phys. Lett.* **B484**, 10–22.

Allen, A. D. (2001). Szilard endorsed nuclear medicine by example. *Phys. Today Lett. Online*, February 2001, p. 1.

Al-Masri, M. S. (1996). Cerenkov counting technique. *J. Radioanal. Nucl. Chem., Articles* **207**, 205–213.

Al-Masri, M. S., and Blackburn, R. (1994). Simultaneous determination of ^{234}Th and ^{228}Th in environmental samples using Cerenkov counting. *Radiochim. Acta* **65**, 133–136.

Al-Masri, M. S., and Blackburn, R. (1995a). Application of Cerenkov radiation for the assay of ^{226}Ra in natural water. *Sci. Total Environ.* **173/174**, 53–59.

Al-Masri, M. S., and Blackburn, R. (1995b). Radiochemical determination of ^{226}Ra using Cerenkov counting. *J. Radioanal. Nucl. Chem., Articles* **195**, 339–344.

Al-Masri, M. S., and Blackburn, R. (1999). Radon-222 and related activities in surface waters of the English Lake District. *Appl. Radiat. Isot.* **50**, 1137–1143.

Alpher, R. A., Follin, J. W., and Herman, R. C. (1953). Physical conditions in the initial stages of the expanding universe. *Phys. Rev.* **92**, 1347–1361.

Alpher, R. A., and Herman, R. C. (1949). Remarks on the evolution of an expanding universe. *Phys. Rev.* **75**(7), 1089–1095.

Alpher, R. A., and Herman, R. C. (1988). Reflections on early "big bang" cosmology. *Phys. Today* (August) **41**(8), 24–34.

Amaldi, E., D'Agostino, O., Fermi, E., Pontecorvo, B., Rasetti, F., and Segre, E. (1935). Artificial radioactivity produced by neutron bombardment. II. *Proc. R. Soc. Lond.* **A149**(868), 522–558.

Amaldi, E., and Fermi, E. (1935). Absorption of slow neutrons. *Ricerca Sci.* **6**, 344–347.

Amaldi, E., and Fermi, E. (1936a). Groups of slow neutrons. *Ricerca Sci.* **7**(1), 310–313.

Amaldi, E., and Fermi, E. (1936b). Diffusion of slow neutrons. *Ricerca Sci.* **7**(1), 393–395.

Amaldi, E., Fermi, E., and Rasetti, F. (1937). Artificial generator of neutrons. *Ricerca Sci.* **8**(2), 40–43.

AMS Collaboration (2000). Cosmic protons. *Phys. Lett.* **B490**, 27–35.

Anderson, C. D. (1930). Space distribution of x-ray photoelectrons ejected from the K and L atomic energy levels. *Phys. Rev.* **35**, 1139–1145.

Anderson, C. D. (1932). The apparent existence of easily detectable positives. *Science* **76**, 238.

Anderson, C. D. (1933a). Energy loss and scattering of cosmic-ray particles. *Phys. Rev.* **43**, 381.

Anderson, C. D. (1933b). The positive electron. *Phys. Rev.* **43**, 491–498.

Anderson, C. D. (1933c). Free positive electrons resulting from the impacts upon atomic nuclei of the photons from ThC. *Science* **77**, 432.

Anderson, C. D. (1936). The production and properties of positrons. Nobel Lecture, December 13, 1939. In: *Nobel Lectures, Physics, 1922–1941*, (1965), Elsevier, Amsterdam.

Anderson, C. D., Adams, R. V., and Rau, R. R. (1947a). On the mass and the disintegration products of the mesotron. *Phys. Rev.* **72**, 724–727.

Anderson, C. D., Millikan, R. A., Neddermeyer, S. H., and Pickering, W. (1934). The mechanism of cosmic-ray counter action. *Phys. Rev.* **45**, 352–363.

Anderson, C. D., and Neddermeyer, S. H. (1933). Positrons from gamma rays. *Phys. Rev.* **43**, 1034–1035.

Anderson, C. D., and Neddermeyer, S. H. (1936). Cloud chamber observations of cosmic rays at 4300 meters elevation and near sea-level. *Phys. Rev.* **50**, 263–271.

Anderson, E. C. (1949). Natural Radiocarbon. Ph.D. Thesis. University of Chicago, Chicago, IL.

Anderson, E. C., Libby, W. F., Weinhouse, S., Reid, A. F., Kirshenbaum, A. D., and Gosse, A. V. (1947a). Radiocarbon from cosmic radiation. *Science* **105**, 576–577.

Anderson, E. C., Libby, W. F., Weinhouse, S., Reid, A. F., Kirshenbaum, A. D., and Gosse, A. V. (1947b). Natural radiocarbon from cosmic radiation. *Phys. Rev.* **72**, 931–936.

Anderson, H. H., Sørensen, H., and Vadja, P. (1969). Excitation potentials and shell corrections for the elements $Z_2 = 20$ to $Z_2 = 30$. *Phys. Rev.* **180**, 373–380.

Anderson, H. L., Fermi, E., and Hanstein, H. B. (1939a). Production of neutrons in uranium bombarded by neutrons. *Phys. Rev.* **55**, 797–798.

Anderson, H. L., Fermi, E., and Szilard, L. (1939b). Neutron production and absorption in uranium. *Phys. Rev.* **56**, 284–286.

Andreo, P., Seuntjens, J. P., and Podgorsak, E. B. (2005). Calibration of photon and electron beams. In: *Review of Radiation Oncology Physics: A Handbook for Teachers and Students* (ed. E. B. Podgorsak), International Atomic Energy Agency (IAEA), Vienna, pp. 301–355.

Anonymous. (1975). Obituary. Sir George Thomson. *Nature* **257**, 726.

Argonne National Laboratory. (1963). *Reactor Physics Constants*. ANL-5800, 2nd edn., United States Atomic Energy Commission, Washington, DC.

Arnold, J. R., and Libby, W. F. (1949). Age determinations by radiocarbon content: checks with samples of known age. *Science* **110**, 678–680.

Arrhenius, S. (1922). Presentation Speech for the Nobel Prize in Physics 1921 awarded to Albert Einstein. December 10, 1922. In: *Nobel lectures, Physics 1901–1921*, 1967. Elsevier, Amsterdam.

Arzumanov, S., Bondarenko, L., Chernyavsky, S., Drexel, W., Fomin, A., Geltenbort, P., Morozov, V., Panin, Yu., Pendlebury, J., and Schreckenbach, K. (2000). Neutron lifetime measured by monitored storing of untra-cold neutrons. *Nucl. Instrum. Methods Phys. Res. A* **440**, 511–516.

Atkinson, R. d'E., and Houtermans, F. G. (1929). Aufbaumöglichkeit in Sterren. *Zeits. Physik* **54**, 656–665.

Attas, M., and Abushady, I. (1997). Cerenkov Viewing Device for Spent Fuel Verification at Light Water Reactors. IAEA Inspector Training Manual. ©IAEA, Vienna. Atomic Energy of Canada, Ltd., Pinawa, Manitoba.

Attas, M., Chen, J. D., and Young, G. J. (1990). A Cherenkov viewing device for used-fuel verification. *Nucl. Instrum. Methods Phys. Res. A* **299**, 88–93.

Attas, M., Chen, J. D., and Young, G. J. (1992). An ultraviolet imager for nuclear safeguards inspectors. In: Proceedings of the International Meeting on Electron Tubes and Image Intensifiers. San Hose, CA, Feb. 10–12, 1992. *Proc. SPIE* **1655**, 50–57.

Auger, P. (1923). Secondary β rays produced in a gas by X-rays. *Compt. Rend.* **177**, 169–171.

Auger, P. (1925a). Sur les rayons β secondaires produits dans un gaz par des rayons-X. *Compt. Rend.* **180**, 65–68.

Auger, P. (1925b). Sur l'effect photoélectrque compose. *J. Phys. Radium* **6**, 205–208.

Auger, P., Ehrenfest, P., Maze, R., Daudin, J., and Fréon, R. A. (1939). Extensive cosmic-ray showers *Rev. Mod. Phys.* 11, 288–291.

Auger, P., and Maze, R. (1938). Les grandes gerbes cosmiques de l'atmosphere. *C. R. Acad. Sci. Ser. B* **207**, 228–230.

Auger, P., and Maze, R. (1939). Extension et pouvoir pénétrant des grandes grebes de rayons cosmiques. *Compt. Rend.* **208**, 1641–1643.

Auger, P., Maze, R., and Grivet-Meyer, T. (1938). Heavy electrons in cosmic ray showers originating in the atmosphere. *Compt. Rend.* **206**, 1721–1723.

Baas, H. W. (2004). *Neutron Activation Analysis of Inhomogeneous Samples*. Delft University Press, pp. 174.

Bacon, G. E. (1969). *Neutron Physics*. Wykeham Publications, London.

Bailey, J., Borer, K., Combley, F., *et al.* (1977). On muon lifetimes and time dilation. *Nature* **268**, 301–305.

Bailey, J., Borer, K., Combley, F., *et al.* (1979). Final report of the CERN muon storage ring including the anomalous magnetic moment and the electric dipole moment of the muon, and the direct test of relativistic time dilation. *Nucl. Phys. B* **150**, 1–75.

Balasubramanian, P. S. (1997). Anodically oxidized aluminum layer as a useful substrate for the fabrication of ^{147}Pm sources for beta-ray thickness gauges. *J. Radioanal. Nucl. Chem.* **223**(1–2), 79–81.

Balasubramanian, P. S. (1998). A simple procedure for the fabrication of high activity beta-radiation sources of ^{147}Pm for use in beta-ray thickness gauges. *J. Radioanal. Nucl. Chem.* **229** (1–2), 157–160.

Bambynek, W., Crasemann, B., Fink, R. W., Freund, H. U., Mark, H., Swift, C. D., Price, R. E., and Rao, P. V. (1972). X-ray fluorescence yields, Auger, and Coster-Kronig transition probabilities. *Rev. Mod. Phys.* **44**, 716–813.

Barany, A. (Ed.) (1999). *The Nobel Prizes 1999 Les Prix Nobel: Nobel Prizes, Presentations, Biographies, and Lectures*. Almquiest & Wiksell International, Stockholm.

Barkas, W. H. (Ed.) (1963). *Nuclear Research Emulsions: Pure and Applied Physics*, Academic Press, London, pp. 518.

Bartusiak, M. (1996). The woman behind the bomb. *Washington Post*, Sunday, March 17, 1996.

Basini, G., *et al.* (1999). The flux of cosmic-ray antiprotons from 3.7 to 24 GeV. Proceedings of 26th International Cosmic Ray Conference, Salt Lake City, **3**, 101.

Bassham, J. A., Benson, A. A., and Calvin, M. (1950). The path of carbon in photosynthesis. *J. Biol. Chem.* **185**, 781–787.

Basu, S. K. (2003). *Hand Book of Oceanograohy*. Global Vision Publishing, Delhi, India.

Battista, J., and van Dyk, J. (2002). London celebrates fifty years of cobalt-60 radiotherapy. *Can. Med. Phys. Newsl.* **48**(1), 15.

Beach, A. S., *et al.* (2001). Measurement of the cosmic-ray antiproton-to-proton abundance ratio between 4 and 50 GeV. *Phys. Rev. Lett.* **87**, 271101.

Becquerel, A. H. (1896a). On the rays emitted by phosphorescence. *C. R. Acad. Sci. Paris* **122**, 420.

Becquerel, A. H. (1896b). On the invisible rays emitted by phosphorescent bodies. *C. R. Acad. Sci. Paris* **122**, 501.

Becquerel, A. H. (1901). The radioactivity of matter. *Nature* **63**, 396–398.

Becquerel, A. H. (1903). On radioactivity, a new property of matter. Nobel Lecture, December 11, 1903. In: "Nobel Lectures in Physics, 1901–1970." Vol. 1, *Nobel Lectures, Physics, 1901–1921*, (1967), (eds. Science Incorporated Elsevier), Elsevier, Amsterdam.

Behnken, H. (1927). Die Absolutbestimmung der Dosisinheit "1 Röntgen" inder Physicalisch-Technischen Reichsanstalt. *Strahlentherapie* **26**, 79–100.

Belcher, E. H. (1953). The luminescence of irradiated transparent media and the Cerenkov effect. I. The luminescence of aqueous solutions of radioactive isotopes. *Proc. Royal Soc. Lond. Ser. A* **216**(1124), 90–102.

Bellotti, R., *et al.* (1996). Measurement of the negative muon spectrum between 0.3 and 40 GeV/*c* in the atmosphere. *Phys. Rev.* **D53**, 35.

Bellotti, R., *et al.* (1999). Balloon measurements of cosmic-ray muon spectra in the atmosphere along with those of primary protons and helium nuclei over midlatitude. *Phys. Rev.* **D60**, 052002.

Bem, E. M., Bem, H., and Reimschussel, W. (1978). The use of Cherenkov radiation in the measurement of cadmium-115 m in blood and other tissues. *Bull. Environ. Contam. Toxicol.* **19**, 677–683.

Benson, A. A., Kawauchi, S., Hayes, P., and Calvin, M. (1952). The path of carbon in photosynthesis. *J. Amer. Chem. Soc.* **74**, 4477–4482.

Berger, M. J., and Hubbell, J. H. (1997). Photon attenuation coefficients. In: *Handbook of Chemistry and Physics*, 77th edn. (eds. D. R. Lide and H. P. R. Frederikse), CRC Press, Boca Raton, FL, pp. 10-250–10-254.

Berger, M. J., and Seltzer, S. M. (1982). Stopping powers and ranges of electrons and positrons. *Nat. Bureau Standards Publ.* NBSIR 82-2550, pp. 168.

Berry, M. (1998). Paul Dirac: the purist soul in physics. *Phys. World*, February 1998, p. 36.

Bethe, H. A. (1933). Quantenmechanik der Ein- und Zwei-Electronen-Probleme. In: *Handbuch der Physik*, (eds. H. Geiger and K. Scheel), 2nd edn., Vol. 24, Part I, Springer, Berlin.

Bethe, H. A. (1937). Nuclear physics II. Nuclear dynamics, theoretical. *Rev. Mod. Phys.* **9**, 69–244.

Bethe, H. A. (1939). Energy production in stars. *Phys. Rev.* **55**, 434–456.

Bethe, H. A. (1942). Energy production in stars. *Am. Sci.* **30**(4), 243–264.

Bethe, H. A. (1967). Energy production in stars. Nobel Lecture on December 11, 1967, Stockholm. In: *Nobel Lectures, Physics, 1942–1962*, (1964), (eds. Nobel Foundation Staff), Elsevier, Amsterdam.

Bethe, H. A. (1968). Energy production in stars. *Physics Today* **21**(9), 36–44.

Bethe, H. A. (2000). The German uranium project. *Phys. Today* **53**(7), 34–36.

Bethe, H. A., and Ashkin, J. (1953). Passage of radiations through matter. In: *Experimental Nuclear Physics* (ed. E. Segré), Vol. 1, J. Wiley, New York, NY.

Bethe, H. A., and Bacher, R. F. (1936). Nuclear physics I. Stationary states of nuclei. *Rev. Mod. Phys.* **8**, 82–229.

Bethe, H. A., Bacher, R. F., and M. Stanley Livingston (1986). *Basic Bethe: Seminal Articles on Nuclear Physics, 1936–1937*, American Institute of Physics, New York, NY.

Bethe, H. A., and Critchfield, C. L. (1938). The formation of deuterons by proton combination. *Phys. Rev.* **54**, 248–254.

Bethe, H. A., and Peierls, R. E. (1934). The neutrino. *Nature* **133**, 532.

Bethe, H. A., and Placzec, G. (1937). Resonance effects in nuclear processes. *Phys. Rev.* **51**, 450–484.

Biersack, H.-J., and Freeman, L. M. (eds.) (2007). *Nuclear Medicine Concise*. Springer Verlag, pp. 380.

Billington, D. (1992). *Radioisotopes (Introduction to Biotechniques)*, BIOS Scientific, pp. 168.

Birattari, C., Bonardi, M., Groppi, F., and Gini, L. (2001). Review of cyclotron production and quality control of "high specific activity" radionuclides for biomedical, biological, industrial, and environmental applications at INFN-LASA. In: "Cyclotrons and Their Applications 2001". Proceedings of Sixteenth International Conference (ed. F. Marti), American Institute of Physics, Melville, NY.

Bird, D. J., *et al.* (1995). Detection of a cosmic ray with measured energy well beyond the expected spectral cutoff due to cosmic microwave radiation. *Astrophys. J.* **441**, 144–150.

Birge, R. T. (1940). Account of Professor Lawrence's Work. Speech given at Presentaion Ceremony of Nobel Prize, Physics to Ernest O. Lawrence on February 29, 1940 at The University of California, Berkeley. In: *Nobel Lectures, Physics, 1922–1941*, (1965), Elsevier, Amsterdam.

Bjorklund, R., Crandall, W. E., Moyer, B. J., and York, H. F. (1950). High energy photons from proton-nucleon collisions. *Phys. Rev.* **77**, 213–218.

Blackburn, R., and Al-Masri, M. S. (1993). Determination of radon-222 and radium-226 in water samples by Cerenkov counting. *Analyst* **118**, 873–876.

Blackburn, R., and Al-Masri, M. S. (1994). Determination of uranium by liquid scintillation and Cerenkov counting. *Analyst* **119**, 465–468.

Blackett, P. M. S. (1922). On the analysis of alpha-ray photographs. *Proc. R. Soc. Lond.* **102A**, 294–318.

Blackett, P. M. S. (1923). The study of forked α-ray tracks. *Proc. R. Soc. Lond.* **103A**, 62–80.

Blackett, P. M. S. (1925). The ejection of protons from nitrogen nuclei, photographed by the Wilson method. *Proc. R. Soc. Lond.* **107A**, 349–360.

Blackett, P. M. S. (1927). An automatic cloud chamber for the rapid production of alpha ray photographs. *J. Sci. Instrum.* **4**(14), 433–439.

Blackett, P. M. S. (1929). On the design and use of a double camera for photographing artificial disintegrations. *Proc. R. Soc. Lond.* **123A**, 613–629.

Blackett, P. M. S. (1934). On the technique of the counter controlled cloud chamber. *Proc. R. Soc. Lond.* **146A**, 281–289.

Blackett, P. M. S. (1936). The measurement of the energy of cosmic rays. I. The electro-magnet and cloud chamber. *Proc. R. Soc. Lond.* **154A**, 564–573.

Blackett, P. M. S. (1948a). Cloud chamber researches in nuclear physics and cosmic radiation. Nobel Lecture on December 13, 1948. In: *Nobel Lectures, Physics, 1942–1962.*, (1964), (eds. Nobel Foundation Staff), Elsevier, Amsterdam.

Blackett, P. M. S. (1948b). *Military and Political Consequences of Atomic Energy,*. Turnstile Press, London.

Blackett, P. M. S. (1954). *Lectures on Rock Magnetism*. The Weizmann Press, Tel Aviv, Israel.

Blackett, P. M. S. (1962). *Studies of War*, Oliver and Boyd, London.

Blackett, P. M. S. (1971). *Science, Technology and Aid in Developing Countries*, Edinburgh University Press, Edinburgh.

Blackett, P. M. S., and Brode, R. B. (1936). The measurement of the energy of cosmic rays. II. The curvature measurements and the energy spectrum. *Proc. R. Soc. Lond.* **154A**, 573–587.

Blackett, P. M. S., and Champion, F. C. (1931). The scattering of slow alpha particles by helium. *Proc. R. Soc. Lond.* **130A**, 380–388.

Blackett, P. M. S., and Hudson, E. P. (1927). The elasticity of the collisions of alpha particles with hydrogen nuclei. *Proc. R. Soc. Lond.* **117A**, 124–130.

Blackett, P. M. S., and Lees, D. S. (1932a). Investigations with a Wilson Chamber. I. On the photography of artificial disintegration collisions. *Proc. R. Soc. Lond.* **136A**, 325–338.

Blackett, P. M. S., and Lees, D. S. (1932b). Further investigations with a Wilson Chamber. II. The range and velocity of recoil atoms. *Proc. R. Soc. Lond.* **134A**, 658–671.

Blackett, P. M. S., and Lees, D. S. (1932c). Further investigations with a Wilson Chamber. III. The accuracy of the angle determination. *Proc. R. Soc. Lond.* **136A**, 338–348.

Blackett, P. M. S., and Occhialini, G. P. S. (1932). Photography of penetrating corpuscular radiation. *Nature* **130**, 363.

Blackett, P. M. S., and Occhialini, G. P. S. (1933). Some photographs of the tracks of penetrating radiation. *Proc. R. Soc. Lond.* **139A**, 699–726.

Blais, J. S., and Marshall, W. D. (1988). Determination of lead-210 in admixture with bismuth-210 and polonium-210 in quenched samples by liquid scintillation counting. *Anal. Chem.* **60**, 1851–1855.

Blasko, J. C., Grimm, P. D., Radge, H., and Schumacher, D. (1998). In: *Prostate Cancer* (eds. M. S. Ernstoff, J. A. Heaney, and R. E. Peschel), Blackwell Science, Cambridge, MA, p. 137.

Blau, M. (1938). Photographic tracks from cosmic rays. *Nature* **142**, 613.

Blau, M. (1956). Hyperfragments and slow K-mesons in stars produced by 3 BeV protons. *Phys. Rev.* **102**, 495–501.

Blau, M., and Coulton, M. (1954). Inelastic scattering of 500-MeV negative pions in emulsion nuclei. *Phys. Rev.* **96**, 150–160.

Blau, M., Coulton, M., and Smith, J. E. (1953). Meson production by 500 MeV negative pions. *Phys. Rev.* **92**, 516.

Blau, M., and De Felice, J. A. (1948). Development of thick emulsions by a two-bath method. *Phys. Rev.* **74**, 1198.

Blau, M., and Oliver, A. R. (1956). Interaction of 750 MeV pi-mesons with emulsion nuclei. *Phys. Rev.* **102**, 489–494.

Blau, M., and Wambacher, H. (1935). Über die Empfindlichkeit desensibilisierter photographischer Schichten in Abhängigkeit vom Luftsauerstoff und von der Konzentration der Desensibilatoren. *Mitteilungen des Institutes für Radiumforschung Nr. 367. Akad. Der Wissenschaften Wien, Mathematisch-naturwissenschaftliche Klasse, Sitzungsberichte,* **Abteilung IIa, Band 144**, Heft 1-10, S. 403.

Blau, M., and Wambacher, H. (1937). Disintegration processes by cosmic rays with the simultaneous emission of several heavy particles. *Nature* **140**(October 2), 585.

Boezio, M., *et al.* (1999). The cosmic-ray proton and helium spectra between 0.2 and 200 GeV. *Astrophys. J.* **518**, 457.

Boezio, M., *et al.* (2000a). The cosmic-ray electron and positron spectra measured at 1 AU during solar minimum activity. *Astrophys. J.* **532**, 653.

Boezio, M., *et al.* (2000b). Measurements of the flux of atmospheric muons with the CAPRICE94 apparatus. *Phys. Rev.* **D62**, 032007.

Bohr, N. (1913). On the constitution of atoms and molecules. *Philos. Mag.* **26**(6), 1–25.

Bohr, N. (1914). On the spectrum of hydrogen (Address to the Physical Society of Copenhagen, Dec. 20, 1913). *Physik Tidsskrift* **12**, 97, translated by A.D. Udden "The Theory of Spectra and Atomic Constitution—Three Essays 1922." In: "Autobiography of Science" 1950. (eds. F. R. Moulton and J. J. Schifferes), Doubleday, New York, NY.

Bohr, N. (1921a). Atomic structure. *Nature* **107**, 104–107.

Bohr, N. (1921b). Atomic structure. *Nature* **108**, 208–209.

Bohr, N. (1922). The structure of the atom. Nobel Lecture, December 11, 1922. In: *Nobel Lectures, Physics 1922–1941*, (1965), Elsevier, Amsterdam.

Bohr, N. (1928). The quantum postulate and the recent development of atomic theory. *Nature* **121**, 580–591.

Bohr, N. (1936). Neutron capture and nuclear constitution. *Nature* **137**, 344–348.

Bohr, N. (1939). Disintegration of heavy nuclei. *Nature* **143**, 330.

Bohr, N. (1950). Open letter to the United Nations. *Science* **112**(2897), 1–6.

Bohr, N., Heisenberg, W., and Jordan, P. (1926). Quantum mechanics II. *Z. Phys.* **35**(8/9), 557–615.

Bohr, N., and Lindhard, J. (1954). Electron capture and loss by heavy ions penetrating through matter. *Kgl. Danske Videnskab. Selskab, Mat.-Fys. Medd.* **28**(7), 1–30.

Bolotovski, B. M., Vavilov, Y. N., and Kirkin, A. N. (1998). Sergei Ivanovich Vavilov, the man and the scientist: a view from the threshold of the 21st century. *Uspekhi Fizicheskikh Nauk Russ. Acad. Sci.* **41**(5), 487–504.

Bolotovski, B. M., Vavilov, Y. N., and Schmeleva, A. P. (2004). Sergei Vavilov: Luminary of Russian Physics. *Cern Courier*, 12 November 2004, Institute of Physics, London.

Boltz, C. L. (1970). *Ernest Rutherford*, Heron Books, Norwich, UK, p. 371.

Bonardi, M. L., Birattari, C., Groppi, F., Gini, L., and Mainardi, H. S. C. (2004). Cyclotron production and quality control of "high specific activity" radionuclides in "no-carrier added" form for radioanalytical applications in the life sciences. *J. Radioanal. Nucl. Chem.* **250**(3), 419–425.

Bonbardieri, E., Buscombe, J., Lucignani, G., and Schober, O. (Eds.) (2007). Advances in Nuclear Oncology. Informa Healthcare, p. 600.

Bone, A. (2005). The Russell-Einstein Manifest and the Origins of Pugwash. Eric Fawcett Memorial Lecture to Canadian Pugwash and Science for Peace. Joint Forum of the Canadian Pugwash Group and Science for Peace, October 1, 2005, p. 11.

Bonn, J., Bornshein, L., Bornshein, B., Fickinger, L., Flatt, B., Kraus, Ch., Otten, E. W., Schnall, J. P., Ulrich, H., Weinheimer, Ch., Kazachenka, O., and Kovalik, A. (2005). The Mainz neutrino mass experiment. In: Proceedings of the 19th International Conference on Neutrino Physics and Astrophysics (Neutrino-2000), Sudbury, Canada, June 16–21, 2000, Elsevier, Amsterdam.

Bonner, T. W., and Brubaker, W. M. (1936a). The disintegration of nitrogen by neutrons. *Phys. Rev.* **49**, 223–229.

Bonner, T. W., and Brubaker, W. M. (1936b). The disintegration of nitrogen by slow neutrons. *Phys. Rev.* **49**, 778.

Borisenko, V. E., and Ossicini, S. (2004). *What is What in the Nanoworld: A Handbook on Nanoscience and Nanotechnology*, Wiley-VCH Verlag GmbH & Co. KgaA, Weinheim, Germany.

Born, M. (1926a). The quantum mechanics of the impact process. *Z. Phys.* **37**(12), 863–867.

Born, M. (1926b). Quantum mechanics in impact process. *Z. Phys.* **38**(11/12), 803–840.

Born, M. (1926c). Zur Wellenmechanik der Stossvorgänge. *Göttingen Nachr. Math. Phys. Kl.*, 146–160.

Born, M. (1926d). The adiabaten principle in the quantum mechanics. *Z. Phys.* **40**(3/4), 167–192.

Born, M. (1954). The statistical interpretation of quantum mechanics. In: *Nobel Lectures, Physics, 1942–1962*, (1964), (eds. Nobel Foundation Staff), Elsevier, Amsterdam.

Born, M. (1969). *Atomic Physics*, Blackie and Son, Ltd., Glasgow, Scotland, p. 36–37.

Born, M. (1978). *My Life: Recollections of a Nobel Laureate*, 1st U.S. edn., Schribner, NY, p. 308.

Born, M., and Fock, V. (1928). Bewas des Adiabatensatzes. *Z. Phys.* **51**, 165–180.

Born, M., Heisenberg, W., and Jordan, P. (1926). Quantum mechanics II. *Z. Phys.* **35**(8/9), 557–615.

Born, M., and Oppenheimer, R. (1927). Quantum theory of molecules. *Ann. Phys.* **84**(20), 457–484.

Bornemann, A., Aschwer, U., and Mutterlose, J. (2003). The impact of calcerous nannofossils on the pelagic carbonate accumulation across the Jurassic-Cretaceous boundary interval. *Palaeogeogr. Palaeoclimatol. Palaeoecol.* **199**, 187–228.

Bothe, W. (1926). The interconnection between elementary radiation activities. *Z. Phys.* **37**, 547–567.

Bothe, W. (1954). The coincidence method. Nobel Lecture. In: *Nobel Lectures, Physics, 1942–1962*, (1964), (eds. Nobel Foundation Staff), Elsevier, Amsterdam.

Bothe, W., and Becker, H. (1930). Künstliche Erregung von Kern-Strahlen. *Z. Phys.* **66**, 289.

Bothe, W., and Geiger, H. (1924). Über das Wesen des Compton effecks. *Z. Phys.* **26**, 44.

Bowen, I. S., Millikan, R. A., Korff, S. A., and Neher, H. V. (1936). The latitude effect in cosmic rays at altitudes of up to 29,000 feet. *Phys. Rev.* **50**, 579–581.

Bowen, I. S., Millikan, R. A., and Neher, H. V. (1934). A very high altitude survey of the effect of latitude upon cosmic ray intensities and an attempt at a general interpretation of cosmic ray phenomena. *Phys. Rev.* **46**, 641–652.

Bowen, I. S., Millikan, R. A., and Neher, H. V. (1937). The influence of the earth's magnetic field on cosmic-ray intensities up to the top of the atmosphere. *Phys. Rev.* **52**, 80–88.

Bradt, H. L., and Peters, B. (1948). Investigation of the primary cosmic radiation with nuclear photographic emulsions. *Phys. Rev.* **74**, 1828–1837.

Bragg, W. L. (1913). The diffraction of short electromagnetic waves by a crystal. *Proc. Camb. Philos. Soc.* **17**, 43–57.

Bragg, W. L. (1914a). The analysis of crystals by the x-ray spectrometer. *Proc. R. Soc. Lond., Ser. A* **89**, 468–489.

Bragg, W. L. (1914b). *X-rays and Crystaline Structure*, Morse King Publishers, p. 802.

Bragg, W. L. (1922). The diffraction of x-rays by crystals. Nobel Lecture, September 6, 1922. In: *Nobel Lectures, Physics, 1901–1921*, (1967), Elsevier, Amsterdam.

Brajnik, D., Korpar, S., Medin, G., Starič, M., and Stanovnik, A. (1994). Measurement of ^{90}Sr activity with Cherenkov radiation in a silica aerogel. *Nucl. Instrum. Methods Phys. Res. A* **353**, 217–221.

Brajnik, D., Medin, G., Stanovnik, A., and Starič, M. (1995). Determination of high energy β-emitters with a Ge spectrometer or Cherenkov radiation in a silica aerogel. *Sci. Total Environ.* **173/174**, 225–230.

Brasch, A., Lange, F., Waly, A., Banks, T. E., Chalmers, T. A., Szilard, L., and Hopwood, F. L. (1934). Liberation of neutrons from beryllium by x-rays: radioactivity induced by means of electron tubes. *Nature* **134**, 880.

Brown, J. L., Glaser, D. A., and Perl, M. L. (1956). Liquid xenon bubble chambers. *Phys. Rev.* **102**, 586–589.

Brown, L. M. (2000). Book review: The discovery of antimatter: the autobiography of Carl David Anderson, the youngest man to win the Nobel Prize. *Phys. Today* **53**(10), 81.

Browne, E., Firestone, R. B., and Shirley, V. S. (1986). *Table of Radioactive Isotopes*, Wiley, New York, NY.

Bryant, P. J. (1992). A brief history and review of accelerators. In: *Proceedings, General Accelerator Physics*, Vol. I, CERN, Geneva, pp. 1–16.

Buchtela, K. (2003). Gas ionization detectors. pp. 123–178, In: *Handbook of Radioactivity Analysis*, 2nd edn. (ed. M. F. L'Annunziata), Elsevier, Amsterdam, p. 1273.

Buchtela, K., and Tschurlovits, M. (1975). A new method for determination of ^{89}Sr, ^{90}Sr, and ^{90}Y in aqueous solution with the aid of liquid scintillation counting. *Int. J. Appl. Radiat. Isot.* **26**, 333–338.

Burcham. W. E., and Goldhaber, M. (1936). The disintegration of nitrogen by slow neutrons. *Proc. Camb. Philos. Soc.* **32**, 632–636.

Burcham, W. E., and Jobes, M. (1994). *Nuclear and Particle Physics*, Longman Scientific & Technical, Essex.

Burden, D. L., and Hieftje, G. M. (1998). Cerenkov radiation as a UV and visible light source for time-resolved fluorescence. *Anal. Chem.* **70**(16), 3426–3433.

Buttlar, H. V. (1968). *Nuclear Physics, an Introduction*, Academic Press, New York, NY.

Byers, N. (2000). *Marietta Blau*, University of California, Los Angeles website http://cwp.library.ucla.edu.

Byrne, J. (1994). *Neutrons, Nuclei and Matter: An Exploration of the Physics of Slow Neutrons*, Institute of Physics, London.

Caffee, M. W., Finkel, R. C., Nimz, G. J., and Borchers, J. (1992). Isotopic composition of chlorine in groundwater from the Wawona basin, Yosmite National Park. *EOS Trans. Am. Geophys.l Union* **73**, 173.

Callahan, A. C., *et al.* (1966). Measurement of the $K^+_{\mu 3}$ decay parameters. *Phys. Rev.* **150**, 1153–1164.

Calvin, M. (1949). The path of carbon in photosynthesis. VI. *J. Chem. Educ.* **26**, 623–657.

Calvin, M. (1953). The path of carbon in photosynthesis. *Chem. Eng. News* **31**, 1622–1625.

Calvin, M., Bassham, J. A., and Benson, A. A. (1950). Chemical transformations of carbon in photosynthesis. *Fed. Proc.* **9**(2), 524–534.

Calvin, M., and Benson, A. A. (1948). The path of carbon in photosynthesis. *Science* **107**, 476–480.

Camerini, U., Muirhead, H., Powell, C. F., and Ritson, M. (1948). Observations on slow mesons of the cosmic radiation. *Nature* **162**, 433–438.

Canberra Nuclear. (1996). Neutron detection and counting. In: *Canberra Nuclear Instruments Catalog*, 9th edn., Canberra Industries, Meriden, CT, pp. 37–39.

Carlson, A. G., Hooper, J. E., and King, D. T. (1950). Nuclear transmutations produced by cosmic-ray particles of great energy—Part V. The neutral meson. *Philos. Mag.* **41**, 701.

Carmon, B., and Dyer, A. (1987). Čerenkov spectroscopic assay of fission isotopes. II. Čerenkov counting of ^{106}Ru on UV-colour-quenched solutions containing other β-emitters. *J. Radioanal. Nucl. Chem. Articles* **109**, 229–236.

Caroll, J., and Lerche, I. (2003). *Sedimentary Processes: Quantification using Radionuclides (Radioactivity in the Environment)*, Elsevier, Amsterdam, p. 282.

Casnati, E., Baraldi, C., Boccaccio, P., Bonifazzi, C., Singh, B., and Tartari, A. (1998). The effect of delta rays on the ionometric dosimetry of proton beams. *Phys. Med. Biol.* **43**, 547–558.

Cathcart, B. (2004). *The Fly in the Cathedral: How a Group of Cambridge Scientists Won the International Race to Split the Atom*, Farrar, Straus and Giroux, New York, NY, p. 320.

CERN Courier. (2004). *Bookshelf: Marietta Blau – Sterne der Zertrümmerung*. January/February 2004, Vol. 44 (No. 1).

Chadwick, J. (1932a). Possible existence of a neutron (received Feb. 27, 1932). *Nature* **129**, 312.

Chadwick, J. (1932b). The existence of a neutron (received May 10, 1932). *Proc. R. Soc. Lond. Ser. A* **136**, 692–708.

Chadwick, J. (1935). The neutron and its properties. Nobel Lecture, December 12, 1935. In: *Nobel Lectures, Physics 1922–1941*, (1965), Elsevier, Amsterdam.

Chadwick, J., and Goldhaber, M. (1934). A "nuclear photo-effect": Disintegration of the diplon by γ-rays. *Nature* **134**, 237–238.

Chamberlain, O. (1959). The Early Antiproton Work, Nobel Lecture, December 11, 1959. In: *Nobel Lectures, Physics, 1942–1962*, (1964), (eds. Nobel Foundation Staff), Elsevier, Amsterdam.

Chamberlain, O., Chupp, W. W., Ekspong, A. G., Goldhaber, G., Goldhaber, S., Lofgren, E. J., Segrè, E., Wiegand, C., Amaldi, E., Baroni, G., Castagnoli, C., Franzinetti, C., and Manfredini, A. (1956). *Phys. Rev.* **102**, 921.

Chamberlain, O., Segrè, E., Wiegand, C., and Ypsilantis, T. (1955). Observation of antiprotons. *Phys. Rev.* **100**. 947–951.

Chang, T.-M., Chen, S.-C., King, J.-Y., and Wang, S.-J. (1996). Rapid and accurate determination of $^{89/90}$Sr in radioactive samples by Cerenkov counting. *J. Radioanal. Nucl. Chem. Articles* **204**, 339–347.

Charpak, G. (1969). Particle detection by gas discharges. *J. Phys. (Paris) Colloq.* **2**, 86–96.

Charpak, G. (1970). Evolution of the automatic spark chambers. *Annu. Rev. Nucl. Sci.* **20**, 195–254.

Charpak, G., Bouchlier, R., Bressani, T., Favier, J., and Zupancic, C. (1968). *Nucl. Instrum. Methods* **62**, 262–268.

Charpak, G., Rahm, D., and Steiner, M. (1970). Some developments in the operation of multiwire proportional chambers. *Nucl. Instrum. Methods* **80**, 14–34.

Charpak, G., and Sauli, F. (1978). The multistep avalanche chamber: A new high-rate, high-accuracy gaseous detector. *Phys. Lett.* **78B**, 523–528.

Charpak, G., and Sauli, F. (1979). Multiwire proportional chambers and drift chambers. *Nucl. Instrum. Methods* **162**, 405–428.

Chase, G. D., and Rabinowitz, J. L. (1968). *Principles of Radioisotope Methodology*, 3rd edn., Burgess Publishing Company, Minneapolis, MN, pp. 140–143.

Chaudhuri, A., and Glaser, D. A. (1991). Metastable motion anisotropy. *Vis. Neurosci.* **7**(4), 397–407.

Chaundy, R. (2002). Paul Dirac: The unsung genius. *In Depth News, BBC Newsmakers,* Friday, July 5, 2002 (http://news.bbc.co.uk).

Cherenkov, E. (2005). Pavel Cheerenkov—a person and a physicist—through the eyes of his daughter. *Nucl. Instrum. Methods Phys. Res.* **A553**, 1–8.

Cherenkov, P. A. (1934a). Visible light from clear liquids under the action of gamma radiation. *C.R. Dokl. Akad. Nauk, SSSR* **2**(8), 451–454.

Cherenkov, P. A. (1934b). Sichtbares Leichten von Reinen Flüssigkeiten unter der Einwirkung von γ-Strahlen. *C.R. Dokl. Akad. Nauk, SSSR* **2**(8), 455–457.

Cherenkov, P.A.(1936). Die Wirkung eines Magnetfeldes auf das durch Gamma-Strahlen Hervorgerfene sichtbare Leuchten der Flüssigkeiten. *Dokl. Akad. Nauk, SSSR* **3**(9), 413–416.

Cherenkov, P. A. (1937a). Sichtbares Leuchten der Reinen Flüssigkeiten unter der Einwirkung von Harten β-Strahlen. *Dokl. Akad. Nauk, SSSR* **14**(3), 101–105.

Cherenkov, P. A. (1937b). Winkelverteilung der Intensität des Leuchtens, das durch γ-Strahlen in Reinen Flüssigkeiten Hervorgerufen wird. *Dokl. Akad. Nauk, SSSR* **14**(3), 105–107.

Cherenkov, P. A. (1937c). Visible radiation produced by electrons moving in a medium with velocities exceeding that of light. *Phys. Rev.* **52**, 378–379.

Cherenkov, P. A. (1937d). Visible radiation of pure liquids under the action of fast electrons. *Byull. Akad. Nauk, SSSR* **4–5**, 455–492.

Cherenkov, P. A. (1938a). Spatial distribution of visible radiation produced by fast electrons. *Dokl. Akad. Nauk, SSSR* **21**(7), 319–321.

Cherenkov, P. A. (1938b). The spectrum of visible radiation produced by fast electrons. *Dokl. Akad. Nauk, SSSR* **20**(9), 651–655.

Cherenkov, P. A. (1938c). Absolute output of radiation caused by electrons moving within a medium with super-light velocity. *Dokl. Akad. Nauk, SSSR* **21**(8), 116–121.

Cherenkov, P. A. (1958). Nobel Lecture: Radiation of particles moving at a velocity exceeding that of light, and some of the possibilities for their use in experimental physics. Stockholm, December 11, 1958. In: *Nobel Lectures, Physics, 1942–1962,* (1964), (eds. Nobel Foundation Staff), Elsevier, Amsterdam.

Cherenkov, P. A. (1986). At the threshold of discovery. *Nucl. Instrum. Methods Phys. Res.* **A248**, 1–4.

Chmielewski, A. G., and Haji-Saeid, M. (2004). Radiation technologies: past, present and future. *Radiat. Phys. Chem.* **71**, 17–21.

Chourasia, A. R., and Chopra, D. R. (1997). Auger electron spectroscopy. In: *Handbook of Instrumental Techniques for Analytical Chemistry* (ed. F. A. Settle) Prentice Hall, Upper Saddle River, NJ, pp. 791–808.

Christian, P. E., and Waterstram, K. (2007). Nuclear medicine and PET/CT Technology and Techniques. Mosby, p. 688.

Clapp, T. G., Titus, K. J., Olson, L. H., and Dorrity, J. L. (1995). The on-line inspection of sewn seams. National Textile Center Annual Report (August), 221–230.

Clark, R. W. (1970). *Einstein: The Life and Times*, Avon Books, New York, NY, p. 681.

Claus, R., Seidel, S., Sulak, I., Bionta, R., Blewitt, G., *et al.* (1987). A waveshifter light collector for a water Cherenkov detector. *Nucl. Instrum. Methods Phys. Res. A* **261**, 540–542.

Clay, J. (1928). Penetrating radiation. *Proc. R. Acad. Amst.* **31**, 1091.

Clay, J. (1934). Dutch cosmic ray expedition, 1933. *Physica Vol. I:* 363–382.

Clem, J. M., and Evenson, P. A. (2002). Positron abundance in galactic cosmic rays. *Astrophys. J.* **568**, 216–219.

Cobelli, C., Foster, D., and Toffolo, G. (2006). *Tracer Kinetics in Biomedical research: From Data to Model*. Springer, pp. 468.

Cockcroft, J. D. (1951). Experiments on the interaction of high-speed nucleons with atomic nuclei. Nobel Lecture, December 11, 1951, Stockholm, Sweden. In: *Nobel Lectures, Physics, 1942–1962*, (1964), (eds. Nobel Foundation Staff), Elsevier, Amsterdam.

Cockcroft, J. D., Gilbert, C. W., and Walton, E. T. S. (1934a). Production of induced radioactivity by high velocity protons. *Nature* **133**, 328.

Cockcroft, J. D., Gilbert, C. W., and Walton, E. T. S. (1934b). Experiments with high velocity positive ions. IV. The production if induced radioactivity by high velocity protons and diplons. *Proc. R. Soc. Lond.* **148A**(863), 225–240.

Cockcroft, J. D., and Lewis, W. B. (1936a). Experiments with high velocity positive ions. V. Further experiments on the disintegration of boron. *Proc. R. Soc. Lond.* **154A**(881), 246–261.

Cockcroft, J. D., and Lewis, W. B. (1936b). Experiments with high velocity positive ions. VI. The disintegration of carbon, nitrogen, and oxygen by deuterons. *Proc. R. Soc. Lond.* **154A**(881), 261–279.

Cockcroft, J. D., and Walton, E. T. S. (1930). Experiments with high velocity positive ions. *Proc. R. Soc. Lond.* **129A**(811), 477–489.

Cockcroft, J. D., and Walton, E. T. S. (1932a). Artificial production of fast protons. *Nature* **129**, 242.

Cockcroft, J. D., and Walton, E. T. S. (1932b). Disintegration of lithium by swift protons. *Nature* **129**, 649.

Cockcroft, J. D., and Walton, E. T. S. (1932c). Experiments with high velocity positive ions. (I) Further developments in the method of obtaining high velocity positive ions. *Proc. R. Soc. Lond.* **136A**(830), 619–630.

Cockcroft, J. D., and Walton, E. T. S. (1932d). Experiments with high velocity positive ions. (II) The disintegration of elements by high velocity protons. *Proc. R. Soc. Lond.* **137A**(831), 229–242.

Cockcroft, J. D., and Walton, E. T. S. (1933). Disintegration of light elements by fast protons. *Nature* **131**, 23.

Cockcroft, J. D., and Walton, E. T. S. (1934). Experiments with high velocity positive ions. III. The disintegration of lithium, boron, and carbon by heavy hydrogen ions. *Proc. R. Soc. Lond.* **144A**(853), 704–720.

Collins, G. B., and Reiling, V. G. (1938). Čerenkov radiation. *Phys. Rev.* **54**, 499–503.

Compton, A. H. (1923a). A quantum theory of the scattering of x-rays by light elements. *Phys. Rev.* **21**(5), 483–502.

Compton, A. H. (1923b). The spectrum of scattered x-rays. *Phys. Rev.* **22**(5), 409–413.

Compton, A. H. (1927). X-rays as a branch of optics. Nobel Lecture, December 12, 1927. In: *Nobel Lectures, Physics, 1922–1941*, (1965), Elsevier, Amsterdam.

Compton, A. H. (1932). Variation of the cosmic rays with latitude. *Phys. Rev.* **41**, 111–113.

Compton, A. H. (1933). A geographic study of cosmic rays. *Phys. Rev.* **43**, 387–404.

Compton, A. H., and Simon, A. W. (1925a). Measurements of β-rays associated with scattered x-rays. *Phys. Rev.* **25**, 306–314.

Compton, A. H., and Simon, A. W. (1925b). Directed quanta of scattered x-rays. *Phys. Rev.* **26**, 289–299.

Cook, G. J. R., Maisey, M. N., Britton, K. E., and Chengazi, V. (2007). *Clinical Nuclear Medicine*, 4th edn., A Hoddar Arnold Publication, p. 896.

Cook, G. T., Passo, Jr., C. J., and Carter, B. (2003). Environmental liquid scintillation counting. In: *Handbook of Radioactivity Analysis*, 2nd ed. (ed. M. F. L'Annunziata), Elsevier, Amsterdam, p. 537–607.

Cook, I., Dolan, T., Gordon, C., Gulden, W., Marbach, G., Moshonas, K., Ohira, S., and Petti, D. (2001). Fusion reactor safety. Report on the 7th IAEA Technical Committee Meeting held at Cannes, France, 13–16 June 2000. *Nucl. Fusion* **40**(11), 1535–1750.

Cork, B., Lambertson, G. R., Piccioni, O., and Wenzel, W. A. (1956). Antineutrons produced from antiprotons in charge-exchange collisions. *Phys. Rev.* **104**, 1193–1197.

Coursey, B. M., Calhoun, J. M., and Cessna, J. T. (1993). Radioassays of yttrium-90 used in nuclear medicine. *Nucl. Med. Biol.* **20**(5), 693–700.

Coursey, B. M., and Nath, R. (2000). Radionuclide therapy. *Phys. Today Online* (April), p. 9.

Coutu, S., *et al.* (2000). Energy spectra, altitude profiles and charge ratios of atmospheric muons. *Phys. Rev.* **D62**, 032001.

Cowan, Jr., C. L. (1964). Anatomy of an experiment: An account of the discovery of the neutrino. *Smithsonion Institution Annual Report:* 409.

Cowan, Jr., C. L., Reines, F., Harrison, F. B., Anderson, E. C., and Hayes, F. N. (1953). Large liquid scintillation detectors. *Phys. Rev.* **90**(3), 493–494.

Cowan, Jr., C. L., Reines, F., Harrison, F. B., Kruse, H. W., and McGuire, A. D. (1956). Detection of the free neutrino: a confirmation. *Science* **124**, 103–104.

Cronin, J. W. (1999). Cosmic rays: the most energetic particles in the universe. *Rev. Mod. Phys.* **71**(2), S165–S172.

Crookes, W. (1903a). The emanation of Radium. *Proc. R. Soc.* Lond. **A71**, 405–408.

Crookes, W. (1903b). Certain properties of the emanation of Radium. *Chem. News* **87**, 241.

Crow, J. F. (2004). Addendum: Leo Szilard 1898–1964. Addendum to Maas, W. (2004). Leo Szilard. A personal remembrance. *Genetics* **167**, 555–558.

Cucinotta, F. A., Nikjoo, H., and Goodhead, D. T. (1998). The effects of delta rays on the number of particle-track traversals per cell in laboratory and space exposures. *Radiat. Res.* **150**, 115–119.

Curie, E. (1937). *Madame Curie: A Biography*, Country Life Press, Garden City, NY, p. 448.

Currie, L. A. (2004). The remarkable meterological history of radiocarbon dating. *J. Res. Natl. Inst. Stand. Technol.* **109**, 185–217.

Curie, M. (1911). Radium and the new concepts in chemistry. Nobel Lecture on December 11, 1911. In: *Nobel Prize Lectures in Chemistry, 1901–1970.* (1967). Nobel Foundation Staff, Eds., Vol. 1, 1901–1921, Vol. 2, 1922–1941, Vol. 3, 1942–1962. Elsevier, Amsterdam.

Curie, P. (1905). Radioactive substances, especially radium. Nobel Lecture on June 6, 1905. In: *Nobel Lectures in Physics, 1901–1970.* Vol. 1, *Nobel Lectures, Physics, 1901–1921*, (1967), (eds. Science Incorporated Elsevier), Elsevier, Amsterdam.

Dahlgren, E. W. (1911). Presentation Speech of Nobel Prize physics to Wilhelm Wien. In: *Nobel Lectures in Physics, 1901–1970*, Vol. 1—*Nobel Lectures, Physics, 1901–1921*, (1967), (eds. Science Incorporated Elsevier), Elsevier, Amsterdam.

Dalitz, R. H. (Ed.) (1995). *The Collected Works of P. A. M. Dirac: Volume I: 1924–1948*, Cambridge University Press, p. 1334.

Dannen, G. (1997). The Einstein-Szilard refrigerators. *Sci. Am.*, (Jan), **276**(1), 90–95.

Dannen, G. (1998). Leo Szilard the inventor. In: *Leo Szilard Centenary Volume*. (ed. George Marx), Proceedings of the Leo Szilard Centenary, Eötvös University, Budapest, February 9, 1998. Eötvös Physical Society, Budapest.

Dannen, G. (2001). Szilard's inventions patently halted. *Phys. Today*, March 2001.

Das, A., and Ferbel, T. (1994). *Introduction to Nuclear and Particle Physics*, Wiley, New York, NY.

Davisson, C. (1937). The discovery of electron waves. Nobel Lecture, December 13, 1937. In: *Nobel Lectures 1922–1941*. Elsevier, Amsterdam.

Davisson, C., and Germer, L. H. (1920). The emission of electrons from oxide-coated filaments under positive bombardment. *Phys. Rev.* **15**, 330–332.

Davisson, C., and Germer, L. H. (1927a). The scattering of electrons by a single crystal of nickel. *Nature* **119**, 558–560.

Davisson, C., and Germer, L. H. (1927b). Diffraction of electrons by a crystal of nickel. *Phys. Rev.* **30**, 705–741.

Davisson, C., and Germer, L. H. (1928). Diffraction of electrons by a single layer of atoms. *Phys. Rev.* **31**, 135.

Davisson, C., and Kunsman, C. H. (1922). The secondary electron emission from nickel. *Phys. Rev.* **20**, 110.

Davisson, C., and Kunsman, C. H. (1922). The scattering of electrons by aluminum. *Phys. Rev.* **19**, 534–535.

Davisson, C., and Kunsman, C. H. (1923). The scattering of low speed electrons by platinum and magnesium. *Phys. Rev.* **22**, 242–258.

Davisson, C., and Pidgeon, H. A. (1920). The emission of electrons from oxide-coated filaments. *Phys. Rev.* **15**, 553–555.

Dean, J. A. (1995). *Analytical Chemistry Handbook,* McGraw-Hill, New York, NY.

de Broglie, L. (1923a). Waves and Quanta. *Nature* **112**, 540.

de Broglie, L. (1923b). Ondes et quanta. *Compt. Rend.* **177**, 507–510.

de Broglie, L. (1923c). Quanta de lumière, diffraction et interférence. *Compt. Rend.* **177**, 548–550.

de Broglie, L. (1924). Rescherches sur la théorie des quantas. Thesis, Paris.

de Broglie, L. (1925). Rescherches sur la théorie des quantas. *Ann. Phys. (Paris)* **3**(10), 22–128.

Dee, P. I., and Walton, E. T. S. (1933). A photographic investigation of the transmutation of lithium and boron by protons and of lithium by ions of the heavy isotope of hydrogen. *Proc. R. Soc. Lond., A* **141**(845), 733–742.

de Hevesy, G. (1944). Some applications of isotopic indicators. Nobel Lecture, December 12, 1944. In: *Nobel Lectures, Chemistry 1942–1962.* (1964), Elsevier, Amsterdam.

de Hevesy, G., and Levi, H. (1936a). Action of slow neutrons on rare-earth elements. *Nature (London)* **137**, 185.

de Hevesy, G., and Levi, H. (1936b). The action of neutrons on rare-earth elements. *Kgl. Danske Vidensk. Selsk. Math-Fys. Medd.* **14**(5), 1–3.

de Hevesy, G., and Paneth, F. (1913a). Die Löslichkeit des Bleisulfids und Bleichromats. (RaD). *Z. Anorg. Chem.* **82**, 223.

de Hevesy, G., and Paneth, F. (1913b). RaD als Indikator des Bleis. *Z. Anorg. Chem.* **82**, 322.

Derenzo, S. E. (1969). Measurement of the low-energy end of the μ^+ decay spectrum. *Phys. Rev.* **181**, 1854–1866.

Dicke, R. H., Peebles, P. J. E., Roll, P. G., and Wilkinson, D. T. (1965). Cosmic black-body radiation. *Astrophys. J.* **142**, 414–419.

Dirac, P. A. M. (1928a). The quantum theory of the electron. *Proc. R.l Soc. Lond. A* **117**, 610–612.

Dirac, P. A. M. (1928b). The quantum theory of the electron, Part II. *Proc. R. Soc. Lond., A* **118**, 351–361.

Dirac, P. A. M. (1933). Theory of electrons and positrons. Nobel Lecture, December 12, 1933. In: *Nobel Lectures, Physics, 1922–1941*, (1965), Elsevier, Amsterdam.

Dolan, T. J. (1982). *Fusion Research: Principles, Experience and Technology*, Pergamon Press, New York, NY.

Domanus, J. C. (Ed.) (2004). *Practical Neutron Radiography*, Springer, p. 292.

Donahue, R. J., and Fassó, A. (2002). Radioactivity and radiation protection. *In* Hagiwara *et al.*, (Eds.). *Phys. Rev.* **D66**, 01001-1.

Drury, L. O'C. (1983). An introduction to the theory of diffuse shock acceleration of energetic particles in tenuous plasmas. *Rep. Prog. Phys.* **46**, 973–1027.

Dunne, P. (2002). A reappraisal of the mechanism of pion exchange and its implications for the teaching of particle physics. *Phys. Educ.* **37**(3), 211–222.

DuVernois, M. A., *et al.* (2000). Cosmic-ray electrons and positrons from 1 to 100 GeV: measurement with HEAT and their interpretation. *Astrophys. J.* **559**, 296–303.

Eary, J. F., and Brenner, W. (Eds.) (2007). *Nuclear Medicine Therapy*, Informa Healthcare, London, p. 500.

Eddington, A. S. (1926). Bakerian Lecture—Diffuse matter in interstellar space. *Proc. R. Soc. London, Ser. A* **111**, 424–456.

Ehman, W. D., and Vance, D. E. (1991). *Radiochemistry and Nuclear Methods of Analysis*, JWiley, New York, NY.

Einstein, A. (1905). Über einen die Erzeugung und Verwandlung des Lichtes betreffenden heuristischen Gesichtspunkt. *Annalen der Physik, Leipzig* **17**, 132–148.

Einstein, A. (1911). Über den Einfluß der Schwerkraft auf die Ausbreitung des Lichtes. *Annalen der Physik, Leipzig* **35**, 898–908.

Einstein, A. (1915). Erklärung der Perihelbewegung des Merkur aus der allgemeinem Relativitätstheorie. *SPAW* **1915**, 831–839.

Einstein, A. (1916). Die Grundlage der allgemeinen Relativitätstheorie. *Annalen der Physik, Leipzig* **49**, 769–822.

Einstein, A. (1923). Fundamental ideas and problems of the theory of relativity. Nobel Lecture delivered to the Nordic Assembly of Naturalists at Gothenburg, July 11, 1923. In: *Nobel Lectures, Physics, 1901–1921*, (1967), Elsevier, Amsterdam.

Einstein, A. (1931). The world as I see it. In: *Living Philosophies*, Simon and Schuster, New York, NY, pp. 3–7.

ElBaradei, M. (2004). Nuclear power, an evolving scenario. *IAEA Bull.* **46**(1), 4–8.

El-Dessouky, H. T., and Ettouney, H. M. (2002). *Fundamentals of Salt Water Desalination*, Elsevier, Amsterdam, p. 690.

Elkin, L. O. (2003). Rosalind Franklin and the double helix. *Phys. Today* (March), **56**(3), 42–48.

Elster, J., and Geitel, H. (1903). Über die durch radioactive Emanation erregte scintillierende Phosphoreszenz der Sidot-Blende. *Phys. Z.* **4**, 439–440.

Engelmann, J. J., Ferrando, P., Soutoul, A., Goret, P., and Juliusson, E. (1990). Charge composition and energy spectra of cosmic-ray nuclei for elements from Be to Ni—Results from HEAO-3-C2. *Astron. Astrophys.* **233**, 96–111.

Estermann, I., and Stern, O. (1930). Beugung von Molekularstrahlen. *Z. Phys.* **61**, 95–125.

Evans, R. D. (1955). *The Atomic Nucleus.* 1st Ed. McGraw-Hill, New York, NY.

Evans, R. D. (1972). *The Atomic Nucleus*, McGraw-Hill, New York, NY.

Fajans, K., and Göring, O. H. (1913). Über das Uran X_2- das neue Element der Uranreihe. *Phys. Z.* **14**, 877–884.

Fakley, D. C. (1983). The British mission. *Los Alamos Science,* (Winter/Spring), **7**, 186–189.

Farhataziz, and Rodgers, M. A. J. (1987). *Radiation Chemistry, Principles and Applications*, VCH Publishers, New York, NY.

Feather, N. (1932). The collisions of neutrons with nitrogen nuclei. *Proc. R. Soc. Lond., A* **137**, 229–242.

Feather, N. (1938). Further possibilities for the absorption method of investigating the primary β-particles from radioactive substances. *Proc. Cambridge Philos. Soc.* **34**, 599–611.

Feather, N. (1940). *Lord Rutherford*, Blackie & Son, London.

Fedak, G. (2002). Origin of Marquis Wheat. Eastern Cereal and Oilseed Research Centre, Agriculture and Agri-Food Canada, Ottawa (http://sci.agr.ca/ecorc/au/marquis_e.htm).

Feld, B. T., and Weiss Szilard, G. (Eds.) (1972). *The Collected Works of Leo Szilard: Scientific Papers*, MIT Press, Cambridge, p. 642.

Federmann, G. (2003).Viktor Hess und die Entdeckung der Kosmischen Strahlung. Dimplomarbeit zur Erlangung des akademischen Grades Magister der Naturwissenschaften. Institut für Radiumforschung und Kernphysik, Vienna.

Fenyves, E., and Haiman, O. (1969). *The Physical Principles of Nuclear Radiation Measurements*, Academic Press, New York, NY.

Fermi, E. (1934a). Versuch einer theorie der β-Strahlen. *Z. Phys.* **88**, 161–177.

Fermi, E. (1934b). Attempt at a theory of beta rays. *Nuovo Cimento* **11**, 1–21.

Fermi, E. (1934c). Radioactivity induced by neutron bombardment. *Nature* **133**, 757.

Fermi, E. (1934d). Possible production of elements of atomic number higher than 92. *Nature* **133**, 898–899.

Fermi, E. (1938). Artificial radioactivity produced by neutron bombardment. Nobel lecture, December 12, 1938. In: *Nobel Lectures, Physics, 1922–1941*, (1965), Elsevier, Amsterdam.

Fermi, E. (1940a). Reactions produced by neutrons in heavy elements. *Science* **92**, 269–271.

Fermi, E. (1940b). Reactions produced by neutrons in heavy elements. *Nature* **146**, 640–642.

Fermi, E. (1949). On the origin of the cosmic radiation. *Phys. Rev.* **75**(8), 1169–1174.

Fermi, E., and Amaldi, E. (1936). On the absorption and the diffusion of slow neutrons. *Phys. Rev.* **50**, 899–928.

Fermi, E., Amaldi, E., D'Agostino, O., Rasetti, F., and Segre, E. (1934). Artificial radioactivity produced by neutron bombardment. *Proc. R. Soc. Lond., A* **146**(857), 483–500.

Fermi, E., Amaldi, E., and Wick, G. C. (1938). On the albedo of slow neutrons. *Phys. Rev.* **53**, 493.

Filby, R. H. (1995). Isotopic and nuclear analytical techniques in biological systems: A critical study. Part IX. Neutron activation analysis. *Pure Appl. Chem.* **67**(11), 1929–1941.

Firestone, R. B., Shirley, V. S., Baglin, C. M., Frank Chu, S. Y., and Zipkin, J. (1996). *Table of Isotopes*, Vols. I and II, 8th edn., Wiley, New York, NY.

Flammersfeld, A. (1946). Eine Beziehung zwischer Energie und Reichweite für Beta-Strahlen kleiner und mittlerer Energie. *Naturwissenschaften* **33**, 280–281.

Fletcher, H. (1982). My work with Millikan on the oil-drop experiment. *Phys. Today*, June 1982, 43–46.

Fowler, P. H. (1950). Nuclear transmutations produced by cosmic-ray particles of great energy. III. Nature of the slower particles. *Philos. Mag.* **41**, 169–184.

Fox, M. S. (1998). Some recollections and reflections on mutation rates. *Genetics* **148**, 1415–1418.

Frame, P. W. (1997). Tales from the atomic age. Fermi strikes gold and Hevesy invents neutron activation analysis. *Health Phys. Newsl.*, April .

Franck, J. (1926). Transformations of kinetic energy of free electrons into excited energy of atoms by impacts. Nobel Lecture, December 11, 1926. In: *Nobel Lectures. Physics 1922–1941*. Elsevier, Amsterdam.

Franck, J. (1935). Remarks on photosynthesis. *Chem. Rev.* **17**, 433–438.

Franck, J., and Hertz, G. (1914a). Über Zusammenstösse zwischen Elektronen und den Molekülen des Quecksilberdampfes und die Ionisierungsspannung desselben. *Verh. Phys. Ges. Berlin* **15**, 457–467.

Franck, J., and Hertz, G. (1914b). Über die Erregung der Quecksilberresonanzlinie 253.6p.p. durch Elektronentösse. *Verh. Phys. Ges. Berlin* **15**, 512–517.

Franck, J., and Herzfeld, K. E. (1941). Contribution to a theory of photosynthesis. *J. Phys. Chem.* **45**, 978–1025.

Franck, J., Hughes, D. J., Nickson, J. J., Rabinowitch, E., Seaborg, G. T., Stearns, J. C., and Szilard, L. (1945). The Franck Report. Report of the Committee on Political and Social Problems, Manhattan Project, Metallurgical Laboratory, University of Chicago, June 11, 1945. U.S. National Archives, Record Group 77, Manhattan Engineer District Records, Harrison-Bundy File, Folder #76, Washington, DC.

Franck, J., and Rosenberg, J. L. (1964). A theory of light utilization in plant photosynthesis. *J. Theor. Biol.* **7**, 276–301.

Frank, I., and Tamm, I. (1937). Coherent visible radiation of fast electrons passing through matter. *Dokl. Akad. Nauk, SSSR* **14**(3), 109–114.

Frank, I. M. (1944). Interference phenomena in the case of Cherenkov radiation. *C. R. Dokl. Akad. Nauk, SSSR* **42**(8), 341–344.

Frank, I. M. (1958). Optics of light sources in refractive media. Nobel Lecture, December 11, 1958. In *Nobel Lectures, Physics, 1942–1962*, (1964). Nobel Foundation Staff, Eds., Elsevier, Amsterdam.

Frank, I. M. (1979). *Sergei Ivanovich Vavilov: Ocherki I Vospominaniia*, Vol. 1, Nauka, Moscow, p. 296.

Frank, I. M. (1981). *Sergei Ivanovich Vavilov: Ocherki I Vospominaniia*, Vol. 2, Nauka, Moscow, p. 350.

Frank, I. M. (1988). *Izluchenie Vavilova-Cherenkova: Voprosy Teorii (Vavilov-Cherenkov Radiation: Theoretical Aspects)*, Nauka, Moscow, p. 285.

Frank, I. M. (1991). *Sergei Ivanovich Vavilov: Ocherki I Vospominaniia*, Vol. 3, Nauka, Moscow, p. 374.

Freier, P., Lofgren, E. J., Ney, E. P., Oppenheimer, F., Bradt, H. L., and Peters, B. (1948). Evidence for heavy nuclei in the primary cosmic radiation. *Phys. Rev.* **74**, 213–217.

Friedlander, G., Kennedy, J. W., and Miller, J. M. (1964). *Nuclear and Radiochemistry*, 2nd edn., Wiley, New York, NY.

Friedrich, W., Knipping, P., and von Laue, M. (1912). Interferenz-Erscheinungen bei Röntgenstrahlen. *Sitzunsberichte der Königlich Bayerischen Akademie der Wissenschaften* 303–322.

Frisch, O. (1978). Lise Meitner, nuclear pioneer. *New Sci.t* **80**(1128), 426–428.

Fritz, P., Kraus, H. J., Mühlnickel, W., Hammer, U., Dölken, W., Engel-Riedel, W., Chemaissani, A., and Stoelben, E. (2006). Stereotactic, single-dose irradiation of stage I non-small cell lung cancer and lung metastases. *Radiat. Oncol.* **1**, 30.

Gaisser, T. K. (2000). Origin of cosmic radiation. *Astron. AIP Conf. Proc.* **558** (2001), 27–42.

Gaisser, T. K., and Maurer, R. H. (1973). Cosmic $\bar{p}$ production in interstellar pp collisions. *Phys. Rev. Lett.* **30**, 1264–1267.

Gaisser, T. K., and Schaefer, R. K. (1992). Cosmic-ray secondary antiprotons: a closer look. *Astrophys. J.* **394**, 174–184.

Gaisser, T. K., and Stanev, T. (2002). Cosmic rays. *In* Hagiwara, *et al.*, (Eds.) "The Review of Particle Physics". *Physical Review* **D66**, 23.1–23.19 (University of California, Lawrence Berkeley Laboratories).

Galison, P. L. (1997a). Marietta Blau: Between Nazis and Nuclei. pp. 146–159. In: *Image and Logic: A Material Culture of Microphysics*. University of Chicago Press, p. 982.

Galison, P. L. (1997b). Marietta Blau: Between Nazis and nuclei. *Phys. Today* (November), 42–48.

Gamow, G. (1948). Evolution of the universe. *Nature* **162**, 680–682.

Gamow, G., and Houtermans, F. (1929). Quantenmechik der radioactive Kerns. *Zeitschrift fuer Physik* **57**(7–8), 496–509.

Gautier, M. A. (Ed.) (1995). Health and environmental chemistry: Analytical techniques, data management, and quality assurance. Manual LA-10300-M, Vol. III, UC-907, pp. WR190-1-WR190-16. Los Alamos National Laboratory, Los Alamos, NM.

Gautreau, R., and Savin, W. (1999). *Theory and Problems in Modern Physics*, 2nd edn., McGraw-Hill, New York, NY.

Geiger, H. (1908). The scattering of the alpha particles by matter. *Proc. R. Soc. Lond., A* **81**, 174–177.

Geiger, H., and Marsden, E. (1909). On a diffuse reflection of the α-particles. *Proc. R. Soc. Lond. Ser. A* **82**, 495–500.

Gerward, L. (1999). Paul Villard and his discovery of gamma rays. *Phys. Perspect.* **1**, 367–383.

Gerward, L., and Rassat, A. (2000). Paul Villard's discovery of gamma rays – A centenary. *C. R. Acad. Sci., Paris* **t. 1, Ser. IV**, 965–973.

Gibbs, W. W. (1998). A massive discovery. *Sci. Am.* **279**(2), 18–20.

Gibson, J. A. B., and Piesch, E. (1985). Neutron Monitoring for Radiological Protection. Technical Report Series No. 252, International Atomic Energy Agency, Vienna.

Giese, R., Gildemeister, O., Paul, W., and Schuster, G. (1970). A high resolution Cherenkov chamber. *Nucl. Instrum. Methods* **88**, 83–92.

Ginzburg, V. L. (2001). *The Physics of a Lifetime: Reflections on the Problems and Personalities of 20th Century Physics*, Springer, Berlin, pp. 396–402.

Glaser, D. A. (1952). Some effects of ionizing radiation on the formation of bubbles in liquids. *Phys. Rev.* **87**, 665.

Glaser, D. A. (1953). Bubble chamber tracks of penetrating cosmic-ray particles. *Phys. Rev.* **91**, 762–765.

Glaser, D. A. (1960). Elementary Particles and Bubble Chambers, Nobel Lecture, December 12, 1960. In: *Nobel Lectures, Physics, 1942–1962*, (1964), (eds. Nobel Foundation Staff), Elsevier, Amsterdam.

Glaser, D. A. (1994). Invention of the bubble chamber and subsequent events. *Nucl. Phys. B* (Proc. Suppl.) **36**, 3–18.

Glaser, D. A., and Barch, D. (1999). Motion detection and characterization by an excitable membrane: The Bow Wave Model. *Neurocomputing* **26–27**, 137–146.

Glaser, D. A., and Rahm, D. C. (1955). Characteristics of bubble chambers. *Phys. Rev.* **97**, 474–479.

Glaser, D. A., Rahm, D. C., and Dodd, C. (1956). Bubble counting for the determination of the velocities of charged particles in bubble chambers. *Phys. Rev.* **102**, 1653–1658.

Glässel, P. (1999). The limits of the ring image Cherenkov technique. *Nucl. Instrum Methods Phys. Res. A* **433**, 17–23.

Glasser, R. G., Seeman, N., and Stiller, B. (1961). Mean lifetime of the neutral pion. *Phys. Rev.* **123**, 1014–1020.

Glendenin, L. E. (1948). Determination of the energy of beta particles and photons by absorption. *Nucleonics* **2**, 12–32.

Golden, R. L., *et al.* (1979). Evidence for the existence of cosmic-ray antiprotons. *Phys. Rev. Lett.* **43**, 1196–1199.

Golden, R. L., Mauger, B. G., Nunn, S., and Horan, S. (1984). Energy dependence of the $\bar{p}/p$ ratio in cosmic rays. *Astrophys. Lett.* **24**, 75–83.

Goldsmith, B. (2004). *Obsessive Genius: The Inner World of Marie Curie (Great Discoveries)*, W.W. Norton & Co., 320 pp.

Goldstein, D. (2001). In defense of Robert Andrews Millikan. *Am. Sci.*, **89**(1)(Jan./Feb), 54.

Goldstein, G. R. (2001). A review essay on "Lise Meitner and the Dawn of the Nuclear Age" by Patricia Rife. *Peace Change J. Peace Res.* **26**, 95.

Gonfiantini, R., Fröhlich, K., Araguás-Araguás, L., and Rozanski, K. (1998). Isotopes in groundwater hydrology, pp. 203–245. In: *Isotope Tracers in Catchment Hydrology*, (eds. C. Kendall and J. J. McDonnell), Elsevier, Amsterdam, pp 839.

Gorelink, G. E. (1995). Moscow Physics, 1937. In: *Tragicheskie Sud'by* (Tragic Destinies). Nauka, Moscow, pp. 54–75.

Govorkov, B. B. (2005). Cherenkov detectors in Cherenkov's laboratory. *Nucl. Instrum. Methods Phys. Res.* **A553**, 9–17.

Granqvist, G. (1915). Presentation Speech for the Nobel Prize in Physics 1914. In: *Nobel Lectures, Physics 1901–1921*, (1967), Elsevier, Amsterdam.

Gratta, G., and Wang, Y. F. (1999). Towards two-threshold, real-time solar neutrino detectors. *Nucl. Instrum. Methods Phys. Res. A* **438**, 317–321.

Grau Carles, A., and Grau Malonda, A. (1995). Radionuclide standardization by Cherenkov counting. *Appl. Radiat. Isot.* **46**, 799–803.

Grau Carles, A., and Grau Malonda, A. (1996). "Applicación de al Radiación Cherenkov a la Metrología de Radionucleidos." Editorial CIEMAT, Avda. Complutense, 22, 28040 Madrid, p. 140.

Grau Malonda, A. (1999). *Free parameter Models in Liquid Scintillation Counting*, Editorial CIEMAT, Madrid, pp. 146.

Grau Malonda, A., and Grau Carles, A. (1998a). The anisotropy coefficient in Cerenkov counting. Proceedings of ICRM 1997. National Institute of Standards and Technology, Gaithersburg, MD.

Grau Malonda, A., and Grau Carles, A. (1998b). The anisotropy coefficient in Cerenkov counting. *Appl. Radiat. Isot.* **49**(9–11), 1049–1053.

Grau Malonda, A., and Grau Carles, A. (2000). Standardization of electron-capture radionuclides by liquid scintillation counting. *Appl. Radiat. Isot.* **52**, 657–662.

Grau Malonda, A., and Grau Carles, A. (2002). Half-life determination of ^{40}K by LSC. *Appl. Radiat. Isot.* **56**, 153–156.

Green, J. H., and Maddock, A. G. (1949). (n, γ) recoil effects in potassium chromate and dichromate. *Nature* **164**, 788–789.

Gruhn, C. R., and Ogle, W. (1980). In: *Liquid Scintillation Counting, Recent Applications and Developments* (eds. C. T. Peng, D. L. Horrocks and E. L. Alpen) Vol. 1, Academic Press, New York and London, pp. 357–374.

Grünewald, Th. (1984). Food preservation. pp. 271–301. In: *Isotopes and Radiation in Agricultural Sciences*, Vol. 2, *Animals, Plants, Food and the Environment*, (eds. M. F. L'Annunziata and J. O. Legg), Academic Press, London, p. 356.

Gullstrand, A. (1923). Presentation Speech, The Nobel Prize in Physics 1923. In: *Nobel Lectures, Physics 1922–1941*, (1965), Elsevier, Amsterdam.

Gullstrand, A. (1925). Presentation Speech, The Nobel Prize in Physics 1924. In: *Nobel Lectures, Physics 1922–1941*, (1965), Elsevier, Amsterdam.

Gurney, R. W., and Mott, N. F. (1938). The theory of the photolysis of silver bromide and the photographic latent image. *Proc. R. Soc. Lond., A* **164**, 151–167.

Hahn, O., and Meitner, L. (1918). Die Muttersubstanz des Actiniums, eines neues radioactives Element von langer Lebendauer. *Phys. Z.* **19**, 208–218.

Hahn, O., and Strassmann, F. (1939a). Über den Nachweis und das Verhalten der bei Bestrahlung des Urans mittels Neutronen entstehenden Erdalkalimetalle. *Naturwissenschaften* **27**, 11–15.

Hahn, O., and Strassmann, F. (1939b). Nachweis der Entstehung activer bariumisotope aus Uran und Thorium durch Neutronenbestrahlung; nachweis weiterer aktiver Bruchtucke bei der Uranspaltung. *Naturwissenschaften* **27**, 89–95.

Hall, B. (1995). A tribute to Hans Albrecht Bethe. *Cornell Magazine*, August–September Issue.

Halpern, A. (1988). *Schaum's 3000 Solved Problems in Physics*, McGraw-Hill, New York, NY.

Halpern, L. (1993). Marietta Blau (1894–1970). In: *Women in Chemistry and Physics: A Bibliographic Sourcebook* (eds. L. S. Grinsten, R. K. Rose and M. H. Rafailovich) Greenwood Press, Westport, CT, p. 736.

Hammer, G. P., Zeeb, H., Tveten, U., and Blettner, M. (2000). Comparing different methods of estimating cosmic radiation exposure of airline personnel. *Radiat. Environ. Biophys.* **39**, 227–231.

Harbottle, G. (1954). Szilard-Chalmers reaction in crystalline compounds of chromium, radiochemical analysis. *J. Chem. Phys.* **22**, 1083.

Harms, A. A., and Wyman, D. R. (2005). *Mathematics and Physics of Neutron Radiography*, Springer, p. 184.

Hatch, M. D., and Slack, C. R. (1966). Photosynthesis by sugarcane leaves. A new carboxylation reaction and the pathway of sugar formation. *Biochem. J.* **101**, 103–111.

Hawking, S. W. (1988). *A Brief History of Time*, Bantam Press, London.

Hawryluk, R. J., Adler, H., Alling, P., Ancher, C., Anderson, H., Anderson, J. L., Ascroft, D., Barnes, C. W., and Barnes, G. (1994). Confinement and heating of a deuterium-tritium plasma. *Phys. Rev. Lett.* **72**, 3530–3533.

Hayashida, N., *et al.* (1994). Observation of a very energetic cosmic ray well beyond the predicted 2.7 K cutoff of the primary energy spectrum. *Phys. Rev. Lett.* **73**, 3491–3494.

Heaviside, O. (1888). The electromagnetic effects of a moving charge. *The Electrician* **22**, 147–148.

Heaviside, O. (1889). On the electromagnetic effects due to the motion of electricity through a dielectric. *Philos. Mag.* **27**, 324–339.

Heaviside, O. (1892). *Electrical Papers*, Vol. II, Macmillan, London, pp. 494–497.

Heaviside, O. (1899). Note on the motion of a charged body at a speed equal to or greater than that of light. In: *Electromagnetic Theory*, Vol. 2, Electrician Printing and Publishing Co., London, pp. 533–536. Reprinted in 2001 by Elibron Classics, Adamant Media Corp., Boston, p. 592.

Heaviside, O. (1912a). Theory of the steady rectilinear motion of a point-charge or 'electron' through the ether when $u < v$ and when $u > v$. In: *Electromagnetic Theory*, Vol. 3, Electrician Printing and Publishing Co., London, pp. 25–28. Reprinted in 2001 by Elibron Classics, Adamant Media Corp., Boston, p. 556.

Heaviside, O. (1912b). The steady rectilinear motion in its own line of a terminated electrified line when $u > v$, and interpretation of the impure conical wave following an electron. In: *Electromagnetic Theory*, Vol. 3, Electrician Printing and Publishing Co., London pp. 30–33. Reprinted in 2001 by Elibron Classics, Adamant Media Corp., Boston, MA, p. 556.

Heilbron, J. L. (2003). *Ernest Rutherford and the Explosion of Atoms*, Oxford University Press, New York, NY, p. 139.

Heisenberg, W. (1926a). Quantum mechanics. *Naturwissen.* **14**, 989–994.

Heisenberg, W. (1926b). Multi-body problem and resonance in the quantum mechanics. *Z. Phys.* **38**(6/7), 411–426.

Heisenberg, W. (1926c). Fluctuation appearances and quantum mechanics. *Z. Phys.* **40**(7), 501–506.

Heisenberg, W. (1927). Über den anschaulichen Inhalt der quantentheoretischen Kinematik und Mechanik. *Z. Phys.* **43**, 172–198.

Heisenberg, W. (1929). The development of the quantum theory 1918–1928. *Naturwissen.* **17**, 490–496.

Heisenberg, W. (1932a). Über den Bau der Atomkerne. I. *Z. Phys.* **77**, 1–11.

Heisenberg, W. (1932b). Über den Bau der Atomkerne. II. *Z. Phys.* **78**, 156–168.

Heisenberg, W. (1933a). Development of Quantum Mechanics. Nobel Lecture, December 11, 1933. In: *Nobel Lectures, Physics 1922–1941*, (1965), Elsevier, Amsterdam.

Heisenberg, W. (1933b). Werner Heisenberg—Biography. In: *Nobel Lectures. Physics 1922–1941.* (1965), Elsevier, Amsterdam.

Heisenberg, W. (1933c). Über den Bau der Atomkerne. III. *Z. Physik* **80**, 587–596.

Heisenberg, W. (1946). Über die Arbeiten zur technischen Ausnutzung der Atomkernenergie in Deutschland. *Naturwissen.* **33**(11), 325–329.

Hertz, G. (1926). The results of the electron-impact tests in the light of Bohr's theory of atoms. Nobel Lecture, December 11, 1926. In: *Nobel Lectures. Physics 1922–1941.* Elsevier, Amsterdam.

Hertz, G. (1933). Manufacture of pure heavy hydrogen isotope by diffusion. *Naturwissenschaften* **21**, 884–885.

Hertz, H. (1887). Über einen Einfluss des ultravioletten Lichtes auf die elektrische Entladung. (An effect of ultraviolet light on electrical discharge). *Sitz.-ber. Berl. Akad., 9th June*, also *Wied. Ann Physik, Vol. 31.*

Hertz, H. (1892). Über den Durchgang von Kathodenstrahlen durch dünne Metallschichten. (The passage of cathode rays through thin metal layers). *Wied. Ann. Phys.* **45**, 88.

Hess, V. F. (1912). Über Beobachtungen der durchdringenden Strahlung bei sieben Freiballonfahrten. *Phys. Z.* **13**, 1084–1091.

Hess, V. F. (1932). The cosmic ray observatory at the Hafelekar. *Terr. Mag. Atmos. Elect.* September 1932, p. 399.

Hess, V. F. (1936). Unsolved problems in physics: tasks for the immediate future in cosmic rays studies. Nobel Lecture, December 12, 1939. In: *Nobel Lectures, Physics, 1922–1941*, (1965), Elsevier, Amsterdam.

Hess, V. F. (1939). The significance of variations in cosmic ray intensity and their relation to solar, earthmagnetic and atmospheric phenomena. *Rev. Mod. Phys.* **11**, 153–157.

Hess, V. F. (1940). The discovery of cosmic radiation. *Thought* (Fordham Quarterly) **15**, 225–236.

Hess, V. F., and Demmelmair, A. (1937). World-wide effect in cosmic ray intensity, as observed during a recent magnetic storm. *Nature* **140**, 316–317.

Hess, V. F., Demmelmair, A., and Steinmaurer, R. (1938a). Relations between terrestrial magnetism and cosmic ray intensity. *Wiener Sitz. Ber.* IIa. **147**, 89.

Hess, V. F., Steinmaurer, R., and Demmelmair, A. (1938b). Cosmic rays and the Aurora of January 25–26. *Nature* **141**, 686–687.

Hetherington, E. L., Sorby, P. J., and Camakaris, J. (1986). The preparation of high specific activity copper-64 for medical diagnosis. *Int. J. Rad. Appl. Instrum.* **37**(12), 1242–1243.

Hildebrand, R. H., and Nagle, D. E. (1953). Operation of a Glaser bubble chamber with liquid hydrogen. *Phys. Rev.* **92**, 517–518.

Hof, M., *et al.* (1996). Measurement of cosmic-ray antiprotons from 3.7 to 19 GeV. *Astrophys. J. Lett.* **467**, L33.

Holden, N. E. (1997a). Table of isotopes. In: *CRC Handbook of Chemistry and Physics* (ed. D. R. Lide), 77th edn., CRC Press, Boca Raton, FL, pp. 11-38–11-143.

Holden, N. E. (1997b). Neutron scattering and absorption properties. In: *CRC Handbook of Chemistry and Physics* (ed. D. R. Lide), 77th edn., CRC Press, Boca Raton, FL, pp. 11-144–11-158.

Hubbell, J. H. (1969). Photon Cross Sections, Attenuation Coefficients, and Energy Absorption Coefficients from 10 keV to 100 GeV. NSRDS-NBS 29, Natl. Stand. Ref. Data Ser., National Bureau of Standards (U.S.), p. 80.

Hubbell, J. H. (1991). Faster than a speeding photon. *Curr. Contents* **34**(August 26), 10.

Huber, G., Passler, G., Wendt, K., Kratz, J. V., and Trautmann, N. (2003). Radioisotope mass spectrometry. P. 799–843, In: *Handbook of Radioactivity Analysis*, (ed. Michael F. L'Annunziata) 2nd edn., Elsevier, Amsterdam, p. 1273.

Hulthén, L. (1978). Presentation Speech for the Nobel Prize in Physics 1978, December 8, 1978, Stockholm. In: *Nobel Lectures, Physics 1971–1980*, (ed. S. Lundqvist), World Scientific Publishing Co., Singapore.

IAEA. (2000). *Nuclear Techniques in Integrated Plant Nutrient, Water and Soil Management*, IAEA, Vienna, p. 479.

IAEA. (2001). *Introduction of Nuclear Desalination: A Guidebook*, IAEA, Vienna, p. 281.

IAEA. (2002). *Assessment of Soil Phosphorus Status and Management of Phosphatic Fertilizers to Optimize Crop Production*, IAEA, Vienna, p. 473.

IAEA. (2004a). *Isotope Hydrology and Integrated Water Resources Management*, IAEA-CSP-23, IAEA, Vienna, p. 487.

IAEA. (2004b). *Radiation Processing of Polysaccharides for Health Care Applications*, International Cooperative Agreement. IAEA, Vienna, p. 39.

IAEA. (2005a). *World Survey of Activities in Controlled Fusion Research*, 10th edn. International Atomic Energy Agency, Vienna.

IAEA. (2005b). *Nuclear Technology Review 2005*, International Atomic Energy Agency, Vienna.

IAEA. (2005c). *Gamma Irradiators for Radiation Processing*, IAEA, Vienna, p. 40.

IAEA. (2005d). *Optimization of the Coupling of Nuclear Reactors and Desalination Systems*, IAEA, Vienna.

IAEA. (2005e). *Nutrient and Water Management Practices for Increasing Crop Production in Rainfed Arid/Semi-arid Areas*, IAEA, Vienna, p. 221.

ICRP. (1991). *Publication 60, 1990 Recommendation of the International Commission on Radiological Protection* Pergamon Press.

Ironside, R. (1928). The diffraction of cathode rays by thin films of copper, silver and tin. *Proc. R. Soc. Lond., A* **119**, 668–673.

Ising, G. (1924). Prinzip einer Methode zur Herstellung von Kanalstrahlen hoher Voltzahl. *Arkiv för Matematik, Astronomi och Fysik* **18**(30), 1–4.

Isotope Products Laboratories. (1995). Californium-252 fission foils and neutron sources. In: *Radiation Sources for Research, Industry and Environmental Applications*, Isotope Products Laboratories, Burbank, CA, p. 55.

Janni, J. F. (1982). Proton range-energy tables, 1 keV–10 GeV. *At. Data Nucl. Data Tables* **27**, 147–339.

Jelley, J. V. (1958). *Čerenkov Radiation and its Applications*, Pergamon Press, New York, NY, pp. 296.

Jelley, J. V. (1962). Čerenkov radiation: its origin, properties and applications. *Contemp. Phys.* **3**, 45–47.

JET Team. (1994). Fusion energy production from a deuterium-tritium plasma in the JET tokamak. *Nucl. Fusion* **32**, 187–201.

Jia, W., and Ehrhardt, G. J. (1997). Enhancing the specific activity of ^{186}Re using an inorganic Szilard-Chalmers process. *Radiochim. Acta* **79**(1–4), 131–136.

Joliot-Curie, I. (1931). Sur l'excitation des rayons gamma nucléaires du bore par les particules alpha. Énergie quantique du rayonnement gamma du polonium. *C. R. Acad. Sci.* (Paris) **193S**, 1415–1417.

Joliot-Curie, I., and Joliot, F. (1932). Èmission de protons de grande vitesse par les substances hydrogénées sous l'influence des rayons γ trés pénétrants. *Comptes Rendus* (Paris) **194S**, 273–275.

Joliot-Curie, I., and Joliot, F. (1933). Contribution à l'étude des électrons positifs. *Comptes Rendus* (Paris) **196S**, 1105–1107.

Joliot-Curie, I., and Joliot, F. (1934a). Artificial production of a new kind of radioelement. *Nature* **133**, 201.

Joliot-Curie, I., and Joliot, F. (1934b). Un nouveau type of radioactivité. *Comptes Rendus* **198S**, 254–256.

Joliot-Curie, I., and Joliot, F. (1934c). Séparation chimique des nouveaux radioéléments émetteurs d'électrons positifs. *Comptes Rendus* **198S**, 559–561.

Joliot-Curie, I., and Savitch, P. (1938a). Sur les radioelement de période 3,5 heures formé dans l'uranium irradié par les neutrons. *Comptes Rendus* **206S**, 906–908.

Joliot-Curie, I., and Savitch, P. (1938b). Sur la nature du radioélements de période 3,5 heures formé dans l'uranium irradié par les neutrons. *Comptes Rendus* **206S**, 1643–1644.

Joliot-Curie, I., and Savitch, P. (1939). Sur les radioéléments formés dans l'uranium et le thorium irradiés par les neutrons. *Comptes Rendus* **208S**, 343–346.

Joliot, F., and Joliot-Curie, I. (1935). Chemical evidence of the transmutation of elements. Nobel Lecture on December 12, 1935. In: *Nobel Prize Lectures in Chemistry, 1901–1970.* (1967). Nobel Foundation Staff, Eds., Vol. 1, 1901–1921, Vol. 2, 1922–1941, Vol. 3, 1942–1962. Elsevier, Amsterdam.

Joram, C. (2002). The evolution of the Cherenkov imaging technique in high energy physics. *Nucl. Phys. B (Proc. Suppl.)* **109B**, 153–161.

Jungk, R. (1956). *Heller als tausend Sonnen*, Scherz & Goverts, p. 368.

Jungk, R. (1957). *Staekere end Tusind Sole: Atomforskernes Skaebne*, Jespersen og Die, p. 309.

Jungk, R. (1958). *Hotter Than a Thousand Suns: A Personal History of the Atomic Scientists*, Harcourt Brace, New York, NY, p. 369.

Kaiser, T. R. (1974). Heaviside radiation. *Nature* **274** (8 Feb.), 400–401.

Kallman, H. (1950). Scintillation counting with solutions. *Phys. Rev.* **78**(5), 621–622.

Kalmus, G. E., and Kernan, A. (1967). Experimental study of $K^+ \rightarrow \pi^0 + e^+ + \nu$ decay. *Phys. Rev.* **159**, 1187–1194.

Karelin, Y. A., Gordeev, Y. N., Karasev, V. I., Radchenko, V. M., Schimbarev, Y. V., and Kuznetsov, R. A. (1997). Californium-252 neutron sources. *Appl. Radiat. Isot.* **48**(10–12), 1563–1566.

Kawauchi, S., Benson, A. A., Hayes, P., and Calvin, M. (1952). The path of carbon in photosynthesis. *J. Am. Chem. Soc.* **74**, 4477–4482.

Kearns, E., Kajita, T., and Totsuka, Y. (1999). Detecting massive neutrinos. *Sci. Am.* **281**(2), 64–71.

Keisch, B. (2003). *The Atomic Fingerprint: Neutron Activation Analysis*, University Press of the Pacific, Honolulu, HI, p. 64.

Kendall, C., and McDonnell, J. J. (Eds.) (1998). *Isotope Tracers in Catchment Hydrology*, Elsevier, Amsterdam, p. 839.

Kenrick, F. B., Gilbert, C. S., and Wismer, K. L. (1924). The superheating of liquids. *J. Phys. Chem.* **28**, 1297–1307.

Kesterbaum, D. (1998). Neutrinos throw their weight around. *Science* **281**(5383), 1594–1595.

Kirby, M. (2003). IFORS' operational research hall of fame, Patrick Maynard Stuart Blackett. *Int. Trans. Opt. Res.* **10**, 405–407.

Klassen, W. (2003). Edward F. Knipling: titan and driving force in ecologically selective area-wide pest management. *J. Am. Mosq. Control Assoc.* **19**(1), 94–103.

Kloft, W. J. (1984). Entomology. pp. 51–103. In: *Isotopes and Radiation in Agricultural Sciences*, Vol. 2, *Animals, Plants, Food and the Environment*. (eds. M. F. L'Annunziata and J. O. Legg), Academic Press, London.

Knapp, A. K., Abrams, M. D., and Hulbert, L. C. (1985). An evaluation of beta attenuation for estimating aboveground biomass in a tallgrass praire. *J. Range Manage.* **38**(6), 556–558.

Knipling, E. F. (1955). Possibilities of insect control or eradication through the use of sexually sterile males. *J. Econ. Entomol.* **48**(4), 459–462.

Knipling, E. F. (1957). Controlled eradication of screw-worm fly by atomic radiation. *Sci. Mon.* **85**(4), 195–202.

Knipling, E. F. (1959). Sterile-male method of population control. *Science* **130**(3380), 902–904.

Knipling, E. F. (1960a). Use of insects for their own destruction. *J. Econ. Entomol.* **53**(3), 415–420.

Knipling, E. F. (1960b). The eradication of the screwworm fly. *Sci. Am.* **203**(4), 54–61.

Koch, L. (1995). Radioactivity and fission energy. *Radiochim. Acta* **70/71**, 397–402.

Koppers, A. A. P., Staudigel, H., and Duncan, R. A. (2003). High-resolution Ar-40/Ar-39 dating of the oldest oceanic basement basalts in the western Pacific basin. *Geochem. Geophys. Geosyst.* **4**(11), 8914.

Koppers, A. A. P., Staudigel, H., and Wijbrans, J. R. (2000). Dating crystalline groundmass separates of altered Cretaceous seamount basalts by the Ar-40/Ar-39 incremental heating technique. *Chem. Geol.* **166**(1–2), 139–158.

Korff, S. A. (1940). On the contribution to the ionization at sea level produced by the neutrons in the cosmic radiation. *Terr. Mag. Atmos. Elect.* **45**(June 1940), 133–134.

Korff, S. A., and Danforth, W. E. (1939). Neutron measurements with boron-trifluoride counters. *Phys. Rev.* **55**, 980.

Kortschak, H. P., and Hartt, C. E. (1966). Effects of varied conditions on carbon dioxide fixation in sugarcane leaves. *Naturwissenschaften* **53**, 253.

Kortschak, H. P., Hartt, C. E., and Burr, G. O. (1965). Carbon dioxide fixation in sugarcane leaves. *Plant Physiol.* **40**, 209–213.

Kostroun, V. O., Chen, M. S., and Crasemann, B. (1971). Atomic radiation transition probabilities to the $1s$ state and theoretical K-shell fluorescence yields. *Phys. Rev.* **A3**, 533–545.

Kragh, H. (1990). *Dirac: A Scientific Biography*, Cambridge University Press, Cambridge, p. 399.

Krane, K. S. (1988). *Introductory Nuclear Physics*, Wiley, New York, NY.

Križan, P. (2001). Recent progress in Čerenkov counters. *IEEE Trans. Nucl. Sci.* **48**(4), 941–949.

Križan, P., *et al.* (2001). The performance of the HERA-B RICH at high track densities. *Nucl. Instrum. Methods Phys. Res. A* **471**, 30–34.

Krönig, B., and Friedrich, W. (1918). *Physikalische und Biologische Grundlagen der Strahlentherapie*, Urban & Scharzenberg, Berlin.

Kudo, H. (1995). Radioactivity and fusion energy. *Radiochim. Acta* **70/71**, 403–412.

Kulcsar, F., Teherani, D., and Altmann, H. (1982). Study of the spectrum of Cherenkov light. *J. Radioanal. Chem.* **68**, 161–168.

Kullander, S. (2005). Accelerators and Nobel Laureates. Pages 30. http://nobelprize.org/physics/articles/kullander/index.html.

Kumar, T., and Glaser, D. A. (1993). Initial performance, learning, and observer variability for hyperacuity tasks. *Vision Res.* **33**, 2287–2300.

Kumar, T., and Glaser, D. A. (1995). Depth discrimination of a crowded line is better when it is more luminant than the lines crowding it. *Vision Res.* **35**, 657–666.

Kuribara, M. (1994). Spent fuel burnup estimation by Cerenkov glow intensity measurement. *IEEE Trans. Nucl. Sci.* **41**(5), 1736–1739.

Kullander, S. (2005). Accelerators and Nobel Laureates. Pages 30. http://nobelprize.org/physics/articles/kullander/index.html.

Kumar, T., and Glaser, D. A. (1993). Initial performance, learning, and observer variability for hyperacuity tasks. *Vision Res.* **33**, 2287–2300.

Kumar, T., and Glaser, D. A. (1995). Depth discrimination of a crowded line is better when it is more luminant than the lines crowding it. *Vision Res.* **35**, 657–666.

Kuribara, M. (1994). Spent fuel burnup estimation by Cerenkov glow intensity measurement. *IEEE Trans. Nucl. Sci.* **41**(5), 1736–1739.

Kuribara, M., and Nemoto, K. (1994). Development of new U.V. — I.I. Cerenkov viewing device. *IEEE Trans. Nucl. Sci.* **41**(1), 331–335.

Kurie, F. N. D. (1934). A new mode of disintegration produced by neutrons. *Phys. Rev.* **45**, 904–905.

Kushita, K. N., and Du, J. (1998). Radioactivity measurements of ^{188}Re by Cherenkov counting. *Appl. Radiat. Isot.* **49**(9–11), 1069–1072.

Kyle, R. A., and Shampo, M. A. (2002). Rosalyn Yalow—Pioneer in nuclear medicine. *Mayo Clin. Proc.* **77**, 4.

L'Annunziata, M. F. (1971). Birth of a unique parent-daughter relation: secular equilibrium. *J. Chem. Educ.* **48**, 700–703.

L'Annunziata, M. F. (1984). Agricultural biochemistry: reaction mechanisms and pathways in biosynthesis. In: *Isotopes and Radiation in Agricultural Sciences* (eds. M. F. L'Annunziata and J. O. Legg), Vol. 2, Academic Press, London, pp. 105–182.

L'Annunziata, M. F. (1987). *Radionuclide Tracers, Their Detection and Measurement*, Academic Press, London, pp. 505.

L'Annunziata, M. F. (2003a). Liquid scintillation analysis: principles and practice. In: *Handbook of Radioactivity Analysis* (ed. M. F. L'Annunziata), 2nd edn., Elsevier, Amsterdam, pp. 347–535.

L'Annunziata, M. F. (2003b). Cherenkov counting. In: *Handbook of Radioactivity Analysis* (ed. M. F. L'Annunziata), 2nd edn, , Elsevier, Amsterdam, pp. 719–797.

L'Annunziata, M. F. (2003c). Solid scintillation analysis. In: *Handbook of Radioactivity Analysis* (ed. M. F. L'Annunziata), 2nd edn., Elsevier, Amsterdam, pp. 845–988.

L'Annunziata, M. F. (Ed.) (2003d). *Handbook of Radioactivity Analysis*, 2nd edn., Elsevier, Amsterdam, pp. 1273.

L'Annunziata, M. F., Gonzalez, J., and Olivares, L. A. (1977). Microbial epimerization of *myo*-inositol to *chiro*-inositol in soil. *Soil Sci. Soc. Am. J.* **41**, 733–736.

L'Annunziata, M. F., and Legg, J. O., Eds. (1984a). *Isotopes and Radiation in Agricultural Sciences, Vol. 1. Soil-Plant-Water Relationships*, Academic Press, London, pp. 292.

L'Annunziata, M. F., and Legg, J. O., Eds. (1984b). *Isotopes and Radiation in Agricultural Sciences, Vol. 2. Animals, Plants, Food and the Environment*, Academic Press, London, pp. 356.

L'Annunziata, M. F., and Passo Jr., C. J. (2002). Cherenkov counting of yttrium-90 in the dry state; correlations with phosphorus-32 Cherenkov counting data. *Appl. Radiat. Isot.* **56**, 907–916.

Lagemann, R. T. (1977). New light on old rays: N rays. *Am. J. Phys.* **45**(3), 281–284.

Larijani, B., Woscholski, R., and Rosser, C. A. (Ed.) (2006). *Chemical Biology: Applications and Techniques*, Wiley, New York, NY, p. 272.

Lattes, C. M. G., Occhialini, G. P. S., and Powell, C. F. (1947). Observation on the tracks of slow mesons in photographic emulsions. *Nature* **160**, 453–456.

Lauer, W. C. (Ed.) (2006). *Desalination of Seawater and Brackish Water*, American Waterworks Association, p. 780.

Lawrence, E. O. (1935). Transmutation of sodium by deuterons. *Phys. Rev.* **47**, 17–27.

Lawrence, E. O. (1951). "The Evolution of the Cyclotron", Nobel Lecture given on December 11, 1951 in Stockholm, Sweden. In: *Nobel Lectures, Physics, 1922–1941*, (1965), Elsevier, Amsterdam.

Lawrence, E. O., Alvarez, A. W., Brobeck, W. M., Cooksey, D., Corson, D. R., McMillan, E. M., Salisbury, W. W., and Thornton, R. L. (1939). Initial performance of the 60-inch cyclotron of the William H. Crocker Laboratory, University of California. *Phys. Rev.* **56**, 124–125.

Lawrence, E. O., and Cooksey, D. (1936). On the apparatus for the multiple acceleration of light ions to high speeds. *Phys. Rev.* **50**, 1131–1140.

Lawrence, E. O., and Edlefsen, N. E. (1930). The production of high-speed protons without the use of high voltages. *Science* **72**, 376–377.

Lawrence, E. O., and Livingston, M. S. (1931a). The production of high-speed protons without the use of high voltages. *Phys. Rev.* **37**, 1707.

Lawrence, E. O., and Livingston, M. S. (1931b). The production of high-speed protons without the use of high voltages. *Phys. Rev.* **38**, 834.

Lawrence, E. O., and Livingston, M. S. (1932). The production of high-speed light ions without the use of high voltages. *Phys. Rev.* **40**, 19–35.

Lawrence, E. O., and Livingston, M. S. (1934). The multiple acceleration of ions to very high speeds. *Phys. Rev.* **45**, 608–612.

Lawson, C. G., and Krause, C. (2004). Documenting history: minutes of the New Piles Committee meetings. *Nucl. News* (November), 36–38.

Lee, J., and Samios, N. P. (1959). Search for rare decay modes of the μ^+ meson. *Phys. Rev. Lett.* **3**, 55–56.

Lee, Z., Sodee, D. B., Resnick, M., and MacLennan, G. T. (2005). Multimodal and three-dimensional imaging of prostate cancer. *Comput. Med. Imag. Graph.* **29**, 477–486.

Lee, T. D., and Yang, C. N. (1956). Question of parity conservation in weak interactions. *Phys. Rev.* **104**, 254–258.

Leighton, R. B., Anderson, C. D., and Seriff, A. J. (1949). The energy spectrum of the decay particles and the mass and spin of the mesotron. *Phys. Rev.* **75**, 1432–1437.

Leighton, R. B., Wanlass, S. D., and Anderson, C. D. (1953). The decay of V^0 particles. *Phys. Rev.* **89**, 148–177.

Lemaître, G. (1927). Un universe homogène de masse constante et de rayon croissant relevant compte de la vitesse radiale des nébuleuses extragalactiques. *Ann. Soc. Sci. Bruxelles* **A47**, 41–49.

Lemaître, G. (1931a). A homogeneous universe of constant mass and increasing radius accounting for the radial velocity of extragalactic nebulae. *Mon. Not. R. Astron. Soc.* **91**(1), 483–490.

Lemaître, G. (1931b). The evolution of the universe. *Nature* (Suppl., Oct. 24, 1931,) **128** (No. 3234), 704–706.

Lenard, P. (1894). Über Kathodenstrahlen in Gasen von atmosphärischem Druck und im äussersten Vakuum. (Cathode rays in gases at atmospheric pressure and in the highest vacuum). *Wied. Ann. Phys.* **51**, 225.

Lenard, P. (1899). Erzeugung von Kathodenstrahlen durch ultraviolettes Licht. (Production of cathode rays by ultraviolet light). *Sitz.-ber. Kaiserl. Akad. Wiss. Wien, 19th October,* also *Ann. Phys., Vol. 2,* 359.

Lenard, P. (1902). Über die lichtelektrische Wirkung. *Ann. Phys.* **8**, 149–198.

Levi, H. (1985). *George de Hevesy: Life and Work*, Hilger, Bristol, p. 147.

Levine, I. N. (1999). *Quantum Chemistry*, 5th edn., Prentice Hall, Upper Saddle River, NJ, p. 739.

Libby, W. F. (1955). *Radiocarbon Dating*, 2nd edn., University of Chicago Press, Chicago, p. 175.

Libby, W. F. (1960). Radiocarbon dating. Nobel Lecture given in Stockholm on December 12, 1960. In: *Nobel Lectures, Chemistry, 1942–1962.* (1964), (eds. Nobel Foundation Staff), Elsevier, Amsterdam.

Libby, W. F. (1967). History of radiocarbon dating. In: "Radioactive Dating and Methods of Low-Level Counting. Proceedings of a Symposium, Monaco", 1967. pp. 3–25. International Atomic Energy Agency, Vienna.

Lide, D. R. (1997). *CRC Handbook of Chemistry and Physics*, 77th edn., CRC Press, Boca Raton, FL.

Lide, D. R. (2001). *CRC Handbook of Chemistry and Physics*, 81st edn., CRC Press, Boca Raton, FL, pp. 10–220.

Lilley, J. (2001). *Nuclear Physics, Principles and Applications*, Wiley, Ltd., West Sussex, p. 393.

Lindh, A. E. (1950). Presentation Speech for the Nobel Prize in Physics 1950. In: *Nobel Lectures, Physics, 1942–1962,* (1964), (eds. Nobel Foundation Staff), Elsevier, Amsterdam.

Lindhard, J., and Scharff, M. (1960). Recent developments in the theory of stopping power. I. Principles of the statistical method. In: *Penetration of Charged Particles in Matter*, National Academy of Sciences-National Research Council, Publication No. 752, p. 49.

Lindhard, J., and Scharff, M. (1961). Energy dissipation by ions in the keV region. *Phys. Rev.* **124**, 128–130.

Lindsey, C. S., Lindblad, T., Waldemark, K., and Hildingsson, L. (1999). Nuclear fuel assembly assessment project and image categorization. Proceedings of the Ninth Workshop on Virtual Intelligence, Dynamic Neural Networks. Royal Institute of Technology, KTH, Stockholm, Sweden, June 22–26, 1998. *Proc. SPIE* **3728**, 491–499.

Linsley, J. (1963). Evidence for a primary cosmic-ray particle with energy 10^{20} eV. *Phys. Rev. Lett.* **10**, 146–148.

Lionaes, A. (1970). Presentation Speech, The Nobel Peace Prize 1970, In: *Nobel Lectures, Peace, 1951–1970*, 1972, (ed. F. W. Heberman), Elsevier, Amsterdam.

Livingston, M. Stanley, and Bethe, H. A. (1937). Nuclear physics III. Nuclear dynamics, experimental. *Rev. Mod. Phys.* **9**, 245–390.

Lobashev, V. M., Aseev, V. A., Belasev, A. I., Berlev, A. I., Geraskin, E. V., Golubev, A. A., Golubev, N. A., Kazachenko, O. V., Kuznetsev, Yu. E., Ostroumov, R. P., Ryvkis, L. A., Stern, B. E., Titov, N. A., Zadorozhny, S. V., and Zakharov, Yu. I. (2005). Neutrino mass anomaly in the tritium beta-spectrum. In: Proceedings the XIX International Conference on Neutrino Physics and Astrophysics (Neutrino-2000), Sudbury, Canada, June 16–21, 2000, Elsevier, Amsterdam.

Loeber, C. R. (2002). *Building the Bombs: A History of the Nuclear Weapons Complex*, Sandia National Laboratories, Albuquerque, NM, p. 262.

Los Alamos Science. (1997). The Reines-Cowan Experiments, 1953–1956: Detecting the Poltergeist Number 25.

Lovell, B. (1975). Blackett, Patrick Maynard Stuart, Baron Blackett of Chelsea. *Biographical Memoirs of the Fellows of the Royal Society*, Vol. 21, pp.1–115, Royal Society, London, p. 564.

Lowenthal, G., and Airey, P. (2001). *Practical Applications of Radioactivity and Nuclear Radiations*, Cambridge University Press, MA, p. 366.

Luft, R. (1977). Presentation Speech. The Nobel Prize in Physiology or Medicine 1977. In: *Nobel Lectures, Physiology and Medicine 1971–1980.* (ed. J. Lindsten), 1992. World Scientific Publishing Co., Singapore.

Maas, W. (2004). Leo Szilard. A personal remembrance. *Genetics* **167**, 555–558.

Magaritz, M., Kaufman, A., Paul, M., Boaretto, E., and Holos, G. (1990). A new method to determine regional evapotranspiration. *Water Resour. Res.* **26**, 1759–1762.

Mallet, L. (1926). Luminescence of water and organic substances subjected to gamma radiation. *C. R. Acad. Sci. (Paris)* **183**, 274–275.

Mallet, L. (1928). Spectral study of the luminescence of water and carbon disulfide under gamma radiation. *C. R. Acad. Sci. (Paris)* **187**, 222–223.

Mallet, L. (1929). The ultra-violet radiation of substances subjected to γ-rays. *C. R. Acad. Sci. (Paris)* **188**, 445–447.

Mann, W. B. (1978). A Handbook of Radioactivity Measurements Procedures. National Council on Radiation Protection and Measurements. CCRP Report No. 58, Washington, DC.

Mapleston, P. (1997). Film thickness gauges meet market needs for quality, cost. *Mod. Plastics* **74**, 73–76.

Marburger III, J. H. (2001). "Enrico Fermi's Impact on Science". Address given at the Italian Embassy, Washington, D.C., Centennial Celebration of the Birth of Enrico Fermi, November 27, 2001.

Marques-Netto, A., and Abbe, J. C. (1975). Szilard-Chalmers effects in halfnium chelates. *J. Inorg. Nucl. Chem.* **37**, 2235–2238.

Marshall, J. (1952). Particle counting by Cerenkov radiation. *Phys. Rev.* **86**, 685–693.

Martin, R. C., Knauer, J. B., and Balo, P. A. (2000). Production, distribution and applications of Californium-252 neutron sources. *Appl. Radiat. Isot.* **53**, 785–792.

Martin, R. C., Laxon, R. R., Miller, J. H., Wierzbicki, J. G., Rivard, M. J., and Marsh, D. L. (1997). Development of high-activity 252Cf sources for neutron brachytherapy. *Appl. Radiat. Isot.* **48**(10–12), 1567–1570.

Masefield, J. (2004). Reflections on the evolution and current status of the radiation industry. *Radiat. Phys. Chem.* **71**, 8–15.

Mather, J. C., and Boslough, J. (1996). *The Very First Light: The True Story of the Scientific Journey Back to the Dawn of the Universe*, Basic Books, New York, NY, pp. 41–42, 328 pp.

Matsumoto, T., Korpar, S., Adachi, I., Fratina, S., Iijima, T., Ishibashi, R., Kawai, H., Križan, P., Ogawa, S., Pestotnok, R., Saitoh, S., Seki, T., Sumiyoshi, T., Suzuki, K., Tabata, T., Uchida, Y., and Unno, Y. (2004). Studies of proximity focusing RICH with an aerogel radiator using flat-panel multi-anode PMTs (Hamamatsu H8500). *Nucl. Instrum. Methods Phys. Res. A* **521**, 367–377.

McDougall, I., and Harrison, T. M. (1999). *Geochronology and Thermochronology by the $^{40}Ar/^{39}Ar$ Method*, Oxford University Press, New York, NY, p. 269.

McLane, V., Dunford, C. L., and Rose, P. F. (1988). *"Neutron Cross Sections"*, Vol. 2, *"Neutron Cross Section Curves"*, Academic Press, San Diego, CA.

McLean, A. (2002). The most versatile physicist of his generation. *Science* **296**(5565), April 5, 2002, 49–50.

McMillan, E. M. (1945). The synchrotron – a proposed high energy accelerator. *Phys. Rev.* **68**, 143–144.

McQuarrie, D. A. (1983). *Quantum Chemistry*, University Science Books, Mill Valley, p. 517.

Mederski, H. J. (1961). Determination of internal water by beta gauging technique. *Soil Sci.* **92**, 143–146.

Mederski, H. J., and Alles, W. (1968). Beta gauging leaf water status: influence of changing leaf characteristics. *Plant Physiol.* **43**, 470–472.

Meitner, L. (1923). Das beta-Strahlenspektrum von UX_1 und seine Deutung. *Z. Phys.* **17**, 54–66.

Meitner, L., and Frisch, O. R. (1939). Disintegration of uranium by neutrons: a new type of nuclear reaction. *Nature* **143**, 239–240 (February 11, 1939).

Meitner, L., and Philipp, K. (1933). Die bei Neutronenenregung auftretenden Elektronenbahnen. *Naturwiss* **21**, 286–287.

Meitner, L., Strassmann, F., and Hahn, O. (1938). Künstliche Emwandlungsprozesse bei Bestrahlung des Thoriums mit Neutronen; Auftreten isomer Reihen durch Abspaltung von α-Strahlen. *Z. Phys.* **109**, 538–552.

Mehlhorn, W. (1998). 70 years of Auger spectroscopy, a historical perspective. *J. Electron Spectrosc. Rel. Phenom.* **93**(1), 1–15.

Menn, W., *et al.* (2000). The absolue flux of protons and helium at the top of the atmosphere using IMAX. *Astrophys. J.* **533**, 281–297.

Menzel, R. G., and Smith, S. J. (1984). Soil fertility and plant nutrition. pp. 1–34. In: *Isotopes and Radiation in Agricultural Sciences. Vol. 1 Soil-Plant-Water Relationships*, (M. F. L'Annunziata and J. O. Legg, Eds.), Academic Press, London, p. 292.

Mettler, F. A., and Guiberteau, M. J. (2005). *Essentials of Nuclear Medicine Imaging*, Saunders, Philadelphia, p. 592.

Meyer, D. I., Perl, M. L., and Glaser, D. A. (1957). Scattering of K^+ mesons by protons. *Phys. Rev.* **107**, 279–282.

Michael Lederer, C., and Shirley, V. S. (Eds.) (1978). *Table of Isotopes*, 7th edn. Wiley, New York, NY.

Michaudon, A. (2000). From alchemy to atoms. The making of plutonium. *Los Alamos Science* **26**, 62–73.

Midbon, M. (2000). A day without yesterday: Georges Lemaître & the Big Bang. *Commonweal*, March 24, 18–19.

Miley, G. H., and Sved, J. (1997). The IEC—A plasma-target-based neutron source. *Appl. Radiat. Isot.* **48**(10–12), 1557–1561.

Millikan, R. A. (1911). The isolation of an ion, a precision measurement of its charge, and the correction of Stoke's law. *Phys. Rev.* (Series I) **32**, 349–397.

Millikan, R. A. (1913). On the elementary electrical charge and the Avogadro constant. *Phys. Rev.* **2** (Series 2), 109–143.

Millikan, R. A. (1914). A direct determination of "*h*". *Phys. Rev.* **4**, 73–75.

Millikan, R. A. (1916a). Einstein's photoelectric equation and contact electromotive force. *Phys. Rev.* **7**, 18–32.

Millikan, R. A. (1916b). A direct photoelectric determination of Planck's "*h*". *Phys. Rev.* **7**, 355–388.

Millikan, R. A. (1921). The distinction between intrinsic and spurious contact E.M.F.S. and the question of the absorption of radiation by metals in quanta. *Phys. Rev.* **18**, 236–244.

Millikan, R. A. (1924). The Electron and the Light-Quant from the Experimental Point of View. Nobel Lecture, May 23, 1924. In: *Nobel Lectures, Physics 1922–1941*, (1965), Elsevier, Amsterdam.

Millikan, R. A. (1950). *The Autobiography of Robert A. Millikan*, Prentice Hall, New Jersey, p. 311.

Millikan, R. A., and Bowen, I. S. (1926). High frequency rays of cosmic origin I. Sounding ballon observations at extreme altitudes. *Phys. Rev.* **27**, 353–361.

Millikan, R. A., and Cameron, G. H. (1928). High altitude tests on the graphical, directional, and spectral distributions of cosmic rays. *Phys. Rev.* **31**, 163–173.

Millikan, R. A., and Neher, H. V. (1935). The equatorial latitude effect in cosmic rays. *Phys. Rev.* **47**, 205–208.

Mitchell, J. W., Barbier, L. M., Christian, E. R., Krizmanic, J. F., Krombel, K., Ormes, J. F., Streitmatter, R. E., Labrador, A. W., Davis, A. J., Mewaldt, R. A., Schindler, S. M., Golden, R. L., Stochaj, S. J., Webber, W. R., Menn, W., Hof, M., Reimer, O., Simon, M., and Rasmussen, I. L. (1996). Measurement of 0.25–3.2 GeV antiprotons in the cosmic radiation. *Phys. Rev. Lett.* **76**(17), 3057–3060.

Molins, R. A. (Ed.) (2001). *Food Irradiation: Principles and Applications*, Wiley, New York, NY.

Molnar, G. (2004). *Handbook of Prompt Gamma Activation Analysis: with Neutron Beams*, Springer, p. 423.

Mook, W. G., and van der Plicht, J. (1999). Reporting ^{14}C activities and concentrations. *Radiocarbon* **41**, 227–239.

Moore, W. (1989). *Schrödinger, Life and Thought*, Cambridge University Press, MA, pp. 479–480.

Mora, G. (2003). Temperature and salinity changes in the Caribbean sea during glacial/interglacial intervals as inferred from elemental and isotopic data. Geological Society of America Annual Meeting, Seattle, November 2–5, 2003.

Moritz, C. (Ed.) (1963). Victor Hess. In: *Current Biography Yearbook 1963*, H.W. Wilson Company, New York, NY, pp. 180–182.

Morozov, V. A., Churin, I. N., and Morozova, N. V. (1998). Nuclear experimental techniques—three-dimensional delayed coincidence single-crystal scintillation time spectrometer. *Instrum. Exp. Tech.* **41**(5), 609.

Morrison, J. W. (1960). Marquis wheat – a triumph of scientific endeavor. *Agric. Hist.* **34**(4), 182–188.

Morrissey, R. F., and Herring, C. M. (2002). Radiation sterilization: past, present and future. *Radiat. Phys. Chem.* **63**, 217–221.

Moscow News (2006). Father of Russian Genetics and Selection of Plants. *Moscow News*, No. 15, Tuesday, 2 May 2006.

Moseley, H. G. J. (1913). The high frequency spectra of the elements. *Philos. Mag.* **26**(1), 1024–1034.

Moseley, H. G. J. (1914). The high frequency spectra of the elements. *Philos. Mag.* **27**(2), 703–713.

Mozumder, A., Chatterjee, A., and Magee, J. L. (1968). Theory of radiation chemistry. IX. Mode and structure of heavy particle tracks in water. *Am. Chem. Soc. Adv. Chem. Ser.* **1**, 27.

Mundy, J. N., Rothman, S. J., Fluss, M. J., and Smedskjaer, L. C. (Eds.) (1983). *Methods in Experimental Physics*, Vol. 21, Academic Press, New York, NY, p. 504.

Murin, A. N., and Nefedov, V. D. (1955). Concentration of artificial radioactive isotopes of Groups IV and V by method of recoil atoms. *Primenenie Mechenykh Atomov v Anal. Khim., Akad. Nauk S.S.S.R.*, 75–78. UCRL Trans.

Murray, R. L. (1993). *Nuclear Energy. An Introduction to the Concepts, Systems, and Applications of Nuclear Processes*, 4th edn. Pergamon Press, Oxford, p. 437.

Murray, R. L. (2001). *Nuclear Energy. An Introduction to the Concepts, Systems, and Applications of Nuclear Processes*, 5th edn. Butterworth-Heinemann, Boston, MA, p. 490.

Mutterlose, J., Bornemann, A., and Herrie, J. O. (2005). Mesozoic calcareous nanofossils – state of the art. *Palaeentologische Zeitschrift* **79**, 113–133.

Nagano, M., and Watson, A. A. (2000). Observations and implications of the untrahigh-energy cosmic rays. *Rev. Mod. Phys.* **72**, 689–732.

Nakahata, M. (2000). Neutrinos underground. *Science* **289**(5482), 1155–1156.

Nakayama, F. S., and Ehrler, W. L. (1964). Beta ray gauging technique for measuring leaf water content changes and moisture status of plants. *Plant Physiol.* **39**, 95–98.

National Academy of Sciences (1992). Oceanography in the Next Decade: Building New Partnerships. National Academy Press, Washington, D.C., pp. 202.

Navarro, N., Grau Carles, A., Alvarez, A., Salvador, S., and Gomez, V. (1997). Standardization of ^{234}Th by Cerenkov counting. *Appl. Radiat. Isot.* **48**(7), 949–952.

Neddermeyer, S. H., and Anderson, C. D. (1937). Note on the nature of cosmic-ray particles. *Phys. Rev.* **51**, 884–890.

Neddermeyer, S. H., and Anderson, C. D. (1938). Cosmic-ray particles of intermediate mass. *Phys. Rev.* **53**, 88–89.

Nielsen, D. R., and Cassel, D. K. (1984). Soil water management. pp. 35–65. In: *Isotopes and Radiation in Agricultural Sciences. Vol. 1 "Soil-Plant-Water Relationships"* (eds. M. F. L'Annunziata and J. O. Legg), Academic Press, London, p. 292.

Nimz, G. J. (1998). Lithogenic and cosmogenic tracers in catchment hydrology. pp. 247–289. In: *Isotope Tracers in Catchment Hydrology* (eds. C. Kendall and J. J. McDonnell) Elsevier, Amsterdam, p. 839.

Nimz, G. J., Caffee, M. W., and Borchers, J. W. (1993). Extremely low ^{36}Cl.Cl values in deep ground water at Wawona, Yosemite National Park, California: evidence for rapid upwelling of deep crustal waters? *EOS Trans., Am. Geophys. Union* **74**, 582.

Nimz, G. J., Moore, J. N., and Kassameyer, P. W. (1997). ^{36}Cl.Cl ratios in geothermal systems: preliminary measurements from the Coso field. *Geothermal Res. Council Trans.* **21**, 211–217.

Nobel Prize Lectures, Chemistry, 1901–1970, (1967), (eds. Nobel Foundation Staff), Vol. 1, 1901–1921, Vol. 2, 1922–1941, Vol. 3, 1942–1962. Elsevier, Amsterdam.

Nobel Lectures, Chemistry, 1942–1962, (1964), (eds. Nobel Foundation Staff), Elsevier, Amsterdam.

Nobel Lectures, Physics, 1901–1970. Vol. 1, In: *Nobel Lectures, Physics, 1901–1921*, (1967). (eds. Science Incorporated Elsevier), Elsevier, Amsterdam.

Nobel Lectures, Physics, 1922–1941. (1965), Elsevier, Amsterdam.

Nobel Lectures, Physics, 1942–1962, (1964), (eds. Nobel Foundation Staff), Elsevier, Amsterdam.

Noor, A., Zakir, M., Rasyid, B., Maming, and L'Annunziata, M. F. (1996). Cerenkov and liquid scintillation analysis of the triple label ^{86}Rb-^{35}S-^{33}P. *Appl. Radiat. Isot.* **47**, 659–668.

Nordling, C. (1994). Presentation Speech, The Nobel Prize in Physics 1994. In: *Nobel Lectures, Physics 1991–1995*, (ed. G. Ekspong), 1997, World Scientific Publishing Co., Singapore.

Northcliffe, L. C. (1963). Passage of heavy ions through matter. *Ann. Rev. Nucl. Sci.* **13**, 67–102.

Nour, S., Burnett, W. C., and Horwitz, E. P. (2002). ^{234}Th analysis of marine sediments via extraction chromatography and liquid scintillation counting. *Appl. Radiat. Isot.* **57**, 235–241.

Novick, A., and Szilard, L. (1950). Experiments with the chemostat on spontaneous mutations of bacteria. *Proc. Natl. Acad. Sci. USA* **36**, 708–719.

Novick, A., and Szilard, L. (1951). Genetic mechanisms in bacteria and bacterial viruses. I. Experiments on spontaneous and chemically induced mutations of bacteria growing in the chemostat. *Cold Spring Harbor Symposia on Qualitative Biology* **16**, 337–343.

Novick, A., and Szilard, L. (1952). Anti-mutagens. *Nature* **170**, 926–927.

NRL. (1998). The exposure of New Zealand aircrew to cosmic radiation. National Radiation Laboratory, Information Sheet 19, February 1998, p. 6, Christchurch, New Zealand.

Nuclear Technology Review. (2004). *International Atomic Energy Agency*, Vienna, p. 94.

Nuclear Technology Review Update. (2005). *International Atomic Energy Agency*, Vienna, p. 36.

Nuclear Technology Review. (2006). *International Atomic Energy Agency*, Vienna, p. 127.

Nye, M. J. (2004). *Blackett, Physics, War, and Politics in the Twentieth Century*, Harvard University Press, Cambridge, MA, p. 249.

Obregewitsch, R. P., Rolston, D. E., Nielsen, D. R., and Nakayama, F. S. (1975). Estimating relative leaf water content with a simple beta gauge calibration. *Agron. J.* **67**, 729–732.

Occhialini, G. P. S., and Powell, C. F. (1947). Nuclear disintegrations produced by slow charged particles of small mass. *Nature* **159**, 186–190.

Occhialini, G. P. S., and Powell, C. F. (1948). Observations on the production of mesons by cosmic radiation. *Nature* **162**, 168–173.

Ogando, J. (1993). Nuclear web gauging keeps pace with processor needs. *Plastics Technol.* **39**, 46–49.

Ohm, H., Liang, M., Molner, G., and Sistemich, K. (1990). Delayed-coincidence measurement of subnanosecond lifetimes in fission fragments. In: "The Spectroscopy of Heavy Nuclei 1989: Proceedings of the International Conference on the Spectroscopy of Heavy Nuclei, Agia Pelagia, Crete," June 25–July 1, 1989. *Institute Physics Conference Series*. No. 105, 323–328. Hilger, Bristol.

O'Leary, G. J., and Incerti, M. (1993). A field comparison of three neutron moisture meters. *Aust. J. Exp. Agric.* **33**, 59–69.

Ongena, J., and van Oost, G. (2004). Energy for future centuries. Will fusion be an inexhaustible, safe and clean energy source? *Fusion Sci. Technol.* **45**, 3–14.

Oseen, C. W. (1926). Presentation Speech for The Nobel Prize in Physics 1925 on December 10, 1926. In: *Nobel Lectures. Physics 1922–1941* (1965), Elsevier, Amsterdam.

Panofsky, W. K. H., Aamodt, R. L., and Hadley, J. (1951). The gamma-ray spectrum resulting from capture of negative π^- mesons in hydrogen and deuterium. *Phys. Rev.* **81**, 565–574.

Parker, R. P., and Elrick, R. H. (1970). Cerenkov counting as a means of assaying β-emitting radionuclides. In: The Current Status of Liquid Scintillation Counting (ed. E. D. Bransome, Jr.), Grune and Stratton, New York, NY, pp. 110–122.

Pasachoff, N. (1996). *Marie Curie and the Science of Radioactivity*, Oxford University Press, New York, NY, p. 112.

Pasachoff, N. (2005). *Ernest Rutherford: Father of Nuclear Science*, Enslow Publishers, Berkeley Heights, NJ, p. 128.

Passo, C. J., Jr., and Cook, G. T. (1994). *Handbook of Environmental Liquid Scintillation Spectrometry. A Compilation of Theory and Methods*, Packard Instrument Company, Meriden, CT.

Patel, S. B. (1991). *Nuclear Physics, an Introduction*, Wiley, New York, NY, p. 346.

Paul, M., Kaufman, A., Magaritz, M., Fink, D., Henning, W., Kaim, R., Kutschera, W., and Meirav, O. (1986). A new ^{36}Cl hydrologic model and ^{36}Cl systematics in the Jordan River/Dead Sea system. *Nature* **321**, 511–515.

Pauli, W. (1924). Zur Frage der theoretischen Deutung der Satelliten eineger Spektrallinien und ihrer Beeinflussung durch magnetische Felder. *Naturwiss* **12**, 741–743.

Pauli, W. (1925). Über den Zusammenhang des Abschlusses der Electronengruppen in Atom der Komplexstruktur der Spectren. *Z. Phys.* **31**, 765–783.

Pauli, W. (1946). Exclusion principle and quantum mechanics. Nobel Lecture, December 13, 1946. In: *Nobel Lectures, Physics, 1942–1962*, (1964), (eds. Nobel Foundation Staff)., Elsevier, Amsterdam.

Paul, W., and Steinwedel, H. (1955). Interaction of electrons with matter. In: "Beta- and Gamma-ray Spectroscopy" (ed. K. Siegbahn) North-Holland, Amsterdam.

Penzias, A. A., and Wilson, R. W. (1965). A measurement of excess antenna temperature at 4080 Mc/s. *Astrophys. J.* **142**, 419–421.

Perkins, D. H. (1947). Nuclear disintegrations by meson capture. *Nature* **159**, 126.

Perlmutter, A. (2000). *Marietta Blau's Work after World War II*. Manuscript Biography No. MB 2001-251, American Institute of Physics, College Park, Maryland, p. 44.

Perrino, R., *et al.* (2001). Performances in the aerogel threshold Cherenkov counter for the Jefferson Lab Hall A spectrometers in the 1–4 GeV/c momentum range. *Nucl. Instrum. Methods Phys. Res. A* **457**, 571–580.

Pestotnik, R., Križan, P., Korpar, S., Bračko, M., Starič, M., and Stanovik, A. (2002). Investigation of ^{90}Sr detection with Cherenkov radiation in silica aerogels. In: "Proceedings of IEEE Symposium on Nuclear Science and Medical Imaging including Nuclear Power Systems." Nov. 4–9, 2001, San Diego, IEEE, Piscataway, NJ.

Piccioni, O. (1950). On the secondary particles of local penetrating showers. *Phys. Rev.* **77**, 6–10.

Pichlmaier, A., Butterworth, J., Geltenbort, P., Nagel, H., Nesvizhevsky, V., Neumaier, S., Schreckenbach, K., Steichele, E., and Varlamov, V. (2000). MAMBO II: neutron lifetime measurement with storage of ultra-cold neutrons. *Nucl. Instrum., Methods Phys. Res. A* **440**, 517–521.

Pickering, W. H. (1991). *Biography: Carl David Anderson, September 3, 1905—January 11, 1991.* California Institute of Technology, Institute Archives, Pasadena, p.16.

Pinto da Cunha, J., Neves, F., and Lopes, M. I. (2000). On the reconstruction of Cherenkov rings from aerogel radiators. *Nucl. Instrum. Methods Phys. Res. A* **452**, 401–421.

Pleijel, H. (1933). Presentation Speech, December 10, 1933. The Nobel Prize in Physics 1932 and 1933. In: *Nobel Lectures, Physics 1922–1941*, (1965), Elsevier, Amsterdam.

Pleijel, H. (1935). Presentation Speech, December 10, 1935. The Nobel Prize in Physics 1935. In: *Nobel Lectures, Physics 1922–1941*, (1965), Elsevier, Amsterdam.

Pleijel, H. (1936). Presentation Speech, December 10, 1936. The Nobel Prize in Physics 1936. In: *Nobel Lectures, Physics 1922–1941*, (1965), Elsevier, Amsterdam.

Povinec, P., and Sanchez-Cabeza, J. A. (Eds.) (2006). International Conference on Isotopes and Environmental Studies, Vol. 8: Aquatic Forum 2004, 25–29 October, Monaco (Radioactivity In the Environment). Elsevier, Amsterdam, p. 660.

Powell, C. F. (1950). The cosmic radiation. Nobel Lecture at Stockholm, Sweden on December 11, 1950. In: *Nobel Lectures, Physics, 1942–1962*, (1964), (eds. Nobel Foundation Staff), Elsevier, Amsterdam.

Powell, C. F. (1987). *Cecil Powell, Fragments of Autobiography*, University of Bristol, UK, p. 28.

Powell, C. F., and Rosenblum, S. (1948). A new method for the determination of the mass of mesons. *Nature* **161**, 473–475.

Powsner, R. A., and Powsner, E. R. (2006). *Essentials of Nuclear Medicine Physics*, 2nd edn., Blackwell Publishing, p. 224.

Pullen, B. P. (1986). Cerenkov counting of ^{40}K in KCl using a liquid scintillation spectrometer. *J. Chem. Educ.* **63**, 971.

Quinn, S. (1996). *Marie Curie: A Life*, Addison Wesley, NY, p. 509.

Rabinowitch, E. (1964). Editorial. Bell Atomic Scientists. October, 16–20.

Rachinhas, P. J. B. M., Simões, P. C. P. S., Lopes, J. A. M., Dias, T. H. V. T., Morgado, R. E., dos Santos, J. M. F., Stauffer, A. D., and Conde, C. A. N. (2000). Simulation and experimental results for the detection of conversion electrons with gas proportional scintillation counters. *Nucl. Instrum. Methods Phys. Res. A* **441**, 468–478.

Radchenko, V. M., Ryabinin, M. A., Andreytchuk, N. N., Gavrilov, V. D., and Karelin, Ye A. (2000). Curium-248 standard neutron source. *Appl. Radiat. Isot.* **53**, 833–835.

Ramesh, A., and Subramanian, M. S. (1997). Trace determination of tellurium using liquid scintillation and Cerenkov counting. *Analyst* **122**, 1605–1610.

Ramsay, W., and Soddy, F. (1903). Experiments in radioactivity, and the production of helium from radium. *Proc. R. Soc. Lond.* **72**, 204–207.

Rant, J., Milič, Z., Istenič, J., Knific, T., Lengar, I., and Rant, A. (2006). Neutron radiography examination of objects belonging to the cultural heritage. *Appl. Radiat. Isot.* **64**, 7–12.

Reid, A. (1928). The diffraction of cathode rays by thin celluloid films. *Proc. R. Soc. Lond.* **A119**, 663–668.

Reines, F. (1960). Neutrino interactions. *Annu. Rev. Nucl. Sci.* **10**, 1–26.

Reines, F. (1963). Neutrinos, old and new. *Science* **141**, 778–783.

Reines, F. (1979). The early days of experimental neutrino physics. *Science* **203**, 11–16.

Reines, F. (1982). Neutrinos to 1960. Personal recollections. *J. Phys.* **43**(supplement to 12), C8–C237.

Reines, F. (1994). 40 years of neutrino physics. *Prog. Part. Nucl. Phys.* **32**, 1–12.

Reines, F. (1995). The neutrino: from poltergeist to particle. Nobel Lecture, December 8, 1995. In: *The Nobel Prizes 1999 Les Prix Nobel: Nobel Prizes, Presentations, Biographies, and Lectures* (ed. A. Barany), Almquiest & Wiksell International, Stockholm.

Reines, F., and Cowen, Jr., C. L., (1953). Detection of the free neutrino. *Phys. Rev. Lett.* **92**, 330.

Reines, F., and Cowen, Jr., C. L. (1956). The neutrino. *Nature* **178**, 446–449.

Reines, F., and Cowen, Jr., C. L. (1957). Neutrino physics. *Phys. Today* **10**(8), 12–18.

Reines, F., and Cowen, Jr., C. L., Harrison, F. B., McGuire, A. D., and Kruse, H. W. (1960). Detection of the free antineutrino. *Phys. Rev.* **117**(1), 159–173.

Reynolds, G. T., Harrison, F. B., and Salvini, G. (1950). Liquid scintillation counters. *Phys. Rev.* **78**(4), 488.

Rhodes, R. (1986). *The Making of the Atomic Bomb*, Touchstone/Simon & Schuster, New York, NY, p. 886.

Rhodes, R. (1999). Atomic Physicist, Enrico Fermi. *Time*, March 28, 1999.

Rife, P. (1999). *Lise Meitner and the Dawn of the Nuclear Age*, Birkhäuser, Boston, MA, p. 432.

Ritson, D. M. (1961). *Techniques in High Energy Physics*, Interscience, New York, NY.

Roberts, A. (1960). A new type of Čerenkov detector for the accurate measurement of particle velocity and direction. *Nucl. Instrum. Methods* **9**, 55–60.

Robson, J. M. (1950a). Radioactive decay of the neutron. *Phys. Rev.* **77**, 747A.

Robson, J. M. (1950b). Radioactive decay of the neutron. *Phys. Rev.* **78**, 311–312.

Rochester, G. D., and Butler, C. C. (1947). Evidence for the existence of new unstable elementary particles. *Nature* **160**, 855–857.

Roentgen, W. C. (1895). Über eine neue Art von Strahlen. *Sitzungsberichte der Würzburger Physik-medic Gesellschaft.* **9**, 132–141.

Roentgen, W. C. (1896). On a new kind of rays. *Nature* **53**(1369), 274–276.

Rohrlich, F., and Carlson, B. C. (1954). Positron-electron differences in energy loss and multiple scattering. *Phys. Rev.* **93**, 38–44.

Rosenberg, J. L. (2004). The contributions of James Franck to photosynthesis research: a tribute. *Photosynth. Res.* **80**, 71–76.

Rosner, R., and Strohmaier, B. (2003). *Marietta Blau—Sterne der Zertrümmerung. Biographie einer Wegbereiterin der modernen Teilchenphysik.* Böhlau Verlag, Vienna, p. 192.

Rossi, B., and Hall, D. B. (1941). Variation of the rate of decay of mesotrons with momentum. *Phys. Rev.* **59**, 223–228.

Roy, R. R., and Reed, R. D. (1968). *Interactions of Photons and Leptons with Matter*, Academic Press, New York, NY.

Rucker, T. L. (1991). Calculational method for the resolution of ^{90}Sr and ^{89}Sr counts from Cerenkov and liquid scintillation counting. In: *Liquid Scintillation Counting and Organic Scintillators* (eds. H. Ross, J. E. Noakes, and J. D. Spaulding), pp. 529–535. Lewis Publishers, Chelsea, MI.

Rutherford, E. (1899). Uranium radiation and the electrical conduction produced by it. *Philos. Mag. Ser.* **5, 47**, 109–163.

Rutherford, E. (1900). A radioactive substance emitted from thorium compounds. *Philos. Mag. Ser.* **5, 49**, 1–14.

Rutherford, E. (1903). The magnetic and electric deviation of the easily absorbed rays from radium. *Philos. Mag. Ser.* **6, 5**, 177–187.

Rutherford, E. (1906). Retardation of the alpha particle from radium in passing through matter. *Philos. Mag. Ser.* **6, 12**, 134–146.

Rutherford, E. (1911). The scattering of α and β particles by matter and the structure of the atom. *Phil. Mag.,* **21**(6), 669–688.

Rutherford, E. (1913). The structure of the atom. *Nature* **92**(2302), 423.

Rutherford, E. (1919). Collision of α-particles with light atoms. *Nature (London)* **103**, 415–418.

Rutherford, E. (1920). Bakerian Lecture: nuclear constitution of atoms. *Proc. R. Soc. Lond.* **97A**, 374–401.

Rutherford, E. (1929). "Annual Address to the Royal Society." *Proc. R. Soc. Lond.* **122A**, 1–23.

Rutherford, E. (1936). The development of the theory of atomic structure. In: *Background to Modern Science* (eds. J. Needham and W. Pagel), Macmillan Company, New York, NY, pp. 61–74.

Rutherford, E., and Soddy, F. (1902). The cause and nature of radioactivity. *Philos. Mag.* **4**, 370–396.

Samios, N. P., Plano, R., Prodell, A., Schwartz, M., and Steinberger, J. (1962). Parity of the neutral pion and the decay $\pi^{0} \rightarrow 2e^{+} + 2e^{-}$. *Phys. Rev.* **126**, 1844–1849.

Sanuki, T., *et al.* (2000). Precise measurement of cosmic-ray proton and helium spectra with the BESS spectrometer. *Astrophys. J.* **545**, 1135–1142.

Scarpitta, S. C., and Fisenne, I. M. (1996). Cerenkov counting as a compliment to liquid scintillation counting. *Appl. Radiat. Isot.* **47**, 795–800.

Schoenberger, J., Rozeboom, S., Wirthgen-Beyer, E., and Eilles, C. (2004). Evaluation of the clinical value of bone metabolic parameters for the screening of osseous metastases compared to bone scintiigraphy. *BMC Nucl. Med.* **4**, 3.

Schoenborn, B. P. (2006). How single hydrogen atoms came into view. Forty years devoted to neutron techniques for structural biology. *Los Alamos Sci.* **30**, 196–203.

Schrödinger, E. (1915a). Notice regarding capillary pressure in gas blasts. *Ann. Phys.* **46**(3), 413–418.

Schrödinger, E. (1915b). The theory of drop and rise tests on Brownian motion particles. *Phys. Z.* **16**, 289–295.

Schrödinger, E. (1917a). The acoustics of the atmosphere. *Phys. Z.* **18**, 445–453.

Schrödinger, E. (1917b). Addendum to my essay: on the acoustics of the atmosphere. *Phys. Z.* **18**, 567–574.

Schrödinger, E. (1921). Isotopia and Gibbs paradox. *Z. Phys.* **5**, 163–166.

Schrödinger, E. (1926a). Quantisierung als Eigenwertproblem I. *Ann. Phys.* **79**(4), 361–376.

Schrödinger, E. (1926b). Quantisierung als Eigenwertproblem II. *Ann. Phys.* **79**(6), 489–527.

Schrödinger, E. (1926c). Über das Verhaltnis der Heisenberg-Born-Jordan'schen Quantenmechanik zu der meinen. *Ann. Phys.* **79**(8), 734–756.

Schrödinger, E. (1926d). Quantisierung als Eigenwertproblem III. *Ann. Phys.* **80**(13), 437–490.

Schrödinger, E. (1926e). Quantisierung als Eigenwertproblem IV. *Ann. Phys.* **81**(18), 109–143.

Schrödinger, E. (1926f). The constant crossover of micro- to macro molecules. *Natirwissenschaften* **14**, 664–666.

Schrödinger, E. (1926g). An undulatory theory of the mechanics of atoms and molecules. *Phys. Rev.* **28**(6), 1049–1070.

Schrödinger, E. (1927). Energieaustausch nach der Wellenmechanik. *Ann. Phys.* **83**(15), 956–968.

Schultz, H. (2005). Nobel Efforts. *Nat. Geographic Mag.* May, p. 1.

Schulz, H. D., and Zabel, M. (2006). *Marine Geochemistry*, Springer, p. 574.

Schwarzschild, A. (1963). A survey of the latest developments in delayed coincidence measurements. *Nucl. Instrum. Methods* **21**, 1–16.

ScienceWeek (2001). History of Physics: Millikan's Oil Drops, Electron Charge, and Cooked Data. February 23, 2001, Vol. 5, No. 8.

Seelig, C. (1954). *A. Einstein. Ideas and Opinions, based on Mein Weltbild*, Bonzana Books, New York, NY, pp. 8–11.

Segrè, E. (1959). "Properties of Nucleons", Nobel Lecture, December 11, 1959. In: *Nobel Lectures, Physics, 1942–1962*, (1964), (eds. Nobel Foundation Staff), Elsevier, Amsterdam.

Segrè, E. (1968). *Nuclei and Particles*, W. A. Benjamin, New York, NY.

Seltzer, S. M., and Berger, M. J. (1982a). Evaluation of the collision stopping power of elements and compounds for electrons and positrons. *Int. J. Appl. Radiat. Isot.* **33**, 1189–1218.

Seltzer, S. M., and Berger, M. J. (1982b). Procedure for calculating the radiation stopping power for electrons. *Int. J. Appl. Radiat. Isot.* **33**, 1219–1226.

Seltzer, S. M., and Berger, M. J. (1984). Improved procedure for calculating the collision stopping power of elements and compounds for electrons and positrons. *Int. J. Appl. Radiat. Isot.* **35**(7), 665–676.

Sequinot, J., and Ypsilantis, T. (1994). A historical survey of ring imaging Cherenkov counters. *Nucl. Instrum. Methods Phys. Res. A* **343**, 1–29.

Serway, R. A., Moses, C. J., and Moyer, C. A. (1997). *Modern Physics*, 2nd edn., Harcourt College Publishers, New York, NY.

Sheffield, J. (2001). The future of fusion. *Nucl. Instrum. Methods Phys. Res. A* **464**, 33–37.

Shu, F., Ramakrishnan, V., and Schoenborn, B. P. (2000). Enhanced visibility of hydrogen atoms by neutron crystallography on fully deuterated myoglobin. *Proc. Natl. Acad. Sci. USA* **97**(8), 3872–3877.

Siegbahn, K. (1958). Presentation Speech, Nobel Prize in Physics, Stockholm, December 11, 1958. In: *Nobel Lectures, Physics, 1942–1962*, (1964), (eds. Nobel Foundation Staff), Elsevier, Amsterdam.

Siegbahn, K. (1960). Presentation Speech, Nobel Prize in Physics, Stockholm, December 12, 1960. In: *Nobel Lectures, Physics, 1942–1962*, (1964), (eds. Nobel Foundation Staff), Elsevier, Amsterdam.

Sime, R. L. (1996). *Lise Meitner: A Life in Physics*, University of California Press, Berkeley, CA, p. 526.

Simpson, J. A. (1948). The latitude dependence of neutron densities in the atmosphere as a function of altitude. *Phys. Rev.* **73**, 1389–1391.

Simpson, J. A. (1951). Neutrons produced in the atmosphere by cosmic radiations. *Phys. Rev.* **83**, 1175–1188.

Simpson, J. A. (1983). Elemental and isotopic composition of the galactic cosmic rays. *Annu. Rev. Nucl. Part. Sci.* **33**, 323–381.

Simpson, J. A. (2001). The cosmic radiation. In: *The Century of Space Science* (eds. J. A. M. Bleeker, J. Geiss, and M. C. E. Huber, Eds.), Vol. 1, Kluwer Academic Publishers, Dordrecht, The Netherlands, pp. 117–151.

Simpson, J. A., Fonger, W., and Treiman, S. B. (1953). Cosmic ray intensity-time variations and their origin: neutron intensity variation method and meteorological factors. *Phys. Rev.* **90**, 934–950.

Simpson, J. A., and Garcia-Munoz, M. (1988). Cosmic-ray lifetime in the galaxy: Experimental results and models. *Space Sci. Rev.* **46**, 205–224.

Singh, B. P., and Kumar, B. (2005). *Isotopes in Hydrology, Hydrogeology, and Water Resources*, Alpha Science, p. 250.

Skottsberg, C. (1949). Banquet Speech, Nobel Banquet, Stockholm, December 10, 1949. In: *Les Prix Nobel en 1949* (ed. Arne Holmberg), Nobel Foundation, Stockholm.

Slater, R. J. (Ed.) (2002). *Radioisotopes in Biology*, 2nd edn., Oxford University Press, Oxford, p. 328.

Smoot, G. F., and Scott, D. (2002). Cosmic background radiation. pp. 22.1–22.14. *In* Hagiwara *et al.* (2002). *Phys. Rev.* **D66**, 010001-1.

Snell, A. H., and Miller, L. C. (1948). On the radioactive decay of the neutron. *Phys. Rev.* **74**, 1217–1218.

Snell, A. H., *et al.* (1950). Radioactive decay of the neutron. *Phys. Rev.* **78**, 310–311.

Snow, W. M., Chowdhuri, Z., Dewey, M. S., Fei, X., Gilliam, D. M., Greene, G. L., Nico, J. S., Sørensen, H., and Andersen, H. H. (1973). Stopping power of Al, Cu, Ag, Au, Pb, and U for 5–18 MeV protons and deuterons. *Phys. Rev.* **8B**, 1854–1863.

Snow, W. M., Chowdhuri, Z., Dewey, M. S., Fei, X., Gilliam, D. M., Greene, G. L., Nico, J. S., and Wietfeldt, F. E. (2000). A measurement of the neutron lifetime by counting trapped protons. *Nucl. Instrum. Methods Phys. Res., Sect. A* **440**, 528–534.

Soddy, F. (1913a). Inter-atomic charge. *Nature* **92**, 399–400.

Soddy, F. (1913b). Radioactivity. *Chem. Soc. Ann. Rep.* **10**, 262–288.

Soddy, F. (1913c). The radio-elements and the Periodic Law. *Chem. News* **107**, 97–99.

Soddy, F. (1922). The origins of the conceptions of isotopes. Nobel Lecture, December 12, 1922. In: *Nobel Lectures, Chemistry 1901–1921* (1966), Elsevier, Amsterdam.

Soddy, F., and Cranston, J. A. (1918). The parent of actinium. *Nature* **100**, 498–499.

Sowerby, B. D. (1971). Čerenkov detectors for low-energy gamma-rays. *Nucl. Instrum. Methods* **97**, 145–149.

Spano, H., and Kahn, M. (1952). Enrichment of tin activity through the Szilard-Chalmers separation. *J. Am. Chem. Soc.* **74**, 568–569.

Spinks, J. W. T., and Woods, R. J. (1990). *An Introduction to Radiation Chemistry*, 3rd edn., Wiley, New York, NY.

Steere, A. R. (2005). "A Timeline of Major Particle Accelerators". Masters of Science in Physics Thesis, Michigan State University, Department of Physics and Astronomy, East Lansing, MI.

Steigman, G. (1977). Secondary antiprotons: a valuable cosmic-ray probe. *Astrophys. J.* **217**, L131–L133.

Strachan, J. D., Adler, H., Barnes, C. W., Barnes, G., *et al.* (1994). Fusion power production from TFTR plasmas fueled with deuterium and tritium. *Phys. Rev. Lett.* **72**, 3526–3529.

Street, J. C., and Stevenson, E. G. (1937). New evidence for the existence of a particle of mass intermediate between the proton and electron. *Phys. Rev.* **51**, 1005.

Strömholm, D., and Svedberg, T. (1909a). Untersuchungen über die Chemie der radioactiven Grundstoffe. I. *Z. anorg. Chem.* **61**, 338–346.

Strömholm, D., and Svedberg, T. (1909b). Untersuchungen über die Chemie der radioactiven Grundstoffe. II. *Z. anorg. Chem.* **63**, 197–206.

Stuiver, M., and Polach, H. A. (1977). Reporting ^{14}C data. *Radiocarbon* **19**, 355–363.

Suess, H. E. (1955). Radiocarbon content in modern wood. *Science* **122**, 415–417.

Sundaresan, M. K. (2001). *Handbook of Particle Physics*, CRC Press, Boca Raton, FL, p. 446.

Szilard, L. (1925). Über die Ausdehnung der Phänomenologischen Thermodynamik auf die Schwankungserscheinungen. *Z. Phys.* **32**, 753–788.

Szilard, L. (1929). Über die Entropieverminderung in einem thermodynamischen System bei Eingriffen intelligenter Wessen. *Z. Phys.* **53**, 840–856.

Szilard, L., and Chalmers, T. A. (1934a). Chemical separation of the radioactive element from its bombarded isotope in the Fermi effect. *Nature* **134**, 462.

Szilard, L., and Chalmers, T. A. (1934b). Detection of neutrons liberated from beryllium by gamma rays: a new technique for inducing radioactivity. *Nature* **134**, 494–495.

Szilard, L., and Chalmers, T. A. (1935). Radioactivity induced by neutrons. *Nature* **135**, 98.

Tabata, T., Ito, R., and Okabe, S. (1972). Generalized semiempirical equations for the extrapolated range of electrons. *Nucl. Instrum. Methods.* **103**, 85–91.

Tait, W. H. (1980). *Radiation Detection*, Butterworths, London.

Takiue, M., Fujii, H., and Aburai, T. (1993). Reliability of activity determined by Cerenkov measurements in a liquid scintillation counter. In: *Liquid Scintillation Spectrometry 1992* (eds. J. E. Noakes, F. Schönhofer and H. A. Polach) *Radiocarbon*, pp. 69–73.

Takiue, M., Natake, T., Fujii, H., and Aburai, T. (1996). Accuracy of Cerenkov measurements using a liquid scintillation spectrometer. *Appl. Radiat. Isot.* **47**, 123–126.

Takiue, M., Yoshizawa, Y., and Fujii, H. (2004). Cerenkov counting of low-energy beta-emitters using a new ceramic with high refractive index. *Appl. Radiat. Isot.* **61**, 1335–1337.

Tamm, I. E. (1958). General characteristics of radiations emitted by systems moving with super-light velocities with some applications to plasma physics. Nobel Lecture, December 11, 1958. In: *Nobel Lectures, Physics, 1942–1962*, (1964), (eds. Nobel Foundation Staff), Elsevier, Amsterdam.

Tamm, I., and Ivanenko, D. (1934). Exchange forces between neutrons and protons and Fermi's theory. *Nature* **133**, 981–982.

Taylor, C. C. W. (1999). *The Atomists: Leucippus and Democritus*, University of Toronto Press, Toronto.

Taylor, L. S., Tubiana, M., Wyckoff, H. O., Allisy, A., Boag, J. W., Chamberlain, R. H., Cowan, E. P., Ellis, F., Fowler, J. F., Fränz, H., Gauwerky, F., Greening, J. R., Johns, H. E., Lidén, K., Morgan, R. H., Petrov, V. A., Rossi, H. H., and Tsuya, A. (1970). "Linear Energy Transfer." ICRU Report 16. International Commission on Radiation Units and Measurements, Washington, D.C.

Taylor, R. E. (2000). Fifty years of radiocarbon dating. *Am. Sci.* **88**, 60–67.

Telegdi, V. L. (2000). Szilard as inventor: accelerators and more. *Phys. Today* **53** (10), October 2000, 25.

Thomson, G. P. (1927). The diffraction of cathode rays by thin film of platinum. *Nature* **120**, 802.

Thomson, G. P. (1928). Experiments in the diffraction of cathode rays. *Proc. R. Soc. Lond. A* **119**, 600–609.

Thomson, G. P. (1929). The crystal structure of nickel films. *Nature* **123**, 912.

Thomson, G. P. (1935). Electron diffraction as a method of research. *Nature* **135**, 492–495.

Thomson, G. P. (1938). Nobel Lecture. "Electronic Waves", June 7, 1938. In: *Nobel Lectures, Physics 1922–1941*, (1965), Elsevier, Amsterdam.

Thomson, G. P., and Reid, A. (1927). Diffraction of cathode rays by a thin film. *Nature* **119**, 890–895.

Thomson, J. J. (1897). Cathode rays. *Philos. Mag.* **44**, 293–316.

Titus, K. J, Clapp, T. G., and Zhu, Z. (1997). A preliminary investigation of a beta-particle transmission gauge for seam quality determination. *Textile Res. J.* **67**, 23–24.

Treves, S. T. (Ed.) (2006). *Pediatric Nuclear medicine/PET*, 3rd edn., Springer, p. 542.

Tsoulfanidis, N. (1995). *Measurement and Detection of Radiation*, 2nd edn. Taylor and Francis, Washington, DC.

Tsybin, A. S. (1997). New physical possibilities in compact neutron sources. *Appl. Radiat. Isot.* **48**(10–12), 1577–1583.

Tucker, A. (2005). Obituary—Hans Bethe. *The Guardian*, March 8, 2005.

Tumul'kan, A. D. (1991). Typical calibration curves for beta thickness gauges. *Meas. Tech.* **34**(1), 24.

Turner, J. E. (1995). *Atoms, Radiation and Radiation Protection*, 2nd edn. Wiley, New York, NY.

Tyapkin, A. A. (1974). The first theoretical prediction of the radiation discovered by Vavilov and Čerenkov. *Sov. Phys. Usp.* **17**, 288.

UCSD. (2005). The Register of Leo Szilard Papers 1898–1998, MSS 0032. Mandeville Special Collections Library, Geisel Library, University of California, San Diego.

UIC. (2005). *Nuclear Fusion Power*. UIC Nuclear Issues Briefing Paper No. 69, June 2005, p. 5. Uranium Information Centre, Ltd. Melbourne, Australia.

Upham, L. V., and Englert, D. F. (2003). Radionuclide imaging. In: *Handbook of Radioactivity Analysis*, 2nd edn. (ed. M. F. L'Annunziata) Elsevier, Amsterdam, pp. 1063–1127.

USDA-ARS. (2006). http://www.screwworm.ars.usda.gov/HISTORY1.htm

U.S. News & World Report. (August 15, 1960). Leo Szilard, Interview: President Truman Did Not Understand, pp. 68–71.

Vavilov, S. I. (1934). On the possible causes of blue γ-glow of liquids. *C.R. Dokl. Akad. Nauk, SSSR* **2**, 457–459.

Vavilov, S. I., and Levshin, V. L. (1926). The relation between fluorescence and phosphorescence in solid and liquid media. *Z. Phys.* **35**, 920–936.

Vavilov, Yu. N. (2002). The work and the life of S. I. Vavilov. CERN Seminar 21 November 2002. *CERN Wkly Bull. Issue* **46**, 11 November 2002.

Va'vra, J. (2000). Particle identification methods in high energy physics. *Nucl. Instrum. Methods Phys. Res. A* **453**, 262–278.

Veksler, V. I. (1944). A new method of accelerating relativistic particles. *C. R. Dokl. Akad. Nauk, SSSR* **43**(8), 329–331.

Veksler, V. I. (1945). A new method of accelerating relativistic particles. *J. Phys. USSR* **9**, 153–158.

Veksler, V. I. (1946). Concerning some new methods of acceleration of relativistic particles. *Phys. Rev.* **69**, 244.

Villard, P. (1894a). On the carbonic hydrate and the composition of gas hydrates. *C. R. Acad. Sci. Paris* **119**, 368–371.

Villard, P. (1894b). Experimental study of gas hydrates. *Ann. Chim. Phys.* **11**(7), 353–360.

Villard, P. (1896). Combinaison de l'argon avec l'eau. *C. R. Acad. Sci., Paris* **123**, 377–379.

Villard, P. (1899). Sur l'action chimique des rayons X. *C. R. Acad. Sci., Paris* **129**, 882–883.

Villard, P. (1900a). On the chemical action of the x-rays. *Philos. Mag.* **49**, 244.

Villard, P. (1900b). Sur la réflexion et la refraction des rayons cathodiques et des rayons déviables du radium. *C. R. Acad. Sci., Paris* **130**, 1010–1012.

Villard, P. (1900c). Sur le rayonnement du radium. *C. R. Acad. Sci., Paris* **130**, 1178–1179.

Villard, P. (1900d). Rayonnement du radium. *Séances de la Société Française de Physique* p. 45–46.

Villard, P. (1908). Instruments de mesure á lecture directe pour les rayons x. Sunstitution de la méthode électrométrique aux autres methods de mesure en radiology. Scleromètre et quantimètre. *Arch. d'électricité Médicale* **16**, 692–699.

Volkman, J. K. (Ed.) (2006). *Marine Organic Matter: Biomarkers, Isotopes and DNA*, Springer, p. 374.

von Laue, M. (1912). Eine quantitative Prüfung der Theorie für die Interferenzerscheinungen bei Röntgenstrahlen. *Sitzunsberichte der Königlich Bayerischen Akademie der Wissenschaften*, 363–373.

von Laue, M. (1913). Concerning the detection of x-ray interferences, Nobel Lecture, November 12, 1915. In: *Nobel Lectures, Physics 1901–1921*, (1967), Elsevier, Amsterdam.

van Loon, R., and van Tihhelen, R. (2004). Radiation dosimetry in medical exposure. A short historical overview. *Ann. Assoc. Belge Radioprot.* **29**(2), 163–174.

von Weizsäcker, C. F. (1937). Über Elementumwandlungen im Innern Sterne. *Phys. Z.* **38**, 176.

Waddington, C. J. (Ed.) (1988). *Cosmic Abundances of Matter*, A.I.P. Conf. Proceedings No. 183, p. 111.

Waller, I. (1949). Presentation Speech, December 12, 1949. Nobel Prize in Physics 1949. In: *Nobel Lectures, Physics, 1942–1962*, (1964), (eds. Nobel Foundation Staff), Elsevier, Amsterdam.

Walton, E. T. S. (1951). The artificial production of fast particles. Nobel Lecture, December 11, 1951. In: *Nobel Lectures, Physics, 1942–1962*, (1964), (eds. Nobel Foundation Staff), Elsevier, Amsterdam.

Webber, W. R., and Potgieter, M. S. (1989). The artificial production of nuclear gamma-radiation. *Astrophys. J.* **344**, 779.

Webster, H. C. (1932). The artificial production of nuclear γ-radiation. *Proc. R. Soc. Lond., A* **136**, 428–453.

Weise, W. L., and Martin, G. A, (1989). *A Physicists Desk Reference*, American Institute of Physics, New York, NY, p. 94.

Weiss, R. J. (Ed.) (1999). *The Discovery of Anti-matter: The Autobiography of Carl David Anderson, the Youngest Man to Win the Nobel Prize*, World Scientific, River Edge, NJ, p. 144.

Wideröe, R. (1928). Über ein neues Prinzip zur Herstellung hoher Spannungen. *Archiv für Elektrotechnik* **21**, 387–406.

Wiebe, L. I., Helus, F., and Maier-Borst, W. (1978). Čerenkov counting and Čerenkov scintillation counting with high refractive index organic ligands using a liquid scintillation counter. *Int. J. Appl. Radiat. Isot.* **29**, 391–394.

Wiebe, L. I., McQuarrie, S. A., Ediss, C., Maier-Borst, W., and Helus, F. (1980). Liquid scintillation counting of radionuclides emitting high-energy beta radiation. *J. Radioanal. Chem.* **60**, 385–394.

Wietfeldt, F. E. (2000). A measurement of the neutron lifetime by counting trapped protons. *Nucl. Instrum. Methods Phys. Res. A* **440**, 528–534.

Wien, W. (1893). Eine neue Beziehung der Strahlung schwarzer Korper zum zweiten Hauprsatz der Warmtheorie. *Sitzbericht der Akad. Wissenshaft* **9**, 55–62.

Wigner, E. (1933). On the mass defect of helium. *Phys. Rev.* **43**, 252–257.

Williams, J. L., and Dunn, T. S. (1979). Radiation sources-gamma. *Radiat. Phys. Chem.* **14**, 185–201.

Williamson, L. M. (1998). Transfusion associated graft versus host disease and its prevention. *Heart* **80**, 211–212.

Wilson, C. T. R. (1911). On the method of making visible the paths of ionizing particles through a gas. *Proc. R. Soc. Lond., A* **85**, 285–288.

Wilson, C. T. R. (1912). On the expansion apparatus for making visible the tracks of ionizing particles in gases and some results obtained by its use. *Proc. R. Soc. Lond., Ser. A* **87**, 277–291.

Wilson, C. T. R. (1923). Investigations on x-rays and β-rays by the cloud method I.—x-rays. *Proc. R. Soc. Lond., Ser. A* **104**, 1–24.

Wilson, C. T. R. (1927). On the cloud method of making visible ions and the tracks of ionizing particles. Nobel Lecture, December 12, 1927. In: *Nobel Lectures, Physics 1922–1941*, Elsevier, Amsterdam.

Wilson, E. J. N. (2001). *An Introduction to Particle Accelerators*, Oxford University Press, p. 272.

Woan, G. (2000). *The Cambridge Handbook of Physics Formulas*, Cambridge University Press, Cambridge.

Wolfe, R. R. (1992). *Radioactive and Stable Isotope Tracers in Biomedicine: Principles and Practice of Kinetic Analysis*, Wiley-Liss, New York, NY, p. 480.

Wolfe, R. R., and Chinkes, D. L. (2004). *Isotope Tracers in Metabolic Research: Principles and Practice of Kinetic Analysis*, 2nd edn., Wiley-Liss, New York, NY, p. 488.

Wood, J. G. (1954). Bubble tracks in a hydrogen-filled Glaser chamber. *Phys. Rev.* **94**, 731.

Wooldridge, D. E. (1936). The Separation of Gaseous Isotopes by Diffusion. Doctoral Dissertation, California Institute of Technology, Pasadena, California.

Wooldridge, D. E., and Smythe, W. R. (1936). The separation of gaseous isotopes by diffusion. *Phys. Rev.* **50**, 233–237.

Wyckoff, H. O., and Henderson, J. E. (1943). The spatial asymmetry of Cerenkov radiation as a function of electron energy. *Phys. Rev.* **64**(1–2), 1–6.

Wynn-Williams, C. E. (1931). The use of thyratrons for high-speed automatic counting of physical phenomena. *Proc. R. Soc. Lond., A* **132**, 295–310.

Wynn-Williams, C. E. (1932). A thyratron scale-of-two counter. *Proc. R. Soc. Lond., A* **136**, 312–324.

Yalow, R. S. (1977). Radioimmunoassay: A probe for fine structure of biologic systems. Nobel Lecture, December 8, 1977. In: *Nobel Lectures, Physiology and Medicine 1971–1980*, (1992), (ed. J. Lindsten). World Scientific Publishing, Singapore.

Yalow, R. S., and Berson, S. A. (1959). Assay of plasma insulin in human subjects by immunological methods. *Nature* **184**, 1648–1649.

Yi, C. Y., Han, H. S., Jun, J. S., and Chai, H. S. (1999). Mass attenuation coefficients of β^+-particles. *Appl. Radiat. Isot.* **51**, 217–227.

Ypsilantis, T., and Sequinot, J. (1994). Theory of ring imaging Cherenkov counters. *Nucl. Instrum. Methods Phys. Res. A* **343**, 30–51.

Yukawa, H. (1935). On the interaction of elementary particles. I. *Proc. Phys.-Math. Soc. Japan* **17**, 48–57.

Yukawa, H. (1949). Meson theory in its developments. Nobel Lecture, December 12, 1949. In: *Nobel Lectures, Physics, 1942–1962*, (1964), (eds. Nobel Foundation Staff), Elsevier, Amsterdam.

Zahn, U. (1967a). A study of the recoil behavior of ^{56}Mn atoms in dilute solutions of manganese carbonyl compounds. *Radiochim. Acta* **7**, 170–175.

Zahn, U. (1967b). Recoil reactions in crystalline ^{56}Mn carbonyls. *Radiochim. Acta* **8**, 177–178.

Zeeman, P. (1897). The effect of magnetism on the nature of light emitted by a substance. *Nature* **55**(1424), 347.

Zeisler, S. K., and Weber, K. (1998). Szilard-Chalmers effect in holmium complexes. *J. Radioanal. Nucl. Chem.* **227**, 105–109.

Ziessman, H. A., O'Malley, J. P., and Thrall, J. H. (2005). *Nuclear Medicine: The Requisites*, 3rd edn., Mosby, Philadelphia, p. 704.

Zinn, W. H., and Szilard, L. (1939). Emission of neutrons by uranium. *Phys. Rev.* **56**, 619–624.

Index

A number, *see also* Mass number
definition, 71
Absorption cross sections, neutron,
279–287
Accelerators
as neutron source, 268–269
isotope production with, 33, 503–504
radiotherapy with, 9
Accelerator, linear
Cockcroft and Walton and, 506–511
development of, 506–508
Ising and, 238, 498–499, 506
principle of, 498–499, 506–508
Wideröe and, 237–238, 498–499,
500, 506
Activity, radionuclide
definition, 549
radionuclide mass equivalents,
549–552
units for, 551
Alpha-particles
constitution of, 71
energies of, 71–75
fluorescence by, 76
identification of, 59
interactions with matter, 75–78
linear energy transfer of, 138–139
nuclear scattering of, 76–78
origin of, 59, 61
properties of, 56–58
ranges of, 12–20, 78–84
ranges in air, 125
ranges in water, 140
specific ionization of, 20
stopping power of, 133
Americium-241
decay energy calculations of, 71–74
decay scheme, 71, 73
Anderson, Carl D.
discovery of positron by, 325–326,
350–354
kaon measurements by, 358–359

Millikan and, 351–352
Neddermeyer and, 326, 353–357
muon studies by, 357
observation of pair production by, 354
Wilson cloud chamber and, 352–354
Annihilation
electron/positron, 7, 154, 195–196
neutron/antineutron, 384, 393
proton/antiproton, 384–386, 393–394
Anode, 49–50
Anticathode, 50
Antineutrino *see also* Neutrino
beta decay and, 100–102
Antiparticles
Dirac and, 325–327
Antiproton
discovery of, 326
observation of, 384–386, 393
prediction of, 326
Argon-37
decay scheme of, 200
Argon-40
isotope dating with, 32
Atomic bomb
Szilard patent for, 236, 239–240
Atomic number
definition, 71
Moseley and, 174–175
Atomic pile, 104, 245–246
Atomic radius, 76
Auger, Pierre Victor, 394–397
Auger effect, 394
Auger electrons
definition of, 394
discovery of, 228, 344, 394
energies of, 345, 395–396
fluorescence yield and, 345–346
following electron capture, 128
following internal conversion, 341
following x-ray emission, 198
origins of, 344–345, 394–396
Auger electron spectroscopy, 396

Barkla, Charles Glover, 175–177
Becquerel, Henri, 50–52
Becquerel unit, definition, 551
Beria, Lavrenty, 427, 434, 436
Berson, Solomon, 22–23
Beta decay
    Fermi, 100–102
    Pauli and, 108–109
    reverse, 112
Beta-particles
    absorption and transmission, 129–132
    energy spectra of, 100, 120–121
    interaction with matter, 123, 132–140
    linear energy transfer of, 138–139
    origin of, 61, 68, 119–120
    properties of, 56–58, 119–121
    ranges of, 123–125
    ranges in water, 140
    stopping power of, 133–137
Bethe, Hans A., 512–518
Big Bang
    CMB and, 414–415, 419–420
    Georges Lemaître and, 413–414
    hypothesis for, 419–420
    primordial explosion theory of, 413–414
Binding energy, nuclear
    definition, 264, 270, 274
    deuterium, 222–224
    fission product, 265–266
    per nucleon, 264–266, 273–274
    Uranium-236, 264–265
Blackbody radiation, 142–144, 415–422
Blackett, Patrick M.S.
    discovery of positrons and, 366–367
    nuclear collision measurements by,
        360–361
    Occhialini and, 355, 363–367
    pair production and, 355
    proof of nuclear transmutation by, 360–363
Blau, Marietta, 373–374
Blix, Hans, 15–16
Blondlot, René R., 445
Bohr, Aage, 295
Bohr, Niels
    atomic nucleus and, 290
    atomic theory of, 289–310
    Franck and, 305
    Manhattan Project and, 295–296
    nuclear nonproliferation and, 296, 298

Open Letter to the UN, 297–298
    Oppenheimer and, 296
    Planck and, 289–290, 305
    quantum electron characterization by,
        290–295
    Roosevelt and, 296–297
    Rutherford and, 289–290
    Rydberg formula and, 290–292
Borlaug, Norman, 24–26
Born, Max
    electron probability densities and, 323–324
    Schrödinger and, 323–324
Boron-12
    decay scheme of, 126
Brachytherapy, 10
Bragg's Law, 168–169, 172
Bragg, W. H. and W. L., 166–171
Breeder reactor, 239, 246
Bremmstrahlung, 49–50, 198–200
    from accelerators, 269
    energy loss by, 135–137
    internal, 128
Brockhouse, Bertram N., 33–34
Brownian motion, 153
Bubble chamber
    antineutron annihilation in, 393
    antiproton annihilation in, 393, 394
    development of, 387–390
    kaon tracks in, 390–392
    muon tracks in, 391
    pair production tracks in, 392
    particle velocity measurements in, 390
    pion tracks in, 390
Burkart, Werner, 16, 21, 27–30

Cadmium-109(Silver-109m)
    decay scheme of, 341–342
Calcium-45
    decay scheme of, 120
Californium-252
    spontaneous fission, 261–262
Calvin, Melvin, 12–13
Cancer
    diagnosis and treatment, 2–10
Carbon-14
    biological research with, 12–15
    dating with, 519–528
    decay scheme, 102, 120
Cathode, 49–50

Cathode ray tube, 48–50, 67
Cathode rays, 48, 67
  Lenard window and, 68–69
Cerenkov, *see* Cherenkov
Cerium-144
  decay scheme of, 192
Chadwick, James,
  neutron discovery by, 217–224, 260
  neutron mass and, 220–224
  nuclear binding energy and, 222–224
Chain reaction, nuclear, 242–246
Chamberlain, Owen, 326, 384–386
Chart of the Nuclides, 127
Cherenkov, Pavel A.
  adolescence, 438
  discovery of radiation, 429, 439–441, 465
  parent's persecution, 438–439
  rejection by *Nature*, 445–446
  research findings, 440–457
  student years, 439
  Vavilov. S. and, 429, 439, 441
Cherenkov effect, *see* Cherenkov radiation
Cherenkov radiation
  asymmetric property of, 441–449, 460–461,
    471–475
  discovery of, 429, 439–441
  electrons and beta-particles and, 465,
    469–474, 477–478, 480, 482–485,
    487, 491
  energy output of, 452–456, 461–462,
    475–476
  Frank and, 427, 429–430, 436, 441, 443,
    452–453, 458–463, 466
  gamma radiation and, 482–484
  index of refraction and, 443–444, 448–450,
    460–461, 468–471
  kaons and, 486–487, 490–492
  Mallet and, 429, 466
  Marie Curie and, 429, 466
  muons and, 470
  origin of, 196, 430, 441, 465–467
  particle energy and, 443–444, 448–450,
    460–461, 468–471
  particle energy discrimination and, 457
  particle identification with, 475, 485–492
  particle speed and momentum and, 457
  photon intensity and, 462, 476–480
  photon pulse duration, 480–482
  pions and, 485, 487, 490–492

prediction by Heavyside, 465–466
protons and, 470, 485, 487, 490–491
radionuclide analysis by, 457, 492–495
ring imaging, 486–490
Tamm and, 427, 429–430, 436, 441, 443,
  452–453, 458–463, 466
wavelength spectrum of, 196, 450–452, 475
Chlorine-36
  decay scheme of, 120
  water resource studies with, 29
Cloud chamber, Wilson
  annihilation radiation tracks in, 354
  development of, 90–92
  first particle tracks in, 91–94
  positron discovery with, 353–354
Cobalt-58
  decay scheme of, 125
Cobalt-60
  decay scheme of, 9
  insect pest control with, 16
  radiation processing with, 39–41
  radiotherapy with, 8–9
Cockcroft, John D., 504–511
Coincidence counting
  delayed, 113
Compton, Arthur H., 158–163
  x-ray scattering and, 202–205
Compton edge, 204–205
Compton Effect, 159–162, 202–205
Controlled thermonuclear reactors, 269
Cormack, Allan M., 4
Cosmic Microwave Background
  discovery of, 414–415
  measurement of, 415–417
  prediction of, 412–414
  spectrum, 416–417
  temperature equivalent, 415–422
Cosmic radiation, *see also* Cosmic radiation,
    air showers of,
  beryllium isotopes in, 400
  classification of, 399–400
  composition of, 400–401
  Compton and, 162
  definition, 399
  discovery of, 347–349, 399
  dose from, 424–426
  energies of, 402, 405–406
  intensities of, 400–402, 404–406
  kaon detection in, 358–359

Cosmic radiation (*Continued*)
  latitude effects on, 400–401, 426
  microwave background radiation, 412–421
  Millikan and, 186
  muon detection in, 356–357, 380–381
  origins of, 412
  pion detection in, 379–381
  positron discovery in, 352–354
  primary, 400
  relativistic speeds of, 402–404
  secondary, 400
  underground, 411–412
Cosmic radiation, showers of,
  Auger and, 396–397
  characterization of, 399, 409
  definition, 399, 406
  discovery of, 394, 396
  intensities vs. altitude, 399, 409–410, 425
  measurements of, 397
  mesons in, 406–407, 409
  origin of, 406
  particle energies in, 397
  underground, 411–412
Coulombic barrier
  atomic electrons and, 103
Cowan Jr., Clyde, 110–117
Crick, Francis, 171
Critical mass, U-235
  Peierls and, 243
  Szilard and, 240
Crooke's tube, 48, 67
Crookes, William, 76
Curie, Marie
  death of, 55
  Nobel Prizes and, 53–54
  recognition of, 55
Curie, Pierre,
  death of, 54
Curie, Marie and Pierre
  nuclear medicine and, 2, 8, 53
  polonium and, 53
  radium and, 53
  thorium and, 53
  uranium and, 53
Curie unit, definition, 63, 551
Cyclotron
  invention of, 2
  isotope production by, 2, 33
  worldwide use of, 33

Cyclotron
  applications, 504
  development of, 236–238, 500–504
  invention of, 499–500
  principle of, 500–502

Dating, Carbon-14
  applications of, 527–528
  conceptualization of, 520–521
  low-background counting in, 524–526
  principle of, 519–526
  Suess effect and, 528
Dating, isotope, *see also* Dating, Carbon-14
  Argon-40 and, 32
Davisson, Clinton
  de Broglie matter-wave and, 329–331
  electron elastic scattering and, 329
  electron diffraction and, 329–332
  Germer and, 329–332
de Broglie wavelength, 257–259
de Broglie, Louis
  matter-waves and, 146–149
  matter-wave duality and, 33, 309,
    329–331
  electron diffraction and, 148–149
Decay constant, radionuclide, 532–534
Decay energy
  calculations of, 71–72
  definition, 72
Decay, radionuclide
  branching, 549–550
  complex schemes of, 548–550
  decay constant in, 532–534
  equations for, 532–535, 538–549
  half-life and, 529–530
  out of equilibrium, 547–548
  secular equilibrium in, 539–544
  transient equilibrium in, 544–547
de Hevesy, George
  Levi and, 37
  neutron activation analysis and, 37
  Paneth and, 11
  radiotracer technique and, 10–12
  Rutherford and, 10
Democritus, 47–48
Desalinization, 43–44
Deuterium, *see also* Deuterons
  discovery of, 12
  gamma-ray disintegration of, 222–223

Deuterons
   linear energy transfer of, 138–139
   ranges in water, 140
   stopping power of, 133
Diffraction
   electron, 329–339
   x-ray, 166–171
Dirac, Paul A. M.,
   antiparticles and, 325–327
   prediction of annihilation, 326
   prediction of antiproton, 326
   prediction of positron, 325–326
   quantum theory of, 325–326
Dose, radiation
   definition, 421
   sources of, 424
   units of measurement, 421–423
   weighting factors of, 423–424

Eddington, Arthur
   Einstein and, 155
Einstein, Albert
   Brownian motion and, 153
   general theory of relativity and, 155
   mass–energy equivalence and, 71, 122,
      153–154, 509, 517
   perihelion of Mercury and, 156
   photoelectric effect and, 69–70, 151–152
   photon and, 33, 69, 151–152, 162, 181–185,
      309
   quantum theory and, 151–152
   Roosevelt and, 157, 242–243
   Russell and, 158
   Russell-Einstein Manifesto and, 158
   solar light deflection and, 155–156
   special theory of relativity and, 153
   Szilard and, 157, 237, 239, 242–243
Einstein-Planck equation, 144, 188,
      202–203
Elastic scattering, neutron, 277–278
ElBaradei, Mohamed, 298–300
Electron beam
   radiation processing with, 39–40
Electron capture, 128–129, 196–197, 342
Electron charge
   Fletcher and, 186
   Millikan and, 180–185
Electron diffraction,
   de Broglie and, 148–149

Electrons, atomic, *see also* Electrons
   binding energy of, 201
   energy level transitions, 290–295
   orbitals of, 106–108, 317–322
   quantum characterization of, 290–295
   quantum levels of, 106–108
   quantum numbers of, 106–107
   subshells of, 106–108
Electrons
   Bragg's Law and, 332, 336–338
   diffraction of, 148–149, 329–339
   discovery of, 62, 67–68
   elastic scattering of, 329
   in cathode ray tube, 48–50
   internal conversion, 341–344
   matter-wave duality, 329–331, 333–334
   photoelectric effect and, 215
   wavelengths of, 147–149
Electromagnetic interactions, 100
Electroscope, gold leaf, 227, 348–349
Electroweak force, 100
Endocurietherapy, 10
Epimerization
   carbon-14 and, 15
Exclusion Principle, Pauli, 106–108

FAO/IAEA, 18, 27
Fermi, Enrico
   atomic bomb and, 104–105
   atomic pile and, 245–246
   beta decay and, 100–102
   first nuclear reactor and, 104
   neutrino and, 100
   neutron-induced nuclear reactions and,
      102–103
   nuclear chain reaction and, 104–105
   radioisotope production and, 102–103
   Segré and, 104–105
   Szilard and, 104, 242, 244, 246
Fissile materials, 268
Fission, nuclear
   Byrnes and, 246
   discovery of, 4, 224, 229–235
   energy liberated from, 231–232, 243–245
   Hahn and, 229–236
   Meitner and, 229–235
   neutron induced, 41–43, 104–105, 262–268
   Peierls and, 243
   Plutonium-239, 267–268

Fission, nuclear (*Continued*)
    Roosevelt and, 242–243, 246
    spontaneous, 261–262
    Szilard and, 240, 242–243, 245–246
    Uranium-233, 267–268
    Uranium-235, 262–266
Fissionable isotopes, 268
Fletcher, Harvey, 186
Fluorescence
    alpha particles and, 76
    discovery of, 76
Fluorescence yield, 345–346
Franck-Hertz experiment, 305–309
Franck, James
    Bohr and, 303
    energy quantum absorption and, 305–309
    Franck Report and, 303–305
    Hertz and, 305–309
    Manhattan Project and, 303
    Nazi Germany and, 303
    Nobel Prize with Bohr and Planck, 305
    photosynthesis and, 305
Franck Report, 247, 303–305
Frank, Il'ja M., 458–463
Franklin, Rosalind, 171
Frequency, radiation, 190
Fusion, nuclear
    compact devices for, 272
    deuterium-deuterium, 270–272
    deuterium-tritium, 270–272, 274–275
    energy from, 270–272, 274–276
    environmental impact of, 272, 275–276
    fuel for, 272, 274–275
    future energy source, 44–45
    neutron source, 272
    reactor safety of, 276

Gabor, Dennis, 238
Gamma-radiation
    absorption edges of, 214–215
    attenuation, 201–214
    discovery of, 56–57
    dual nature of, 187–191
    energies of, 191
    following alpha emission, 73–75
    linear energy transfer of, 138–139
    origins of, 61, 187
    properties of, 56–58, 187–191
Gamma-ray scattering

Compton and, 159–162
Gaseous diffusion, 301–302
Geiger, Hans, 60–63
Germer, Lester, 329–332
Ginzburg, Vitaly, 436–437, 463
Glaser, Donald A., 386–394

Hafnium
    discovery of, 229
Half-life
    applications of, 534–535
    definition, 529–530
    determination of, 531, 535–538
    Rutherford and, 529–530, 532
    Soddy and, 532
Half-value thickness, 208–212
Hahn, Otto
    Meitner and, 226–236
    nuclear fission and, 229–236
    Strassmann and, 228–231, 233–235
Heavyside, Oliver, 465–466
Heisenberg, Werner
    atomic bomb and, 315–316
    Bethe and, 316
    Bohr and, 314–316
    Uncertainty Principle and, 311–314
    wave-particle duality and, 314–315
Heisenberg Uncertainty Principle
    concepts of, 311–314
    meson prediction and, 369–370
Hertz, Gustav
    energy quantum absorption and, 305–309
    Franck and 305–309
    isotope enrichment and, 301–302
    USSR and, 302
Hertz, Heinrich, 68–69, 151, 187
Hess, Victor F., 347–350
Hodgkin, Dorothy, 171
Hounsfield, Godfrey N., 4
Hydrogels, 38
Hydrogen-1, *see* Protons
Hydrogen-2, *see* Deuterium
Hydrogen-3, *see* Tritium
Hyperons, 385

IAEA, 16, 21, 27–30, 38, 42, 44
IAEA, NNP, 298–300
Inelastic scattering, neutron, 278–279
Internal conversion, 341–344

Iodine-125
  brachytherapy with, 10
  radioimmunoassay with, 23–24
  receptor binding assays with, 30–31
Ionization
  alpha particles and, 75–76
  energy loss by, 135–137
Iron-55
  decay scheme of, 200
Ising, Gustaf, 238, 498–499, 506
Isomeric transition, 194
Isotope dilution, 19
Isotope enrichment, 301–302
Isotopes
  discovery of, 87
ITER, 272, 276

Joliot-Curie, Irène and Frédéric
  Marie Curie and, 97–98
  neutron radiation and, 95
  nuclear fission and, 98
  nuclear medicine and, 99
  positron emission and, 95–98
  radioisotope synthesis and, 95–98
  Savitch and, 98

K capture, see Electron capture
Kaons
  bubble chamber observations of, 390, 392
  decay of, 358–359
  properties of, 358
Kapitsa, Pyotr L., 434–436
Kendrew, John, 171
Kinetic energy, particle, 254–256
Knipling, Edward F., 17–18

Lattes, Césare, 375, 378–379
Lawrence, Ernest, 2, 497–504
Lebedev, Pyotr N., 427–430
Lemaître, Georges
  Big Bang theory of, 413–414
  Eddington and, 413
  Einstein and, 414
Lenard, Philipp, 68–70
Levi, Hilde, 37
Libby, Willard F., 518–528
Linear attenuation coefficient, 208–214
Linear energy transfer, 132, 137–140
Lifetime, neutron, 288

Lorentz, Hendrick A., 64–65
  Einstein and, 65
Lorentz-Einstein transformations, 146
Lorentz transformations
  muon lifetime dilation and, 65
Lysenko, Trofin, 432–434

Mallet, Lucien, 429–430
Manhattan Project, 104–105, 240, 243, 246–247
Marine resources, 29–32
Marsden, Ernest, 76
Mass attenuation coefficient, 209, 211–214
Mass number
  definition, 71
Mass thickness, 79–83
Matter-waves, de Broglie, 146–149
MAUD Committee, 338–339
Maxwell, James Clerk, 64, 187
Meitner, Lisa
  Auger electrons and, 228
  at Kaiser Wilhelm Institute, 226–229
  at University of Vienna, 225
  Einstein and, 226, 229
  Frisch and, 226–227, 230–231
  Hahn and, 226–236
  Nobel Prize and, 234
  nuclear fission and, 229–235
  Planck and, 226
  Strassmann and, 228, 233–235
Mesons
  discovery of, 368
  prediction of, 368–370
  properties, 369, 371
Millikan, Robert A., 180–186, 351–352
Momentum, particle, 189–191
Moseley, Henry G. J., 172–175
Muons
  bubble chamber measurements of, 391
  decay of, 357–358, 380, 409
  properties of, 356, 380
  relativistic time dilations of, 410–411
  sea-level abundances of, 409
  tracks in nuclear emulsions, 380
  underground intensities of, 411–412

Neddermeyer, Seth, 326, 353–357
Negatrons, see also Beta particles
  annihilation of, 195–196
  creation of, 205–207

Negatrons (*Continued*)
  origin, 119–120, 126–127
  stopping power of, 133–137
Negatron emission
  Fermi and, 100–102
Neutrinos
  beta decay and, 100–102
  Cowan and, 110–116
  detection of, 112–116
  Fermi and, 100–102, 121–122
  following electron capture, 128
  mass of, 122
  origin, 119–121
  Pauli and, 100, 108–109, 115–116, 120–121
  properties of, 112, 119, 122
  Reines and, 110–116
Neutron activation analysis, 36–37
Neutron attenuation, 281–287
Neutron capture, 279–280
Neutron decay, 287–288
Neutron diffraction, 34–35
Neutron radiation
  attenuation of, 281–287
  capture, 279–280
  capture cross sections of, 279–287
  classification, 253–255
  decay of, 287–288
  discovery of, 95, 217–224, 260
  elastic scattering of, 277–278
  fission induced by, 262–268, 273–276, 279, 281
  inelastic scattering of, 278–279
  interactions with matter, 276–281
  nonelastic reactions, 280–281
  origins of, 253
  properties of, 253–259
  range of, 281–285
  sources of, 260–276
  velocity, 255–257, 259
  wavelength, 255, 257–258
Neutron radiography, 35–36
Neutrons, *See also* Neutron radiation
  activation analysis with, 36–37
  Chadwick and, 217–224
  diffraction and scattering of, 33–35
  decay of, 102
  discovery of, 95, 217–224
  Fermi and, 102–104
  isotope production with, 102–103

  Joliot-Curie's and, 218, 229–230
  mass measurements of, 220–224
  moisture measurement with, 20–21
  radiography with, 35–36
  Rutherford and, 217–218
  thermal neutron capture, 103
New Piles Committee, 42, 246
Nitrogen-12
  decay scheme of, 126
Nitrogen-13
  decay scheme, 102
Nitrogen-15
  fertilizer studies with, 19–20
Nonelastic reactions, neutron, 280–281
Nuclear chain reaction
  Fermi and, 104–105
Nuclear disarmament
  Pugwash Conferences for, 248, 251–252
  Russell-Einstein Manifesto and, 248–251
Nuclear emulsions
  cosmic-ray studies with, 373–384
  nuclear collisions in, 378–379
  particle identification with, 382
  stars in, 378, 384
  technique of, 374–375
Nuclear medicine, 2–10
Nuclear nonproliferation
  Bohr and, 297–298
  ElBaradei and, 298
  Franck and 303–305
  IAEA and, 298–300
Nuclear power
  desalinization and, 43–44
  worldwide energy from, 41–43
Nuclear reactor
  applications of, 104
  Fermi and Szilard design of, 104
  patent for first, 246
  worldwide numbers of, 104
Nuclear radius, 78
Nuclear reactions
  alpha-particle induced, 260–261
  neutron-induced, 102–104
Nuclear reactor
  Fermi and, 41–42
Nuclear resonance effect
  neutron capture and, 103
Nuclear stability
  N/Z radios and, 127–128

Nuclear transmutations
    Cockcroft and Walton and, 508–510
    Lawrence and, 503–504
    Rutherford and, 61–62, 498, 505–506
Nucleus, atomic
    Bohr and, 230–231
    liquid-drop model of, 230–233, 244, 295
    Meitner and, 230
N/Z ratios
    nuclear stability and, 127–128

Occhialini, Giuseppe, 355, 363–368, 370, 373,
    375–376
Oil-drop experiment, 181–184, 186
Oppenheimer, J. Robert, 104, 296
Oxygen-18
    marine resource studies with, 32

Pair production, 153–154, 205–207, 213–215
    discovery of, 354
Palladium-103
    brachytherapy with, 10
Paneth, Frederick, 11
Pauli, Wolfgang
    beta decay and, 108–109
    Cowan and, 115–116
    Einstein and, 106
    Exclusion Principle and, 106–108
    neutrino and, 108–109, 115–116
    Reines and, 115–116
Penzias, Arno, 414–415
Perutz, Max, 171
Phosphorus-32
    decay scheme of, 120, 194–195
    fertilizer studies with, 19
Photoelectric effect
    discovery of, 68–69, 151
    Einstein and, 69–70, 151–152, 181, 185,
        187–188
    Einstein equation for, 201–202
    electron binding energies and, 215
    Hallwachs and, 151
    Hertz and, 68–69, 151
    Lenard and, 68–70, 151, 188
    Millikan and, 181, 185
Photographic emulsions, *see* Nuclear
    emulsions
Photon
    discovery of, 69, 151–152, 187

    Einstein and, 181, 185, 187–188
    Millikan and, 181, 185
    properties of, 188–191
Photoneutron sources, 268
Photonuclear reactions, 268
Photosynthesis, 13–14
Pierre Auger Observatory, 397
Pions
    bubble chamber tracks of, 390, 408
    cloud chamber observations of, 358–359
    decay schemes of, 380–381, 407–408
    properties of, 371
    tracks in nuclear emulsion, 380
Planck, Max
    black-body radiation and, 142
    quantum theory of, 143–144, 188, 309
Planck constant, 142–144, 188
Planck-Einstein equation *see* Einstein-Planck
    equation
Plant breeding, 24–27
Plutonium
    spontaneous fission of, 262
Plutonium-239
    preparation, 267
Positrons, *see also* Beta particles
    Anderson and, 366–367
    annihilation of, 195–196, 366
    Blackett and Occialini and, 366–367
    creation of, 205–207
    discovery of, 350–354, 366–367
    origins of, 124–127, 356
    pair production and, 365
    prediction of, 325–326
    stopping power of, 133
Positron emission
    electron capture and, 129
    Fermi and, 100–102
    Joliot-Curies and, 95–98
Powell, Cecil F.
    Lattes and, 375, 379, 381–382
    Occhialini and, 375–376
    pion measurements and, 379–380
    Rutherford and, 372–373
    Wilson and, 372–373
Protactinium
    discovery of, 228–229
Protons
    discovery of, 62
    linear energy transfer of, 138–139

Protons (*Continued*)
  ranges in water, 140
  stopping power of, 133
Pugwash Conferences, 248, 251–252

Quantum level, 106–108
Quantum mechanics, 310, 312,
  314–315
Quantum number, 106–107
Quantum theory
  atomic electron, 106–108
  de Broglie and, 146–149
  Einstein and, 144, 146, 188
  Planck and, 143–144, 151

Radioactivity
  definition, 1
  discovery of, 2, 51
  sources of, 1
Radiation
  deleterious effects of, 53–54
  food preservation with, 38–40
  industrial processing with, 37–41
  insect pest control and, 16–18, 39–40
  ionization by, 51–52
  medical sterilization with, 38–41
  medicine and, 2–10, 32–33
  neutron, 20–21
  plant breeding and, 24, 26–27
  rubber curing with, 39–41
Radiation processing, 37–41
Radiochemistry
  birth of, 60, 97
Radioimmunoassay, 22–24
Radioisotopes
  agricultural production and, 15–27
  animal production and, 22–24
  biosciences and, 10–15
  industrial applications, 32–41
  medicine and, 2–10, 33
  peaceful applications of, 5–45
  production of, 32–33, 102–103
Radiotherapy, 8–10
Ramsay, William, 86
Range
  alpha-particle, 78–84, 125
  beta-particle, 123–125
  gamma- and x-ray, 207–214
  neutron, 281–285

Reactor, nuclear
  Fermi-Szilard patent for, 246
  isotope production and, 33
  research reactors worldwide, 33
Reactor, nuclear power
Receptor binding assay, 30–32
Recoil energy, 72–74
Reines, Frederick, 110–117
Relativistic mass, particle, 189, 254
Relativistic speed, particle, 256–257
Relativistic wavelength, particle, 257–258
Relativity
  general theory of, 155
  special theory of, 153
Rest energy, particle, 189, 254, 257–259
Rest mass, particle, 189, 254
Reverse beta decay, 112
Ring Imaging Cherenkov counters,
  486–490
Rotblat, Joseph, 251
Röntgen, Wilhelm C., 48–50
Röntgenogram, 49
Roosevelt, Franklin Delano, 157
Rubidium-86
  decay scheme of, 192
Russel, Bertrand
  Einstein and, 158
Russell-Einstein Manifesto, 158, 248–251
Rutherford, Ernest
  α-, β-, γ-ray nomenclature by, 57–59
  α-, β-, γ-ray properties and, 57–60
  α-particle deflection and, 60–61
  α-particle identification and, 59
  α-particle scattering and, 76–78
  accelerator development and, 504–506
  atomic nucleus and, 61
  atom splitting and, 61–62
  Ci unit and, 63
  Geiger and, 60–63, 76
  ionization detectors and, 63
  isotope decay laws and, 63, 85–86,
  529–530, 532
  isotope production and, 2, 61–62
  Marsden and, 76
  nuclear radius calculations of, 77–78
  nuclear transmutations and, 61–62, 498,
  505–506
  Powell and, 62
  proton discovery by, 62

Rutherford, Ernest (*Continued*)
  radioisotope half-life and, 63
  radioisotope decay constant and, 63
Rutherford unit, definition, 551
Rydberg formula, 290–292
Rydberg, Janne, 290

Savitch, Paul, 98
Schmidt, Gerhard, 53
Schrödinger, Erwin
  Bohr and, 317, 321
  Born and, 317
  de Broglie and, 317
  electron orbitals and, 317–322
  Heisenberg and, 321
  matter-wave duality and, 317
  Nazi Germany and, 321
  Pauli and, 318, 321
  radioactive decay and, 316
  wave mechanics and, 317–321
Scintillation, liquid, 112
Scintillation proximity assay, 31–32
Segrè, Emilio, 104–105, 326, 384–386,
    393
Shull, Clifford G., 33–34
Siegbahn, Manne, 177–179
Sodium-22
  decay schemes of, 129, 193
Soddy, Frederick
  isotope decay laws and, 86, 532
  isotope displacement law of, 86–89
  isotope nomenclature by, 87
  isotope periodic law of, 86–89
  Ramsay and, 86
  Rutherford and, 85–86
Specific ionization, 84
Stalin, Josef, 427, 429, 431–437
Stellar energy
  discovery of source, 513
  energy yields in, 516–518
  nuclear reactions in, 514–515
  reaction lifetimes in, 518
Sterile Insect Technique, 16–18
Stokes, George, 183
Stokes Law, 183–184
Stopping power, 132–137
Strassmann, Fritz, 228–231, 233–235
Strontium-87
  water resource studies with, 29

Strontium-89
  decay scheme of, 120
Sulfur-35
  decay scheme of, 120
Surface density, 79
Szilard-Chalmers effect, 240–242
Szilard, Leo
  atomic bomb and, 236, 239, 242–243, 246
  breeder reactor and, 239
  cyclotron and, 236
  electron microscope and, 236, 238
  Einstein and, 157, 237, 239, 242, 243
  Fermi and, 41–42, 104, 242, 244, 246
  first reactor design and, 104
  Gabor and, 238
  nuclear chain reaction and, 104
  nuclear disarmament and, 246–248, 251
  nuclear fission and, 240, 242–243,
      245–246
  nuclear medicine and, 252
  Pugwash Conferences and, 248,
      251–252
  Roosevelt and, 246
  von Laue and, 237
  Zinn and, 41, 242

Tamm, Igor I., 458–464
Technetium-99m
  medicine and, 5–6
Thomson, George Paget
  de Broglie matter-wave and, 333–334
  electron diffraction and, 333–339
Thomson, Joseph J.
  electron discovery by, 67–68
  electron properties and, 68
  Rutherford and, 66–67
Thorium-232, decay chain, 550
Threshold Cherenkov counters, 485
Time-of-propagation Cherenkov counters,
    490–492
Tin-119m
  decay scheme of, 194
Tokamak, 272
Tomography
  computed, 3–6
  positron emission, 6–7
Total attenuation coefficient, 212–214
Tracers, isotope
  agricultural research with, 15–24

Tracers, isotope (*Continued*)
  animal health studies with, 22–24
  biological research with, 10–15
  discovery of, 10–11
  fertilizer use efficiency studies with, 18–20
  marine resource studies with, 29–32
  water resource studies with, 27–29
Transition energy, 74
Tritium
  decay scheme of, 120
  receptor binding assays with, 30–31

Uncertainty Principle, 311–314
United Nations,
  Bohr and, 297–298
  ElBaradei and, 298–300
Uranium-233
  preparation, 268
Uranium-235
  fission, 230–235, 242–245
  fission energy from, 262–267, 274–275
  neutron-induced fission of, 262–266
Uranium-236
  binding energy of, 264–266
  decay of, 263–266
  fission energy from, 263–266, 274–275
Urey, Harold C., 12, 246–247

Vanadium-49
  decay scheme of, 200
Vavilov–Cherenkov radiation, *see* Cherenkov
    radiation
Vavilov, Nicolai I., 427–428, 432–433
Vavilov, Sergei I.
  Cherenkov and, 429
  discovery of Cherenkov radiation, 429
  Frank and, 429–430, 434–435, 441, 463
  Ginzburg and, 436–437
  leader of Russian physics, 430–431
  research on luminescence, 428
  Sakharov and, 433–434
  Stalin's regime and, 427, 429–437
  Tamm and, 429–430, 431–432, 441
  Vavilov, N. and, 432–433
  Veksler and, 432
Veksler, Vladimir I., 431–432, 457
Velocity, particle, 254–257
Villard, Paul U.
  Becquerel and, 57

  gamma radiation and, 55–57
  radiation dosimetry and, 58
  Rutherford and, 57–58
von Laue, Max, 163–166
von Wetzsäcker, C. F., 157

Walton, Ernest T. S., 504–511
Wambacher, Hertha, 373–374
Water resources, 27–29
Water use efficiency, 20–21
Watson, James, 171
Wavelength, particle, 253–255, 257–259
Wavelength, radiation, 189–190
Wave mechanics, 317–321
Weak interaction, nuclear, 100
Wedgewood, Thomas, 142
Whole-body counter, 116–117
Wideröe, Rolf, 237–238, 498–500, 506
Wien, Wilhelm, 142, 418–419, 422
Wilson, C. T. R., 89–94
Wilson cloud chamber
  automation of, 363–365
  development of, 90–92
  Compton effect and, 160–162
  first particle tracks in, 92–94
  kaons observed with, 359
  muons observed with, 356–357
  nuclear collision measurements with, 360–361
  pair production observed with, 351–354
  pions observed with, 359
  transmutation observations with, 362–363
Wilson, Robert, 414–415

x-radiation
  absorption edges of, 214–215
  artificial, 200–201
  attenuation, 201–205
  computed tomography with, 3–5
  discovery of, 3, 48
  dual nature of, 187–191
  early tubes for, 49–50
  energies, 394, 396
  following electron capture, 128
  linear energy transfer of, 138–139
  origins of, 3, 49–50, 172–174, 176–177,
      187, 196–201, 394, 396
  properties of, 48–50, 164–165, 176–177,
      187–191, 196–198
  radiography with, 35–36

x-ray crystallography, 171
x-ray diffraction
  Bragg, W. H. and W. L. and, 166–171
  crystal structure and, 164–165, 168–171
  von Laue and, 163–165
x-ray emission spectra
  Barkla and, 176
  Moseley and, 173–174
x-ray fluorescence
  Barkla and, 176
  definition, 396
x-ray radiography, 35–36
x-ray scattering
  Compton and, 159–162

x-ray spectrometer, 169
x-ray spectroscopy
  Moseley and, 172–174
  Siegbahn and, 178–179

Yalow, Rosalyn, 22–23
Yukawa, Hideki, 367–371

Zeeman, Pieter, 65–66, 188
Zinc-65
  decay schemes of, 129
Zinn, Walter, 242
Z number, *see also* Atomic number
  definition, 71